Leonhard Euler

C·1

Leonhard Euler

Mathematical Genius in the Enlightenment

Ronald S. Calinger

Princeton University Press
Princeton and Oxford

Published by Princeton University Press, 41 William Street, Princeton, New Jersey 08540
In the United Kingdom: Princeton University Press, 6 Oxford Street, Woodstock, Oxfordshire OX20 1TW
press.princeton.edu

Jacket art: Detail of 19th century engraving of Leonhard Euler, private collection. Image

Library of Congress Cataloging-in-Publication Data

Calinger, Ronald.
Leonhard Euler : mathematical genius in the Enlightenment / Ronald S. Calinger.
pages cm
Includes bibliographical references and index.
ISBN 978-0-691-11927-4 (hardcover : alk. paper) 1. Euler, Leonhard, 1707–1783. 2. Mathematicians—Germany—Biography. 3. Mathematicians—Russia (Federation)—Biography. 4. Mathematicians—Switzerland—Biography. 5. Physicists—Germany—Biography. 6. Physicists—Russia (Federation)—Biography. 7. Physicists—Switzerland—Biography. 8. Mathematics—History—18th century. I. Title.
QA29.E8C35 2015
510.92--dc23
[B]

2014045172

British Library Cataloging-in-Publication Data is available

This book has been composed in Baskerville 10 Pro.

Printed on acid-free paper. ∞

Printed in the United States of America

1 3 5 7 9 10 8 6 4 2

Contents

Preface

For his profound and extensive contributions across pure and applied mathematics, Leonhard Euler (1707–83) ranks among the four greatest mathematicians of all time, the other three being Archimedes, Isaac Newton, and Carl Friedrich Gauss. Euler is a principal figure in applying calculus to create the modern mathematical sciences of celestial mechanics, analytical mechanics, elasticity, and optics. In the twentieth century, six concise biographies of him were produced, beginning in 1927 with Louis du Pasquier's *Léonard Euler et ses amis* (Leonhard Euler and his friends) and two years later Otto Spiess's *Leonhard Euler*. In 1961 Vadim V. Kotek published a short biography in Russian, while in 1982 Adolf P. Yushkevich brought out another in English and Rüdiger Thiele a narrative in German. In 1995 Emil A. Fellmann supplied a text in German of fewer than two hundred pages. In 2007, the tercentenary of Euler's birth, Fellmann's text was translated into English by Erika and Walter Gautschi, and Philippe Henry published *Leonhard Euler: "incomparable géomètre."* Yet in no language has a treatment of his life and work appeared that is full in length and scope.

This book attempts to offer the first detailed and comprehensive account in the context of Euler's life, research, computations, and professional interactions that centers on his achievements in calculus and analytical mechanics. The growing body in print of primary sources by Euler, many long inaccessible, together with the secondary literature on him and his research, are making this possible. Central to the first effort is the near completion of the more than eighty large volumes of Euler's *Opera omnia* (Collected works). Series 1, on mathematics, comprises twenty-nine volumes; series 2, on mechanics and astronomy, comprises thirty-one volumes; and series 3 comprises twelve volumes on physics and *varia*. Series 4A will have eight volumes with annotated versions of his massive correspondence, and a planned Internet database provisionally called Euler Heritage will make accessible his remaining manuscript catalog, including his twelve notebooks totaling four thousand pages. It has taken more than a century to near the finish of the *Opera omnia*. The catalog of Euler's complex writings by Gustaf Eneström appears in *Die Schriften Eulers chronologische nach den Jahren geordnet, in denen sie verfasst worden sind* (The Writings of Euler chronologically arranged, according to the year they were published), which was published in the *Jahresbericht der Deutschen Mathematiker-Vereinigung* (Annual Report of the German Mathematical Society) from 1910 to 1913. Eneström lists 866 of Euler's publications, which include eighteen books. Today each has a number indicating its ordered place from the time of appearance annually in the Eneström index.[1] Thus, E88 stands for "Nova theoria lucis & colorum," which was printed in 1746 coming after his first 87 publications. The Euler Archive, directed by Dominic Klyve, Lee Stemkoski, and Erik Tou has made the originals of almost all of these writings available on the Internet.[2] One result of the scope and depth of Euler's research is that most scholarship on him is fragmentary, centering on one or another particular subject. The thorough investigation of Euler's correspondence and the subjects and reliability of his notebooks is at an early stage.

Euler publications and letters span five languages. Most are in Latin and French, some are in German and Russian, and Euler himself translated one book from English into German. Recent Euler scholarship, especially that published by the Mathematical Association of America, includes many articles and books in English. Research on Euler requires expertise ideally in at least the first three of these five languages and collaboration among scholars who together address all of them. The bibliography in the present work indicates that competence in Italian, Spanish, Chinese, and Japanese also helps.

The range, depth, and volume of Euler's work make it highly unlikely that any one scholar can master all the fields that he pursued. This comprehensive biography examines the known principal areas of Euler's work. In describing, explaining, and summarizing what Euler achieved, the present book is a scientific biography, but it is not a scientific treatise exploring his central concepts at length. An exhaustive treatment of Euler's accomplishments is not yet possible, for there is still much to learn about them. Instead, this book presents a synoptic study of the full scope of his research. It stresses the discovery of new information about contributions, the character of his colleagues and rivals, and the sources of problems. A future book might have a team of specialists with expertise from different fields write sections of essentially an anthology and be overseen by an editor to strengthen a coherent perspective. This initial comprehensive biography does not offer a social, cultural, political, or economic narrative.

By paying greater attention to Euler's correspondence and academic records than did earlier concise biographies, this volume hopes to begin to bring out Euler's distinct personality, to remove myths (about, for example, his dealings with Denis Diderot), and to note his relations with other great Enlightenment figures, particularly Jean-Baptiste le Rond d'Alembert and Frederick II. Each of these topics remains underdocumented and underexamined.

Two other vital sources exist: the minutes of the Petersburg and Berlin Academies of Sciences, along with the anniversary or jubilee volumes for dates marking multiples of a full or half century since Euler's birth or death—that is, 1907, 1957, 1983, and 2007. From 1897 to 1911, the minutes of the earlier Petersburg Academy of Sciences, mostly in Latin and French, were reprinted in four volumes; they were later annotated in the *Chronicles of the Russian Academy of Sciences* in another four volumes (2000–2004). The *Trudy* (Works), volume 17 (1962), describes all Euler documents in the Archives of the Russian Academy of Sciences. The present-day Berlin-Brandenburg Academy of Sciences and Humanities lacks minutes from the eighteenth century but in 1957 issued extracts, its *Registres.* These too merit more examination. The Euler jubilee anniversary volumes published in Basel, Berlin, and Saint Petersburg, along with recent critical studies published in Washington, DC, have added crucial information about Euler's life and work. An excellent account of all these primary and secondary texts is Gleb K. Mikhaĭlov's "Euleriana: A Short Bibliographical Note."[3]

In the investigation of the intellectual, social, and personal odyssey of Leonhard Euler, this biography follows a chronological order, and its chapters are broken down into sections that follow episodes from his life and research.[4] Of the settings, cultures, and circumstances that influence his studies,[5] the institutional is the foremost, involving royal financing, programs, and administration at the Berlin and the Petersburg academies. These offered libraries, opportunities for publication, the safety of the laboratory, foreign connections, and a measure of freedom.[6] The Royal Academy of Sciences in Paris, with its prestigious annual prizes, set out important topics and promoted competitions across Europe.

The sections of this biography emphasize a few significant problems, innovative and daring computations, and proofs. These draw upon basic concepts; new methods; disciplinary intuition; symbols such as e, π, and i that Euler invented or made standard; and guesswork. Readers who do not wish to review the technical steps in computations and proofs may proceed directly to results. The time divisions for chapters largely follow Euler's places of residence, major developments in his career, or times of war. The first Saint Petersburg period was when he laid the groundwork of his research; during his Berlin years, he attained the summit of his career when he presided over mathematical research and the transformation of the old geometric exact sciences into the modern mathematical sciences; his second Saint Petersburg stay, according to his grandson Paul Heinrich von

Fuss, was a time of "prodigious activity" when the degree of Euler's writing increased.[7] Although the preparation of many of Euler's books and articles came years before their actual dates of publication, this biography will date them according to when they were printed but occasionally also give the times of composition. The book generally follows the Gregorian calendar.

Biographies in the mathematical sciences have open and protean boundaries. In the development of mathematics, two principal interpretations exist: the Platonic, in which the subject advances independently of time and setting by the force of its interior logic; and the external, in which cultural, intellectual, political, social, and institutional forces shape and support the subject's growth or impede it.[8] But no such simple dichotomy alone can capture the diversity and nuances of a vigorous discipline. Prominent among other sources are aesthetics or beauty and beginning studies with the final proofs, computations, and other results of mathematicians.[9] Mathematics is the most certain and exact of all the sciences, and studies of its history benefit greatly from an interpretation that centers on the close and critical reading of original written sources. The rigorous examination of these materials can elucidate the methods that guided Euler's research and can identify persistent problems, the pattern of his breakthroughs, his orderly arrangement of fields, and the few errors that escaped his methods and near flawless intuition in mathematics.

Since Ernst Mach's *Die Mechanik in ihrer Entwicklung historisch-kritisch dargestellt* (The science of mechanics: a critical and historical account of its development), first published in 1883, it was often believed that Newton's *Principia mathematica* had provided the complete framework for classical mechanics, with the eighteenth century adding little. But René Dugas, Craig Fraser, Walter Habicht, Thomas Hankins, Gleb K. Mikhaĭlov, István Szabo, Stephen Timoshenko, Clifford Truesdell, Eduard Winter, and others have brought out the vigorous, inventive research during the Enlightenment. Upon Newton's *Principia mathematica* Euler and his competitors built a coherent foundation for classical mechanics, proposed novel procedures for solving problems, and successfully applied differential calculus.[10]

Truesdell, a master of six languages, including Greek and Latin, skillfully turned to a critical inquiry into Euler's writings and especially reestablished the significance of Euler's work in theoretical physics, editing five volumes of the *Opera omnia*.[11] For these he provided exemplary historical and scientific introductions on rational mechanics, elasticity, and fluid mechanics. Truesdell has also written *Essays in the History of Mechanics*, published in 1968, and founded the journal *Archive for History of Exact Sciences*.[12]

Although the importance of what was called the progress of the sciences to the Enlightenment is recognized, the history of science scarcely appears in general histories of the period or biographies of Enlightenment leaders associated with Euler. Thorough studies of his correspondence and much of his *Opera omnia* could illuminate points in common with the French Enlightenment and ideas distinctive to Berlin and Saint Petersburg. Neither Tim Blanning's impressive *The Pursuit of Glory: Europe 1648–1815* (2007) nor Robert Massie's *Catherine the Great: Portrait of a Woman* (2011) mention Euler, while David Fraser's *Frederick the Great* (2000) has only one sentence on him.[13] Thus, the present study of the life of Euler and the importance of his research seeks to remove a major gap in the history of science and of the Enlightenment.

Acknowledgments

At the time of my doctoral studies at the University of Chicago, I had considered writing a full-length biography of Euler. Saunders Mac Lane, my mathematics adviser, thought it not yet possible and recommended beginning with an investigation into Euler's research in calculus and mechanics. For many years Mac Lane, historian of science Allen G. Debus, and I continued discussions on the subject.

Over the past decade, various sources offered assistance. A grant from the Hitachi Foundation partly supported research on Euler. In 2002 the Euler Society was founded; it has developed under the guidance of Robert Bradley, Ed Sandifer, Dominic Klyve, Erik Tou, Lee Stemkoski, Fred Rickey, John Glaus, and Mary Ann McLoughlin. Ed, Rob, Dominic, and Sandro Caparrini have adroitly reviewed and added clarity to selected chapters of this book. John Glaus translated Euler's correspondence with Johann Kaspar Wettstein, d'Alembert's letters to Euler, and Voltaire's *L'histoire du docteur Akakia,* which ridiculed in the 1750s the ideas and character of Pierre Maupertuis, the president of the Royal Prussian Academy of Sciences in Berlin. In 2007 and 2009, with Ronald Brashear and Lilla Vekerdy as the heads of special collections, respectively, the Dibner Library of the History of Science and Technology at the Smithsonian Institution awarded me resident scholarships. Kirsten van der Veen helped with book searches. Leslie K. Overstreet, curator of the Joseph F. Cullman 3rd Library of Natural History, explained early bookmaking techniques and provided excellent quarters during the period the Dibner Library was closed for renovation.

Others have helped this project move forward. Martin Mattmüller, director of the Bernoulli-Euler Zentrum with the Euler Archive at the University of Basel sent information on Euler's Swiss origins, while he and Gleb K. Mikhaĭlov made incisive critiques about the content and style of portions of the present volume. Elena Polyakhova of Saint Petersburg State University composed abstracts in English of articles from the Russian Academy of Sciences Euler Conference in 2007, and Alexey Lopatukhin translated into English portions of the Kotek biography. Located in Euler's last residence, Saint Petersburg's School 27, which emphasizes literature and languages, provided a marvelous tour that included posters by students about Euler's life and work. Ioan James of the University of Oxford shared perspectives on Euler's health and on graph theory. Daniele Struppa of Chapman University and Klaus Fischer of George Mason University commented on Euler's early work, and Bonnie Fors of Heidelberg University encouraged the project. At the University of Maryland–College Park, the Engineering and Physical Sciences Library provided Euler's *Opera omnia*, while senior reference librarian Jim Miller gave vital assistance with computer searches; the McKeldin Library offered guest privileges. At the Catholic University of America, history department chair Robert A. Schneider provided release time for research, Dean L. R. Poos supported the project, Robert Hand looked at some initial translations from Latin, and library reference specialist Anne B. Lesher and staff members obtained through interlibrary loans works difficult to find. Katya Mouris and Martin R. Waldman helped format an early version of the bibliography while they were doctoral candidates in history at the Catholic University of America.

At Princeton University Press, executive editor Vickie Kearn has been stellar in encouragement and guidance, while Quinn Fusting and Nathan Carr have been most helpful in completing the project. From start to finish, Thomas R. West, formerly of the Catholic University of America, has carefully read each chapter and deftly appraised style. Nonetheless, mistakes that remain in the book are mine alone.

My special thanks go to family members. Recently Tobias Enke looked in Berlin for letters written by Johann Albrecht Euler, and Hubi "HK" Koizar discussed science fiction and astronomy. During the entire project my son John and daughter Anne have had great patience, and my wife Betty has, among many contributions, improved computer searches and manuscript programs.

Author's Notes

All English translations in the text and notes are, unless otherwise noted, the author's own or from the Euler Archive.

The names of the various science academies mentioned in this volume were subject to change over time as the academies underwent various reorganizations and renovations. Frequently they are referred to in the text with shortened names in English (the Berlin Academy; the Paris Academy; and the Petersburg Academy also referred to as the Russian Academy). Following are the names of the societies, with common English translations and applicable years:

The Berlin Academy

1700–44 Societas Regia Scientiarum (Royal Society of Science)

1744–1810 Académie Royale des Sciences et Belles-Lettres de Prusse (Royal Prussian Academy of Sciences and Belle-Lettres)

The Paris Academy

1666–1793 Académie Royale des Sciences de Paris (Paris Royal Academy of Sciences)

The Petersburg Academy

1725–1803 Academia Scientiarum Imperialis Petropolitanae (Petersburg Imperial Academy of Sciences)

1803–36 Académie Impériale des Sciences (Imperial Academy of Sciences)

Leonhard Euler

Introduction

In the 1720s the European Enlightenment began. This was to be an age of growth in all aspects of life at the time, including state centralization, industrialization, an expansion in overseas empires, population growth, larger military forces, increases in literacy, and expanded learning, notably with a passion for criticism and for major advances in mathematics and the natural sciences.[1] Historians generally date this period as lasting to 1789, the "short eighteenth century." The previous century had brought what has been called the Scientific Revolution, in which mathematics and celestial mechanics were the sciences par excellence. Would they continue to be paramount and their subjects transformed? Where might this occur? How, and why? By late in the century a general assessment came from outside the groups of mathematicians. In the preface to his *Critique of Pure Reason*, published in 1781 with a second edition in 1787, Immanuel Kant expressed the belief that new ideas in mathematics and the natural sciences testified best to the depth of Enlightenment thought.[2] What had happened in the years prior to justify this assessment? The year 1727 marked an important historical moment when, shortly after Isaac Newton died, Leonhard Euler began his distinguished career in Saint Petersburg.

In mathematics at the start of the Enlightenment, many expected few major new achievements or fundamental innovations. The seventeenth century—when most of the field's practitioners came from the aristocracy or from positions in medicine, law, or religion—was considered a golden age of mathematics; at midcentury René Descartes and Pierre de Fermat had each separately created what we now call analytic geometry, and the period culminated in the beginnings of differential calculus in the "method of fluxions" of Newton and the work of Gottfried Wilhelm Leibniz. Many thought that little of general importance was left to pursue.[3] But other scholars anticipated instead a fecund era not only in calculus, including the creation of its core branches, but also across mathematics in theory and application. Above all, the extensive research and writings of Leonhard Euler were to ensure that all of this would occur.

Driven by enormous energy, a passion for mathematics and the exact sciences, a commitment to building a strong institutional base for these fields, and an insistent defense of Reform Christianity, Euler diligently pursued an immense research, computational, and writing program across pure and applied mathematics and related technologies from his days in Basel onward except during a few bouts of severe fever. In calculus alone he provided hundreds of discoveries and proofs, along with many fearless computations to simplify and clarify techniques for differential calculus, infinite series, and integral calculus; he was the principal inventor of the core branches of differential equations, together with a semigeometric analytic form and later the analytic calculus of variations. In hundreds of articles and a calculus trilogy starting with the two-volume *Introductio in analysin infinitorum* (Introduction to analysis of the infinite, E101 and E102, 1748),[4] Euler identified foundations; methodically arranged, elaborated, and transmitted calculus; and set out the initial program for calculus's development. As a primary result of his studies, analysis displaced synthetic Euclidean geometry from its two-millennium primacy in mathematics and was the exemplar for reason in the *esprit géométrique* of the period. In pure mathematics Euler did more: he substantially advanced number theory and made headway in algebra, combinatorics, graph theory, probability, topology, and geometry, which included pioneering the differential geometry of surfaces. Drawn deeply also to the exact sciences of mechanics, optics, and astronomy, Euler made contributions across applied mathematics that were unparalleled in combined scope and depth.

Not since the second-century Alexandrian astronomer Claudius Ptolemy had a single geometer so dominated all branches of the exact sciences. Euler was the first to systematically apply calculus to rational mechanics, beginning in 1736 with his two-volume *Mechanica sive motus scientia analytica exposita* (Mechanics of the science of motion set forth analytically, E15 and E16). Prior to the time of William Rowan Hamilton, it was Euler, not Newton, who formulated most of the differential equations in mechanics. Euler founded continuum mechanics and in both print and correspondence led a talented group of competitors and rivals, including Daniel Bernoulli, Alexis Claude Clairaut, Jean-Baptiste le Rond d'Alembert, and Joseph-Louis Lagrange, in transforming into the modern mathematical sciences based on calculus what were then called the mixed mathematics of mechanics, geometric astronomy,[5] optics, dioptrics, acoustics, pneumatics, and games of chance—along with two major fields of physics, physical astronomy and cosmology.[6] Building upon Clairaut's research, Euler proved in theoretical astronomy that Newton's inverse-square law of gravitation by itself accounts for all lunar motion, a major confirmation of Newtonian dynamics. In the mid-eighteenth century Euler was unique in creating a mathematical language for the exact sciences that would stand for the next two centuries.

During the Enlightenment, Euler was crucial in helping to build European reputations for the new royal academies of sciences in Berlin and Saint Petersburg. Together with the academy in Paris and the Royal Society of London, they now surpassed the universities in scientific research. At the academies in Saint Petersburg and Berlin, Euler interacted beyond mathematics and the exact sciences in culture, economics, law, politics, religion, and society.[7] The transmission and refinement of his work often relied on correspondence, mostly transmitted through the postal service. Three of these academies were in the capitals of the rising powers of the eighteenth century at a time when France remained dominant. Frederick II of Prussia and Catherine II of imperial Russia assigned Euler royal tasks vital to the growth of commerce, trade, exploration, empire, and the centralizing state; these duties included developing a more exact astronomy, cartography, and geodesy. At the same time the goal was to advance the technologies of artillery, shipbuilding, bridge construction, and instrument building—especially that of clocks, thermometers, microscopes, and telescopes, all of which were essential to making discoveries in the sciences.[8] In Berlin Euler criticized Wolffian philosophy before Kant, argued for improving science education on the university level, and devised plans for state lotteries and pensions. A devout Protestant, he defended traditional religion against the skepticism of the *Encyclopédistes* and freethinkers, but he did not experience a crisis of conscience. In the dispute over the principle of least action, Euler gave to Pierre Maupertuis credit that Euler himself deserved as he questioned the new public sphere that was growing in Europe. In journals and other print, the new public challenged old standards set by state officials, censors, churches, and universities.[9] Euler's influence and the Enlightenment were to extend beyond Europe: he became one of the first foreign members of the American Academy of Arts and Sciences.

Chapter 1

The Swiss Years: 1707 to April 1727

Leonhard Euler was born on 15 April 1707 in Basel, the third largest city in the Swiss Confederation.[1] He was the first child of Basler Evangelical-Reformed minister Paul III Euler and his wife Margaretha (née Brucker). On 17 April, the Sunday before Easter, Leonhard was baptized in Saint Martin's Church as his father had been. The church's baptismal registry for the years 1663–1762 lists for the child three godparents, who were not relatives but city officials. He was named Leonhard after one of these godparents, city privy councilor Leonhard Respinger, a close friend of the family. Exactly where in Basel Leonhard Euler was born is not known. Usually women gave birth at home, assisted by a midwife; pediatricians, gynecologists, and most pharmaceutical medicines were as yet unavailable. Many babies born in winter suffered from respiratory illnesses, and children born in summer from gastric problems, but the infant Euler was spared both. Whether his birth occurred at his parents' home or possibly that of his maternal grandparents, the residence was likely located in Saint Martin's Quarter, the picturesque center of Basel. Since this part of the city was extensively renovated later, it is doubtful that the house remains standing.

"Das alte ehrwürdige Basel" (Worthy Old Basel)

In 1727, the year Euler left, Basel was a damp and misty city at the farthest point of navigation on the Rhine; it was the wealthiest in the German-speaking Swiss areas and known for its learning and piety.[2] The city was the capital of the Canton of Basel, one of the thirteen cantons that constituted the Swiss Confederation, having joined in 1501 mainly for reasons of self-defense. It had separated from the Holy Roman Empire, in which it had held the status of *Reichsstadt* (free imperial city). In Reformation Europe urban population growth usually depended upon immigration. Except during a few periods of large losses from plagues, Basel had a

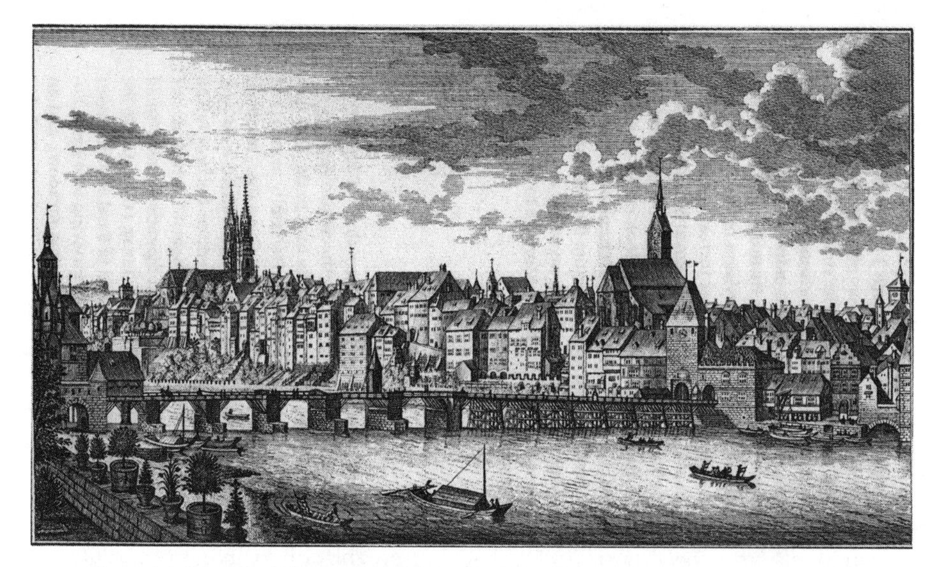

Figure 1.1. Basel in 1761 as viewed from the Rhine River. A copperplate engraving by David Herrligberger (1697–1777), following a pen-and-ink drawing by Emanuel Büchel (1705–1775). Saint Martin's is the church on the right.

restrictive policy on immigration and citizenship; as a result its population had decreased from about twenty thousand in 1500 to seventeen thousand in the early 1700s. Max Weber's thesis that Protestant religions rose in parallel with capitalism and the modern sciences could look to Basel for reinforcement.[3] Following its rejection of the Catholic faith in the 1520s, the city became a great center of northern humanism, publication, and the book trade, principally through the efforts of its intellectually inquisitive silversmiths, ribbon and fine paper makers, printers, and engravers. The printing presses developed there produced major works in the new sciences, publishing in 1566 the second edition of Nicolaus Copernicus's *De revolutionibus orbium coelestium* (On the revolutions of the heavenly spheres).

The year 1522 saw the beginning of the period known as the *Herrschaft*, when the city council came under the oversight of the two most powerful groups in the city, the artisans in the eighteen *Zünfte* (roughly, trade guilds organized by the merchants and manufacturers) and the *Gnädige Herren* ("gracious lords," consisting of the old nobility and the church establishment). Income distribution was quite uneven. The political leaders of the city, the *Ratsherren*, or senators, were the members of the council. The trade guilds opposed the *liberum commercium*, or freedom of trade across boundaries, which the *Herren* promoted. With city warehouses to hold finished goods—particularly silk ribbons, *passementerie* (trimmings), and, to a lesser degree, fine printed cottons—the merchants made Basel a hub of what was termed the *putting-out system* of manufacturing. Taking advantage of cheap labor in the countryside and acting essentially as middlemen, they provided rural weavers raw materials and the most advanced looms available. The weavers completed the textiles and took them to the new merchant warehouses in Basel, where the merchants bought the finished textiles at meager piecework prices. They were thus able to avoid the restrictions of the city's guilds, and this greatly increased productivity and lowered costs. At the time, silk ribbons were a far more important item in the garment industry than they are today; the fortunes of the *Bändelherren,* or ribbon lords, became the foundation for a private banking system in the region. As a result, Basel had transformed itself from a medieval guild city into a commercial, manufacturing, and banking center.

While its sister city of Zurich had Ulrich Zwingli as its hero and Geneva had John Calvin, early sixteenth-century Basel had the Christian humanists Oecolampadius (Johannes Heussgen) and Desiderius Erasmus.[4] At the time scholars customarily took as pseudonyms classical versions of their original appellations or names based on admired concepts; thus Oecolampadius, of Greek origin, which means "house shine." Oecolampadius had studied at Tübingen, Stuttgart, and Heidelberg, absorbing the theories of Johannes Reuchlin, Philipp Melanchthon, and Wolfgang Capito. In 1522 he moved to Basel as vicar of Saint Martin's Church and, as a talented philologist of Greek, Hebrew, and Latin, was a proofreader for a translation of patristic literature. The next year he also became a reader of scripture at the university. Oecolampadius was among the first advocates in Europe for the movement away from Latin to the vernacular in churches and academia. In church services he preached in German, while another university faculty member in Basel, the cantankerous Paracelsus,[5] defied academic tradition by lecturing in German.

The most abrasive of the reformers transforming natural philosophy, Paracelsus also sought to overturn Aristotelian philosophy and replace it with Christian Neoplatonism and with Hermetic mysticism.[6] Using these he worked to uncover truths in the two so-called divine books: the Bible and, metaphorically, the book of nature. For this enterprise he employed what was known as *magia naturalis* (natural magic) and cosmic sympathy, the latter being the influence of the stars on human life and also known as macrocosm-microcosm interpretation. In chemical science, Paracelsus proposed that his three principles—ideal forms of salt, sulfur, and mercury—were not matter but fundamental states of it, and he accepted quantification, though he adopted sidereal mathematics to indicate when to give medicines.

Paracelsus objected to mathematical abstraction as a way of understanding chemical science and based his research consistently on observation and experiment. His thought was a mixture of the mystical and the evolving modern scientific during this time when the meaning of science was shifting. He attacked the dominant Galenic and Arabic herbal medicine traditions retained from antiquity and the Middle Ages and argued for the superiority of medical chemistry, especially the use of mercury.

In the early 1500s the printer and publisher Johannes Froben asked Oecolampadius to assist Erasmus in translating the New Testament. From 1522 on, Oecolampadius—who, like Froben, protested ecclesiastical abuses in the Catholic Church—demanded in Basel a complete reformation, a total Protestant break with the church. A series of riots ensued, with mobs seeking to remove images from churches. Fearing that change of any sort might escalate to revolution in concert with the peasant revolt that was sweeping the central and southern German states and the Alpine region of imperial Austria in 1524 and 1525, the city council hesitated to act. But in 1529 it forced out the Catholic bishop and instituted a new Protestant church order.

From the time of the Reformation, Basel offered a place of refuge for well-to-do religious dissidents from northern Italian city-states, France, Alsace, Holland, the Spanish Netherlands, and neighboring German-speaking lands. The government in Basel did not accept the poor and the indigent; only immigrants having the *kunstreiche und wohlhabende* (skills and capital) to expand commerce and craft in their adopted home were welcomed as citizens, albeit for a substantial fee. Most of the new Basler *Herren* were widely traveled, spoke several languages, and were shrewd businessmen. Basler merchants bought, sold, and speculated in every major European market; they had contacts that by the early eighteenth century reached to Amsterdam, Bordeaux, Le Havre, London, Paris, Saint Petersburg, and Vienna. This network provided critical information about commercial conditions and opportunities.

During the seventeenth century the Euler family members were Baslers, citizens of the city-republic of Basel, the *regio basiliensis* (as Latinists termed it). The regions of the Swiss Confederation were highly independent; no federal Swiss state existed before the nineteenth century, and loyalties were to cities and cantons. Its geographic, economic, linguistic, and cultural connections made cosmopolitan Basel more a European city than a Swiss one.

Lineage and Early Childhood

Over time the Euler family name has had a variety of spellings, with Öwler and Äweler in the fourteenth and fifteenth centuries and Öwbler, Ewbler, and Ouwler in the sixteenth.[7] Since at least 1287, when the first written record of an earlier form of the name Euler appeared, family members had lived on the far side of what is today known as the northern Swiss Alps, in Lindau on the Bodensee (Lake Constance) in Bavaria.[8] But a continuous residence for them did not begin there until 1458. The family bore a double name, today often written as Euler-Schölpi. The name Euler refers not to an owl but to *Äule*, meaning a modest wet field or swampy meadow. In German regions, an *Äuler* (pronounced "oyler") was a land proprietor. This suggests that some of Euler's ancestors were landowners and tillers of the soil. *Au*, shorthand for *Äuler*, is part of the name of many small towns, such as Lindau, Dessau, and Nassau. The belief that the family name comes from the word for small pottery from Roman times (*olla* in Latin) transformed into the middle-high German *Aul* and then to *Aulner* and *Iller* is incorrect. Schölpi, which derives from the words *schelb* and *schief*, related to *schielen*, signifies squint-eyed, cross-eyed, or crooked. (Figuratively it can also refer to a small-time cheat or rascal.) The epithet, which became the second part of the family name and had been spelled in different ways, seems to suggest that the Eulers had a susceptibility to an eye malady.

As a *Reichsstadt* in the Holy Roman Empire, Lindau had for a long time enjoyed close economic, political, diplomatic, and religious ties with Basel, and the two are two hundred kilometers apart. In 1594 Hans-Georg (or Jörg) Euler, the great-great-grandfather of Leonhard, moved to Basel and became a full citizen at the age of twenty-one. The reasons for his move are unclear, but the death of his mother was possibly a factor. There were also earlier family contacts, among them the Basel shoemaker Martin Frisschmann, who died before 1582. Hans-Georg had trained to become a brush- and combmaker. As the youngest son he had to seek a new profession, because his oldest brother would succeed to his father's position in the guild, and only one place was available. In Basel Hans-Georg pursued the combmaking occupation and became a tradesman belonging to the hospitality or public house guild. He now dropped the second part of the family's double name and was to become the scion of its Basel branch. Hans-Georg married twice, fathering nine children in the first marriage and six in the second.[9] Only four of his eleven sons would continue the family name: Remund II, Hans-Georg II, Paul I, and Johann Jakob.[10] Hans-Georg the elder lived to be ninety. In Lindau most of his ancestors had been small landowners and vintners. Hans-Georg's oldest son, Remund II, broke with him in 1631, when the father would not surrender to him his master's position in the guild. Remund, whose father-in-law was a book dealer, may have been the first Euler in Basel to send his sons to school. Not Paul I but Hans Georg II the Younger was the great-grandfather of Leonhard Euler; Hans's son Paul II was Leonhard's grandfather.

Most male members of the first three Basler generations engaged in brush- and combmaking. None achieved distinction beyond the confines of his community, but each built a modest financial base. In the fourth generation in Basel, two of the fourteen male cousins became combmakers, but it was now possible for five of the others, including Paul III, to study to be Basler Evangelical-Reformed ministers.

The Basler Evangelical-Reformed Church adhered to a blend of Protestant beliefs, distinct from the Calvinist, Lutheran, and Zwinglian faiths, which in the eighteenth century developed into Pietism. It was thus to be a dissident current in Protestant faiths that often conflicted with state churches. Although he was not a prominent theologian, as other founding reformers were, Oecolampadius excelled in public disputations, and he briefly assisted Zwingli. He was the first to want congregations, not an episcopal hierarchy or clergy, to constitute a secular magistracy representing their church—a position that initially distressed the Basel city council and Martin Luther, but won through the efforts of Calvin in Geneva. Oecolampadius professed the primacy of the Gospel and defended the view held by Zwingli and Luther of the metaphorical nature of the Eucharist. Rather than stress salvation through Christian faith alone, as Luther did, Oecolampadius emphasized the working of the Holy Spirit. Nor did he examine in depth the concept of predestination. He agreed with Zwingli, who was called the Cicero of his age, on esteeming humane learning but in one respect went further, rejecting Zwingli's ban on organ music in church services. Oecolampadius also stressed the education of youth; children had to receive Sunday school teaching from the *Wochenkinderlehre* (weekly children's instruction) booklet. By the time of Leonhard Euler, religious education would be based on the catechism of the Basler theologian Samuel Werenfels, the *Nachtmahlbüchlein* (Nighttime pamphlet), which included prayers and referred to those who had come before in Holy Communion and had left a rich legacy.

Paul III Euler, the father of Leonhard, bore the same name as his father. The name Paul II Euler appears in the registers of students matriculating into the University of Basel in 1654, but no record of his studies has survived. Like his forebears, Paul II was a brush- and combmaker. In 1669 he married Anna-Maria Gassner, the daughter of a pastry baker, who had emigrated from Vöcklabruck in Upper Austria near Lake Constance; the next year Paul III was born.

In 1685 Paul III moved away from the artisanal profession of his father, entering the University of Basel for general studies in the philosophical faculty preparatory to a major in Protestant theology. The university had existed since 1460. Two years earlier its founder, a lowly clerk in Basel, had been elected the humanistic Pope Pius II.[11]

During his first semesters Paul III had to study mathematics. In October 1688 he defended, in a debate chaired by Jakob (Jacques) Bernoulli, Bernoulli's thesis "Positiones mathematicae de rationibus et proportionibus" (Mathematical positions on ratios and proportions), which contained fifteen theorems and postulates. The thesis was published in 1688 and again in the posthumous collection edited by Jakob Bernoulli's nephew Nikolaus II in 1744. It begins with algebra, rather than geometry, as basic to mathematics and stresses general methods; both were approaches Paul III was subsequently to convey to his young son. By 1769 the subject of this debate would be a section (or chapter) 3 of Leonhard Euler's *Vollständige Anleitung zur Algebra* (Complete elements of algebra).[12] In addition to Jakob Bernoulli, Paul likely knew his younger brother, fellow student Johann I Bernoulli.

After defending his senior thesis, Paul III Euler turned to theology and in 1688 received his master of arts degree. In 1693 at the age of twenty-three, he completed theological studies at the University of Basel, becoming Sacri Ministerii Candidatus, the requisite for entry into the ministry. Yet, perhaps because of an excess of pastors, he did not take up his first position until 1701, at the Basel orphanage; two years later he became pastor at Saint Jakob's Church on the River Birs. That position was an improvement, but both posts came with only a meager salary and no parsonage. The small Saint Jakob's Church still exists today.

Possibly owing to a lack of income, Paul did not marry Margaretha Brucker, the daughter of the local hospital's minister, Johann Heinrich Brucker, until April 1706. About her we know almost nothing. In the Basler Protestant faith, the ability to read was essential, especially to study the Bible. Margaretha was thus taught to read, and she studied Greek and Roman classics. While Paul Euler was a *homo novus*, a new man lacking a pedigree of notable academic predecessors, his wife's family stemmed from a distinguished line of humanistic scholars. These included the Latinist Celio Secondo Curione (d. 1569), the jurist Bernhard Brand (d. 1594), the Hebraist Johannes Buxtorf (d. 1629), and members of the learned Zwinger dynasty, including Theodore and his son Jakob (d. 1588). A year after Paul and Margaretha's marriage, the first of their four children was born.

Leonhard Euler was to spend his beginning youth not in Basel but in the nearby Swiss countryside. In early June 1708 Bonifacius Burckhardt, the pastor at Saint Martin's Church in nearby Riehen, died after a three-year illness following a stroke. Three weeks later Paul III Euler was named pastor of Riehen-Bettingen, a position that he held until his death in March 1745. During his pastorate the congregation developed Riehen Pietism, stressing not the main current in Protestant thought, obedience to orthodox theology, but the Christian inner life of rebirth, brotherly love, and living belief.[13] The small community, the collegia pietatis, centered on Bible study and group prayer. Along with the improvement of catechism came church music, prayer, and mission—especially service on behalf of youth expressed through the establishment of schools and orphanages. To supplement his nominal income as pastor, Paul taught lessons at the religious school for children. That was part of a pastor's normal duties. Until his death he was active in multiple pastoral duties and was loved by his congregation.

Riehen was a small village located near the valley of the Wiese, a tributary of the Rhine, about 5.4 kilometers northeast of Basel and halfway to Lörrach. Located closer to the smaller Wiese, it is today a residential suburb of Basel. Supported by a more temperate, sub-Mediterranean climate, the region had rich vegetation appearing as early as February and lasting through the fall. It was especially known for the striking white blossoms of its cherry trees and the leaves of gold and red on a variety of grapes in the vineyards, as well as for its plum trees and wines. It is here that Leonhard Euler probably developed his interest in plants and trees, as evidenced later in Berlin and Charlottenburg in his gardening and the task he assumed of ordering trees for the Royal Prussian Academy of Sciences in Berlin. In 1693, Riehen and Bettingen combined had about fourteen hundred inhabitants and two hundred households.[14]

In November 1708, when Paul was installed as pastor, the Euler family, including Paul's mother Anna Maria, moved to Riehen. Domestic conditions were modest: the family of six lived in a cramped two-room parsonage .[15] One room served as a study, the other as general living quarters. The parsonage exists to the present day; it was enlarged in 1712 for the Euler family and further expanded with the addition of a third floor in 1851.

Leonhard grew up with two younger sisters; Anna Maria was born in 1708 and Maria Magdalena in 1711.[16] His paternal grandmother remained with the family until her death in 1712. Most likely there were other births and infant mortality. The fourth surviving child, Johann Heinrich, named for his deceased maternal grandfather, was born in 1719 after Leonhard departed for studies at the Gymnasium in Basel; he would later live with

Leonhard in Saint Petersburg. The walk from Riehen to Basel took about an hour;[17] as Euler grew, he likely accompanied his mother on visits to relatives, and perhaps he occasionally went on foot from the village to school in the city.

An eighteenth-century contemporary told a rather charming story, first put in print by Johann Wernhard Herzog in his *Adumbratio eruditorum basiliensium* (Sketch of Basel scholars) of 1780. Clearly completely fabricated, it claims to involve Euler at about age four, and is not unlike generic stories of curious boys that can be found in folklore. As the story goes, in the late afternoon Leonhard's parents gave him the task of collecting eggs. When he had been gone for hours and failed to appear for dinner, they became worried. After searching, they found him in the corner of the chicken coop. He had observed the hens laying eggs and sitting on them to hatch chicks. So he supposedly gathered from the nests the eggs for the day, placed them in the corner of the coop, and was sitting next to them. Asked what he was doing, he replied that he wanted to produce young chickens. The *Adumbratio* included outlines of the lives and achievements of Euler and his sons;[18] in correspondence with his father in 1781, Nicholas Fuss, who was then living with Euler, related that Herzog's work also included other made-up "facts."

On the character of Euler his biographers agree. A talented child, he was open, cheerful, and sociable. Fuss's eulogy declares that the simplicity of rural life and the example of his parents contributed greatly to the forthright nature and generally even disposition for which the adult Euler was known.[19] His *bürgerliche* (lesser bourgeois) family was not wanting materially, and instilled in him the values of family closeness, individual freedom, and attention to financial security. These were the virtues of a free burgher in the remarkable Swiss Confederation in which—while disparities in wealth were great—no absolute monarch supported by the nobility ruled atop a people arranged in a hierarchical social pyramid. Even after he left Basel and went to Saint Petersburg and Berlin, Euler conducted himself as a middle-order professional rather than a courtier. This was crucial in the transition during the eighteenth century from the vagaries of patronage toward a professional status for the sciences.

Occasionally young Euler could erupt in an argument, but when he calmed down he laughed at his own behavior. On one issue for him there was no room for debate—the question of Christian belief. From his parents, grandmothers, and reform Protestantism in Riehen and Basel, Euler imbibed religious convictions from which he never wavered. This explains the intensity of his opposition to such critics of traditional religion as the Wolffians as well as freethinkers and French Encyclopedists, notably in his apologetic *Rettung der göttlichen Offenbahrung* (Rescue of divine revelation, E92) of 1747.

At this time it was not only the nobility that had coats of arms or escutcheons; so did the lesser orders of society. It was probably only Leonhard, not the entire family, who was later awarded an insignia of a unicorn in a field of blue.[20] He also had several others, including insignia bearing his initials under a crown, with a shield in the background reflecting the initials; a mythical beast; and the unicorn. The coats of arms were important for signet rings used to seal letters.

Leonhard Euler's parents were his first teachers. Although it is unlikely that she was versed in the Greek and Latin of the humanistic tradition, his mother, Margaretha, is thought to have instructed him in beginning reading.[21] The elementary education that his father gave included mathematics. Like his teacher Jakob Bernoulli, Paul taught his young son mathematics not as an isolated discipline but as underlying all natural knowledge, thickly interrelated with other fields. To teach Leonhard the first principles of mathematics, Paul selected an algebra rather than a geometry textbook: the leading sixteenth-century German cossist Michael Stifel's second edition (in 1553) of the Silesian Christoph Rudolff's *Coss* of 1525. (The German word *Coss* is the equivalent of the Italian *cosa*, "thing," here roughly indicating "something unknown."[22]) Euler was to have a copy of the 1553 edition into the 1740s that was possibly a 1615 reprint. German cossists had undertaken to explain arithmetical laws in some generality, to solve equations still largely in verbal form, and to apply appropriate methods to resolve such problems as finding square roots. While Rudolff improved upon notation, using for example the plus (+) and minus (−) signs, he still represented powers to nine by special symbols, such as *z* for *zensus* (2) and *c* for *cubus* (3), without indicating the relationship x^2 and x^3.

Stifel's enlarged second edition of Rudolff's *Coss*, the first comprehensive text on algebra in German, included many annotations from Stifel, who had expanded the two-part text from 208 to 484 pages. His additions included replacing a number of rules for special cases with a general method. At Leonhard's age, the Stifel book was extremely difficult. The twelve chapters of part 1 cover materials to be mastered before studying algebra, including place-value notation, the basic arithmetical operations, root extraction, and summation formulas for arithmetic and geometric progressions, along with the three types of roots of equations that the book recognizes—rational, irrational, and communicant (the third of these being two roots sharing a common rational factor). In his *Arithmetica Integra* of 1544 Stifel, who was skilled in computing with irrational numbers, had stated that "an irrational number is not a true number"; he operated with them, but did not offer an exact definition.[23] Part 2 of the *Coss* gives rules for solving first- and second-degree equations and then sets out 434 problems. It closes with three cubic equations but not their solutions, which were unknown to Rudolff. These three equations were intended to promote further research on their solution. Euler's short, incomplete autobiography, dictated in 1767 to his oldest son, recalls how he diligently studied the entire *Coss* for several years before he entered the Gymnasium in Basel, which was a Latin school, and making progress in solving its problems. Possibly a few solutions were added to his copy by his father or others. Still, at just seven years of age, only an exceptional child could have hoped to make it through such a difficult text.

After Simon Stevin and François Viète's new notation and John Napier's invention of logarithms surpassed its knowledge, Rudolff's book had limited use. Young Euler seems to have been demonstrating himself to be a child prodigy, although the term was not applied to him as it was later in the century to Wolfgang Amadeus Mozart in music and Carl Friedrich Gauss in mathematics. Yet while young Leonhard's problem solving was limited to elementary algebra, Paul Euler must have recognized his son's unusual aptitude for mathematics more generally, though he possibly considered it mainly a diversion. Despite having introduced mathematical reflection to Leonhard's fertile mind, he wanted his son to study Evangelical-Reformed theology and become a rural pastor.

Formal Education in Basel

With Leonhard's required instructional preparation completed, his parents—seeing that he had extraordinary intellectual ability—sent him to Basel, perhaps as early as the age of eight, where he lodged at the home of his widowed maternal grandmother, Maria Magdalena Brucker-Faber. This eliminated the need for daily travel from Riehen. Initially this was probably only during the week. Euler must have been delighted to be out of the cramped one-room family living quarters and to have a room to himself, or perhaps with a cousin. A few Brucker grandchildren likely also resided with their grandmother; the parents would likely have contributed eggs, vegetables and fruits from their gardens, and meats.

By the second decade of the 1700s Basel was long past its golden period, but the work of Jakob and Johann Bernoulli nonetheless made it a center for scientific and mathematical research. Reform Protestant faiths and the *vernünftige Orthodoxie* (enlightened orthodoxy) movement that combined Calvinism, biblical humanism, and Erasmian criticism were familiar to many of its well-educated citizens. At the start of the seventeenth century, the Calvinist Bernoulli family of merchants had moved from Amsterdam to Antwerp. During the Thirty Years' War, they then had to flee Antwerp to avoid persecution by Catholic Spanish forces, and the Bernoullis reached Basel by 1622. After Louis XIV of France revoked in 1685 the Edict of Nantes supporting religious toleration, the influx of Calvinist immigrants to Basel again briefly increased. Yet even though these immigrants helped to revitalize the city, it banned immigration after 1706 except for that of a few very wealthy individuals.

Euler began his formal education at the Basel Gymnasium. The cultural and intellectual energies of Basel seem to have bypassed the school. Documents of the time depict it as being in a pitiful state; students could learn little more than the Latin language and selections from its classical literature as well as optional Greek. Teaching was conventional; the pedagogical reforms of Johann Pestalozzi and the increase of mathematics and sciences in the curriculum promoted by Christoph Bernoulli were not to come until early in the nineteenth century.[24] In Euler's day, the rod was not spared: schoolboys were sometimes beaten or knocked about until they bled.[25] At other times the teacher might find himself dragged by the hair by an enraged father making an unannounced visit to the class. Fistfights broke out in the classroom. There is no reason to suppose that Euler's intelligence and demeanor saved him from routine indignities and brutalities.

Many parents engaged private tutors for further instruction. Most of these were students from the university, and many of them combined enthusiasm for learning with a sensitivity toward students. In 1715 Paul Euler hired as a private tutor Johannes (Johann Jakob) Burckhardt, a twenty-four-year-old theologian with a tolerable background in mathematics. At the time Burckhardt, who would become pastor in Kleinhüningen in 1721 and Oltingen in 1732, was supporting Johann Bernoulli in arguments with Brook Taylor and other members of the Royal Society of London over which was superior, Gottfried Wilhelm Leibniz's differential calculus or Isaac Newton's method of fluxions. This was also the time of the controversy over whether Leibniz or Newton deserved priority for the invention of calculus. Bernoulli publicly refused to engage in the priority dispute but was deeply involved to the point of faking evidence. In 1713 Leibniz had released a letter from Bernoulli, without giving his name, that argued against Newton's interpretation of the higher differential. Bernoulli despised the "scurvy English" and "English buffoons" for accusing Leibniz of plagiarism. Afterward Bernoulli demonstrated that for finding ballistic curves Leibnizian differential calculus was better than

Figure 1.2. The University of Basel in the seventeenth century. The numeral 1 indicates the lower college.

Newton's fluxions. Burckhardt must have influenced Euler's education, but exactly how is not clear. Reporting his death, Daniel Bernoulli, who became Euler's closest friend for many years, referred to him simply as "magni Euleri praeceptor in mathematicis" (teacher of the great Euler in mathematics).[26] Young Euler displayed diverse interests; his concentration on mathematics still lay in the future. Yet to have taught mathematics to Euler remains Burckhardt's greatest achievement.

The University of Basel in the Seventeenth Century

After completing his gymnasium education, in October 1720 Leonhard registered at the University of Basel for courses in the philosophical faculty, which covered fields of learning outside recognized professions. It was the equivalent of the modern secondary school.[27] The new student was thirteen. To enter a university at that age was not unusual for the time, and admission was a civic right of the citizens of Basel. Long in decline, the university was little more than a benevolent institution for townsmen. Its deterioration had culminated in the election of professors by lots, a system initially instituted to reduce corruption in the selection process and give candidates from nonpatrician clans a better chance at employment. But it was still often difficult for the best candidates to win, and through this lottery system able younger candidates could be passed over in favor of older, less-qualified ones. Having enrolled more than a thousand students a year in the early seventeenth century, by the start of the eighteenth the university's higher faculties—religion, medicine, and law—matriculated twenty to forty students per year. In 1720 the higher faculties numbered only nineteen professors and a little over a hundred students.[28] No foreign medical students matriculated, and not enough law students (generally under six) attended to necessitate the scheduling of bar examinations. Except for Johann Bernoulli, who had succeeded his brother Jakob in 1705, the university's badly underpaid faculty was mediocre and instruction was poor.[29] Bernoulli single-handedly made the university a European center for mathematics. In general, the professors and the university officials, who with few exceptions were viewed as less dogmatic, sought to inculcate religious precepts and morals as a matter of state policy. At the university, the seat of learning in the generally open-minded Basel, deviation from orthodoxy led to dismissal.

The philosophical faculty imparted a general education to a student before he chose a specialty for a higher degree. Through hard work and an astonishing memory, Euler mastered all of his subjects. During his first two years he was enrolled in Johann Bernoulli's class for beginners in geometry as well as practical and theoretical arithmetic. Before his fellow students, the fourteen-year-old gave a speech in Latin titled "Declamatio: De arithmetica et geometria" (Rhetoric: On arithmetic and geometry) in which he presented a treatment of the two branches of arithmetic, the practical and the theoretical. In practical applications and the fine arts, he commended the superiority of geometry.

Euler spent two years earning his *prima laurea*, roughly equivalent to a bachelor's degree, receiving it in June 1722 at the age of fifteen after presenting in a public speech in Latin, the *ad lectiones publicas*, his undergraduate thesis, "De temperantia," extolling moderation. In conversations and intellectual debates Euler began to display to a wider public his substantial learning and his command of Latin. As a result, the precocious youth was appointed three times in 1722 to exchange views publicly about papers, twice by candidates for a professorship at the university, in logic in January and in jurisprudence rooted in the history of Roman law in November. The candidate authors of the first paper, "Positiones logicae . . . ," are listed as J. J. Battier, J. Burckhardt, and J. R. Iselin.[30] The third instance was for a student earning a master's degree.

Euler was to remember for life the subjects that he studied in the classical humanities, and his correspondence suggests that until he became blind in his later years he continued reading in them.

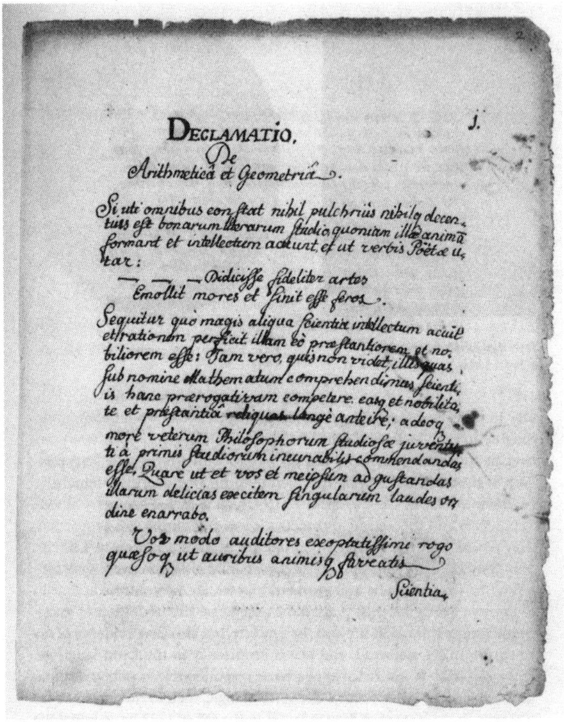
Declamatio. 1.
De
Arithmetica et Geometria.

Si uti omnibus constat nihil pulchrius nihilq decentius est bonarum literarum studio, quoniam illa animū formant et intellectum acuunt, et ut verbis Poëtae utar:

— — — Didicisse fideliter artes
Emollit mores et sinit esse feros.

Sequitur quo magis aliqua scientia intellectum acuit et rationem perficit illam eo praestantiorem et nobiliorem esse: Jam vero, quis non videt, illis quas sub nomine Mathematum comprehendimus scientiis hanc praerogativam competere, easq et nobilitate et praestantia reliquas longè antecire; adeoq more veterum Philosophorum studiosae juventuti a primis studiorum incunabilis commendandas esse. Quare ut et vos et meipsum ad gustandas illarum delicias excitem, singularum laudes ordine enarrabo.

Vos modo auditores exoptatissimi rogo quaesoq ut auribus animisq faveatis.

Scientia,

Figure 1.3. The first page of fourteen-year old Euler's Latin speech "Declamatio de Arithmetica et Geometria" presented before other university students.

While he advanced rapidly in lower-level mathematics courses, Euler did not quickly gain the attention of his teacher Johann Bernoulli. Engaged in substantial research, disputes, and school reform in Basel, the fractious Bernoulli, witty when at ease but as a younger adult known for his Flemish pugnacity, devoted little time to students in undergraduate classes, though he did offer public lectures for beginners each workday at two in the afternoon. The course topic for 1720–21 was geometry; for 1721–22, theoretical and practical arithmetic; for 1722–24, selected subjects in the applications of geometry; and for 1724–25, astronomy.

A report on Bernoulli as a teacher in these courses comes from later in the summer of 1733, when Johann Jakob Ritter of Bern and Johann Samuel König enrolled in his course on theoretical and practical arithmetic. Apparently it had changed little from Euler's time; although Ritter and König were not novices in mathematics, Ritter relied upon the assistance of König in order to follow Bernoulli's concise lectures.[31] In order to benefit most from Bernoulli's class, students needed some grounding in algebra,[32] yet as Bernoulli grew older he displayed an increasing unwillingness to teach courses in elementary algebra and geometry, offering them "with the greatest resentment"; he instead spent his time immersed in the new calculus. Yet even the painful onset of gout did not keep him from his teaching. According to Ritter, Bernoulli "could entertain a crowd in class with his clever ideas" and "was very generous, often giving needy auditors funds from his own lecture fees."[33] Confronted by Bernoulli's way of teaching, Euler probably tended to skip algebra and geometry lectures soon after the first few and deplored the poor level of course offerings. His reaction to dry lectures was not unlike that of Charles Darwin, Albert Einstein, and Bertrand Russell.[34]

In the autumn of 1723 Euler passed the examination of the philosophical faculty for the *Magister Artium* (master of arts) degree, and he officially received it on 8 June 1724 at the age of seventeen. His classmate Johann II (Jean II) Bernoulli, who was three years younger, also earned his master's degree. At the graduation session on that day, Euler gave a public lecture in Latin comparing the natural philosophy of René Descartes with that of Isaac Newton and indicating the consequences of each. It was probably in consultation with Burckhardt and Johann I Bernoulli that he had chosen this important and timely topic for his master's lecture. An intense rivalry, centering at the Royal Academy of Sciences in Paris, was newly under way between the Cartesians and the Newtonians for supremacy in the sciences in Continental Europe. Against criticism from both Catholics and Protestants that its animate automatons and mechanistic philosophy led to fatalism, Cartesian science had come, late in the seventeenth century, to dominate at European universities, supplanting varieties of Aristotelian thought.

Descartes's *Discours de la méthode* of 1637 set forth a rationalism, founded in radical, exhaustive doubt, that he intended as a program for pruning away unfounded beliefs and discovering new and reliable knowledge. As a truth invulnerable to doubt, he posited "Je pense, donc je suis," which was expressed in Latin as "Cogito, [ergo] sum" and is generally translated as "I think, therefore I am." From that one assurance, Descartes held that he could proceed to expand the range of factual knowledge, achieving results that would stand up to even the most severe doubting. His new epistemology had revolutionary implications; the Cartesian rational method surpassed the Aristotelian syllogism designed as a means of confirming knowledge. Descartes freely used deduction but drew little from mathematics.

Descartes's *Le monde*, with its section titled *Traité de l'homme*, was completed by 1633, and his *Principia philosophiae* followed in 1644; together the two works gave a comprehensive account of his mechanistic natural philosophy. In his universe, the essence of matter was extension, and the fundamental phenomenon was matter in motion. He posited that motion imparted by God breaks matter into three gradations: agglomerates of the largest size form the planets; the middle size comprises liquids, fire, and gases, including the atmosphere; and the smallest, the most minute pieces found in the interstices, are components of heat and light. These most subtle pieces fill the ether throughout space, making his universe a plenum. His corpuscles of matter differed from the atoms of the ancient Greeks by being infinitely divisible. Rejecting the occult qualities of the scholastics, Descartes held that a change in motion could be caused only by a direct mechanical contact between bodies, which demands establishing the laws of collision. He maintained that the quantity of matter multiplied by velocity, his measure of force or momentum, was conserved. Central to his science is the theory of impact. Descartes's refusal to accept the vacuum, and action at a distance, put him at odds with the mechanics of Galileo Galilei and Johannes Kepler. Largely because he denied any void, Descartes rejected absolute time and space. Like his predecessors, he separated terrestrial from celestial physics. In his theory, vortices—whirlpools of ether—of different sizes and speeds filled the heavens, accounting for celestial motions including, with difficulty, Kepler's three laws of planetary motion. In his books Descartes's physics was largely qualitative, but his correspondence, particularly with Marin Mersenne, was directed toward an emerging mathematical physics.[35] His main illustrations for his method in the *Discours* had been mathematical in dioptrics and meteorology as well as employing analytic geometry. Within Descartes's natural philosophy, dualism meant that mind and matter were separate, only connected in the pineal gland.

Newtonian science is based primarily on the *Principia mathematica* of 1687, along with the *Opticks* of 1704. Newton's *Principia* gave his three laws of motion and incorporated into a general dynamics—a science that he systematized—Kepler's three planetary laws and Galileo's law of free fall. For the force that was in opposition to Christiaan Huygens's centrifugal force (pulling away from the center), Newton coined the word *centripetal* (moving toward the center). While Aristotelians defined rest as the only natural state, Newton added the preservation of motion in a straight line; he rejected the Cartesian concepts that space and time are relative, positing instead that they are absolute. Newton's *Principia* was the first work to unify on a theoretical level celestial and terrestrial physics, which he accomplished under the inverse-square law of gravitational attraction, and his dynamics provided the first physical basis for Copernican astronomy. Although the *Principia* made use of a Euclidean geometrical format, Newton employed his new fluxional calculus to reach some results.[36] In methodology, he and John Locke propounded critical empiricism. Newton's *Principia* marks the apex of what has been called the Scientific Revolution.

Newton's *Opticks* gave a corpuscular theory of light, accepted the existence of the vacuum, and posited that the universe is almost entirely a void. In the theory of matter, Newton accepted atomism from the ancients, holding that matter is ultimately composed of hard, indivisible, and passive atoms. The *Opticks* had almost as great an impact as had the *Principia*. After the 1720s, Newton's theories of optics dominated in western Europe. His "Tractatus de quadratura curvarum" (Treatise on the quadrature of curves), one of two appendixes to the *Opticks*, presented an exposition of the fluxional calculus that Newton had invented but that lacked satisfactory foundations; these were developed by others over the next two centuries. Newton had set aside the use of infinitesimals for fluxions. Two essential terms of his new calculus were *fluent* and *fluxion*. Having a strong sense of continuous motion and time, Newton assumed that such motion propagates mathematical quantities analogous to a moving point tracing a curve. Each of the flowing quantities or variables he called a *fluent*. The velocity or rate of change of a fluent is the *fluxion* of the fluent and is represented by $\dot{x}$. Acting over an infinitely brief time or moment o, it is $o\dot{x}$. This is Newton's dot notation for differentiation. The fluxion is the original term for what is today called a derivative and remains dependent on time. Euler probably had not yet examined the higher mathematics of the "Tractatus" and did not know Newton's extensive studies of alchemy.

His master's lecture located Euler within the chief current in the development of eighteenth-century science—the diffusion, confirmation or criticism, and mathematical articulation of Newton's dynamics and optics; he was to spend the rest of his life contributing to these subjects. In the 1720s Newtonian science was encountering opposition on the Continent, primarily at the Paris Academy. There was a continuing debate over whether attraction was an occult quality or a general law of nature, and the precision of Newtonian dynamics was challenged on the tides, the shape of the planet Earth, the orbits of comets, planetary and lunar motion, and fluid dynamics. Until the mid-1730s Paris Academy astronomer Jacques Cassini's geodetic measurements were said to refute Newton's proof of the shape of Earth. Cassini, Johann Bernoulli, and the Cartesians now questioned and found fault with Newtonian science, impeding its general acceptance for at least a decade. But on the Continent the Dutch physicist Willem 'sGravesande, the author of the two-volume

Physices elementa mathematica, experimentis confirmata: Sive introductio ad philosophiam Newtonianam (Mathematical elements of natural philosophy, confirmed by experiments; or an introduction to Newtonian philosophy, 1720–21), was essential to making the University of Leiden in Holland a citadel of Newtonian science.

Euler's master's lecture did not review the methodology, mathematics, and natural philosophy of Leibniz; the comparison of Cartesian with Newtonian science was in itself an imposing subject. But in universities in German-speaking Europe, Leibnizian thought was to be the main competitor to that of Newton. Johann I Bernoulli had likely begun to convey Leibnizian doctrines, two of which Euler described in his "Dissertatio physica de sono" (Dissertation on the physics of sound). Euler's response to Leibniz's thought was to be selective, accepting some aspects and rejecting others. Likely he wanted more time to develop a view that partly differed from the position of Johann Bernoulli and accorded with his own religious beliefs.

Leonhard's increasing attention to mathematics and natural philosophy did not please his father, who obliged him to register in the theology faculty in October 1723 in preparation for taking holy orders. In the theology division Euler studied under Samuel Werenfels, a professor of theology and the dean of the school, and Samuel Battier, who taught Greek. Werenfels had also been among Paul's professors. In 1696 Johann Wettstein, a professor of Greek, had moved to the law faculty, so he did not teach Euler; his nephew or younger cousin, Johann Kaspar Wettstein, was to become a friend of the adult Euler through correspondence, but the two never met. The theology curriculum included Protestant theology and classical humanities but no longer Hebrew and ancient Greek. Euler was later to confess in his autobiography that he did not make much progress in any of these subjects. He now displayed his eidetic memory by reciting long passages from Virgil's *Aeneid*, which contains more than ninety-five hundred verses and which he knew entirely by heart. Even at the age of seventy, he could cite the beginning and closing words on each page of the text he had read as a young man.

The theology curriculum allowed Euler to continue to study mathematics, an opportunity that he eagerly seized. Despite his discouragements in his early mathematics classes, Euler's interest in the subject and in theoretical natural knowledge in general had deepened, and his "Declamatio" expressed that; Euler's failure to advance in other subjects may be traced to his turning "most of my time to mathematical studies." He felt fortunate to continue Saturday meetings with the stern Johann Bernoulli. Since Leibniz was dead and Newton old and less active, Bernoulli was the leading mathematical preceptor in all of Europe. The previous year Euler's friend and classmate Johann II, the youngest Bernoulli son, had helped gain him access to a possible tutorial with the difficult Johann I.

If his autobiography is accurate, Euler had begun meeting with the elder Bernoulli before October 1723. Bernoulli gave able students private lessons, *privatissima*, in mathematics and physics on a sliding scale; for those who could not afford to pay, the course was free. But Bernoulli was busy, and flatly refused to give Euler such private lessons. Instead he advised Euler "to start reading some more difficult mathematical books and work through them on my own as diligently as I could, and if I came across some obstacle or difficulty, I could visit him every Saturday afternoon."[37] Bernoulli, who must have negotiated these Saturday tutorials, would then show him how to overcome the unsolved problems. These tutorials, which covered astronomy and physics along with mathematics, were most fruitful. They examined conceptual roots along with problems and exercises.

While a complete account of the readings for the tutorials is not available, a plausible reconstruction of some principal titles could be deduced from Euler's conviction, which he shared with the Bernoullis, that whereas the universe is God's creation, fundamental connections exist between theology and the mathematical sciences. This belief and later research projects suggest that Euler examined such classics as the second edition of Copernicus's *De revolutionibus orbium coelestium* from 1566 as well as Johannes Kepler's *Astronomia nova* of 1610 and Galileo's *Dialogo sopra i due massimi sistemi del mondo* (Dialogue concerning the two chief world systems), published in Florence in 1632. Euler's master's lecture shows that he was studying Descartes's *Principia philosophiae*, which contains the author's mechanistic physics, and *La géométrie,* which presents Descartes's invention of analytical geometry, as given in the fourth of Franz van Schooten's Latin editions, *Geometria*, or Jakob Bernoulli's edition published in 1695. He possibly also read Jacques Rohault's masterful *Traité de physique*, the major physics text of the late seventeenth century, which appeared in 1671. While arbitrating between Aristotle and Descartes, Rohault's *Traité* accepts and explains the Cartesian mechanistic philosophy and its laws. For the science of Newton, Euler must have read the *Opticks* with its appendix on the method of fluxions and the second edition of the *Principia mathematica*, which appeared in 1713. By increasing the value of resistance by three halves, Newton had—in proposition 10, book 2—arrived at the correct answer, three-fourths of the amount given in the first edition.[38]

The interests of Euler's teacher, together with Basler publications in the mathematical sciences and notes to his first articles supplied by editors of his *Opera omnia*, make it probable that he read a range of works relating to the new calculus and its applications, beginning with the *Analyse des infiniment petits pour l'intelligence des lignes courbes* (Analysis of the infinitely small in order to understand curves), published in 1696 in Paris, with a second edition following in 1715.[39] The *Analyse des infiniment petits*, the first textbook on differential calculus in print, is an influential introduction to the Leibnizian version of it. Though issued anonymously and sometimes attributed to Guillaume-François-Antoine, Marquis de L'Hôpital, it was based mainly on the lectures of Johann I Bernoulli. Euler possibly examined two texts of Pierre Varignon, a correspondent, disciple, and close friend of Bernoulli: the *Projet d'une nouvelle méchanique* (Project of a new mechanics), published in 1687, and *Nouvelles conjectures sur la pesanteur* (New conjectures on gravity) from 1690. Varignon's two-volume *Nouvelle méchanique ou statique* (New mechanics or statistics) and *Éclaircissemens sur l'analyse des infiniment petits* (Explanations of the analysis of the infinitely small), both from 1725, may also have been available. His *Éclaircissemens*, a favorable commentary on L'Hôpital's *Analyse*, joined with advocates of the new calculus in their battle with the partisans of Euclidean synthetic geometry. More precise than L'Hôpital, Varignon defended the new calculus, defining the differential as a variable and warning that the use of infinite series must consider the remainder term. Euler must have studied Jakob Bernoulli's articles on the theory of infinite series, published from 1682 to 1704 and reprinted in 1713; his *Ars conjectandi* (The art of conjecturing) on probability, with a preface by Nikolaus I, published posthumously in 1713; and Jakob Hermann's *Phoronomia, sive de viribus et motibus corporum solidorum et fluidorum* (Phoronomis, or the forces and motions of bodies solid and fluid) of 1716. Other books to which Euler referred are John Wallis's *Arithmetica infinitorum* (1656), volume 2 of his *Opera Mathematica* (1693), and Brook Taylor's *Methodus incrementorum directa et inversa* (Method of direct and inverse increments, 1715), which Bernoulli criticized.

No confirming evidence exists on whether Euler saw Johann Bernoulli's pioneering article on what would become the calculus of variations for the Paris Academy in 1718. Bernoulli recognized that many of his earlier attempts to solve isoperimetric problems had been unsuccessful. He did not at first see that the number of free variables needed to solve these problems depends upon the number of side conditions, with one more free variable than their total. He had one variable, which suffices only when there is no side condition. But the isoperimetric problem has one side condition, which requires two free variables, as his late brother Jakob argued. For this reason, Johann initially failed to solve the problem. The emerging field was quickly to gain Euler's attention.

In order "not to bother [his teacher] unnecessarily," Euler expended his full energy to reduce to as few as possible his questions about the concepts, methods of solution, and challenging problems introduced in the readings.[40] This was a very small number. Efforts to solve the many problems, some of which Euler found intractable, did not exhaust him, and he recognized that in the search for solutions the methods then prevailing offered him promising routes and deterred him from unproductive approaches. Bernoulli must have posed a few queries to be sure that Euler had mastered the materials. Whenever Bernoulli "resolved one obstacle for me," noted Euler, "ten others disappeared right away, which is certainly the way to make happy advances in the mathematical sciences."[41] He benefited immensely from this demanding tutorial, which deepened his commitment to the mathematical sciences. In it Bernoulli discovered his student's genius. Throughout his career Euler would continue to believe that for a bright, untiring student, working from master writings in a tutorial with an able teacher was "a splendid way to succeed in learning mathematical subjects." His eulogist Nicholas Fuss called it "eine herrliche Methode" that had exceptional results, at least with so talented a student as Euler, who possessed a vast genius bound with a nearly inexhaustible diligence.[42]

Euler found the tutorial work exhilarating, and appreciated the freedom of the study and the close association in the sciences with Johann II; he felt less of a connection with the other two brothers, Daniel and Nikolaus II, however.[43] Devoting most of his time to mathematical studies after 1723, Euler had decided that he wanted to become a mathematician and natural philosopher or physicist rather than a rural Evangelical-Reformed pastor. To make this case to his father, he asked the help of Johann I Bernoulli, who possibly traveled to Riehen in 1725 to meet with his college friend Paul III Euler, though there is no actual record of this. Bernoulli, now nearly sixty, had adjudged his eighteen-year-old student to be a genius who—alone among his contemporaries, including the Bernoulli sons—was worthy to be his successor. In an article in the supplemental volume 9 of Leipzig's *Acta eruditorum* in the late 1720s and republished in volume 2 of his *Opera omnia* in 1742, Bernoulli described Euler as possessing the highest acumen, together with the mental agility and ingenuity, to be able to penetrate the most profound secrets of higher mathematics.[44] In the hypothetical meeting Bernoulli is believed to have argued that young Euler was not cut out to be a rural pastor. While likely concerned about employment for his genial son, Paul Euler accepted Bernoulli's request for Leonhard's shift out of theology. Although there were no graduate programs in mathematics and the exact sciences, once his father's consent was obtained Euler redoubled his efforts in their study. No record exists of his father's response to Leonhard's change, but it seems he was supportive of his son.

The shift in courses was not a break with religion; Euler did not set unfettered reason against revealed religion, as did the Enlightenment French *philosophes* Denis Diderot and Voltaire. As they applied Newtonian dynamics and the new calculus to reduce imprecision in describing physical phenomena, the partisans of reason largely desacralized the world. Euler, like Kepler, Newton, and the Bernoullis, assumed that intimate connections existed between religion and the mathematical sciences. From his college days on, both reason and religious faith inspired Euler's research.

In 1725, after recovering from a near fatal fever, Daniel Bernoulli returned to Basel from Venice and announced his intention to accept a post in the new Petersburg Academy of Sciences. His highly regarded *Exercitationes mathematicae* (Mathematical exercises) had appeared in print the previous year, and he now won the Royal Academy of Sciences' Prix de Paris for a paper on perfecting clepsydras for use as more accurate timepieces on ships. The three Bernoulli brothers became closer friends with Euler, and perhaps through discussions with the father and sons Euler completed a remarkable first "notebook" projecting an ambitious research program for himself in mechanics and hydrodynamics. Through their clarity, frankness, and mature genius, these notes compare favorably with Jakob Bernoulli's ponderous *Meditationes* and Gauss's terse *Notizenjournal*, whose 121 entries are almost entirely limited to his research to age twenty-four. Among the problems that Euler now studied was the outflow of water from a vessel. This early work was lost until 1965,[45] but it and later significant contributions supporting Daniel Bernoulli in founding hydrodynamics are now available.

While declining a position in Saint Petersburg because of his age, Johann I had helped persuade Christian Wolff and academy officials to offer it to his son Daniel. But he neatly indicated that unless the nearly inseparable Daniel and Nikolaus II were invited and could travel together, neither would go. In June 1725 Wolff wrote to the man who became the first president of the Petersburg Academy, Laurentius (Lavrentii) Blumentrost, the royal physician to Peter the Great, "Ich praefirire diesen [Daniel] seinem Bruder Nicolao, weil er dem Vater näher kommt, als der andere" (I prefer this one [Daniel] to his brother Nikolaus, because he comes nearer to his father than the other).[46] Daniel was offered an annual pension of six hundred rubles and free housing, along with sufficient wood and candles for heating and lighting, but he wanted to also include a position for his brother. In response to these entreaties and confusion over which brother had been initially invited, academy president Blumentrost proposed to raise Daniel's salary to eight hundred rubles and include a place for Nikolaus II; the brothers accepted the new offer. Euler noted in his diary that he was pleased that the brothers were leaving for the academy and he hoped to go with them.[47] He apparently asked them to recommend him to fill the first suitable vacancy there, and they promised to do so.

Initial Publications and the Search for a Position

By the age of seventeen in 1724, after completing his university studies but earning no degree, Euler remained in Basel to seek employment. Apparently he continued to live with his maternal grandmother; as his selection of a graduate major indicates, there may have been some tension with his father. In the early eighteenth century, conditions for beginning a career in mathematics were difficult; it was not yet a profession with regular positions, and it was struggling for recognition in the academic world. The Swiss had no journals dedicated to the field, and there were almost none elsewhere in Europe, though Swiss geometers could submit papers to Leipzig's more general scholarly *Acta eruditorum*.[48] The Paris Academy's *Mémoires de l'Académie Royale des Sciences*, the *Journal des Sçavans,* and the *Philosophical Transactions* of the Royal Society of London treated the sciences in general. The financial risk made publishers reluctant to publish articles on theoretical mathematics. Swiss universities had few chairs in mathematics and salaries were poor; they were graduating more students in the mathematical sciences than could be hired in the field. The Baslers were exporting this talent chiefly to northern Italian cities and the new Petersburg Academy of Sciences. The lottery system of choosing professors in Basel was another problem, often making it difficult for the best candidates to win. Daniel Bernoulli, for example, had been rejected three times in applications for a professorship there. Before departing from the city, Euler had begun to make his scholarly reputation by publishing two articles and an essay in addition to having another essay in press.

At the age of eighteen Euler had written a three-page paper titled "Constructio linearum isochronarum in medio quocunque resistente" (Construction of isochronal curves in any forms of resistant media, E01), which appeared in 1726 in *Acta eruditorum* as his first publication.[49] He may have written another paper earlier, but that seems unlikely. The "Constructio" was in Latin, as would be four-fifths of his publications. Obtaining isochronal curves (for the fastest descent) in a resistant fluid was a difficult mathematical problem; long competitions over isoperimetric problems and ballistics, in which Johann I Bernoulli participated, generated the search for these. To reach the lowest point on an isochronous curve, a weighted body sliding along it forced by gravity must take equal time of descent regardless of where the object begins on the curve. Euler began by observing that in a nonresistant medium the ordinary cycloid is an isochrone or tautochrone, which requires finding the curve a weighted particle follows to reach a given point of the curve in equal time no matter what the starting point is. During that era, the catenary or hanging chain was also being studied as a means of finding the lowest center of gravity.

In his background studies Euler apparently drew upon proposition 17 of Brook Taylor's *Methodus incrementorum directa et inversa*, possibly the first edition from 1715, which solves an isoperimetric problem. He likely also drew on Johann I Bernoulli's article of 1718 for the Paris Academy, which demonstrates that among curves of the same length the cycloid suffices to answer the brachistochrone problem. That is, it determines the curve of quickest descent by a weighted particle, free from friction and resistance, falling in a gravitational field from one point to another not directly under it in a vacuum. In *Acta eruditorum* for June 1696, Bernoulli had initiated the challenge to mathematicians to solve the brachistochrone problem. He and more generally his brother Jakob, as well as Leibniz, Newton, and L'Hôpital, had each solved it. Galileo had conjectured in the *Discorsi* of 1638 that the solution is a semicircle; but the cycloid is the correct curve, which Leibniz called the apple of Eve. The cycloid was generally known as the Helen of Geometers, the answer to several challenge problems. Through his study of the pendulum problem, for example, Huygens had found that a pendulum bob traversing a cycloid takes exactly the same time to complete swings however small or large the amplitude. Thus, the cycloid was called the tautochrone. When Johann I Bernoulli received a solution in a paper without the name of the writer, he knew by its authority and skill that the author was Newton "ex ungue leonem" ("as the lion is recognized by its claw"). "You will be astonished," wrote Bernoulli in *Acta eruditorum* for May 1697, "when I say that this same cycloid, the tautochrone of Huygens, is the brachistochrone we are seeking." He called the cycloid "the fateful curve of the seventeenth century."[50]

Thoroughly scrutinizing Henri Sully's system of the pendulum tautochrone, Euler's "Constructio linearum isochronarum" confirms that its solution is a cycloid. This seems surprising, for the brachistochrone curve involves the least time, while an isochronal curve involves equal time. In any case, Euler did not yet have a formula for it. Newton, he observed, had shown the isochrone to be a cycloid. Euler's paper generalizes, shortens, and improves upon Bernoulli's resolution; it refers to Huygens and Newton on the nature of the cycloid and shows that a particle acting under uniform gravity in this case gives a brachistochrone. Others who surpassed Bernoulli usually aroused the master's jealousy, but Euler remained his favorite, and Bernoulli praised the higher mathematical dexterity and ingenuity of the young student.[51] Euler's first paper contains one of his rare errors, posing and solving the wrong differential equation. He discovered that mistake and corrected it within two years, giving the right differential equation in the article "Curva tautochrona in fluido resistentiam faciente secundum quadrata celeritatum" (Tautochrone curves in a fluid offering resistance according to the square of the speed, E13) in 1735 in the *Commentarii Academiae Scientiarum Imperialis Petropolitanae*, the journal of the Petersburg Academy.[52] His first paper cites but leaves for several years in the future a problem raised by Bernoulli on the brachistochrone in a resistant medium.[53]

In this and other work of determining extremals for isoperimetric problems, the Bernoulli brothers along with Taylor and Euler pioneered the calculus of variations in its semigeometric phase. In 1700 and 1701 Jakob had employed differential equations to attack isoperimetric problems, finding that they require a second degree of freedom. Only after Jakob died in 1705 did Johann accept the requirement. Apparently he set this work aside for nearly a decade, for not until the Paris Academy paper of 1718 did he elaborate his isoperimetric condition in which the abscissa is the independent variable. After 1715 Johann was also engaged in a quarrel with the British over ballistics problems. In the article "Exertatio geometrica de traiectoriis orthogonalibus" (Geometric probe of orthagonal trajectories) in *Acta eruditorum* for 1720, his favorite son Nikolaus II had proposed the problem of determining reciprocal trajectories—that is, constructing curves and their reflections, which fall between two parallel axes and constantly cut each other at right angles. Johann Bernoulli, Euler explained, had given him this problem principally as a brain teaser for the new calculus rather than as a search for physical applications. Johann bested Henry Pemberton, a Briton who at that time remained anonymous, by devising a better and simpler general geometric solution for arriving at curves that satisfy the condition that the trajectories must be reciprocal. The reciprocal curve is the reflection intersecting with the original curve, either at right angles or more generally at a constant angle.

Euler's second published paper, "Methodus inveniendi trajectorias reciprocas algebraicas" (A method for finding algebraic reciprocal trajectories, E03), is on the construction of such trajectories derived from algebraic curves.[54] His mentor, "the celebrated master Johann Bernoulli, my teacher and patron" had assigned this as a homework problem.[55] These trajectories are curves, which are reflections between parallel lines, also known as parallel axes. They cut each other at right angles, and they grow steeper in one direction as their angle, not slope, increases and become less steep in the other. Euler's article appeared in *Acta eruditorum* in September 1727, but this paper had undoubtedly been completed before he departed from Basel. Finding reciprocal trajectories seems to have been a challenge problem mostly involving the new calculus rather than physical application. The main aim was to obtain trajectories that could be expressed in a simple form. Euler quickly and acutely found solutions written as arc length formulas. Once he had these rectifiable algebraic curves, he verified that trajectories constructed from them were reciprocal. The lack of modern notation and Euler's failure to explain all of his assumptions detract from the article. Three more papers continued to examine the problem of reciprocal trajectories, including instances in which it could be expressed in the less complicated polynomial functions, bettering Euler's presentation each time and exhausting the subject, which would soon be forgotten.[56] He lacked the notation needed to present his work more clearly. Since more algebraic functions exist for this problem, it was easier to resolve the algebraic case. Euler also resolved questions of the still anonymous Pemberton about inverse tangents. Bernoulli praised his student's "most fortunate talents" and "the ease and adroitness with which he penetrated the most secret fields of higher mathematics."[57]

In 1727 Euler competed for the Royal Academy of Sciences' annual Prix de Paris.[58] The nautical problem posed was "what is the best manner to mast [sailing] ships, with respect to the situation, number, and height of masts?" The goal was to maximize speed.

In 1715 Jean Baptiste Rouillé de Meslay had provided funds for that academy to create prizes in two categories: theory and application. The academy was to propose problems to the republic of letters generally in the form of questions that had eluded solution and then award the best answers.[59] The greater prize, given in even years for theory, was to be for the best treatise on astronomy, matter theory, mechanics, optics, or physics; it carried a handsome financial award of at least 2,500 livres. The other prize, given in odd years, was for discoveries that more accurately determined longitude, which was a famous problem, or provided useful information for improving navigation and ship construction; the winner of this prize received 2,000 livres. The financial awards were roughly a Petersburg academician's annual salary, or one-tenth to one-fifth of the annual income of a prosperous bourgeois in Paris.[60] No prize was solely in mathematics, but the competitions indirectly broadened the field's range. Occasionally a question was continued and a double prize awarded two years later; in a few instances the academy split the prize. Usually it published only the winning essay, but sometimes it included another explaining or expanding upon it; *accessit* or honorable mention papers were almost never printed. Each year the Paris Academy assembled a prize report, and the reports were subsequently published in its prize volumes, *Pièces qui ont remporté le prix de l'académie royale des sciences de Paris* (Papers that have won the prize of royal academy of sciences in Paris).[61] To prevent the selection of favored authors, the process did not allow names to appear on the papers. A sealed envelope accompanied each paper, and the two had matching numbers. Only after the winning paper was selected did the academy director publicly open the appropriate envelope and announce the name. From the initial competition for it in 1720 through most of the eighteenth century, the Prix de Paris was the most distinguished scientific award in Europe.

After moving as a child to his grandmother's, Euler had observed the boats—mainly ferry boats, freighters, and canoes—at the nearby dock and on the Rhine. But in 1727 when he completed "Meditationes super problemate nautico, de implantatione malorum, quae proxime accessere" (Thoughts about a nautical problem on the positioning of masts, E04),[62] Euler had yet to see a large oceangoing ship. The rule for writing papers was cosmopolitan: they could be in French or Latin, the ubiquitous language of learning in Europe.[63] Since Euler had not yet learned French, he wrote his paper in Latin. The "Meditationes" followed Johann Bernoulli's *Théorie de la manoeuvre des vaisseaux* (Theory of the maneuver of ships) of 1714 and specifically his solution of the velaria problem, which required determining the shape of a sail under the pressure of wind. Its solution took a second-order differential equation.[64] Euler generalized and shortened Bernoulli's solution by reducing sailing to problems of statics and dynamics; his paper reflects a powerful intuition as regards physics.

Occasionally the prize topic was designed for a favored scholar to win. For several years the royal hydrographer and geodesist Pierre Bouguer, France's foremost authority on all nautical matters and a leading natural philosopher, had been composing a treatise on finding the optimal way to set up masts on a ship. He submitted the entire manuscript, which won first prize—the first of three for him—and the treatise would quickly become the standard publication on the topic. The excellence of Bouguer's paper merited its victory, but Euler's entry did what few earlier winning papers had done: it solved a problem.[65] His paper placed third, sharing the *accessit* from the Paris Academy for 1727 with a paper by the French mathematician and mechanist Charles Étienne Louis Camus.[66] There was clearly high regard for the work of the young newcomer, for only about half the annual prize competitions were awarded the *accessit*, and Euler's was one of the few such papers to be published by the academy. Over the next three decades, Euler and Bouguer would become intense rivals in naval science.

When in a first draft of the Paris paper Euler described his own experiments with a ship model, he did not yet express an absolute faith in his theory.[67] In the last section (number 100) of the final version of the paper, he apparently shifted, proudly declaring himself a formalist and following a strictly rational methodology.[68] That draft cites experiments, but Euler did not believe it necessary to confirm his theory empirically, since he based it on certain and impregnable principles of mechanics: "Haud opus esse existimavi istam meam theoriam experientia confirmare, cum integra et ex certissimis et irrepugnabilibus principiis mechanicis deducta, atque adeo de illa dubitari minime possit" (I have judged it not to be necessary to confirm this theory of mine by experiment, for it is sound and has been deduced from the most certain and indisputable principles of mechanics, and on this account there can be scarcely any doubt about it).[69] Throughout his career Euler was to excel in a priori proofs. But after recording astronomical observations, conducting physical experiments, and carrying out water projects in Saint Petersburg that in part addressed possible flooding, he would insist on the necessity of quantitative experience in the form of experiments and systematic observations for confirming scientific theories and refining differential equations.

Upon the death in September 1726 of Johann R. Beck, the professor of physics at Basel, Bernoulli urged Euler to apply to fill the vacancy. Each applicant had to submit a specimen essay, and Euler presented a sixteen-page in-quarto essay on acoustics titled "Dissertatio physica de sono" (Dissertation on the theory of sound, E02).[70] By late 1726 he had decided on its content and wrote to Daniel Bernoulli describing it. In December Bernoulli responded, "Not without pleasure have I learned that you intend to write a dissertation on sound; thus you will show very well how necessary is the joining of physics to higher mathematics. I doubt that anything can be said regarding the speed of sound, since there is no right explanation of its propagation."[71] On the morning of 18 February in the law auditorium, Euler defended the essay in a public disputation.

Written in a clear and direct style that came quickly to be Euler's signature, the essay summarizes without any calculations all existing knowledge about producing and propagating sound, adds some details of his own, and closes with a page containing six brief *annexa*—essentially, appendixes. For the speed of sound, Euler showed a diversity of existing results. Later he would work to obtain more accurate measures for the speeds of sound and

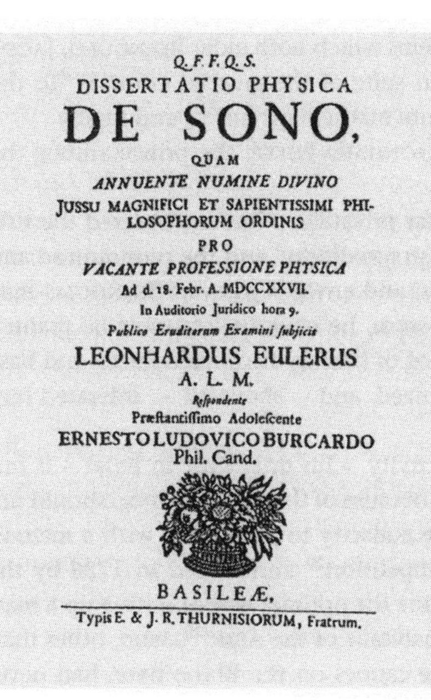

Q. F. F. Q. S.
DISSERTATIO PHYSICA
DE SONO,
QUAM
ANNUENTE NUMINE DIVINO
JUSSU MAGNIFICI ET SAPIENTISSIMI PHILOSOPHORUM ORDINIS
PRO
VACANTE PROFESSIONE PHYSICA
Ad d. 18. Febr. A. MDCCXXVII.
In Auditorio Juridico hora 9.
Publico Eruditorum Examini subjicit
LEONHARDUS EULERUS
A. L. M.
Respondente
Præstantissimo Adolescente
ERNESTO LUDOVICO BURCARDO
Phil. Cand.

BASILEÆ,
Typis E. & J. R. THURNISIORUM, Fratrum.

Figure 1.4. The title page of Euler's Habilitationschrift, "Dissertatio Physica de Sono."

light. The paper showed its author to be an original thinker who was able to identify and organize a synthesis of elements from divergent scientific traditions and transmute them. What Euler extracted from Aristotelian, Cartesian, Leibnizian, and Newtonian thought had to be logically consistent and contribute to the construction of his general theory. Reprinted in 1751 in a collection of important scientific works edited by the Berner anatomist Albrecht von Haller, a professor at the new University of Göttingen, the essay became a classic that was cited for over a hundred years, offering a program for research in acoustics. Subsequently Euler wrote more than fifty papers directly or indirectly related to solving its problems.

The six appendixes grafted to the end of the paper address ideas that Euler was pursuing. They themselves offer sufficient illustration of their author's critical, selective, and independent mind while placing him within the main scientific currents of his time.

The first two depart from the thought of Johann I Bernoulli, which is surprising for any student of his. The opening appendix rejects the Leibnizian doctrine of the preestablished harmony between body and soul in which each is depicted as independent but the two are arranged to act as synchronized clocks. It is not known whether Bernoulli ever discussed this concept, but for religious—not scientific—reasons Euler opposed it. A preestablished harmony, he argued, would limit the freedom of the spirit and so contradict fundamental Christian belief. In later battles with Wolffian freethinkers in Berlin, Euler was to add the argument of Calvinist clerics who condemned the doctrine as harmful to the Christian concept of original sin. Contrary to Cartesian thought, the second appendix asserts that Newton's gravitational attraction gives the most satisfactory explanation of all celestial motions. Although Euler asserted that the mutual attraction of bodies is beyond doubt, like Newton he would search for a mechanical explanation and he later thought that a slight revision might be needed in celestial mechanics.

The third appendix involves a thought experiment: if a shaft is drilled through the center of the Earth, what happens to a stone that falls down through it? Euler was not proposing that the Earth was hollow, an idea first set forth by Edmond Halley in 1681. Newton and others had unsuccessfully grappled with the shaft problem. Euler rejected the commonly accepted theory that the stone would pass beyond the center and reach the other side of Earth, then return to the center again. He had it turn suddenly at the center and rebound along the same path to the starting point on the surface. For this unexpected position he received a literary rap on the knuckles twenty-five years later from Voltaire,[72] who had read "De sono" in 1751 as reprinted in the Haller collection. Euler's corrected explanation of his position would come later, in numbers 49 and 50 of his *Lettres à une Princesse d'Allemagne, sur divers sujets de physique et de philosophie* (Letters to a German princess on diverse subjects of physics and philosophy), published from 1768 to 1772.[73]

The next appendix places Euler within the *vis viva* controversy. Accepting the idea of *vis viva* advanced by Huygens and Leibniz, it gives as the true measure of their force the quantity of matter of moving bodies multiplied by the square of their speeds, mv^2—that is, living force or twice kinetic energy. Quantity of matter was not yet defined as mass; the Cartesians and Newtonians measured force by *quantitas motus* (essentially, momentum, or mv). The *vis viva* controversy was ardently pursued for twenty more years until Jean-Baptiste le Rond d'Alembert in his *Traité de dynamique* of 1743 used two different ways of calculating force and concluded that both quantities are correct. He called the quarrel "une dispute de mots" (a battle of words).[74] Still, not everyone was ready to consign the question to semantics. Measurements of centripetal force, for example, seemed to produce contradictions. Thus the dispute persisted for another decade.

The last two appendixes, like the third, address problems of oscillation. The fifth determines that the speed of a sphere rolling frictionless along an inclined plane compared to another falling perpendicularly is $\sqrt{5}$ to $\sqrt{7}$. The sixth repeats Euler's solution to the velaria problem.

With "De sono" Euler had entered the fields of mechanics and philosophy. Often described wrongly as his doctoral dissertation, it was essentially his *Habilitationsschrift*, a postdoctoral thesis based on independent scholarship of a higher quality than that for a doctorate (which, in fact, he did not have); the *Habilitationsschrift* was a thesis that was required for qualifying and gaining permission to be a professor. In "De sono" Euler embarked on the course of mastering all of physics that was known at the time, a goal that he was to achieve, and he began to display his extraordinary ability to solve special problems computationally.

Although the University of Basel gave "De sono" a good reception and Johann I Bernoulli strongly supported him, Euler's name was not included among the final three selected for the chance drawing by lots after a review of the candidates. The selection committee probably considered him too young, for he was not yet twenty. The most prominent candidate on the list was Jakob Hermann of the Petersburg Academy of Sciences, but neither would he win. Instead the position went to Benedict Stähelin, an alumnus of the Universities of Basel and Paris who knew less physics than anatomy and botany but had a modest reputation for scholarship; he was also a corresponding member of the Paris Academy of Sciences. Although he was selected by lot, the fact that the Stähelins were a patrician family in Basel may have strengthened his candidacy. University positions were becoming sinecures for the sons of Basler *Herren* who did not enter the family business.[75]

Around 1725 Euler began to sketch in notebooks a substantial research program for himself.[76] His first notebook contains his proposal to write a major treatise on all of mechanics and outlines in three sections a book on a new mathematical theory of music composition.[77] For this project he gave the titles of all but one of its fifty-eight chapters and sections. His initial study of the important vibrating string problem also appears in the first notebook. Brook Taylor had published his research on this problem in 1713, and Euler must have consulted with Johann I Bernoulli, who was also studying it.[78] The goal was to discover the center of oscillation in the curve formed by the fundamental mode of a vibrating string. Since a vibrating string is flexible, the initial task was to determine its shape; then Euler needed to integrate the differential equation for its motion. He would complete each of these projects impressively in the decades after his departure from Basel.

Euler's failure to obtain the Basel professorship is generally thought of as fortunate. Although mathematics can be a singularly internal subject, and thereby chiefly separate from its surrounding environment, it cannot be created without mathematicians. The social, institutional, and intellectual setting throws light on the intensity of collaboration and debate, and it is a principal factor in determining what questions and problems will be central. The rejection at Basel forced Euler to move from a small republic. A broader political setting and an institutional base more appropriate to his brilliant research and technological contributions lay ahead.

Leonhard Euler's future began in the Russian colossus. Even before the death of Nikolaus II Bernoulli after a lengthy illness, a position had opened at the Petersburg Academy of Sciences. True to his promise, Daniel Bernoulli, along with his father, recommended Euler to President Blumentrost;[79] Christian Goldbach may also have supported Euler's candidacy, and Euler was selected for a post in physiology; a vacancy in mathematics was not considered at that time. In the summer of 1726, Blumentrost sent a letter inviting Euler to accept the position of adjunct in physiology with an annual pension of two hundred rubles. The European postal service was evolving, and in Russia it was not yet reliable, so the letter was lost. At the end of September Daniel wrote to Euler that Blumentrost had sent a letter of appointment a few months earlier; Daniel explained that he found the pension no match for Euler's worth, but that he believed the president would increase it. While he urged Euler to submit quickly a letter of acceptance, it was not until 9 November that Euler wrote that he was honored to be admitted to the Petersburg Academy of Sciences. His hesitation in responding may be partly explained by his continuing effort to win the physics professorship in Basel. His unstated plan was that if he was not successful in Basel he would travel to Russia when the weather cleared in the spring.

Daniel Bernoulli advised Euler to use the intervening time to get ready for the position by taking courses in anatomy and physiology that the University of Basel offered as preparation for a medical degree. The spring term did not end until June. But on 5 April 1727, three days after enrolling at the university with dean Johann Rudolf Zwinger for beginning medical courses during the summer semester, Euler boarded a boat on the Rhine and left Basel. He would never return. Although he was to retain his Basel citizenship for life, the timing of his leaving seemed to involve more than the excitement of moving from a small republic to the capital city of a great empire; Euler's departure, only ten days before his twentieth birthday, may suggest some disagreement with his father's wishes concerning the beginning of his new career. Yet he maintained toward his father a respect that is reflected in his lifelong familial adherence to the prayers and scriptural readings that the senior Euler favored. Having lived apart from the family for ten years may have made less difficult Euler's decision to leave. While Daniel Bernoulli wanted to return to Basel, Euler was not so inclined, but he nonetheless was to maintain lifelong contact with the city via correspondence, and once he was well established in Saint Petersburg he hired Baslers as household staff.

Chapter 2

"Into the Paradise of Scholars": April 1727 to 1730

Leonhard Euler's second "notebook" details in its first seventeen pages the stations on his seven-week-long journey to Saint Petersburg. It is among the twelve notebooks listed in the inventory of the *Rukopisnye materialy*, published in 1962 (RM 397–408), at the Petersburg Academy Archive.[1] It indicates that at 9:00 a.m. on 5 April 1727, he boarded a boat in Basel that sailed down the Rhine to Mainz.[2] Via stage coach, Euler proceeded through Giessen and Kassel to nearby Frankfurt am Main. At 8:00 a.m. on 11 April he departed Frankfurt, passing through Friedberg and Butzbach before arriving in Marburg in Hesse-Kassel. The next day, as Johann I (Jean I) Bernoulli must have recommended, Euler met Christian Wolff, whose religious views he opposed. An exile from the cameralist University of Halle, Wolff had counseled Czar Peter I, also known as Peter the Great, on the selection of most of the Petersburg Imperial Academy of Sciences' original members.[3]

Euler stayed only briefly at Marburg. He probably next spent a night at Wetzlar before continuing north on land through Hannover to Hamburg, where he received a hundred rubles sent from the Petersburg Academy of Sciences to cover his travel expenses. From there he traveled to the Hanseatic port city of Lübeck.

In a letter sent through Hamburg and dated 20 April, Wolff expressed deep regret that Euler's haste had left insufficient time for a full discussion of various matters or to show Wolff's high regard for the Academy of Sciences and the friendship of Johann I Bernoulli. "You are now entering into the paradise of scholars,"[4] Wolff declared, and he wished that God would grant Euler good health and years of good fortune in Saint Petersburg; he asked Euler to convey his "humble respects to the academy's President," Laurentius (Lavrentii) Blumentrost, and its members "Mr. Bilfinger, Hermann, Daniel Bernoulli, Martini, and Leutmann, and . . . remember me kindly. . . ."[5]

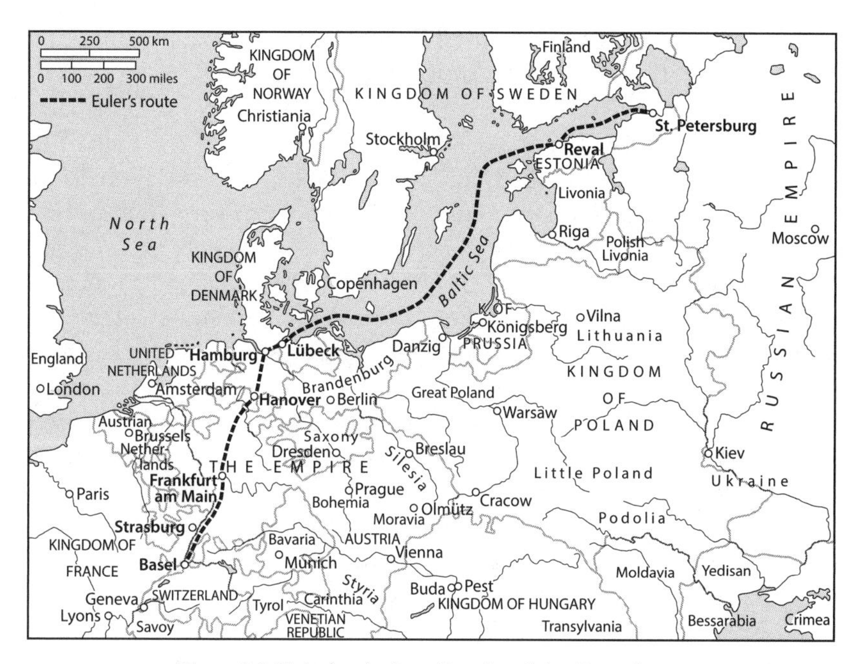

Figure 2.1. Euler's trip from Basel to Saint Petersburg, in a drawing based on a map of Europe from about 1740.

With no ship in Lübeck ready to sail to Saint Petersburg, Euler was forced to take a boat to Rostock. On 30 April, the boat was severely rocked by storms in Lübeck Bay and all passengers, including Euler, became seasick. The boat landed to the east on the Baltic at Wismar before reaching Rostock. From Rostock Euler traveled along the coastal waters of the Baltic to Reval, the modern Tallinn in Estonia. The trip to this point had taken almost four weeks. From Reval Euler crossed the Gulf of Finland to the citadel island of Kotlin, with the city and fortress of Kronstadt that Peter I had taken from the Swedes in 1710; Euler's boat must have landed at its large western or merchant harbor. After a short ferry trip from Kotlin, Euler reached the mainland and proceeded on foot around Saint Petersburg on 24 May 1727.

On his journey, Euler carried a few personal belongings, along with items that Daniel Bernoulli had requested: "fifteen pounds of coffee, one pound of the best green tea, a half dozen bottles of good Danzig brandy, twelve dozen fine tobacco pipes, and several dozen packs of playing cards."[6] The number of pipes suggests that Bernoulli, like Euler later, was a heavy smoker. On this trip Euler possibly traveled with the Basel mechanist and geographer Isaac Bruckner (Brucker).[7]

Founding Saint Petersburg and the Imperial Academy of Sciences

Located on the Gulf of Finland, the easternmost arm of the Baltic Sea, Saint Petersburg was founded on 29 June 1703, the feast day of Saints Peter and Paul, as a crude log-and-earth fort under construction on Hare Island that was intended to guard the mouth of the River Neva. In beginning this city on the periphery of his empire, Peter I aimed to increase his influence on the Baltic region, to provide an opening to central and western Europe, and probably to have a place to test his reforms.[8] It was initially to be a city located on seven islands connected by bridges. In 1706, Peter recruited the Swiss-born architect Domenico Andrea Trezzini to transform the crude fort into the Peter and Paul Fortress. The settlement that grew around it had various names, such as Petropolis, the exact Greek translation of Petersburg, but the name that gained wide acceptance was Saint Petersburg (Sankt Piter Burkh). The czar's commitment to building the metropolitan center, even while he was engaged in the Great Northern War against Sweden from 1700 to 1721, was a daunting task that some other European rulers thought a reckless gambit. A site selected for military and trade reasons rather than its climate and terrain

Figure 2.2. Portrait of Peter the Great (1672–1725) by Jean-Marc Nattier.

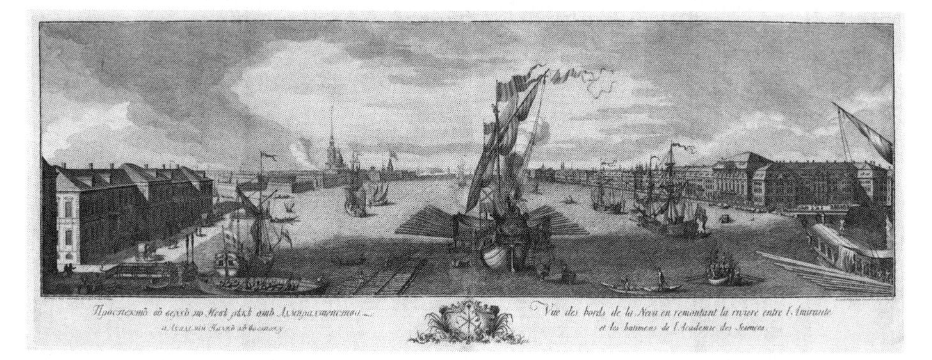

Figure 2.3. A view of Saint Petersburg upstream on the River Neva from the admiralty. An engraving by Yefim Vinogradov, taken from a 1749 drawing by Mikhail Makhayev.

(it had very cold winters, as well as bogs and moss), the city was to suffer from two main dangers: floods and fires. At this time, fireplaces were essential for heating and cooking; difficulties with them and the lack of adequate firefighting equipment produced large fires in most major cities.

Trezzini, the first chief architect, made plans for the city and designed buildings, among them Peter and Paul Cathedral and the Summer Palace. In 1714 the French architect Jean-Baptiste Alexandre Le Blond succeeded Trezzini. The flamboyant style, the theatrical religiosity, the irregular church ornamentation, and the monumental size of the city's major structures characterized the Age of the Baroque. Since it was built on the banks and islands where the Neva has two main tributaries, the Greater and Little Neva, Saint Petersburg soon had an integrated network of canals and bridges for transportation and communication and beautiful water vistas; some called it the Venice of the North. This was almost a cliché, for a few other northern European cities with similar characteristics also gained the accolade, but in fact Peter desired another Amsterdam.

After the defeat of the Swedish monarch Charles XII at the Battle of Poltava in 1709, Saint Petersburg was deemed secure and Peter moved quickly to make it into more than a military outpost;[9] he wanted it to be Russia's capital. While he issued no official decree, the process of moving the governmental bodies that began in 1710 made the city two years later the effective capital. The czar ordered a thousand aristocrats, hundreds of merchants and traders, and two thousands artisans, together with their families, to move there. A large body of soldiers accompanied them. To accelerate the growth of Saint Petersburg, the czar dedicated almost 5 percent of the royal budget to government buildings, and aristocrats had impressive palaces constructed. The population grew from around 8,000 in 1710, to 24,000 in 1717, to approximately 40,000 in 1725, already more than twice the size of Basel.[10] By 1725 the city could boast of being the focal point of Russia's export trade with the West as well as the center of the Enlightenment in Russia and of pursuing a growing fleet modeled after the British Navy. Rapid increases over the next three decades would bring the population of Saint Petersburg to about 150,000.[11]

An essential component in Peter's ambitious modernizing plan was to introduce Russia to the Enlightenment along with Western science and scientific methods; his first visit to the West had stimulated if not initiated his interest in the sciences. In Holland in 1697 he met Nicolaas Hartsoeker, who guided him in observations of the planets and stars; and he visited hospitals and botanical gardens and attended lectures on medicine at the University of Leiden, where he spoke with the physician Hermann Boerhaave. Acquiring and cultivating medical skills was among Peter's major ambitions for an academy. In 1698 he recruited fifty surgeons for the fledgling Russian navy. Later that year he visited the Royal Society of London and probably met with Isaac Newton there.[12]

To the Petrine reforms, which caused upheaval in the ranks of Russia's culturally reactionary church and aristocracy, the Petersburg Academy of Sciences was central. In the eighteenth century chartered, royal scientific institutions were to complement and in research surpass the universities, whose main task was transmitting knowledge rather than advancing it. After 1700 the number of these institutions grew; by 1789 there were about seventy of them.[13] Chief among them were those in London, Paris, Berlin, and Saint Petersburg. Charles II had chartered the Royal Society of London in 1662, and Louis XIV and his minister Jean-Baptiste Colbert had established the Royal Academy of Sciences in Paris in 1666. Primarily through Gottfried Wilhelm Leibniz, the Royal Brandenburg Society of Sciences was begun in Berlin in 1701. Leibniz was the intellectual founder of the Petersburg Academy as well.

Peter named Leibniz a Russian privy counselor of justice, and in 1711 invited him to meet in Karlsbad. The chief assignment was to promote mathematics and the sciences in Russia as well as to encourage the building of a system of public schools and universities. Leibniz's attempt to establish scientific academies across central and eastern Europe would fail in Dresden and Vienna, but the czar accepted Leibniz's proposal for an academy, commencing communication with China to learn about its science and ordering observations of magnetic declination across his empire.[14] The two maintained a correspondence, and in 1712 and 1716 met again. For the last meeting Leibniz prepared several memoranda that included outlining a series of studies to bring Western scientific standards to

Russia. His letters and memoranda recommended a systematic collection of chronicles and linguistic works from ancient and medieval Russian and Byzantine history and ethnography; encouragement of missionary activities from western and central European Christian churches in Russia; what may be roughly defined as a distinctive school with shared faculty productions and transmission of collective knowledge that would include artists and painters; observations of terrestrial magnetism and the incline of the compass to improve navigation; building a chemistry laboratory and an astronomical observatory; a survey of plants, animals, and ores of Russia and its southern neighbors; and the translation into Russian of useful Western books on the sciences and technology, especially industrial technology.[15] One memorandum included a call for founding a "college of sciences."[16] Shortly before his death in 1716, Leibniz urged Peter to accept as a scientific adviser Leibniz's most prominent follower, Christian Wolff.[17] For his contributions to the evolution of the sciences in Russia, Bernard le Bovier de Fontenelle proclaimed in his *éloge* to Leibniz delivered to the Paris Academy, "In the history of science in Russia he can never be forgotten, and his name will be placed alongside that of the Czar."[18]

In the sciences Peter I had predominantly practical and utilitarian purposes, broadly understood, that were tied closely to his mercantilist economy. According to mercantilist theory, states must protect the wealth of their economies through monopolies, restrictions, and tariffs; they need a favorable balance of trade as a means to attract and hoard gold and silver and prohibit their export, all of this falling under the term *bullionism*. Peter saw mechanics, mathematics, astronomy, and chemical science as tools in the construction of ships, canals, and docks and to provide for navigation, improve artillery, develop mining installations, and benefit public health. He wanted to expand Russia's trade and make his empire a European power. Accurate Russian cartography, based on reliable geographic surveys and astronomical observations, was essential for an efficient government, and it offered a source of political unity. Peter recognized the sciences as a compound of theory, experiment, and practical research, and he was open to a wide range of studies. He did "not damn the alchemist who seeks to transform metals into gold nor the student of mechanics who, in his search for perpetual motion, looks for the extraordinary, for he unexpectedly discovers many collateral things."[19] The Russian word for science, *nauka*, which had referred to technical ingenuity and technique, was now taking on a meaning close to that of the German term *Wissenschaft*, which together with the natural and social sciences included history and philosophy. Though first drawn to the practical and utilitarian side of the Enlightenment, the czar also came to accept the restlessly inquisitive spirit in theory that animated utilitarian accomplishments without subordinating itself to their demands. As the understanding of what comprised the natural sciences evolved, alchemy and magic were being displaced. And since the word *mathematics* carried a connotation of magic stemming from Alexandrian antiquity, the term *geometer* was now preferred over *mathematician*.

On his second trip to the West, in 1717, Peter visited the Royal Academy of Sciences in Paris. In addition to being elected an honorary member, he established contacts for gathering information on practical steps preliminary to establishing a science academy, such as building a library and collecting scientific instruments. The next year he wanted to begin to create an academy, but did not yet press the project;[20] it was debated over time both within and outside Russia.[21]

In 1719 the czar asked Blumentrost and the Alsatian librarian Johann Schumacher to learn in detail the functions, organization, and projects of the Royal Society of London and the Paris Academy of Sciences and to make recommendations to him; he also spoke with Iakov Vilimovich Brius (Jacob Daniel Bruce), the son of Scottish immigrants and a friend from childhood, who was the first Russian Newtonian and the president of the Russian College of Mines and Manufactures. Russia's first historiographer, Vasilii Nikititch Tatishchev, a voracious reader of René Descartes, Robert Boyle, Thomas Hobbes, John Locke, Nicolas Malebranche, Gottfried Leibniz, and Wolff, recommended studies in mathematics, geography, and natural history.[22] In 1714 the Royal Society of London had elected as a fellow Prince Aleksandr Menshikov, a supporter of the czar.[23]

Disliking the independence of government that members of the Royal Society enjoyed and the limits to governmental control over its corporate activities, Peter rejected it as a model; he preferred instead the Paris Academy of Sciences, for it was closely allied with the royal government and answered the questions the authorities posed. Through correspondence with Wolff, in the course of which Wolff compared himself to Aristotle, Blumentrost gained knowledge about the Royal Brandenburg Society. The Petersburg Academy, like the Brandenburg Society, was not to be limited to the natural sciences but would include historical and linguistic subjects. Since Russia lacked the corporate bodies and personnel to advance the sciences, Peter extended governmental financial and moral support; following the practices of the Paris Academy, he planned to provide pensions for academicians and funding for useful experiments, buildings, and equipment. For the maintenance of his own academy of sciences, Peter allocated an annual budget of 24,912 rubles collected from customs and licensing fees in several cities on the Baltic.[24]

In the Russian Orthodox Church, the planned academy encountered some antagonism; church hierarchy considered Western science a new religion that would contaminate the Russian soul. An antirational theology branded books and articles on the Western sciences as the work of the Antichrist, and church censors prohibited their publication. Attempting to subordinate the church to the state and to accommodate religious ideology with Continental rationalism and science, Peter issued the Ecclesiastical Regulation in 1721. The devout Russian Orthodox czar did not challenge the principal beliefs of the church but attacked antirationalism as inconsonant with Christianity. Secularizing knowledge and disseminating the sciences were, Peter believed, the sole way to overcome the cultural inertia that the ultraconservative Russo-Byzantine intellectual tradition had imposed on his kingdom.

His regulation had only a minor effect. The Orthodox censors, for example, continued their long-standing opposition to Copernican astronomy. The Russian translation in 1661 of Willem and Johan Blaeu's modest *Atlas novus* (1645), which spoke favorably of Copernicus and heliocentric astronomy, remained unprinted in manuscript form until around 1724.

Meanwhile Wolff, the outspoken professor of mathematics and natural science at the University of Halle, was encountering troubles. Under the leadership of August Hermann Francke, the Lutheran Pietists, who were themselves subjected to attack by orthodox Lutherans, criticized Wolff's *Vernünftige Gedancken von Gott, der Welt, und der Seele des Menschen, auch allen Dingen überhaupt* (Rational thoughts on God, the world, and the soul of man, and all things whatsoever, 1720) for its metaphysics, which they considered mechanistic and feared would lead, like Benedict de Spinoza's metaphysics, to fatalism and atheism.[25] The next year Wolff, in his role as acting highest official of the university, delivered his inaugural lecture, "Oratorio de Sinarum philosophia practica" (On the practical philosophy of the Chinese), which praised Confucian moral precepts and the power of reason to obtain moral concepts in general. This work provoked an open break with the Pietists. In 1722 and 1723, Peter sent emissaries to offer Wolff the vice presidency of the projected Petersburg Academy of Sciences,[26] as Leibniz had advised several years earlier; preferring to remain at a university, Wolff declined all offers. Disillusioned at the poor state of the Royal Brandenburg Society of Sciences and considering universities a prerequisite to science academies, Wolff wrote to Blumentrost in June 1723 urging that instead of an academy the Russians found a university stressing scientific studies.[27] In advancing royal science academies, Euler was closer to Leibniz than was Leibniz's disciple Wolff. The czar, who recognized that his capital's geographical remoteness, harsh winters, and intellectual provinciality would make recruitment difficult, received Wolff's promise to supply a list of scholars more willing to accept nomination to the academy. Wolff took seriously the task of preparing his first list of names, which was submitted in April 1723.[28] As Count Aleksandr Gavrilovich Golovkin noted, Wolff "had in mind only a learned society and demanded great achievements and high reputations of those who wanted to join this institution; he did not seek out scholars who wanted to be only professors."[29] None of Wolff's choices were Russian scholars.

Had Wolff decided to accept a position in Russia, his career there would have been stormy. To demean him in the eyes of the czar Francke had, through his acquaintances in the hierarchy of the Russian Orthodox Church, spread rumors that Wolff leaned toward atheism. But Wolff had on his side archbishop Feofan (Theophan) Prokopovich, one of Russia's foremost intellectuals, who was an ideologist and Peter's director of clerical education. Prokopovich was a liaison with Protestant churches and worked to bring the Enlightenment to Russia. His library included the first of Wolff's four-volume *Institutiones mathematicae*, published between 1713 and 1737, together with works of Francke and Philipp Jacob Spener on Pietism.[30] Among the high sciences, mathematics and astronomy were particularly attractive to Prokopovich, who possessed Johannes Kepler's *Epitome astronomiae Copernicanae* (1618–21) and who made his own telescopic observations.

In 1723 the Lutheran Pietists at Halle gave a new argument against Wolff intended for the Prussian soldier-king Frederick William I, claiming that Wolff's determinism would suggest that authorities did not have enough free will to chose to punish a soldier for a crime or disciplinary offense; the punishment would take place only if it was predetermined. This would suggest an act of God, interpreted by state and religious officials. In the eighteenth century severe punishment, including public floggings and hangings, was considered fundamental to military discipline.

Enraged, Frederick William I gave Wolff forty-eight hours to leave Prussia or be hanged. Wolff fled to Marburg in Hesse-Kassel, whose university quickly renewed its offer of a professorship. The expulsion made him a cause célèbre; he was seen as standing against religious intolerance and arbitrary royal power. During the next five years, enrollment at the University of Marburg increased by 50 percent.

With the basic arrangements for the academy now complete, in January 1724 Peter I had Blumentrost and Schumacher submit a detailed plan for it, which he immediately accepted and sent to the senate. On 2 February he corrected a draft of the status of the academy but did not sign any decree. The academy was to consist of three classes or departments: mathematics and physicomathematics in the related sciences of astronomy, geography, and navigation; the whole of the physical sciences, including experimental and theoretical physics, anatomy, botany, and chemical science; and the humanities, embracing rhetoric, the study of antiquities, ancient and modern history, law, economics, and politics. In the eighteenth century physicomathematics had different meanings. But at the early Petersburg Academy of Sciences, it generally entailed going beyond the older mixed mathematics toward a mathematical physics. Following Wolff, Blumentrost and Schumacher proposed that a university be part of the academy and consist of three faculties: law, medicine, and philosophy. An academic gymnasium was to prepare young men for university studies. The purpose of the new academy was not merely to transmit knowledge but to invent and expand it.

In 1724, when Peter set about recruiting members, he commissioned Tatishchev to go on a trading mission to Sweden to enroll twenty-two Russian students in mining and metallurgy courses in schools and factories there. He was also to obtain members for the academy, but this effort garnered no one.[31] Tatishchev had instructions as well to study the Swedish currency. The persistent czar had Golovkin write on 21 October 1724 to Wolff in exile, and he sent a legation headed by the Russian official Berndisz that once more invited him to come to Saint Petersburg to be vice president of the academy. Again Wolff declined, but he sent a list of eighteen recommended men, all but two of whom would be chosen as founding members.

To make his offers more appealing, Peter provided higher salaries, and he ordered construction along the Neva of a spacious, comfortable building, the Kunstkammer (in Latin, Kunstkamera), to house the academy on Vasilievsky Island. Mostly finished and occupied by 1727, the Kunstkammer contained—in addition to a three-story observatory in its dome—a museum for Peter's collection of monstrosities and curiosities in natural history and ethnography, a library on the second and third floors, an anatomical dissecting theater, a conference room, faculty offices, and service rooms. The academy did not gain jurisdiction over it until 1728, and work finishing it took until 1734—that is, the government did not release its control over the building for a few years. The library began by acquiring the personal library of Peter I, rich in geography and military science, and that of his deceased son Alexis, which was filled with religious books, as well as several other private noble libraries. Peter's collection of technical instruments for the study of physics also went to the academy, forming the nucleus of its Museum Physicum. Technical

Figure 2.4. The Kunstkammer, the location of Euler's office. Image © Dimos/ Shutterstock.com. Image used under license from Shutterstock.com.

workshops in which academicians designed and helped produce optical and precision instruments, along with drafting tools, were auxiliary to the enterprise.

After Peter died on 8 February 1725, his successor and widow Catherine I acted rapidly to make the academy functional. The claim that she granted its members an audience with her on 26 August is unlikely because from the start she made it clear that the institution was under her control, not theirs. At a reception the European nature of the academy was displayed when the Basler mathematician Jakob Hermann greeted the empress in French and Georg Bernhardt Bilfinger did so in German. Catherine I put the finishing touches on Peter's plans by ratifying these as the *Project* of 1724, formally nominating Blumentrost to be president (to the approbation of assembled members) and confirming Schumacher as its first secretary and librarian.[32] When she assumed power Catherine could not write her name; as such, Menshikov acted as regent. Menshikov, beginning as Peter's drinking buddy and then becoming governor of Saint Petersburg, had assisted with Peter's reforms. Throughout the summer and fall of 1725 the academy members arrived in Saint Petersburg. By November, seventeen of the eighteen members—comprising fourteen Germans (seven of them from the University of Tübingen), three Swiss, and one Frenchman—were present; only the German Sinologist and classical historian Gottlob (Theophilus) Siegfried Bayer from Königsberg had not yet arrived.[33] The dual responsibilities in research and the transmission of knowledge included institutional duties outlined in Peter's *Project* of 1724. Members were to meet weekly to present and discuss scientific topics, assembling at 4:00 p.m. on Thursdays and Fridays. Annually there would be three public meetings of the academy with scholars speaking on their sciences, and each member had a joint appointment with the title of professor at the academic university.

Among the academicians at Saint Petersburg who investigated the chief currents of scientific thought, the most distinguished was Hermann, who received the largest pension, at two thousand rubles per year, and held the wise counselor designation, Nestor, of the Basel delegation of three. A student of Jakob Bernoulli and defender of Leibniz's differential calculus, Hermann had on Leibniz's recommendation been elected a member of the Royal Brandenburg Society in 1701 and had taught at the Universities of Padua and Frankfurt an der Oder. His magnum opus, *Phoronomia, sive de viribus et motibus corporum solidorum et fluidorum* (Phoronomis, or the forces and motions of bodies solid and fluid), published in 1716 in Amsterdam, attempted to apply the new calculus to the mechanics of solid bodies and fluids. Primarily because of his age (he was fifty-seven) Hermann received from his colleagues the lofty title of *professor primarius et matheseos sublimioris*. The other Baslers were Daniel Bernoulli, who worked among the first rank of physicists for eight years at the academy, and his brother Nikolaus II Bernoulli, his father Johann's favorite son; their father was named a distinguished foreign member.

The oldest academician (by one year over Hermann) was the country pastor Johann Georg Leutmann, a theologian and archaeologist from Wittenberg. On Hermann's recommendation, Leutmann shifted to the study of practical mechanics, physics, and optics. A skilled technologist, he improved clocks, calligraphy, and glass grinding. The contract of another natural philosopher, Christian Martini, stipulated that he would advance physics according to Wolff's principles; at the fourth meeting of the academy he claimed to have found a way to a perpetuum mobile. Afterward Martini was considered incompetent in physics and moved to logic, metaphysics, and ethics. At the academy's second meeting the Swabian pastor Georg Bernhardt Bilfinger (or Bülfinger, whose twelve fingers were perhaps hereditary), a noted Wolffian philosopher who had also studied mathematics, physics, and botany with Wolff, moved from logic to complement Martini in metaphysics and replace him in experimental and theoretical physics. Bilfinger had introduced Wolff's ideas at the University of Tübingen, which provoked hostility over questions of his orthodoxy and set back his academic career, so he enthusiastically accepted the Russian invitation. The academy resolved that Bilfinger should start lecturing four times a week on physics, using as a text Willem 'sGravesande's *Philosophiae Newtonianae Institutiones*, published in 1723, which drew principally upon Newton's *Principia* and *Opticks* and recommended giving experimental proofs of scientific principles. Schumacher had met 'sGravesande in Leiden and had come to admire his work. Schumacher likely influenced the choice of texts. The French astronomer Joseph-Nicholas Delisle brought from Paris scientific instruments and equipment ordered by Peter I, made the three-story observatory on Vasilievsky Island one of the finest in Europe, and founded the Saint Petersburg astronomical school.

Of the remaining academy members, all of them from German lands, the diplomat and mathematician Christian Goldbach, the son of a pastor from Königsberg, was the only German who came to Saint Petersburg without a recommendation from Wolff or anyone else; in fact, he came uninvited. Citing his knowledge of ballistics and medicine as well as publications in Leipzig's *Acta eruditorum*, in July 1725 Goldbach requested by mail an academic position from Blumentrost. He was at first refused, but eventually gained a five-year appointment after his arrival in September. Since his knowledge of mathematics was still limited,[34] Goldbach was uneasy with the title of professor. He wanted instead to be chief of protocol. After studying law at the University of Groningen in Holland, where he earned a licentiate in 1712, and then a brief stint at the University of Königsberg, Goldbach had undertaken the life of a wandering scholar, traveling widely across Europe; he had been introduced to Leibniz, attended Oxford University in 1713, and gained a rudimentary knowledge of the mathematical work of John Wallis and Isaac Newton. During these and later sojourns, he developed an impressive proficiency in languages. Goldbach next met Nikolaus II and later Daniel Bernoulli in Venice, and began corresponding with them in 1721 and 1723, respectively. The three examined Diophantine problems, Fibonacci numbers, Newton's method of fluxions, and the nonlinear ordinary differential equation to be subsequently named after Jacopo Riccati by Jean-Baptiste le Rond d'Alembert, in modern notation $dy/dx = py^2 + qy + r$, when p, q, and r are given functions of x. Having declined a professorship at the University of Padua, the noble Riccati had through his private studies introduced Newton's fluxions into northern Italian city-states, and as the number of differential equations increased he sought to solve them by replacing special techniques with general rules. He reduced second-order ordinary differential equations to first order and later, with Euler and Alexis Claude Clairaut, developed criteria for integrability. Riccati received considerable attention for being the first to integrate his own nonlinear differential equation for several cases.

Other academy members were Johann Simon Beckenstein in jurisprudence, Michael Bürger in chemistry and practical medicine, Johann Kohl in church history, and in mathematics the philosopher Friedrich Christoph Mayer, who worked with Goldbach and along with Christian Friedrich Gross in moral philosophy was part of the Bilfinger retinue. The outstanding Württemberg anatomist Johann Georg Duvernoy (or Duvernois) had studied medicine in Basel and Paris, and had been Albrecht von Haller's doctoral adviser at the University of Tübingen; he was named professor of anatomy, surgery, and zoology. The Saint Petersburg police force provided him with corpses for study, and he received such exotic animals as an elephant, lions, leopards, and once—so he claimed—an animal hermaphrodite. During the winter of 1726, it is presumed, he gave a public lecture on the elephant and its dissection. Duvernoy brought with him two talented young mathematicians: Josias Weitbrecht, who quickly switched to anatomy and physiology and became an authority on them, and Georg Wolfgang Krafft, who would in time become a close friend and collaborator of Euler's. Weitbrecht and Krafft were appointed adjuncts, as was the historian and geographer Gerhard Friedrich Müller from Würrtemberg. Until 1728 Müller taught Latin, geography, and history at the academic gymnasium.

Not counting Hermann, Leutmann, and Delisle, more than half the new academicians were under the age of thirty. This meant that the institution was not overburdened with a strong hierarchy of seating by ranks, as in the Paris Academy of Sciences, or with stressing experimental philosophy, to the near exclusion of speculation, as was advocated by most senior fellows at the early Royal Society of London. The opportunities for the younger academicians were most promising, and they would write letters declaring rosy expectations.[35]

With the arrival in Saint Petersburg of all but Bayer, on 13 November 1725 the academy held its initial—though unofficial—meeting. For the first time the scientific ideas of Newton and Descartes were formally introduced into Russia in a learned setting, but the discussion was in Latin, not Russian, and no Russians attended the meeting. Cartesian science remained preeminent on the Continent, but Newtonian dynamics and optics were beginning to challenge it; studies of the tides, the nature and orbits of comets, and planetary and lunar motion were under way, with steadily improving telescopes and more accurate surveyor's instruments now available to determine which system was superior. At the meeting, which was essentially a panel discussion, the academy had as its problem the shape of Earth taken from book 3, problem 19 of Newton's *Principia.* Newton had demonstrated that because gravity diminishes at the poles from the diurnal motion of Earth, the planet's circumference had to be larger at the equator and flattened at the poles; it would be not a perfect sphere, but an oblate spheroid. But along with centrifugal force and rectilinear inertia, the Cartesian *tourbillons* or vortices—whirlpools of ether of differing density in the heavens—were supposed to explain celestial motions and project instead an elongated ellipsoid caused by Earth's rotation in the ether. A passage from Voltaire's *Philosophical Letters* of 1734 states the crux of the continuing dispute: "In Paris you picture the earth as shaped like a melon; in London it is flattened on both sides."[36] Resolving the issue took almost another two decades; at that time it was principally members of the Paris Academy who sufficiently gathered globally separated, precise measurements of arcs of meridian. Beginning with a globe shape for Earth, they computed the complete circle for meridians based on the arc measures at different latitudes to determine the size of Earth in each; the arc sizes near the equator and north pole were crucial. Euler would later begin with a fluid Earth and make computations with new differential equations in fluid mechanics to determine the shape more exactly.

Delighted at the arrival with Delisle of more powerful telescopes, at the 13 November meeting the genial Hermann did not oppose Newton's position, though he indicated that the related computations in *Phoronomia* were more accurate, while Bilfinger rejected the shape of Earth as Newton had calculated it. Hermann's interest in Delisle's telescopes extended to another question that intrigued him: whether they could confirm the Cartesian, Newtonian, and Wolffian belief that there was human life on the moon. The debate over extraterrestrial life was long standing; even before Johannes Kepler spoke of men on the moon belief in a plurality of human worlds had existed and had gained increased support in scientific circles through Fontenelle's *Entretiens sur la pluralité des mondes* (Conversations on the plurality of worlds, 1686, commonly referred to as *Mondes* at that time), which marked the high tide of Cartesian thought in the sciences, and Christiaan Huygens's posthumous *Cosmotheoros: sive de Terris Coelestibus, earumque ornatu, conjecturae* (Cosmotheory: or, conjectures concerning the inhabitants of the planets, 1698). Fontenelle's *Mondes* intended to make the discoveries in science intelligible to the small but growing reading public, and it was an inspiration to Huygens.[37] The *Cosmotheoros,* an *apologia* of an eminent man of science, is chiefly a fantasy speculating on the existence of life on other planets in our solar system and around other stars, a notion that seized the popular imagination; it was soon translated from its original Latin into English, French, German, and Russian. Czar Peter I had ordered the Russian translation, completed by Bruce in 1717. The 128-page work touches upon Copernican astronomy, the research of Tycho Brahe and Johannes Kepler, and Descartes's theory of vortices along with Newton's inverse-square law of gravitational attraction for explaining the elliptical orbits of planets and the paths of comets. Huygens advanced a theory of vortices differing from that of Descartes, holding that the planets swim in an ether at rest; the czar may have helped write the introduction to the book. On scientific grounds and the religious belief in the uniqueness of humanity, the dispute over the existence of extraterrestrial life remained popular throughout the Enlightenment.[38]

A Fledgling Camp Divided

On 7 December 1725, Catherine I sent a decree to the Russian senate confirming the establishment of the academy following Peter's plan from 1724; oddly, however, she did not ratify the decree. Catherine named Blumentrost the academy's president; born in Moscow to a German physician in Russian service, he was an excellent choice, because he spoke flawless Russian and read adequate Latin and Greek. Blumentrost had studied medicine at the Universities of Halle, Oxford, and Leiden, had supervised the imperial library, and had been indispensable in planning the academy. His background indicated that he could work well with foreign scholars, and he was influential at the imperial court.

At their first official public meeting on 7 January 1727, the academicians met in a conference room of Baron Peter Pavlovich Shafirov's palace. The renovation of their official building, the palace of Peter I's widowed sister-in-law Princess Praskovea, which was to be for meetings and typography, would not be completed until 1728, nor was the Kunstkammer ready in 1726. (The Princess's palace contained the library and observatory until the Kunstkammer was completed.) At this opening meeting the members discussed the physics theories of Leibniz and Wolff, a subject that was to be a continuing preoccupation among the academicians.

The academy normally met on Tuesdays and Fridays at 4:00 p.m.; these conferences were lively, sometimes heated. Although all members accepted the great potential of the calculus invented independently by Newton and Leibniz, the academy split into opposing factions in mechanics, with neither party uncritically and entirely accepting its own theory.

Cultural nationalism among French and German savants opposed to the British contributed to the quarrel, as did personal intellectual rivalries. Bilfinger headed the Wolffian majority; Hermann, who endorsed much of Leibniz and the work of Jakob (Jacques) and Johann I Bernoulli in Basel, generally opposed Newton and allied with the Wolffians, while Daniel Bernoulli led the Newtonians. The issues were the validity and precision of each science; the animated debate between the partisans of Wolff and the champions of Newton was doubtless in a bewildering jumble of Latin and German dialects.[39] Hermann and Bilfinger resided together, and Goldbach attributed the alliance between them to the persuasiveness of the younger man in discussions that extended beyond the academy. The debate generally showed that physics, previously experimental and speculative, was becoming a more modern mathematical science.

Bilfinger was not new to the polemic and its distinctive character in central Europe and Russia. His *Commentatio hypothetica* of 1723 had supported Leibniz's doctrines of monadology and a preestablished harmony. Addressing the question of how the mind knows the body or matter and thus enables learning, Leibniz had rejected the Cartesian doctrine of occasionalism in which the communication nexus between them was thought to occur in the pineal gland and God alone was responsible for all fundamental causal activity. The notion that all events were a continual miracle, God using mind and body simply as an occasion for cause and effect, Leibniz had found untenable. To explain the concomitance between mind and body he posited instead a divine clockmaker who creates two perfectly synchronized components and then steps aside. Leibniz rejected the idea of God as a perpetually tinkering clockmaker for this contradicts the idea that God is a perfect artificer. Here he was to draw the ire of Newton, who believed in providence or *bienfaisance*. The *Commentatio hypothetica* was put on the Roman inquisition's *Index librorium prohibitorum* (Index of prohibited books), a catalog of books that Catholics were forbidden to read. Bilfinger, who also endorsed Leibniz's view that space, time, and motion are relative, saw himself as the leader in defending and disseminating Leibniz's ideas. He compared his role to that of the English theologian and metaphysician Samuel Clarke, a major supporter of Newton's dynamics, notably in Clarke's addenda to Jacques Rohault's *Traité de physique* and correspondence with Leibniz. Shortly after arriving in Saint Petersburg, Bilfinger wrote to Clarke, politely defending Leibnizian science. By this time Bilfinger had in his *Dilucidationes philosophicae* already begun modifying his Leibnizian views but still opposed Newton's; he shrewdly coined the term *Wolff-Leibnizian succession* to proclaim the durability of Leibnizian thought. Yet Wolff

Figure 2.5. Gottfried Wilhelm Leibniz.

appropriately opposed the equivalent phrase *Leibnizo-Wolffian* that his Halle critics proposed, feeling that his dogmatic rational methodology and science, now in a formative stage, were not simply an extension of Leibniz's ideas; there were to be substantial modifications.

From the late seventeenth century through the Enlightenment, scholars debated whether faith in written revelation or trust in reason was the superior path to truth. Generally the literate community believed in a providential order of nature with periodic divine interventions that, taken together, explained the creation of the physical universe and certain phenomena in it such as comets and earthquakes. It was thought that God continuously supervised the universe, providing occasional infusions of energy and mending it. Leibniz's physicotheology conceived the universe as a natural machine created by God and thereafter operating without need for further divine intervention. The belief in an interceding deity in nature did not diminish in the physical sciences until the development of Newtonian dynamics by the end of the Enlightenment, when Newton's God became indistinguishable from Leibniz's.

At the close of the seventeenth century, Leibniz and the French Huguenot refugee Pierre Bayle had been the principal figures in the dispute between faith and reason. Bayle's *Dictionnaire Historique et Critique* of 1697, considered the summa of Pyrrhonism or skepticism, argues that while both faith and reason must lead to the same truths, in the case of incongruities faith is the higher arbiter. That same year, Leibniz's *De Rerum Originatione* (Ultimate origination of the universe) depicted a Platonic universe in which God geometrizes. His *Théodicée* (Theodicy, 1710) again claims that the laws of nature emanating from the mind of God must be rational; the truths of faith and the truths of reason must agree, but when they do not, reason is supreme in resolving all problems, including dilemmas in theology. The *Theodicy* was published anonymously, a practice common at the time for works on controversial topics, but the reading public knew who the author was.

After Descartes, Leibniz was the foremost contributor to continental rationalism. The concept of unity pervades his work: the sciences are not isolated but part of a universal science, including the mathematical and physical sciences, metaphysics, and jurisprudence. The universal science has a single method of demonstration and of discovery or invention, which is capable of either pursuing deductive certainty or estimating probabilities. Leibniz rejected the methodological break with the past undertaken by Descartes and Francis Bacon. His *philosophia perennis* (the name given by Leibniz scholars to his concept) proposed a new rationalism, a general logic founded on three axiomatic principles: predicate-in-notion, sufficient reason, and contradiction, the last two of which stemmed from Plato's *Timaeus*. Leibniz assumed that the concepts of the subject of every true proposition were more sweeping than the concepts of the predicate; they independently expressed the whole universe and thus contained all possible predicate concepts. Leibniz's principle of predicate-in-notion reversed the traditional Aristotelian syllogism, which sought to confirm the truth of propositions, instead making the mind an engine for discovery. Extending the Latin statement *nihil est sine ratione* (nothing is without a reason), the principle of sufficient reason covered physical causality, relations between antecedent and consequence, and teleology. Leibniz coupled sufficient reason with the principle of greatest perfection, borrowed from Aristotle's *summum bonum* (the highest good) and leading to the metaphysical law of the "the best of all possible worlds." This law did not refer to what the British historian Edward Gibbon would describe in 1781 at the end of the third volume of his *Decline and Fall of the Roman Empire*, that "every age of the world has increased and still increases the real wealth, the happiness, the knowledge, and perhaps the virtue of the human race."[40] Rather, Leibniz posited that in each passing instant God selects the best from an infinite number of possible worlds for gaining the greatest perfection and harmony; perfection requires a maximum of variety and a superior order that fosters this diversity. Elaborated by Johann I Bernoulli and Isaac Newton, sufficient reason became the guiding principle of classical physics: if predicate notions repeat subject notions, propositions are identical; applying appropriate substitutions in proofs and arriving at noncontradiction in a finite number of steps establish consistency and identity.

At the core of Leibniz's metaphysics and science was the distinction between the two levels of truths: the necessary truths of reason and the contingent truths of fact. The necessary or absolute truths of reason applied to the deeper world of the monads and were independent of testimony deriving from the senses, while contingent laws of nature could be discovered by observation, experiment, and induction in the physical world, but they had a rational foundation. Contra Locke, who adopted the Aristotelian view that the mind is a passive *tabula rasa*, a blank tablet upon which the senses write, Leibniz followed Plato in believing that the mind is active and filled with necessary truths. The source of contingent truths is sensation, while necessary truths derive from reflection. Leibniz presented these ideas in his *Nouveaux essais sur l'entendement humain* (New essays concerning human understanding), which was completed in 1704, but he refused to publish it when he learned that Locke had died. To his conscience, an argument against a recently deceased opponent was an unworthy one.

Leibniz's science was founded on the doctrine of monads, his theory of energy underlying matter. Leibniz rejected the hard, indivisible, passive atoms of the ancient Greeks as the ultimate, smallest particle of matter and disagreed with Descartes's equating of matter with extension. These could not resolve Zeno's paradox on the beginning of motion, validate Leibniz's belief in an organic continuum in nature, or bridge the chasm between the metaphysical and physical worlds. The monads are percipient, elastic, geometric points endowed with an entelechy—a primal dynamic. Following Thomas Hobbes, Leibniz assumed that they possess a *conatus* or tendency to motion, which provides for a continuum of motion. The concept of continuity deriving from the medieval law that nature does not leap and the reality of a preestablished harmony were essential to the monadic doctrine: monads are well ordered by the degree of clarity in their rational perception of the world, receding from the absolute clarity—God—to the basest element. Since the monads are windowless, mind and matter are attuned to each other by the parsimonious preestablished harmony. Reason may go awry, however, in examining two labyrinths: the continuum and the human freedom to choose.

Leibniz believed the study of the world at large to fall preeminently to the field of physics, and he began to apply the calculus to solve its problems. He accepted generally the Cartesian mechanistic explanations—impulsion by contact and by the ether in the heavens—and came to reject Newton's concept of gravitational attraction occurring among bodies over a distance without any intervening medium. Since Newton did not give an underlying mechanical explanation of action at a distance, Leibniz charged that attraction belonged to the discredited Scholastic concept of *occult quality*. At the foundation of physical phenomena, Leibniz proposed a principle of conservation. For the correct measure of force, he rejected the Cartesian and Newtonian concept of momentum (quantity of matter multiplied by velocity, mv), and accepted Huygens's *vis viva* (in modern notation roughly twice kinetic energy, mv^2), but he exceeded Huygens in asserting its general conservation. Derived from different old ways for determining *vis viva*, it could also be written in modern notation as $mv^2/2$.[41] This all suggests that Leibniz was groping toward an intuitive conception of the energetic principle, the conservation of energy. His principle of conservation went against the Newtonian belief in occasional divine interventions in the universe. In correspondence with Samuel Clarke, Leibniz found erroneous Newton's view that space, time, and motion are absolute and proposed instead that they are entirely relative and relational, a position difficult to defend before the separate inventions of non-Euclidean geometries. Leibniz rejected Newton's position that space is the sensorium of God, contending in part that God does not need this organ to order the universe.

For its breadth and extensive applications, Leibniz exalted mathematics and made seminal contributions to it. Among these, he posited mathematics to be the universal language of the sciences and to have a dual role: it reduced the imprecision in the sciences and provided a technique for discovering new facts from a body of knowledge. His emphasis on the algorithmic nature of method made him a founder of formalism in mathematics as opposed to intuitionism. Leibniz is chiefly known in the field for being, like Newton, an inventor of the powerful differential calculus that includes the use of infinite series to compute tangent slopes, trigonometric lines and their inverses, and the quadrature of curves. His superior notation includes the integral sign from the first letter of the word *summa* as well as *dy* and *dx*. His research advanced algebraic analysis against the dominant geometric synthesis; this continued the tension between analysis and synthesis that had existed since Euclid.

Leibniz never presented a coherent and comprehensive account of his complete philosophy. He was a baroque philosopher in a baroque world, and most of his extensive collection of letters, articles, and books long went unpublished. Leibniz's only major book printed during his lifetime was the *Theodicy*. His *Discours de métaphysique* of 1686 was not published until the nineteenth century; his *Nouveaux essais sur l'entendement humain* appeared in print in 1765; and his *Monadologia* from 1714 and the Leibniz-Clarke letters from 1715–16 were published posthumously in 1717. Many of Leibniz's private papers were unknown, and his articles in *Acta eruditorum* on calculus were too difficult for readers other than a select few,

Figure 2.6. Christian Wolff.

including Jakob and Johann I Bernoulli;[42] outside their circle and excepting Paris academicians, scholars of this time knew him mainly as the Leibniz of the *Theodicy*. Few within the European republic of letters in the early eighteenth century grasped the full scope and profundity of his contributions to mathematics and natural philosophy.

The task of systematizing Leibniz's thought, transmitting it to a wider public, and articulating or modifying it fell primarily to his disciple Wolff, a father of the German *Aufklärung* (Enlightenment). According to Immanuel Kant, Wolff introduced into German philosophy the spirit of exactness and rigor—that is, *Gründlichkeit* (thoroughness)—in covering the detail of a subject and provided an impetus for clarity and the precision that became the chief pursuit of the *Aufklärung*. Writing most of the forty quarto volumes of his books in German created a vocabulary for German philosophy, and it urged that in instruction and research at German universities the secular language replace Latin. During the 1720s, Wolff's philosophy began to displace the Peripatetic thought of the Wittenberg professor Philipp Melanchthon that had been preponderant at German universities since the sixteenth century. From 1713 to 1725, Wolff's seven volumes of *Vernünftige Gedanken* gave a comprehensive exposition of his thought. Prominent among these were *Vernünftige Gedanken von Gott, der Welt und der Seele des Menschen, auch allen Dingen* (Rational thoughts on God, the world, and the soul of man, and all things whatsoever, 1720), which offered a German metaphysics, and *Vernünftige Gedanken von den Wirkung der Natur* (Rational thoughts on the operation of nature, 1723), which was a German cosmology.

Yet by 1725 two essential Wolff works, both in Latin, had not yet appeared: the *Philosophia prima rationalis sive Logica* (Preliminary discourse on philosophy in general, 1728), giving a systematic account of Wolff's philosophy, *Philosophia prima sive ontologia* (First philosophy or ontology, 1730), and the *Cosmologia generalis* (Universal cosmology, 1731), which describes the visible world as a perfect machine and gives the laws of motion by which it operates.

Wolff altered and vulgarized Leibniz's ideas and made these a foundation for his philosophy. In methodology, Wolff was a strict rationalist, but he centered rationality in the principle of contradiction instead of in sufficient reason and attempted to prove it. In mathematics Wolff stressed Euclidean deductive synthesis over Leibniz's infinitary analysis or calculus; his method was one of neo-Scholastic formalism. In the theory of matter, Wolff replaced Leibniz's inanimate category of monads with passive ultimate corpuscles. His indivisible and nonextended *atomi naturae* followed the laws of mechanics; Wolff's corpuscular theory was thus strictly mechanistic and deterministic, though he retained an element from Leibniz's search for designs in nature constructed to benefit humankind. Wolff accepted a preestablished harmony but revised Leibniz's concepts of relative space and time: space, he posited, is an order of things, whereas time is the order of successive events within a continuous series.

Until the second edition of Kant's *Kritik der reinen Vernunft* (Critique of pure reason) in 1787, the Wolffians dominated philosophy at German universities. At midcentury, Leonhard Euler was to be their major critic in religion, mathematics, and physics. Over time, Wolffian philosophy would prove unsatisfactory.

At the academy's meeting on 5 March 1726, during an opening year that was characterized by disputes that tested the civility and decorum of the members, Bilfinger criticized a report by Daniel Bernoulli. Ten days later Bernoulli found fault with Hermann's *Phoronomia*, pointing out an "erroneous" proof in it and expounding what he thought to be a better version. Departing from his generally calm demeanor, Hermann stoutly defended his result. The debate became so heated that President Blumentrost ordered both parties to calm down and asked Bernoulli to apologize to Hermann. The principal controversy involved Huygens's and Leibniz's law of the conservation of *vis viva*. The argument had already defined the 1724 Paris Academy's Prix competition on the communication of motion in hard bodies as the accurate measure of force. There Johann I Bernoulli submitted a paper that was disqualified for rejecting the existence of ultimate hard bodies and holding hardness to pertain to perfectly elastic bodies under infinite pressure, while the Scottish mathematician Colin Maclaurin won, not only basing physical laws on momentum but also criticizing the use of *vis viva* to resolve the question;[43] Bernoulli was instead awarded an *accessit* (honorable mention).

In the 1726 Paris Academy competition on the perfect or imperfect impact of elastic bodies, Bernoulli resubmitted his paper with an addendum on the physical cause of elasticity; he again took an *accessit*, losing this time to the Cartesian Pierre Mazière, a priest of the Oratory who based the laws of collision of elastic and hard bodies strictly on momentum.[44] The virulence of the March 1726 meeting of the Petersburg Academy of Sciences bespoke the reach and intensity of the *vis viva* controversy there.

At the meeting on 10 May, just as Bilfinger began to read his paper "De viribus corpori moto insitis et illarum mensura" (On the essential forces of bodily motion and their measure), which discounted Huygens's view of equal proportion between actual downward motion and potential upward motion, where force equals mv^2, Daniel Bernoulli angrily exclaimed "Errasti! Errasti!" (False! False!). In the *Journal des Sçavans* in 1669, Huygens had presented his thesis that only the elevating force and not the entire amount of energy is conserved. In this, Bernoulli considered Huygens more accurate than Leibniz.[45] It took three meetings for Bilfinger to complete reading his paper. When Duvernoy tried to lessen the row by changing the subject, Nikolaus II Bernoulli brusquely declared that Duvernoy did not understand the matter under discussion.

At the 14 June meeting, another clash setting Daniel Bernoulli against Hermann and Bilfinger brought a reprimand from the president; at this assembly Hermann read a paper, "De calculo integrali," claiming that Johann I Bernoulli had alone invented integral calculus. Nikolaus II Bernoulli, who knew the work of Newton, declared this account to be biased. The sudden death of Nikolaus on 9 August robbed Daniel of his most potent ally and best friend. At the academy, he succeeded Nikolaus as a professor of mechanics.

In "De mensura virium corporum" (Estimate of the force of bodies), which appeared in the first volume of the *Commentarii* for 1726, Hermann presented a standard defense of the concept of *vis viva*. It began with a review of the contributions by Galileo Galilei, Evangelista Torricelli, Giovanni Alfonso Borelli, Marin Mersenne, Johann I Bernoulli, 'sGravesande, Huygens, Edme Mariotte, Wallis, Christopher Wren, and the Cartesians François de Catelan and Denis Papin, and followed with the position of Newton as expressed in the correspondence between Leibniz and Clarke.[46] Hermann concentrated on proving the validity of the conservation of *vis viva* by examining the collision of a perfectly elastic sphere A, having quantity of matter 1 and velocity 2, with a motionless globe C of quantity of matter 3. The impact repulses A with velocity of

Figures 2.7a–f. a) Laurentius Blumentrost, b) Georg Bernhardt Bilfinger, c) Daniel Bernoulli, d) Nikolaus II Bernoulli, e) Jakob Hermann, and f) Joseph-Nicholas Delisle.

1. If A then collides with a motionless globe D of quantity of matter 1, it communicates the velocity to D and comes to rest. This meant for Hermann that force is 3 + 1, or 4. Thus, in this case, where $m = 1$ and $v = 2$, the amount of force imparted is 4 or mv^2, 1×2^2, and not the Cartesian momentum of mv, 1×2. But Hermann failed to keep straight the direction of the force vectors; the direction reverses at collision. In addition, the total 4 can be 2^2: 2 times 2. The paper proved nothing.

In addition to papers by Hermann, Bilfinger, and Wolff defending the principle of the conservation of *vis viva*,[47] the first volume of the *Commentarii* contained Daniel Bernoulli's article "Examen principiorum mechanicae et demonstrationes geometricae" (Examination of the mechanical principles and geometric demonstrations), which sought to show that Newton's parallelogram theorem of forces could be proven analytically without further assumptions—that it is an a priori or necessary truth and not a contingent one. This theorem holds that a body acted upon simultaneously by two forces that may be represented in magnitude and direction as two sides of a parallelogram will describe a diagonal of this parallelogram. The young Bernoulli's leadership, and the time he spent on elaborating Newton's dynamics—including solving difficult problems irresolvable with geometry but achievable with the new calculus—was to vex his father Johann I, who believed that his son was not only heading a competing camp but encroaching upon his own attainments.

In 1727 the quarrel among Cartesian, Leibnizian, and Newtonian camps regarding mechanics intensified. After Daniel Bernoulli and Leonhard Euler had reported on the water outflow from vessels, Johann I Bernoulli's article "Theoremata de conseruatione virium vivarum" (Theorems of the conservation of living forces) was presented in October to the academy for the second volume of the *Commentarii*; it defended *vis viva*.[48] When Bernoulli supported the Newtonian theory of capillarity by using theorems from book 2 of Newton's *Principia* Bilfinger rejected his argument, and afterward he took issue with the Newtonian insistence that Cartesian vortices could not explain planetary motion.[49] Bilfinger's Prix de Paris paper for 1728, "De causa gravitatis physica generalis physica disquistio experimentalis" (Inquiry into the general physical cause of gravity), published in Paris by Claude Jombert as a forty-page pamphlet, defined vortices as the cause of gravity, and hypothesized that the ether composing them is so rarified that it can penetrate ethers across different vortices without affecting the velocity of each. It is thus the vortices that account for planetary motion.[50] This paper, which also repeated the charge that attraction at a distance was an occult quality, won the Prix de Paris for 1728. It appeared in print one year after the Paris Academy's *Histoire* containing Fontenelle's "Éloge de

M. Newton." The "Éloge" described Newton's dynamics as building upon the work of Kepler and Galileo, and it destroyed the Cartesian system of vortices.[51] Bilfinger's paper was timely in the controversy but was soon forgotten. From England, Clarke worsened the debate by accusing his opponents of trying to besmirch the name of Newton.[52]

At the Petersburg Academy Bilfinger referred to Newton's law of gravitational attraction as a "vulgar hypothesis," and Bernoulli subsequently refused to speak to him. For his part, Bernoulli portrayed the Cartesian hypothesis of vortices as insufficient and not worth saving. Over the next three years, the rift between Bernoulli and Bilfinger widened, each seeking support for his science by conducting experiments on the capillarity of fluids in an array of vessels. The nub of their dispute was about more than gravitational attraction; it extended to the status of the sciences in the world of learning and approaches toward investigating them. Since Bilfinger lacked training in higher mathematics, Bernoulli believed that he could not comprehend anything about physical phenomena. Subsequently Euler also insisted that preparation in higher mathematics was necessary for anybody wanting to make fundamental studies of physical phenomena.

In 1727 the academy's education branch consisted of the gymnasium and the university. Only the gymnasium had a good beginning, and it soon joined the rest in dire straits; it first enjoyed an enrollment of 112 students, coming mainly from foreign families in Russia with a few from the Russian aristocracy. A combination of factors worked against continuing this promising start, however: the nobles wanted a faculty that could speak Russian; they resented the absence of a policy that would discriminate among applicants by social order, disliked the emphasis on the sciences over the humanities, and found frustrating a haphazard curriculum responding to the interests of individual faculty members. All of these factors reduced the numbers of noble sons at the gymnasium. Instead the Russian aristocracy established "estate schools" exclusively for the nobility. These schools, along with the death of Peter I, undercut the development of the relatively open gymnasium. By 1729 enrollment had fallen to seventy-four, and by 1737 it was only nineteen.

The academy's university quickly encountered problems. Lectures had to be in Latin and classes were required to have a minimum number of students. Russia lacked adequate preparatory schools: together with the engineering, artillery, and naval schools in Saint Petersburg, the Moscow school of mathematical and naval sciences stressed scientific subjects, but they were designed to prepare future officers for careers in military strategy, artillery, and fortifications.[53] Weak academically, they gave little training in Latin, and seminaries opposed enrolling their students in a secular university. Eight students from Vienna represented almost the entire student body at the new university.[54] To meet the minimum class size, some academicians attended the lectures of their colleagues. But the decline in enrollment at the university as well as the gymnasium freed academicians of teaching duties and allowed them to devote their full time to scientific research.

In 1726 and 1727, the poet, future diplomat, and prince Antiokh Kantemir was the principal student at the gymnasium and the university. Born in Constantinople and fluent in Greek, Latin, and Russian,[55] he was deeply interested in mathematics and philosophy and thus drawn to the academy. Mayer taught him algebra and geometry, while the Bernoulli brothers taught him in higher mathematics. Kantemir studied physics with Bilfinger, history with Bayer, and philosophy with Gross. By 1730 the university had no students, but in schooling Kantemir it had sent forth an important partisan of Western science.

To enhance its prestige, the academy required its members to publish. From its start it desired to have a modest publications program of its own that included printing its proceedings under the title *Commentarii*. It was to be divided into three classes: *mathematica*, *physica*, and *historica*. The *mathematica* section would contain articles in pure mathematics, at first emphasizing differential equations and infinite series, and in applied mathematics directing infinitary analysis to resolving problems in astronomy and mechanics. The academy began without a printing press, and it needed one with Latin and German typefaces, but the proceedings were nonetheless edited for publication.

The Entrance of Euler

Euler arrived at the academy to begin his distinguished career in May 1727 at the age of twenty; it was eleven years after the death of Leibniz and two months after that of the eighty-four-year-old Newton. Catherine I, the benefactor of the academy, had died after a two-year reign at the age of forty-three a week earlier, and the struggle for succession was under way; Euler arrived on the day that her death was made known. The capital, wrote Euler, experienced "the greatest consternation, yet I had the pleasure of meeting not only Daniel Bernoulli . . . but also . . . Professor Hermann . . . , who gave me every imaginable assistance. My [annual] pay was 300 rubles along with free lodging, [wood for] heat, and [candles for] light, and since my inclination lay altogether and only toward mathematical studies, I was made adjunct in higher mathematics, and the proposal to busy me with medicine was completely dropped."[56] Hermann and Bernoulli, perhaps along with Goldbach to a degree, had arranged for Euler to enter in mathematics. Euler spent his first morning there with Hermann; that evening he sought out Bernoulli. Euler's initial position was listed as *élève* (student), but it was quickly changed to adjunct. Before Euler's arrival, Daniel Bernoulli had successfully negotiated the financial terms, and Euler was free to engage in the meetings of the academy and give papers that went into the *Commentarii*, which started publication the following year.[57] Until his marriage six years later, Euler, Bernoulli, and others lodged at the quarters the academy provided for all the younger, unmarried academicians; the two fruitfully collaborated on some research in mathematics and physics. Despite small and temporary misunderstandings, Euler's presence pleased Daniel Bernoulli, who following the death of his brother Nikolaus had no other close colleague in Saint Petersburg.

Since nearly half the members of the initial academy were in mathematics and mechanics, these were the principal research subjects. As the first eight volumes of the *Commentarii* for the years 1726–41 show, in mathematics the areas were portioned among members: Mayer covered algebra, geometry, and spherical trigonometry, the last of which Euler would later put in its near modern form. Hermann studied integral calculus, differential equations, elliptical conics, and solid loci in a Cartesian framework, while Bernoulli wrote articles on infinite series, recurrent series, and the integration of differential equations. Goldbach considered integration, transformation series, and limits of infinite series. With the addition of Euler, mathematics and mechanics were to become the academy's forte and continued to be until his departure in 1741. Aside from the academicians there were other Baslers in Saint Petersburg; among them was Isaak Bruckner, who had left his position as royal geographer in Paris to be a mechanical artist and manager of mathematical instruments for the Petersburg Academy. Knowledgeable in mechanics, applied mathematics, and geography, he specialized in constructing terrestrial globes and maps. Euler had known him in Basel.

Euler encountered a dynamic science academy enjoying a splendid era that possibly made it the foremost in Europe in mechanics and mathematics. In this setting his genius flowered, steadied by a temperament that Hermann quickly recognized as prudent, objective in scientific discussions and, though impulsive, reluctant to quarrel. For the academy and associated enterprises, Euler flourished in a multitude of tasks. Applying his powerful memorization and talent for languages, he quickly learned to speak and write in Russian, which many other foreign academicians had thus far failed to do; Euler's "Notebook" for 1727 contains notes on Russian grammar. At the conferences of the academy, he lectured on average ten times a year, while others spoke from one to five times. Because academicians had dual roles in research and instruction, for several years Euler also taught classes in mathematics, logic, and physics at the gymnasium and the university; throughout his career he was an enthusiastic and talented teacher.

Euler sought to advance the academy's goal of building an imperial cadre of Russian scientists to fill the institution, an effort that initially met with almost no success. Unlike universities, the academy encouraged its members to engage in a range of research and practical projects. It also charged Euler with many other duties: the tireless young colleague was an examiner for the cadet corps, and he helped develop tests for the commercial college, the office of weights and measures, and the Saint Petersburg tariff system. He participated in the operation of the sawmill and the fireworks of Andrej Konstantinovich Nartov,[58] the head of the academic workshop. Nartov was well-disposed toward Euler and the two became friends. Euler also drew up a plan for a single-cylinder steam engine according to Papin's principle with water boiled and condensed beneath the piston.

On 25 July 1727, Euler presented a paper to the academy for the first time. The topic given him was the motion of water in tubes bent and inclined. In August of that year Bernoulli wrote to his Paduan colleague Giovanni Marchese Poleni about having hit upon a new "theory of the motion of water, which is very general . . . [and] had been sought by the cleverest geometers, but in vain . . . ; but what is even more remarkable is that at the same time this theory was found by a different method by Mr. Euler of Basel, student of my father, who will do him much honor."[59] In September the Russian Privy Council received a report on the work of the academicians and their plans. That Euler, expanding upon this research, presented to the academy the paper "De effluxu aquae" (Of the flow of water) was not mentioned. Among Euler's paper's thirteen topics was a new theory of water outflow from vessels, upright and bent, that was based on experiments; other leading subjects were water motion in fountains and the theory of sounds,[60] and Bernoulli added sections on the principle of the conservation of *vis viva* and the motion of elastic fluids. The two were assigned a series of problems in hydraulics, which were to be prominent in their early studies. Other members were also pursuing what would later be called hydraulics and hydrodynamics.

Euler's extensive publications in the *Commentarii* began in 1727 with its second volume, in which he had three articles: one on reciprocal trajectories in differential geometry, as it is now termed; one on the tautochrone in the field of mechanics, and one on the elasticity of air in the study of physics. Hermann, astonished at the superior genius that Euler displayed in differential geometry, was immediately taken with the young recruit. After his report to the academy in August 1727, Euler generously set aside classical hydrodynamics so as not to compete in his friend Bernoulli's primary field; the two did not collaborate on the subject for the next decade while Bernoulli was preparing his *Hydrodynamica* of 1738, which was essential to the creation of the subject. In their research the two men complemented each other. Bernoulli's thinking was principally in line with physics; when it was possible, he wanted to avoid mathematics, and after obtaining a mathematical method to solve a physical problem usually did not further develop it. In contrast, Euler, who was primarily a geometer in the wide Enlightenment sense of the term, began his study of a physical problem by drawing upon its nature to formulate the particular mathematical algorithms needed to solve it; he then would proceed systematically to advance and generalize these algorithms. As twenty-nine of the seventy-four volumes in the first three series of his *Opera omnia* attest, Euler—unlike Bernoulli—was taken with the speculative hypotheses in mechanics, optics, and analysis as well as proofs in number theory. Their shared research on hydrodynamics, however, made for some acrimony, which may explain why Euler did not engage in research on the subject for almost twenty years.

The unexpected death of Catherine I threw the academy into a time of insecurity. In a palace revolt her chief minister, "the Goliath" Menshikov, fell before the reactionary Old Believer Russian nobility hostile to Petrine reforms and was sent into exile. By September the Dolgoruki clan controlled Catherine's named successor, twelve-year-old Peter II, the grandson of Peter the Great.[61] Considering the academy an intrusion into Russian culture on the part of German, Swiss, and French aliens, failing to understand the importance of scientific inquiry, and believing that it subverted traditional Russian culture and the Orthodox faith guarded by its priests, the Old Believer nobles treated the institution with indifference and sometimes hostility. Governmental ministers among them withheld payments for it, delayed paying salaries, and froze funding for new positions.

It is often believed that Euler, having not yet established his reputation and lacking independent income, sought another reliable funded position. Many older secondary works repeat the claim that on the basis of his Paris Academy *accessit* prize on ships' masts and courses in physiology, he became a medic in the Russian navy. There is no record of this or of rear admiral Pieter de Sievers's alleged praise for his work. The navy, at any rate, was not a small enterprise in Russia. Employing the British model, Peter I had built it from exactly one obsolete vessel to forty-eight major warships comparable to the best in the British fleet, along with 787 minor and auxiliary craft. The navy grew to more than 28,000 men. Russia quickly became the chief naval power on the Baltic Sea, and Saint Petersburg was at the center of an expanding Russian shipbuilding industry during what became the fully transatlantic Enlightenment.

While closest to Bernoulli and soon Goldbach, Euler benefited from working with Hermann. Yet Hermann, while comfortable in his well-paid, prestigious position, wanted to return home to Basel. After failing to gain the professorial chair in ethics in Basel in 1706 and in physics in early 1727 (the latter of which Euler had also sought), he finally succeeded in 1727 in obtaining through the faculty lottery the chair of ethics. Petersburg Academy officials refused to release Hermann from his five-year contract, so the University of Basel allowed a vicar to substitute for him until its conclusion. Euler profited from Hermann's extra years in Saint Petersburg; the two collaborated in studying and better resolving significant calculus problems (though their solution for the brachistochrone was mistaken) and they worked together in developing differential equations for mechanics. Hermann helped introduce Euler to studies of the nature of heat and fire, which—along with the investigation of electricity, magnetism, light, and color—were prominent among topics to which Enlightenment savants in the sciences turned. Hermann and the Bernoullis had questioned the dominant phlogiston theory of combustion formulated by the German physician Georg Ernest Stahl. Phlogiston was supposed to be one of three types of earth comprising all matter, and substances rich in it were said to burn readily; flames were believed to be the whirling of phlogiston as it escaped. In Western Europe the phlogiston theory was to prevail until the 1770s, when the French chemist Antoine Lavoisier discovered oxygen and its role in combustion. But the Petersburg academicians had earlier challenged it; Hermann and the Bernoullis had developed instead a rudimentary mechanical theory of heat in which the movement of the smallest particles of matter produced it. Meanwhile, during the second half of 1727, Goldbach had developed a close friendship with Euler that survived the former's temporary assignment to Moscow; initially he wrote to Daniel Bernoulli, but in October 1729 a steady correspondence with Euler began that lasted until 1764.[62] The bulk of the two hundred letters exchanged with Goldbach delineate Euler's research and results in number theory, and to a lesser degree they deal with differential calculus and the theory of infinite series.

Seeking to push back the modernizing reforms of Peter I, nobles among the Old Believers, the *starovery* or *staroobriadtsy*, returned the capital to Moscow in January 1728. When they moved they took with them the academy's president, Blumentrost, continuing him in the influential position of royal physician. Blumentrost sent a letter to academy members expressing his continuing highest hopes for its success. Goldbach, the conference secretary in deed though not officially, also left to be the czar's tutor. Hermann was ordered to write volumes in the *Abregé de mathématiques*, a textbook on elementary arithmetic for the young monarch. Delisle was ordered to write an introduction to astronomy; while he could include the Copernican system, he treated it cautiously, not wishing at this time to provoke the ire of the Holy Synod. Blumentrost and Goldbach would not return until 1732, when Saint Petersburg again became the capital.

With the departure of Blumentrost the government granted his friend, the despotic and crude Schumacher, nearly unlimited authority over the academy as director of its chancery.[63] Except for being stripped of the position for a year in 1742–43 he served continuously until his retirement in 1759. Coming from Colmar in Alsace, Schumacher had studied in Strasbourg; he joined the service of Peter I, purchasing private libraries and equipment for the czar, and assisted Blumentrost in planning the academy—particularly its library. An assiduous and inventive bureaucrat, Schumacher now applied new methods to control academy members. Life was not secure even at the academy. Once, upon returning from travels, Schumacher found that his private cabinet of goods had been looted. He benefited from free postage, which was a privilege of the academy, but to strengthen his position over the academy Schumacher arranged to have intercepted and reviewed whatever foreign correspondence was sent through the chancery. To maintain control, Schumacher attempted to set leading members against one another; he plotted to put Bilfinger and Bernoulli to arguing, but later efforts to have Euler oppose Bernoulli failed. Schumacher's scheming made senior members Hermann and Bilfinger eager to leave, but by employing tact and diplomacy Euler fended off Schumacher's behavior toward him.

In 1728 the academy, despite attempts by Old Believer nobles to obscure it, lost no members and recruited with a fellowship the nineteen-year-old German chemist, natural historian, and geographer Johann Georg Gmelin, who had studied under Bilfinger and Duvernoy at the University of Tübingen and received a medical degree. That year the academy finally acquired a printing press from Holland and started publishing its proceedings, the *Commentarii*. Its first volume, issued in 1728, consisted of papers delivered two years earlier; that year Euler had submitted a paper on elastic curves, "Solutio problematis de invenienda curva, quam format lamina utcunque elastica insingulus punctis a potentiis quibuscunque sollicitata" (Solution to the problem of finding curves which are formed by an elastic band when a force is applied to a single point, E08),[64] and Bilfinger had won the Prix de Paris on the topic of the cause of gravity. Euler was apparently continuing research begun in Basel on treating several curves in a unified way; he presented differential equations for elastica and some mechanical curves then popular, together with the catenary (a cord hanging between two points having the lowest center of gravity), the velaria (the figure given by a perfectly flexible cord with a uniform normal pressure load), and the lintearia (the shape of a cylindrical cloth when filled with water).

As part of the academy's effort to reach a wider public, Euler gave open lectures on logic and mathematics. At the academy Schumacher also oversaw the founding of the Russian- and German-language newspaper *Sankt Peterburgskie Vedomosti* (Saint Petersburg gazette), issued twice weekly and begun in 1728 to inform literate Russians about current events in Russia and the world; there was also a supplement, the first journal in Russian, the *Primechaniya k Vedomostyam* (Notes to the gazette), which covered various new scientific advances in Russia, including such Western ideas as gravitational attraction. Euler probably wrote some of the anonymous articles for it, and he reviewed for acceptance papers submitted by others, including an article on the quadrature of the circle; Schumacher had recruited Müller to be the editor of the two new periodicals. The initial failure of both journals apparently led to a break in 1731 between Schumacher and Müller, the latter of whom had already become a member of Euler's close circle of friends.

With the academy now possessing a printing press, and with permission to include Copernican astronomy in teaching the young czar Peter II, Delisle, Bernoulli, and Hermann led an effort to spread the heliocentric theory in Russia. They were not the first to do so, but this course of action was dangerous, especially given their public stature. In seventeenth-century Russia even arithmetic and geometry had been branded as "magic" and excluded from school curricula. The first attempt to introduce Copernican heliocentric astronomy into Russia, a manuscript from 1661 of a translation of Johan Blaeu's *Atlas novus*, remained unpublished until around 1724. Thus, the first book in Russian to describe Newtonian cosmology was the translation in 1717 of Huygens's *Cosmotheoros*, which had appeared in French in 1702 under the title *Nouveau traité de la pluralité des mondes*; it was superior to Fontenelle's *Mondes*, for it included Newtonian science. The *Cosmotheoros* began with the bold declaration that "our world is not the center of the universe. . . . [It is] like the other planets and revolves around the sun."[65] Mikhail Avramov, the pious director of the Saint Petersburg printing office, along with other dogmatic Old Believers, considered author and translator Jacob Bruce to be an atheist and the book an invention of the devil.[66] While the Old Believers were persecuted by the Holy Synod founded by Peter, it and censors of the Russian Orthodox Church, possessing a firm hold over publications, rebuffed Delisle, Bernoulli, and Hermann; they prohibited the printing of all articles by Bernoulli and Hermann as well as books on Copernican astronomy and writings by Delisle on the rotation of the Earth, even by the academic press, and attacked science as the work of the Antichrist. Nonetheless, the academy had its relentless champions; especially combative among them was Kantemir, whose *Satira* (Satire I, 1729), circulated in manuscript form, ridiculed obscurantist opponents of Copernican astronomy.

Although belonging to the camp of Delisle, Hermann, Bernoulli, and Kantemir, Euler differed sharply with many leading *philosophes* of the French Enlightenment led by Voltaire, who was a major opponent of the superstitious element in religion. In Euler's Enlightenment stream, as was the case in the British North American colonies that would become the United States, piety and science were not at odds.[67] The Scriptures inspired the devout Euler, who perceived a harmony between written revelation and natural phenomena. He knew, of course, that the sciences had their own major contradictions and paradoxes and likely saw the opposition to Copernican astronomy as such a case in religion; generally avoiding confrontation, he did not challenge the Orthodox clerics at this time. But as the debate over Copernican astronomy continued in Russia during the 1730s, he was to take a more public position regarding it.

In November 1727 Euler had sent a letter to his old teacher Johann I Bernoulli beginning a correspondence of nearly two decades' duration.[68] To July 1730, the two exchanged nine letters; of these, six were from Euler. Most of the correspondence with his mentor was in Latin, which they considered the language of learning. The Latin salutations by Bernoulli show a growing respect for his student, who is addressed in 1728 as "Doctissimo atque ingeniossimo Viro Juveni" (the most learned and ingenious young man of science), and in 1729 as "Viro Clarissimo ac Mathematico longe acutissimo" (the most famous and learned man of mathematics).

Their letters begin with Euler's reference to trajectories, the difficulty of the mathematical investigation of fluid motion, and his dissertation on sound. Only briefly does Euler mention, "By chance [I found myself] struggling with the function $y = (-1)^x$." Determining its graph was "extremely difficult . . . for it is sometimes positive, sometimes negative, and sometimes imaginary, so it seems to me that it does not make a steady line but rather infinitely many points spaced equally as 1 unit on either side of the axis."[69] The function reappears in the second, third, and fifth letters between them. In January 1728 Johann I Bernoulli commented on Euler's investigations of fluid mechanics and the velocity of sound before adding that the sum of the infinite series $y = (-1)^x$ is 1.[70] Bernoulli asked Daniel for a copy of the article by the Swiss engineer Nicolas Fatio de Duillier on centers of oscillation and percussion with projections of cannon shots, and announced that he had sent the Petersburg Academy of Sciences his paper on the conservation of *vis viva*.

In December 1728 Euler revisited the controversial topic of logarithms of negative numbers. In a twelve-letter correspondence in 1712–13, Gottfried Leibniz and Johann I Bernoulli had debated what these were; Leibniz had argued that $\log(-1)$ does not exist, while Bernoulli had held that $\log(-x) = \log(x)$. Euler was disinclined to accept Bernoulli's position, which seemed to him contradictory; initially he thought the logarithms of negative numbers to be imaginary. His calculations led by December 1728 to $\log(-1) = \pi i$, thus refuting both Leibniz and Bernoulli, though without proof.[71] Here the symbol π represents the ratio of the circumference of a circle to its diameter, and i is the imaginary unit $\sqrt{-1}$. Euler had not yet settled on the symbol i for this value. While the number -1 is a real number, its square root takes us abruptly outside. Thus, for the square root operation, the real numbers are not complete. Imaginary numbers, a name given by Descartes in his *Géométrie*, refer to what we today call complex numbers, which are complete under the normal arithmetic operations of addition, subtraction, multiplication, division, or finding any root. When these operations are applied to complex numbers, the answer will always be a complex number. For Euler, who developed them in his imagination, the term *imaginary* was especially appropriate.

In April 1729 Bernoulli promised to answer fully Euler's objection to the claim that $\log(x) = \log(-x)$. If this equality were true, Euler knew, $\log(-x)$ would equal 0, as it does when $x = 1$. Euler found unsatisfactory Bernoulli's argument that $\log(-x)$ is in all cases real, while $\log\sqrt{(-x)}$ is imaginary, but he did not push the subject further since Bernoulli wanted to drop it. This was not the last Euler heard of logarithms of negative numbers, for d'Alembert later independently came to the same conclusion as Bernoulli.

Euler was now also investigating problems in what became two central branches of calculus: differential equations and the classical calculus of variations; he would come to be considered the principal inventor of both. At this point, though, he emphasized differential equations. Possibly as early as 1727, but more likely in 1728–29, Euler wrote an undated thirty-page manuscript, titled "Calculus differentialis,"[72] that outlined ideas to be fleshed out later in his *Introductio in analysin infinitorum* (Introduction to the analysis of the infinite, 2 vols., 1748) and *Institutiones calculi differentialis* (Foundations of differential calculus, 2 vols., 1755). Not to be confused with these books, his manuscript was probably for instruction in the academic university, where he lectured on the subject; he was not yet influential enough to have it published. He may have also considered it an unfinished text and would later see its rules and concepts change. Euler's manuscript is not very satisfactory; its brevity possibly comes from his believing that his students were not adequately prepared. An important step in overcoming what was lacking in the education of students was for the academy to establish a translation office and print Russian versions of important mathematical texts. This would not come about until after Euler left the academy in 1740.

"Calculus differentialis" recognized that developing a theoretical underpinning for the new calculus was under way. A crucial issue that Euler addressed for nearly thirty years following, if not longer, was defining exactly the relationship between infinitesimal, finite, and infinite quantities. The proper handling of infinitesimals in calculus would be a matter of dispute. The infinitely small gained less attention in "Calculus differentialis" than in Euler's later books and was not generally treated, but the manuscript warned against errors that would arise if final, complete expressions incorrectly omitted these quantities. He discussed a first set of rules for removing higher order infinitesimals. Guillaume-François-Antoine, Marquis de L'Hôpital's *Analyse des infiniment petits* (Analysis of the infinitely small, 1696) seems to have been Euler's source of rules for operations with the infinitely small or differential by which a variable continuously changes. He closely followed Johann I Bernoulli's definition from an article for the Paris Academy of Sciences in 1718, whose text opens by describing a function as "One quantity composed somehow from one or a greater number of quantities."[73] Thus, from the start Euler's functions had several variables; he represented these with the last letters of the alphabet, and constants with the beginning letters; this would become standard practice. The phrase "composed somehow" embraces arithmetical operations, inversions, root extractions, the finding of logarithms, and combinations of these. Euler's manuscript classifies functions as either algebraic or transcendental; within the transcendental class Euler put only logarithms and exponentials, not yet trigonometric half chords and chords, as he would later do. Until he made them of equal status, he considered exponentials only as the inverse of logarithms.

The years from 1727–44 were a transitional time for Euler as he moved from geometry to infinitary analysis, partly in developing a general concept of function based on additional operations.[74] This was part of an ongoing dispute over the foundations of calculus. In accepting finite and infinite magnitudes as basic and following an intuitive approach, Euler was traditional. A sound treatment of arithmetical differences came only in the next century with the theory of sets. Throughout his efforts, Euler adapted methods and operations to problems that he faced. His stressing of functions and moving them to the heart of analysis clearly broke with ancient and medieval mathematics and represented the modern. Euler began to provide rules for the calculus of finite differences—rules that had been either mistaken in or largely absent from the writings of his predecessors. Developing a full set of rules would take almost thirty years. He sought an algorithm for applying the calculus of differences to simple classes of functions so that the field could serve as the basis for all of differential calculus. His manuscript "Calculus differentialis" proposes the sign Σ for summation, for differences the symbols Δy, $\Delta^2 y$, and $\Delta^3 y, \ldots$, and for the increases in quantities $a, b, c, \ldots$ the corresponding Greek letters, $\alpha, \beta, \gamma, \ldots$, while his article "De infinitis curvis eiusdem generis" (On infinitely many curves of a given kind, E44), written in 1734 but published six years later, almost accidentally expresses the function of a linear expression $x/a + b$ with parentheses as $f(x/a + b)$. He replaced Johann Bernoulli's f of x with the symbol $f(x)$;[75] it took decades for this notation to gain acceptance.

A synthesis of ad hoc results from his collaboration in 1727 with Johann I Bernoulli and Hermann and his own original ideas underlay the first phase of Euler's contributing several fundamental concepts, computations, and advances to the development of calculus. A second phase after 1740 transformed these into a semigeometric stage of the classical calculus of variations. Variational theory studies extrema; in David Hilbert's modern viewpoint, "Given a set of mathematical objects, $a, b, c, \ldots$ and a relation so that each element a is associated with a real number N_a, then look for the element or elements which have a minimal or maximal associated number . . . if there is any."[76] As elements Euler had curves or functions—generally expressed by definite integrals—usually called functionals; he did not have modern abstract definitions. He had to find associated maxima and minima values, and gave detailed explanations for these. Currently, the more appropriate title for this field is in the narrow sense the calculus of variations, because variations or differentials of functionals involve more than extrema.

In perhaps his most famous homework assignment, Johann I Bernoulli had asked Euler, probably in 1727, to find the general methods and differential equations giving the shortest line between two given points on any surface.[77] Bernoulli had only ad hoc solutions. Such problems in what became the calculus of variations had ancient origins. They have been traced to at least the Dido Isoperimetric Problem of finding the largest area enclosed by a given length for the perimeter. Virgil had stated this problem in the *Aeneid*; in the translation by John Dryden, it appears in book 1. As the story goes, Dido had fled Tyre to begin Carthage. She was given the problem of obtaining the largest possible area enclosed by a curve of given length; the answer is a circular curve. Since Carthage borders on the sea, the original problem may have been to find a circular arc rather than a circle. Although it is not in Euler's own handwriting, a copy of the eight-page manuscript containing his answer has been preserved in Moscow.

Responding to Bernoulli's request, Euler wrote his first published paper in this field, the twelve-page "De linea brevissima in superficie quacunque duo quaelibet puncta iungente" (Concerning the shortest line [or curve] on any surface by which any two given points can be joined together, E09), which appears in volume 3 of the *Commentarii* (1728) and was printed in 1732.[78] Consisting of thirty-six paragraphs, it notes that Johann I Bernoulli had posed for Euler the project of constructing general methods for finding equations for the shortest lines or curves on any surface. The material from the fourteenth to the last paragraph reads like a worksheet showing how Euler reached his conclusions; the reader gets to see him at work.[79] The paper indicates that the shortest line between two points in a plane is a straight line and a great circle in a sphere. Euler derived the two second-order differential equations that furnish minimizing lines or curves on cylinders—conoidal, or cone-shaped—on surfaces, and on surfaces of revolution, providing solutions to equations that are integrable; he also discussed some special cases. "De linea brevissima" thus begins to set out the analytical foundations for the calculus of variations. For convex surfaces, Euler also described what he terms a "mechanical solution" by stretching a string or chord between the two points; another hand wrote on the manuscript the completion date, November 1728. In December, a letter from Bernoulli on geodesic curves again asked Euler to find a general equation giving the shortest curve on a surface.[80] The following February Euler supplied the differential equation being sought, a development that supports the manuscript date given here.[81] Solving these problems led to his founding of the semigeometric stage of the classical calculus of variations.

Euler had another reason for creating the field. In a paper of 1727 for volume 2 of the *Commentarii* titled "Theoria generalis motuum" (General theory of motions), Hermann had given a mechanical solution to problems of shortest distances or times using three-dimensional coordinates and had attempted to improve upon the solutions by Jakob and Johann Bernoulli of the brachistochrone problem; Hermann's solution was incorrect, a fact that Euler did not point out until 1731. Euler thought the problem important, and over the next decades devised new solutions, but he did not yet in 1728 display a continuing interest in inventing the semigeometric stage of the classical calculus of variations. In May 1729, Euler gave three ways to integrate second-order differential equations, while a letter in October gave findings on the tautochrone and isochrone curves with relations between second- and first-order differential equations. In 1730 Euler provided a first-order differential equation for the tautochrone curve.

Even before Euler arrived in Saint Petersburg, Daniel Bernoulli and Christian Goldbach had pursued problems of interpolating two particular sequences, that of the factorial numbers, 1, 2, 6, 24, 120, 720, . . . , which they called the "hypergeometric progression," and the question of partial sums of harmonic series 1, 3/2, 11/6, ... Probably motivated by efforts to develop a differential calculus of fractional order, they sought to generalize factorial $n!$—that is, to provide an expression for nonintegral values of n, but these problems stumped them. Having fractional values of n well defined, Euler consequently interpolated the sequences. Making extensive computations with numerous identities, transformations, and daring integrations and following a process akin to interpolating the factorial progression led him to invent a mathematical object, later known as the gamma function, $\Gamma(n) = (n - 1)!$, which extends the factorial to complex numbers. With this development the twenty-two-year-old Euler quickly solved both problems.[82] In October 1729 Euler wrote to Goldbach, "Most celebrated Sir, I have been thinking about the laws by which a series may be interpolated [that is, the sequence of values falling between its defined whole number sums]. . . . [Daniel] Bernoulli suggested that I write to you" and reported that he had its solution.[83] That solution and his computations of $n!$ for nonintegral values and integrals that produce it led to Euler's invention that year of the gamma function.[84] It appeared in his article "De progressionibus transcendentibus, seu quarum termini generales algebraice dari nequeunt" (On transcendental progressions, or those for which the general term is not given algebraically, E19) and developed it in the sequel "De summatione innumerabilium progressionum" (Concerning the summation of innumerable progressions, E20).[85] In December 1729 Goldbach reviewed the gamma function, and in a letter the next year Euler outlined his integral methods in calculating it; he was roughly a century ahead of his time. Only in the nineteenth and twentieth centuries would the gamma function appear in many formulas and have extensive applications in solving problems in mathematics and mathematical physics—among others, in complex variables, Laplace transforms, probability, and fractional derivatives, the last of which was another discovery that Euler had made at the end of the first of these three papers.

Number theory, its connections with other fields of mathematics, and the aesthetic appeal its abstraction held for him—particularly the beauty he found in Pierre de Fermat's conjectures—began by 1728 to produce in Euler a profound fascination with the subject. This appears to be mainly independent of the Bernoullis, for none after Jakob showed any great interest in number theory. Encouraging the recovery of Fermat's conjectures and obtaining proofs for many of them, contributing to second-degree Diophantine equations, and formulating more important principles for the field than had all of his predecessors combined, Euler vitalized and essentially re-created number theory, preparing the way for Carl Friedrich Gauss. In later deriving without proofs the equivalences of the prime number theorem and the law of quadratic reciprocity—the two fundamental theorems of the field—Euler glimpsed what he called its inner *Herrlichkeit*, or splendor. First, however—by early 1728 and probably at the instigation of Johann I Bernoulli—Euler had turned to investigating the challenging Basel Problem, summing exactly the infinite series of the reciprocals or inverse of square integers $1 + 1/2^2 + 1/3^2 + 1/4^2 + \dots$, now known as $\zeta(2)$; the zeta symbol was to come later from Bernhard Riemann.

The Basel Problem was not new; the Bolognese mathematician Pietro Mengoli had posed it in his *Novae quadraturae arithmeticae* of 1650. Its name comes from Jakob Bernoulli's *Tractatus de seriebus infinitis* (Treatise on infinite series), which in 1689 first made it widely known among mathematicians, though its solution long eluded them. Resolving the Basel Problem was thought as important as Leibniz's summing in 1674 for the arctan x for $x = 1$, obtaining $1 - 1/3 + 1/5 - 1/7 + \dots = \pi/4$.[86] The connection with π was a major achievement. The slow convergence of the infinite series of the reciprocals of the square integers was difficult to overcome. John Wallis's *Arithmetica infinitorum* of 1656 had summed $\zeta(2)$ to three decimal places as 1.645. Through a term-by-term comparison of related series, Jakob Bernoulli confirmed that $\zeta(2)$ must be less than 2. Decomposing the infinite series in the Basel Problem, Bernoulli sought to improve upon Wallis's approximation. Among others, Huygens, Leibniz,[87] and the older Bernoulli brothers, along with Nikolaus I Bernoulli, James Stirling, and Abraham de Moivre, had failed to find an exact solution. In August 1728 Daniel Bernoulli wrote to Goldbach that he had approximated the sum to be "very nearly 8/5." The next January Goldbach computed $\zeta(2) - 1$ more closely, as falling between 16,223 ÷ 25,200 and 30,197 ÷ 46,800, or 0.6437 and 0.6453, respectively. Put into simpler fractions, his upper and lower bounds were $1\ 16/25 = 1.64 < \zeta(2) < 1\ 2/3 = 1.66. \ldots$ Now it was Euler's turn.

By 1729 young Euler had reckoned the sum of the Basel series to approach 1.644924, which is correct to six decimal places. He was thus more accurate than his colleagues and pretty nearly had the exact sum, which he soon found to be $\pi^2/6$. But the sum had not yet been linked to π. To find a more precise answer, he recognized, he could not fruitfully attack the problem directly. Summing the first hundred terms of the reciprocals of the squares gives 1.63498, and the first thousand gives 1.64393. The first figure is accurate only to one decimal place and the second to two. Instead of conducting a direct assault, Euler ingeniously derived an alternative expression for the Basel series. In a 1730 letter to Goldbach, he reported that he had found the functional equation of $\zeta(2) = (\log 2)^2 + \Sigma_1^\infty 1/(n^2 2^{n-1})$.[88] The formula was a breakthrough, for with the term 2^{n-1} in the denominator the new series converges far more rapidly than that of the inverse squares, though still slowly. Such transformations were employed principally to make divergent series with alternate signs and a regular progression turn into sluggishly convergent ones. As done here, the rule could also be applied to a convergent series to make it converge more swiftly. This was important in making summations, and this work suggests recognition of the importance of convergence and the rate of convergence. Notably, it encouraged Euler to compare integrals with infinite series. Responding to inquiries about the one-half place and the three-half place, Euler proceeded to interpolate partial sums of the harmonic series $H_n = 1, 1/2, 1/3, 1/4, 1/5, 1/6, \ldots 1/n$; this gives the total for the first, 1; the second, 3/2; and the third, 11/6. . . . This work deftly joins geometric and logarithmic series, improper integrals, and termwise integration, albeit without proof of its existence, along with his skillful manipulation of symbols, seeking an alternative expression of the series.

Euler's work with Fermat's conjectures, which was among the vital stimuli of his passion for number theory, first began in 1729 when he read Fermat's *Varia opera mathematica* (1679), and perhaps John Wallis's *Commercium epistolicum* (1658). In October 1729 Euler began his correspondence with Goldbach by asking questions leading to the achievement of Euler's own beta and gamma functions with interpolations of infinite series, including a slowly converging series whose product gives the sum of the hypergeometric sequence 1, 2, 6, 24, 120, . . . for factorial numbers $n!$. He had not yet discovered lemmas connecting these series with logarithms. To this point, his reading of Fermat seemed only cursory. Then a postscript to a letter from Goldbach in early December 1729 further prompted Euler's labors in number theory:

> Notae Tibi est Fermatii observatio omnes numeros hujus formulae $2^{2x-1} + 1$, nempe 3, 5, 17, etc. esse primos, quam tamen ipse fatebatur se demonstrare non posse, et post eum nemo, quod sciam, demonstravit.
>
> Is Fermat's conjecture known to you, that all integers $2^{2n-1} + 1$ [for $n = 1, 2, 3, 4, \ldots$], that is 3, 5, 17, etc. are prime numbers, but he did not claim that he could demonstrate it, nor, as far as I know, has anyone else been able to prove it?[89]

Goldbach asserted that by himself he too could not prove this; in his next letter, Euler admitted the same.

After failing over the preceding decade to engage others in number theory—even Daniel Bernoulli, to whom he posed Diophantine problems—Goldbach had turned to Euler.[90] His wish would be fulfilled beyond what he could have imagined; Euler's profound work in number theory would fill four volumes of his *Opera omnia*, and these accomplishments alone are enough to make him significant in the history of mathematics. Responding to a query from Bernard Frénicle de Bessy sent through Mersenne in 1640, Fermat had posited the preceding conjecture; lacking a proof, he asked Wallis and other English mathematicians to provide one. Euler's terse reply to Goldbach in January 1730 gave little hint of what was to come; he remarked that Fermat had found this speculation by induction. Euler confirmed that $2^{2^n} + 1$ is prime for $n = 0, 1, 2, 3$, and 4 and noted that a table of prime numbers extending beyond 100,000 has no case when a number has this form. But he observed that Fermat had not gone to $n = 5$. The closing statement of his letter, "I have been able to discover nothing about the conjecture noticed by Fermat,"[91] suggests that for the moment he considered the problem a distraction. In June of that year Goldbach stated that he had not yet read Fermat's works but urged Euler to do so; Euler had already started on them.

The two men were to correspond over the next thirty-five years, writing almost two hundred letters. While not a leading mathematician, Goldbach was the most influential within Euler's circle in encouraging him to pursue number theory, persistently posing substantial problems that required difficult solutions and keeping him working steadily on the subject. Euler's responses would be among his first achievements.

Shortly after Euler arrived in Saint Petersburg, Delisle had recruited him to collaborate in work at the observatory but not to act as an observer until trained. Astronomy was to gain increasing attention from him.[92] From his first visits to the observatory, he noticed that clocks were running at different rates. He worked on their regulation and took at the start Huygens's design for a pendulum clock as a model for an accurate chronometer. Reliable records of lunar and solar eclipses were important for checking the accuracy of measurements of longitude. The *Sankt Peterburgskie Vedomosti*, reporting the latest information from the Petersburg Academy of Science observatory in August 1728, noted that the entire mathematical department witnessed a complete lunar eclipse. In September the permanent members of the observatory—Delisle, Krafft, Meyer, and the French instrument maker P. Vignon—made more detailed observations of a lunar eclipse. The next month Delisle observed the occultation, or conjunction, of Venus as it disappeared behind the disc of the moon. Having thirteen-, fifteen-, and twenty-three-foot telescopes, he employed the longest in an attempt to minimize optical defects; his chief purpose was to determine whether the moon had an atmosphere. The refraction of light in blues and reds from Venus and the absence of a vaporous surrounding for the moon blocking the view of Venus just before conjunction suggested to Delisle that the moon in fact had no atmosphere. But his Parisian colleague Jacques Louville argued for the existence of a lunar atmosphere, which Euler accepted. In 1729 Euler studied different conjectures on the nature of polar aurora.

As the astronomical findings of Euler, Gmelin, Krafft, and Weitbrecht were reported in the *Primechaniya k Vedomostyam* and became more popular, the academicians formed a society that met every Saturday evening at Schumacher's house to decide on what themes their articles should take up; their topics included the ebb and flow of the tides, finding longitude at sea, and the rise and fall of the River Neva. Euler's article on the tides in the November *Primechaniya k Vedomostyam* generally coincides with the arguments in numbers 62–67 of his *Lettres à une princesse d'Allemagne* (Letters to a German princess). This article also contains his first criticism of Descartes's ideas—in this case on a lunar whirlwind's influence on the tides.[93]

Chapter 3

Departures, and Euler in Love: 1730 to 1734

Late in 1729 the Russian nobility officially announced the engagement of "the boy czar" Peter II to Princess Anastasia Dolgorukaia. But the succession to the throne would again change quickly. Early the next year while preparing for the wedding in Moscow, Peter died suddenly from smallpox; he was not yet fifteen. Peter II ended the line of the Romanovs through male succession. Anna of Courland was the daughter of Ivan, a half-brother of Peter the Great. When she succeeded the young czar, the Petersburg Academy of Sciences began to regain royal favor; she had ascended to the throne with the aid of the reformist Archbishop of Novgorod, Feofan (Theophan) Prokopovich, and the aristocratic poet Antiokh Kantemir. Powerful Russian magnates attempted to restrict her powers but failed, for she was popular with the imperial guard and the educated German lesser nobility. As Anna surrounded herself with foreign nobles, especially Baltic Germans, the Russian boyars lost favor at court, and she continued Peter I's modernizing efforts.

For ten years the German Ernst Johann von Biron (or Biren, or Bühren), the empress's lover, and his clique controlled the government in what became known as the *Bironovschina*. Police persecutions and thousands of executions darkened this period, known as the time of the German Yoke. In the academy, the arrogance of some German members provoked resentment among students and translators, but the material fortunes of the academy improved, and in 1732 Anna returned the capital to Saint Petersburg. One sign of her support for the academy came in 1730: when Jakob Hermann left, Anna awarded him a lifetime pension of two hundred rubles annually.

Hermann returned to the University of Basel to take up the chair of ethics that had been offered him three years earlier. Hermann never attained the position that he desired in mathematics, the post that Johann I (Jean I) Bernoulli continued to hold, and two years after returning to Basel he succumbed to a fever. In 1730 Georg Bernhardt Bilfinger went back to Tübingen; the new duke, Karl of Württemberg, had invited him to be a military and state council official. Shortly before leaving Saint Petersburg, he and Daniel Bernoulli made peace and the rift between them was mended.[1]

In 1730, Daniel Bernoulli succeeded Hermann in the prestigious position of professor of higher mathematics, although with a salary of only one thousand rubles (half what Hermann had received); Leonhard Euler, who was to replace Bilfinger, received a new four-year contract at four hundred rubles a year for the first two and six hundred yearly for the remainder. He was also granted sixty rubles annually for housing, wood for heat and cooking, and candles for light.[2] In 1731 the academy officially named Euler, now twenty-three years old, to fill Bilfinger's professorial chair of physics and continue as a full-time member.

One of Euler's first tasks was to carry out assignments for the naval academy. An early job was to provide (in September 1731) an expert evaluation of lieutenant Stepan Malygin's booklet *Navigation nach der Carte de Reduction* (Navigation according to the small scale map). Euler found it rich in the ways it treated problems, and a "useful [work that] could serve as a textbook."[3] The Petersburg academicians were expected to give a few reports annually at the twice-weekly conferences. Until his departure in 1741, Euler averaged ten reports a year, double the highest number submitted by any other scholar. Clearly Daniel Bernoulli and Euler surpassed their predecessors in scientific achievement, but the academy suffered from its inability to fill positions that opened when senior members, including Hermann and Bilfinger, and later Bernoulli, left. After 1731 a six-year break occurred in Euler's documented correspondence with Johann I Bernoulli, which perhaps suggests his satisfaction in Saint Petersburg, an absorption in research and, after 1733, his shifting to a regular correspondence with Daniel Bernoulli, who had by then returned to Basel. Apparently, however, some letters to Johann I during this period were simply lost.

In 1730 Euler continued his study of anomalous polar auroras. His articles that year in the *Primechaniya k Vedomostyam* (Notes to the gazette) of the Academy of Sciences in Saint Petersburg rejected the Aristotelian explanation, saying that it dealt "more with empty words than with reason and real concerns," as well as the associated Cartesian description, which Euler declared to be "no longer considered of any significance."[4] He also discarded the most popular idea of the time: that the polar lights reflected the eruptions of volcanoes upon the breaking up of ice.[5] Euler examined all explanations for the lights and found none convincing; he proposed further investigation of their frequency in sharply changing weather conditions. These studies had to be anchored in recognizing the "universe for what it truly is . . . a real chemical laboratory." In 1730, through its study of auroras together with initial observations of sunspots, the academy increased its activity in exploring weather, the nature of the atmosphere, and interconnections with solar activity.

At this time a combative facet of Euler's personality began to emerge at the academy; the subject of the dispute was his salary. Throughout his career, he strenuously negotiated his income; in 1731 the genial Euler, who had always been known for his calm disposition, became angry with Laurentius (Lavrentii) Blumentrost and Johann Schumacher.[6] That Johann Georg Gmelin, Georg Wolfgang Krafft, Gerhard Friedrich Müller, and Josias Weitbrecht, who had received lesser salaries previously, were proposed for professorships, each with a salary of four hundred rubles—the same as his—upset Euler. His annual income, moreover, was frozen for two years. On 23 June 1731, Euler complained sharply to Schumacher, "It seems to me that it is very disgraceful for me, that I, who up to now have had more salary than the others, shall now be set equal to them. . . . I think that the number of those who have carried [mathematics] as far as I [have] is pretty small in the whole of Europe, and none of them will come for 1000 rubles."[7]

Euler's seven published articles had already begun to set him apart from the others, but on the basis of these alone, his argument was a bit bold. Because of his salary and other reasons, Euler had some strained relations with Schumacher, who advised Blumentrost not to make any concession to Euler lest he become impudent. By keeping the salary of all five at four hundred rubles, instead of offering Euler the five hundred that he requested, Schumacher won this round. But as Euler's reputation grew he was to become a tougher salary negotiator, successfully arguing for raises in both 1735 and 1741. But in 1731 he avoided public disputes with Schumacher; young Euler attended parties given at the Schumachers' house, and the invitations contributed to the forming of a research circle around Euler.

The articles in the Petersburg Academy of Sciences' *Commentarii* show Euler's emerging primacy. Of the thirteen articles on mathematics in volume 5 (for 1730–31, but not published until 1736), eleven are by Euler and two by Daniel Bernoulli. Krafft, who was central within the research group, grew closer to Euler.

Throughout the 1730s Euler had teaching as a regular duty. The academy's course catalog for 1732 had Joseph-Nicholas Delisle teaching astronomy, Daniel Bernoulli elementary geometry, and Leonhard Euler physics. These were only a few of the available courses. Euler lectured from ten to eleven on Monday, Wednesday, and Thursday, and on Saturday he added experiments demonstrating physical laws. Among the faculty, he alone listed a textbook for his course, Willem 'sGravesande's *Physices elementa mathematica*, an introduction to Newtonian science that had first appeared in two volumes in 1720 and 1721. Euler seems not to have known the author directly. 'sGravesande, the foremost expositor of Newtonian science on the Continent, made Leiden a citadel for it. His physics courses did not, for the most part, include mathematics. (Despite its title, *Physices elementa mathematica*, his text contained scant mathematics.) The courses, which were conducted much like salons, became popular, and among the attendees would eventually be Voltaire. In his lectures using 'sGravesande's book, and in his experiments, Euler must have added mathematics. Until Daniel Bernoulli left Russia in 1733, Euler taught only physics, but he then switched to two basic mathematics courses.

As early as 1730 Daniel Bernoulli had wanted to leave Russia; its harsh climate buffeted his delicate health, and he regretted having come to Saint Petersburg, the city where his brother Nikolaus had died. The censorship, the hostility to the Germans, the constant intrigues of Schumacher attempting to pit others against Bernoulli, and the failure to pay his salary according to his contract upset him, as did the poor financial position of the academy and the political instability during the *Bironovschina*. To counter Bernoulli's desire to leave, the imperial court in 1731 increased his duties and promised a retirement fund equal to half his salary, stating that he could retire whenever he chose. Wishing to keep a friend and talented collaborator in research, Euler urged Bernoulli to stay in Russia. But when University of Basel professor of botany and anatomy Emmanuel König died in 1731 Bernoulli applied for the vacant position and won the lottery competition against Benedict Stähelin and Johann Jakob Huber—the latter a student of Albrecht von Haller. In the autumn of 1732 Daniel's younger brother Johann II (Jean II), a university friend of Euler's, arrived in Saint Petersburg; he was apparently seeking a position at the academy, having already made contributions to the mathematical sciences. Although Euler urged his appointment, Schumacher blocked it; controlling the first Basel trio had been difficult, and a new one would be daunting.

To the great disappointment of Euler, Daniel and Johann II remained in Saint Petersburg only until the summer of 1733. Daniel would not often again achieve the level of research he had during his eight years in Saint Petersburg—six of these in close interaction and amicable competition with Euler. When Daniel Bernoulli departed, the twenty-six-year-old Euler succeeded him as professor of higher mathematics, the premier post in the academy, but he had lost a gifted collaborator and good friend. The failure to fill the more junior vacancy created by his departure would weaken the academy. On their leisurely journey home, the Bernoulli brothers had stopped in Danzig; Stettin; Groningen, which was Daniel's birthplace; Amsterdam; and Paris, where they met (among others) Bernard le Bovier de Fontenelle, Pierre Maupertuis, Alexis Claude Clairaut, Charles Marie de la Condamine, and Jean-Jacques d'Ortous de Mairan of the Paris Academy of Sciences and attended a lecture there. In September Daniel wrote to Euler that in mathematics and mechanics the members of the Paris Academy had not come as far as Daniel in achievements, and he deplored the slowness in publishing articles in the Petersburg Academy's *Commentarii*.[8] It was growing worse: in 1731 the delay in publication went from two to five years. After he left Paris and arrived in Basel in September 1733, Daniel wrote in German to Euler that he felt like "another man, since he can breathe our good Swiss air";[9] the air, of course, was unpolluted by the Russian autocracy.

Before leaving Saint Petersburg, Daniel had completed a paper based on Newtonian dynamics that determined the inclination of planetary orbits relative to the plane of Earth's orbit. He proposed that an atmosphere resembling air rotating about the solar axis caused the inclination to increase toward the solar equator. Submitted to the Paris Academy's Prix competition for 1734, his paper shared the prize with his father's paper, which gave a Cartesian explanation of the inclination of planetary orbits. This outcome was an important indication that the Paris Academy was beginning to shift from Cartesian to Newtonian science. After the splitting of the prize the father became increasingly upset with Daniel's acceptance and elaboration of Newtonian science, along with his work in hydraulics, which engendered a reputation that had grown to the point of challenging his own.

In March 1733, according to the Petersburg Academy's logbook, Euler with a group of geodesists first came to assist at the well-equipped observatory, where he pursued more of the groundwork for his research. Delisle, who headed the observatory, taught Euler astronomy and observational techniques with telescopes and other instruments. Students from France had been trained to aid Delisle in making systematic observations of sunspots, the rotation of Earth, the transits of Mercury, the paths of comets, and the eclipses of Jupiter's moons—in the last case to determine longitude. Volunteers from the academy and the academic gymnasium joined the project. The initial tasks for Euler and the geodesists were to synchronize four clocks, to observe the setting of Jupiter's main moon, and to measure elevations of the upper edge of the sun. It took Euler almost a year to become proficient in making astronomical observations; during this time, he assisted Delisle in assembling the observational data and collaborated with him in making calculations in related celestial mechanics.

In September 1734 Delisle suggested that before beginning regular, independent observations the following month Euler work with a three-foot quadrant. Euler, who had first to compute when the sun crossed the meridian of longitude for Saint Petersburg, provided an estimate of errors in his results. "His own person" now in the observatory, Euler entered numerous and detailed entries into its logbook and wrote articles for the *Primechaniya k Vedomosti*.[10] Data were entered twice daily, between eight and nine in the morning and between two and three in the afternoon; in this work Euler first experienced the centrality of adept and sustained observations to making advances in the physical sciences. Seeking observations for determining longitude, Delisle published on the eclipses of Jupiter's satellites. Engaged in the long-standing effort to draw the first accurate, large-scale map of Russia, which the government desired, he founded the geography department of the academy with the help of Euler. Delisle and his assistants were thus occupied with cartography and geodesy, two subjects in which Euler developed a deep interest.

Courtship and Marriage

After the Bernoullis left, Euler had more time for dating, and in 1733 he was courting Katharina Gsell; being that it was Russia, he probably called her Katya. An improved financial situation permitted him to contemplate marriage that year; in addition to being the premier mathematician at the academy, with a substantial yearly salary of six hundred rubles (with sixty more added for household costs), he reportedly received four hundred rubles from field marshal and count Burkhardt Christoph von Münnich to lecture to the Cadet Corps and provide advice on its inspections. Katharina, who was born in Amsterdam in 1707, the same year as Euler, was the daughter of the Swiss-born artist Georg Gsell, who around 1690 had moved from Saint Gallen to Amsterdam and in 1697 wedded his first wife, the Hessian Marie Gertrud von Loen. They were married until her death in 1713. During his second trip to the West in 1717, Peter the Great had been impressed with the work of Gsell and asked him to travel with the imperial entourage. Peter invited him to become court painter, keeper of the imperial painting gallery, and a teacher at the art academy in Saint Petersburg, which Gsell accepted in 1719, coming to Saint Petersburg with his second wife, Dorothea Maria Henriette, an artist and naturalist. The daughter of the famous naturalist

Maria Sybilla Merian, with whom she had studied insects, butterflies, and flowers in Surinam, Dorothea Maria also had many duties at the painting gallery and art school;[11] she helped arrange the catalog of the art cabinet, did watercolors of its most important rarities, and taught watercolor painting and copper engraving.

From her father and stepmother Katharina Gsell likely received training in art, a subject that interested Euler; his brother and a cousin were to become professional artists. From his friend and confidant Georg Gsell, who was now in his sixties, director Schumacher often tolerated plain truths about the academy and its operation. Euler and Katharina may have met at one of the social gatherings at his house. There he would also meet his future father-in-law, who was a close friend of the Schumachers.

That year's Christmas season was a joyous time for Euler. On 7 January 1734 he married Katharina. To celebrate the wedding, the academy had fireworks and illuminations;[12] following its custom for special events, it tasked its poet Gottlieb Friedrich Wilhelm Juncker with writing commemorative verse.[13] The long poem celebrating the wedding, written in German, begins,

> It seems the polar influence
> Increases and strengthens the Teutonic flames
> And draws together what was formerly separate
> Into a passion of the heart.

Later it asks the question that reflects the general view of Euler at the academy:

> Who would have thought it,
> That our Euler should be in love?
> Day and night he thought constantly
> How he wanted more to calculate numbers,
> His profound learned sense was free.

The poem closes with these words for the wedding couple:

> The hearts join, like the hands
> As . . . [Euler's] calendar says, it makes
> A beautiful end of the year for you.

The marriage seems to have been happy. Before the wedding Euler and Katharina had hunted for a house. Planned in a grid with the nobility at the center, the city had streets proceeding out in parallel by ranks in society through physician, lawyer, trader, and academic. The couple found a comfortable, large house made entirely of wood on a brick foundation; it was in moderately good condition, located on the bank of the Neva River on the tenth line of Vasilievsky Island close to its big avenue. On a plot of land measuring thirty by ten *sazhen* (about one-third of an acre, or fourteen thousand square feet),[14] it had five bedrooms, one kitchen, a covered inner courtyard, two painted stoves, and a fireplace; several rooms were without heat. In the yard was a garden with a few raspberry and currant bushes, and next to the garden a log hut with porches, a shed, a stable, and a wooden shed for storing firewood.[15] The grounds were near the main academy building and Münnich's palace. Katharina set about making the house into a home, and the fireplace decoration and the painting of one bedroom with lime were probably her doing. Apparently Katharina ran the entire household splendidly on her own, freeing Euler's time for his studies.

In May 1734 Euler wrote to his father Paul of his wish for Katharina to become a citizen of Basel. There were strong arguments in her favor: she was of Swiss ancestry, she was married to Euler, and he would pay the required fee, equal to eleven hundred rubles. While recognizing the need for other applicants to pay, he questioned the fees for his wife and complained of the difficulties in the process of naturalization. But at midcentury, Basel—which had for a long time severely limited immigration—denied all applications for citizenship. In a world that praised growth, Basel's elite resisted expansion, intending to defend its civic tradition and its social power. Its merchant oligarchy was deeply entrenched. The rejection of Katharina's application this time must have angered Euler; it probably helped end any thought of his returning to his birthplace.

On 27 November 1734 the couple's first child, Johann Albrecht, was born. In a letter to Müller, Euler proudly announced that "endlich ist auch meine Liebste . . . mit einem jungen Sohn niedergekommen" (finally, my darling has delivered a young son).[16] Euler's son was named after Baron Johann Albrecht von Korff, a *Kammerjunker* or German noble at the imperial court, who in September 1734 had just become president of the academy, a position that he would hold until 1740.[17] Korff made reforms but was unable to stabilize the finances or curb the difficult influence of Schumacher.[18] Johann Albrecht Euler's godfather, Christian Goldbach, was the conference secretary from 1734 to 1742. While these two good friends were at the academy, Euler's position was most secure. Johann Albrecht was the first of thirteen children, but only three Euler sons and two daughters were to survive past early childhood.

In the correspondence by Euler and his associates for the years after 1734 only six references to Katharina have been found thus far. In the

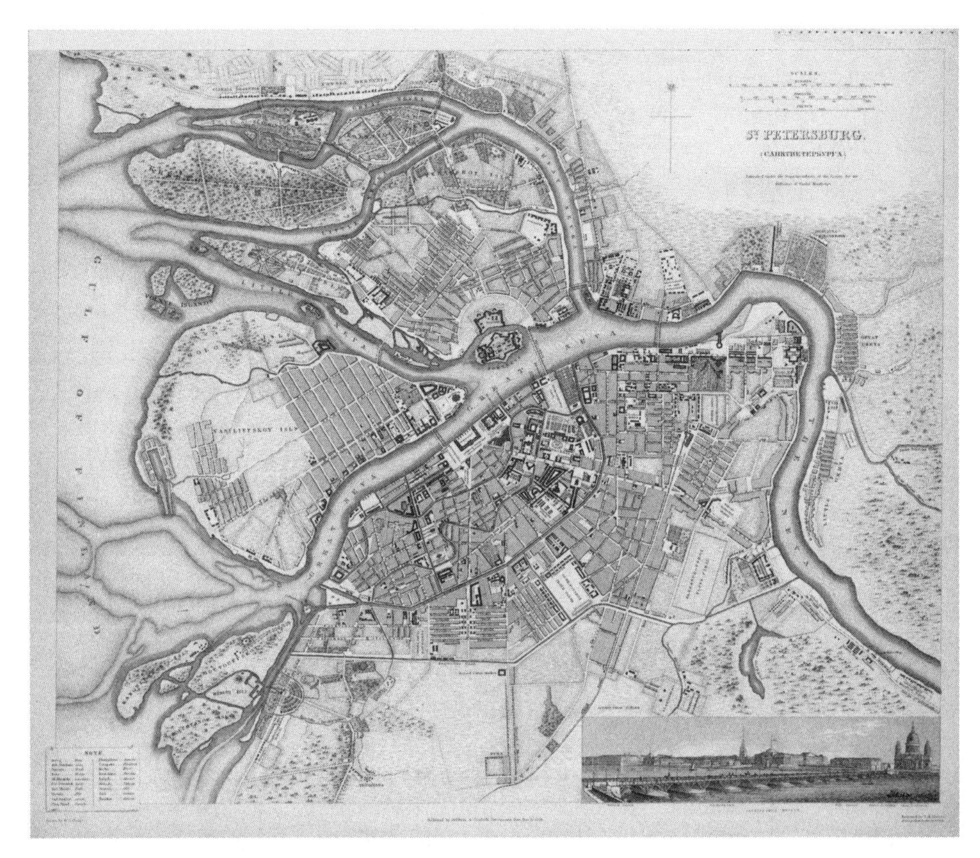

Figure 3.1. Map of Saint Petersburg. On the facing page is a detail of the above map with arrows indicating the general location of Euler's first house, the Kunstkammer, the Admiralty, and the Hermitage. Copyright National Library of Israel.

baroque age, silence about a wife indicated that the marriage was happy, and in this case it was. But even for Euler, who in his personal correspondence often wrote of other families, the later absence of comments about Katharina seems odd.

Groundwork Research and Massive Computations

As his correspondence with Johann I Bernoulli and the eleven articles published in the *Commentarii* in 1730 attest, Euler continued to immerse himself in research and many computations—for example, regarding ways to calculate π. Such intense work was not new to him, and had continued since his last years in Basel. Within the capital of absolutist Russia sycophants flattered the violent Biron. At the same time, hostility

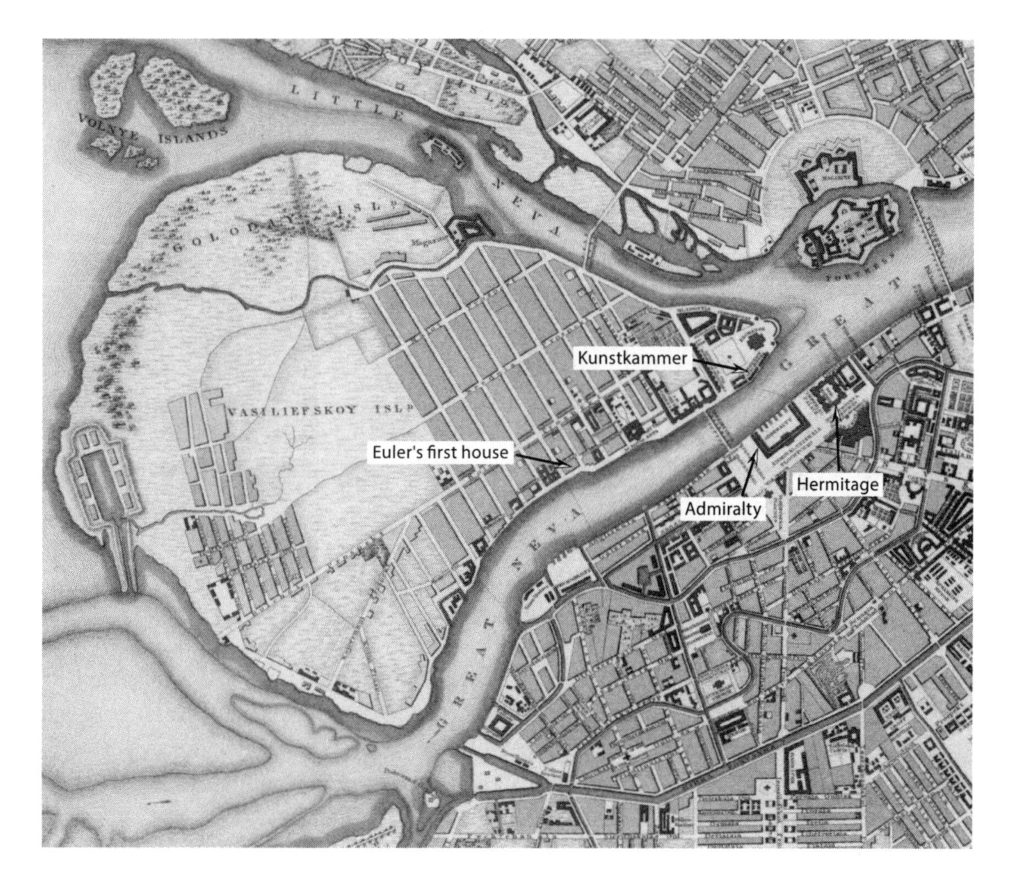

to the academy and the new sciences by the Russian nobility and Orthodox clerics still posed dangers. The political circumstances were an incentive to embrace the considerable academic freedom and reinforce the habit of mind to persevere in research under the shelter of the academy. This and the academy's generous publication program, especially the *Commentarii*, helped Euler become prolific in print. A letter from him to Schumacher in November 1749 praises the institution: "I and all others who have had the good fortune to spend some time in the Russian Imperial Academy of Sciences must admit that we owe all that we are to the advantageous circumstances in which we found ourselves there. For my part, had I not had that splendid opportunity, I should have had to devote myself primarily to some other field of study, in which by all appearances I would have become only a bungler."[19] Beyond Leipzig's *Acta eruditorum*, few scholarly journals in which to publish existed that were independent of the royal science academies. The creation of distinguished scholarly journals devoted to mathematics, such as *Crelles Journal für die reine und angewandte Mathematik* (Crelle's journal of pure and applied mathematics) and the *Journal de Liouville*, would not occur until the next century.[20]

Euler's *Commentarii* articles, correspondence, and notebooks, as well as observations in the first two volumes of the academy's *Vedomosti*, show his research through 1734 ranging widely over algebra, arithmetic, astronomy, ballistics, conic sections, differential geometry, elasticity, infinite series, music theory, number theory, and oscillations while his main field was rational mechanics.[21] Convinced of the unity of the mathematical sciences, Euler set about perfecting each branch. His early research occurred at various preliminary stages by subject area, and not all were of equal importance. The number and intellectual depth of Euler's *Commentarii* articles indicate a concentration on infinitary analysis or differential and integral calculus, along with rational mechanics—the application of calculus to problems of mechanics. His first eleven *Commentarii* articles, published up to 1734, were on differential equations, isoperimetric problems, and rational mechanics. He excelled in each of these fields, and his multiple discoveries and new methods were to transform them. Throughout his career they remained central to his research.

A thorough examination of Euler's twelve surviving mathematical notebooks remains at its early stages today. These comprise about four thousand pages and span his entire career. In the late twentieth century, members of the Petersburg Academy of Sciences and other scholars have studied them, but much basic analysis remains.[22] The exact progression of Euler's early work from one mathematical idea to another is occasionally unclear; his notes often change from subject to subject. A fairly accurate dating is facilitated by the notes generated from his reading of articles related to his research and by letters from fellow geometers and natural philosophers. The notebooks do reveal his persistent paths or strategies to reach solutions and disclose the research and exercises that he was busy with at any given time,[23] and they illuminate his inner life and the workings of his creativity within the sciences. The first six volumes, dating from his Basel, initial Saint Petersburg, and Berlin years, unsystematically address issues across the exact sciences and mathematics—especially differential and integral calculus with emphasis on problems of integration, number theory, and Diophantine analysis. Euler's first notebook sketches show that he planned to investigate the mechanics of fluids and gases, along with the theory of the motion of solid bodies, but he would not publish on these subjects until more than two decades later.

The first of Euler's notebooks, which dates from 1725 to 1727, when he was still studying in the city, is known as the Basel notebook; it contains notes and several research projects. A letter of June 1731 to Johann Bernoulli reports that Euler had nearly completed a short work titled *Tractatus de musica*, but it was not possible to publish a science monograph so early in his career. It took more research after the *Tractatus* to reach his *Tentamen novae theoriae musicae* (An attempt at a new theory of music, E33).[24] Bernoulli's reply in August 1731 criticizes the foundations of music theory, especially the lack of precise ideas of harmony, and hints at displeasure with Euler for devoting so much time to it; he urged Euler to quickly complete that treatise and commit himself to finishing the planned *Mechanica sive motus scientia analytica exposita* (Mechanics of the science of motion set forth analytically).[25]

In the dynamics of his mathematical creation, it was characteristic of Euler to begin his studies by accepting or setting a problem and following it with numerous calculations to reach increasingly close approximate values of sums of infinite series. In this stage of preparation, he was an experimental mathematician and a peerless calculator in both the elementary and the formal sense, the latter with transformations and an abundance of clever substitutions. Euler had a strategy for constantly developing new methods to solve problems; his disciplinary intuition probably served as a check to his creativity or to support some of it. These new methods and Euler's fertile analogies and algorithms provided the degree of accuracy that was especially crucial for problems in physics. In some cases it would take a century for scientists to grasp the proper use of his procedures. Ordinarily Euler continued computations in an area until an obstacle blocked further advance; while never losing interest in a subject, he would apparently set it aside completely for a time, letting his subconscious or the work of others find the solution. During what could be the next two stages, which Jacques Hadamard would later define as *incubation* and *illumination*, Euler probed a mass of combinations and discovered connections. Over time he obtained solutions.[26] Once a breakthrough came in the power or generality of method, Euler returned to the problem to make exhaustive new computations. In his search for deeper interconnections among the branches of mathematics, for example, those in the case of $\zeta(2)$ would be among harmonic series, natural logarithms, and trigonometry. In his research Euler stressed discovery and announced his many results, leaving for the future the establishment of rigorous foundations. (His definition of $\zeta(2)$ and early work on summing it appear in chapter 2.)

Among Euler's first papers written in Saint Petersburg was his "Meditatio in experimenta explosione tormentorum nuper instituta"

(Meditation upon experiments made recently on the firing of a cannon, E853).[27] The military had long urged and supported the study of ballistics, and Johann Bernoulli had worked on problems in it. In August 1727, soon after Euler arrived in Russia, he and other academicians attended trial firing experiments conducted by a Russian general in the ordnance section, building upon an earlier experiment by Delisle in Paris. The goal was to measure accurately the speeds, as transmitted through the air, of the flash and sound of cannon shot, and Daniel Bernoulli provided the main evaluation. Composed after September 1727 for the army, the "Meditatio" sought to compute maxima for firings with pyrotechnic explosions from gunpowder. For shots Euler gave precise measurements in seconds of height, time of flight, and descent, as well as figures for the total distance. On the basis of four firing experiments, Bernoulli had attempted in an article in the second volume of the *Commentarii* to compute what became known as interior ballistics. From the dynamics of the explosion of the gunpowder and afterward a drag proportional to air resistance in the Newtonian law, interior ballistics sought to determine the muzzle velocity, and Bernoulli developed a formula for it. Expanding the firing experiments to seven, Euler improved upon Bernoulli's work. In addition to figuring in air resistance and time, Euler's computations covered the specific gravity and power of pyrotechnic shot. The experiments supported the wave or pulse theory of sound as compressions or rarefactions and his search for a mathematical model of atmospheric phenomena. Various explanations have been given for why Euler never attempted to publish the "Meditatio": perhaps Bernoulli asked him not to do so, and perhaps Euler recognized that errors remained in the protocol for the experiments and in his results.

Though Euler calculated with the decimal logarithms of the *Arithmetica logarithmica* of 1628 by Adriaan Vlacq,[28] his paper was the first to employ the symbol e to denote 2.7182818 . . . and suggest it as the base for natural logarithms.[29] From the start of his career, Euler had recognized either instinctively or from his teacher Johann I Bernoulli the importance of the exponential function e^x. It is likely that Bernoulli introduced him to this function, in which the exponent is the variable and the derivative of the function is the function itself. In his second notebook (folio 35r), which dates from 1727, Euler defined the number e and using series expansions calculated it to twelve decimal places. His first mention of e in print would not appear until the *Mechanica* of 1736, but a letter of 1731 on a differential equation uses it again, defining it as the "number whose hyperbolic logarithm = 1," which is approximately 2.71828. By delineating it as the limit of the binomial expansion $(1 + 1/n)^n$, as n approaches infinity, Isaac Newton had discovered in 1665—and Johann I Bernoulli would discover later—the infinite series, in modern notation, $e = 1 + 1/1! + 1/2! + 1/3! + \ldots + 1/n!$. The factorials in the denominators cause this series to converge rapidly. The sum of its first eleven terms, which end with 1/10!, is 2.718281801.[30] The significance of the exponential function was not yet established, and Euler continued to perfect his computation of it.

His article "De summatione innumerabilium progressionum" (On the summation of innumerable progressions, E20) appeared in volume 4 of the *Commentarii*, was presented to the academy in March 1731, and was published in 1738. In it Euler continued with his effort to sum exactly the Basel Problem;[31] this included improving results from the components of the functional equation noted in chapter 2 of $\zeta(2) = (\log 2)^2 + \sum_{n=1}^{\infty} 1/n^2 2^{n-1}$. Euler computed (log 2) to dozens of decimal places, and found that $(\log 2)^2 = [\sum_{n=1}^{\infty} 1/n2^n]^2$ is approximately 0.480453. Since he knew logarithms quite accurately, it is surprising that he gave this value to only six decimal places. Combining it and the first fourteen terms in his formula $2\sum_{n=1}^{\infty} 1/n^2 2^n$, which is approximately 1.164481, he estimated that $\zeta(2)$ is about 1.644924, which is correct to six decimal places. Reaching this degree of accuracy was extraordinary. Euler proclaimed the answer to be more precise than could be obtained directly by adding more than a thousand terms. It actually would require summing the first 30,000 terms of the original infinite series. From the start of his career, Euler had summed ever more precisely many progressions.[32] He did not yet give the difference between constants and variable quantities. In his treatment of algebraic progressions, his concept of function was at an early stage; he characteristically worked to refine it as he sought to strengthen the foundations of calculus.[33] He no longer treated variables as curved lines but as abstract entities. Regarding $\zeta(2)$, Euler had read Bernard Fontenelle's *Géométrie de l'infini* (1732), which was mistaken about the sum, but probably not yet James Stirling's *Methodus differentialis* (1730), which introduced a method giving the most exact total to that date, 1.644934066, a value that is correct to eight decimal places. By 1734–35 Euler's studies were to produce the long-sought exact answer. Stirling's *Methodus differentialis* covers arctangents, geometric series, the interpolation method of Second Viscount William Brouncker, the infinite product for $2/\pi$ by John Wallis, and the hypergeometric series of factorials. Euler had already begun extending each of these series through interpolations from integers to fractions.

As in the infinite series of James Gregory and Gottfried Wilhelm Leibniz, Euler must have realized from this approximation in 1731 a connection between the total value and π, a symbol that he employed in his first notebook and in the *Mechanica*. By the 1740s it became standard; other geometers generally adopted it. In August 1730 Euler had pointed out to Goldbach the quickly converging series of Friedrich Christoph Mayer for computing π and the existing infinite series for $\pi/2$, $\pi/4$, and $\pi/6$, and he applied his unmatched computational ability to create infinite series for $\pi/8$, $\pi/12$, $\pi^2/36$, and $\pi^4/72$, which equals $1 + 1/2^3(1 + 1/2 + 1/3^3)(1 + 1/2 + 1/3 + \ldots)$. Exhilarated by calculations with infinite series, part of which Euler devised, he carried out with aplomb extensive computations of convergent series. In a few cases he summed the first terms of infinite series until they started to diverge (*quod termini divergere incipient*).[34] Today called asymptotic convergence, it was essentially part of the Euler and Colin Maclaurin's summation formula.[35] Euler's working to solve the Basel Problem would be crucial to his discovery in 1732 of that formula;[36] his computational goal was to cut down and simplify computations from divergent series to finite sums and examine the formula with remainders.

Most likely as a result of his collaboration with Daniel Bernoulli, in 1731 Euler examined a problem in probability now called the Saint Petersburg Paradox. Jakob (Jacques) Bernoulli's *Ars conjectandi* (The art of conjecturing), published posthumously in 1713, had given the first comprehensive theory of probability. Curiously, Euler did not cite it. Daniel's brother Nikolaus II Bernoulli had invented the paradox, and the two brothers had worked on it together. Its name comes from the city where Daniel Bernoulli lived, and the paradox drew on his article "Specimen theoriae novae de mensura sortis" (Exposition of a new theory on the measurement of risk) in volume 5 of the academy's *Commentarii* (for 1730–31, but published in 1738). The Saint Petersburg Paradox describes a gambling game that involves probability and decision theory; a random variable with geometric progressions and related infinite series leads to an infinite expected value or payoff, computed at a certain point when the game stops.[37] One version begins with the house tossing a coin until a head is reached. The probability of tossing n heads in a row is $1/2^n$. On the nth throw and not before, the player is to receive, let us say, two dollars. The game has a value of $\sum_{n=1}^{\infty} 2^n \bullet 1/2^n = \sum_{n=1}^{\infty} 1 = \infty$. But with each round, the player has to put a wager into the pot. The paradox is not the infinite expected value, but that no reasonable person would pay a large sum to be permitted to play the game. Bernoulli's article, translated into English and published in *Econometria* in 1954, distinguishes between the value of classical expectation, mathematical or moral, and that of fortune, physical or moral. It thus offers a foundation for what we now call valuation theory. Instead of offering a geometric mean property for valuing games, Euler and Bernoulli discovered a property later called the marginal utility function that would not be developed further until the late nineteenth century by the economists William Stanley Jevons, Carl Menger, and Marie-Esprit-Léon Walras. Except for unusual situations that Bernoulli alone described, the utility from a small increase in wealth decreases as the wealth possessed increases: the more wealth that you have, the less value you will place on adding another ruble. The utility finds the small increase in wealth to be inversely proportional to previously possessed wealth; the utility function is a logarithmic curve. Euler introduced the concept of status, giving the player's worth after a particular outcome. Rather than appeal to Bernoulli's moral expectation, he concluded that when the outcome in a game is *b* or *c*, its worth is their geometric mean, $\sqrt{bc}$. Instead of imagining a coin toss, Euler thought of rolling a die in seeking to obtain an even or an odd number; for him the game ends when the odd number is achieved. By rolling even numbers, the player wins. Only for infinitely large chances in a game would it be possible to guarantee equality. This is one of many models of declining marginal utility.

Euler did not solve the Saint Petersburg Paradox, but his proposal suggested that games of chance are evil. He did not publish his four-page paper, "Vera aestimatio sortis in ludis" (On the true value of risk in games, E811), perhaps because Bernoulli's "Specimen" was excellent. Euler's article appeared posthumously in 1862.[38] Bernoulli demonstrated in the "Specimen" that the utility function is a logarithmic curve, and he obtained Euler's geometric mean property—two items important in modern economic theory.

From the early 1730s on, Euler investigated the statics of thin elastic ribbons or bands—that is, laminae.[39] In a series of letters after 1733, Daniel Bernoulli posed problems about these for Euler and worked with him until a solution was found. Euler principally sought to uncover the shapes of laminae in equilibrium when they are pressed by various loadings. Missing in earlier studies of elasticity and inelasticity had been the modern ideas of stress and stress analysis, which he approached. Natural philosophers considered an inelastic body a perfectly deformed elastic one. Galileo Galilei, Edme Mariotte, Gottfried Wilhelm Leibniz, Christiaan Huygens, and Pierre Varignon achieved important results in their studies of elasticity—particularly for bending rods. In his *Traité de la percussion ou chocs des corps* (Treatise on percussion or the impact of bodies), published in 1673, the French physicist Mariotte examined the work of Galileo and Huygens and was the first to treat the impact of both bodies comprehensively. But Euler's research stemmed from Jakob Bernoulli's derivation in the 1690s of the differential equations of elastic curves when external forces act upon the ends of the laminae. Euler's first article on elasticity appeared in the

Commentarii in 1732, and he wrote an unpublished piece, likely sometime in the 1730s, on an annulus, a washer-like ring, disturbed from equilibrium and put into motion.[40] This research led subsequently to Euler's buckling formula, giving the maximum load that can be exerted on an elastic column before it bends. To compute the exact answer took time, for Euler had first to develop elliptic integrals, which refer to expressing the arc length of an ellipse as an integral. It gave the solution to one of these integrals. Euler never used the term *elliptic integrals* or presented a unified theory for them, but he studied these integrals for four decades.

In 1732 Euler's "Nova methodus innumerabiles aequationes differentiales secundi gradus reducendi ad aequationes differentiales primi gradus" (A new method by means of which innumerable differential equations of the second degree can be reduced to differential equations of the first degree, E10) was published in the *Commentarii*.[41] Submitted four years earlier, it had important consequences for mechanics as well as mathematics. In the eighteenth century, modern mathematical physics passed in a transition essentially in format from Newton's *Principia mathematica,* with its geometrical-proportional dynamics without symbols and only an introduction to limits,[42] to Joseph Fourier's analytical equations for the theory of heat.[43] Newton had lacked relational laws with equal measures that were not simply proportionalities. Critical in the beginning of the transformation is Euler's "Nova methodus," taking dimensional thinking and concepts beyond mathematics, reducing to the first degree different types of differential equations of the second degree, and introducing in print the application to mechanics of *dimensionum*, the Latin word for "dimensions." All of these algebraic equations must obey homogeneity. Some of Euler's early efforts were crude; for example, he considered weight and mass to be homogeneous. His formulation of differential equations for mechanics, a difficult task, would significantly continue in his *Mechanica*.

The title page of Daniel Bernoulli's *Hydrodynamica, sive de viribus et motibus fluidorum commentarii* (Hydrodynamics, or commentaries on forces and motions of fluids), published in Strasbourg in 1738, asserts that the book was the fruit of ten years of research, which suggests that a draft had been completed before he left Saint Petersburg in 1733.[44] Huygens's conservation of *vis viva* is its guiding principle, and Bernoulli presented the academy with a copy. The *Hydrodynamica*, a difficult text that covers hydraulics, hydrostatics, mechanics, and the science of equilibrium, includes the first successful analysis of the pressure and velocity of a moving fluid. It introduced the term *hydrodynamics* into the vocabulary of science, defining *hydrostatics* as applying to pressure and *hydraulics* as applying to motion. Pressure is the force a fluid asserts on the walls of its enclosing tube or vessel, and one measurement of pressure is the height of a column of stationary fluid. In August 1727 the twenty-year-old Euler had given the academy the results of his own study in hydrodynamics, just two weeks after Bernoulli's. This produced a sensitive situation between the two, but Euler conceded that his more senior colleague should publish them.[45] Bernoulli prepared the draft of the *Hydrodynamica* virtually on his own, but after he returned to Basel the two corresponded about making it ready for publication. Euler repeatedly urged that its new ideas be soon put into general discussion within the wider scientific community. Nonetheless, for nearly a quarter century he stopped research in hydrodynamics.

As his early notebooks reveal, by the early 1730s Euler was already beginning to make original contributions to astronomy by devising important concepts and principles. He now supported freedom of the press in response to a request to publish the translation into Russian of Fontenelle's celebrated popularization *Entretiens sur la pluralité des mondes* (Conversations on the plurality of worlds, 1686, and commonly referred to as *Mondes* at that time), an exchange between a philosopher and a curious woman Fontenelle called Madame the Marquise.[46] In 1730 Kantemir completed a manuscript of his translation and submitted it to the press of the Petersburg Academy.

Orthodox Church censors resolutely opposed Copernican astronomy, which in the first part, or what was called an "evening," of *Mondes* is set forth as most likely the correct system. Fontenelle imagined that his representative (the philosopher) and the Marquise would spend evenings observing the night sky. Thus the academy's bringing Copernican astronomy within the grasp of the literate public was acting essentially in defiance of the Russian censors. While the heliocentric universe was no longer an issue among the scientific community in the West, early in the eighteenth century a few books by clerics in north central and eastern Europe still accepted Claudius Ptolemy's geocentric model. The Russian translation of Huygens's *Cosmotheoros: sive de Terris Coelestibus, eaxumque ornata, conjecturae* (Cosmotheory: or, conjectures concerning the inhabitants of the planets, 1698), another source of the new astronomy, had appeared in print only through the order of Peter I, whose death removed a powerful advocate of the sciences in Russia. Conversations on Fontenelle's "second and third evenings" examined an infinite universe and possible voyages to the moon; the "fourth and fifth evenings," at the end of the 1686 edition, sketch the Cartesian vortex cosmology, which allows no place for a vacuum. (The edition of 1687 added a sixth conversation on the latest discoveries in astronomy.) *Mondes* holds that there is life on the other planets in our solar system, on comets, and on planets circling the myriad of stars.

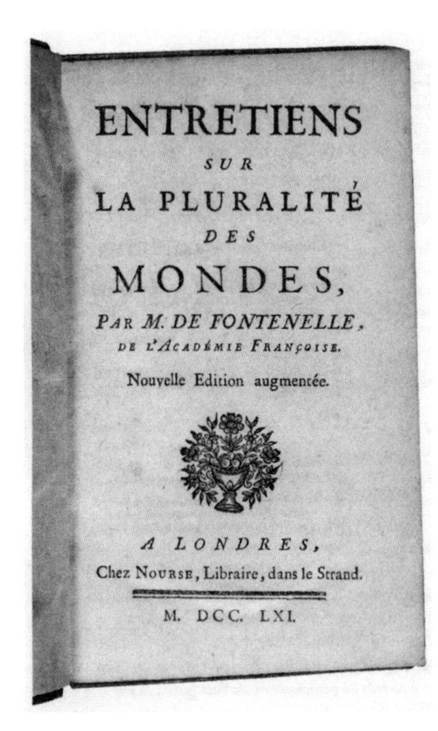

ENTRETIENS
SUR
LA PLURALITÉ
DES
MONDES,
PAR M. DE FONTENELLE,
DE L'ACADÉMIE FRANÇOISE.
Nouvelle Edition augmentée.

A LONDRES,
Chez NOURSE, Libraire, dans le Strand.
M. DCC. LXI.

Figure 3.2., The title page of Bernard le Bovier de Fontenelle's *Entretiens sur la pluralité des mondes,* 1761 edition.

In 1687 the book was put on the Catholic *Index librorium prohibitorum* (Index of prohibited books). To keep his book fresh, Fontenelle provided updates, notably in 1708 and 1742, with data from the most recent astronomical observations. Among other texts on the plurality of worlds was Johannes Kepler's novel *Somnium* (The dream) of 1634, but Fontenelle's *Mondes* (originally published in 1686) became the best known.

A product of the great age of French letters, *Mondes* is written with wit and humor that go far in accounting for its popularity during the Enlightenment. It substantially contributed to the diffusion of the new sciences and helped make them fashionable for the European reading public, largely by creating a genre of literature written for women. It influenced Francesco Algarotti and, it can be conjectured, Euler's later *Lettres à une princesse d'Allemagne* (Letters to a German princess). The introduction of *Mondes* extols Cartesian rational scientific knowledge, lauds Francis Bacon's program of experiments and observations, and is skeptical toward metaphysics. Kantemir's translation of the *Mondes* would enrich Russian scientific terminology.

Meanwhile, Church censors in Russia acted in character, suppressing the publication of *Mondes* throughout the 1730s. Euler's support for printing the translation was fruitless. Protests from Kantemir, Delisle, and—before their departures—Hermann and Bernoulli against restrictions on scientific publications fared no better. The issue of *Mondes* was not Kantemir's initial brush with the censors. His *Satira I* (First satire) of 1729 had attacked the concept of a hereditary aristocracy and criticized religious opposition to Copernican astronomy. The academy's chronicle by Petr Pekarskii reports that educational primers published by the church from 1700 to 1725 attacked everything published outside Russia as heretical. But as is not uncommon among agencies stifling free speech, the Holy Synod was inconsistent. It failed to block a series of articles in 1732 for the *Primechaniya k Vedomosti* in Russian and German, including Euler's polemical "Von der Gestalt der Erden" (On the shape of Earth, E32),[47] it did not prevent speeches by Bernoulli and Delisle at public meetings of the academy defending Newton, and it did not suppress Vasilii Nikitich Tatishchev's *Razgovor dvukh priiatelei o pol'ze nauk I uchilishch* (Conversation of two friends on the utility of science and of schools, 1733), which defended free scientific inquiry.

During the early 1730s Euler also studied Samuel de Fermat's edition with commentary and with notes by his father Pierre of the six known books of the *Arithmetica* by Diophantus of Alexandria, which was based on the Latin translation by Claude Bachet de Méziriac of 1621. Among other authors that Euler later read on number theory were Euclid, François Viète, René Descartes, Christoph Rudolff, Franz van Schooten, Claude Bachet de Méziriac, Bernard Frenicle de Bessy, Philippe de la Hire, Joseph Sauveur, and John Wallis from the past and Christian Wolff and Philipp Naudé from his own time.[48] Reading Fermat's *Varia opera mathematica* in June 1730, Euler came across what he called the "not inelegant theorem" that every natural number could be represented as the sum of four squares.[49] Proving such conjectures, Euler explained in a letter that month to Goldbach in Moscow, could "contribute greatly to the enrichment of analysis . . . and also shed useful insight into other areas of mathematics."[50] His exploration of number theory was becoming more emphatic. His letters to Joseph-Louis, Comte de Lagrange after Goldbach's death in 1764 credit Euler's correspondence with Goldbach for its persistent encouragement in keeping his attention on the subject.

The historical record and a survey of Euler's articles on number theory in the Petersburg Academy of Sciences' *Commentarii* and later the *Novi Commentarii* and *Nova Acta Academiae Scientiarum Imperialis Petropolitanae* confirm that Goldbach was the most important stimulus for Euler's research in the field. To 1756 his correspondence on number theory, almost solely with Goldbach, showed his progress as well as the conventionality of most of Goldbach's ideas; the new insights lay with Euler. Their letters, written in Latin up to 1740, had continued after Goldbach returned to Saint Petersburg in 1732, when the two probably saw each other daily. By posing acute questions and problems without proof, and often misstated rather than giving general results, Goldbach provided an opportunity for Euler to discover proofs and place their discussion on a firmer foundation; he kept Euler steadily focused on the subject.[51] In May–June 1730 he had redrawn Euler's attention to Fermat's numbers. Prime numbers and divisibility—both of them essential to number theory—intrigued Euler, and he was to investigate them for the rest of his career. Refining criteria for prime numbers from Fermat, he computed them up to 10^6 and, for a few, beyond that.

In the commentary on his translation of Diophantus, Bachet had stated and demonstrated the "not inelegant theorem" for integers up to 325 but could not prove it. Possibly through careless reading, Euler failed to realize that Fermat had claimed to prove the theorem with his method of infinite descent.[52] In June 1730, Euler wrote to Goldbach that problems with numbers as the sums of three triangular numbers, along with four nontrivial integral squares and five pentagonal numbers, were blocking his attempts to prove that theorem. The next year he rediscovered Fermat's "outstanding" Little Theorem: if p is a prime number that does not divide the integer a, then p divides the expression $a^{p-1} - 1$.[53] Mathematicians have called this the Little Theorem to separate it from what became Fermat's celebrated Last Theorem.[54] It would take Euler six years to obtain his first proof of the Little Theorem in "Theorematum quorundam ad numeros primos spectantium demonstratio" (A proof of certain theorems concerning prime numbers, E54, written in 1736 but not published until 1741), which was by induction.

After noting in his reply to the postscript of Goldbach from 1729 that Fermat believed that all numbers of form $2^{2^n} + 1$ are prime, Euler's article "Observationes de theoremate quodam Fermatiano, aliisque ad numeros primos spectantibus" (Observations on theorems that Fermat and others examined concerning prime numbers, E26), which he presented to the academy in 1732, discovered that $n = 5$ gives a counterexample. He factored $2^{32} + 1 = 4{,}294{,}967{,}297$ into $641 \times 6{,}700{,}417$,[55] thus disproving Fermat's conjecture.[56] Euler had found that composite Fermat numbers must possess divisors of the form $2^{n+1}k + 1$. Once he recognized this, it was much easier to arrive at the divisor 641, for the case when $n = 5$ and $k = 10$ gives $2^6(10) + 1 = 64(10) + 1$.

Before obtaining this result, Euler had studied Christian Wolff's *Elementa matheseos universae* (Elements of the universal exact sciences, 1730), which presented number theory as a major field of mathematics,[57] a status that it had not previously enjoyed, and examined prime numbers. Following a curious narrative style rather than the deductive theorem-proof format, Euler began by arguing, "If there are prime numbers of the form $a^n + 1$, they must all be included in the form $a^{2n} + 1$."[58] If a is an odd number, then $a^n + 1$ is divisible by 2, while a as an even number can be either prime or composite. Wolff investigated cases in which the Mersenne numbers $2^n - 1$ are prime. He showed that when $n = 37$ and 43, this gives Mersenne numbers. Marin Mersenne had previously asserted (in part erroneously) that $n = 2, 3, 5, 7, 13, 17, 31, 67, 127$, and 257 result in primes.[59] Wolff's list of primes includes $2^{11} - 1 = 2047$.[60] But Euler contradicted Wolff, asserting that Mersenne numbers may be composites for cases in which n is a prime, noting in volume 6 of the *Commentarii*, "I see the celebrated Wolff did not notice [the mistake] in his *Elementa* . . . where he . . . listed $2^{11} - 1 = 2047$ among the primes."[61] Euler noted that $2047 = 23 \times 89$.

At the end of "Observationes de theoremate" Euler listed six theorems. The first is Fermat's Little Theorem, and three are special cases of the Euler-Fermat Theorem, a generalization of Fermat's Little Theorem that Euler would later pursue. Euler could not yet prove any of these. He asserted that Fermat had expressed the Little Theorem "without proof [and obtained it] merely by induction."[62] Nor did he know of papers, now in the University of Hannover Library, containing the proof by Leibniz in 1683. Like Fermat, Euler had proceeded experimentally to his discovery of the theorem. He began with $2^{p-1} - 1$, and it took until 1736 before his first, crude proof of it appeared in print.[63]

Euler's correspondence with Delisle in 1732 continued with composites and primes among the Mersenne numbers, which he related to Euclidean perfect numbers, a topic that his work on Fermat, Descartes, and Frans van Schooten had led him to investigate. He posited that perfect numbers are positive integers whose proper positive divisors less than the number add up to that number, such as $6 = 1 + 2 + 3$.[64] Book 9, section 36 of Euclid's *Elements* states that if $2^n - 1$ is prime, where n is a natural number greater than 1, then the number $2^{n-1}(2^n - 1)$ is perfect. Euler demonstrated that all even perfect numbers fit this formula. The question of odd perfect numbers is still unresolved today. Euler's result made him akin to a collaborator with Euclid, earning for his discovery the name the Euclid-Euler Theorem.[65] Determining when Fermat numbers are composites took another decade.

In corresponding with Goldbach and the Bernoullis in 1731–32 about developing methods to solve differential equations, including those without separated variables, Euler demonstrated that some Riccati equations are solvable.[66] He began with seeking integer solutions to them. The general equation is $a(x)y^2 + b(x)y + c(x) = y'$. After Goldbach provided a small base for them, Euler built a broad foundation and, by ingeniously using clever transformations, substitutions, and continued fractions, obtained solutions. While examining Fermat primes and perhaps perfect numbers, Goldbach asserted that for given numbers, including 1 along with the primes, "each number greater than 2 is an aggregate of three primes" and "each number of the form $4n + 2$ is the sum of two primes of the form $4n + 1$," Goldbach's theorem.[67] He mistakenly believed that $4n + 1$ can be expressed as $cx^2 + y$, whenever c is any divisor of n. Euler disagreed, finding this probably true for x and y rational fractions, but not always when they are integers. He gave as an example $89 \neq 11x^2 + y$.

Beginning with the article "De solutione problematum Diophanteorum per numeros integros" (On the solution of Diophantine problems by integer numbers, E29) presented to the academy in 1733 but published later, in the *Commentarii* for 1738, Euler pursued solutions for second-degree Diophantine equations,[68] indeterminate algebraic equations with several unknowns expressed in successive conditions that require rational solutions. Breaking with the past, Fermat had given the modern meaning to these equations by restricting solutions to integers. They are critical in number theory and for Euler's rectifying of curves. Apparently Euler was not yet aware of the Diophantine equation now known as Fermat's Last Theorem: $x^n + y^n = z^n$ has no positive integer value solutions except for $n = 1$ and 2, along with the uninteresting trivial cases (0, 0, 0) (0, 1, 1), and (1, 0, 1). Stated otherwise, it has no integer solution for powers of x, y, and z higher than 2. In this case the word *theorem* had long been a misnomer. Proving it became the most celebrated problem in mathematics, drawing numerous attempts to solve it, though some of them were false claims. Euler was about to embark on a search for a proof, but it would remain an unsolved problem until the 1990s.

By 1731 Euler had begun to formulate his own theory regarding maximum and minimum curves in what became the calculus of variations. Generally variational theory studies extrema—families of maximum and minimum values for integrals, often called functionals—and from these values determines an optimal curve. Since the variation or differential of a functional involves more than extrema, today the more appropriate term for this research is the calculus of variations in the narrow sense. Euler's colleague Hermann's *Theoria generalis motuum* (General theory of motion, written in 1727 but published in 1732) offered an incorrect solution to the curve of quickest descent in a resistant medium. A letter of Johann Bernoulli to his son Daniel in 1727 had alerted Euler to these extremals and variational problems.[69] In 1731 it was left to Euler at the academy to improve upon and correct Hermann's solution. At this time, when the concepts of analytic functions and of analytic surfaces did not yet exist, Euler provided a simple mechanical solution. But in a letter of September 1736, Daniel Bernoulli was to point out that Euler's solution also was wrong.

In 1732 Euler had an article in the *Commentarii* and in 1734 another that belonged to the origins of the calculus of variations. Euler stated that Hermann had raised the brachistochrone problem in a resistant medium in 1727. His two articles elaborate methods from Hermann's *Phoronomia* together with strategies in Brook Taylor's *Methodus incrementorum* and methods from Johann and especially Jakob Bernoulli. In *Acta eruditorum* for 1697 Jakob had observed that this problem without subsidiary conditions was rather straightforward, while many isoperimetric problems having more restrictions on admissible curves were more interesting. Euler emphasized finding extremal curves for actions of forces, including gravity. Instead of following a systematic course, both articles treated special ad hoc problems.

Maintaining that in the "broadest sense general solutions of isoperimetric problems are possible," Euler's difficult "Problematis isoperimetrici in latissimo sensu accepti solutio generalis" (An account of the solution of isoperimetric problems in the widest sense, E27), presented to the academy in 1732, suggested the future direction of his research in the calculus of variations.[70] The summation written for this article at the beginning of the *Commentarii* was likely not by Euler, for its author seems not to have recognized that answers for the emerging calculus of variations are curves, not numbers. Euler's essay analyzes problems that require only two curve elements. The cycloid and the catenary are curves with the desired properties. The "Problematis isoperimetrici" initiates the author's search for Euler's necessary condition, in which a particular differential must become 0 for a maximum or minimum. Euler's "De linea celerrimi descensus in medio quocunque resistente" (On curves of fastest descent in a resistant medium, E42), probably written in 1729 but saved for submission until 1734, addresses the brachistochrone problem in the case of friction. Throughout his early career, Euler had sustained an interest in providing better solutions for that problem and for motion in a resisting medium. Believing that variational principles manifest themselves in general laws of physics, he was the first to widely apply the methods of

METHODUS
Incrementorum
Directa & Inverſa.

AUCTORE
BROOK TAYLOR, LL.D. &
Regiæ Societatis Secretario.

LONDINI
Typis Pearſonianis: Proſtant apud Gul. Innys ad Inſignia
Principis in Cœmeterio Paulino. MDCCXV.

Figure 3.3. Brook Taylor (left) and the title page of his *Methodus incrementorum directa et inversa,* 1715 (right).

the calculus of variations to physical problems. He viewed the field as a Leibnizian *ars inveniendi*, or method of discovery.

In the early 1730s, Euler's assault on the Basel Problem continued. His "Methodus generalis summandi progressiones" (General methods for summing progressions, E25) in volume 6 of the *Commentarii* for 1732–33, published in 1736, addresses it. The article begins, "I have just last year [in E20] put forth a method of summing innumerable progressions."[71] In a plethora of computations, Euler gave increasingly better approximations. These investigations produced a special case of the calculus tool now known as the Euler-Maclaurin Summation Formula, which for a function gives the relation between its integral and the sum of its values at equally spaced points.[72]

After making initial steps in 1729 toward the task of bringing forward and defining gamma or his constant, Euler had put related summations aside for five years. Gamma, one of the most famous of the mathematical constants, 0.5772156649 . . . , is not to be confused with the more familiar *e*;[73] it has applications in fields including analysis and number theory, and much of its nature—for example, whether it is irrational—remains a mystery. Returning to the subject in "De progressionibus harmonicis observationes" (Observations on harmonic progressions, E43), presented to the academy in 1734, Euler announced his breakthrough related to the gamma constant in the discovery of a fascinating interrelationship between the natural logarithms and n-th partial sum of the harmonic series $H_n = 1 + 1/2 + 1/3 + \ldots + 1/n$ that enabled him to arrive at a more nearly exact value of his constant.[74] These are its two defining elements. The first series is called harmonic because each term is the harmonic mean of the two surrounding terms. As he examined convergent infinite series and partial summing of terms of infinite series, again "until they begin to diverge," the search to identify characteristics of convergent series, conditional and absolute, and achieve exact solutions guided Euler's efforts. Applying his instinct, research on the differences between natural logarithms and divergent "more natural harmonic series combined with geometric ones," and dangerous rearrangements and manipulations of conditionally convergent series that have been described as acrobatic, he found the convergent zeta series for his constant. As he put it,

> Quae series, cum sint convergentes, si proxime summentur prodibit . . .
>
> This series, since each term is convergent taken one after the other, will proceed [that is, $1 + 1/2 + 1/3 + \ldots - 1/i = \log(i + 1) + 0.577218$, approximately. . . .]
>
> Huius igitur quantitatis constantis c valorem deteximus, quippe est C = 0,577218.
>
> Therefore, we have revealed the value of his constant to this accuracy to be $C = 0.577218$.)[75]

This is wrong in the sixth decimal place. Two years later in E47 Euler calculated the numerical value of the constant, defining C to be the limit as n approaches infinity of $(H_n - \ln n)$, or numerically the decimal fraction correct to the first fifteen decimal places, 0.577215664901532. . . . Here ln stands for the French of logarithmic natural—that is, the natural logarithm to base e. Claiming that his number was "worthy of considerable attention,"[76] Euler pursued deeper connections between the divergent harmonic series and natural logarithms, his constant, the formula $\sin x/x$, and related labors on the Basel Problem and the Euler-Maclaurin Summation Formula; this combined work was the beginning of analytic number theory. Euler must have seen that the exact sum of the Basel Problem is near $\pi^2/6$. Euler initially gave no proof of the existence of his constant; his fundamental concern was over whether it was rational or irrational, which to the present day remains unresolved. He believed that his constant derived from another constant or function or that it was the logarithm of another number, but he failed to show this. Euler usually represented his constant by the symbol C, sometimes as O or A. His constant was not yet called gamma, and its significance was not immediately recognized.

Alongside the transcendental numbers π and e, the gamma constant remains one of the most important real constants in analysis and have been called the trinity of mathematics.[77] Euler is closely associated with all three, inventing the first and computing more accurately and making standard the notation for the next two. Despite sustained, serious examination of the gamma constant for more than two centuries by many mathematicians who have shown its beautiful and far-ranging consequences in mathematics and mathematical physics, to the present day its exact nature remains so mysterious that we do not even know whether it is a fraction; it is not known whether the gamma constant is rational, algebraic, or transcendental. In the twentieth century, Godfrey Harold Hardy pledged to cede his Savilian professorship at the University of Oxford to any scholar who could prove that the gamma constant is algebraic.[78]

In the article "Solutio problematis arithmetici de inveniendo numero qui per datos numeros divisus, relinquat data residua" (Solution to the arithmetic problems of finding a number that, divided by given numbers, leaves given remainders, E36), written in 1734 but not published in volume 7 of the *Commentarii* until 1740, Euler independently rediscovered and proved the Chinese Remainder Theorem, which also belongs to number theory.[79] According to a Chinese word puzzle from the first century A.D.,

> Certain things have an unknown number. Dividing that number N by 3 gives the remainder 2, dividing by 5 has the remainder 3, and dividing by 7 has the remainder 2. What is the number of things N?

The solution requires finding an integer N that satisfies three simultaneous linear Diophantine equations. For any integer k, the solution is $N = 338 - 105k$. Euler's rediscovery of the theorem was separate from most of his work and perhaps accidental. It is one of the few theorems that he found and did not take up again later.

The same issue of the *Commentarii* reveals a colleague who exerted a minor influence on Euler in number theory. Krafft had a modest article on the subject "De numeris perfectis" (On perfect numbers) that claimed to prove that all even perfect numbers must follow the form given in book 9, section 36 of Euclid's *Elements*: if $2^n - 1$ is prime, then $2^{n-1}(2^n - 1)$ is perfect. Another article, and "Observationes arithmeticae de septenario" (Arithmetic observations on the septenary), apparently by a relative of Krafft named Johannes, explains how to recognize whether a number is divisible by 7. Mathematicians besides Goldbach in the Petersburg Academy of Sciences encouraged Euler's research in number theory.

Able to complete the many early practical academic assignments quickly and possessing a strong physical constitution, Euler had the time, energy, and freedom in the academy to pursue his research and have the *Commentarii* publish his results. By the end of 1734 he had completed and published or had accepted for publication thirty-five articles. These circumstances went into the "good fortune" that Euler later described to Clairaut regarding the situation in the early years of the academy.[80] But the publication of thirty-five articles did not satisfy Euler; these were to comprise only 4 percent of his total work. He set about increasing his writings.

A valuable source of Euler's scientific research is his correspondence with Daniel Bernoulli. Resuming in 1733, it continued for thirty-five years, except for a break from 1754 to 1767, provoked by what Bernoulli took to be the misbehavior of Euler regarding publications and Bernoulli's relations with the Petersburg and Berlin Academies, including a possible pension from the Petersburg Academy.[81] In 1734 Euler sent three letters to Daniel Bernoulli. His February letter addresses the speed of ships, the Riccati differential equation, a remarkable series for logarithms, the brachistochrone problem in a resistant medium, and tautochrone curves for fluids. His November missive supported Bernoulli's request for a pension from the Petersburg Academy of Sciences. Euler confirmed that he had completed volume one of his *Mechanica* and inquired about the publication of Bernoulli's *Hydrodynamica*. He also referred to his unfulfilled wish to earn a medical degree in Basel, but considered the cost too high. A letter of December expresses Bernoulli's delight that Korff had been named academy president and urged that the prodigy Clairaut from Paris be invited to Saint Petersburg. Bernoulli noted that after reviewing recent general principles of mechanics, he examined small vibrations of elastic rods. Nearly simultaneously he and Euler had devised the differential equation for simple transverse vibrations of the rods, discovering that Leibniz's finding in 1684 of the connection between elastic and acoustic properties of bodies was correct. Their research on special theories of vibratory phenomena aided their advance to general principles of mechanics.

Thus far, to 1735, the academy had assigned Euler only one major state project, in geography and cartography. The Russian monarchy's principal charge and funding for the academy after 1730 was to draw the first accurate, large-scale maps of Kamchatka and neighboring territories, including the western North American coast. In their correspondence, Daniel

Bernoulli requested from Euler information on it and the instruments La Condamine had for measurements of the arc of meridian in Peru.

Successors of Peter the Great wanted to continue his investigation of the economic potential of Siberia, and Empress Anna desired to add splendor to her court. Improvements in clocks, surveying instruments, telescopes, and mathematical techniques (one of which allowed a more accurate synchronization with clocks) were also prompting a rejuvenation of mapping across Europe. In 1730 the Danish navigator and explorer Vitus Jonassen Bering, who had served in the Russian Navy and led the first Kamchatka expedition, proposed the second—or Great Northern—expedition to explore all of Siberia, including its eastern portion of Kamchatka, and to approach the terra incognita of northwestern America.[82] Ivan Kirilov, the high secretary of the Russian senate, supported it; Kirilov himself was preparing a map of the Russian Empire and supported geographical research. A year later Anna accepted the proposal and ordered that detailed plans be made for it. Bering and Russian officials arranged for a nautical and geographical study with geodetic surveys and astronomical observations. But the academy requested more; it wanted to expand the expedition into a general scientific exploration to include recording barometric and thermometric results, painting landscapes, and writing reports and gathering archives on the colonization, ethnography, and history of nearby areas. It also sought to conduct botanical, metallurgical, and zoological studies. At the academy Müller and Gmelin, the latter named professor of chemistry in early 1731, led in advocating this wider project. After falling from favor with Schumacher that year, Müller moved wholly to conducting extensive research on Russian history and collecting primary source materials that were to appear in the *Sammlung Russischer Geschichte* (Collection of Russian history), printed in 1732–33. He also searched for documents and reports for a history of the early academy. (It would not go to press until 1890.) In 1731 Anna accepted the ambitious combined plans of Bering and the academy.

The Russian senate had instructed Delisle to prepare a map of Kamchatka and neighboring trading company areas, which reached to the northwestern American coastlands, indicating graphically uncharted areas to be filled in by the expedition.[83] Completed in 1732, the map was based on faulty information and Delisle's still unfounded concepts of mathematical cartography.[84] An article on the nature of land regions and their history accompanied the map; when they left Saint Petersburg, Bering and the expedition astronomer Louis de l'Isle de la Croyère received copies of both.

Paraphrasing Gottlob (Theophilus) Siegfried Bayer, Peter Lauridsen has called the expedition "the greatest geographical enterprise ever undertaken." The largest group research project of the early academy, it was to give Russia claims to eastern Siberia and what later came to be known as Alaska.[85] In early 1733, the group began departing from Saint Petersburg in detachments, the first of them with over 187 men under the command of Bering. They included a surgeon, a landscape painter, an interpreter, and an instrument maker, together with fourteen bodyguards.[86] Gmelin, Müller, and La Croyère were the initial academy contingent. Speaking on the importance of chemistry in February 1732 and writing in two articles for volume 5 of the *Commentarii*, the increasingly influential Gmelin had urged construction of a chemical laboratory at the academy that came to stress metallurgy.[87] The three academicians received ample instruments; La Croyère, for example, had nine wagons filled with equipment, including two clocks, twenty-seven barometers, and telescopes of five, seven, thirteen, and fifteen feet. The academicians also carried a library of several hundred books, covering the sciences and history, special interests, Latin classics, and light reading.[88] Gmelin and Müller were good choices for the expedition,[89] but not La Croyère, an amiable but unmotivated character who cared little for scientific research and was ill-prepared to be chief astronomer. (His brother Joseph Delisle had obtained the position for him.) In the face of the contempt of his academic associates and the hardships of Siberian living, La Croyère's work deteriorated, and his astronomical observations, mostly the work of his assistants, were to prove worthless. Over its entire course, the expedition included thirteen ships and roughly three thousand men.[90]

As Delisle's interest in the Kamchatka expedition lessened, Euler's association with it increased. He began as the mathematician for the project, assisting Delisle in establishing mathematical cartography working on the academy's geography department; they prepared young astronomers and geodetic practitioners for cartographic and geodetic work. Euler was now moving from student of Delisle to colleague; by late 1734 he independently made astronomical observations needed for constructing meridian tables that were published the following year.

In March 1734, Euler's friend Müller wrote to him from Tobolsk congratulating him on his marriage. At the end of that year, Euler replied.[91] This letter suggests a growing role for Euler: he informed Müller that he would receive frequent reports on developments in Russia and from abroad. On a personal level, Euler reported, "I have purchased a house on the tenth line not far from the Field Marshal [Münnich's] palace. Finally, my dearest wife has on 16 November [27 November on the Gregorian calendar] delivered a young son." In closing the letter, he wished Müller "all pleasure in pursuing this great expedition."[92] Drawing on his ten years to 1743, Müller—who would later become known as the father of Siberian historiography—was to write his magnum opus, the four-volume *Istoriya Sibiri* (History of Siberia), but only the first volume (in 1750) and part of the second appeared in print before his death. Euler encouraged Müller's collection of archives and geographical materials for maps, along with cultural and ethnographic data, such as that on the ceremonies of different Siberian societies; the two men corresponded regularly until 1767.

Chapter 4

Reaching the "Inmost Heart of Mathematics": 1734 to 1740

By 1740, Leonhard Euler had established a European reputation in the mathematical sciences for his publications, scientific correspondence, and entries for the Paris Academy of Science's prestigious annual Prix de Paris. During his first Saint Petersburg stay, Euler published more than fifty articles and books, including thirty-seven essays published in the *Commentarii* after 1733.[1] He continued to concentrate on rational mechanics, together with differential and integral calculus (and especially its application to mechanics), while giving much attention to ship theory, number theory, astronomy, and music theory. Demonstrating a broad range of interests across the mathematical sciences, Euler published preparatory studies in eight other scientific disciplines: algebra, arithmetic, celestial mechanics, physics, tidal theory, variational theory, geometry, and the roots of differential geometry, as well as in pedagogy. Four accomplishments in particular were to bring him European distinction. The most important were the exact solution of the Basel Problem in 1735 and in the next year the publication of his *Mechanica sive motus scientia analytice exposita* (Mechanics of the science of motion set forth analytically), which was reviewed in Paris in 1740. Of lesser consequence were his computing the accurate shape of Earth independently of the Lapland Expedition and sharing the Prix de Paris in 1738 and 1740 as well as winning it alone in 1739. In addition to conducting state and technical projects, Euler's correspondence—especially with Daniel Bernoulli and Giovanni Jacopo Marinoni—seemed to encourage telescopic observations;[2] it is possible that Euler was still likely making such observations. He was skillful in experiments as well. Elements of what may be termed his Eulerian synthesis began to emerge; he drew elements mainly from Isaac Newton's mechanics while avoiding contradiction—for example, accepting the law of attraction but believing that some modification was necessary for celestial dynamics. René Descartes had asserted that an ether filled all space, and Euler accepted this concept. He added his own original insights in the sciences and sought to develop and apply to dynamics and astronomy the Leibniz-Bernoullian calculus. Euler had already begun to question Newton's corpuscular optics but had not yet developed his pulse theory, and he was quietly criticizing Wolffian mathematics and physics as rife with errors.

In his "Éloge de M. Euler" Nicolas, Marquis de Condorcet, spoke of the "impossibility . . . of conveying in detail . . . an accurate idea of the multiplicity of discoveries, new methods of investigation, and ingenious views" contained in the massive publications and correspondence of Euler, an observation that remains true to the present day.[3] By 1735 Euler's writings and ideas were already beyond the scope of any moderate-length biography.

This chapter will investigate chronologically and thematically Euler's major publications in pure and applied mathematics for the years 1734–40, his contributions to practical projects, and the tenor of his life.

In January 1735 the Petersburg Academy of Sciences instructed Euler to compile tables "by which at any given latitude one can determine what the time for that person is, provided only that he possesses some fixed collection of the elevations of the sun."[4] He was being asked to obtain the equation of noon, where the word *equation* would carry the ancient meaning of correction rather than the modern reference to equality. This was an early part of the eighteenth-century torrent of cases linking experience with underlying, evolving mathematical models. After working for two years on tables, Euler proposed the midday correction equation when he was given a table of two separate observations of the sun's identical altitude above the horizon for every degree of deviation between one and eighteen hours from Saint Petersburg; the observations of equal height occurred at 8:21 a.m. and 3:49 p.m. Since the sun tracks not Earth's equator but the ecliptic—that is, the sun's annual path or great circle around the sky—determining the exact latitude requires an adjustment of the original equation. When superimposed, these two planes make an angle of about 23 1/2°; this is known as the obliquity angle, which gives the increment or change, dt, of the sun's declination—in this case, north of the celestial equator in one day. The obliquity of the ecliptic is its angle to the celestial equator. Euler derived the term *correction* between true noon (as seen with a sundial, mean noon from using a clock) and actual noon (depending on equivalent solar altitudes). In this effort, the academy attempted to construct more accurate clocks.

Euler surely knew that book 1, chapter 2 of Claudius Ptolemy's *Almagest* predicted celestial positions using plane triangles and chords on a circle, and that book 3, chapter 6, gave changes in longitude of the sun's position. Perhaps Euler also had access to the Persian mathematician and astronomer Kushyar ibn Labban's *Jami Zij* (Comprehensive astronomical handbook, ca. 1025), which improved upon Ptolemy and added information on Hindu mathematics and medieval Islamic trigonometry.[5] Euler may be seen as Ptolemy's and ibn Labban's heir in that he joined a systematic mathematics with detailed, numerous, and skillfully collected astronomical data. He began by examining the relationship of the sine and cosine of arcs with these values and small deviations. Like his predecessors, Euler appealed rigorously to spherical and plane geometry, trigonometry, and algebra, but he had the advantage over them of recent advances and growing sophistication in all of these and especially "joining them together with the new infinitesimal calculus."[6] He was also already familiar with the post-Copernican celestial mechanics and planetary theory, particularly of Johannes Kepler and Newton.

Euler astonished the Saint Petersburg academicians by completing in three days a task they expected might take three months, an accomplishment that his Basler elegist and grandson-in-law Nicholas Fuss attributed to indefatigable labor. In Euler's paper "Methodus computandi aequationem meridiei" (A method for the improvement or correction of the equation of time, E50), he gave two formulas or equations.[7] The second has the cosine of declination dividing the product of the sine of obliquity, cosine of longitude, and daily change in longitude. We do not know how long it took to devise the correction equation to make the computations more simply and quickly, but it was derived from ideas presented in two anonymous papers in the academy observatory's *Primechaniya k Vedomostyam* (Notes to the gazette) for 1731 that had been coauthored by Leonhard Euler and Georg Wolfgang Krafft. The tables in the manuscript submitted at the academy conference in mid-January 1735 employing the new formula required no more than eight hours a day for three days to finish. Since the paper was to appear in the *Notes*, Euler took it home to make accurate copies. At the conference a week later, Joseph-Nicholas Delisle took copies to study, and thereafter the academy had a single cannon firing daily to mark the moment of the solar zenith. Though submitted in 1735, "Methodus computandi aequationem meridiei" was not published until 1741 in the *Commentarii*.

According to Fuss, who was nearly fifty years younger and did not meet Euler until 1773, then becoming his amanuensis and assistant in 1778, the three days of these astronomical calculations in 1735 were a major factor in Euler's problems with his vision. Yet, for blindness as progressive as Euler's was to be, eyestrain alone is not an adequate explanation. The length of time over which this occurred seems a lapse in the memory of the elderly Euler, who was the source for Fuss via their almost daily conversations. Euler's health did suffer in early 1735, apparently not from exhaustion from work on this problem but from headaches and a life-threatening fever to which harsh weather contributed.

It is not possible today to diagnose the illness related to the fever; one supposition is that it was scrofula, a disease known from antiquity and still existing well into the eighteenth century; it was called the King's evil, after a belief in France and other lands that a royal touch could cure it.[8] Scrofula, today labeled tuberculous adenitis, is a tuberculosis of the lymph glands in the neck, which enlarge and grow firmer while the patient endures fever and chills. After 1735 occasionally Euler suffered from serious fevers, but even with his strong constitution scrofula would have been more debilitating over the years; most likely the illness was something else. That year Euler kept it secret from his parents and colleagues abroad; only after his recovery did he inform Daniel Bernoulli of the illness, who reported the health problem to Euler's parents.

Later in 1735 Euler's sixteen-year-old brother Johann Heinrich, an aspiring painter, arrived in Saint Petersburg; probably with assistance from his father-in-law, Euler had his brother appointed a student or adjunct in the art academy there. Heinrich apparently lived with the Euler family until departing with them from Saint Petersburg in 1741.

Among the duties that Euler—after having few state and technical tasks assigned to him at the academy before 1735—was now charged with was that of being a member of the permanent faculty of the department of geography, which the Supreme Senate approved in August. That year the Senate also appointed him department supervisor, a position that increased his pay to twelve hundred rubles. Euler's main responsibility was to oversee the Second Kamchatka Expedition, which would last to 1743, in part to prepare the first accurate land maps of the Russian Empire and its regions through determining latitude and longitude astronomically and gathering better geodetic measurements. The expedition was the academy's principal and most heavily funded project and Euler's chief assignment. In this enterprise he began by assisting Delisle and interacting with Christian Nikolaus Winsheim. In 1737 Euler computed ephemerides and drew up uniform instructions for the expedition's geodesists as well as for the Russian scholar Vasilii Nikitich Tatishchev, who was studying economic possibilities in the Ural Mountains.

At the end of August 1735, Euler appeared for the first time in the geography department of the academy. The introductory protocol read, "Mr. Professor Euler has been charged [by president Johann Albrecht von Korff] to aid Mr. Delisle in composing the general map of Russia [to its European borders] and has come today in order to be an adjunct for us."[9]

The official appointment from the Senate came a week later, on 2 September, when Euler met with Delisle, who explained his map projection to fit a curved surface on a flat sheet. Improving upon Ptolemy, Delisle was developing an equidistant conic projection with constant parallel placement along all meridians the same scale; the latitude and longitude provided the coordinates of a point, and Euler agreed that this projection was needed for precision. He generally worked in the department daily but did gain permission from Delisle to carry out some work at home. The task was enormous and painstaking; detailed records exist taken from departmental minutes written in French from 1735 to 1747, but their examination, begun by Nina I. Nevskaya, remains far from complete.

In November 1735, Delisle, Euler, and their associates were invited to visit the office of engineers in the chancellery of the main administration of artillery and fortification to view the engineers' maps, and in early December they did so. Having begun work on a good general map of the Russian Empire, Euler wrote to academy president Korff that preparing it would be difficult and take a long time; it required comparing the components of a large number of special maps and assuring their accuracy through reliable geodetic measures, sophisticated astronomical observations, and determinations of longitude not previously available in Russia. In January 1736, Korff sent an order that Euler's map be completed as soon as possible; the minutes of the geography department show that "since . . . Professor Euler who, together with Professor Delisle, has agreed to the preparation of the map of Russia's boundaries requested by the Supreme Senate, Mr. Delisle [returned to] him the map yesterday with a note begging him . . . to complete [it]."[10] The Second Kamchatka Expedition would provide many of the materials required for accuracy. Euler started with a study of special maps of the empire's districts and built upon them to prepare accurate ones of the nineteen provinces, then combined these in drawing the most complete and reliable map of the empire.[11] In September 1736 he and Delisle completed their first map of the European borders of Russia. Euler assigned his student Vasilii Adodurov, an aristocrat from Novgorod and one of his first gymnasium students, to enter the Russian nomenclature because his own knowledge of Russian was still not sufficient. He planned to place maps of the nineteen provinces into a Russian atlas with a general map of the empire.

From 1736 on Euler apparently labored mainly with the newly appointed astronomer Gottfried Wilhelm Heinsius on astronomical calculations; Heinsius became one of his closest friends. Applying Kamchatka observations and their better measurements, along with their own computations, Euler and Heinsius painstakingly corrected land maps almost daily. During his second Saint Petersburg stay in 1777 and 1778 he would publish three articles on cartography, including his new map projection to minimize distortion.

In the mid-1730s another project of the geography department was a comparative analysis of different calendars from ancient and eighteenth-century China, Japan, southern India, and the other areas of the east. Gottlob (Theophilus) Siegfried Bayer, Delisle, Krafft, and Vasily Kirillovich Trediakovskii participated along with Euler. Delisle believed that this research could inspire the creativity that was essential to making discoveries in astronomy.

In 1736 Euler began to study the solar year in the traditional Sanskrit calendrics of the Hindus, collaborating with Bayer, who had apparently heightened his interest in ancient texts. While Bayer had come to Saint Petersburg in 1726, it is not clear how the two men met; it may have been through Christian Goldbach and Delisle. (Bayer lived in Delisle's house for a brief period.) Bayer, who did not understand technical matters in theoretical astronomy computations, checked occasionally with Euler. This work was a part of comparative chronology, a field that Bayer helped the sixteenth-century antiquarian Joseph Justus Scaliger to have founded. This work apparently increased Euler's interest in ancient sources, and his findings appeared as an appendix, "De Indorum anno solari astronomico" (On the astronomical solar year of the Indians, E18) to Bayer's *Historia regni Graecorium bactriani* (History of the kingdom of the [ancient] Bactrian Greeks, 1738).[12] In part Euler's comparison finds the Indian year slightly longer than the Gregorian year—essentially by twenty-three minutes. In the preface to his book, Bayer wrote, "Leonhard Euler, the noted mathematician and most closely linked to me by the ties of collegiality, has now shown to some extent how welcome these [discoveries] must be to everyone."[13]

The Basel Problem and the *Mechanica*

During the early eighteenth century, the search for the exact sum of the Basel Problem, the series of the inverse squares of the positive integers, had assumed a mystique not unlike the quest for a proof of Pierre de Fermat's Last Theorem that would come in the twentieth century. Its resolution promised to match in importance Gottfried Wilhelm Leibniz's striking discovery sixty years earlier that the sum of the infinite alternating series $1 - 1/3 + 1/5 - 1/7 + 1/9 - \ldots$ is $\pi/4$. By 1668 the Scottish mathematician James Gregory had the equivalent of the arctan or $\tan^{-1}$ series—that is, an inverse trigonometric series—but apparently did not make this computation; the series is today known as Gregory's or Leibniz's.

Widespread scrutiny of the Basel Problem for more than half a century by leading geometers—who were hampered by its slowly converging infinite series—had produced ever closer approximations, such as those of James Stirling, but had failed to find an exact solution. In 1735 Euler was continuing his attempt to sum $\zeta(2)$, which he had pursued since 1728. Having discovered his summation formula by 1732, in a paper read in October of that year he used but a few terms to sum the Basel Problem to twenty decimal places, an extraordinary achievement. This result and his study of series for π might not be suggestive to many mathematicians, but apparently Euler saw that the answer to the problem was near to $\pi^2/6$; Still, he initially believed it unlikely that he would make any further advance with the problem that had previously stumped him so. A letter in December calmly began, "So much work has been done on the series [$\zeta(2)$] that it hardly seems likely that anything new about them may still turn up. . . . I too, in spite of repeated efforts, could achieve nothing more than approximate values . . .," but the tone quickly changed: "Now, however, quite unexpectedly, I have found an elegant formula for the sum of the infinite series $1 + 1/4 + 1/9 + 1/16 + \ldots$, which depends upon the quadrature of the circle [that is, upon π] . . . I have found that six times the sum of this series equals the square of the circumference of a circle whose diameter is 1."[14] While this statement is oblique geometrically, in modern notation Euler had the marvelous formula $\sum_{n=1}^{\infty} 1/n^2 = \pi^2/6$. The previously intractable problem had fallen before Euler's unmatched ability in elementary and formal computation, including masterful substitutions and interpolations, his skill in mathematical integration, and his unequaled intuition in mathematics, all applied to recently derived series for trigonometric chords and π. The discovery seems to have evoked a joy not unlike that of Archimedes of Syracuse in his "eureka moment."

In a paper read at the academy, also in December 1735, Euler announced his sensational discovery.[15] His article "De summis serierum reciprocarum" (On the sums of series of reciprocals, E41), published in the *Commentarii* for 1734–35, solved the Basel Problem by exactly summing $\zeta(2)$ in three different ways, though with unsatisfactory procedures and with a proof having gaps generally unrecognized at the time.[16]

In the best known of his three computations of $\zeta(2)$, Euler recklessly applied Newton's results on summing powers of roots of finite algebraic equations and transcendental functions, such as $1 - \sin x/x$. To work with these, he made two modest yet bold assumptions, policed by his splendid instinct for warning as regards errors.[17] From his study of algebraic

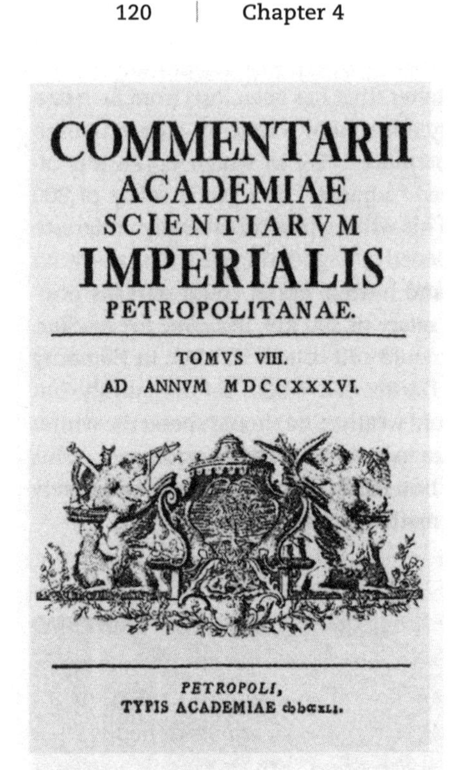
COMMENTARII
ACADEMIAE
SCIENTIARVM
IMPERIALIS
PETROPOLITANAE.

TOMVS VIII.
AD ANNVM MDCCXXXVI.

PETROPOLI,
TYPIS ACADEMIAE cIↃIↃCCXLI.

Figure 4.1. The title page of volume 8 of the *Commentarii Academiae scientiarum Imperialis Petropolitanae.*

theorems, Euler asserted that for an nth degree polynomial, $P(x)$, with nonzero roots $a_1, a_2, a_3, \ldots$ it is possible to factor $P(x)$, where $P(x) = 0$, into $P(x) = (1 - x/a_1)(1 - x/a_2)(1 - x/a_3) \ldots (1 - x/a_n)\ldots.$ Substituting 0's for x's in the factored polynomial gives the answer of 1. $P(0) = 1$ is a crucial condition upon which the factorization hinges. In another leap of faith, Euler daringly extended to infinite polynomials this rule for finite polynomials, assuming that the infinitely many roots of transcendentals could be factored likewise;[18] he could not yet prove it, but considered it a natural extension. In modern symbolism, Euler began his proof with

$$P(x) = 1 - x^2/3! + x^4/5! - x^6/7! + x^8/9! - \ldots$$

Next he needed $sinx$, which equals $x - x^3/3! + x^5/5! - x^7/7! + \ldots$ So he multiplied $P(x)$ by x/x, giving him $(x - x^3/3! + x^5/5! - x^7/7! + x^9/9! - \ldots)/x$, where the numerator is Brook Taylor's series expansion of $sinx$. When $sinx = 0$ for $x = \pm k\pi$ and for k a real number, $P(x)$ is zero. Euler represented $P(x)$ as "an infinite product."

$$P(x) = [(1 - x/\pi)(1 + x/\pi)][(1 - x/2\pi)(1 + 2\pi) \ldots] = [1 - x^2/\pi^2][1 - x^2/4\pi^2][1 - x^2/9\pi^2] \ldots .$$

Expanding this infinite product and setting it equal to the above equations gave

$$-1/3! = -(1/\pi^2 + 1/4\pi^2 + 1/9\pi^2 + 1/16\pi^2 + \ldots)$$
$$-1/6 = -1/\pi^2 (1 + 1/4 + 1/9 + 1/16 + \ldots)$$

The numerical coefficient is $-1/3! = -1/6$. Dividing both sides by $-1/\pi^2$ gives the desired series and the exact summation $1 + 1/4 + 1/9 + 1/16 + \ldots = \pi^2/3! = \pi^2/6$. The Basel problem was thus solved, QED.

Euler recognized the shakiness of his computational procedures and spent the next eight years searching for a better proof.[19] Often a pathfinder like Leibniz and Newton, Euler would arrive at solutions without subsequent requisite tools for proofs; even so, almost all of his results and formulas have withstood tests of rigor. At the time, however, he refused to withhold his many discoveries from publication. Only after Euler—in the nineteenth century—would attention among mathematicians turn more to rigor.

Feverish with excitement, Euler was not satisfied to stop with $\zeta(2)$. To reach the results given in "De summis serierum reciprocarum" he applied "multo labore" and a known method—probably indeterminate coefficients—to extend his procedure by 1736 to compute the zeta function for the even integers up to 12. The coefficients were more difficult to obtain, but he correctly gave $\zeta(4) = \pi^4/90$, $\zeta(6) = \pi^6/945$, $\zeta(8) = \pi^8/9450$, and $\zeta(10) = \pi^{10}/93555$, up to $\zeta(12) = 691\pi^{12}/(6825 \times 93555)$.[20] Fermat had anticipated these coefficients, while pages 96 and 97 of Jakob (Jacques) Bernoulli's posthumous *Ars conjectandi* (The art of conjecturing, 1713) had defined the generating function leading to them, the equivalent of the Taylor expansion of $1/(e^x - 1)$. By 1735 Euler had computed rational multiple coefficients of π^2, π^4, π^6, π^8, . . . , seeking to solve the Basel Problem, and discovered that they were related to the numbers Jakob Bernoulli had described.[21] Bernoulli's priority was not immediately recognized. In his *Miscellanea analytica* of 1730, Abraham de Moivre had named these numbers *Bernoullian*, a term that Euler eagerly accepted.[22]

A generating function for Bernoulli numbers with the Taylor expansion of $1/(e^x - 1)$, together with the Euler-Maclaurin Summation Formula, which generalized the work of Jakob Bernoulli, made it easier to obtain his numbers, $B_1 = 1/6$, $B_2 = 1/30$, $B_4 = 1/42$, $B_6 = 1/30$, $B_8 = 5/666$, and $B_{10} = 691/2730$, the initial Bernoulli numbers. That the prime 691 appears in both B_{10} and the coefficient for $\zeta(10)$ probably suggested to Euler an interconnection. He was now also bringing out the central role of the Taylor series in differential calculus. In 1738 Maclaurin independently arrived at the summation formula—hence its shared name—and Maclaurin immediately employed it to compute $\zeta(2)$ to twenty decimal places.[23] It made integration easier in obtaining the numbers.

Two properties of Bernoulli numbers are that $B_n = 0$ for all odd $n \geq 3$, and that the even subscripts B_{2k} have alternating signs.[24] Not until his article on certain series, "De seriebus quibusbam considerationes" (E130), published in the *Commentarii* in 1750, did Euler prove the formula $\zeta(2n) = (2^{2n-1}/(2n)!) \mid B_{2n} \mid \pi^{2n}$, with B_{2n} the Bernoulli numbers. Euler did not yet realize their importance to number theory,[25] and not until chapter 2, part 5, of his *Institutiones calculi differentialis* (Foundations of Differential Calculus) of 1755 did he start to bring out their significance, which would emerge only in the next century. Remarkably silent about 3 and other odd integer exponents, Euler and his correspondents recognized some brashness in his reaching his findings without proofs—as the Latin proverb proclaims, *audaces fortuna iuvat* (fortune favors the bold.) His "De summis serierum reciprocarum" (On the sums of series of reciprocals, E41) does more than solve the Basel Problem; it also shows Euler's interest in approximating the value of π, a problem that he often revisited.

Delays in the publication of the *Commentarii* meant that information about Euler's discoveries often first reached western and central Europe prior to the published proceedings of the Petersburg Academy of Sciences. Since volume 7 of the *Commentarii*, containing "De summis serierum reciprocarum," was not published until 1740, news of his exact result spread relatively slowly in 1736 and 1737 through mail to and by friends and a growing number of correspondents, among them James Stirling in Edinburgh and Giovanni Marchese Poleni in Padua.[26] Euler's summation provoked a lively discussion among geometers while its lack of a rigorous foundation brought criticism. When in a letter of April 1737 Johann I (Jean I) Bernoulli received news of the solution, he is said to have exclaimed, "Utinam frater superstes effet!" (If only my brother [Jakob] were alive!) He inquired about a proof and a well-founded computational procedure and set about finding them himself. The timing of this exchange suggests that the solution to the Basel Problem furthered the development of Euler's distinguished reputation, which his *Mechanica* in particular had launched.

During the eighteenth century, the study of mechanics lay at the center of research in the mathematical sciences; it and astronomy were then considered the sciences par excellence. To establish repute, every natural philosopher and geometer of note sought to make advances in mechanics. Though the Cartesians and Wolffians competed on the Continent until midcentury, Newton's dynamics in the *Principia mathematica* dominated this discourse by the 1740s.[27] One historical interpretation has mechanics developing according to two laws: Apollo's Arrow and the land of the Ancients' Inferno.[28] Apollo's Arrow is a metaphor for a growing collection of means for solving particular theoretical and practical questions and problems as well as enabling progress in the field. Often natural philosophers and geometers knew the correct answer to a problem or gathered more or less imprecise data with improved instruments such as telescopes, pendulum chronometers, barometers, and surveying equipment; their task was to provide a method or the mathematics needed to reach the correct result. The Ancients' Inferno represents the intelligence required and the slow and tortuous development of Newtonian mechanics and Euler's originality and synthesis, along with the endlessness of the pursuit of science.[29] Ambiguities, confusion, and errors, together with an array of accomplishments, were among the factors affecting the pace of development.

After making detailed plans for future treatises on point dynamics and fluid flow theory in his Basel notebook, prepared between the ages of eighteen and twenty, Euler had set about the work that led to his *Mechanica*. This record indicates that mechanics rather than mathematics was his first major and continuing interest. From 1725 to 1730 he completed a dozen articles on the subject and two drafts of a tract on motion under central forces.[30] Unable to solve the problem of a body forced to rotate around two perpendicular axes and afterward freed, Euler dropped his initial work on the *Mechanica* around 1730. His penultimate draft was divided into three sections: motion from forces acting on a freely moving "infinitely small body that may be treated as a point" (a free mass point); motion from forces acting on the constrained motion of a mass point; and the motion of rigid bodies that are driven by any forces, including the center of oscillating motion of rigid rods. These bodies exist in the absence of a resisting medium, that is, a vacuum, or a resisting medium along geodesics on a given surface. Today a mass point is defined as a geometric or point particle assigned a finite mass. The third section proposes the theorem of the moment of momentum.

Euler found the existing principles of mechanics insufficient to account for the motion of rigid bodies. Together with Johann I and Daniel Bernoulli, he explored a series of scattered problems, many of which he had not anticipated, and to solve them employed special devices to discover differential equations for them. At first neither Euler nor the Bernoullis went much beyond the findings of Christiaan Huygens on oscillation. It took Euler another quarter century to develop a general theory of the motion of rigid bodies, including the nonplanar case; he did not let this delay his announcement and progress with his grand program in mechanics, however; he deleted the third section from the planned *Mechanica* and concentrated on clarifying and revising the first two.

On the path to rational mechanics, the principal question facing Euler was how to reorganize mechanics by applying the *analytic method*, along with how to formulate the fundamental, general principles at its base.[31] The analytic method, which refers to his systematic employment of the terms and symbols of differential and integral calculus, was for Euler a heuristic mode of thought and a means to an end rather than simply an algorithmic approach. Behind the synthetic geometric format of Newton's *Principia*, including conic sections and quadratures, Euler saw a host of methods: infinitesimals, geometric limit procedures, interpolation techniques, and infinite series. Still, Newton had not devised the partial differential equations of motion. Euler was more indebted to the Continental geometers, led by the Bernoullis, who in an ad hoc fashion were applying differential calculus, then barely half a century old, to achieve precise solutions of problems of motion, its generations, and its alteration—including instantaneous acceleration—in physics that neither analytic geometry nor the ancient synthetic method of Euclidean geometry could attain. Breaking decisively with those two geometries, Euler sought to express Newtonian mechanics with Leibnizian differential equations and the partial differential equations that the Bernoullis had introduced, and he added many more of the latter. He differed from Leibniz, however, in evolving his concept of function by examining formulas and the relations between quantities rather than curves; he thus formalized the subject. By 1734 Euler had returned to writing the *Mechanica* with a singleness of purpose and soon completed the first volume. By then he had written for the *Commentarii* nine articles on mechanics and eleven on solving its problems; he was to include a summary of these in the *Mechanica*.

In 1736 the Petersburg Academy, which had promised funding for the publication of most of Euler's future books, issued his 980-page *Mechanica sive motus scientia analytice exposita* (Mechanics, or the science of motion set forth analytically, E15 and E16) in two volumes. These were each too large to appear in the *Commentarii* but were, as noted on the title page, supplements to it.[32] Their title states Euler's central concept, though he briefly considered *Dynamica* as an alternative title. Johann I Bernoulli wrote in a letter that he thought *Dynamica* had merit,[33] but Euler quickly settled on *Mechanica* and dedicated these volumes to his "illustrious and most excellent" friend, academy president Korff.

The preface to the *Mechanica* emphasizes its importance:

> [I]f analysis is needed anywhere, then it is certainly in mechanics. Although the reader can convince himself of the truth of the exhibited

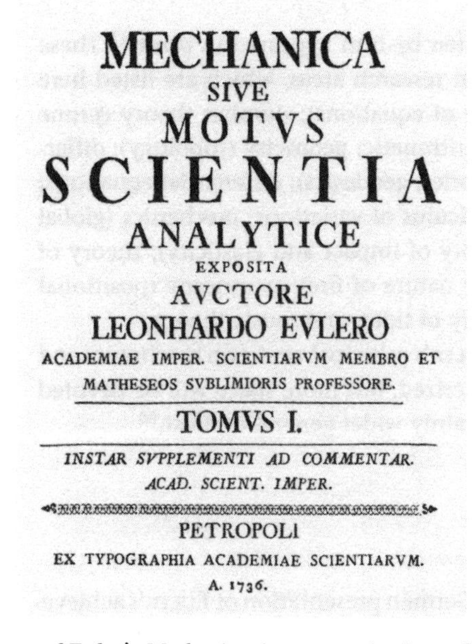
MECHANICA
SIVE
MOTVS
SCIENTIA
ANALYTICE
EXPOSITA
AVCTORE
LEONHARDO EVLERO
ACADEMIAE IMPER. SCIENTIARVM MEMBRO ET
MATHESEOS SVBLIMIORIS PROFESSORE.
TOMVS I.
INSTAR SVPPLEMENTI AD COMMENTAR.
ACAD. SCIENT. IMPER.
PETROPOLI
EX TYPOGRAPHIA ACADEMIAE SCIENTIARVM.
A. 1736.

Figure 4.2. Title page of Euler's *Mechanica sive motus scientia analytice exposita*, 1736.

propositions, he does not acquire a sufficiently clear and accurate understanding of them, so that if those questions be ever so slightly changed he will not be able to answer them independently unless he turns to analysis and solves the same propositions using analytic methods. This in fact happened to me when I began to familiarize myself with Newton's *Principia* and Hermann's *Phoronomia*; although it seemed to me that I clearly understood the solutions of many of the problems, I was nevertheless unable to solve problems differing slightly from them. But then I tried, as far as I was able, to distinguish the analysis [hidden] in the synthetic method and to my own ends rework systematically those same propositions, as a result of which I understood the problem much better. . . . I expounded . . . [the propositions anew] using a systematic and unified method and reordered them more conveniently.[34]

In putting forward simple and uniform analytic methods and logically arranging fundamental propositions and new definitions, thus making them intelligible, Euler had begun conceptually to organize infinitary analysis and was the first to apply it systematically to dynamics.[35] This was a major feature of the *Mechanica*: Euler wanted to demonstrate that the proofs of the laws of mechanics are certain and necessary.[36] He accepted Newton's dynamics, but his research program was not simply an extension of it; he was also to incorporate elements from the Cartesian and Leibnizian sciences and his own thought. In the preface to the *Mechanica* Euler declared that he had included "both what I have found in the writings of others on the motion of bodies and what I myself have thought out in my ruminations."[37] Among his sources were Archimedes on statics; Galileo Galilei on kinematics, and Huygens, Newton, Johann I Bernoulli, and Jakob Hermann on dynamics. During his early Saint Petersburg years, he had become familiar with many of the writings of Galileo, Huygens, and Newton; he also referred to the second volume of Christian Wolff's *Elementa matheseos universae* of 1733, in part an exposition on statics and elementary mechanics. (Euler did not comment on its aerodynamics and hydraulics.) Clearly Newton's *Principia mathematica* and Hermann's *Phoronomia* most influenced him. For its systematic application of infinitary methods, the *Mechanica*'s preface confidently concludes, "[Qui] in analysi tam finitorum quam infinitorum satis fuerit exercitatus, is mira facilitate omnia intelligere atque sine ulla manuductione integrum hoc opus perlegere queat" (He who has enough practice in finite and infinite analysis will understand everything with wonderful ease and be able to read through this entire work without help). On the Continent, Euler was assuming the lead in applying infinitary analysis to mechanics. By 1750 analysis surpassed geometric synthesis.

Among the many novelties in the *Mechanica* is the precision it gives to the concept of bodies, which Newton had employed only vaguely, giving it at least three disparate meanings. Seeing that in general Newton's laws correctly applied to masses reducible to points, Euler introduced the concept of mass points. The conclusions in his punctual mechanics apply only to punctiform—and not extended—bodies. Analogous to his treatment of motion in a plane, he introduced rectangular Cartesian coordinates in tracing the trajectories of mass points, explicitly decomposing the paths along the axes of forces acting on them. The binormal, principal normal, and tangent give the axes. It is mistakenly believed that Newton had shown this, but Newton's geometric method had only intrinsic coordinates. Euler asserted that motion is continuous, so we know each point in the space covered in a trajectory. This builds upon Leibniz's principle of continuity. From its initial to its final place, Euler demonstrated moreover that the straight-line path of a mass point to the center of force is the shortest. To Cartesian coordinates he added directed magnitudes or vectors at different points. Euler now also employed vectors, which had previously applied only to static forces, for changes in speed and direction; their addition lies at the heart of the Enlightenment concept of space-time.

The concept of force was crucial to Euler's mechanics, and he treated it as an external entity to a body causing change in motion. His physical model was still incomplete and would generate criticisms; in the preface to his *Traité de dynamique* (1743), Jean-Baptiste le Rond d'Alembert described as vague and obscure purely mathematical explanations of motion without the inclusion of external observations. In "Recherches sur l'origine des forces" (E181, 1750) Euler would provide the absolute impenetrability of matter as the previously missing element in his explanation of force.[38] Both Newton and d'Alembert had accepted this concept, but Euler went further in examining the relation between this property and impact phenomena.

In the *Mechanica*'s book 2, proposition 20, Euler proved Newton's second law: "The change of motion is proportional to the motive force impressed; and is made in the direction of the right line in which that force is impressed." Newton had offered a dichotomy. His second law treated motion in two forms: in modern symbols $F = \Delta(mv)$ for $F = ma$, according to whether the force is acting continuously or discontinuously. Before Euler's time, that law was often auxiliary in studies of motion and mostly applied to one degree of freedom.[39] Euler's proof reduced the ambiguity and argued that Newton's second law was an underlying principle of motion; among geometers and natural philosophers, Euler was prominent in developing it to apply to many degrees of freedom. He did not examine two bodies subject to the same forces, but his study of Kepler's planetary laws in book 1, chapter 5 of the *Mechanica* suggests that he could reduce these problems to determining the motion of a single body. Working only with mass points, he did not require Newton's third law on opposite and equal reaction.

Occasionally histories have misinterpreted Euler's *Mechanica*. In the preface to his *Mechanique analitique* of 1788 Joseph-Louis, Comte de Lagrange gave Euler's work qualified praise, terming it "le premier grand ouvrage où l'Analyse ait été appliquée à la science du mouvement" (the first great book in which [infinitary] analysis has been applied to the science of motion).[40] This assertion is not exactly correct; instead, the *Mechanica* was the first large book to apply only infinitary analysis to mechanics, for Hermann's *Phoronomia* had a mixture of analysis and geometry. Also incorrect is the occasional belief, which is not from Lagrange, that the *Mechanica* simply translates Newton's *Principia mathematica* into the language of infinitary analysis. Euler's section on mass points moved by a force toward a fixed center brilliantly reformulates analytically the corresponding part of Newton's *Principia* and somewhat introduces Euler's later works in celestial mechanics. Limited to the rectilinear and curvilinear motion of mass points, together with a few chapters on motion in resisting media, Euler's *Mechanica* has a much narrower scope than the *Principia* and consolidates only a small portion of rational mechanics.

Its preface announced that the *Mechanica* would be the first of Euler's books in which he would apply "his complete intelligence to determining the motion of finite bodies." The general scholium to section 98 of chapter 1 promised that since partial differential equations were lacking for finite bodies and existing principles were unsatisfactory for their general treatment, he would, in a highly ambitious program, turn in succession to resolving computationally the motion of bodies of finite magnitude that are rigid, flexible, elastic, subject to impacts with one another involving compression and extension, and fluid[41]—a task that would have exhausted most scientific geniuses. But through his brilliance, combined with his nearly untiring work habits and strong physical constitution, Euler successfully completed this work over the next twenty years. In addition to this achievement in mechanics, which alone would have made him famous in the annals of science, he was driven to cultivate all the other mathematical sciences and make significant contributions to each.

Spirited criticism of the *Mechanica*, especially for errors and lack of analytical sophistication, came from Cartesian natural philosophers as well as its chief opponent, the English mechanist Benjamin Robins, discoverer of the ballistic pendulum. In 1739 he published a small book, *Remarks on Mr. Euler's Treatise of Motion, Dr. Smith's Complete System of Opticks, and Dr. Jurin's Essay upon Distinct and Indistinct Vision*, the first third of which comprised "Remarks on Mr. Leonard Euler's Treatise Entitled *Mechanica*." Proceeding with what Euler's editor Paul Stäckel would in the twentieth century call "impudent arrogance," Robins faulted the metaphysical and philosophical grounding of the *Mechanica* for lacking an experimental base, pointing out several computational errors and what he deemed were bizarre assertions.[42]

Robins was no stranger to controversy. Against Bishop George Berkeley's critique of the foundations of calculus with evanescent increments, Robins had in 1735 defended Newton's method of fluxions based instead on prime and ultimate ratios—or, in essence, limits. In his *Remarks* Robins, like the Cartesians, believed it possible to reach some of the *Mechanica*'s results by replacing the numerous differential equations with simpler rational arguments; this might reduce errors. He found Euler's efforts to prove the laws of mechanics "contrived" and wanted all laws checked via experiments. He asserted that the third proposition, "that in any unequal motion the least elements of the space described may be passed over with a uniform motion," was not universally true. Euler had "obscure suspicions of the fallacy of the proposition but . . . applies it without scruple."[43] Euler had only one half of a correct sine value in a triangle depending on force and motion; there were "capital errors arising from an unwarranted adherence to the third proposition." The "7th, 8th, and 9th propositions relating to the continuation of a body in its state either of rest, or of equable motion in a right line," maintained Robins, stemmed from experiments and not metaphysical notions of bodies and motion.[44]

Robins found Euler's demonstrations of propositions "pretended" and his computational results half what they should be.[45] He faulted Euler's proof on uniform motion in proposition 15, called the proof of proposition 18 "absolutely inconclusive," and found all equations for proposition 21 "false."[46] To introduce a new idea or problem numerical in nature, including those with differential equations, Euler used theorems. In the *Mechanica*'s chapter 5, on curvilinear motion, Robins claimed that every proposition depended on an erroneous equation. He found mistakes in propositions 43–103, and asserted that the last chapter was filled with unnecessary computation and prolix.

Robins criticized Euler for his "blind submission to his computations."[47] Euler recognized what might lead to mistakes, but that did not stop him when he deemed something was a step toward advancing the field. Throughout his career, Euler constantly sought to perfect thought and did not refrain from publishing, even for reasons of absolute rigor. Most of Robins's criticisms dealt with minor points, his dislike of Leibniz's calculus, and his opposition to metaphysics in general rather than to Euler's work itself. Robins's sharpest criticism was of Euler's theory of centripetal force, that when a mass point moving rectilinearly into the center of a force reaches that center it will remain there, for the counterforces will be equal and thus cancel it out. Without explaining how he arrived at his conclusion, Robins argued that Newton's law of gravitational attraction required the point to oscillate like a pendulum between the center and a return point.

Besides being significant in the development of mechanics, the *Mechanica* is notable in the history of mathematics. Through applying first-order ordinary differential equations and inventing many of the second order, Euler was participating especially with the Bernoulli family in founding differential equations as a core branch of calculus. Corollary 832 in the *Mechanica*'s volume 2 places Euler among the first to introduce partial differential equations of functions with two independent variables; a few years earlier Nikolaus Bernoulli had worked on these, recasting the trigonometric lines—given as the traditional Ptolemaic chords and half chords—into numerical ratios and particularly working with the tangent. Euler was searching for deeper connections among the three elementary transcendental functions: the exponential, the Napierian logarithmic, and now the trigonometric. The *Mechanica* also touches upon cyclometric functions, which measure revolutions of a wheel, and elements of what would later become the calculus of variations, differential geometry, and the theory of geodesics. Attentive to a better symbolism, Euler adopted π in computing the area of a circle, extensively used e for the base of natural logarithms, and introduced his summation sign Σ. In section 184 of volume 1, the sign $A{\cdot}x$ represents the function of arctan x; in the second volume it stands for secant x, while $At{\cdot}x$ becomes the sign for arctan x. The continuing appearance of these symbols in Euler's many influential publications quickly made them conventional.

Throughout early Enlightenment Europe, most of the scientific community quickly recognized the *Mechanica* as a landmark in the history of physics and knew that a genius of the highest magnitude had arrived. The initial review, published in 1737 in the journal *Bibliothéque germanique* in Amsterdam, was by the Calvinist pastor and philosopher Jean-Henri-Samuel Formey, the leader of a scholarly circle in Berlin. Formey praised Euler for employing the analytical method both to prove established truths and to lead to generalizations and new theorems. Johann I Bernoulli, who wrote the next review, indicated the growing repute of Euler in 1737 by addressing him in a letter as the "most famous and wisest man of science Leonhard Euler." His review appeared in Leipzig's *Nova acta eruditorum* the next year.[48] Bernoulli, whose enthusiasm recalled that of Leibniz for Hermann's *Phoronomia*, lauded the genius and acumen of Euler: "Until now no book has appeared that so splendidly reached the inmost heart of mechanics. It has brought out in abundance sublime and hidden things. . . . The whole book is analytical, so that no synthetic ballast bores the reader."[49] Another commendation in 1738 came from Wolff, who cited Euler's application of differential calculus for the utility and ease in reading that it brought to the entire book. In his *Remarks* Robins indicated a rising reputation for Euler, who for several recent articles "has been celebrated in the various literary journals of different countries with high encomiums."[50]

The Königsberg Bridges and More Foundational Work in Mathematics

In March 1736 Karl Ehler, the mayor of Danzig (now Gdansk), a city eighty miles from Königsberg, imparted to Euler his thoughts on a recreational puzzle about the seven bridges of Königsberg. It was part of their ongoing correspondence, which covered such items as artillery, real and imaginary numbers, and the rectification of curves. Ehler called the bridges problem "an outstanding example of the calculus of position." Euler had already solved it. The city of Königsberg in East Prussia (now

Kaliningrad in Russia) comprises four sections. At the center is an island in the Pregel River, and in Euler's time seven bridges spanning the river connected the island with the three other sections. The question was whether someone could pass over the bridges in a connected walk, crossing each bridge once, and return to the same spot. The puzzle itself, unrelated to Euler's mathematical research, was among several problems that he addressed only once. While Leibniz and Wolff posed problems of this type, Euler seems to have learned of them from Johann I Bernoulli.[51] Finding the Königsberg Bridge Problem simple, Euler solved it negatively—not with mathematics but with reason alone. The article "Solutio problematis ad geometriam situs pertinentis" (Solution of a problem relating to the geometry of position, E53, 1735) gives his conclusion. Submitted the next year for volume 8 of the *Commentarii* (see figure 4.1 on page 120), it was not published until 1741 in an issue containing thirteen mathematical articles—two by Daniel Bernoulli and eleven by Euler.[52]

"Solutio problematis" contains no graphs but is considered the first work in graph theory. The requisite type of graphs to represent the possible Königsberg walk under the given conditions did not appear until the mid-nineteenth century. Euler divided this paper into twenty-one numbered paragraphs. After paragraph 3 rejects as unworkable any attempt to solve the problem by checking all possible paths, the paper considers the transit entrances to land regions rather than the crossing of bridges. Paragraphs 4–8 simplify the puzzle by determining the relation between the number of bridges connecting a land mass and the number of times represented by capital letters in the regions as shown in figure 4.3. The journey around Königsberg then became a string of capital letters—for instance, "DCA"—would carry us from the right-most region D across a bridge to region C and then across another to the island A. Euler noted that, if a region has an odd number of ridges (k), then the letter of that region must appear $(k + 1)/2$ times in the string of capital letters that represent the entire journey. From this he showed that no path could cross each bridge exactly once. Since five bridges connect the island to the city's other regions, the frequency will be $(5 + 1) \div 2 = 3$. The frequency for b, c, and d, there being three bridges, is $(3 + 1) \div 2 = 2$. The sum of these frequencies is nine, but the sum for a path crossing each of the seven bridges only once is eight; the Königsberg tour under the given conditions is thus impossible; the problem has no solution.

Among the Enlightenment contributors to differential and integral calculus—whether in organization, discoveries, or articulation—Euler was to be the foremost; he was making analysis the ascendant field of mathematics. From 1736 on, and aided by correspondence with Daniel

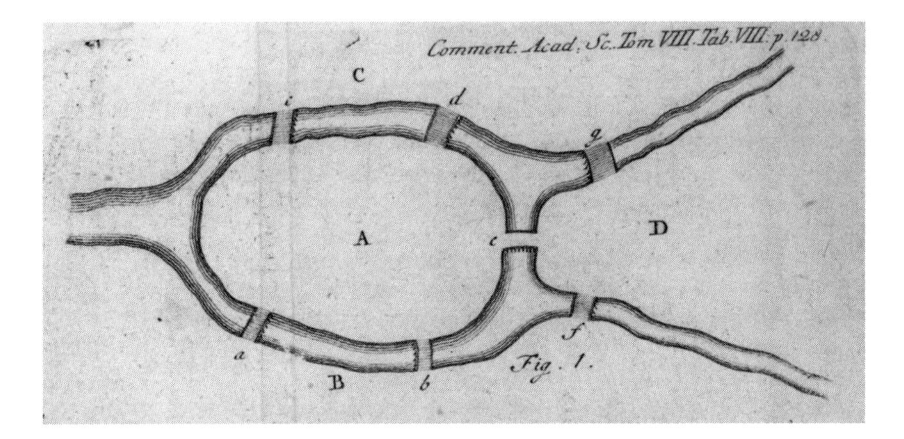

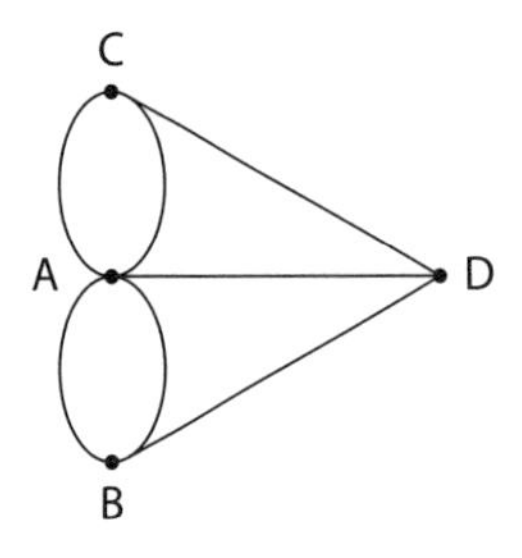

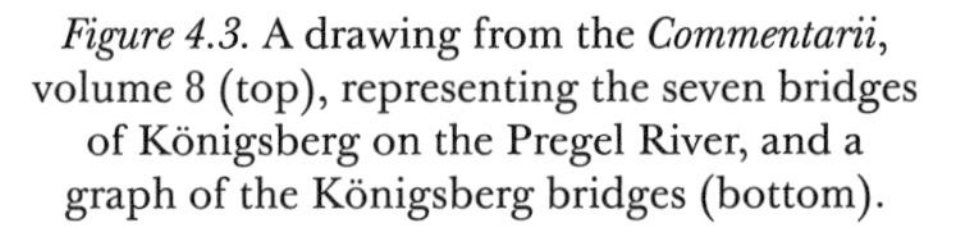
Figure 4.3. A drawing from the *Commentarii*, volume 8 (top), representing the seven bridges of Königsberg on the Pregel River, and a graph of the Königsberg bridges (bottom).

and Johann I Bernoulli, Christian Goldbach, James Stirling, and soon Alexis Claude Clairaut, Euler continued to put studies of infinitary analysis at the center of his research. He identified its core principles, and in his fearless and massive calculations perfected computational methods. This work was preparatory to his *Introductio in analysin infinitorum* (Introduction to the analysis of the infinite, E101 and E102) of 1748. This groundwork rested chiefly upon the writings of Jakob and Johann I Bernoulli, John Wallis, Newton, Brook Taylor, Gregory of Saint Vincent, and possibly Pierre Varignon, as well as articles in Leipzig's *Acta eruditorum* and the *Philosophical Transactions* of the Royal Society of London. Euler was transforming the understanding of the three elementary classes of transcendental functions—the logarithmic, the exponential, and the trigonometric—and these were still only vaguely understood. Descartes had divided curves into two classes: geometric and mechanical; Leibniz called these algebraic and transcendental, the latter meaning insoluble by algebraic equations. Euler accepted this division, as had Johann I Bernoulli. In Saint Petersburg Euler also discovered—independent of Maclaurin—the integral test for convergence.

By discarding the Ptolemaic chords and half chords drawn on a circle and making them numerical ratios, Euler was recasting trigonometry. In early modern circle squaring, the French lawyer and mathematician François Viète had discovered the remarkable infinite product for $\pi/2 = 1/(\sqrt{1/2} - \sqrt{(1/2 + 1/2\sqrt{1/2 \ldots})})$, which appeared in his *Variorum* in 1593, and by 1656, by ingenious interpolations, John Wallis produced another in his *Arithmetica infinitorum*: $4/\pi = 3/2 \cdot 3/4 \cdot 5/4 \cdot 5/6 \cdot 7/6. \ldots$ Both thus used integers exclusively, while in 1671 James Gregory gave the power series expansion $\arctan x = x - x^3/3 + x^5/5 - x^7/7 + \ldots$, from which Leibniz three years later discovered for $x = 1$ or $1/1 - 1/3 + 1/5 - 1/7 + \ldots = \pi/4$; Huygens believed that this achievement would make young Leibniz "forever famous among geometers."[53] The slow convergence of these series means that their numerical evaluation was not easy. Since approximations of π existed to as many as thirty-five decimal places, the chief purpose of this work was not practicality. Rather, it suggested that π, defined in connection to the circle, can be expressed precisely by use of integers only. This increased the search to demonstrate whether it is a terminating expansion of the ratio of two integers—that is, a rational or irrational number. The proof that π is a nonterminating, nonrepeating expansion, not only irrational but transcendental, lay in the future. These many computations of π and similar recent results by others, along with Euler's summing of zeta numbers, helped to show that trigonometric formulas connected with the circle could be represented solely by integers. Euler was the first to establish this method. Continuing in this vein, he made trigonometry part of analysis and provided it with stronger foundations.

Extending in the mid-1730s the results of John Wallis and Jakob Bernoulli on the progressions of factorial numbers, Euler interpolated and determined the general nth term of the sequence 1, 2, 6, 24, 120, . . . by using the infinite product

$$1 \cdot 2^n/(1+n) \cdot 2^{1-n}3^n/(2+n) \cdot 3^{1-n}4^n/(3+n) \ldots .$$

Even before that time, as early as his first letter to Goldbach in 1729, he had introduced two new integrals.[54] One is for positive integers n:

$$\int_0^1 x^e(1-x)^n dx = n!/(e+1)(e+2) \ldots (e+n+1).$$

The other is for nearly all integers n:

$$\int_0^1 (-\ln x)^n dx = n!$$

These integrals and the zeta function became the chief nonelementary transcendental functions of the eighteenth century. In his *Institutionum calculi integralis* (1768–70) Euler would refine the two integrals. In the second of his three-volume *Exercices de calcul integral* (1811–17), the French mathematician Adrien-Marie Legendre denoted these as the first and second Eulerian integrals. Legendre and Carl Friedrich Gauss named the second integral the gamma function, and in 1839 Jacques Binet designated the first the beta function.

By 1736 Euler, who knew that the exponential function $e^x = \lim_{n\to\infty} (1 + x/n)^n$, was the first to employ the Taylor series to compute the value of the number $e = 1/0! + 1/1! + 1/2! + 1/3! + \ldots$ (given here in modern factorial notation); he later found it accurate to nine decimal places, 2.718281828. He knew that in the exponential function e^x, the exponent is the variable, and the derivative of the function is the function itself. One of his letters written in 1737 repeats the notation for it. Why Euler chose e is not known; seemingly this was not, as some have posited, because it was the first letter of his name. Perhaps it was because e is the first letter of the Latin word *exponens*. Since a, b, c, and d were already employed as mathematical symbols, e was the first available letter.

In 1737 Euler made notable discoveries regarding e. For computing it, he formulated an infinite continued fraction:

$$e = 2 + \cfrac{1}{1 + \cfrac{1}{2 + \cfrac{2}{3 + \cfrac{3}{4 + \cfrac{4}{}}}}}$$

It and another infinite continued fraction for $(e + 1)/(e - 1)$ were among the many Euler obtained involving e and π. He also proved that every finite continued fraction gives a rational number, while every nonterminating or infinite continued fraction produces an irrational number. Proceeding from the above expression, Euler proved that both e and e^2 are irrational. He also worked toward a proof that π is irrational, but Johann Heinrich Lambert would be the first to make such a rigorous proof. Pietro Cataldi had invented continued fractions in 1613 and Second Viscount William Brouncker employed them to confirm to ten decimal places the value of $\pi/2$ from Wallis's *Arithmetica infinitorum* of 1656, which contains Wallis's discovery that the value is $(2/1)(2/3)(4/3)(4/5)(6/5)(6/7)\ldots$. Until that time they had been neglected. Brouncker further employed continued fractions to solve Diophantine equations, but Euler was the first to make them prominent.

Building on these results and computations of e, the function $\sin(x)/x$, and associated logarithms, in 1737 Euler obtained a form of his famous identity $e^{ix} = \cos x + i \sin x$, the cardinal formula of analytical trigonometry. His reading of volumes of the *Philosophical Transactions* and the correspondence between Johann Bernoulli and Abraham de Moivre suggest that Euler knew of the pioneering work on this by Roger Cotes and Moivre. The *Philosophical Transactions* for 1714 and Cotes's posthumous *Harmonia mensuram* of 1722 contain Cotes's equivalent of

$$\log(\cos x + i \sin x) = ix$$

though not in this notation. In addition, Moivre's *Miscellanea analytica* of 1730 implies but does not explicitly state the famous theorem

$$(\cos x + i \sin x)^n = \cos nx + i \sin nx$$

now known as De Moivre's Formula. From this, Euler's identity can be simply derived, but in his initial computations with e he apparently found—independent of Cotes and Moivre—his result:

$$(\cos z + \sqrt{-1} \sin z)^n = \cos nz + \sqrt{-1} \sin nz$$

Number theory continued to be Euler's passion and a wellspring of challenging problems; its higher degrees of abstraction attracted him. Among the circle of scholars he met or corresponded regularly with were Goldbach and Krafft, both of whom particularly discussed number theory with him, and by 1736 Euler was inventing ways to prove its theorems by introducing the concepts, definitions, and methods required to complete these theorems; he assiduously tried to consolidate the methods.

After beginning with studies of $2^{p-1} - 1$, in 1736 Euler stated (without modern notation) Fermat's Little Theorem: if p denotes a prime number, then in modern notation $a^{p-1} \equiv 1 (\mathrm{mod}\ p)$, unless a is divisible by p—that is, a and p must be relatively prime.[55] This means that $a^{p-1} - 1$ is always divisible by p. In "Theorematum quorundam ad numeros primos spectantium demonstratio" (A proof of certain theorems regarding prime numbers, E54) Euler gave his first of four proofs of the theorem, a clumsy additive one based on mathematical induction on the notation a in the formula $a^p - 1$ with p prime and $a^p - 1$ always divisible by p, unless p divides a.[56] His proof employs the binomial expansion of $(1 + 1)^{p-1} - 1$, the subtraction of consecutive binomial coefficients and their divisibility, and an appropriate rearrangement of terms. Subsequent proofs were devised in many ways; probably only the Pythagorean Theorem has more proofs. Euler's "Theorematum" article was not published until 1741 in the *Commentarii*.

After noting in 1732 Fermat's conjecture that if $n + 1$ is a prime and divides neither a nor b, then $a^n - b^n$ is divisible by $n + 1$, Euler offered his proof in the essay "Variae observationes circa series infinitas" (Various observations about infinite series, E72; submitted to the *Commentarii* in 1737 but not published until 1741).[57] Its introductory paragraph credits Goldbach with introducing him to what we today know as Bernhard Riemann's zeta function and proves new results about it. E54 and E72 relate the distribution of prime numbers to properties of the zeta function, find the density of reciprocals of primes to log log n, and suggest a proof that there are infinitely many primes.[58] Since cryptographers produced nearly unbreakable military codes with them, prime numbers and their decoding had from earlier on been of considerable interest to governments. Primes steadily become sparser, seemingly decreasing in number. Euler probably checked whether they follow a regular pattern, but they do not.

"Variae observationes" showed Euler's enthusiasm for computations and displayed well his skill and originality. He proved that the sum of the infinite series 1/3 + 1/7 + 1/8 + 1/15 + 1/24 + 1/26 + 1/31, + 1/35 + . . . is 1; this includes the sum of reciprocals of numbers one less than a perfect power. In April 1729 Goldbach had posed this problem to Daniel Bernoulli. Euler apparently did not know about it until Goldbach sent him the problem in a letter (now lost) in 1737; Euler thereupon quickly solved it.[59] His "Variae observationes" also introduces his famous product decomposition formula p for the set of primes,

$$\prod_{p \in P} (1 - p^{-s})^{-1} = \sum_{n=1}^{\infty} 1/n^s$$

Multiplying the right side of the equation yields $\sum_{n=1}^{\infty} n^{-s} = \zeta(s)$. When $s = 1$, $\zeta(1)$ is the harmonic series, which diverges to ∞. By applying the divergence of the harmonic series to the occurrence of primes, Euler proved indirectly their infinitude, a fact known since antiquity. The corresponding product must have infinitely many factors. Analytic number theory is thus said to have begun with Euler's product formula. He demonstrated further that the series of reciprocals of primes diverges, and his study of the zeta function helped move him toward the prime number theorem, which concerns the distribution of such numbers.

In June 1736 Euler sent James Stirling his zeta summations and construction of infinite series for logarithms of integers based on the harmonic series.[60] In March 1738, Daniel Bernoulli wrote to Euler that Stirling, in his book *Methodus differentialis*, had been the first to solve the Basel Problem.

Yet while Stirling had obtained the most accurate approximation to eight decimal places, Euler wrote to Bernoulli in April of that year that he had made the connection to π two years before Stirling and was the first to find an exact summation. His earliest equation for computing that sum was

$$\sin x = x - x^3/1 \cdot 2 \cdot 3 + x^5/1 \cdot 2 \cdot 3 \cdot 4 \cdot 5$$

The equation is false. It gives only the first three terms of an infinite series.

Finally, in June 1739, Stirling responded to Euler's letter on logarithmic series. Motivated by his attempts to sum the zeta series and similar problems, Euler had by 1732 invented the Euler-Maclaurin Summation Formula, submitting it that year for publication in the *Commentarii* under the title "Methodus generalis summandi progressiones" (A general method for summing series, E25). Three years later Euler gave a full exposition of that procedure.[61] But the article was not printed until 1738 (in volume 6 of the *Commentarii*)—the year Maclaurin independently made his discovery, which was scheduled to appear in his *Treatise of Functions* in 1742.[62] In his response, Stirling compared the methods of the two men; he also asked Euler to submit his independent findings for the *Philosophical Transactions* and to allow Euler's name to be entered for election to the Royal Society in London. Since Stirling had taken two years to answer Euler's letter, he feared that Euler might engage in a priority dispute, but Euler was characteristically gracious rather than proprietary. Relieved, Stirling wrote to Maclaurin in October 1738 that Euler "is under no uneasiness about your having fallen on the same theorem with him, because both his statement and its demonstration were publicly read in the academy about four years ago. . . ."[63]

In 1738 Euler made two important contributions to number theory. In March Daniel Bernoulli inquired about Bernard Frenicle de Bessy's *Traité de triangles*, which the Paris Academy had reprinted in its *Mémoires* for 1729 (volume 5); this volume had probably just arrived in Saint Petersburg. In an article for the *Commentarii* for 1738, Euler treated in detail equations $x^4 \pm y^4 = z^2$ and $x^4 \pm y^4 = 2z^2$. For the first time he addressed what is called Fermat's Last (or Great) Theorem: for $n > 2$, there are no positive integers x, y, and z, such that $x^n + y^n = z^n$, and in April he informed Daniel Bernoulli that he had proved it for $n = 4$. While Euler had earlier known the *Varia opera* (Various works) and *Commercium epistolicum* (Exchange of letters) of Fermat, his work in number theory in 1738 suggests a recent encounter with Fermat's *Diophantus* of 1670; this probably set off a search through French correspondents for more documents by Fermat. Only in 1748 and 1753 would Euler return to this conjecture on the part of Fermat;

he called it "a very beautiful theorem." In 1753 he stated that he could prove Fermat's Last Theorem for $n = 3$ and $n = 4$, but not beyond that.[64]

Before 1738 Euler had investigated separate isoperimetric problems and methods in three articles. In "Solutio problematis cuiusdam a celeb. Dan. Bernoulllio propositi" (The solution of a certain problem proposed by the celebrated Daniel Bernoulli, E99) in the *Commentarii* for that year, he embarked on more systematic variational studies.[65] The opening sections of "Solutio problematis," which praised the Bernoullis, Hermann, and Taylor for their acuteness, classified variational problems according to their number of side conditions. Euler had studied the associated tautochrone, isochrone, and brachistochrone curves in detail, and he found that in isoperimetric problems they and related integrals become maxima and minima. Seeking formulas, he moved toward exhausting geometric techniques in the emerging calculus of variations. His goal after 1738 was to transform those techniques by devising differential equations and demonstrating the utility and full range of application of the new field, becoming the first mathematician to do so. He did err, however, in treating problems with integrand variables connected by differential equations. Principally because he had other books to complete or put into print, Euler's progress in variational studies was slowed for two years.

After 1735 the number of technical projects and duties that the academy assigned to Euler grew; the projects included research on such topics as magnetism and ship construction, which were vital to the status of Russia as a rising sea power on the Baltic. Euler served on several technological committees that were testing and improving balances, clocks, fire pumps, machines, saws, and scales. Spurred by Jan Andrej (Johann Andreas) von Segner's water-powered machine, in about 1737 he began to experiment on this and to develop a water turbine. Two centuries later, in 1944, Jakob Ackeret would build a turbine following Euler's precepts that attained the remarkable efficiency of 71 percent out of a total possible of slightly over 80 percent. Euler also gave public lectures on logic and advances in mathematics. To introduce the sciences to the general public, he sent popular essays to the Petersburg Academy's *Primechaniya k Vedomostyam*, and he reviewed papers of others, including an essay on the quadrature of a circle.[66] These supplements were published from 1729 to 1741 in French and German simultaneously. In German they appeared under the title *Anmerckungen über die Zeitungen*, which was a weekly journal; it appears that Euler anonymously wrote several articles for it; this is currently being investigated. His responsibilities at the Petersburg Academy went so far as ordering ink and paper for its printing press, and in 1738 he was appointed to the academy's

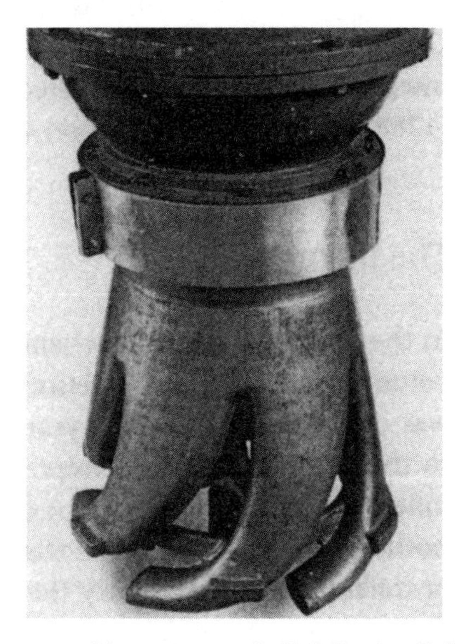

Figure 4.4. A water turbine prototype built following Euler's proposals.

commission on weights and measures. For volume 10 of the *Commentarii* he wrote "Disquisitio de bilancibus" (An inquiry into balances, E93), describing a perfect balance, the structure of balances, and conditions for equilibrium and stability in them, and "De machinarum tam simplicium quam compositorum usu maxime lucroso" (On the most profitable application of simple as well as composite machines, E96), examining motion in simple and complex machines, introducing differential equations for it, and defining equilibrium attained in machines between active forces and resistances. Both articles were submitted in 1738 but the volume was not published until 1747.[67]

Although most of Euler's publications were theoretical, he had a strong interest in practical questions and contributed extensively to mixed mathematics (now called applied mathematics) and to engineering mechanics. In the late 1730s the court of Empress Anna and the Russian and German intelligentsia in Saint Petersburg believed that what mathematicians had achieved in mechanics and astronomy they were obliged to do with numbers for military science and machine construction. They also desired canons for a theory of beauty in the fine arts (aesthetics and the sublime), which became important in the later Enlightenment, especially in the writings of Johann Joachim Winckelmann and Immanuel Kant in German-speaking Europe and Edmund Burke in England. Aesthetics were about the only enterprise that Euler neglected to pursue at length, and even in that field he made minor contributions as regards musical composition.

Scientia navalis, Polemics, and the Prix de Paris

Upon completing the *Mechanica*, Euler turned intently in 1737 to finishing the second of his volumes on the subject. This included, during his first Saint Petersburg period, his two-volume *Scientia navalis* (E110 and E111) on the construction of ships and problems of their propulsion and navigation. He was the first geometer to undertake the project of making floating navigation a complete science; he had encountered the subject in 1735 in an article by the general commissioner of the French fleet, Cesar Marie de la Croix, in the *Mémoires de Trévoux*. Previously geometers had studied only elements of navigation and nautical astronomy, while the essays on mechanical principles submitted to a contest on the speed and maneuvering of ships around 1690 by Christiaan Huygens and lieutenant general Bernard Renau d'Eliçagaray were imperfect.[68] Euler worked on portions of the book in 1735 and completed most of it in 1738, but not until 1740 did it officially become a state project, commissioned by the Russian admiralty.

In his program to develop rational mechanics, *Scientia navalis* was Euler's second milestone work. Outstanding in theoretical and applied mechanics, it applied variational principles in its first volume, making Euler the first to mathematize ship propulsion with a degree of perfection. He set down the principles of hydrostatics with clarity and systematically ordered them, thereby establishing hydrostatics as a branch of science. To the present this work retains its importance. Euler's investigation of the equilibrium of floating bodies in volume 1 provided a way to return stability to a computed value. Effects of external forces upon floating bodies and the resistance of water to their motion posed difficult problems, and Euler corrected many false conclusions about ship motion. Through fluid pressure he defined an ideal fluid. *Scientia navalis* introduces the concepts of centroid and metacenter separate from center of gravity. Volume 2 stresses applications, which allowed Euler to improve his general theory, particularly for ships. Among its topics are the rocking and oscillation of ships, the force of wind on sails, the effect of oars and rudders, and the masting of sailing ships. As Euler formulated the differential equations of ship motion he determined the optimal design of well-constructed ships, providing a foundation for the theory of naval architecture; general principles alone had been insufficient to determine construction. The search for utmost stability countered that for speed; the two had to be reconciled by deciding which properties of each a ship required. Euler also began to develop the kinetics and dynamics of rigid bodies; by then he had established the principle of angular momentum regarding a fixed axis and introduced and defined the terms *moment of inertia* and, speculatively, *principle axes of inertia*.

In part Euler's close attention to practical design and problems of ship handling in *Scientia navalis* came from experiments with ship models and, from 1736 to 1740, correspondence with the Danish naval officer Friedrich Weggersløff concerning the stability of moving ships.[69] Euler found the formulation of abstract laws to describe ship motion extraordinarily difficult, and he appreciated Weggersløff's reports for their worthwhile examples and explanations, but he believed that to correct errors in important existing theories many more observations were needed.

Euler's work on the *Scientia navalis* runs contrary to the notion that he relied almost solely on theoretical mathematics and ignored experience. Any reading confined to his published articles in the *Commentarii* to 1741, Benjamin Robins's criticism of his *Mechanica*, and Condorcet's "Éloge," which treats Euler as purely an analyst, would be misleading. Above all, he sought to express physical problems in mathematical terms and was sure of his computational approximations, instinct, and analogies, but *Scientia navalis* demonstrates the importance of experience and more exact observations to stimulate and guide his devising of difficult partial differential equations and his confirming of theories.

Although virtually completed in 1738, *Scientia navalis* was not published for eleven years. The delay in its publication may be traced in part to turbulent Russian political circumstances after 1740 and strained imperial budgets, but it is more likely that it resulted from Euler's move to Berlin in 1741 and Johann Schumacher's reluctance to recommend a stipend for Euler and funding for the printing. The delay in the publication of *Scientia navalis* meant that Euler lost priority among many of his constituents in ship theory because Pierre Bouguer published his *Traité du navire* in 1746.

During the mid-eighteenth century, the transmission, confirmation, and elaboration of Newton's dynamics and optics, together with the successful application of differential calculus to them, were the chief developments in the physical sciences. The reception of Newton's dynamics included responses to the inverse-square law of mutual gravitational attraction, the principle of universal gravitation and its mutual action, and the principle of action at a distance.[70] Since the concept of gravitational attraction presented in Newton's *Principia mathematica* was more an intellectual than an experimental construct, natural philosophers, astronomers, and geometers busied themselves attempting to uphold or deny it with quantitative measurements, experiments, and telescopic observations that became increasingly accurate with new technology. While during the 1720s and 1730s Newton's gravitational theory had more support than any other hypothesis, most geometers and astronomers stood against it, and each proposed a law of his own; three basic camps now arose in response to Newton's dynamics. The British and a few Continental scholars like Willem 'sGravesande and Daniel Bernoulli generally accepted Newton's dynamics. On the Continent, the Cartesians—who at the time dominated the Paris Academy—strongly opposed it. In the intense controversy between Newtonians and Cartesians, Bernard le Bovier de Fontenelle and the astronomer Jacques Cassini led the partisans of Cartesian science. In central Europe the Wolffians followed Leibniz in questioning the existence and cause of Newton's attraction and his theories on the measure of force. Euler belonged to none of these three camps; in dynamics he was closest to the Continental Newtonians who allowed that a slight corrective factor might be needed to modify the law of the inverse square in celestial mechanics. Five subjects that Newton had begun to explore—the shape of Earth, the theory of sound, the tides, lunar motion, and the nature and orbits of comets—were the vital tests for determining which of the two sciences, Cartesian or Newtonian, more accurately described the phenomenon quantitatively.

While it was agreed in astronomy that Earth has a round shape, two major views arose on the matter. In book 3, propositions 18 and 19 of his *Principia mathematica* Newton had argued that Earth is a spheroid flattened at the poles and with a bulge at the equator. He based that conclusion on gravity, along with its diminution toward the poles occasioned by Earth's diurnal revolution, centrifugal force, and a uniform density of matter. To complete that perspective Newton added a crude hydrostatics that assumed a fluid spheroid in equilibrium. He crucially utilized the telescopic observations and study of pendulum clocks by the French astronomer Jean Richer in 1672–73 on Cayenne Island to confirm his computation of Earth's shape.[71] Newton also gave telescopic observations of Jupiter, including those of Gian Domenico Cassini in 1691, and maintained that if Earth were not an oblate spheroid the equatorial region would be underwater. His finding of Earth's shape provoked a sometimes angry controversy that lasted more than half a century.[72] Why the fluid spheroid should be in equilibrium was quickly questioned; that all laws of force projecting a spheroid Earth are virtually the same added confusion. Since attraction had no mechanical cause, the law of force became the heart of the controversy for the Cartesians.

During the 1730s the position of the Cartesians was eroding at the Paris Academy. Jacques Cassini, using measurements of arcs of meridian to the north and south of Paris, provided in 1718 the second major interpretation that Earth rotating in the ether is *allongé*, a spindle elongated at the poles; this blocked Delisle's plan in the early 1720s for a polar geodetic expedition. Delisle knew that measurements more geographically separated were needed, and justifying their concept of the shape of Earth was critical to the Cartesians. With scathing criticism, the editor Elie de Joncourt claimed in his *Journal histoire de la république des lettres* for 1732–33 that a margin of error of twenty seconds in measuring terrestrial arcs made it possible to prove or disprove Cassini's conclusion. Cassini's cartographic data, he observed, were six times finer than what was possible with existing instruments, and Cassini had rejected figures contradicting his theory. The margin of error from the instruments placed great doubt in Cassini's conclusions.

An array of factors, including the dominance of the Paris Academy over geographical studies in France; the rejuvenation of mapping there in the 1730s; the ferocity of the dispute, which engaged strong personalities; and a spate of papers on geodesy prompted a push to measure arcs of meridian in latitudes near the north pole and the equator to conclusively determine the shape of Earth. These would also be a good test case for resolving a dispute among craftsmen over which new surveying equipment was superior. Cassini's measuring of an arc of the great circle perpendicular to the Paris meridian in 1733–34 did not put an end to the discussion. The French monarchy and the Paris Academy funded geodetic expeditions to Peru (under Charles de la Condamine and Pierre Bouguer, 1735–44) and to Swedish Lapland (under Pierre Maupertuis and Clairaut, 1736–37). Both expeditions were equipped with the latest instruments, mainly for the purpose of measuring an arc of meridian of one degree in far different latitudes: inside the Arctic circle in Lapland and near the equator in Peru. Maupertuis and Clairaut wanted to confirm the correctness of Newton's shape of Earth as *aplati* (flattened), not *allongé* (prolate or ellipsoid).

In May 1736 the French and Swedish expedition left Dunkirk for the Gulf of Bothnia. Its fieldwork was based in the small military outpost town of Torneå in the north of Sweden. Battling swarms of insects in summer and bitter cold in winter, the expedition made astronomical observations to determine the end points of latitudes under consideration and a series of cartographic surveying triangulations to measure the ground length of an arc of meridian. Partly with the hope of forestalling future wrangles over the reliability of the expedition data, the new Graham zenith sector was introduced, which Swedish astronomer Anders Celsius had spent most of 1735 gaining experience with in London.

Maupertuis also brought to Lapland the abbé Reginald Outhier, who was respected for his surveying and mapmaking skills. To corroborate their results and lower inaccuracies, the members of the expedition gathered pendulum data with new clocks showing centrifugal force to be weaker near the pole. To his surprise, Maupertuis's data indicated that Earth is flatter at the pole than Newton had thought. He returned quickly to Paris in August 1737, declaring the precision of his expedition's telescopic observations and geodetic measurements. To correct his observation of the motion of stars, Maupertuis even incorporated James Bradley's discovery in the 1720s of stellar aberration, which is caused by the finite velocity of light and the motion of Earth in space. Although this was negligible, it demonstrated Maupertuis's search for a higher level of precision.

While Maupertuis's claims for the achievement of the Lapland Expedition were dubious and the accuracy of his results made in the Arctic terrain was later shown to be poor, he gained support from many craftsmen for the soundness of his geodetic methods and instruments.[73] But the general response at the Paris Academy was frosty at best. The Cartesians, especially Jacques Cassini, did not accept the accuracy of Maupertuis's checks for his data, while their ally Johann I Bernoulli cited Clairaut's myopia. Irritated, Maupertuis went briefly to Cirey to visit Émilie, Marquise du Châtelet, who lauded the exactitude of the expedition's measurements and acclaimed its leader as Sir Isaac Maupertuis.[74] And Voltaire waxed enthusiastic:

> Herós de la physique, Argonautes nouveaux
> Qui franchissez les monts, qui traversée les eaux,
> Dont la travail immense et l'éxacte mesure
> De la terre étonée ont fixé la figure.
>
> Heroes of physics, new Argonauts
> Who cross mountains, who traverse seas,
> Whose immense labor and exact measurements
> The shape of Earth have determined.[75]

Apollonius of Rhodes, the librarian of the ancient Alexandrian Museum, had retold the Argonaut myth of heroes in search of truths and treasure within the framework of quest and adventure amid treachery and intrigue. The Lapland data provided a major triumph for Newtonian science in Paris, but they did not silence the Cartesians.

Daniel Bernoulli declared in a letter of March 1738 that the findings of the Lapland Expedition were definitive, and a month later Euler wrote for the German community a series of seven articles in the Saint Petersburg newspapers, generally titled "Von der Gestalt der Erden" (On the shape of earth, E46), addressing the debate.[76] Euler's articles mark somewhat of a deviation from his avoidance of public disputes in the sciences. He had closely followed the Lapland Expedition, he noted, and judged it crucial for geography and for expanding knowledge of the deepest principles of natural science. He praised the precision of its telescopic observations and geodetic measurements of an arc of meridian near the north pole. These confirmed that Earth was not a perfect sphere but more like an orange, flattened at the poles; the oscillation of a pendulum near the equator had already shown centrifugal force to be greater there. A number of telescopic observations revealing the orange-like shape of Jupiter were also suggestive. Apparently Euler calculated and repeated in several articles in the *Sankt Peterburgskie Vedomosti* (Saint Petersburg gazette or news) and its *Primechaniya k Vedomostyam* that the shape of Earth is more like that of an orange than a lengthy melon; the existence of these articles has not yet been confirmed. By 1738 he had completed "De attractione corporum sphaeroidico-ellipticorum" (On the attraction of spherico-elliptical bodies, E97), another important contribution to the science on the shape of Earth.

Maupertuis started corresponding with Euler in May 1738, declaring his admiration for the *Mechanica* and sending a copy of his own book, *La figure de la terre* (The figure of the Earth), which was partly a travelogue of the expedition to Lapland. Responding to Maupertuis's letter, Euler expressed the belief that the Lapland data were the most accurate so far, having been made with the best instruments available, but alone insufficient to give the shape of Earth—a position that was similar in rigor to Clairaut's. Telescopes were not yet accurate to less than five seconds, a measure that Euler required for a conclusive proof, and he awaited the return of the Peru Expedition.

In a December 1738 article Euler proposed computing the polar flattening by assuming a diminution of the radius of the polar circle by 1/289. He correctly lessened the estimate that put the flattening at 1/233 to 1/234 and Newton's at 1/229 to 1/230. Euler discussed differing measures of an arc of meridian by Jean Picard, the Cassinis, and Pieter van Musschenbroek. Through his own computations Euler introduced a second assumption: that the internal density of Earth was not homogeneous but variable. While in Lapland, Clairaut had decided similarly to apply hydrodynamics and calculus in the determination of Earth's shape. Euler made it clear that he had not found an error in Newton, whose dynamics were correct; he continued to accept Newton's inverse-square law of attraction but was considering a slight modification. Colin Maclaurin retained Newton's position holding the radius of the polar circle to be 1/229 less than that of the equator. Euler closely followed the continuing research on the shape of Earth, and this partly involved calculating the equilibrium of a fluid spheroid. As Euler was to inform Daniel Bernoulli in September 1740, he admired two articles on the polar flattening by Clairaut that had been published in the *Philosophical Transactions*. The most accurate result on the radius of the polar circle, a lesser flattening of 1/300, was to appear in Clairaut's *Théorie de la figure de la terre* in 1743.

Euler was the chief Enlightenment opponent of Newton's corpuscular theory of light—a qualitative description that Euler rejected as imperfect in its foundations. In a 1737 letter Johann I Bernoulli asserted that he was seeking to synthesize Newton's corpuscular theory with Huygens's wave theory, but ultimately his and others' attempts at synthesis did not succeed. Euler—building on his powerful intuition in physics and holding a strict analogy between the propagation of sound in air and light in ether (which was believed to fill the universe)—proposed his formative theory of light. He assumed that air was composed of infinitely small but springy particles compressed together and that sound was a pressure wave transmitted through it. Not unlike Cartesians Nicolas Malebranche and Jakob Hermann, with their concept of pressure waves, Euler proposed that pressure vibrations or pulses within the medium of elastic ether propagated light. His letter of September 1740 to Daniel Bernoulli stated that Newton's explanations of reflection and refraction were unclear. Yet while Euler investigated optics at the end of his first Saint Petersburg stay, his mature optics—shorn of the ideas of Malebranche—did not appear until he had relocated to Berlin.

The taciturn Euler had earlier refrained from joining the argumentative Daniel Bernoulli in attacking the Wolffian philosophy. But in a letter of August 1736 to Karl Ehler in Danzig Euler gently opened criticism of Wolff's science in the *Philosophia prima sive ontologia* (First philosophy or cosmology), *Cosmologia generalis* (Universal cosmology), and the theory of positive and negative infinity presented in the latest edition of Wolff's *Elementa matheseos universae* (Elements of general mathematics).[77] A letter in February 1737 to Ehler noted regrets that no satisfactory book on integral calculus existed and announced Euler's plans to embark on such a project. While he had not yet read Wolff's forthcoming *Theologia naturalis* (Natural theology), he assumed that it would be no better than the *Cosmologia generalis*, which had many computational errors. At the academy meeting on 16 May 1738, Euler—apparently for the first time—publicly disputed Wolff's thought, submitting a notice now lost with a critique of faults in the *Cosmologia generalis*. A letter of 10 July to Georg Bernhardt

Bilfinger in Tübingen asked for comments on the notice but suggested that it was preliminary and should not be circulated to Wolff. In another letter to Bilfinger in November 1738 Euler especially questioned Wolff's reduction of primal or elemental substance to animate monads in the *Ontologia* and *Cosmologia*, as implied in the discussion of infinitesimals with Ehler;[78] monads, he argued, contradicted the law of inertia in physics, and the acceptance of eternal monads as a basis for cosmology, he believed, would lead to atheism. This appears to be Euler's earliest criticism involving implications of Wolff's science for religion. He reproved as mistaken the Wolffian rational method that reduced the importance of Leibniz's principle of sufficient reason, and he found their employment of the principle otherwise inadequate.

Another debate was over James Jurin's vigorous criticism of the principle of the conservation of *vis viva*. Initially Jurin had opposed Bilfinger at the academy, but after 1735 he primarily assailed Johann I and Daniel Bernoulli. Their quarrels appeared in articles in Leipzig's *Acta eruditorum*, and in a letter to Euler in April 1737 Johann I mentioned his ongoing dispute with Jurin. In 1741 Émilie du Châtelet reignited the debate when in the appendix to her *Institutiones de physique* she supported Hermann's arguments for the conservation of *vis viva* and challenged contrary views presented by the Cartesian Jean-Jacques d'Ortous de Mairan in his "Dissertation sur les forces motrices des corps" (Dissertation on the motive forces of bodies).

Euler's progress toward center stage in the sciences in Europe, which was among an efflorescence of his achievements in 1738, began with sharing the annual Prix de Paris from the Paris Academy of Sciences. After receiving an *accessit* (honorable mention) in 1727, Euler had submitted eighteen essays to the Prix de Paris competitions, sometimes anonymously. He won alone or shared these prizes and was the laureate an astonishing twelve times, including once under the name of his son Johann Albrecht in 1760, besides taking another *accessit* in 1741. He is also sometimes credited with writing for Johann Albrecht in 1761 the winning paper on the ballast of ships and ways of obtaining greater speeds. Euler's fifteen Prix awards (he won the prize itself twelve times and the *accessit* three times) is as yet unmatched; his closest competitor, Daniel Bernoulli, won a total of ten awards. In competing for the Paris prizes, which were offered in theoretical and applied subjects in alternate years, Euler engaged extensively in both categories. He was to receive five prizes for applied papers on navigation and shipbuilding.

In the prize competition for 1738 on the nature and properties of fire, Euler and two others won. Along with air, electricity, heat, and magnetism, fire was a crucial research subject of mid-eighteenth-century natural philosophers; they considered it the most volatile of the four Aristotelian elements—earth, air, fire, and water. In central Europe lengthy studies were undertaken of the combustion and calcination associated with mining as well as investigations into Paracelsus's three principles—ideal forms of salt, sulfur, and mercury—concerning the most basic elements. These texts led to the theory espoused by German physician Georg Ernest Stahl that substances rich in something then called phlogiston burned readily. But this theory lacked adequate phenomenological descriptions and quantitative laws. To create a richer theoretical framework and identify better what to measure in the various phenomena of fire, natural philosophers began to address the crucial issue of its weight; notable among these was Musschenbroek, whose *Elementa physicae* was published in 1734. Musschenbroek rejected the idea of innate bodies and called for accurate observations and careful experiments, conjecturing that fire had weight. He gave as an example the heating of iron: as the iron expanded the increase in the volume of displaced air produced an upthrust that exactly compensated for that weight. Others thought that fire possessed no weight, and the inability to properly weigh phlogiston was raising doubts about its existence.

The methods and analysis in the Prix de Paris papers for 1738 were diverse. In "Dissertatio de igne, in qua ejus natura et proprietates explicantur" (Dissertation on fire, in which its nature and properties are explained, E34) Euler proposed a model for fire as resulting from the bursting of tiny glass-like balls of highly compressed air located within the pores of bodies; the phenomenon of heat, he posited, similarly involved motion of the smallest particles.[79] This made it possible to describe both fire and heat through the laws of mechanics without any appeal to the old, discredited occult qualities. Euler gave a formula for a law applying to both heat and fire, but did not explain how he had derived it. Of the two other winners, the Jesuit Louis-Antoine de Lozeran du Fesc rejected the idea that fire was an element, preferring instead to explain it by a theory of vortices, and Jean-Antoine de Créquy, Comte de Canaples, found fire to be caused by the motion in countercurrents of an ethereal fluid.

As interesting as the winners are the two who received *accessits*: Voltaire and Émilie du Châtelet, both of whom had submitted papers anonymously. In a most unusual decision, their papers were published with those of the winners. Voltaire was upset at not winning the prize and blamed the outcome on a Cartesian dominance at the Paris Academy. Since his *Lettres philosophiques* (also known as *Letters concerning the English Nation*, 1733 and 1734), he had been a partisan of Newton and opposed the Cartesian imaginary vortices of ether; the *Lettres* gave the first published account of an apple falling on Newton in a garden. In 1738 Voltaire's *Elémens de la philosophie de Newton* (Elements of Newton's philosophy)—a work primarily dealing with optics and intended to bring Newton's science to the general reading public—appeared in Paris. The reviewer for the *Journal des Trévoux* observed that while few Parisians had previously known Newton, now "all Paris resounds with Newton, all Paris stutters Newton."[80] But Voltaire's simple division of the academy into two camps, the Cartesian and the Newtonian, failed to recognize the complexities of arguments and the diversity within each group. Before writing his essay on the particulate nature of fire for the Prix de Paris, Voltaire had visited Leiden and conducted experiments in Cirey. His essay lacked originality, and its evidence regarding the weight of fire particles was derivative, coming from the research of 'sGravesande and especially Musschenbroek, and this fact was not lost on the judges. In scientific research, what the English physicist Charles Percy Snow later described as two cultures—one of literary intellectuals and the other of scientists—seemed to exist.[81] For Voltaire, who possessed an essentially literary mind and was missing a strong education in mathematics and the sciences, the sciences were becoming too complex; he was unable to make major advances in them. Although his essay was published anonymously, the Paris Academy's description of its author as "one of our premier poets" (*un de nos premiers poetes*) left little doubt about who it was.

In 1738 Euler had a valuable agent in Paris when Antiokh Kantemir was named ambassador to France, a post that he held until 1744. While ambassador to Great Britain from 1731 to 1738, Kantemir had grown interested in British science and studied it. Among the works that he read in London were selected mathematical papers and *The Chronology of Ancient Kingdoms* by Newton, along with James Gregory's *Manual of Geography* and John Desagulier's *Course of Experimental Philosophy* on Newtonian science. As a result Kantemir shifted his support from Descartes to Newton. He met Hans Sloane, Newton's successor as president of the Royal Society of London, and had Sloane elected as a foreign member of the Petersburg Academy of Sciences.

The publication by the academy of the series of articles titled "Von der Gestalt der Erden" in 1738, under the protection of the Archbishop of Novgorod, Feofan (Theophan) Prokopovich, a staunch defender of the academy in the Holy Synod, improved the climate for Copernican astronomy in the Russian capital.[82] These articles, most of which Euler wrote, supported Copernicus and Newton but did not criticize or belittle the Orthodox Church's position that God had created the whole universe for the inhabitants of Earth; the censors found this acceptable.

Although no longer a Cartesian, Kantemir saw the translation of Fontenelle's *Entretiens sur la pluralité des mondes* (Conversations on the plurality of worlds, 1686) through the academic press in 1740, insisting that the publication date given on the title page be 1730, the year of his original submissions.

In Paris, Kantemir acted as a relatively modern Marin Mersenne;[83] besides promptly delivering publications of the Paris Academy to Saint Petersburg, he provided scientific savants in Paris—such as the academicians Georges-Louis Leclerc, Comte de Buffon, Clairaut, and Mairan, as well as Émilie du Châtelet—with reports and printed works of the Petersburg Academy of Sciences. In 1738 he had Maupertuis named an honorary member. The next year, at the urging of Émilie du Châtelet, he translated into Russian two books by the Venetian count Francesco Algarotti: *Dialogue sur l'Optique de Newton* and the popularization *Il Newtonianismo per la dame* (Newtonianism for the lady), a work also mostly on optics that had been published in 1737. At Kantemir's urging, Algarotti traveled to Russia in 1739 to meet Empress Anna;[84] Euler apparently met Algarotti during the visit.

In the eighteenth century, finding evolutes (curves generated by a taut string, attached to a point, that is pulled along the tangent to that point) and their inverse, involutes (spiral or inward curves), was an important mathematical problem with applications in navigation, mechanics, and clocks. After developing the pendulum clock in 1656, describing it in his first *Horologium* (1658) and expanding upon this in his principal *Horologium oscillatorium sive de motu* (Theory and design of the pendulum clock, 1659), Huygens had studied this problem as it related to the oscillations of a pendulum. The pendulum problem was to become part of the unfolding conflict between geometric and algebraic descriptions of curves. In chapter 5 of Guillaume-François-Antoine, Marquis de L'Hôpital's *Analyse*, Johann I Bernoulli had turned this from a geometric problem in special cases to a problem in differential calculus.[85] Answering a question posed by Krafft, in August 1739 Euler submitted "Investigatio curvarum quae evolutae sui similes producunt" (Investigation of curves that produce evolutes that are similar to themselves, E129), which generalized Bernoulli's solution.[86] This was a developing stage in Euler's turning differential calculus from the study of curves toward operations with functions.

Pedagogy and Music Theory

Euler enjoyed teaching and undertook it with enthusiasm. The director of the Petersburg Academy required leading members to write elementary textbooks and handbooks, and Euler was assigned to write an arithmetic text in German for the academy's gymnasium. He was pleased to do so, and the result was *Einleitung zur Rechen-kunst* (Introduction to arithmetic, E17) in two parts.[87] After composing most of the first part in 1735, Euler finished and published it in 1738; the second part appeared in print two years later. At the time German—even with the great Lutheran Bible as part of its history—was still in the process of becoming a literary language, not yet unified in grammar and spelling; Wolff (in Marburg) and later Johann Christoph Gottsched (in Leipzig) undertook formulation of the principles and rules for its grammar.[88] Particularly rigorous as regards sentence structure, Euler made the *Einleitung* easily readable; he also made small refinements in orthography, replacing the letter *y* in such words as *seyen* and *drey* with *i*, or *seien* and *drei*, while *teusch* became *deutsch*. He simplified symbols, reducing the mathematical constants *ck* to *k* and *tz* to *z*, and he designated null by the word *Ziffern* rather than *Cyphren*.

The *Einleitung* defines arithmetic as a science that examines the nature and property of numbers and gives rules sufficient to solve problems in practical life and work. After explaining place-value notation in base 10, Euler discussed the four basic arithmetical operations: addition, subtraction, multiplication, and division. For each he gave numerical examples and word problems involving whole numbers, fractions, and mixed numbers. Word problems incorporated basic knowledge from geometry, physics, commerce, and astronomy, and this must have made them more stimulating. The text also emphasizes the conversion of the many units of weight and money.

The section on division, an operation that Euler stressed in his study of number theory, includes problems with remainders and the method for finding the greatest common divisor of two numbers, the Euclidean algorithm. Proceeding step by step, Euler found the greatest common divisor of 2,904 and 1,578. First he subtracted 1,578 from 2,904 to obtain 1,326, which he divided into 1,578, leaving a remainder of 252. The process was repeated until he reached an answer without a remainder. The greatest common divisor is 6.[89] Noting in one multiplication problem that a great circle is divided into 360 degrees and that 105 *Werste* (one *Werst* is about .66 miles) make up a degree, the *Einleitung* asks, how many *Werste* are there in the circumference of Earth? Since each degree has 105 *Werste* and Earth has 360 degrees, the number of *Werste* in its circumference is 105 multiplied by 360, or 37,800 *Werste*. In reviewing each arithmetical operation with fractions, Euler computed practical monetary totals—for example, from denominations of rubles, gewirs, and kopeks.[90] This elementary text did not critically and vertically develop problems, nor did decimals, proportions, or radicals appear in parts 1 or 2.

According to Fuss's eulogy of Euler, music was one of the few relaxations that Euler allowed himself. This suggests that he possibly played the clavier often and may have deepened his schema for music theory during moments of pleasure from the harmony that he achieved in playing compositions. Amid chords he likely calculated proportions. Harmony is central to most Western music, entailing a movement or progression among chords. The clavier was at this time becoming more complex; it had delicate tones and was intended for performance in a room. The larger harpsichord, with richer tones and meant for concert halls, Euler knew well and listened to, but no record exists of his playing it. His Basel notebook outlines a book on music theory. His correspondence with Johann I Bernoulli reveals that subject as also being an assignment from the academy. By May 1731 Euler had nearly completed a short treatise on music theory, its first proofs bearing the title *Tractatus de musica*. It was not possible, however, to publish a science monograph so early in his career. Following the *Tractatus,* he worked diligently to perfect his ideas by adding new insights and conclusions. The result was a more mature work, the 263-page *Tentamen novae theoriae musicae ex certissimis harmoniae principiis dilucide exposita* (An attempt at a new theory of music composition exposed in full clarity according to the most certain principles of harmony, E33),[91] which the press of the Petersburg Academy published in 1739. Of the trilogy of books from Euler's first Saint Petersburg period it was the second text in applied mathematics, readied after finishing the *Scientia navalis* two years earlier; it is not simply an extension of the first work.

In December 1738 Euler informed Johann I Bernoulli of his forthcoming book on music theory based on acoustical, aesthetic, and physical regularities: "At the beginning of next year my memoir on music, which I had written a few years ago, will now go into print, in which, so I think, are revealed the true and inherent principles of harmony."[92] In March Bernoulli excused himself from the discussion, writing "in music I am not very well versed, and with the foundations of this science I am not sufficiently familiar that I could judge your discoveries in this area," but he added, "what you touch on in your letter . . . seems . . . really outstanding."[93] Johann I wanted to concentrate on hydraulics; regarding priority over it, he was about to clash with his son Daniel.[94] Johann did not return to the topic of music, but Daniel took it up; in 1739 he questioned Euler's belief that "the general term $2^n \cdot 3^m \cdot 5^p$ covers almost all pleasantly perceived tones." He doubted that musicians would accept Euler, for they did not insist on perfect harmony; they would prefer to follow geometric progressions, which have simple proportions and give tones exactly enough.[95] In March Euler replied that his book was nearly through the press and that his general term was not simply an observation but that it accurately agreed with the latest temperate study of keys and tones.[96]

In the late baroque period Euler was not the first to propose a new theory of the tonal structure and aesthetics of music. Independently, the French composer Jean Philippe Rameau had already published in 1722 his systematic theory of composition, entitled *Traité de l'harmonie réduite à ses principes naturels* (Treatise on harmony reduced to its natural principles). Rameau derived the principles of harmony from laws of acoustics and investigations with strings, and he anchored tonality in the tonic, dominant, and subdominant chords. He displaced the medieval counterpoint, which had a hierarchy of melodic relations among parts for voices and entailed two or more harmonious notes. The usual technique for these relations had been imitation. Because the Parisian music community believed that music theorists could not be good composers, Rameau remained obscure for a time, but upon being adjudged a creation of genius, his *Traité de l'harmonie* quickly brought him recognition;[97] he would come to be known as the Isaac Newton of music.[98] Rameau's *Nouveau systéme de musique* of 1726 appealed to overtones and applied basic mathematics only when associated with observations. His system of harmony worked for lower overtones but not higher ones. Euler opposed it. Throughout his career Rameau had critics, especially, at midcentury, Jean-Jacques Rousseau and the Encyclopedists. Among his defenses were to be *Demonstration du principe de l'harmonie* (1750) and *Nouvelles réflexions de M. Rameau* (1752).

Still conceiving of music as a mathematical science, Euler continued a tradition dating from antiquity, when the Pythagoreans, building upon their doctrine that "everything is a number," discovered empirically the numerical ratios of musical intervals for producing harmonious sounds. In these studies the early Pythagoreans employed stringed instruments, but they may have originated in Pythagoras's hearing of different blacksmith hammers. The Pythagorean system of tones and rhythm followed the same mathematical laws that operate in the physical world. The Pythagorean quadrivium, a name that was adopted in the Latin Middle Ages, had been made fundamental in classical Greek schools by Plato and Aristotle; it continued to be dominant in the later Hellenistic period. Arithmetic, geometry, *harmonia* (music), and *astrologia* (astronomy) are its subjects, and all were considered branches of mathematics. Book 3 of Plato's *Republic* argues for the importance of music in education.

In addition to providing entertainment, music was believed to discipline the mind, and most composers accepted its doctrine that the affections it arouses aid the control of the emotions. In antiquity two schools of music emerged: the Pythagoreans stressed mathematical theory without giving adequate attention to hearing, while their critic Aristoxenus of Tarentum, a disciple of Aristotle and the author of *Elementa harmonica* (Elements of harmonics) and *Elementa rhythmica* (Elements of rhythmics), did the reverse. In the second century A.D., Ptolemy, the author of the astronomical masterwork *Almagest*, which significantly advanced all known mathematical sciences, attempted in his three-book *Harmonics* to build a mathematically satisfactory middle ground between the two schools of music. His work seems to have been a model for young Euler.

Ptolemy's *Harmonics* provided material for the Enlightenment work in music; Kepler, whose astronomy Euler admired, had read a poor Latin translation of Ptolemy's treatise made before 1600. From Ptolemy's belief that musical ratios exist in the heavens and the soul, the harmony of the spheres followed as a basic concept in astronomy. Explaining planetary motion by harmonic theory and its mathematical ratios intrigued Kepler, who based harmony on geometry rather than arithmetic. Kepler's research in harmonics preceded his discovery of his three planetary laws, the first being the elliptical orbits of planets. Also within Euler's reach was a new edition of Ptolemy's *Harmonics* prepared by John Wallis in 1682. In book 1, chapters 2, 5, and 6 treat the Pythagoreans, and chapters 9 and 12 discuss Aristoxenus. Other seventeenth-century sources for Euler's *Tentamen*

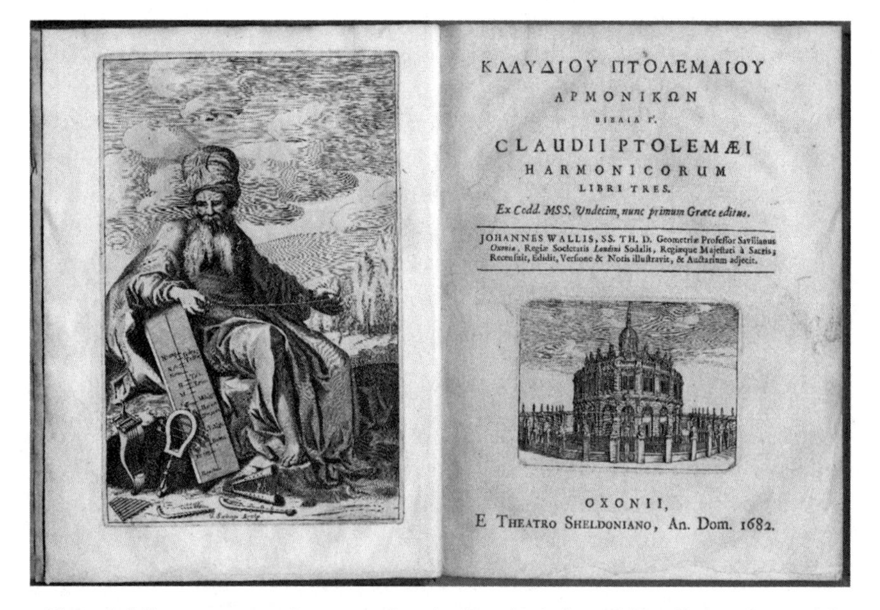

ΚΛΑΥΔΙΟΥ ΠΤΟΛΕΜΑΙΟΥ
ΑΡΜΟΝΙΚΩΝ
ΒΙΒΛΙΑ Γ'.
CLAUDII PTOLEMÆI
HARMONICORUM
LIBRI TRES.
Ex Codd. MSS. Undecim, nunc primum Græce editus.

JOHANNES WALLIS, SS. TH. D. Geometriæ Professor Savilianus *Oxoniæ*, Regiæ Societatis *Londini* Sodalis, Regiæque Majestati à Sacris; Recensuit, Edidit, Versione & Notis illustravit, & Auctarium adjecit.

OXONII,
E THEATRO SHELDONIANO, An. Dom. 1682.

Figure 4.5. Late seventeenth-century frontispiece depiction of Claudius Ptolemy and the title page for the 1682 edition of his *Harmonics*, which Euler probably read.

include sections of the *Discorsi* of Galileo Galilei on acoustics and the pendulum, the *Harmonie universelle* of Mersenne, the *Compendium musicae* of Descartes, the popular *Musurgia universalis* of 1646 of Athanasius Kircher (Fuldensis), and writings of Leibniz "as amateur" exploring the range of consonance in music. The last three influenced Euler's number theory conception of degrees of consonance.

In revising his *Tentamen* Euler examined relevant portions of Newton's *Principia mathematica* and musical experiments that Joseph Sauveur—a mathematician at Paris's College Royal and a tutor in Louis XIV's court—made from 1700 to 1716. Sauveur, whose results appeared in the Paris Academy's *Mémoires*, along with the organ builder Pierre-François Deslandes, investigated the frequency and pitch of organ pipes. Using the rates of vibration of the higher notes, they measured the speed of sound. In not restricting his study of music to harmonics, Sauveur developed a new field that he called *acoustique*. Book 2, section 7 of Newton's *Principia* analyzes sound as pressure waves through a compressible medium and recognizes the finite velocity of sound as frequency multiplied times wavelength. Frequency is $v \, \sigma \, 1/l\sqrt{F}/\sigma$, where l is length, F tension, and σ cross-sectional area. The first edition of the *Principia* in 1687 had the velocity of sound as 968 feet per second. After Newton reviewed frequency measures by Sauveur in five-foot organ pipes and also those by William Derham, the second edition in 1713 increased the velocity to 1,020 feet per second, which is his closest approximation. Section 6 of Euler's *Tentamen* gives only the numerical result of his computations, an approximation of 1,100 Rhenish [and Danish] feet per second.[99] The Rhenish foot defined in chapter 1 of "Dissertatio physica de sono" (Dissertation on the theory of sound, E02) was later shown to be 313.835 millimeters; converting these gives 1,134.8 feet per second. Euler was pleased to arrive at a result larger than Newton's, and correctly believed it to be more nearly exact. (He was slightly high; the speed of sound has since been calculated at 1,125.8 feet per second.) Euler's letter of August 1737 to Johann I Bernoulli defends his computational technique and result.[100]

That letter also objects to Newton's method for calculating the speed of light. Euler declared that light had the greatest velocity known in his time, and its speed became for him a long-standing interest. He compared the speed of sound to that of light by the difference in time between seeing a bolt of lightning and hearing the thunder. Descartes had earlier held that the velocity of light was instantaneous, but Galileo believed it to be finite. Through observations of the appearance of eclipses of Jupiter's moons, the Danish astronomer Ole Römer had in the 1670s scientifically established the value as approximately 140,000 miles per second, compared to the current value of 186,282 miles. Through measurements with optics Euler later arrived at a closer but somewhat high approximation of twelve million English miles per minute or 200,000 miles per second.[101]

In the *Tentamen* Euler sought to provide a new music theory, one that was "part of mathematics and [would] deduce in an orderly manner *ex certissimus harmonicae principiis* [from the most certain principles of harmony] everything which can make pleasing a fitting together and mingling of tones."[102] His music theory had a dual foundation: from the sciences came an exact knowledge of sound and from metaphysics came the method of how the relations of the frequency of vibrations in the air by harmony please—or by disharmony displease—auditory perceptions. To decide all this Euler appealed to reason, experience, and adept computations. The *Tentamen* demonstrates his pursuit of systematic order, his skillful crossing of disciplinary boundaries, and his search for deeper connections among fields—in this case acoustics, optics, and vibration theory. Vibratory motion, he concluded, was the cause of sounds and hearing, with sound transmitted by pulses in the air.

Chapter 1 of the *Tentamen* summarizes the principles and a structure of acoustics that expands upon Euler's "De Sono." Section 9 presents his formula for a vibrating string's frequency: $355/113 = \sqrt{3166n/a}$, where n is the ratio of a string's stretching weight to its normal weight and a gives the string's length. The value $355/113 = 3.1415929$ is accurate to six decimal places of π; in the late fifth century AD, the Chinese engineer and astronomer Zu Chongzhi had discovered this ratio. In section 34, Euler employed the Archimedean approximation of 22/7.

Essential to Euler's notion of order in music theory was basing the tonal scale upon numerical ratios of frequency. By preferring to pursue mathematical computations rather than philosophical analysis for foundations, he separated himself from his contemporaries; no one had taken music theory as far mathematically as had Euler. Seventeenth-century music theorists had shown pitch to be proportional to frequency: musical intervals were frequency intervals, the inverse of length ratios. Euler thus rejected Aristoxenus, who wanted no numerical ratios in music theory. Euler found the work of the Pythagoreans and Aristoxenus to be filled with confusion and errors, and he warned that Pythagorean principles alone were insufficient and, when their limits were unrecognized, a potential source of errors. He carefully built upon Pythagorean and Ptolemaic laws, claiming an agreement between the ancients and his modern thought.

Toward scientific authority Euler displayed a critical, skeptical attitude that was sometimes playful. Vibrations are heard as tones,[103] and in diatonic systems succeeding tones give the individual terms in the next octave. The ratio of frequencies of two sounds given in small integers from the simplest order of two notes in unison (1:1) proceeds through the fundamental chords or consonance of the octave, 2:1, to the fifth, 3:2, where one of the numbers is not 1, to the fourth, 4:3, and so forth. Euler saw these as blows upon the tympanum of the ear conveyed through the intermediary of the air.

One of Euler's two supports for the metaphysical foundation for music in a text filled with mathematical formulas and notes is the degree of consonance or pleasantness among tones, his *gradus theoria*. He cautioned, however, that there were no clear borders between the notes C and D. He agreed with the ancients that the simpler the ratios of vibrations of the air the more consonant and pleasing the intervals would be. As had Ptolemy and Descartes, he calculated these tonal scales by intervals using the small prime numbers of the harmonic triad, consisting of the musical integers 2, 3, and 5, together with the "exponent of the Ptolemaic tonal system," $2^n \cdot 3^3 \cdot 5$. N = 6 he called the exponent of accord, and he allowed for higher octaves with multiples and powers of these integers. From these Euler constructed a finite number of genera. He differed with his contemporaries in music, who employed only 3 and 5. Instead he gave 2 an essential role. He disagreed with them, for example, that the pitch 8 is a combination of 7 + 1 or 9 − 1. Rather, he maintained that it is 2^3, the third octave of basic vibration 1. So far he confirmed the findings of Leibniz, who held that the musical world had not yet learned to count beyond 5. Measuring lengths, weights, and tensions of two strings for a given musical note, Euler calculated their absolute frequencies or vibrations per second. But he provided no experimental confirmation of the frequencies. His computations draw upon analogic reasoning and include difficult logarithms and continued fractions.

Chapter 4 gives a first approximation of ratios in the progression from the octave to the fourth to the fifth. Multiplying two smaller semitones comes to .30103001/.1760903, or log 2/log 1.5. Euler expressed this as the continued fraction

$$1 + \cfrac{1}{1 + \cfrac{1}{2 + \cfrac{1}{3 + \dots}}}$$

Chapters 10 and 13 warn that attempts to introduce any number besides the triad (2, 3, and 5) are difficult and have produced numerous errors.[104]

In book 1, sections 19, 28, and 32 of the *Tentamen* Euler mostly followed Ptolemy. To reach one or more octaves and arrange all consonances by degrees, he multiplied by a second power of consonance: 1; 3, 2^2; 5; $3 \cdot 5$; $3^2 \cdot 5$; $3^3 \cdot 5$; . . . . Each octave of the diatonic-chromatic system of his time includes twelve tones. The octave for the first degree contains the relation of C:E:G:B, which makes a major seventh chord. Euler later computed it to be the proportion 36:45:54:64. The least common multiple or exponent of 8,640 is $2^6 \cdot 3^3 \cdot 5$, which Euler named the exponent of accord. This puts related figures into the seventeenth degree of consonance. How high a sound could be, Euler discovered, depended upon the rapidity of tones, and he investigated limits of hearing. In the development of vibration theory and continuum mechanics, the measurement of audible vibrational frequencies to their upper and lower limits was critical.

A second essential support for the foundation of music theory, along with his *gradus theoria*, was Euler's substitution theory concerning proper hearing as basic for explaining music. Section 20 of chapter 10 and section 19 of chapter 13 warn of the great difficulty for computing harmony in applying 7, the next prime number after the harmonic triad. Euler declared that attempts had produced numerous errors. He found it difficult to introduce 7, "since the consonances sound too hard." He asserted that 7 neither joined with 2 or 3 and added to 5 would produce pleasant sounds. For the moment, he did not clearly explain a dominant seventh chord on the tonal triad and treated it as purely a conjecture. But in the 1760s, and in a paper in 1773, he accepted it. The seventh leading note, 7:4, needed examination. Mersenne's *Harmonice universelle* of 1636, which invoked 7 and employed analogies between music and other physical phenomena, probably influenced Euler, and Leibniz's extension to 7 of the basics for music most certainly did.[105] Euler's drawings in the *Tentamen* and his later *Lettres à une Princesse d'Allemagne* (Letters to a German princess) are considered the origins of Venn diagrams in mathematics.

Good practice and his principles, Euler thought, might bring music to its highest perfection. Maintaining that theory alone cannot attain the requisite pleasure, he also intended his *Tentamen* to help musical instrument and machine makers improve their handiwork. Even Euler considered music theory a dry and tiresome subject, one in which it was difficult to maintain interest. Composers, musicians, and savants did not agree that all music and related feelings could be reduced to mathematics and Enlightenment reason, yet it is not accurate to describe Euler's work in music as purely mathematical. The *Tentamen* employs mathematics as a

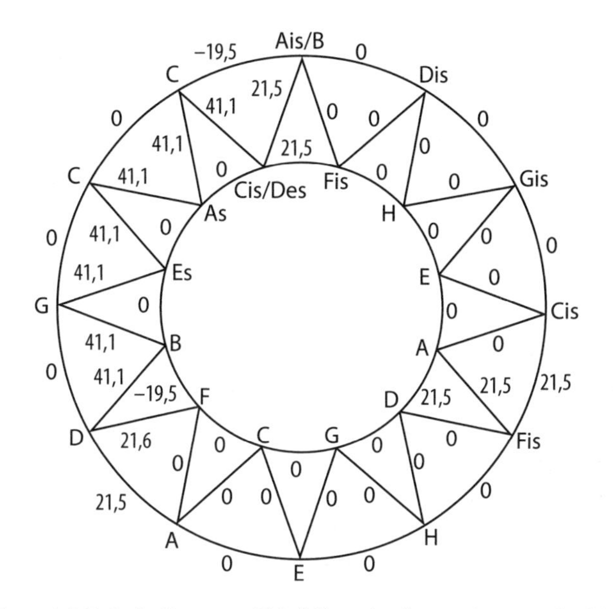

Figure 4.6. Euler's diagram of his "diatonic-chromatic genus," which has twelve absolutely pure and twelve discordant notes.

language to describe musical phenomena, while the work's metaphysics also brings into play fundamental acoustical, aesthetic, and physical regularities.

In November 1740 and January 1741, Daniel Bernoulli sustained a skeptical stance toward the *Tentamen*. A letter from January reads, "I with Mr. Pfaff, who is an excellent musician, decided . . . to have my piano tuned according to your prescribed manner. But he doubts that this will produce a good effect and, he said, one should not pay attention solely to the harmony, especially when one has to deal with imperceptible differences."[106] No further correspondence addresses this matter, so the outcome is not known.

At midcentury Euler's *Tentamen* had many friends and many doubters.[107] A notable critic was Rameau, while the Wolffian Lorenz Mizler at Leipzig University provided, in volume 3 of his *Musikalische Bibliothek* (Musical library, 1752) a commentary of more than a hundred pages. The *Musical Library*, a monthly that lasted to 1754, represents the beginnings of a formal journal reviewing important ancient and modern works in music. In it Mizler spoke mainly as an amateur and a lover of the field;[108] he had only briefly studied composition with Johann Sebastian Bach. Strongly influenced by the philosophy of Wolff and the style of Johann Christoph Gottsched, his teacher at Leipzig, Mizler held that music and poetry were sister fields and that music was a philosophical field. He recognized the prominence of Euler in the sciences, referring to him as "the famous Mr. Euler." Both Euler's partisans and critics, Mizler held, gave incomplete accounts. Mizler, who also stressed tones and consonance, examined vibrations through the air, in the ear, and in instruments with pipes and strings. To grasp beauty in music's nature and its relations with sorrow and joy, he claimed, Euler should have begun with Wolffian metaphysics. Mizler reviewed numerical relations from the Pythagoreans and the improvements by Euler, finding some typographical errors in computations, and held that further advances were needed. Unlike Euler, Mizler in his table of intervals for harmonic sounds lacked any prime numbers. In the study of nature, he agreed on the centrality of number, weight, and motion. An outsider to music, Euler had a deductive, mathematical tonal system that contrasted with Rameau's concept of harmony emphasizing three vibrations. Euler's system was not tempered by appeals to modern musical masterworks or current musical practice, and Rameau opposed it.

While the many new ideas in the *Tentamen* elicited substantial discussion, the work encountered sharp criticism that soon led to its neglect. The principal criticism given was that the work was "for musicians too [advanced in its] mathematical [computations] and for mathematicians too musical." For musicians, it was inadequate, its presentation sometimes seeming backward to the way they played their instruments, its complex computations of chords devolving into an excess of overtones or partial sounds. It lacked instruction. Most immediate criticism came from music theorists of empirical bent. This would be especially true in Berlin, where the musicologist Friedrich Wilhelm Marpurg,[109] a dominant figure at the court of Frederick II, countenanced no opposition to his empirical approach. Euler wrote to Goldbach that he was disappointed that all of his critics left out the core of his music theoretical conceptions.

Daniel Bernoulli and Family

In 1739 and 1740 Euler continued corresponding with Daniel Bernoulli, and the letters indicate the breadth of Euler's scientific studies. Recurring topics were elasticity, flood theory (which was a vital subject in Saint Petersburg), hydraulics, isoperimetrics, and maximum and minimum values for integrals, along with lunar motion, the tides, Newtonian science and the Cartesian leader Cassini, and finances. Euler explored small transverse vibrations in an elastic bar and how to integrate related linear differential equations with constant coefficients. He read Alexis Fontaine and

Clairaut from Paris but still lacked a way for making computations on such integration. At least partly this seems due to Euler's own emphasis on empirical mathematics, in which results of careful experiments guide the formulation of method. In a letter of May 1739 he reported that for the most part he had proved and found accurate the computations in Bernoulli's *Hydrodynamica* except for one error.[110] Among other topics were the thermometers of Maupertuis and Delisle, the election of Maupertuis as a foreign member of the Petersburg Academy of Sciences, the effect of the length of pendula on their swinging, computations with partial differential equations having two variables, the interpolation and summation of infinite series, Musschenbroek's law of magnetic attraction, Euler's manuscript for *Scientia navalis*, and the influence of the sun and the moon on tides. Euler found Newton's explanation for the reflection and refraction of light to be unclear; in December 1738 he completed a manuscript in Latin on astronomical refraction and sent it to Delisle. Proceeding from a mass of astronomical observations, Euler developed formulas and checked data to construct tables of refraction; he enclosed a copy of a table by Bouguer that differed in places from his own observational findings. In May 1739 Delisle returned the manuscript, having found errors in it and stating that improvements needed to be made before publication. Euler probably revised the paper, but it remained unpublished, perhaps amid the unfolding turmoil in Russia. He soon proposed an alternative pulse theory in optics. Daniel Bernoulli observed that in their correspondence Euler preferred writing in German, remaining most comfortable in his native tongue.[111] But Euler also wrote in Latin and soon, like Bernoulli, extensively employed French.

In the years 1736–40, Euler mourned the deaths of three small daughters, suffered from a serious health problem, and was troubled by the billeting of Russian soldiers on his property. Nonetheless, his achievements in research and writing continued uninterrupted—and, in fact, accelerated.

Euler and his wife Katharina adhered to the eighteenth-century Christian dictum that the chief purpose of marriage was to multiply; in Saint Petersburg, they had a child every year or two. They did not, however, escape the sorrow of high infant and child mortality of the time. The three children born after Johann Albrecht died quickly: Anna Margaretha, born in June 1736, survived less than a month; at almost two years of age Maria Gertrud passed away in May 1739; and Anna Elizabeth died two weeks after her birth in November 1739. During the two years of Euler's worst fevers no children were born, and over the years five children survived.

In 1736 Paul Euler attempted to draw his son Leonhard home to Switzerland for a visit at which he was to be named godfather to the child of his sister Marie Magdalena. Marie, married to pastor Johann Jacob Nörbel, had grown up with Leonhard and the two were close. Unable to return, Euler allowed another to represent him. His father proudly entered his son's name in the baptismal register as "[Herr Magister] Leonhard Euler, Prof. Math. Sublimioris of the Petersburg Imperial Academy."[112]

In 1737 the painter and drawing instructor at the Saint Petersburg art gymnasium, Johann Georg Brucker, who came from Holstein and was probably a cousin, completed the oldest known portrait of Euler. The following March Daniel Bernoulli reported that he had received the very good portraits of Euler and his wife.[113] The portraits, which were intended for Euler's parents, are now both lost, and what happened to them is not known. One unlikely conjecture, however, is that Katharina's father, then in his seventies, had painted them rather than Brucker. An engraving by twenty-year-old Vassili Sokolow that was rediscovered in the mid-twentieth century in Leningrad preserved the portrait of Euler; the engraving is based on Brucker's painting, and depicts a happy, confident, and seemingly playful individual. If it was completed after 15 April 1737,

Figure 4.7. Earliest known portrait of Euler, 1737, from the mezzotint copy by Vassili Sokolow of the lost portrait by Johann Georg Brucker.

Euler would have been thirty years old, and no major problem with his eyes was yet apparent in the engraving, but that was soon to change.

In late summer of 1738 Euler experienced his second major health crisis since moving to Saint Petersburg. A dangerously high fever and strong infection caused an abscess in his right eye. It was probably inadequate hygiene that worsened the condition and damaged his vision. Fuss's "Lobrede auf Herrn von Leonhard Euler" (Eulogy in memory of Leonhard Euler, 1783) incorrectly attributes the fever to three days of intense astronomical calculations to meet a deadline in computing the latitude of Saint Petersburg, which actually occurred in 1735. An abscess brought on by the fever of 1738 alone, the "Lobrede" reports, led to the total loss of sight in Euler's right eye and began a steady course of problems leading to his near blindness by 1767.[114] The notion of a constant deterioration in Euler's sight from 1738 via a three-stage pathogenesis, in which excessive eye strain brought on a fever that led to blindness, is also doubtful. More likely the process transpired in four stages, typical in high stress cases involving eyesight, in which great stress leads to a high fever, which together with an infection produces an eye abscess that contributes to the evolution of blindness; there are nonetheless occasional partial remissions in one's failing sight. Euler's dangerous fevers, starting in 1735, support this interpretation. His failure to tell his parents and correspondents of the course of his illness perhaps came of a desire not to reveal that convalescence from these fevers required lengthy, persistent nursing.

Assuming that the depiction of his eyes has not been altered, two portraits of Euler provide valuable information on his medical history. The Brucker portrait in 1737 displays no serious eye malady. Euler's loss of vision in the right eye thus must have come later. In a pastel portrait from 1753, his Basler colleague Emanuel Handmann gave close attention to the eyes, detailing problems with the upper lid and a condition of strabismus in the right eye. That the right eye does not appear totally blind suggests a partial, temporary remission of some vision loss. A strong dark pupil in the left eye indicates that it was sound. A subsequent cataract, rather than a constant deterioration beginning in 1738, must have most harmed that eye.

In the eighteenth century, European monarchs drew power to themselves as they centralized their states and created standing armies, which generally contained an element of social misfits, criminals, and vagrants. Imperial Russia, which was extremely agricultural and suffered from high rates of illiteracy, raised most of its soldiers by conscripting serfs from rural areas. Rising military costs from wars against Sweden, France, and the Ottoman Empire, combined with chronic financial shortages in the imperial treasury, meant that funds for sufficient barracks for the growing army in Russia did not exist; this resulted in the billeting of many soldiers in private homes. In 1738 eight soldiers stayed on the Euler property; they apparently were quartered not in his house but in a large shed with windows and a stove in his yard. The soldiers could be crude, rowdy, and dishonest, and their heavy smoking discolored the building's interior. For Euler, his wife, their four-year-old son Johann Albrecht, a daughter less than two years old, and Euler's brother Heinrich, this was an imposition. Euler strongly opposed billeting and went repeatedly to the imperial government to complain, but to no avail. Freedom from billeting was a term that he later negotiated with the officials of Frederick II in Berlin.

Chapter 5

Life Becomes Rather Dangerous: 1740 to August 1741

In 1740 Leonhard Euler's accomplishments were growing and his reputation increasing, especially in France, where his *Mechanica sive motus scientia analytice exposita* (Mechanics of the science of motion set forth analytically) received a highly favorable review in the influential Parisian Jesuit journal *Mémoires de Trévoux*, and he won his third consecutive annual Prix de Paris. The reviewer credited Euler's *Mechanica* with founding modern mechanics and declared that the older mechanics, to which Galileo Galilei had contributed so much, was only statics. The latter part of this assertion is odd, for Galileo had pioneered the mathematical study of the motion of bodies or kinematics. The reviewer also praised Euler for removing the difficult geometric impediments and abstruse calculations from Isaac Newton's *Principia mathematica*. While Euler was already well known in the Paris Academy of Sciences, this review brought him for the first time to the attention of the broader Parisian scientific community. The *Mechanica* and winning the Prix de Paris were essential elements in catapulting Euler into the first rank of Europe's geometers, and the *Mechanica*'s analytical approach has dominated in physics ever since.

Another Paris Prize, a Textbook, and Book Sales

Along with three others—Daniel Bernoulli, the Jesuit mathematician Antoine Cavalleri, and Colin Maclaurin—Euler won the Prix de Paris in 1740 for his paper "Inquisitio physicia in causam fluxus et refluxus maris" (An investigation of the physical causes of the ebb and flow of the sea, E57). Rarely had the Paris Academy seen such brilliant competition. Originally Euler's submission was lost in the post, and when Antiokh Kantemir in Paris and the Petersburg Academy of Sciences learned this, they successfully petitioned the Royal Academy of Sciences in Paris to accept a copy from Saint Petersburg.[1] In a letter in March 1740, Euler thanked Kantemir for making sure that his paper was not late for the competition.[2]

For over a decade the Petersburg academicians, including Euler, had been examining the tides. Initially Euler studied the Cartesian concept that pressure from the lunar whirlwind caused the tides but rejected it and the entire Cartesian lunar theory for predicting low water levels at times when the water was generally high. Instead he accepted Johannes Kepler's idea of an attractive force. Continuing his own work on the tides, Euler completed an article on observing their ebb and flow that appeared, without anything indicating the author, as parts 9 and 10 for the *Primechaniya k Vedomostyam* (Notes to the gazette, 1740), printed by the academy.[3] These also appeared in Russian and in German as *Anmerkungen über die Zeitungen*. While Bernoulli's Prix de Paris submission followed strictly Newtonian principles, Euler's began with a theory of vortices and explained the tides with the sea as an equilibrium figure and the water level as depending on the horizontal disturbing force. He explained the tides as influenced solely by the force of the moon and sun on the sea, with the force of the moon being four times that of the sun. Euler's theory of tides brought together the fundamental dynamics of the motion of the tides, their inertia, and alterations in them produced by these two forces acting as oscillations. As he developed imposing integrations to calculate these forces he initially considered water inertia to be inconsequential but had to include it to explain reciprocal motion. Euler's theory of tides moved beyond the earlier Newtonian theory of statics. While differing in approach, Euler and Bernoulli reached the same conclusions; both men likely began with precise results and worked backward to find explanations for them. The two also agreed on a few points, including those bearing on the case of tides located under the polar ice caps. Euler's system accurately predicted the tides. In the controversy between Newtonians and Cartesians at the Paris Academy, the year 1740 represented a turning point: afterward no Cartesian paper won the annual Prix de Paris, though a modest debate continued there for another decade.

Bernoulli, who in correcting Newton put at 2:5 the ratio of the forces of the sun and moon on the ebb and flow of tides, chided Euler in a letter in April 1740 for still questioning the extent to which Newton's attraction accounted for celestial motions. Studies of the pendulum and the shape of Earth, he argued, made Newton's superiority clear. In June, to confirm the remarkable precision of Newton's dynamics, Bernoulli invoked a range of authorities: Newton, Alexis Claude Clairaut, and Colin Maclaurin. He also cited data from the recent Lapland Expedition and a comparison of the measure of a degree of meridian from the polar circle with those from Paris and Amiens accurately made by Jean Picard in 1670 and Dominic Cassini in 1701. That comparison supported the Newtonian position that

Earth is flattened at the poles. In September Euler responded that he was pursuing flood theory, which was crucial in Saint Petersburg, which often flooded; he pointed out the contrary judgment of Jacques Cassini on the shape of Earth based on conflicting data and also mentioned his own developing eye problem.

During his hectic schedule at the Petersburg Academy in 1740, Euler signed a contract to supply his manuscript on shipbuilding for a stipend of twelve hundred rubles, but the money was not forthcoming. For the geography section, Euler continued to pore over maps; he also computed more astronomical tables and ephemerides and called for new telescopic observations made with the latest instruments. There was not yet a research team with assistants to aid Euler, but he worked with Gottfried Wilhelm Heinsius in astronomy. At a meeting of the academy in March 1740 he presented an article for volume 12 of the *Commentarii* titled "Emendatione tabularum astronomicum per loca planetarum geocentrica" (Emendation to astronomical tables for locating the geocenters of planets, E131), which would be published in 1750.[4] The article examines the difficulties in computing planetary orbits precisely; Euler asserted that three observations can provide only a first approximation of them; for an exact accounting of entire orbits, he knew that a greater number of observations covering daily positions were required. He declared the heliocentric system completely established and the pursuit of the geocentric thus useless. Orthodox clerics still opposed Copernican astronomy; they accepted the word *geliotsentricheskiy* (heliocentric) as an addition to the Russian language but criticized its meaning. The heliocentric system still needed theoretical additions; these would be vital for resolving anomalies arising from such matters as lunar motion and the orbit of Mercury.

During 1740 the second part of Euler's *Einleitung zur Rechen-kunst* (Introduction to arithmetic—art of computing, E17) was also published, and the first part was translated into Russian by Vasilii Adodurov, the only Russian made a member of the early academy, having entered as an adjunct in mathematics in 1733. The second part of the *Einleitung* covers addition and subtraction; multiplication and division for currency conversion; computation of wills; distribution of pensions monthly or yearly; comparisons of apothecary weights from a list of those employed in more than fifty European cities; and appreciation of time divisions—all matters of interest to young nobles. Euler gave a long set of currencies from Russia and sixteen other European realms, converted these through both multiplication and division, and reduced answers to their simplest form. The Russian currency, for example, had a *ruble* (equal to one hundred kopeks), a *pultin* (fifty kopecks), and a *griwen* (ten kopeks). The English had pounds, shillings, pence, and farthings, and there were the Prussian *Thaler*, *Groschen*, and *Dreyer*. In one problem Euler by three divisions and additions simplified the sum of 511 rubles, 926 griwen, 1,732 kopeks, and 53 polushks to 621 rubles, 0 griwen, and 5 kopeks.[5] Another problem asked how many minutes there are in four weeks, five days, fourteen hours, and thirty-six minutes; Euler showed how to compute the answer: 48,936 minutes. Given that the Julian year was 365 days, 5 hours, 48 minutes, 57 seconds, and 12 milliseconds, another problem asked what number divided into the year would give exactly one day. He went through the steps in finding the dividend of 365 + 52,343/216,000.[6] His showing of dexterous procedures or tricks—for example, for multiplying by 5/24 by substituting (1/3 − 1/8), which requires dividing by 3 and 8 and subtracting—suggests a move to a higher level of arithmetic.

Having completed the *Tentamen novae theoriae musicae ex certissimis harmoniae principiis dilucide exposita* (An attempt at a new theory of music composition exposed in full clarity according to the most certain principles of harmony, E33) and the *Einleitung zur Rechen-kunst*, Euler began in about 1740 to write *Methodus inveniendi lineas curvas maximi minimive proprietate gaudentes* (A method of finding curves that show some property of maximum or minimum, E65), which concluded a second stage of his research in founding a semigeometric calculus of variations.[7] The exact dating of the second stage of this research is unclear. Probably Euler's fourth and last paper in Saint Petersburg on the subject, which appeared in print in 1741, was written about 1736.[8] As was noted in chapter 3, a formative stage of his work on the calculus of variations had occurred in the early 1730s. Euler's discovery of his mistake with integrand variables in this and the previous paper apparently set off a period of intense research on variational mathematics that produced greatly improved methods. As his correspondence with Clairaut and a letter of March 1746 to Pierre Maupertuis noted, he had completed the entire manuscript (except for its two appendixes on elasticity) by the spring of 1741, but the book was not published until 1744.

In September 1740 Clairaut began corresponding with Euler. They were to exchange sixty-one letters that were friendly, respectful, and without disagreement. This set of letters is one of the most important sources for the history of science in the eighteenth century as well as the research of the two men. In 1733, seven years earlier, Daniel Bernoulli had first brought Euler's attention to the precocious French genius Clairaut; Euler followed his achievements and in 1735 recommended Clairaut for a post at the Petersburg Academy. At the start, their letters were to deal with topics in infinitary analysis. In 1739 Bernoulli forwarded to Euler Clairaut's request for difficult problems to test his approach to integrating differential equations and his comments on the theorems of Alexis Fontaine on differentiation as well as an inquiry into where Euler had published on these matters. Clairaut sent a letter in September 1740, and Euler responded a month later, connecting the questions raised on his integrations with prior results of Nikolaus Bernoulli and Jakob Hermann on orthogonal trajectories, noted that the second volume of his *Mechanica* contained difficult integrations, and praised Clairaut's work on the shape of Earth. Upon receiving Euler's reply, Clairaut quickly generalized to multiple variable analogues his integration of differential expressions. In a paper for the Paris Academy of Sciences in 1740, he asserted that their correspondence had made the works of Euler better known to him.

In addition to engaging in his own prolific research, Euler was active in most operations of the Petersburg Academy, even ordering paper and ink for the printing press and, from at least the time he became a full professor, assisting in editing.

Health, Interregnum Dangers, and Prussian Negotiations

Not all was well in Saint Petersburg in 1740. Because the astronomical observations Louis de l'Isle de la Croyère had sent from the Second Kamchatka Expedition were nearly worthless, early that year Joseph-Nicholas Delisle traveled to Siberia to join his brother and make his own observations. He left all of the work of the geography department in Euler's and Heinsius's hands. In July the imperial senate ordered the department to establish the reliability of its map materials and to give the earliest possible date it could expect a general map of Russia. Heinsius, especially, felt that more geodetic measurements were necessary, and Euler immersed himself in this huge task. The work was extremely difficult for him; in an August 1740 letter to Christian Goldbach, the now presbyopic Euler offered a powerful health reason for modifying this assignment. Lapsing from his typical Latin into German, he claimed that more than his reading and writing his painstaking labors on land maps had strained his eyes and were worsening his eyesight problem, which he traced back as beginning in the summer of 1738. "Geography is fatal to me," he wrote. "You know that I have lost an eye and [the other] currently may be in some danger. When a new batch of charts was sent to me this morning to examine, I felt a mild pain."[9] It seems that Euler also had trouble declining new official assignments, and that he wanted his friend Goldbach to intercede to save him from having more cartographic work. Alarmed, Goldbach responded the same day. In a letter of September 1740 Euler reported to Philipp Naudé in Berlin that his right eye was almost unusable and his left eye was deteriorating.

Through the summer of 1740, Euler's family was happily settled in Saint Petersburg. His brother Johann Heinrich continued to live with him and studied art and painting, and a second son, Karl Johann, was born to Leonhard and Katharina on July 15. Though he would always treasure the solid bourgeois, essentially republican culture of his native Basel, Euler felt that good fortune had brought him to imperial Russia, where an absolutist rule provided him a home for his family and freedom and the facilities in which to pursue his research.

Even in the midst of Euler's health problems, he was strongly recruited by the Prussian monarch Frederick II, a twenty-eight-year-old who had ascended the Prussian throne at the end of May 1740. The invitation was extended as part of the royal court's effort to turn the nearly moribund Royal Prussian Society of Sciences into a leading scientific center. In a letter of 14 June, Frederick instructed his close friend and ambassador in Saint Petersburg, the Saxon count Ulrich Friedrich von Suhm, to offer the thirty-three-year-old Euler a position in a renovated and reinvigorated academy of sciences and belles-lettres in Berlin. Suhm—who had translated writings of Christian Wolff into French and had helped Frederick as crown prince to understand them—was a strong advocate of a new science academy in Berlin. Probably from Jean-Henri-Samuel Formey's circle, the Paris Academy prizes, and information from Naudé, the Prussian Society's number theorist, the monarch knew of Euler's European reputation. Even though he referred to Euler as "the grand algebraist," Frederick offered him only a relatively modest salary of 1,000 to 1,200 Reichsthaler, while extending 1,600 and later 3,000 to Maupertuis; subsequently the monarch proposed 12,000 Reichsthaler for Jean-Baptiste le Rond d'Alembert and 20,000 for Voltaire.[10]

Since the Petersburg Academy wanted to pay Euler twelve hundred rubles, the equivalent of sixteen hundred Reichsthaler, he declined to accept the Prussian position.[11] Count Suhm was upset that the commoner Euler wanted to make a counterproposal, but on 15 July Frederick ordered him to continue the discussion.[12] Suhm, who was of fragile health, agreed to convey Euler's request for a salary matching the sixteen hundred Reichsthaler and up to five hundred more for moving expenses, but he died en route to Berlin in November. This salary was much higher than the average for junior members of the Brandenburg-Prussian Society, around three hundred Reichsthaler. When Johann I (Jean I)

Bernoulli—who received 262 Basel pounds annually in addition to a supply of wine and grain—learned the amount of Euler's pension he called it "a truly royal salary."[13]

Beyond the pension request, Euler must have been reluctant to leave a secure position in Saint Petersburg for an unsure post in a kingdom ruled by an untested monarch who had ascended to the Brandenburg-Prussian throne only two months earlier. But in October 1740, soon after Frederick's bargaining began, the Russian empress Anna died; in the tumultuous factional struggle for control of the government, Euler's more tranquil existence in Saint Petersburg came under threat, and life in Russia grew dangerous.

Shortly before her death, Anna had adopted her great nephew, the two-month-old grand duke, Ivan VI, and named him her successor; Anna's chief minister, Ernst Johann von Biron, was to be regent. Biron, the most hated man in Russia because of his cruelty, so bullied Anna and her husband that they considered fleeing the country. After Anna confided this plan to the field marshal and count Burkhardt Christoph von Münnich, he conspired with others to remove Biron and earn the gratitude of the new czar's mother, Anna Leopoldovna, who was from Braunschweig. General Christoph Manstein and the palace regiment participated in a late nighttime arrest of Biron at his house, and the following coup d'etat ousted the very powerful minister of foreign affairs, Count Andrei Ostermann, and made Anna Leopoldovna regent.[14] Refusing to have Biron tortured or put on trial with possible quartering, she banished him to Siberia.

By the end of November, Anna Leopoldovna feared a plot against her. The Russian nobles considered her weak and asked Elizabeth, the young, attractive, and charming but disorganized daughter of Peter the Great, to take over. Amid the turmoil the German noble courtiers and advisers, whom Empress Anna had gathered to her throne, fell from power, and the time of a terrible foreign domination in Russia was ending. The capital seethed with hostility toward Germans and toward other foreigners.

In addition to being a prominent center for the sciences, largely on the strength of Euler's accomplishments, the Petersburg Academy was given several duties by the imperial court; its members were tasked with composing odes for state festivities, preparing fireworks, and forecasting the weather. Euler received the ticklish assignment of casting a horoscope for Ivan VI, who was deposed within a year and shortly thereafter imprisoned despite being still an infant. Since this was an astronomical and astrological assignment, Euler passed it on to his friend Krafft, who during Empress Anna's court had been well known for preparing horoscopes.

One matter that almost all academicians had strictly avoided was becoming involved in imperial court politics; Euler knew the peril, and the menace arising from the existing political maelstrom was a major factor in his relocating to Berlin.

In his brief, unfinished autobiography and a letter of July 1763 to Gerhard Friedrich Müller, when the discussion of his return to Russia was well under way, Euler indicated his reasons for leaving Saint Petersburg after residing there for fourteen years. In 1740 life became awkward and unsettled there; for years Euler's wife Katharina could not think of that time without recalling fear. The death of her father in 1740 also severed for Katharina a tie with the city. On 15 February 1741, apparently partly at her urging, Euler met with the new Prussian ambassador, the aging Baron Axel von Mardefeld, to inquire about his counterproposal to Suhm. Mardefeld informed him that the king had received and accepted it, including the payment of travel costs; Euler thus agreed to move to Berlin. In March he announced to his friend Müller, who was in the Siberian town of Tobolsk, his decision to go to Prussia and enter what was becoming the renewed and reorganized Royal Prussian Academy of Sciences and Belles-Lettres. He noted that of the foreigners invited, only Maupertuis had accepted, while Wolff would return to the University of Halle. In another letter in April, Euler revealed the urgency of his decision, confessing that he had thought nothing less likely than that he would have to establish himself in another place.[15]

Euler quickly informed his parents in Riehen and the Bernoullis in Basel about his planned move to Berlin. In letters dated 28 January and 20 September 1741 Daniel Bernoulli, Euler's closest friend, complimented him for his appointment in Berlin and noted that, along with their father Johann I, Daniel and Johann II (Jean II) had also been invited to Berlin but they had misgivings; Johann II probably wanted to accept, but the situation was not yet resolved.[16] It is possible that Frederick II wanted first to draw Euler to Berlin as a magnet to recruit the Bernoullis; but Daniel did not want to accept a position there, for with Euler as a member the academy already had the highest level of mathematician. He added that he feared that the First Silesian War could stop the whole project. On September 20, he wrote that "dass die Wissenschaften und der Krieg incompatibel mit einander seyen" (it seems that the sciences and war are incompatible).[17] The First Silesian War had taken Frederick's attention away from the academy. Daniel stated that Johann I did not know what his sons would decide but hoped that as soon as the war, which was being fought under the young and pregnant Maria Theresa, neared its end, they would move to Berlin. That was not to occur. In April 1743 Daniel Bernoulli continued his response. He accepted that "in einer Akademie muss einigermassen eine Subordination seyn, als wie in dem Militärstande" (in an academy one must to some extent be a subordinate, as in the military).[18] He thought it best to have a small number of scholars of superior ability direct others to find practical uses for new knowledge and counted himself among those spirits. But Frederick was to be engaged in war and would not have time for the academy.

Maupertuis, who was located near Paris, apparently did not know Daniel's brother Johann II, and Daniel would remain in Basel. The response of the Bernoulli family to Euler's new Prussian position, then, was mixed, but Euler seems to have disregarded the warnings of the younger Bernoullis. Berlin, with a population that by 1740 reached 70,000, had an energetic French Huguenot colony that helped make it a social oasis from the military; Euler must have deemed it a safer place to live than Saint Petersburg.

Writing in Latin, Euler's seventy-three-year-old teacher Johann I Bernoulli composed a letter dated 18 February 1740 that, for the uncertainty of knowing the address in Saint Petersburg, was not sent until 1 September 1741, when it was instead forwarded to Berlin. At this time Bernoulli added a postscript: "With joy have I heard that you have been invited in the name of the Prussian king to organize a new academy in Berlin. May God assist you in your enterprise and be with you on your journey. . . . [I]f I were twenty years younger, I would join you in the blink of an eye." He was pleased that Euler would now be much closer to Basel, and he greatly hoped that Euler would make a trip to his home city to visit his parents. This would provide Bernoulli the opportunity to see Euler once more before he died, and he declared it his most ardent desire: "Nihil jam reliquum est hac vice, quam ut Te, Amice exoptatissime, quamvis absentem, animo exosculer, donec id, si Superis placet, coram facere possim. Vale, iterumque vale." (Nothing now remains, most longed for friend, other than to kiss you, although absent, in spirit, until if it pleases the higher powers, I can do it in your presence. Farewell, again and again farewell.)[19] In another letter in October he added to advanced age his asthma, gout, and responsibilities to the University of Basel as reasons for staying put;[20] Bernoulli still held out hope that Euler would succeed him at the University of Basel.

Euler had first to obtain release from his Russian contract through the academy's director, Johann Schumacher, and opposition from Schumacher, followed by Euler's illness, delayed Euler's leaving Saint Petersburg. He kept an almost daily journal from 15 February to 10 March of his stubborn efforts to gain permission to depart. Schumacher responded that he was under contract and could not leave Russia for a year. Euler explained his situation to the powerful Mardefeld, who in a note to fellow noble Andrei Ostermann and a visit to Karl von Brevern stressed Frederick's desire. In February and March Euler wrote three letters to Brevern. In the first two letters, of 23 February and 5 March, Euler claimed that the difficult climate in Saint Petersburg might ruin his health and that his geography tasks could destroy what was left of his eyesight. The second letter also expressed his great satisfaction to be in Brevern's service at the Petersburg Academy of Sciences, but repeated that the harsh climate made it pressing that he relocate to a milder region. In Berlin as well as Saint Petersburg, Euler's letters raised some alarm. He considered but rejected a proposal that he agree to return to Russia in a year, a stipulation that he thought cowardly. In a letter to Müller in early March he stated that while Frederick II was engaged in the conquest of Silesia, "I await now daily the latest orders of His Majesty and [the acceptance of] my resignation in order to undertake my trip."[21] After Euler received no answer, he again went directly to Brevern, who on 10 March assured him of the passage of a cabinet resolution granting his release. In May Brevern succeeded Korff as academy president. Ultimately Euler left with the good wishes of the Petersburg Academy, and he pledged to continue to collaborate as an honorary member with an annual salary of 200 rubles (about 270 Reichsthaler).

In June Euler asked Schumacher, apparently unsuccessfully, not to hold back part of the money agreed to in his last contract, and in September he reviewed from Berlin the final terms and promises in his negotiations with Schumacher, requesting payment of the promised annual salary due him as an honorary member. Such a stipend was not simply a matter of altruism: Euler was to purchase books and scientific instruments for the academy, write reports on the sciences in western and central Europe, and tutor Russian students who resided with the Euler family, besides submitting a stream of papers for the *Commentarii*, for which he had become the de facto editor.

In early June 1741 Euler and his wife packed the family belongings. On 19 June the Euler family—Leonhard, Katharina, their sons Johann Albrecht and Karl, and Leonhard's brother, Johann Heinrich—boarded a ship to begin their journey from Saint Petersburg to Berlin; Euler left with deep regrets. After a rough, three-week passage on the Baltic, the family arrived in Stettin; in a letter to Goldbach, Euler announced that he was the only family member who had not suffered from seasickness.[22] Stettin conferred various honors on Euler, and there he also attended a gymnasium disputation. Using a carriage and a freight wagon, the Eulers left Stettin on 22 July, arriving in Berlin three days later.

In a letter in September Euler thanked Schumacher for his friendship, and a letter from October mentioned his work for the academy as well as his joy at news of the Russian victory over Sweden and the sale of his Russian house; he meticulously noted that the house must be repaired before winter. His August letter to Goldbach had noted that while the chancellery architect estimated its value at four hundred rubles, Schumacher had paid only a paltry hundred rubles for it.[23] Now Euler was settled in Berlin, and ready to begin the work of the new Royal Prussian Academy.

Chapter 6

A Call to Berlin: August 1741 to 1744

"Ex Oriente Lux": Toward a Frederician Era for the Sciences

Long before he ascended the Prussian throne on 31 May 1740, the future Frederick II wanted to found a vibrant royal science academy in Berlin comparable to institutions in Paris, London, and Saint Petersburg. He saw it as an instrument for building industry and the power of his state and for bringing intellectual distinction to Berlin. A letter of 6 July 1737 from Frederick to Voltaire speaks of Louis XIV's establishing the Paris Academy of Sciences, gathering astronomers pledged to discover stars, and attracting innovative botanists and physicians.[1] The Paris Academy and the Académie française would serve as Frederick's guides. In the early stages of correspondence between Voltaire and the young crown prince—which would last forty years—the sciences were a continuing topic. When Frederick first wrote in August 1736, Voltaire was forty-two and famous; Frederick was twenty-four. Voltaire's letter of 20 August to Frederick mentioned the research of Isaac Newton, Gottfried Wilhelm Leibniz, Robert Boyle, John Locke, and their partisans. That December Frederick sent as a gift a treatise by Christian Wolff, whom he called "the most celebrated philosopher of our day." Voltaire considered the treatise nonsense and did not study it, but his paramour Émilie, Marquise du Châtelet did, wanting to learn more of Leibnizian and Wolffian philosophies. Voltaire was becoming a Continental champion of Newtonian science. Regarding the Frenchman's nearly completed popular exposition *Éléments de la philosophie de Newton* the Prussian monarch wrote in December 1736 and again in March 1737 that he awaited the book "with great impatience"; it was published in 1738. Frederick's eagerness to make a name for himself and his country in the sciences was manifest in his letter to Voltaire in July 1737, which reviewed the history of the Royal Prussian Society of Sciences in Berlin from its establishment on Sunday, 11 July 1700.

Building upon the model of the Paris Academy, Leibniz had planned the Prussian society and received financial support from the well-educated patroness Queen Sophie Charlotte, whom he had known in Hanover. Another founder of the society was Daniel Ernst Jablonski, the ecumenical Calvinist court preacher. The society consisted of four classes: physics, medicine, and chemistry; mathematics, astronomy, and mechanics; German language and history; and literature, stressing oriental writings for missionaries spreading the Gospel to the east. Under Leibniz, its name was to change from the Electoral Brandenburg Society of Sciences to the Royal Prussian Society of Sciences; this followed the new higher status of the realm in moving from an electorate to a kingdom.

Frederick's correspondence does not bring out the importance of the society's observatory, which he planned to replace.[2] The effort of the Corpus Evangelicorum to improve the calendar in Protestant lands,[3] particularly by abolishing the Julian model, argued for its quick building, and records gathered from observatories in France, England, and China aided the calendric reform project. Pastors and astronomers were also eager to confront the claims of astrology. To fund the society as a whole, moreover, the Prussian court had followed the earlier suggestion of Leibniz by allocating all of the revenue from the privilege of the royal Prussian calendar monopoly, though that proved inadequate.

Uncertain financing and the lack of support outside the French community in Berlin, then a town of thirty thousand, delayed the inauguration of the society until 1711. Under the military king Frederick William I, who ruled from 1711 to 1740, it declined, for he did not protect it and in fact disdained it; in 1731 he named the royal jester at court its vice president. Royal contempt, refusal to provide funding sufficient for the maintenance of the observatory and hops garden, and an absence of the freedom enjoyed by the Paris Academy meant that except for adding a medical and surgical section in 1724 the Prussian society languished, and save for its research on phosphorus it was nearly moribund by the late 1730s, as Frederick II would later lament to Voltaire.[4] Under Frederick William I, for example, the forty articles in the last two volumes of the society's occasional journal *Miscellanea Berolinensia*, in 1737 and 1740, included fourteen on the mathematical sciences, and these were not significant.

The original intent of the Royal Prussian Society of Sciences had been to make Berlin the capital of Protestant science and, in northern Europe, of historical apologetics. On 6 June 1740 the new monarch, Frederick II, ordered the Prussian state minister Adam Otto von Viereck, the society's protector or chief official, to prepare a report on its condition. On 9 June, Viereck responded that the underfunded and poorly managed society had not achieved European distinction, and two days later Frederick broke the constraints on the institution. He resolved to end the odious reference to its members as "collected royal buffoons" that Frederick William I had employed; he restored to Philipp Naudé, professor of mathematics, an annuity of two hundred Reichsthaler; and he extended his favor and protection to the society's members.[5] The king urged these members to be more active and proclaimed that an era in learning was about to begin in Berlin.

Frederick envisioned a rejuvenated, reorganized, and expanded academy of sciences and liberal arts as the hub of the cultural and intellectual community in the city. It was essential to making Berlin what he dreamed it could be, an Athens on the Spree River.[6] In the sciences, Frederick wanted utility to have precedence over theory. The leading members of the old society, such as the philologist and zoologist Johann Leonhard Frisch; the French theologian, chronologist, and mathematician Alphonse des Vignoles; and its theologian founder Jablonski, its president since 1731, were elderly, and Frederick set out to recruit scholarly celebrities from outside Prussia. As possible sources, he and the high nobility rejected scholastic thought as pedantic and Protestant theology as dated. In north German states the most dynamic element in intellectual life was Wolffian philosophy, not Cartesian mechanical thought or the English critical empiricism of Boyle, Locke, and Newton. The French high culture of the Parisian salons, debating the merits of the Cartesian and Newtonian sciences, also influenced the high nobility and the French community in Berlin. In Paris the advocates of each science took up specific components but not necessarily the complete system. Later celebrated as a hero of German nationalism, the king was both Francophile and Francophone, and awkward in the language of his own country, and he searched among the French for academicians.

The recruitment of scholars of the highest caliber to conduct important pure and applied research and to lead the new science academy, all with the idea of gaining prestige for Frederick's court, began with the selection of a president. Ideally he wanted the president to be French and of noble descent; he should possess experience in mobilizing resources and carrying out major projects in the sciences. Frederick wished the president to be a lively conversationalist who mixed easily in noble company and who could be a close friend. But above all, the president was to have a distinguished reputation for scholarship that extended beyond the scientific community. Apparently, the monarch's first choice for the presidency was not a noble but Voltaire—based on his fame.[7] Up to 1740, however, Voltaire rebuffed invitations from Frederick to visit Berlin. The next French savant who met all of Frederick's criteria was Pierre Maupertuis.

After studying in Basel under the aging Johann I (Jean I) Bernoulli in 1729–30 and becoming a protégé, the nobleman Maupertuis had first gained attention in Paris with his *Discours sur les differentes figures des astres avec une exposition des systems de MM. Descartes et Newton* (Discourse on the various shapes of stars, with an exposition of the systems of Mssrs. Descartes and Newton), which was published in Paris in 1732.[8] Maupertuis did not join with Jacques Cassini, Bernard le Bovier Fontenelle, and Jean-Jacques d'Ortous de Mairan in opposing Newton's concept of action at a distance over empty space. Nor did he accept Bernoulli's attempt to develop a synthesis that held the inverse-square law to be explainable by fluid mechanics and thus to agree with the theory of vortices. Although professing to be an agnostic regarding Newton's theory of attraction, Maupertuis compared it favorably to the Cartesian notion of vortices, which he, like Newton, found incompatible with Johannes Kepler's laws of planetary motion. Maupertuis was the first to discover a case of the operation of attraction outside the solar system between two rotating stars. In Paris he became the center of the spirited circle at the Café Gradot, fostering connections among savants to converse with in the absence of hierarchical aristocratic constraints. In the 1720s there were 380 cafés in Paris; some of them were known for meeting circles that were exclusively male, but Maupertuis insisted that Émilie du Châtelet be allowed to join the discussion. To strengthen his position in the republic of letters he actively participated in salons, such as that of Madame Marie-Thérèse Rodet Geoffrin. In 1735 the Royal Prussian Society elected him an *associé étranger* (foreign associate). But Maupertuis had little reason to be connected with a society not among the first rank of scientific institutions, and he did not respond until after he returned in 1737 from the famous Lapland Expedition.

From 1736 on, Frederick apparently had another criterion for the society: he wanted a copresidency, and hoped that Wolff would be the other officeholder. In May 1740 Frederick, having read Wolff in French translations, declared him "the teacher of princes,"[9] and on 6 June he got the provost of the University of Halle, pastor Johann Gustav Reinbeck, author of a Wolffian tract, to begin trying to draw Wolff into royal service.

Wolff's expulsion from the University of Halle by Frederick William I in 1723 had made him a cause célèbre. His Latin texts, including *Cosmologia generalis*, published in 1731, and *Theologia naturalis*, published in two volumes from 1736 to 1737, brought him the title *Praeceptor germaniae* among his admirers. The previously dominant Scholastic philosophy of University of Wittenberg professor Philipp Melanchthon had already been weakened in the sixteenth century by criticisms from adherents of Petrus Ramus and in the seventeenth century by attacks launched by the

Cartesians and particularly Samuel von Pufendorf, Christian Thomasius, Walter von Tschirnhaus, and Leibniz. But the works of these four men were insufficient to overthrow Melanchthon's peripatetic thought. By 1740 Wolff had succeeded in dethroning it and providing the German world with a new school of philosophy. The next year the prestigious Académie française named him a foreign member, the first German so honored since Leibniz. Wolff possessed in full the scholarly stature Frederick was seeking in his academicians. In 1736 a royal Prussian commission found no dangerous errors in his writings, and the next year the elderly Frederick William I invited him, at Reinbeck's urging, to return to Halle or Frankfurt,[10] but Wolff rejected the offer. In dedicating the first volume of his *Jus naturae et gentium* (Law of nature and nations) in 1740 to the crown prince, he gave hope that he might return under a new monarch.

On 7 June, a day after the newly crowned Frederick II contacted him, Reinbeck wrote to Wolff asking him to enter royal service. On 15 June Wolff declined to come to Berlin, citing privately to Reinbeck several obstacles. The Berlin winters, he stated, were too cold for him; he had a severe problem with his feet even on cool summer days, and would thus have to remain inside for most of the winter. That the language of the academy would probably be French posed another difficulty; Wolff read French but could not speak it fluently. He resented Frederick's preference for the French, and wondered whether materials in Latin and German, including his own, would get adequate attention. He also disapproved of the king's invitation to Maupertuis, who spoke only French and would represent Newtonian science, and Wolff did not wish to share the presidency; he knew that as a commoner he would not be as close to the king at the palace as would be the French noble.[11] In his open response he cited only the Berlin winters, writing that he would come to Berlin only if the king ordered him. Wolff's larger distaste for Berlin accorded with that of many scholars. The city was a military outpost in provincial Prussia. For the sake of commerce with people of differing faiths it practiced religious toleration, but there was little freedom in the city. Until the end of the Seven Years' War in 1763, few potential Royal Prussian Academy of Sciences members chose to relocate there.

A skilled administrator, Frederick persisted into the summer trying to recruit Wolff. In late June the king had Reinbeck offer a two thousand Reichsthaler salary and restated his intention to have an academy of distinction with a new observatory.[12] Wolff repeated on 29 June his preference for university life and on 29 July added—privately to Reinbeck—his desire not to have to teach and raise military cadets in Berlin, which he considered a demotion.[13] Reinbeck believed that Wolff could still be enticed to

Berlin; he asked the king to comment on the freedom of research in the academy, raise Wolff's salary to three thousand Reichsthaler, which Maupertuis was offered, and to provide up to one thousand Reichsthaler for transportation—particularly for moving Wolff's library and instruments. Frederick, eager to get the academy built quickly, was meanwhile courting Maupertuis. On 14 July the king had written to the French savant of science and former lieutenant in the French musketeers, inviting him to inject the French Enlightenment into Prussian culture.[14] Frederick looked to Maupertuis to create and lead the academy and, rather than vacillate any further, he granted Wolff's request to be reappointed to his old position of professor of philosophy at the University of Halle as well as its vice chancellor and offered him a two thousand Reichsthaler income. On 10 August, Wolff accepted; in December he made a triumphant return to Halle as "a professor for the human race." Soon after arriving in Berlin during the next summer, Leonhard Euler was to become his chief critic.

Amid the negotiations with Wolff, the twenty-eight-year-old Frederick II wrote to Voltaire in June 1740 that among his first royal decisions were to enlarge the army and to "establish the foundations of our new science academy."[15] The military growth was for the sake of *aggrandizement*, the common term at the time for the monarchical business of territorial expansion. Prussia was among the three royal states becoming major European military powers in the eighteenth century, the other two being Russia and Britain. The king wanted to acquire the adjacent imperial Austrian duchy of Silesia, with its 1.5 million people, its richness of natural resources, and its linen industry. Silesia was also desirable for its strategic location, and a conquest he thought crucial to making Prussia a power in Europe. The speed of Frederick's invasion of Silesia in December was to surprise Europe's rulers and diplomats. The republic of letters opposed it but greeted with enthusiasm Frederick's decision to establish a prominent science academy. Voltaire, who had corrected and seen through the press the monarch's treatise *AntiMachiavel*, hailed him as the Marcus Aurelius of the north.

Apparently not anticipating difficulty in recruiting leading scholars, Frederick happily asserted in a letter in June to Voltaire, "I have acquired [the services of] Wolff, Maupertuis, and [the Venetian count Francesco] Algarotti. I am waiting to hear from [the Dutch physicist Willem] 'sGravesande, [the Parisian inventor Jacques de] Vaucanson, and Euler."[16] But his letter was in large part wishful thinking: Wolff preferred to return to Halle, and Maupertuis had consented to visit Berlin to plan the academy but not to remain there. Count Algarotti would occasionally visit Frederick's court in Berlin and later in Potsdam, but rarely interacted in academic affairs with commoners in Berlin. Perhaps Frederick calculated that getting Voltaire to believe the recruitment a fait accompli might tempt the French *philosophe* to join in a triumvirate with Wolff and Maupertuis to head the academy. If so, he badly misjudged Voltaire. 'sGravesande and Vaucanson declined the invitation, as would many others. Only Euler came to Berlin, and the condition in which he found the Royal Prussian Society and the observatory dismayed him.[17]

The attempt to recruit Maupertuis continued. In June 1740, the king extended a flattering invitation, stating, "My heart and my inclination aroused in me, from the moment I assumed the throne, the desire to have you here, so that you should give the Berlin Academy the form that only you can give it. So come and graft the slip of the sciences onto the seedling, that it may flourish."[18] The invitation became the talk of Parisian salons and was published in August by the *Gazette d'Utrecht*. Afterward Frederick met with Maupertuis in Wesel in early September, while Voltaire saw the king near Cleves. Maupertuis was pleased that Voltaire, whom he now considered a competitor for the king's favor, had not yet gone to Berlin. In

Figure 6.1. Pierre-Louis Moreau de Maupertuis.

late September Maupertuis traveled to the royal court in Berlin; Algarotti was also there. In Paris considerable gossip arose over Frederick's role for Maupertuis. Madame Françoise de Graffiny, who saw him frequently, attested that he was in Berlin "to set up an academy." Even so, prospects for it were unclear. Maupertuis wrote to the younger Johann II (Jean II) Bernoulli that he simply "wanted to see what direction the academy would take."[19] Frederick offered Maupertuis a lavish salary of twelve thousand livres. No matter how beneficial the offer, Maupertuis refused to settle permanently in Berlin; he desired only an honorific position with the possibility of returning occasionally. In Berlin Maupertuis met Prussia's minister for state and war, Pomeranian noble Kaspar Wilhelm von Borcke, who was also one of the first translators of Shakespeare into German. In 1745 Maupertuis would marry Borcke's close relative Eleanor, a lady in waiting at court.[20] But Maupertuis found the king directing his primary attention to Silesia; the academy would have to wait.

In November, Voltaire joined Algarotti and Maupertuis in Berlin, but during neither this visit nor any of his others was he active in planning the academy. The king, infatuated with the charming Algarotti, had spent the summer traveling with him; this is likely what led Voltaire, in a letter to Maupertuis, to refer to Frederick as an "amiable whore." As his letters of 15 November and 17 December to the younger Johann Bernoulli show, Maupertuis remained in Berlin but had refused to agree to settle there permanently. Not wanting to cut ties with France, he was negotiating with the French royal ministers and had given his word to return to Paris. On 2 December, Voltaire departed before the invasion of Silesia, which he opposed.

The exchanges by mail between Maupertuis and Euler had begun in May 1738, when Maupertuis sent comments on Euler's *Mechanica sive motus scientia analytice exposita* (Mechanics of the science of motion set forth analytically) and promised to forward his own book on the figure of Earth. In February 1740 Euler thanked Maupertuis for dispatching to him a collection of three Prix de Paris 1738 papers on the nature of fire and two more on the same subject—one by Émilie du Châtelet and the other by Voltaire. He was especially eager to see the work of Madame du Châtelet. In a letter of August 1740, Maupertuis wrote to inform Euler that he had shared the winning position in the Prix de Paris competition on the tides. Euler wrote enthusiastically in October of Maupertuis's data on the shape of Earth and later that year suggested that the two might collaborate in research at the planned Berlin academy. In January 1741, Euler informed Maupertuis that he had received Frederick's invitation to come to Berlin and wanted to accept it.

The Battle of Mollwitz in April 1741 was to eliminate direct collaboration in research for another four years, however. As a former French cavalry officer, Maupertuis had agreed to join Frederick at the front. At Mollwitz in Upper Silesia, 21,600 Prussian troops faced 19,000 Austrians.[21] In a day of battle in heavy snow, the Austrian cavalry charges demoralized the Prussians and drove them from the field. They seemed near defeat, and Frederick was advised to leave the battle to avoid capture. According to one account, he and a small coterie departed to find lodgings for the night in the town of Oppeln, only to learn later that the steady, disciplined Prussian infantry had prevailed.

The Austrians, however, having anticipated his destination, had a squadron of fifty of their hussars waiting for him; when Frederick and his party arrived at Oppeln late in the evening, the town gate was closed. Recognizing a trap, the king turned his horse and escaped before the Austrian hussars could open the gate and give pursuit, but the Austrians were able to overtake Frederick's slower companions. Among them was Maupertuis, whose horse is reported to have bolted. He was stripped of his possessions and clothing and, had an Austrian officer who spoke French not taken pity on him, would probably have been killed. Officers on opposing sides were closer to each other than to the peasants, artisans, and mercenaries of their own armies. The Austrian officer took Maupertuis to the general, Count Wilhelm Reinhard von Neipperg, who recognized him and provided him with a new outfit and gold coins for compensation. Instead of being carried to a Hungarian fortress, he persuaded his captors to take him to the Habsburg court in the imperial capital of Vienna. There he charmed the monarch, Archduchess Maria Theresa, who treated him as an honored guest.

After a short stay in Vienna, Maupertuis returned to Berlin to draw up plans for the renovated academy and asked Frederick's approval for recruiting and other practical matters. Because he was occupied with the war Frederick did not respond, so Maupertuis traveled on to Paris to continue his scientific research, leaving Berlin before Euler arrived and giving no indication of when he might return. In 1742 he was named director of the Royal Academy of Sciences in Paris. Supported by Charles-Louis de Montesquieu and Count Jean Maurepas, he was also elected the next year to be one of the forty immortals of the Académie française;[22] to its strengths in poetry, literature, and rhetoric, he added mathematics. Although Maupertuis encountered acerbic opposition from Jacques Cassini for his Newtonian position on the shape of Earth, he was now at the peak of his career.

Figure 6.2. (Top) Map of Mollwitz and Oppeln. (Bottom) Map of their location in Europe.

The Arrival of the Grand Algebraist

On 25 July 1741, the Eulers arrived in Berlin from Stettin. Apparently Heinrich departed on his previously planned trip to Paris and Italian cities to continue his study of art. The Euler family moved into the Barboness house at the Potsdam Bridge near Unter den Linden. To help Euler make contacts in Berlin, the French ambassador in Saint Petersburg had written three letters of introduction to Prussian nobles, while Joseph-Nicholas Delisle added two—to Johann Wilhelm Wagner, the Prussian royal astronomer, and to Christina Kirch, the daughter of the late astronomer Gottfried Kirch.[23] But the letter that Euler had awaited from the monarch since March did not arrive until 4 September when, engaged in the First Silesian War, Frederick wrote a brief note of welcome from his camp at Reichenbach in Silesia. The king was delighted to know that Euler was pleased to be in Berlin, and acknowledged Euler's importance to his plans for a new science academy, ordering the general directory to pay him the annual salary of sixteen hundred Reichsthaler.[24] Frederick asserted that should Euler still need anything he had only to await the king's return to Berlin, and added that he was looking forward to meeting Euler.[25]

Although the Prussian state minister Philippe Joseph de Jariges, the permanent secretary of the Royal Prussian Society, interceded on Euler's behalf, neither the salary nor the meeting with the monarch was quickly forthcoming. This did not deter Euler's initial enthusiasm to be in Berlin, however; a letter of 9 September to Christian Goldbach requesting plans for the academic buildings in Saint Petersburg reports Frederick's intention to construct a new edifice for his academy.[26] Nonetheless, in December Euler noted that Frederick had postponed its establishment. Through the last quarter of 1741, Euler and his family had to live on credit in Berlin, awaiting his official appointment and first salary payment as well as the promised five hundred Reichsthaler for moving expenses (though the receipts he submitted for moving costs exceeded that amount). Fully engaged in the Silesian Wars and not attending any academy meetings, the king apparently would not meet Euler until September 1749. In a court that valued French wit over intellectual distinction and viewed the purpose of science as its utility to the state, Euler was not popular. Still, in a letter to Voltaire celebrating Euler's arrival, Frederick wrote, "You can see that war has not deadened my taste for the arts."[27]

Also taking note of Euler's coming to Berlin were Voltaire and Maupertuis. Writing to Maupertuis in 1741, Voltaire—perhaps thinking of the mounting war costs facing Prussia—asked, "Is he going to run an academy at a discount?"[28] Maupertuis sent to Euler's friend Johann II Bernoulli a puzzlingly caustic message; regarding the negotiations for the Berlin position, he remarked that "Euler has conducted himself singularly throughout these," but in the strictest confidence he proceeded, "he is an extraordinary and quite annoying person . . . who enjoys getting involved in everything. . . . How can he be such a great mathematician and be so lame in philosophy, where he hardly measures up?"[29] This was the general response to Euler among the French *philosophes* as it would be later from Joseph-Louis, Comte de Lagrange; Euler's commitment to religion and his philosophy, addressing metaphysics and the high sciences, clashed with radical elements of the French Enlightenment.

Nicolas, Marquis de Condorcet's "Éloge de M. Euler" relates an episode usually cited to demonstrate Euler's reticence to engage in the show of court life in Berlin, which very possibly expressed in part a good Swiss bourgeois distaste for it. Probably soon after New Year's Day 1742, Condorcet's anecdote relates, the queen mother, Sophia-Dorothea, invited Euler to a gathering of her salon at the royal palace in Berlin to which nobles, literary figures, French scholars, and Calvinist pastors were invited to engage in enlightened discourse. The daughter of George I of England, Sophia-Dorothea as crown princess in Hanover had known Leibniz. But her efforts to draw Euler into the spirited conversation failed. He responded only to queries in monosyllables. The exasperated queen chided him, asking, "Why do you not wish to speak with me?" Euler, who remembered the brutality of the *Bironovschina* period in Russia (see chapter 3), responded, "Madame, it is because I have just come from a country where a man's words can get him hanged."[30]

Euler was referring to reality in czarist Russia where—since his arrival in Berlin—politics had become more dangerous. The Russian palace revolution in November 1741 replaced Anna Leopoldovna (the mother of the infant czar Ivan VI) with Elizabeth (the daughter of Peter the Great), who had the support of the palace regiments. Anna was sent into exile under heavy guard, and the law courts sentenced her chief supporters to death. They condemned Andrei Ostermann, who had headed the foreign office, to be broken on the wheel and then beheaded; Count Burkhardt Christoph von Münnich of the army was to have his hands and then his head cut off. But after the humiliation of being put on the scaffold, both were spared at the last moment and sent into exile in Siberia. The academician Christian Friedrich Gross was not so fortunate; having spoken on the wrong side during the throne revolt, he was arrested and sentenced to hang on New Year's Day 1742. Apparently Euler thought for a time that the sentence had been carried out, but a letter of February 1742 from Gottfried Wilhelm Heinsius informed him of Gross's suicide following the sentencing.

That year Euler was getting settled in Berlin. In October he wrote to Goldbach that he had just purchased for two thousand Reichsthaler "ein artiges Haus" (a nice house) with a large garden. The original asking price had been twice that amount; Euler bought it from a Mademoiselle Mirabel when she married the brigadier Charles de Baudan.[31] The location delighted Euler. The house was well situated between the Friedrichstrasse and the Dorotheenstadtstrasse, near where the king had decided to construct a new palace, the science academy, and the observatory. The situation could not have been more advantageous. Remembering his experiences in Saint Petersburg, Euler persistently requested from Frederick's officials the privilege of being exempt from any military billeting whatsoever, and the king graciously agreed. [32]

Euler did not occupy the house until Michaelmas, 29 September 1743, nearly a year after the purchase. Before moving into the house, Euler made various necessary repairs costing several hundred Reichsthaler. The address from the Calendar for the Royal Capital City for 1743 lists the Eulers as living on Baerenstrasse (later Behrenstrasse); the current address is 21 Behrenstrasse. The house was two stories tall in the eighteenth century (a third floor was added in the nineteenth), and it exhibited a warm domesticity. Possessing a religious frame of mind, Euler never forgot the religious practices he had observed in his father's house and closely followed them. Every evening after dinner Euler gathered his children, servants, and students lodging at his house for a domestic devotion; there were biblical readings and sometimes explanations and discussions. Before bed, he also often read passages from the Bible or scriptural stories for the children. Euler's property included the land of 20 Behrenstrasse, where the Eulers cultivated a large garden. Among the vegetables most common in such plots were tomatoes, potatoes, and strawberries, and the family likely shared any surplus with neighbors. For the next twenty-three years, Euler was to reside at the Behrenstrasse address, where a small circle of friends, students, and colleagues gathered around him.[33]

In 1742 Euler, who had a passion for teaching, was short of funds from the Prussian Society, as money was being drained into the First Silesian War; he thus sought permission from Frederick to give lessons in mathematics to the sons of the Duchess of Württemberg at her home. Writing from his Znaym camp in a letter in March 1742, the king "with pleasure" authorized it.[34] The duchess was an important contact; in the winter of 1741–42 Voltaire had encouraged Jean-Baptiste de Boyer, Marquis d'Argens, a major Provençal critic of intolerance and the Catholic Church, to relocate in Berlin;[35] D'Argens came with his close friend and protector, the duchess. He became the royal chamberlain, and in 1743 he helped found the Nouvelle Société littéraire in Berlin and joined in the advocacy of a new science academy.

Euler had also agreed to tutor and teach the mathematical sciences to students selected by the Petersburg Academy of Sciences whom, until 1755, the Russian government supplied with meager and inadequate monthly stipends.[36] The students, most of them sons of Russian nobles, boarded at Euler's Behrenstrasse house; although the Euler family continued to grow the house had plenty of room for these students. For a year beginning in the summer of 1743, the sixteen-year-old count Kirill Gregor'evich Razumovskij, who had journeyed west "to acquire habits of civilized societies," resided there along with Grigorij Nikolaevich Teplov. Euler gave them mathematical instruction in the Russian language, and he reported to Maupertuis that the students "lived happily there."[37] But Teplov gave a different account; accustomed to large mansions, he reported that Razumovskij and he had to leave in April 1744 "parce que sa maison est trop petite pour nous (because his house is too cramped for us)."[38] In his letter Teplov continued that Euler's house was "also too far away from the residence of Count Mikhail Bestuzhev-Ryumin," the Russian envoy in Prussia, "but we continue using Mr. Euler's lessons."[39] From 10:00 to 11:00 a.m. on weekdays, Euler taught mathematics, astronomy, and physics to these students and others, including Karl Eugen, Duke of Württemberg and later, for four years, Louis Bertrand, a student from Geneva.

By 1743 the Euler family had grown to six. Joining Johann Albrecht and Karl Johann were Katharina Helene, born in November 1741, and Christoph in 1743; both survived beyond childhood. The children were a source of happiness for Euler, but more losses from the high infant mortality rate of the time were soon to come.

The New Royal Prussian Academy of Sciences

From the time of his arrival in Berlin during the first Silesian War, Euler not only pursued his research but strove to reorganize the structure, operations, finance, and publications of the Royal Prussian Society. At the Petersburg Academy he had gained administrative experience in managing and editing its journal, the *Commentarii*, and in helping to plan and prepare instructions for the Second Kamchatka Expedition across eastern Siberia.

On 6 September 1742, at its first meeting of the autumn session, which was held in the observatory, the mathematical class of the Royal Prussian Society formally welcomed according to protocol "the famous Professor of Mathematics Mr. Euler, whom his majesty has called from the Petersburg Imperial Academy." Greetings also went to another new member: Euler's close ally Johann Nathanael Lieberkühn,[40] a physician, astronomer, and natural philosopher who judged the society's French members and its men of belles-lettres to be producing useless work. Director Vignoles and the members pledged counsel, assistance, and friendship in working for the betterment of the institution. Euler, who was soon to choose Naudé's successor, had completed seven papers since arriving in Berlin and at these early meetings of the society he read them all. In one indication of the continuing fertility of his research in Berlin, he submitted the first five for volume 7 of the society's *Miscellanea Berolinensia*. With the addition of these five of them it came to 242 pages, long enough to make its publication possible. As society protocols and space limitations required, the last two pieces, on properties of conic sections and on linear differential equations, were to appear instead in a separate research volume. Euler was placed in charge of publishing volume 7 of the *Miscellanea*, which took a year.

On 7 September, the day after the welcoming ceremony, Euler wrote to the society secretary Jariges with detailed recommendations on how to "aid astronomy, which has fallen into such a sad state." Recognizing that gathering astronomical data was crucial for compiling the encyclopedic almanacs in nine regional varieties that provided almost all of the society's revenues, he included a request for funds to appoint as astronomer Johann Kies of the University of Tübingen.[41] In November 1742 the funds came through.

In 1742 Euler and the Royal Prussian Society of Sciences enjoyed favorable conditions. His letter of 10 April to Gerhard Friedrich Müller in Saint Petersburg reported that he was living in Berlin in peace, unaffected by the war, receiving his salary, and depending only on the king.[42] Though the First Silesian War had eroded the royal treasury, after the Peace of Breslau with imperial Austria in June 1742 Frederick released twenty thousand Reichsthaler to the society; this revenue was intended to make it more than comparable in resources to the royal science academies in Paris and Saint Petersburg.

But funding was a continuing problem, and the income from the Prussian almanac monopoly remained insufficient. Early in 1740 readers rejected the new academic calendars, preferring weather forecasts and the predictions of the old ones. But David Köhler, who had run the sales since 1738, increased the number of calendars in production; an order in December 1740 and the following January of 6,750 calendars by the only bookseller carrying them in Frankfurt-am-Oder suggests healthy sales.[43] In 1741, after the conquest of Silesia, a governmental protocol had proposed publishing a new Silesian almanac to increase the income of the Prussian Society. The main problem was not with the total calendar revenue; it was an open secret that Köhler skimmed 25 to 50 percent of it for himself. By making himself indispensable to the operation of the academy he avoided removal, while Euler seems to have been unaware of the amount of the losses. Ingratiating toward Euler since his arrival, Köhler paid his salary punctually from 1742 on, thereby earning his support. But the calendars still needed further refinements regarding sales, and Euler became an energetic advocate for increasing their yield.

Impatient that the new academy was so slow to take form, in a letter of January 1743 to the king Euler reckoned that the revenues of the society from the sales of better calendars for Silesia and Prussia should exceed twelve thousand Reichsthaler. While the almanac funds were kept a state secret, Euler assumed that they could be as high as twenty thousand Reichsthaler. Even the lower sum was great enough to found an academy on the same financial footing as its fellow institutions in Paris and Saint Petersburg. These funds were to allow Euler to recruit a sufficient number of able savants in the sciences, who in short time would theoretically surpass the Parisian academicians in scholarly distinction. At the Petersburg Academy of Sciences, Euler had independently recorded astronomical observations for Delisle. For improving the computation of ephemerides to refine the Prussian almanacs, he now introduced the methods that he had developed in Saint Petersburg and streamlined production. His contributions increased the almanac income that year from ten thousand to thirteen thousand Reichsthaler. Some booksellers and printers were failing to charge tax on almanacs, which deprived the crown of these funds.[44] The academy's *General Register* was to list many such violations, and local officials in Prussian lands had to police the collection of taxes from the almanac monopoly, which was not an easy task.

While the total income was nevertheless substantial and exceeded the funding for the Petersburg Academy, in a letter of 21 January 1743 from Charlottenburg Frederick rejected Euler's "ideas on pretended funding," insisting that his revenue account using "abstract calculations out of the grandeur of algebra" violated the ordinary rules of arithmetic; otherwise, he posited, Euler could not have obtained "the imagined grand revenues from the Silesian almanacs."[45] Instead Frederick foresaw debits. Euler, the king believed, should stop being deluded about the academy's income and how to raise the finances needed mainly to reward the good works of its members. At this point Frederick's response did not refer to an academy, and this was probably because many of the scholars whom he believed necessary for beginning the new institution—above all Maupertuis—were not yet available to him. Determining income from the almanac monopoly was to be a source of friction with the monarch until Euler's departure from Berlin years later.

Euler did not retreat from the king's cutting letter as the hierarchical order of society might have caused him to do. Quickly replying to the king on 24 January 1743, he stated that the entire funding for the society from almanacs would soon suffice to establish the new academy and that Lieberkühn might better show the financial soundness of his projection.[46] Having the new academy was his "infinite desire," Euler claimed. In his previous letter Frederick had approved Euler's recommendation to hire his Swiss compatriot and engraver Johann Karl Hedlinger to make medals and work on finances for the academy. Euler informed the king that he had offered a contract to Hedlinger, who was working in Sweden. He also confidently announced, "I have discovered the physical causes of the effects of magnetism, which perfectly explains all related phenomena" and he asked permission to submit his fourth paper, an essay on how best to construct inclination compasses to make possible the most precise observations of the inclination of magnetized needles on sea or land, to the annual Prix de Paris competition for 1743.[47]

That paper included Euler's results from the study of the physical cause of magnetism in which he had mounted a magnetic needle on a pivot.[48] Euler's Prix de Paris paper rejected Roger Cotes's position that gravity was universal in all bodies, a view that had been erroneously ascribed to Euler. As Daniel Bernoulli later noted, the academy wanted to perfect inclination compasses to make them more useful in navigation; it was thought possible to determine longitude through the intersections of different inclination and declination curves.[49] In his *Lettres à une princesse d'Allemagne* (Letters to a German princess, E343, E344, and E417) Euler was to continue this magnetic and longitudinal study.

On 29 January 1743, Frederick extended best wishes in authorizing Euler's entry of his essay into the competition, and its capture of the Prix de Paris for that year brought Berlin the kind of glory that the king had been seeking. The topic of magnetism had been given for the prize competition in 1742, but no paper won that year.[50] While the prize went to Euler in 1744, the official award was not made until two years later. Publication of this essay on magnetism came in the *Réceuil des pièces qui ont remporté le prix de l'Académie Royale des Sciences* (Collection of papers that have won the prize of the Royal Academy of Sciences) for 1744. Euler's letter of 24 January had included a request that the monarch employ his brother Johann Heinrich as a painter. The king urged that Johann Heinrich exhibit his art, and in a marginal note stated that he would see what he could do to help, but a position was never found.

In 1743 Frederick clearly differed from Euler over how swiftly to establish the academy. As he was in his research, Euler was restless to get on with the project. But the king had reasons to delay; for the society's founding Frederick wanted to have pomp and circumstance similar to that which had attended the beginnings of the Royal Academy of Sciences in Paris under Louis XIV and Jean-Baptiste Colbert, and he hoped for support for literary in addition to scientific works. The rapid deterioration of his state treasury to about 150,000 crowns and the strain of the fight against imperial Austria, for which the French were lending him little assistance, had prompted Frederick—in his own wording, "prudently"—to leave the First Silesian War with a temporary victory in the Peace of Breslau in 1742.[51] The next year and a half he was uneasy, sharply increasing the military by adding nine field and seven garrison battalions along with twenty squadrons of light cavalry. War was his priority, for a permanent acquisition of Silesia would place Prussia among the major powers of Europe. Frederick recognized 1743 as the midpoint of what in fact was a single long war, and the approaching Second Silesian War gained most of his attention. The king was also hindered over the selection of the "arcanists"—that is, the directors, protector, and secretary of the academy. At the time, Viereck held all of these posts. And until he could bring Maupertuis from Paris to be its president, Frederick would not proceed with plans for the academy.

Euler possessed the skill and experience to organize and recruit members for a distinguished new academy. In April 1743 he informed Daniel Bernoulli of his intention to seek "a few superior geniuses who discern the nexus of the sciences and real applications, study it, and discover more practical uses of the sciences."[52] Even in an academy subordinate to the military, Euler observed, utilitarian discoveries came of an illuminating spirit. But Frederick did not consider higher mathematics the primary font for practical invention. To inform nobles, officials, and scholars not versed in mathematics, Euler later wrote a brief article titled "Commentatio de matheseos sublimioris utilitate" (Commentary on the utility of higher mathematics, E790), which clearly and comprehensively argued that higher mathematics, like elementary mathematics, was a greatly utilitarian science.[53] He held it to be indispensable to the advancement of mechanics, astronomy, hydraulics, hydrostatics, artillery with a greeting appended from the king, optics, physics, and physiology.[54] In the article he pointed out the exactness in the mathematics expressed in Kepler's discovery of elliptical planetary orbits given in *Astronomia nova* and supported in Newton's *Principia mathematica*.[55] Scientific progress, he insisted, required explorations with higher analysis, and limits placed upon mathematical research restrained developments in many fields. Because this article was written primarily for nonmathematicians, it is curious that he chose Latin as its language, especially given that he had to convince the Francophone king, the *literati* from Paris, and the doubtless unpersuaded French-speaking German nobility. Higher analysis, he contended, sharpened the mind and led to the securing of truths. He knew that his opponents would seek in his paper materials for dispute, and his abrupt reference at the end of his paper to them as "the enemies of mathematics" may have been, upon reflection, a reason he did not submit it for publication. The article was not translated into French or published until seventy years after his death.

Euler informed Goldbach in December 1742 that the preparation of volume 7 of the *Miscellanea Berolinensia* was "progressing quite well"; a work, dedicated to Frederick, it was published in 1743 upon the king's return from the First Silesian War. The volume contained a mélange of good papers—on chemistry by Johann Heinrich Pott, and on astronomy and mathematics by Naudé and Alexis Claude Clairaut's father, Jean-Baptiste, in Paris. Euler had evaluated all of them. As he had told Goldbach, "fast die ganze classis mathematica von mir kommt" (almost all of [the articles] of the mathematical class are by me.)[56] The volume includes his papers addressing the first five of the seven significant problems that he had posed at the academy in 1742: determining the orbit of the comet observed that year, theorems concerning reducing certain integral formulas to the quadrature of circles, the finding of definite integrals that assign the value determined after the integration of the variable quantity, the summation of series of reciprocals generated from powers of natural numbers, the integration of differential equations of higher degrees, common properties of conic sections corresponding to a family of other curves, and solving a special differential equation.[57] This was the first of four lists of problems that Euler issued during his career, and this group was given not as a challenge but as a selection of major ideas that he was examining. That the essay by Jean-Baptiste Clairaut is in French rather than Latin makes the *Miscellanea* a transitional work preceding the language shift at the new academy; including those two essays also represented an important advance toward such a shift. Euler saw through the press this final issue of *Miscellanea Berolinensia*, and its publication made for a graceful point in the demise of the old Royal Prussian Society.

In 1742 and 1743 a clash between Johann I and Daniel Bernoulli saddened Euler. In Basel the splitting of the Prix de Paris of 1734 between the two Bernoullis on the subject of the inclination of planetary orbits, along with Daniel's work in hydraulics, which Johann I also addressed, had added jealousy and bitterness to a strained relationship between father and son. Their competition degenerated into Johann I's publishing in his *Hydraulica* and *Opera omnia* of 1742 materials plagiarized from Daniel's *Hydrodynamica*. Earlier Daniel had agreed that his father, despite suffering from gout and asthma, should expand upon the *Hydrodynamica*. But

Johann I now argued for his own priority, claiming to have completed his manuscripts a decade earlier; to the world he presented his son as stealing his ideas. Daniel protested, but he was devoted to his father and had no desire to dispute him publicly.

The affair weighed heavily upon Daniel, who endured it with resignation. In October 1742 he wrote to Euler of Georg Bernhardt Bilfinger's asserting that Daniel had taken everything from his father and contributed nothing himself. The dispute produced such disdain for his studies up to that point, Daniel complained, that he would liked "to have learned a shoemaker's trade rather than that of a mathematician."[58] Bernoulli's letter of September 1743 to Euler asserted that the *Hydrodynamica* was the fruit of ten years of his own research and that he owed not one iota to his father for it, a fact Euler knew well. The letter dealt principally, however, with such scientific matters as the motion of several bodies in a rotating tube and extremals in mechanics; it also reported that Euler's teacher Johannes (Johann Jakob) Burckhardt had recently died.[59] The dispiriting news from Basel did not distract Euler from his institution-building efforts in Berlin, but it likely delayed his research on the foundations of fluid mechanics until after the death of Johann I Bernoulli.

By the summer of 1743 Euler, together with French savants in Berlin and high nobles in Frederick's court—above all his military mentor and favorite field marshal, Count Samuel von Schmettau—were dissatisfied with the lack of progress toward a new academy of sciences, which had by now been a subject of discussion for three years. Schmettau, whose mother was the daughter of the pastor to the French enclave in Berlin, possessed a military education that included the sciences. He had met Fontenelle of the Paris Academy, and after serving for thirty years in Sicily, Corsica, Poland, and Sweden, accepted Frederick's invitation to return from imperial Austria for the rank of field marshal and to oversee what would roughly be an interregnum.

On 1 August, Borcke and d'Argens joined with Schmettau to found the Nouvelle Société littéraire, intended as a basis for the academy. The new society had three classes: mathematics, literature, and physics—essentially the same divisions as in the old Royal Prussian Society of Sciences. Initially it operated without new protocols. The statutes stressed its intent to cultivate what was useful and more than an entertaining diversion; shunning "pure speculation and mathematical calculation," it would "instruct and perfect the mind" in history, literature, mathematics, natural history, astronomy, and the other sciences. Its original members were divided into two categories: twenty were ordinary and sixteen honorary or aristocratic. Half of the ordinary or working members were French

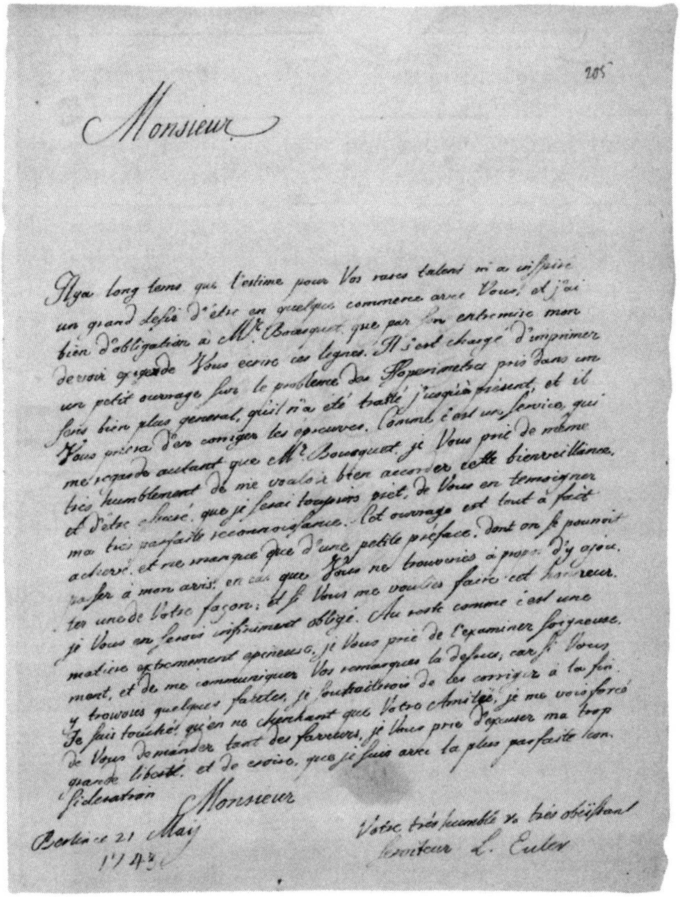

205

Monsieur

Il y a long tems que l'estime pour Vos rares talens m'a inspiré
un grand desir d'être en quelque commerce avec Vous, et j'ai
bien d'obligation à Mr. Bousquet, que par son entremise mon
devoir m'engage de Vous ecrire ces lignes. Il s'est chargé d'imprimer
un petit ouvrage sur le probleme des Isoperimetres pris dans un
sens bien plus general, qu'il n'a été traité jusqu'à present, et il
Vous priera d'en corriger les epreuves. Comme c'est un service, qui
me regarde autant que Mr. Bousquet, je Vous prie de même
tres humblement de me vouloir bien accorder cette bienveillance,
et d'être assuré, que je serai toujours prêt de Vous en temoigner
ma tres parfaite reconnoissance. Cet ouvrage est tout à fait
achevé, et ne manque que d'une petite preface, dont on se pourroit
passer à mon avis; en cas que Vous ne trouveriez à propos d'y ajou-
ter une de Votre façon; et si Vous me voulez faire cet honneur,
je Vous en serois infiniment obligé. Au reste comme c'est une
matiere extremement epineuse, je Vous prie de l'examiner soigneuse-
ment, et de me communiquer Vos remarques la dessus, car si Vous
y trouvez quelques fautes, je souhaiterois de les corriger à la fin.
Je suis touché, qu'en ne cherchant que Votre Amitié, je me vois forcé
de Vous demander tant des faveurs, je Vous prie d'excuser ma trop
grande liberté, et de croire, que je suis avec la plus parfaite consideration

Monsieur

Berlin ce 21 May
1743

Votre très humble et très obeïssant
serviteur L. Euler

Figure 6.3. Euler's letter of 21 May 1743.

Huguenot or of French extraction, and society sessions were in French. Following the statutes of the Paris Academy of Sciences, the Nouvelle Société littéraire rejected the old hierarchical authority and committed itself to freedom and democratic elections.[60] The director and vice director were to be elected semiannually, and the society planned to have twenty-one weekly meetings, the first on 1 August at Schmettau's residence. Frederick gave limited support by providing a room in his Berlin palace. On 8 October Voltaire attended a session, adding distinction to the efforts. At that meeting, Euler conducted some physics experiments.

When Cardinal André-Hercule de Fleury died the French Academy did not elect Voltaire to replace him. The angry *philosophe* accepted

Frederick's blandishments to return to Berlin. Still disapproving of the monarch's military conquests, Voltaire explained that he was paying his respects to a protector of the arts. If his paramour Émilie du Châtelet opposed his trip, Voltaire wrote, he would sacrifice Minerva for Apollo, an allusion Frederick reveled in. But the relationship between the two remained difficult; Frederick considered forging a letter to ruin Voltaire's relations with Louis XV, while Voltaire wanted to be a French spy in Prussia, and the Prussian monarch thought him possibly to be on a mission for the Court of Versailles.

During the autumn of 1743, Voltaire and Madame du Châtelet traveled to Halle, where Wolff was now university chancellor. After 1740, when Madame du Châtelet had anonymously published her *Institutions de physique*, Wolff's reputation had continued to grow. She defended Leibniz's organicism and his concept of living force, *vis viva*, and its conservation. Johann Samuel König, introduced to Leibnizian natural philosophy in the late 1730s in Basel and studying under Wolff in Marburg, had recently tutored her in it, but with the exception of embracing the disputed Leibnizian principle of the conservation of living forces, she remained generally a Newtonian. König helped her revise her treatise, which Wolff admired. From 1741 to 1750 Jean-Henri-Samuel Formey in Berlin was publishing the six volumes of his modest, supportive *La Belle Wolfienne*, which in his detailed attention to personal relations, quarrels, and collaboration reads like a novel about the Wolffians. The series confidently proclaims the victory of Wolff over the Scholastics, Pierre Bayle, and John Locke.

In 1743 the king was trying to draw together Voltaire and Wolff, envisioning them in collaboration with Maupertuis as the trio he needed to lead the new academy. Sometime during the fall, the king and Voltaire were traveling through Halle. But Voltaire claimed that his health was too bad to visit Wolff, now elevated to the nobility as a baron.[61] Despite misgivings that Voltaire was a materialist and skeptic, the chancellor came to the *philosophe* and found him agreeable and jolly. In November 1743, the Royal Society of London elected Voltaire as a member.

The Nouvelle Société littéraire continued to assemble for weekly meetings through 16 January 1744. Minutes of the meetings have been preserved, but the official record breaks with no minutes extant for six more meetings to March 1745.

In pursuing the sciences at the Nouvelle Société littéraire, Euler worked with the chemists Andreas Sigismund Marggraf and Johann Heinrich Pott, the physician Christian Friedrich Ludolff, and the natural philosopher Johann Theodor Eller. To hasten the creation of a new academy, the Nouvelle Société littéraire scheduled its meetings at the homes of nobles at the same time as sessions of the Royal Prussian Society. By the autumn of 1743, no one thought well of the old science society, and Schmettau as well as Euler wanted it terminated, but since that society had funds in its treasury and a meeting place at the old observatory, it survived for a time.

Euler, who was unhappy with the inactivity at the Royal Prussian Society, joined the Nouvelle Société littéraire but had no administrative role in it. He was also not selected for the statute commission or named the director of the mathematics class; Viereck asked the king to confirm Euler for that directorship, but it went to native Berliner Abraham von Humbert, who was an engineer.[62] Probably at his own request, Euler was not elected a deputy or secretary of a class. His correspondence with Russian colleagues indicated that he did not hold the Nouvelle Société littéraire in high regard, but he nevertheless worked diligently to promote high scientific standards at it. Avoiding simple speculation and complex mathematical calculation, at ten meetings he lectured or read papers, most of them on physics, as well as numerous items of correspondence from scholars across Europe.[63] Over the next two months he read a letter from Heinsius in Saint Petersburg covering observations of comets made in 1742 by the Jesuits in Beijing, and another from Heinsius on the shape or phases of the rings of Saturn. Both Delisle and Heinsius corresponded with the Jesuits in China, and in September Euler compared observations of comets made in Beijing, Paris, and Saint Petersburg. In November he read the observations of the transit of Mercury across the sun, reported in two letters by Martin Knutzen, the University of Königsberg professor of logic and metaphysics.

Frederick was displeased with the beginnings of the Nouvelle Société littéraire; in a letter from a miltary camp to d'Argens in June 1743, he noted that he was engaged in serious affairs that demanded all of his attention and counseled delay in founding it.[64] In the fall, when he returned to Berlin for a few weeks, he reiterated that the royal scepter should initiate the new academy. But pressure grew quickly on Frederick from his court. A letter of 19 October from Euler also reminded the king of his "gracious promise" to establish the institution and recommended that Daniel Bernoulli be appointed to the academy and given a small salary to work from Basel on mathematics and astronomy.[65] Noting that he knew of no more capable scholar, Euler argued that a return of Heinsius to the German states would be vital to developing a distinguished academy. Frederick granted neither proposal, and did not act on Euler's request for payment of three hundred Reichsthaler for moving expenses from Saint Petersburg in 1741. The expenses actually totaled almost five hundred Reichsthaler, including a rebate of two hundred Euler paid to the Petersburg Academy of Sciences for salary remitted to him for time beyond his departure. But during a respite between the first two Silesian wars, the king responded to the demands for creation of the institution, working to retain control over its formation.

On 2 November Frederick instructed Viereck to review funding needs and privileges. Schmettau wanted the old Royal Prussian Society and the Nouvelle Société littéraire joined in a new Royal Academy of Sciences and Belles-Lettres of Prussia. That month Vignoles, the ninety-five-year-old director of the mathematics class of the Royal Prussian Society, expressed a wish to retire and recommended that Euler succeed him, but Euler declared that he would not accept that directorship unless the two societies were united. Such a fusion was not a simple task, however: the secrecy of the noble curators in handling their economic affairs had generated discontent; the observatory of the old society was antiquated; and many members, fearing that they would be dropped in new elections, did not want to surrender their funds or traditional prerogatives. Frederick did not warmly receive Euler's joining in the recommendation that Köhler, for his experience and ability, continue to manage the sale of the almanacs; the king found fault with the past overseeing of the almanacs by Köhler, whom he knew was suspected of skimming profits.

On 9 November the king appointed a royal commission, headed by Viereck, to obtain proposals for combining the two institutions into a completely new academy, to disseminate these for discussion, and to submit a final recommendation to him. The ten members of the commission consisted of three government ministers, three honorary and two ordinary members from the Literary Society, and two members from the older Prussian Society; Euler was, conspicuously, not included. In the cabinet order of 13 November to Schmettau, a commission member, the king accepted the idea of unifying the two societies.

The commission meetings on 22 and 29 November discussed the king's intentions. Did he wish to restore the old society or set up something new? Schmettau wanted a new start, and the 6 December meeting was informed of Frederick's support for Schmettau's position.[66] The king wanted the academy to assemble the best intellects, keep to religious freedom, and introduce the French Enlightenment to his archconservative Prussia and its capital.

Within the various proposals and statutes submitted, the structure and administration of the academy generated the most debate. The commission did not concur with Frederick's desire to put imperial and Prussian history in a special position, and it wanted medicine and the German language to have separate statuses. Not proceeding through regular channels, Euler happily presented to Viereck in November a proposal consisting of sixteen propositions that would bring high distinction in the sciences and prevent the academy "from becoming a place of institutionalized honor and ceremony."[67] Drawing upon procedures of the Paris and Petersburg academies as well as the Nouvelle Académie française and a refinement in the structure of the Royal Prussian Society of Sciences,[68] he defined four small classes or departments: mathematics, medicine without physics, a new class of philosophy that included physics, and the two old philology classes collapsed into one; the four classes were to have a total of between twenty and twenty-four regular ("ordinary") members. Shifting physics from the medical to the philosophical class meant increasing attention to experimental and theoretical physics; Euler faulted Wolffian metaphysics for its opposition to this, along with its version of the monadic doctrine and its reduction of the importance of the principle of sufficient reason. He urged the academy to hire secretaries; have postage-free correspondence, as was the case in Saint Petersburg; circulate papers before a meeting so they could be studied in advance; and publish an annual volume of the best papers. As he saw it his principal proposition of the sixteen was his twelfth: ordinary members must be designated with great care in each class, the selection based solely on ability and achievement. This was the practice at the Royal Academy in Paris. The role of the noble curators heading each division was simply to present the names to the king. The reputation of the academy would rest only on the contributions of each of its ordinary members.

Although the commission did not name Euler's proposal and publicly ignored it, the statutes for the new royal academy of sciences are in fact indebted to him. The commission slightly modified his four classes by removing medicine, giving prominence to physics, and ranking mathematics, philosophy, and philology behind it;[69] Euler's realignment of classes thus satisfied neither the king nor Schmettau. Shortly after his arrival, Maupertuis would also alter the proposal.

Europe's Mathematician, Whom Others Wished to Emulate

Among the fields of academic study, which during the Enlightenment had not yet reached the separateness of modern academic specialization, the mathematical sciences along with critical history were attaining the greatest development, and the exemplar of the mathematical sciences was

Euler. In his twenty-five years in Berlin, he completed or wrote more than 380 articles and books (about 280 of these published before he left the city) whose combined depth, originality, difficulty, range, and sheer numbers make his achievement unmatched in the history of the mathematical sciences.

Up to 1744, when the First Silesian War occupied the king, Euler had few Prussian state projects or administrative chores. In a letter to Clairaut in 1742 he proclaimed that in Berlin "je jouis cependant d'un parfait répos" (I enjoy meanwhile a perfect rest).[70] Such time allowed him to immerse himself in extensive and profound research in the mathematical sciences, and Euler continued to contribute substantially to fields pursued in Saint Petersburg, among others. In pure mathematics, infinitary analysis and number theory, together with rational mechanics, remained central to his labors. In applied mathematics Euler expanded his major topics by adding to astronomy the subjects of ballistics and optics, and he began in more detail to investigate magnetism and electricity, which were generating the most excitement within the sciences. In this work he was advancing a second phase of what Clairaut called "l'epoque d'une grande revolution dans la Physique" that had commenced with Newton's *Principia*.[71] Euler's program in mechanics treated elastic, flexible, fluid, and interacting bodies,[72] and prior to the work of William Rowan Hamilton it was Euler, rather than Newton, who formulated most of the basic differential equations for rational mechanics, engineering mechanics, and celestial mechanics.[73] To guide his creation of differential equations for celestial mechanics he pursued more exact observations from more powerful and finely graduated telescopes and improved sextants and chronometers; Delisle had prepared him well in astronomical observational techniques. In making calculus the basis of mechanics and theoretical astronomy Euler—more than his competitors—was transforming these fields into the modern mathematical sciences, the chief achievement of the Enlightenment in the sciences.

Between 1741 and 1744 Euler completed one book and nearly finished four others, but none was yet published. He had eight articles in print, nine more that had been submitted, and in his extant correspondence had about 240 letters. A change of residence did not slow his exceptional accomplishment. While Euler now wrote between five and ten articles a year, many awaited the publishing of the Berlin Academy *Mémoires de l'Académie Royale des sciences et des belles-lettres de Berlin*, which was to begin in late 1745.

Euler's correspondence fell into three categories—scientific, personal, and administrative—that reveal an important circulator of scientific information, known then as an intelligencer. Amid a time of war, the number of letters he wrote annually grew sharply, from 47 in 1740 to 105 in 1743. Because there was no franking privilege for the Berlin academicians, Euler had to pay for his own postage; a letter of August 1755 to Goldbach declared that of his postage fees, none gave him greater joy than those for the letters to his academician friend, and noted that his yearly cost for correspondence was two hundred Reichsthaler.[74] This was not a small amount, for the annual salary of a junior academician was three hundred Reichsthaler. Many of Euler's letters on the sciences are like modern research articles, and in turn elicit from correspondents the latest information on scientific discoveries; thus, his correspondence basically doubles his contributions to the sciences. The letters included personal information on health, families, travels, marriages, births, and deaths, and also illuminated Euler's ongoing close connection with the operation, publishing, and research of the Petersburg Academy of Sciences.

From the time of his arrival in Berlin, Euler was making a host of discoveries in infinitary analysis, accelerating its articulation, and forging three of its core branches—infinite series, differential equations, and the calculus of variations. These were rapid outgrowths of the field. Initially Euler emphasized isoperimetrics or the fledgling calculus of variations. After briefly setting aside work on variational theory, in 1741 in Saint Petersburg he had determined that in articles published earlier that same year and in 1738 he had erred in treating problems with integrand variables connected by differential equations. This recognition had occurred during a breakthrough in computational methods that prompted him to make a second phase of massive calculations of different trajectory curves using older methods and powerful new techniques to obtain as many solutions as possible.[75] As David Hilbert would later do, Euler considered problems to be the heart of mathematics.

By the spring of 1741 at the latest, Euler had finished in Saint Petersburg a draft of *Methodus inveniendi lineas curvas maximi minimive proprietate gaudentes* (A method of finding curves that show some property of maximum or minimum, E65), though without the two appendixes that would come later. In May 1743 the publisher Marcus-Michael Bousquet visited Berlin to present the king a copy of Johann I Bernoulli's *Opera omnia*; Bousquet was impressed with the work of Euler, who handed him the main body of the *Methodus inveniendi* manuscript. That month Euler wrote to the Genevan mathematician Gabriel Cramer, asking him to proofread and correct for Bousquet that small book written on isoperimetric problems. Cramer was known to Euler in part for his commentaries and annotations to Bernoulli's *Opera*. Finding the manuscript admirable, Cramer agreed. Daniel Bernoulli, attempting at the time to determine maxima and minima of elastic curves, recommended to Euler the addition to the *Methodus inveniendi* text of two appendixes on elasticity. Among all curves of a stated length that had tangents at the ends, Euler would minimize the integral of an element of the arc length divided by the radius of curvature squared. By the end of that summer Euler had completed the appendixes, but they were not sent to Bousquet in Lausanne until December.

At the same time as the near completion of the entire manuscript for *Methodus inveniendi* (1741–43) and his overseeing of its printing (1743–44), Euler pursued a second phase of massive computations in infinitary analysis, making a host of discoveries preparatory to his two-volume *Introductio in analysin infinitorum* (Introduction to the analysis of the infinite). Apparently a breakthrough in his perfecting computational methods in 1741 prompted this. His work in subsequent stages to perfect the computing of infinite series, along with elementary transcendental functions, illustrated Euler's pattern of research: he never abandoned a major research subject, but returned to it relentlessly in order to deepen and broaden on each revisiting his understanding of it. Daring and occasionally reckless in his computations, he was safeguarded by his sure instinct. Euler's procedure was to make exhaustive calculations until the limits of existing procedures blocked him from going further. Then he set aside his research in the field for a few years, during which others might reach breakthroughs or ideas could progress in his mind, perhaps subconsciously, while he was working on problems not necessarily related. Once a breakthrough was achieved he returned to the field with powerful new techniques for making extensive computations. Euler possessed an eye for what was essential and unifying. Characteristically he proceeded from simple to ever more complicated examples in his search to uncover underlying connections within branches of a mathematical field or among fields. He sought to make computations simpler and more exact; even when a topic seemed resolved he did not stop but continued to search for what he called more direct and more natural proofs.[76]

By deriving sums for coefficients in the zeta series from sums of Taylor expansions until they begin to diverge rapidly, and by determining the remainder guided by the aid of his potent instinct, by 1741 Euler streamlined summing zeta functions for even integers less than 26.[77] The sums of zeta series of even integers to 16 he calculated to eighteen decimal places. His "Démonstration de la somme de cette suite $1 + 1/4 + 1/9 + 1/16 + \ldots$" for the *Journal littéraire d'Allemagne* two years later contains his first satisfactory proof for zeta functions.[78] Since the *Journal* was popular with those who frequented the salons, he may have been trying to reach a wide readership. But it was not until 1750 that he would solve the zeta problem for all even integers.

Using Taylor expansions, clever substitutions, continued fractions from Second Viscount William Brouncker, and methods from John Wallis's *Arithmetica infinitorum* for finding the area of a circle, plus analogies, conjectures, and his near unerring honed instinct, Euler computed the sums of many infinite series, especially for π and e, almost always more precisely than his predecessors. After employing continued fractions in 1737 to find that $e = 2.718281845904$, he proposed at the latest in 1744 a power series for the logarithm that approximates it accurate to seven decimal places.[79] Using the binomial theorem, Euler expanded $(1 + x)^{1/i}$.[80] In sections 114–16 of his *Introductio* he would treat the exponential function. When not dealing with a finite expression, he resolved continued fractions with integral equations. Earlier Euler had studied Abraham de Moivre's formula $(\cos x + i \sin x)^n = \cos nx + i \sin nx$, though not yet with the i symbolism. He now returned to it, strengthening the connection among ordinary trigonometric functions with both exponential and logarithmic types, while extending their domains to the complex numbers. By 1744 he had perfected his derivation of $\cos x + i \sin x = e^{ix}$ and obtained $2\cos x = e^{2x} + e^{-2x}$. From the first follows $e^{i\pi} = -1$. Euler did not have the modern form of his identity, $e^{i\pi} + 1 = 0$; this would not be revealed until the early nineteenth century.[81]

Defining natural logarithms as the inverse of the exponential function meant that $\ln(-1) = i\pi$; that is, the logarithm of -1 is an imaginary number. Euler and Jean-Baptiste le Rond d'Alembert were soon to debate the nature of logarithms of negative numbers. Euler also derived the expressions $(e^x + e^{-x})/2$ and $(e^x - e^{-x})/2$. This arose from his search for infinite product representations of the sine and cosine functions. He did not employ the term *hyperbolic functions* or make notations for them; the first is today cosh x. Euler's results stemming from his revisiting Moivre's formula were probably his concluding work on the *Introductio*; he took delight in making and perfecting the extensive computations for it and was peerless in them. In the history of mathematics, only Carl Friedrich Gauss compares in importance to him in performing them. In May 1743 Euler had signed a contract with Bousquet to publish the *Introductio*,[82] but he did not submit the entire manuscript until a year later. Bousquet wanted instead to begin with the *Scientia navalis*, but Euler awaited word from the Petersburg Academy of Sciences on funding its printing.

During his first three years in Berlin, Euler's contributions to number theory flowed without interruption. While the Royal Prussian Society emphasized application, Euler insisted that number theory was not distant from infinitary analysis or repugnant to it; on a deeper level, he posited, the two were closely related. Euler's view that number theory was a significant discipline, on equal footing with other fields of mathematics, was not widely shared among mathematicians; Clairaut, Daniel Bernoulli, and many mathematicians of the time considered it more an amusing diversion. Throughout his career, Euler enjoyed returning to number theory to state conjectures precisely and attempt proofs.[83]

In September 1740, when Euler was still in Saint Petersburg, Naudé had asked from Berlin, "How many ways can the number 50 be written as the sum of seven different positive integers?"[84] This letter initiated Euler's interest in partition numbers: determining the number of ways that a positive integer could be represented as a sum of whole numbers. In a series of articles he was to pursue the subject intermittently over the next nearly thirty years. Almost alone this work founded the theory of partitions. In April 1741 Euler responded to Naudé with the paper "Observatione analyticae variae de combinatione" (Various analytical observations about combinations, E158), which was published in 1751 in volume 13 of the *Commentarii*.[85] In it he examined relations comparing a series with the sum of its powers and of its products. He obtained the answer 522, describing it as "a most perfect solution of Naudé's problem." Euler would later improve his method and add detail in chapter 16, section 3.3.1, of the first volume of his *Introductio*. As he wrote to Goldbach in September 1741, he had long known that for m and n positive integers, $4mn - m - 1$ could not be a square, but it took him a year to prove this. That year he also completed one of four of his proofs of Pierre de Fermat's Little Theorem, which in modern symbols is $a^{p-1} \equiv 1 (\mathrm{mod}\, p)$, where p is a prime, a is relatively prime to p, and the exponent is $p - 1$. In this case, p divides the difference $a^{p-1} - 1$. From Euler's proof using Fermat's Little Theorem, it follows that $a^2 + b^2$, when a is prime to b, does not have any prime divisors of the form $4n + 1$. This statement opened the first phase of Euler's seven-year project to investigate sums of two squares and prove all of Fermat's statements about them.[86]

After believing in the 1730s that Fermat had lacked proofs for many of his conjectures, in a letter to Clairaut in April 1742 Euler expressed regret for the loss of Fermat's proofs of these, noting that during the previous fourteen years he had been able to prove only a small portion of them. He inquired whether some unpublished papers of Fermat with proofs existed and urged a search in France to recover them, adding, "Ce seroit un grand avantage . . . si l'on publioit ces démonstrations, peut être que les papiers de ce grand homme se trouvent encore quelque part" (It would be an excellent thing if these proofs were published; maybe the manuscripts of the great man can still be found somewhere).[87] Euler confidently noted that

"I am far from having exhausted the topic. . . . Innumerably many splendid properties of numbers remain still to be discovered."[88] Number lovers and number theorists were enthusiastically discussing Fermat's statements, and Euler became Fermat's successor. But number theory was one area of mathematics on which he and Clairaut diverged, and his attempts to draw Clairaut into its study were to no avail. In his response to the April letter, Clairaut declared that he had "never heard of Fermat's theories, nor do I know what happened to his papers." The study of problems in number theory, he asserted, was "not fashionable and is said to be dry."[89] Clairaut found their study little more than a means of "exercising the mind," and described as extremely subtle but unnecessary Euler's discovery of a method for determining whether a large number is prime. Euler indicated that the Royal Prussian Society in Berlin was also investigating whether a perpetuum mobile was possible. Euler's restraint regarding it prompted Clairaut to respond in July 1742 that it was indeed impossible.

In 1741 Euler had written to Goldbach that he was exploring the "curious properties" of prime divisors. Proceeding from his research on Mersenne numbers and Fermat's divisors, he worked diligently, systematically, and intermittently to bring out the prime divisors of binary quadratic forms $mX^2 + nY^2$, where n may be a positive or negative integer, and he studied solutions of Diophantine equations; he examined sixteen positive and eighteen negative values for n. He sent his results to the Petersburg Academy, which published them in the *Commentarii* for 1750. His guesses on the arrangement of these numbers and his research on the sums of two squares would lead him slowly to his greatest discovery in

Figure 6.4. Alexis Claude Clairaut.

number theory, a clear statement of the complete law of quadratic reciprocity, which is one of number theory's two fundamental laws.[90] When p and q are distinct odd primes, it addresses whether the pair of quadratic congruences $x^2 \equiv q(\text{mod } p)$ and $x^2 \equiv p(\text{mod } q)$ are both solvable or only one of them is. The first preceding symbols indicate that $x^2 - q$ is divisible by p. Congruences, then, involve divisibility. In a letter to Goldbach in August 1742, Euler gave a special case of the law of quadratic reciprocity. (Conjectures on the other basic law in number theory, the prime number theorem, still lay a half century in the future.) While primes appear quite irregularly among the positive integers, this theorem shows their distribution by giving an approximation of how many primes are less than n or equal to it.

In 1742 Goldbach proposed that every integer greater than five was the sum of three prime numbers; such numbers were among Euler's major interests. The stronger binary form of the Goldbach conjecture was that every even number greater than two was the sum of two prime numbers. The conjecture remains unproven. Mathematicians have explored differing ways of writing a number n as the sum of two primes, now known as the Goldbach Partition, which must consider the order of those primes.

The same year as Goldbach's proposal concerning integers, Euler's research in rational mechanics was to engage leading geometers in Basel and Paris who addressed and solved an intriguing problem: the motion of a heavy straight tube rotating about a fixed point on a horizontal plane under the influence of gravity. The year before, Euler had posed the problem to Johann I Bernoulli. By March 1743 Bernoulli solved it in the form given by König, which did not take into account the influence of gravity. It differed from that which Euler sought: a solution applicable to celestial mechanics. Clairaut and Daniel Bernoulli were also sent the problem. Using the same methods in developing differential equations that considered changes in succeeding infinitesimal instants of time, *dt*, Clairaut, d'Alembert, and the Bernoullis independently arrived at basically equivalent solutions. Daniel Bernoulli had the rod and a sphere move along an infinitesimal line with the sphere independent of constraints in the second instant *dt* but arriving simultaneously at a given point. For this he had to add external pressures. Through the application of what amounts to the conservation of angular momentum and *vis viva*, Clairaut arrived at his solution, which appeared in his "Sur quelques principes," submitted to the Paris Academy's *Mémoires* in 1742 but not printed until 1745. Problem 2 of d'Alembert's *Traité de dynamique* of 1743 has a similar resolution. Euler was just becoming aware of d'Alembert's work; in a letter in December 1743, Daniel Bernoulli introduced Euler to the brilliant young French mathematician. Clairaut's second infinitesimal instant *dt* varied, while that of d'Alembert was constant. Euler's distinctive solution included probably the first statement of the concept of a moment of inertia and one of the first of the conservation of momentum.[91] He began by solving the problem of a tube and enclosed body after force has acted in the first *dt*. He derived a series of equations and made skillful substitutions in these in solving for the second infinitesimal instant, which he had as an integral constant. He proved the conservation of the *vis viva* of the whole system and extended the problem to the case when the tube contains three bodies.

Observations by Delisle and Heinsius in Saint Petersburg along with Cassini the elder and Clairaut in Paris of the comet visible there in March 1742 seem to have elevated Euler's interest in the study of the orbits of comets, their tails, and celestial mechanics in general. The same comet was seen for eight days in Berlin. European astronomers still understood only poorly the nature of comets and their paths, and Euler was seeking information on complex questions involving them: Did they belong to the solar system? What force produced their motion? Attempts to compute their orbits remained daunting, for inclined to any angle of the ecliptic plane they travel in highly eccentric orbits regardless of their direction. In their polemic with the Newtonians, the Cartesians held comets to be transitory, while Newtonians considered them permanent. Each investigated whether Newtonian dynamics or Cartesian vortices most accurately describe their orbits.

In June 1742 Delisle wrote of the difficulties he had determining the actual course of the comet that March, and thanked Euler for helping him obtain a copy of Berlin astronomer Christfried Kirch's observations.[92] In addition to commenting on how difficult it was to compute the eccentricities in the orbits of comets, Delisle included a pamphlet that his own sister had sent from Paris containing Clairaut's response to James Bradley's theory of the aberration of the light of fixed stars. Euler considered the discovery of aberration significant for observations and for establishing the Copernican system in astronomy. That same year Delisle reportedly sent his observations of the comet and Jupiter's satellites, which Euler compared with satellite records of Gian Cassini from 1655.[93] Euler's letter to Heinsius in Saint Petersburg in August 1742 requests the chartings by the academy, which were scheduled to be published soon.

Euler wanted to compare the Saint Petersburg results with earlier calculations of the orbits of comets by the Germans, French, and British. He knew that Johann Kies, in not using accurate enough logarithms, had calculated as elliptic the orbits of the comets of 1729 and 1730, while Pierre Bouguer mistakenly found them to be hyperbolic. Euler discovered an error in Edmond Halley's computations, and examined Delisle's reckoning that the paths of comets were close to parabolic. Quarreling with the director Delisle, Heinsius was not visiting the observatory, so the publication of their shared work on comets was delayed. In September 1742 Clairaut had promised to forward his observations, and from his research on fluid mechanics and polar flattening of Earth he gave results that drew upon the work of Daniel Bernoulli. In February 1743, Euler reported on Clairaut's determination—based on hydrostatic principles—of the modified Newtonian shape of Earth as a spheroid flattened at the poles rather than a Cartesian spindle, a prolate ellipsoid. As Euler had projected in 1738, the flattening is less than the Newtonians believed; Earth had an orange rather than a tomato shape. In April Clairaut sent him a copy of his recently published book, *Théorie de la figure de la terre*. Euler's response has been lost, but he considered Clairaut's work to be exceptional. For Euler the argument over Earth's shape was still to be resolved; the Paris Academy's expedition to Peru under Charles Marie de la Condamine, sent to measure an arc of meridian to indicate whether the Newtonian bulge at the equator existed, had yet to return and Euler awaited its data. Clairaut's observations of comets were not sent until December. Heinsius only communicated early in 1744 his writings on the comet of 1742; Euler believed these to be among the most important on the topic.

After asking Delisle, Heinsius, and Clairaut for their sightings of the comet of 1742, Euler had requested three more for the comet of 1743. He anticipated that comparing these good observations from different places would enable him to map comet paths more precisely. He wanted more accurate observations of comets, to be obtained by applying calculus to precisely compute all points in their orbits. He credited Delisle with the principal achievements of the early observatory. After receiving the observations of the comet of 1743, Euler found the paths of both comets to be nearly parabolic ellipses.

In 1742 Euler had another source of information on comets and their paths. Maupertuis published anonymously in Paris a short, popularized book, *Lettre sur la comète*, which neatly ventured across scientific discourse into cultivated literature. Many people saw comets as wonders or portents and thus a source for prophesying, while noble women thought of them as curiosities. Maupertuis, who followed the style of Fontenelle in addressing his tract to women, removed comets from the province of the astrologer and placed them under the rule of physical laws. Cartesian vortices, he believed, could not account for the supposedly very elliptical orbits of comets, and even Newtonian dynamics left unresolved the computation of these orbits. But Maupertuis's cosmology was Newtonian.[94] Euler explained the extreme improbability of a comet's crashing into Earth, a possible event that was exciting worried speculation. The Scriptures, he thought, suggested that this would not occur, which agreed with his scientific studies.

Euler believed exact lunar tables to be the best instrument for proving the validity of Newton's mechanics, which held the law of gravitational attraction to be the explanation for all celestial motions. These provided a case for the application of the three-body problem. But to develop the differential calculus for theoretical astronomy Euler required more accurate and extensive telescopic records of the motion of comets and the moon. He thus began his study of the three-body problem through its application rather than its theory.

At a meeting of the Prussian Society in February 1743, Euler proposed that Clairaut be elected a foreign member. At the following meeting this was agreed upon, and Euler also read his paper "Dissertation sur la lumière et les couleurs." He spoke in French, not Latin, so that the Prussian ministers present could understand; after the fact Naudé corrected the French in the paper. The essay discusses Euler's continuing research on a new medium theory of light and colors, including his discovery through experience of an "admirable harmony" between the two.[95] He concentrated on the physical characteristics of light rather than "pure mathematical speculation and calculus."[96] Having rejected in Saint Petersburg Newton's theory of light as corpuscular, he disagreed with the general Parisian acceptance of Newtonian optics. Knowing the importance of the French language at the academy and the cachet it gave a candidate for president, Euler worked to improve his mastery of it. The translation from German to French of his first paper for the *Mémoires*, which appeared in print at the end of 1745, was in his own hand.[97]

Frederick's Prussian court pressed the connection between military armaments and higher mathematics. The monarch and his General Directory wished especially to make their gunnery officers more competent. Artillery had long been part of applied mathematics, particularly the study of ballistic curves as well as the strength of metals in cannons and the firepower and elasticity of explosives, and the search for accurate descriptions of ballistic curves had drawn important early modern geometers. In 1546 Niccolò Fontana Tartaglia had published in *Quesiti et inventione diverse* his discovery of parabolic trajectories, and Galileo's *Two New Sciences* of 1638 had given a nascent ballistic theory neglecting air resistance and therefore holding true only for mortar shots of low velocity.[98] After writing by late 1727 a short piece on the testing of guns, Euler had continued to investigate the subject; in his study of artillery fire he read widely on fortifications by such authors as Nicholas François Blondel; Jean Errard; Hermann Landsberg;

Count Raimondo von Montecuccoli; Blaise François, Comte de Pagan; and Sébastien Le Prestre, Marquis de Vauban.[99] Computing the curved paths that mortar missiles traced in the air or in any fluid still posed a difficult challenge, although Euler credited Newton with solving it. Among other authors he read on the topic were Huygens, Johann I Bernoulli, and Daniel Bernoulli; he also studied Jakob Hermann's *Phoronomia* and Brook Taylor's *Propositiones* of 1721. These recognized that natural philosophers underestimated the resistance of air to projectiles of higher speed. Air resistance continued to be a major problem, and Johann I Bernoulli investigated whether it is proportional to v or to v^2. To this point no mathematician had yet provided the differential equations needed for practical artillery—a subject that Euler's *Methodus inveniendi lineas curvas* began to touch upon by investigating curves that apply to ballistics.

In 1742, as a good way of displaying the practical application of higher mathematics to the court, Euler apparently proposed an additional worthwhile task, translating into German Benjamin Robins's recently published *New Principles of Gunnery*, a 150-page book that Euler considered the best work on the subject. Ballistics had long been part of applied mathematics, and Euler's concentration on it met with an enthusiastic response from the king and his General Directory.

Prior to writing his *New Principles of Gunnery* Robins had severely criticized Euler's *Mechanica*, calling its author nothing more than a "calculating machine."[100] This attack, which stemmed from a rivalry between the British and the partisans of Bernoulli in mathematics, did not color Euler's judgment. In his "Eulogy in Memory of Leonhard Euler," Nicholas Fuss explained that "while giving Robins his due [for the *New Principles*], Euler in modest fashion corrected his errors, thus avenging Robins' disparagement of the *Mechanica* by making his book famous."[101] In a letter in December 1743, Daniel Bernoulli briefly commented on Robins's new gunnery book, which Euler admired. It is generally easier to translate from a foreign language than into one, but Euler's knowledge of English was sufficiently limited that he may have had assistance with the translation of Robins's work into his *Neue Grundsätze der Artillerie* (E77), which appeared in print in 1745.[102]

Relations with the Petersburg Academy of Sciences

Alongside his range of activities in Berlin and among scholars in central and western Europe, Euler still closely collaborated with the Petersburg Academy. His title, foreign associate, reflected neither the academy's response to him nor the magnitude of his labors on behalf of it. Nicolas de Condorcet remarked that Russian officials did not consider Euler a foreigner, and Fuss correctly wrote that he was member *à tous les titres* (in full status). Euler submitted papers for the *Commentarii* and its newer series after 1747, the *Novi Commentarii.*[103] Despite sending only two papers in 1742 and none in 1744 or 1745, Euler published with the Petersburg Academy of Sciences a total of ninety-six articles written in Berlin.[104] In general, these were more theoretical than his Berlin papers. Since the handful of publishers in Saint Petersburg took little notice of writings by academicians, Euler sought to find publishers for them in other cities, including Leipzig and Amsterdam. In Leipzig, then a major European book market, the dealer Jacob Schuster sold publications of the Petersburg Academy.

Euler also purchased books, instruments, and scientific materials, including a copper printing press, for the Petersburg Academy;[105] He sent information on research in the West, evaluated the studies of academy members, and recommended scholars to fill vacancies. While in philosophy Wolff was the principal disciple of Leibniz, by building up European scientific centers in Berlin and Saint Petersburg Euler was Leibniz's institutional successor.

Between 1741 and 1757, Euler exchanged 299 letters with Johann Schumacher of the Petersburg Academy. Under Empress Elizabeth, Schumacher's authority in the academy slightly diminished, but he continued to be the leading power in it, a good source of information on changes it was undergoing, and effectively Euler's boss there. The extent of Euler's correspondence and its scientific detail sent to an official little interested in the sciences suggests that he was attempting through Schumacher to guide the academy's research program.[106]

Euler's letter of September 1741 expressed gratitude for the "great affection and friendship" shown toward him in Saint Petersburg, but noted that in Berlin he had more opportunities for research than he would have had in Russia.[107] The next April he reported that being situated at the center of the German states he could more easily correspond with able people across central Europe in such cities as Bern, Dresden, Göttingen, Halle, and Leipzig. That month Schumacher thanked Euler for his letters, related that Goldbach had left the academy to join the College of Foreign Affairs in Moscow, and inquired about Euler's progress on the *Scientia navalis* and the *Introductio in analysis* [*sic*] *infinitorum,*[108] indicating that the Petersburg Academy would pay for their publication. In the spring of 1743, Euler informed the astronomer Christian Nikolaus Winsheim that *Scientia navalis* had been completed.[109]

Regarding his annual salary, Euler was tenacious. The financial condition of the academy was poor, but Schumacher reported in April 1742 that Euler's salary had been approved. In sending it the next month he nudged Euler to forward articles. The next year Empress Elizabeth provided no budget line for salaries to honorary members, but funds were nonetheless raised for Euler. As was typical, Euler sent greetings from his family to Schumacher's family, his wife, his brother Jakob, the academy architect, and Schumacher's brother-in-law Johann Conrad Henninger.[110]

The professional relations between Euler and Schumacher were proper but strained. Euler had avoided political quarrels between faculty and the chancellery official that had led many foreign members to depart. His immunity to professional assault grew with his reputation. His neutrality irked some members, but privately he was disturbed by Schumacher's attempts to undercut members and pit them against one another. In a famous case in 1743 Schumacher attacked Mikhail Lomonosov, having him imprisoned for eight months. Nevertheless, Lomonosov was named professor of chemistry two years later and read his paper "Meditationes de caloris et frigoris causa" (Reflections on the reasons for heat and cold), which rejected the dominant phlogiston theory of combustion.

Foremost among the many topics in the natural sciences that Euler covered in his correspondence with the Saint Petersburg academicians was astronomy, and he attempted to turn the academy more toward its study. He worked at devising differential equations to compute more precisely the orbits of comets, meanwhile developing his second lunar theory, which he knew had practical and theoretical significance.[111] Euler's skillful and extensive application of the calculus had phenomenal success; it supported his rejection of the notion that metaphysics and the sciences had separate truths that occasionally contradicted each other, and made that notion the basis for the modern exact sciences. In formulating and refining differential equations for astronomy, Euler supported the record keeping of improved observations gained with better instruments, and he was critical of mathematicians and natural philosophers who did not keep such records.

In April 1737 he had sent two letters to the Basel booksellers Johann and Ludwig Brandmüller requesting information on books of interest to him. They agreed to send him many books for his own use or for sale by consignment in the academy bookstore. After Euler left for Berlin Ludwig Brandmüller made repeated requests to the academy official Sigismund Preisser for an inventory of the books sold and unsold, but he received no reply. So he pleaded with Euler for help. In response Euler wrote to Georg Wolfgang Krafft and Heinsius about the matter. In February 1743 he asked Nikolaj Golovin, the president of the admiralty college, for clarification on the book sales. In June and July he corresponded with state councilor Andrej Konstantinovich Nartov, congratulating him for attaining the distinguished rank of head of the academic chancellery, asking about the pension owed him as an honorary member of the academy, and urging Nartov to order the bookstore to prepare a report detailing the number of the Brandmüller books sold and how many remained. Euler reported on his own sales and the payments he had sent to Basel; the remaining books he had turned over to the bookstore in care of Johann Albrecht von Korff. Apparently Nartov agreed to the request, thus putting an end to the problem—two years after Euler left Russia.

Chapter 7

"The Happiest Man in the World": 1744 to 1746

On 23 January 1744, the eve of Frederick II's thirty-second birthday, his royal commission approved the statutes founding the Royal Prussian Academy of Sciences. It was a spectacular occasion; a meeting on the monarch's birthday at his palace in Berlin conveyed his commitment to the new academy, which was portrayed as a rebirth of the old society. Beyond serving as the royal living residence, the palace of more than seven hundred rooms and salons was the seat of many branches of the Prussian government, including the treasury, and the center for the monarch's birthday festival week. Rather than convene in the old observatory, as had the Royal Prussian Society, or in the private homes of nobles, the new academy would meet in redecorated rooms above the royal stables. The long planned but incomplete academy building was to be part of a *Forum Fredericianum* made up of the royal opera house south of Unter den Linden, a new royal palace to the north, and the academy to the west of the opera.[1] A new observatory tower was planned to rise above the stables, but astronomers soon complained that the warmth and fumes rising from the stables clouded their instruments. Directors of all classes from the Royal Prussian Society were reappointed, except for Alphonse des Vignoles in mathematics, whom Abraham von Humbert replaced; here the king followed his selection for the Nouvelle Société littéraire. Leonhard Euler's recent refusal to become director of the Royal Prussian Society's mathematics class unless it and the Nouvelle Société littéraire were joined may have angered the king, who passed over him for that post in the academy.

Renovation, Prizes, and Leadership

Disturbed by the pace at which the academy was taking shape, Euler intensified his efforts to hasten the process. He continued to give close attention to astronomy, and on 29 January he urged the quick construction of a replacement of the old observatory. In a February meeting Count

Samuel von Schmettau, the acting president of the academy during its first three months, reported on the necessity of rebuilding the observatory, acquiring the latest instruments and models, and assembling a library.[2] As astronomers he appointed David Naudé, the son of Philipp, and Augustin Nathanael Grischow, whose father was also named Augustin; both Philipp and the elder Augustin had been prominent in the Prussian Society. Both sons had experience at the old observatory, and Grischow had studied in Paris and Saint Petersburg. At Euler's request, the president paid David Schumacher, the calculator of the Prussian calendar, one hundred Reichsthaler for continuing the *Manfredini ephemerides,* which the Italian geometer and astronomer Eustachio Manfredi had published since 1715 in Bologna. Ephemerides—astronomical tables of positions of a celestial body over specified intervals throughout the year—were used by navigators to determine longitude and by astronomers to follow the course of such celestial objects as comets. Schmettau agreed with Gottfried Wilhelm Leibniz's earlier proposal that the royal monopoly over the Prussian state almanac be the primary source of revenues for an academy. Because of his capacity and experience he recommended that David Köhler, who had recently conducted an accurate study of calendar debits, continue to manage and distribute the state almanacs, a project that Euler supported; Köhler was given precise instructions on the use of funds. In March Schmettau and the king had the academy investigate establishing as a source of revenue a new press to print the almanac, its *Opera*, and translations of small Protestant religious books. Euler led in bettering astronomical and mathematical bases for the preparation of calendars for Prussia and elsewhere. An attempt by the academy in 1744 to force booksellers and printers to collect taxes on its almanacs would have little success.

In February 1744, Euler wrote to Joseph-Nicholas Delisle expressing his relief at the founding of the new academy but adding, "I was very much mistaken when I thought they would put the new academy on the same footing as that of Paris. The thing is done. We have joined into one body the old and new society [soon officially] under the name of an Academy of Sciences and Belles-Lettres." Euler recognized that the institution faced financial problems; only a few members had pensions, and the almanac revenue could not fully fund adequate equipment for the observatory. Euler lamented to Delisle the poor condition of the old Berlin observatory, noting especially that its telescopes were inadequate for checking the observations of the transit of Mercury that Martin Knutzen was making in Königsberg. Bad weather slowed the academy's research, and from a comet briefly visible from Berlin it was not possible to obtain enough accurate observational data to test Knutzen's new method for computing the paths of comets; that method required four exact observations in order to determine the orbits. Throughout 1744, Knutzen sent his observations of comets and the moon with Newtonian telescopes, which he claimed were more accurate. On the basis of observations of comets that Johannes Hevelius had made in 1652 and Gian Domenico Cassini in 1698, Euler came to believe that they appeared in a forty-five-year cycle. He knew that Britain had more precise telescopes, and the Berlin observatory lacked instruments exact enough to develop systematic, reliable astronomical records needed for him to devise and refine differential equations for celestial motions. While his mathematically formulated theories and celestial mechanics were based on physical principles, in the project on the shape of Earth and many others Euler's mathematics was empirical.[3]

In a letter to Christian Goldbach in July 1744, Euler called Alexis Claude Clairaut's *Théorie de la figure de la terre* "indeed an incomparable work . . . where profound and difficult questions . . . are treated with pleasant, light methods, following which the author makes the most sublime matters completely clear and intelligible."[4] He cautiously awaited the return of the Paris Academy of Science's geodetic expedition to Peru under Charles Marie de la Condamine to support the Newtonian calculation that there must be a bulge at the equator. Sent to a section of Peru that is today Ecuador, the expedition faced difficult conditions, including substantial ranges of temperature and heights, but made excellent measurements of an arc of meridian. Pierre Bouguer returned from the expedition to Paris in 1744, but La Condamine, delayed in Cayenne for five months because he was unable to find a ship to France, did not arrive until the next year with his data confirming the equatorial bulge.[5] The measurements of a degree of an arc of meridian made by the Lapland and Peru expeditions taken together challenged the Cartesian position and the authority of the Paris Observatory astronomers headed by Jacques Cassini, and the question of the shape of Earth was now essentially resolved. Euler, who acquired La Condamine's description of his trip to Peru and Bouguer's account of their measurements, sent both to Delisle in March 1746.

The new academy accepted the practice in the Nouvelle Société littéraire of having a yearly prize competition to examine an issue in the natural sciences important throughout Europe. The model was the Paris Academy's Prix de Paris, which carried an award of fifty ducats, as would the new Berlin prize. The subjects chosen for the Berlin prize competitions were indicative of what awakened curiosity within the educated public in Prussia and across the Continent at that time. The topic in 1744, electricity (which was then conceived of as a subtle fluid), addressed a field of research that had become popular in Europe and North America. It was thought that electrical phenomena would most likely demonstrate the interatomic forces of attraction and repulsion that Isaac Newton had defined. Berlin academicians experimented with rotating glass tubes that generated shocks or fire when touched.[6] In May 1745 Euler announced, on his own initiative, the granting of the prize to the Kassel physicist and treasury official Jakob Sigismund, Freiherr von Waitz for his paper "Abhandlung von der Electricität und deren Ursachen" (Treatise on electricity and its causes).[7] In Saint Petersburg Georg Wilhelm Richmann began his writings on electricity in 1744–45, while at the Berlin Academy of Sciences Christian Friedrich Ludolff ignited gaseous spirits with an electrified iron wand. The next year Ewald Georg von Kleist of Köslin, fascinated by electrical flares, constructed the Leiden jar, which collected a powerful electrical force in the bottom of a bottle. This contradicted the belief, prevalent at the time, that electricity could traverse thick glass. Pieter van Musschenbroek explained an experiment similar to Kleist's exciting discovery and now gave the jar its name.

Prompted by members of the Royal General Directory formerly belonging to the Nouvelle Société littéraire, the new academy had proceeded in early 1744 to address questions concerning faith and reason, immortality and free will, and the clash of materialism with idealism. Between Calvinist ministers in the French community in Berlin and partisans of the German stream of the Enlightenment as well as among the *Aufklärer* (Enlighteners) themselves a rift in what became their world outlook was growing. The teachings of Benedict de Spinoza provoked the most vehement discussions; Reform clerics considered Spinoza a promoter of atheism, whereas Calvinist clergy and noble leaders of the Nouvelle Société littéraire generally associated Cartesian mechanistic science with Spinoza's teachings and considered Wolffian philosophy among movements incompatible with full spirituality. In March 1744 academy secretary Philippe Joseph de Jariges tried to refute Spinoza's thought.[8] Pietistic charges of atheism against Christian Wolff from twenty-five years earlier in Halle were revived. Attempting to bring strains of the radical French Encyclopedists into the academy, Jean-Henri-Samuel Formey presented in April a paper on rights and injustices in connection with the natural law of Hugo Grotius and Samuel von Pufendorf. Formey, the pastor of the influential Französische Friedrichstadtkirche (French Reformed Church in Friedrich City), which Euler attended in the *Gendarmenmarkt*, supported the Wolffian philosophy to which the minister had converted in 1739. Church services were conducted in French.

Figure 7.1. The Französischer Dom in Berlin was built from 1780 to 1785 adjacent to the Hugenot Französische Friedrichstadtkirche (French Reformed Church). Euler attended the Friedrichstadtkirche.

Central to this quarrel in which science stood in varying relations with faith was the question of what was the least component of matter. The Nouvelle Société littéraire had embraced the monadic doctrine of Leibniz as modified by Wolff. Monads, in Leibniz's later description of the physical universe, are geometric points of energy; composed of these points, matter is infinitely divisible. In the north German states, Wolffian philosophy exercised its greatest influence in Berlin, where, according to Nicholas Fuss, the reading public then spoke of nothing but the monadic doctrine and its metaphysical foundation.[9] On 4 and 18 June, and on 19 September, Euler spoke out against Wolff's science; he was determined to keep its core doctrine from becoming dominant. Euler warned that the concept of animate monads on which Leibniz had based his cosmology would lead to atheism. He did not accept the belief on the part of Leibniz and Wolff that reason was the basis of all knowledge and the highest court for resolving issues of faith. Like Pierre Bayle, Euler argued instead that faith and reason are equals and that reason was not the basis for knowledge provided by revelation. While in his criticism of the monadic doctrine many of Euler's initial arguments on the theory of matter were scientific, his underlying purpose was to defend traditional Protestant theology against what he considered Wolffian freethinking and determinism.

In June 1744 Euler delivered his first paper to the Berlin Academy, "Recherches physiques sur la natur des moindres parties de la matiere" (Physical research on the nature of the smallest parts of matter, E91), but it did not appear in print until 1746, in his *Opuscula varii argumenti*.[10] The paper praised the microscope for opening the study of the infinite and diverse least parts of matter, which Euler labeled as molecules; the subject, he asserted, belongs to physics. To confirm the existence of molecules he presented Newton's demonstration that the mass of a body is proportional to its inertia and that all molecules have the same density and specific gravity; inertia, he asserted, replaces the two forces that Leibniz and Wolff had attributed to monads. Euler looked for similarities and differences between total force and Leibniz's living force. The paper presented a very subtle matter, fluid and elastic, comprising the ether that fills the physical universe; this was a view that differed from that of Newton and was sometimes mistakenly associated with the Cartesian concepts of corpuscles and the plenum rather than Euler's pursuit of fluid mechanics and the beginnings of field theory. While the divisibility of matter is incomparably great, the paper maintained, it is finite and the least elements of matter are indivisible. Euler was about to revise sharply his theory of matter, however. Along with the telescope he called the microscope crucial in making possible further scientific discoveries.

In September 1744 Euler argued that the fundamental attribute of the smallest constituent of material substance is impenetrability, which the monads lacked. That November he rejected as a basis for mechanics the indivisible and immaterial Wolffian monads endowed with perception. The next year the academy set the monadic hypothesis as the prize topic for 1747, presumably for the philosophy class to conduct. The general cause of winds, the topic for 1746 under the direction of the mathematics class, was a temporary distraction from this greatly controversial issue. The placement meant that Euler would be the primary judge in choosing the winning paper.

After the Second Silesian War began in May 1744, Frederick II, moving from one military camp to another, had almost no dealings with the academy until January 1745, when following the death of Philipp Naudé the curators proposed allocating to Johann Nathanael Lieberkühn the two hundred Reichsthaler from Naudé's retirement. In a handwritten note, the king responded with an emphatic "No!" and proposed that Euler identify an able scholar in Russia capable of succeeding Naudé.[11] Euler apparently thought that the leadership of the academy, and particularly the presidency, was not yet assured; he seemed to believe that in the absence of Pierre Maupertuis he might be eligible for the post.[12] As the foremost mathematician in Europe and winner of several Paris Academy prizes, he considered himself a logical candidate, but while creating a better future for the sciences, Euler occasionally lived in that future and not within the aristocratic restrictions of his time. His translation into French of his first article for the opening volume of the *Mémoires de l'Académie Royale des sciences et des belles-lettres de Berlin* was probably intended to support his case for the presidency. But the Swiss burgher Euler was not an enlightened *philosophe*, witty in conversation, well-versed in French—the attributes that Frederick sought in a prospective president.

In 1745 Maupertuis's most popular book, *Venus physique*, on the life sciences, appeared in print. Maupertuis had opponents in Paris, and to protect himself from censors he published it anonymously. The debate over the book's authorship only increased its readership, however. In the controversy over embryology Maupertuis defended epigenesis, which held that organisms develop from both parents rather than being preformed. Critics charged that this led to materialism and pantheism; they also questioned Maupertuis's morality for studying the female role in generation—as well as his readiness to leave Paris for Berlin. But since July 1744, when he had sent Euler a copy of his *Astronomique nautique* on navigation that accepted Newtonian attraction, Maupertuis was not in contact with the academy or the king. Only after Frederick's victory at the furious Battle of Hohenfriedberg in June 1745, which was disastrous for the Austrian and Saxon Army, did Maupertuis write that he had permission from the Paris Academy and would return to Berlin. On 4 July Frederick received the message, and a week later he urged Maupertuis to come to Berlin as quickly as possible.[13] Up to this point Maupertuis had conducted research and negotiated allegiances and privileges from Paris while Euler was directly involved in the formation of an important academy. In a letter to Euler from July 1745, Daniel Bernoulli praised Maupertuis on his impending arrival in Berlin and observed that Frederick's entire court would accept the nobleman and distinguished scholar Maupertuis, thus assuring his elevation to the presidency; he predicted that Maupertuis's leadership would bring an expansion in support for the academy.[14] Maupertuis's return to Paris in August ended for the moment any possibility of Euler's appointment to the position.

Between July and the end of the Second Silesian War with the signing of the Dresden Treaty at Christmastime 1745, Maupertuis sent Frederick sixteen letters. The title Maupertuis's Academy has described accurately his autocratic ways and his court politics at Potsdam, which should not be misconstrued to indicate that he was the primary source organizing the institution and the scholar conducting the foremost research in Berlin.[15]

Even before Maupertuis arrived, Frederick approved his salary of three thousand Reichsthaler. At the academy meeting on 22 September 1745, the monarch welcomed him and instructed him to proceed with electing directors along with preparing and managing the budget. But Maupertuis was not yet president and lacked the authority of the royal ministers; he introduced himself to the nobility residing in Berlin but soon moved also to the royal court at Potsdam. His marriage to Kaspar Wilhelm von Borcke's close relative Eleanor in October aided his integration into Berlin's high society and the Prussian aristocracy. The marriage outside the Catholic Church made Maupertuis uneasy, but he and his wife seem to have had good relations, although they lived apart during many of his frequent trips. Activities at court occupied Maupertuis while Euler, never gaining admission to the noble circles in Berlin and Potsdam, had more time for research.

In the absence of Maupertuis until July, and deprived of the full measure of Euler's productivity while he struggled late that year with a severe fever and a period of recovery lasting into December, the new academy was slow to develop in 1745. News in 1744 of the death of both Euler's father in March at age seventy-five in Riehen and of his maternal grandmother Maria Magdalena Brucker-Faber (with whom he had lived for most of his Swiss years) saddened him, but he did not return to Basel.[16] Euler remained as active as his sickness permitted him. In July 1745 Goldbach wrote from Saint Petersburg, congratulating him on becoming director of the mathematics class,[17] an appointment that was still unofficial. Euler's primary organizational project now was to press for building a new observatory equipped with the latest instruments. Astronomical observations, many of them contradictory, were being made across Europe and Asia; their unresolved differences and complications arising from irregularities in the motion of celestial bodies puzzled many geometers. Notably, adequate mathematical methods to solve coupled second-order differential equations and equations of motion, in part to determine the resulting forces, did not yet exist to reach a well-founded mathematical theory in celestial mechanics. And few dioptrical rules existed for the construction of better telescopes, a subject that he soon addressed. Euler considered the observatory vital to research and discoveries in the mathematics class; he was encouraged by the findings of Delisle, whom he wanted to draw to Berlin.

Although Frederick wanted to honor all of Maupertuis's requests, the exact authority of the president along with the protocol and organization of the new academy had yet to be resolved. In October 1745, when Maupertuis asked to design and obtain the academy's regulations, Frederick declined, since he was away from Berlin; much of his time was taken completing the Treaty of Dresden, which again confirmed Silesia as a Prussian territory. Frederick was embarrassed that until he returned he would be unable to deal with the academy and Maupertuis's position, which for the time being was only that of director. Still, when the academy met on 6 December 1745, the king had Maupertuis named as its president but failed to clarify the reach of the new appointee's authority among the Prussian nobility. While Frederick did not interfere with individual research, he had earlier prescribed that all articles in the academy's *Histoire de l'Académie des Sciences et de Belles-Lettres de Berlin, avec les Mémoires pour la même année* (History of the Academy of Sciences and Belle-Lettres in Berlin, with the Memoirs for the same year) be in French. His reasons were several: Frederick wanted the Berlin Academy of Sciences to interact with discussions that were happening in the Parisian salons and to win their plaudits; he supported the powerful older French Calvinist colony in Berlin, which had defended him against near execution at Cüstrin in 1730 when he was crown prince;[18] and he backed the efforts of the growing number of savants within that community and the prospects of their learned journals, such as the *Bibliotheque critique*. In elevating French, Frederick supported a practice of the Nouvelle Société littéraire and the course that Maupertuis and Formey advocated.

Although the Paris Academy had switched from Latin to the vernacular French as far back as 1699, Latin was still the chief language of learning in Europe, and the Berlin Academy's shift to French surprised many in the republic of letters outside Berlin and caused bewilderment among older academicians. Maupertuis desired a world language; Latin might do, but he preferred French as a living speech, noting that it was the second language of the reading public across Europe, while Latin was a dead tongue more helpful to poets and writers than to natural philosophers. French was in frequent use in pathbreaking scientific publications. French, Maupertuis believed, was attaining a perfection and subtlety expressive of a *bel esprit*, a man of wit.[19] Formey, who had been appointed historiographer in July 1745 and was thus a member of the scientific directory, likened the role of French in learning to that of Greek in the age of Cicero and asserted that the use of French would add precision to scholarly discussions.[20] He announced that the *Histoire* would publish in French all of the best works on the sciences in the German states and Great Britain; he began translating articles written in Latin and German and made arrangements for the continuation of that work. Authors in the *Histoire* or *Mémoires* section could submit a draft of their papers in another language but had to arrange for their translation into French, which was normally a difficult task. Since Maupertuis did not know German and the king preferred French, academy discussions were also to be in French. The monarch never participated in academy meetings himself, but he did expect princes and court nobility to attend public sessions.

Investigating the Fabric of the Universe

Driven by extraordinary genius, insatiable curiosity, and great energy, in 1744 and 1745 Euler had three books and thirteen articles published. He also amassed two hundred pieces of correspondence to and from friends and colleagues; these he reviewed for the academy as regards important findings in mathematics and the natural sciences. In April 1744, he informed Goldbach that once the printing of the *Methodus inveniendi lineas curvas maximi minimive proprietate gaudentes* (The method of finding curves that show some property of maximum or minimum, E65) was completed, the publisher Marcus-Michael Bousquet would turn to Euler's *Introductio in analysin infinitorum* (Introduction to the analysis of the infinite). By July he still had not yet heard about the status of the *Methodus inveniendi*, but he had sent the publisher the complete manuscript for the *Introductio*, in which he resolved many difficult problems insoluble with algebra or geometry. He described the book "as a prodromus to analysin infinitorum"— preliminary to his later texts on differential and integral calculus;[21] its publication would take almost four years. Bousquet wanted Gabriel Cramer to supervise the publication of the *Introductio* after the printing of the *Methodus inveniendi*, but in September Cramer informed Euler that he was too busy and wanted to continue writing a treatise on the same subject that book 2 of the *Introductio* covered; Bousquet instead chose Johann Castillon. In December Euler wrote to Cramer thanking him for the corrections he had made to his *Methodus inveniendi*.

The 320-page *Methodus inveniendi* in quarto format was the first book Euler published in the 1740s. A landmark treatise, it appeared in print in September 1744. This was fast, for Euler had submitted its appendixes only the previous December. The study made him the principal creator of the first stage of a new branch of mathematics, the classical calculus of variations, which in the paths of motion sought to determine maximal or minimal lengths of plane curves, if any existed, and pursued extreme values for integrals (often named functionals). Its first section asserts that "since the fabric of the universe is most perfect, and is the work of a most Wise Creator, nothing whatsoever takes place in the universe in which some relation of maximum and minimum does not appear."[22] A

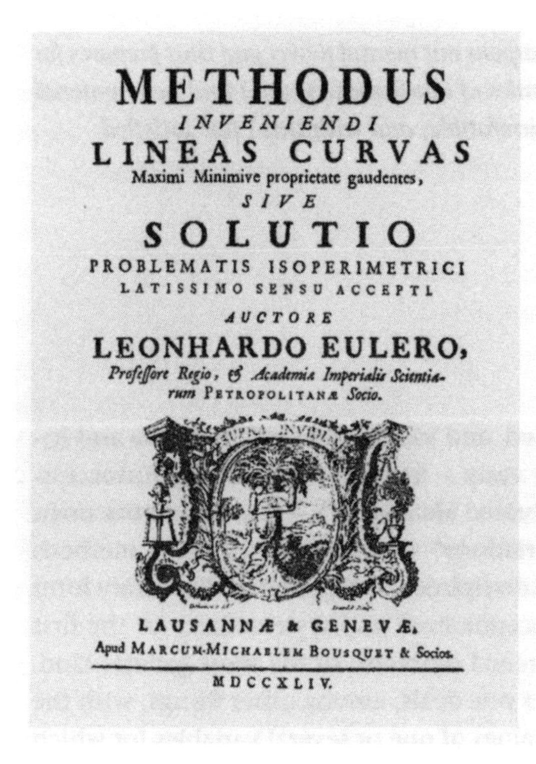
METHODUS
INVENIENDI
LINEAS CURVAS
Maximi Minimive proprietate gaudentes,
SIVE
SOLUTIO
PROBLEMATIS ISOPERIMETRICI
LATISSIMO SENSU ACCEPTI.
AUCTORE
LEONHARDO EULERO,
Professore Regio, & Academiæ Imperialis Scientiarum PETROPOLITANÆ *Socio.*

LAUSANNÆ & GENEVÆ,
Apud MARCUM-MICHAELEM BOUSQUET & Socios.
MDCCXLIV.

Figure 7.2. The title page of Euler's *Methodus inveniendi,* 1744.

letter of December 1745 to Maupertuis repeats Euler's conclusion from the *Methodus inveniendi* that "in the natural course of movements there is a constant maximum or minimum, and I have determined . . . that all trajectory curves, and all bodies drawn toward a fixed center or mutually drawn together have been so described."[23] The baroque title of the work derives from Euler's perception of the new field as a Leibnizian *ars inveniendi*, or method of discovery.[24] Its twentieth-century editor Constantin Carathéodory called it "one of the most beautiful mathematical works ever written."[25]

Through skillful organization in arranging more than a hundred increasingly complex problems in eleven categories and providing new direction and ideas in the *Methodus inveniendi*, Euler replaced the previous ad hoc procedures for special case problems in the formative stage of the calculus of variations, instead offering standard differential equations for general solutions and giving techniques for reaching these equations. His work impressively extended and refined that of Brook Taylor, Jakob

Hermann, and Jakob and Johann I (Jean I) Bernoulli, and was the culmination of their efforts; Euler's success where they had failed in creating the new field magnified his reputation.[26] His methods were closest to Jakob (Jacques) Bernoulli's in that they require two degrees of freedom to extremalize or optimize a curve. Euler's attention to curves, including the tautochrone, isochrone, and brachistochrone, and his use of isoperimetry, a subject popular in the late seventeenth century, kept the new field largely geometric. This was one of the several occasions on which Euler searched for a different solution for the brachistochrone problem. Chapter 3 of the *Methodus invenendi* generalizes this early optimization problem, essentially finding sets of extremals, and section 45 provides the most elegant solution up to 1744, correcting Hermann's solution of the brachistochrone problem in a resisting medium. Chapter 2 contains the vital innovation, the Euler differential equation or the first necessary condition for extremals (now known as the Euler-Lagrange Equation, it is the basic equation in the modern calculus of variations), and chapter 4 takes up the problem of its invariance, but Euler did not recognize that the equation was insufficient to guarantee an extremum.[27] As was typical for Euler, he gave some hundred examples to illustrate its results.

In "De curvis elasticis," the first of two significant appendixes to the *Methodus inveniendi* that Daniel Bernoulli had proposed in a series of letters, Euler—over sixty-six pages with ninety-seven problems—presented the earliest study in print to employ the calculus of variations to solve problems in the theory of elastic curves and surfaces. It is thus the initial general tract on the mathematical theory of elasticity. To obtain his equations, Euler employed the methods of final causes and efficient causes; these came initially from Aristotle, and among others Leibniz had recently studied them. Final causes are teleological, giving the purpose or design of something and contrast with mechanical explanations, which Euler believed to draw upon existing variational principles. Closest to the modern definition of the term *cause*, efficient causes probe the properties of matter and mechanics explaining phenomena. Euler computed the shape of elastica from the forces of efficient causes, and checked to confirm that both approaches led to the same answer; without an appeal to both, the best explanations might not be reached. In his inventory of problems, Euler enumerated nine species of elastic curves and explained how elastic bands bend and oscillate.

The appendix's topics include the problem of the vibrating membrane, at the same time that Daniel Bernoulli was investigating the simpler vibrating string. Euler's buckling formula first appears here;[28] it determines the maximum critical load, now called the first *elastostatic* eigenvalue, which an ideal, slender, long rod pinned at its ends can carry before it

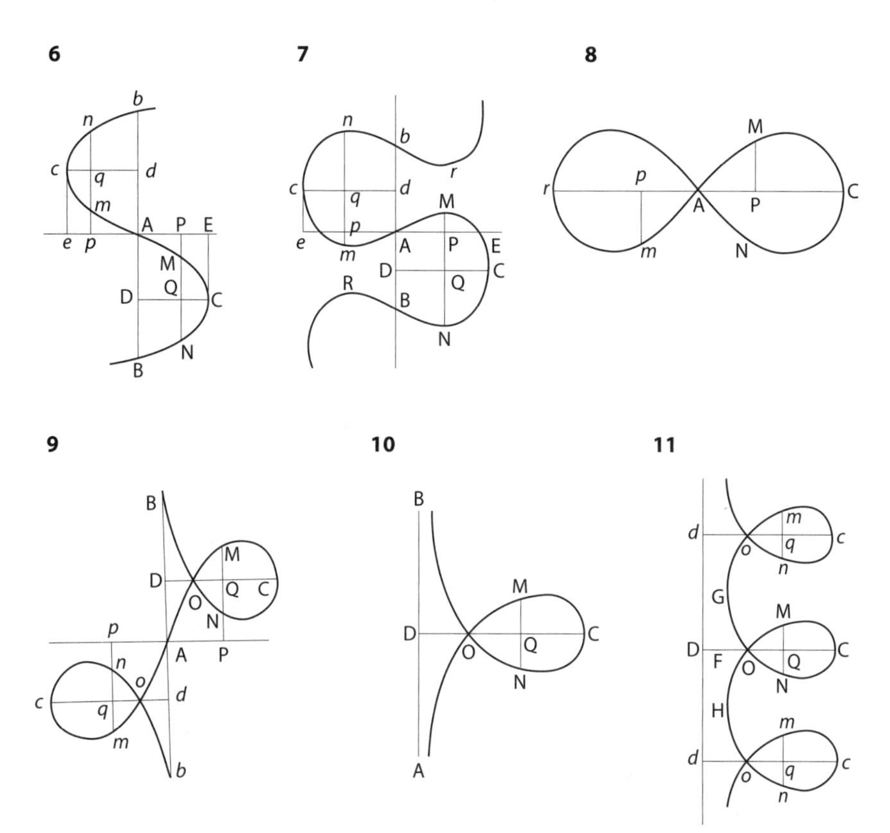

Figure 7.3. Euler's sketches and two earlier basic curves (6 and 7) for "De curvis elasticis."

buckles. The critical axial load applied at its center of gravity needed to bend the rod depends upon the stiffness of its material and how the rod is supported at its ends, and it is proportional to the inverse square of the length of the rod. Euler also computed *elastokinetic* eigenvalues, eigenfrequencies of oscillations of the rod's transversal, and associated eigenfunctions, giving the shapes of a deformed rod.

The other appendix, the ten-page "De motu projectorum in medio non resistente, per methodum maximorum ac minimorum determinando" (On the motion of bodies in a nonresisting medium determined by the method of maxima and minima), introduces a general form of the principle of least action and experimentally determines absolute elasticity.[29] His letters to Daniel Bernoulli show that by late 1738 Euler had mastered that principle.

Euler was now devoting more time to astronomy. At midcentury, when the initial transformation of celestial mechanics into a modern exact science was paramount in the mathematical sciences, its chief unanswered question was whether Newton's inverse-square law of gravitational attraction could exactly describe all astronomical motion. Observational and theoretical astronomers as well as geometers charted the precession of equinoxes; perturbations in the movement of planets and their satellites; inequalities in the mean motion of the two largest planets, Jupiter and Saturn; and the most urgent problem, the motion of the moon. The search was not easy; most of these calculations involved the difficult three-body problem, supposedly the only problem to have given Newton headaches. But this did not apply to the precession of equinoxes, which was a mechanical problem because it lacked the principle of angular momentum. There were conflicting and inaccurate astronomical observations that would slightly slow the development of celestial mechanics for some, but the accuracy attained in the lunar theories of Clairaut, Euler, and Johann Tobias Mayer came rapidly.

Figure 7.4. The frontispiece, with the solar system and comets, from Euler's *Theoria motuum planetarum et cometarum,* 1744.

The year 1744 saw the publication of Euler's second Berlin book, *Theoria motuum planetarum et cometarum* (Theory of the motions of planets and comets, E66), his first on astronomy.[30] This smaller, 187-page treatise was printed anonymously after another comet was sighted on 18 January, the fourth sighting since 13 December. Not one of Euler's chief works, it was among the few translated into German; Johann von Paccassi's translation was published in Vienna. Euler also penned two anonymous supplements that soon followed its publication.[31]

The *Theoria motuum* begins with the orbit of Mercury and makes a table of the interpolations from the radius vector to compute eccentricities more closely. Euler calculated the interpolations with logarithms, trigonometric functions, and fourth-degree roots. He developed the first differential equations for computing every point in the orbits of Earth and Mars. Thereby he illustrated new methods for examining planetary perturbations. Using the latest telescopic observations, Euler differentiated comets from fixed stars and described the research of the time on both of them. While the mass of comets was still unknown, he found their orbits to be nearly parabolic ellipses and sought to devise functions for computing each course element, including anomalies. A much elongated ellipse was assumed if comets were permanent. But Euler found their path closer to a parabola. Comparing accurate observations of the comet of 1742, constructing variations of similar ellipses and parabolas, and employing records concerning comets in 1743 and 1744, Euler revised and improved upon Jacques Cassini's method for computing the time at which a comet reaches perihelion, its nearest point to the sun. After gathering the latest observations, Euler concentrated on formulating new differential equations that better traced the paths of comets. The application of his equations and formulas provided the most accurate computation of most points in the orbits of planets and comets to that time. But the *Theoria motuum* also cites unresolved problems in constructing a precise, theoretical account for comets' entire orbits; Euler recommended caution toward the belief that comets foretell the wrath of God, and rejected the notion—even though they could come close enough to do so—that they would ever destroy Earth, citing the Bible as denying that this could happen.

This work was formulated entirely as a two-body problem and not the classical three-body problem that had drawn Euler's interest as early as 1730.[32] Until the nineteenth century, the book was fundamental for calculating the orbits of planets and comets. It posed severe difficulties for the publisher Ambrose Haude in Berlin. A small but difficult work, it contains errors in the printing of computations and of some formulas; through footnotes, Euler attempted to correct many of these.

Among Euler's articles in 1744 was the popular "Beantwortung verschiedener Fragen über die Beschaffenheit, Bewegung und Würckung der Cometen" (Answer to different questions concerning the nature, motion, and operation of comets, E67). Holding comets to be permanent rather than transient bodies, he insisted that they required further research that must include their tails, which he had depicted as dust particles driven from their cores by rays of the sun. In June he read a letter from Clairaut in Paris describing observations of comets made by Clairaut and Jacques Cassini. Euler's paper on new astronomical tables, "Novae et correctae tabulae ad loca lunae" (New and correct tables for computing the location of the moon, E76), published after October 1745, considered for the first time the perturbations of the sun along with adding an overlooked correction to Johannes Kepler's calculation of the position of the sun. The status of comets, Euler reiterated, was not yet resolved. This was a year after Daniel Bernoulli had sent Euler his improved equations for calculating the paths of comets based on Newtonian dynamics; Bernoulli's letters for 1744 and 1745 also reported on the observations and computing of the orbits of planets made by the young mathematician Philippe de Chéseaux in Lausanne, the possible influence of a comet on Mercury, and the transit of Venus, the so-called holy grail of astronomy. Since the orbital plane of Venus is inclined with respect to the ecliptic, the planet does not annually pass directly between Earth and the disc of the sun with the effect of blocking the sun. Instead this phenomenon occurs in pairs about eight years apart, the pairings separated by more than a century; in the eighteenth century the dates were 1761 and 1769. In 1744–45, Euler developed new statistical methods to handle redundant data. These methods were more crucial than increasingly accurate observations and measurements of the transit to compute it more precisely. To make these observations astronomers employed better telescopes and chronometers as well as instruments with sighting wires. They now obtained the mean solar parallax and, from this, calculated for the first time in Earth measurements the exact distance of Earth from the sun.

In 1745 Empress Elizabeth's allocation of 6,000 rubles for a new observatory heartened Euler. He had—independently of Clairaut and Jean-Baptiste le Rond d'Alembert—been investigating lunar motion, and had made corrections to Gottfried Wilhelm Heinsius's lunar tables. In December 1745 Euler sent the improved tables to Saint Petersburg and to Heinsius, who had moved from the Petersburg Academy to the University of Leipzig the previous year. In these tables Euler computed with differential equations the perihelion and the inclination (obliquity) of the ecliptic. Applying new statistical methods and physical principles that he devised

Neue Grundsätze
der
ARTILLERIE
enthaltend
die Bestimmung der Gewalt des Pulvers
nebst
einer Untersuchung
über den Unterscheid des Wiederstands der Luft in schnellen und
langsamen Bewegungen

aus dem Englischen des Hrn. Benjamin Robins
übersetzt und mit den nöthigen Erläuterungen und
vielen Anmerkungen versehen
von
Leonhard Euler
Königlichem Professor in Berlin.

SAPERE AUDE.

Berlin bey A. Haude
Königl. und der Academie der Wissenschaften
privil. Buchhändler. 1745.

Figure 7.5. The title page of Euler's *Neue Grundsätze der Artillerie,* 1745.

in 1744–45, along with superior skill in the calculus of trigonometric functions and newly derived inequalities and improved observations, Euler was able to compute "Tabulae astronomicae solis & lunae" (Solar and [new] lunar astronomical tables, E87); this treatise, with its improvements for lunar motion, was published in 1746 as part of his *Opuscula varii argumenti*, part 1.

Euler's third book in Berlin, *Neue Grundsätze der Artillerie* (E77), was published in 1745. In this edited version of Benjamin Robins's *New Principles of Gunnery* of 1742, Euler added many explanatory annotations, criticisms and corrections of analytical errors, appendixes, and materials on explosives and artillery that were missing in the original. The result was a book of 720 pages, almost five times longer than Robins's original.[33] Where Robins's book was simply a "little budget of rules, experiments, and guesses,"[34] Euler's *Neue Grundsätze der Artillerie*—although imperfect—was a beginning toward transforming the collection of separate rules, thus making it the initial scientific treatise on gunnery.

In this project Euler began by stressing practical details and numerical computations rather than guiding principles. The problem of the curved paths of projectiles still remained a challenge that he mistakenly believed Newton had solved. Like Newton, he investigated resistance to projectiles in the air or any fluid, including rarefied fluids that do not exist. Johann I (Jean I) Bernoulli had tried to find whether it was to v or to v^2 that the air resistance was proportional, and Taylor underestimated it at higher speeds. (That rarefied fluids are absent from our physical world did not prevent Euler from making possible a case for his theory as well as computations for it. Among mathematicians, ventures into the purely theoretical are not uncommon.) Still lacking a general formula for air resistance, Euler computed its effect through a combination of rules; from this work he found that only at higher velocities does back pressure not exist on a projectile. At lower speeds, his techniques underestimated air resistance. In examining how different temperatures influence the compression of air, Euler developed what are called thermal equations of state. For gaseous substances, these gave for an enclosed space the variable volume and varying pressure recorded at different temperatures; Euler recognized that there might be more than fifty variables in the equations. Euler's study was a matter more of mathematics and speculative natural philosophy.

With the *Neue Grundsätze* and a series of four articles to 1753, Euler developed the first accurate differential equations for ballistic motion in the atmosphere. Taking delight in these equations, Euler placed within calculus his computations for ballistic curves. In better arranging the field, he divided trajectories into families and integrated them as combinations of muzzle velocity, angle of elevation, and air resistance. His simpler approximation theories in ballistics were an advance, but did not compare in difficulty with the attempt to devise closer approximations for the problem in astronomy of three mutually gravitating bodies.

While its concentration was on internal ballistics, the *Neue Grundsätze* also depended on chemical science; it opened with a study of the nature of air and fire and a theory of fluids, and referred to Daniel Bernoulli's "incomparable *Hydrodynamica*." Euler's study of ballistics—for which, in September 1745, Johann I Bernoulli wrote to him making critical observations—benefited from the cooperation of artillery field marshal Count Samuel von Schmettau, and it was to Schmettau's advantage that the project fitted into Euler's larger research program. Publishing it as one of his few books in German rather than Latin or French shows that it was intended mainly to improve the competency of lower military officers. Euler had written extensively on the subject in recent years; this suggests ongoing pressure for the project from the royal court. For over a half century, no work in artillery superior to the *Neue Grundsätze* appeared.

In addition to making extensive applications, Euler continued to set forth the foundations and rules for differential calculus. By mid-1744, for example, in what would appear in book 1, chapter 6 of his *Introductio*, he defined in detail—and more neatly than had previously been done—the logarithm function as the inverse of the exponential; he made this definition popular for the first time in history. But his account of the logarithm function was incomplete, applying only to positive real numbers. After noting in paragraph 103 of chapter 6 that the logarithm of a negative number is an imaginary number—that is, a complex number—in book 2, chapter 21 he cited paradoxes that made it difficult to extend these to the set of all complex numbers. As his correspondence with d'Alembert indicates, in two years he had removed these impediments. As 1746 approached, Euler must have been intensely engaged in developing for differential calculus underlying concepts that remained to be discovered in the provision of clear definitions and fundamental principles. Among other subjects, his studies examined infinite series, differentials, the distinction between "absolutely nothing and a special order of quantities infinitely small," and the character and properties of algebraic and transcendental functions; the results were published in his *Institutiones calculi differentialis* (E212) of 1755.

In the mid-1740s his research on Pierre de Fermat's conjectures moved Euler slowly toward the law of quadratic reciprocity in number theory. He continued to be drawn to the depth of the reasoning in number theory and saw it as an important field in mathematics; it was a major component in his efforts to reach across the whole of mathematics. Sometimes overlooked, his "Theoremata circa divisores numerorum in hac forma $paa \pm qbb$ contentorum" (Theorems about divisors of numbers contained in the form $paa \pm qbb$, E164),[35] his tenth paper on number theory, was published in 1751 (in the volume for 1744–46 of the Petersburg Academy's *Commentarii*). Building upon Goldbach's letter of 28 August 1742, its results relate to factors in the law of quadratic reciprocity. Euler gave fifty-nine statements that he called theorems, offering them without proofs, and he arrived at them apparently by inductive reference. Over the next decade, he provided the missing proofs. Some theorems were simple, others took his great genius to grasp. After the first theorem, Euler gave all except for one in groups of three. The second insists that all prime divisors have this form; and the third is the contrapositive of the first. These theorems give Euler's criterion for when integers a and odd prime p have quadratic residue modulo p or quadratic nonresidue modulo p. Quadratic residues are perfect squares, modulo p. He also probed properties of sums of two squares. In this research Euler demonstrated great computational skills and he identified patterns in the vast array of numbers. That half of his eventual ninety-six papers on number theory were published only posthumously suggests that the subject had not yet been highly regarded in mathematics.

From 1744 to 1746 Euler was increasingly interested in optics. During his career he would spend thirty years investigating its nature and developing differential equations that could be applied in making achromatic objectives and multiple lenses for the improvement of telescopes.[36] He wanted to provide a new theory of light and colors that explained all optical phenomena most clearly and avoided the difficulties hampering some other theories. At the academy he presented in 1744 his "Dissertation sur la lumière et les couleurs" (Dissertation on light and colors)—that is, his "Pensées"—in which he repeated that he found Newton's corpuscular theory of light unsatisfactory; it was a position he had held for a decade, but he did not yet treat in detail his new pulse theory. Afterward he worked diligently on his "Nova theoria lucis & colorum" (New theory of light and colors, E88), which was published in 1746.

Contacts with the Petersburg Academy of Sciences

Even years after leaving Saint Petersburg, Euler corresponded with Petersburg academicians on an array of scientific subjects (prominent among them was astronomy); he also recommended candidates to replace members who had left. Among Euler's recommendations for membership in the Russian Academy were Ludwig Wenz from Basel (to replace Heinsius in astronomy), Johann Samuel König from Bern (to replace Georg Wolfgang Krafft in mathematics), and Jean-Philippe Loys de Chésaux (to replace Delisle in astronomy), but none of the three was accepted. A letter in September 1744 to Johann Schumacher conveys Euler's pleasure in separate visits that August from Heinsius and Krafft,[37] who were now at the University of Leipzig and the University of Tübingen, respectively. On 23 June Heinsius had arrived in Riga while on his way to Berlin. When news reached the Wittenberg mathematician Johann Friedrich Weidler, who was trying to hire Heinsius, he sent Euler a note asking him to pass on an enclosed invitation to Heinsius to stop in Wittenberg.[38]

In Berlin Euler instructed a generation of Russian mathematicians, natural philosophers, and astronomers. They were probably included, as many as nine members at a time, in dinner table gatherings at the Eulers' home in the early phase of what has been called the *Tischgesellschaft* (table society). Notable among the scholars from Russia was Stepan Jakolevič Rumovskij, who lived with the Euler family in 1744–45 and again in 1754–56,[39] writing of his first stay that Euler had "made our visit in Berlin extremely pleasant and useful."[40] The mathematicians Semjon Kirillovič Kotel'nikov and Mikhail Ivanovič Sofronov, as well as the astronomer Nikita Ivanovich Popov, studied with Euler while there. Euler's son Johann Albrecht was his most gifted student; the same age as Rumovskij, he probably took some instruction alongside the Russian youth.[41] The subjects that Euler taught his son included Latin, geometry, and mathematics; Johann Albrecht made such rapid progress in mathematics that at the age of twenty he was to be named a member of the Berlin Academy. Through Euler's instruction, these five men greatly improved their knowledge and computational skills, especially in calculus and mechanics. Subsequently, the four Russians were to become leading representatives of the sciences and of enlightened thought in their country.

Although by 1735 James Jurin had turned his attack on the Leibnizian principle of the conservation of *vis viva* away from the Petersburg Academy of Sciences to Johann I and Daniel Bernoulli, in 1744 Jurin sent to that institution a belligerent forty-eight-page pamphlet titled *De conservatione virium vivarum dissertatio* (Dissertation on the conservation of vis viva) using the pseudonym Phileleutheri Londinensis (Freedom Lover of London). Jurin expressed a hostility shared by many British mathematicians toward their Continental colleagues that was growing out of the controversy over whether Leibniz or Newton deserved priority for the invention of calculus; he added a cultural nationalistic character to this dispute by particularly opposing "some Swiss philosophers"—that is, the Bernoullis—and declaring that "many, especially in England, detest [their hypotheses]." Jurin was known at the Petersburg Academy, having opposed Georg Bilfinger on the subject of capillarity in 1730. At the academy conference on 21 September 1744, Schumacher officially communicated the pamphlet's contents. A vehement exchange followed there between Richmann and the physiologist Josias Weitbrecht. Richmann's negative critique in support of Leibnizian ideas prompted Weitbrecht to defend Jurin in an article, "Notae ad observationes Richmannis in Phileleutheri Londinensis dissertationem de conservatione virium vivarum" (Notes to the observations of Richmann in Phileleutheri Londinensis's dissertation on the conservations of vis viva).[42] Richmann admitted that he did not understand many of Phileleutheri's ideas. When the dispute persisted, the conference on 4 March 1745 decided to send the materials by Jurin, Richmann, and Weitbrecht to Euler, the scholar most learned on the subject. In November, Petersburg Academy secretary Winsheim asked Euler to be a referee between Weitbrecht and

Richmann in determining which arguments had more merit, and thereby to resolve an unpleasant internal dispute.[43]

On 10 December Euler responded to Winsheim; his article "De la force de percussion et de sa véritable mesure" (Percussion and its true measurement, E82) for the first volume of the Berlin Academy's *Mémoires* had criticized the principle of the conservation of *vis viva*. Euler contended that experience disproved it, but also argued that the criticism of the anonymous London author, whom he knew to be Jurin, was shoddy. Euler's letter to Winsheim began with a clear synopsis of the debate over Phileleutheri's article, also posing Richmann against Weitbrecht and Jurin. After referring to Phileleutheri's extraordinary cunning, Euler pointed out his errors on inelastic collisions, general mechanics, fluid mechanics, and the theory of the composite pendulum. Johann I Bernoulli's correct solutions in hydraulics and Daniel Bernoulli's in hydrodynamics, Euler found, refuted the views of Jurin, and he maintained that Richmann's rebuttal of Jurin failed to uncover enough of the British polemicist's errors.[44]

The reading of Euler's letter in Latin at the meeting of the Petersburg Academy of Sciences on 10 January 1746 ended the debate, though Richman responded with a panegyric on the "glorious Leibniz." Each side recognized that it had handled the debate badly, and the authors asked that their papers in this dispute be removed from the academy records and returned to them. The archives, however, obtained these materials and would shortly publish them in a Russian translation. The general quarrel was not finished; from 1750 to 1753 it moved principally to Berlin in the conflict between Maupertuis and König over the principle of least action.

In the mid-1740s, Euler sent the Russian Academy of Sciences descriptions of experiments with electrical wands; the academy was similarly testing heat theory and the force of electricity. Richmann, who in 1745 had succeeded Krafft as professor of physics and director of the physical laboratory at the academy, was studying electricity. One of his projects was investigating with Mikhail Lomonosov the dangerous effects of lightning. He did not limit himself to Leibniz's rationalism and insisted on reliable experiments and observations to establish exact quantitative, phenomenological laws. In January 1745 Euler asked Schumacher to send copies of the Russian and German editions of his *Einleitung zur Rechenkunst* (An introduction to arithmetic, E17), and in March he requested some of his letters and publications that had remained in Saint Petersburg. Exploration and land mapping by the Russian Academy continued to draw his close attention.

In 1745 without the approval of the academic conference, the *Atlas Russicus* (Russian atlas) was published in Russian and Latin, with the title also given in French and German. It was the first atlas produced of Russia, with a general map of the Russian Empire, thirteen smaller maps of its European regions and bordering lands, and six maps of Siberia. These expanded upon the cartography of Ivan Kirilov. They especially drew upon the many sophisticated astronomical observations of Delisle to determine longitude, together with accurate geodetic measurements and the strenuous efforts of Heinsius and Euler that brought out true geographic positions. Much of this came from scientific expeditions. The making of improved maps had been a chief assignment of the academy from its start. The atlas was a major force in the development of geography in Russia and interested cartographers, explorers, and some governments in other lands.

Home, Chess, and the King

The arrival of Charlotte in July 1744 had brought the size of the Euler family to seven; she was the last child to survive to adulthood. Euler's happy home life was attended by health problems. Together with an intermittent deterioration of his vision, he continued to suffer occasionally from fevers that at times became life-threatening. One of the worst came at the end of 1745, a time of intense work for Euler in founding the Royal Prussian Academy of Sciences as well as stress arising during his close study of land maps.

In Berlin Euler had few relaxations. Playing the clavier was the foremost, and he also invited composers to come to his house to give recitals of their new works. He was also becoming a talented chess player; chess was popular in Berlin, he wrote to Goldbach in Moscow, but most people there played it poorly. The extent of interest in chess in Berlin was shown in 1748, when Frederick II played Voltaire; the game was probably on a board with sixty-four squares, eight by eight, as was used in Paris. (There were also at the time a ten by ten board and another that was fourteen by ten.) To improve his game, Euler found a skilled Jewish player and for some time took lessons from him. He advanced so quickly that he was soon able to get the better of his teacher in most of their matches.

To 1746 Euler enjoyed royal favor in Prussia, though not warm relations with Frederick II; as long as the monarch was engaged in the first two Silesian Wars against imperial Austria, he interfered little with Euler and his research. Euler's claims in letters to the Petersburg Academy of Sciences notwithstanding, it appears that he was not overburdened with state projects, and life in Berlin pleased him. Perhaps recalling his interactions with Schumacher in Saint Petersburg, he had announced in 1742 that he "lived completely independently, receiving my pension entirely and depending on no bureaucrat." In a letter written in January 1746, more than a year after submitting the manuscript for the *Introductio in analysin infinitorum* to the publisher, Euler remarked to his friend and fellow Basler Johann Kaspar Wettstein, the chaplain and librarian to the Prince of Wales, "After having left the Petersburg Imperial Academy, I have every reason to be satisfied with my lot. . . . I can do just what I wish [in my research] and no one expects anything from me. The king calls me his professor [*mon professeur*] and I think I am the happiest man in the world."[45] But Euler and Frederick had still not met; the king had been on military campaigns and would soon reside in Potsdam, while the commoner Euler was in Berlin. His relations with Frederick and the Berlin Academy were quickly to turn rougher.

Chapter 8

The Apogee Years, I: 1746 to 1748

During the period of tranquility that Prussia enjoyed in the late 1740s, which were also the early years of the trilingual Royal Academy of Sciences and Belles-Lettres, the European Enlightenment entered its high cultural phase. Paris flourished as the headquarters of advanced thought, and *philosophes* set about applying critical reason to fixed beliefs, seeking freedom of thought and action from absolutist royal courts, the church, and other traditional institutions; they attacked religious fanaticism as well as its attendant repressions and cruelty. Through pamphlets, broadsheets, newspapers, encyclopedias, and satirical books the *philosophes* addressed a growing reading public. Salons and academies entered the fray. Among the period's canonical texts were Charles-Louis de Montesquieu's *De l'esprit des lois* (On the spirit of laws), published in 1748; Voltaire's satire *Zadig* (1747), ridiculing tyranny and quackery, and four years later *Le siècle de Louis XIV*, a leading early text in modern intellectual history, printed in Berlin to add prestige to Frederick II. Most noteworthy was Denis Diderot's monumental twenty-eight-volume *Encyclopédie*, appearing between 1751 and 1772;[1] both a reference work and a giant pamphlet, it consisted of seventeen volumes of text and eleven of plates on tools and manufacturing processes of the time.[2] This bold work, part of which was coedited by Jean-Baptiste le Rond d'Alembert, was intended to organize a compendium of all new and previously attained knowledge and to "change the general way of thinking."[3]

The title *Encyclopédie: Dictionnaire raisonné des sciences, des arts et des métiers* (Encyclopedia: Systematic dictionary of the sciences, arts, and crafts) and its frontispiece give prominence to the sciences, especially to pure and mixed mathematics along with physicomathematics, but the century's progress in the sciences lagged behind that in other fields. In addition to confirming Newtonian dynamics on the tides, the shape of Earth, and the orbits of comets, the moon, and the planets, geometers as well as astronomers and natural philosophers devised and successfully applied differential equations to mechanics and astronomy, transforming them into

Figure 8.1. The frontispiece to the first volume of Denis Diderot's *Encyclopédie*, 1751. The description of the frontispiece reads as follows:

> "Beneath a Temple of Ionic Architecture, Sanctuary of Truth, we see Truth wrapped in a veil, radiant with a light that parts the clouds and disperses them.
>
> On the right of Truth, Reason and Philosophy are engaged, the one in lifting the veil from Truth, the other in pulling it away.
>
> At her feet, kneeling Theology, receives her light from on high.
>
> Following the line of figures, we see grouped on the same side Memory, and Ancient and Modern History; History is writing the annals, and Time serves as a support for her.
>
> Grouped below are Geometry, Astronomy, and Physics.
>
> The figures below this group show Optics, Botany, Chemistry, and Agriculture.
>
> At the bottom are several Arts and Professions that originate from the Sciences.
>
> On the left of Truth, we see Imagination, who is preparing to adorn and crown Truth.
>
> Beneath Imagination, the Artist has placed the different genres of Poetry: Epic, Dramatic, Satiric, and Pastoral.
>
> Next come the other Arts of Imitation: Music, Painting, Sculpture, and Architecture."
>
> (Diderot, 1967, p. viii.)

modern exact sciences. Among the savants themselves and wider society, these accomplishments helped gain confidence in the power of reason.

At midcentury Leonhard Euler was at the peak of his career. Johann I (Jean I) Bernoulli had saluted him as "[t]he incomparable L. Euler, the prince among mathematicians" in 1745, and Henri Poincaré's later description of him as the "god of mathematics" attests to his supremacy in the mathematical sciences. Euler continued to center his research on making seminal contributions to differential and integral calculus and rational mechanics, and producing substantial advances in astronomy, hydrodynamics, and geometrical optics; the state projects of Frederick II required attention especially to hydraulics, cartography, lotteries, and turbines.[4] At midcentury, when d'Alembert and Alexis Claude Clairaut in Paris, Euler in Berlin, Colin Maclaurin in Scotland, and Daniel Bernoulli in Basel dominated the physical sciences, Euler was their presiding genius. Pierre Maupertuis in Berlin, Mikhail Lomonosov in Saint Petersburg, and Gabriel Cramer in Geneva played lesser but still significant parts. In the late 1740s competition among these savants became increasingly intense, and usually it did not pit one against all others. Bernoulli, for example, disputed d'Alembert while Euler initially refrained from doing so, and Bernoulli, Clairaut, and Euler occasionally faced d'Alembert as a common opponent. Their spirited rivalries would produce a break between Euler and d'Alembert, and Euler also apparently had strained relations with Clairaut and Bernoulli. Extending his influence beyond the academy, Euler also criticized Wolffian philosophy and science in Prussia during the German *Aufklärung* (Enlightenment) just before Immanuel Kant's critical period.

The Start of the New Royal Academy

As 1746 began, the positions of Maupertuis and Euler, the two members most important to the new academy of sciences, which was in part a reorganization of its predecessor, remained to be clarified and a new constitution for it had to be completed. Keenly aware of the poor condition of the academy, believing the sciences there were "in a state of collapse and humility" and recognizing the awkwardness and frailty of his position of foreign outsider, Maupertuis argued that only a strong president with powers invested explicitly in him by the king could honorably improve the situation under what he called "the reign of Augustus." Maupertuis, sensitive to rank and status within the king's court, requested on 5 January that Frederick remedy what he called the "contradictions and disaffections" within the chain of command in the academy administration.

Especially missing were specific details about the powers of the president. Maupertuis refused to be subordinate to noble minister-curators and high military officers or to have a lower rank than state ministers, and he wanted to control salaries and all other fiscal matters. His authority, he thought, was not yet clearly spelled out. The king ordered him to prepare statutes that drew upon the model of the Paris Academy of Sciences; surely Maupertuis consulted Euler, but the statutes he wrote went directly counter to the partly democratic model that Euler favored and the Nouvelle Societé littéraire had embodied.

On 3 February the academy elected Euler to replace Abraham von Humbert as director of the mathematics department. The office, like that of all other directors as well as curators and the president, was for life. It had taken Euler five years to attain this position, which had been the reason for his original recruitment. As director of the mathematics department he had to lead the new observatory and manage its construction (which would not be completed until 1752). It was being built above the royal stables, damaged by fire a decade earlier. The foul odors, the astronomers were to complain, hampered their work, and the moisture damaged their instruments. As a member of the academy's directory Euler served on the editorial board of the *Mémoires*, selecting and editing articles, and he supervised the library. Euler was also to prepare the calendars, almanacs, and geographical maps, which were the main sources of funding for the academy, and to arrange for their distribution. Together with Johann Gottlieb Gleditsch he also supervised the botanical gardens. Aiding Frederick's efforts to invigorate Prussian industry, the academy managed the large mulberry plantation outside Berlin. Gleditsch tested ways to cultivate silkworms and trees, and in order to beautify the area around the Berlin Academy, Euler ordered vibrant trees to line its wide avenues.[5]

Although on 3 March the academy inaugurated Maupertuis as president for life, Frederick had not yet signed the official document appointing him. A week later the statutes of the academy, the *Reglement*, appeared and were scheduled for reading at its general assembly meeting on 2 June. The eighth and thirteenth statutes made Maupertuis an autocrat in Prussia within the sciences. As president he was set above all others in the academy of whatever civil or military rank, even marshal and count Samuel von Schmettau and the three minister-curators, and he had final say in managing all economic business. He could set salaries, abolish small positions that he thought lacked merit, and control every other fiscal decision, all of which preempted the chief function of the curators. Nothing was to be done "except by his orders." In an added handwritten amendment to the statutes, his *lettres patentes*, Frederick gave exceptional powers to Maupertuis, likening him to a general who though only of gentry status "command[ed] dukes and princes in an army, without anyone's taking offense." He could command them also in the academy. In a letter in June 1746 to Daniel Bernoulli, Maupertuis's father René Moreau de Maupertuis boasted of his son's new position as unprecedented in the republic of letters, "a thing . . . never . . . seen before," and his son found the king's rulings "so flattering to me that they make me blush."[6]

At the general assembly meeting of the academy on 2 June, the new constitution was read. It separated the academy into four departments that slightly modified the divisions of 1744: experimental philosophy under Johann Theodor Eller, and behind it in rank mathematics directed by Euler, speculative philosophy with Johann Philipp Heinius as its head, and belles-lettres under pastor Jakob Elsner and Jean-Baptiste de Boyer, Marquis d'Argens, all of whom were elected for life. Together chemistry, anatomy, botany, and all natural sciences founded on experience comprised what was known as experimental philosophy. Since all of these had been classified under physics, the major change was in its title. To the geometric, mechanical, and astronomical fields of old mathematics the new regulations added algebra and all sciences that sought abstract theories and numerical bases. The most important change in organization was giving structural form and substance to the speculative philosophy department, the title that Maupertuis assigned it. He had found the old society's philosophy department inadequate in metaphysics, which he believed was essential to scientific research. That accorded with Gottfried Wilhelm Leibniz's view of metaphysics; Jean-Henri-Samuel Formey, describing it as "the mother and queen of the sciences . . . the source of evidence and certitude in our knowledge,"[7] depicted Christian Wolff as employing it to overthrow the Philippist Scholastics and build a new culture of science. Changing the name to speculative philosophy, Maupertuis required that the strict formal logic of metaphysics and theology be used to construct a dogmatic rational philosophy but cautioned against claiming that this research had the same degree of certainty as theoretical mathematics. The speculative philosophy department studied the supreme being, the human spirit, and the smallest component of matter, and it no longer excluded physics, as had the old philosophy department. Appearing now in each of the other three departments, as Euler wanted, were belles-lettres, which replaced philology. Modern academic specialization did not yet exist; belief persisted in a discoverable unity of all knowledge. Embracing antiquities, history, and languages but not inscriptions and medals, belles-lettres were listed occasionally as the historical and philological section of study.

The academy was modest in size. As regular members, each department had three salaried members, three associates, and three veteran members. In addition, there were sixteen noble honorary members and an open number of foreign associates selected solely on the basis of their scientific achievement. Generally the honorary members and French émigrés met separately from the rest. By 1747 these two groups of the institution were in Potsdam. Residing mainly in Berlin with his wife and having living quarters at the Sanssouci palace in Potsdam enabled Maupertuis to keep close to both groups. Plenary sessions were held every Thursday. Through the 1760s, most of these had an attendance of fourteen to twenty members. The academy was to continue awarding an annual prize, but in a departure from earlier practice the academicians were not allowed to submit papers to it.

By ceding his chair, the presiding noble-curator Kaspar Wilhelm von Borcke on 2 June reluctantly turned over authority in the academy to his in-law Maupertuis. The king had reviewed the statutes describing Maupertuis's powers, and on 12 May he informed Maupertuis of his acceptance of the new regulation; Frederick also sent the certificate of appointment for the president. Still, the curators were unhappy with the change; one resigned, while Borcke withdrew from the academy for a time. In a letter in July 1746 to Maupertuis, Euler reported Borcke's persisting unhappiness.[8]

For the winning essay in the Royal Prussian Academy's first prize competition, a committee headed by Euler and including Eller from experimental philosophy and Augustin Nathanael Grischow, Abraham von Humbert, Johann Kies, and Johann Nathanael Lieberkühn from mathematics selected d'Alembert's paper in French titled "Réflexions sur la cause générale des vents" (Reflections on the general cause of the winds, printed in 1747), the first work generally to apply partial differential equations in mathematical physics. This was important to the development of infinitary analysis. While ordinary differential equations have only one independent variable, partial differential equations extend calculus to many variables. The politically astute d'Alembert had written in his dedication (then known as a motto) for the paper a flowery compliment to Frederick II, who "distinguished by a [military] victory [in Silesia] in the eyes of the world, held out the branch of peace."[9] D'Alembert's attribution of wind patterns to effects of tides on the atmosphere was accurate, though he underestimated the impact of heat. In March Euler had informed Christian Goldbach that this paper was "sufficiently above all the others in the importance of its considerations to be worthy of the prize."[10] Euler was later to use his superior computational skills to improve upon the essay's equations and perfect d'Alembert's techniques, an early instance of his refinement of the work of others. The awarding at its first meeting of its annual prize for 1746 to d'Alembert (see figure 8.3, page 258), who was relatively young but a rapidly rising member of the Royal Academy of Sciences in Paris, and the fact that he was chosen by his good friend Maupertuis to be a foreign member (and was then elected unanimously) further enhanced the Royal Prussian Academy's stature.

One competitor was unhappy with the decision; Daniel Bernoulli had believed that Euler had all but promised him the prize. The papers had to be anonymous, but in revealing to Euler the motto on his paper Bernoulli intended to ensure that Euler would pick him as the winner. He was outraged to lose to the "impertinent and puerile youngster" and judged d'Alembert's paper weak and the decision absurd; his admiration for d'Alembert's dynamics did not extend to the Frenchman's ideas in fluid mechanics. To 1750 the prize decision would somewhat strain relations between Euler and Bernoulli, his best friend, who also partly blamed Maupertuis. Finding him two-faced, Bernoulli wrote to Euler, "when Maupertuis is talking to d'Alembert, then you and Mssr. Clairaut are only the gods of minor men, while d'Alembert is raised to an Apollo from whom all knowledge flows as from a true source, but when he talks to me he says [d'Alembert] is the joke of everybody because of his mechanics."

Moving quickly after 2 June to win greater prestige, at its second meeting a week later the Royal Prussian Academy of Science and Belles-Lettres unanimously elected as foreign associates Charles Marie de la Condamine and Voltaire. Initially unimpressed with Formey, the likely candidate for standing secretary who lacked the exuberance and wit valued at Frederick's court, Maupertuis at this session gained approval for an offer to his friend Johann II (Jean II) Bernoulli to fill that post, but Bernoulli was uninterested.[11] Neither was Maupertuis to be successful in a future invitation to Johann II to be director of the department in experimental philosophy (physics). Through his diligence, learning, and French Calvinist leadership, Formey was shortly to be appointed secretary.

The Royal Prussian Academy had separate protocols for the election of new members and foreign associates; both required being nominated by Maupertuis and approved by the king. The usual protocol involved academicians speaking with Maupertuis directly about a name for the list of candidates that he submitted to the assembly. After eight days, the votes of all members were collected, and a candidate winning a majority gained the post. Euler considered more honorable an alternate protocol that gave greater deference to candidates: Maupertuis would present a list of nominees to the assembly, which voted by acclamation. Under both procedures

Maupertuis always got all candidates that he wanted. Except in a few cases of French *philosophes*, Maupertuis sought Euler's recommendations for ordinary members and foreign associates. His advice was decisive.

On 23 June the academy gained further standing when Maupertuis on behalf of Frederick announced that the king had proclaimed himself the protector of the institution.[12] Since the monarch had refused to be the protector earlier under the curators, Maupertuis took this development as testifying to the king's personal respect for him. Although Frederick attended no meetings, he liked to assert as late as 1753 that he was an *Académicien* who submitted historical and philosophical papers and occasionally eulogies. These appeared anonymously, for the king's name could not be given on these, but readers knew who the author was. All of these actions served to give him credentials for being a philosopher king and burnished his reputation. Maupertuis enjoyed his full confidence and favor; Frederick's letters called its Catholic president "the pope of our academy" and confirmed Maupertuis's dominance there. That Frederick expected princes and court nobility to attend public sessions was yet another way of promoting the status of the academy. The institution was to remain close to the king, celebrating his birthday yearly and designing medals to commemorate his reforms and military victories. The discoveries of Euler, the presidency of the energetic Maupertuis, the new constitution, the strong support of the monarch, and the awarding of its 1746 prize to the respected d'Alembert all placed the Royal Prussian Academy prominently alongside the three other leading royal science institutions in Paris, London, and Saint Petersburg.

On 30 June the academy approved a distinguished array of foreign associates, including Britain's royal astronomer James Bradley; the antiquarian Martin Folkes, president of the Royal Society in London; the imperial Austrian astronomer Giovanni Jacopo Marinoni; Charles-Louis de Montesquieu of the Royal Academy of Sciences in Paris; the Utrecht mathematician Pieter van Musschenbroek; Uppsala's famous botanist Carol Linnaeus; and the Scottish mathematician James Stirling. Maupertuis, who had proposed this foreign associate list in consultation with Euler, was absent for the election, having received news before 30 June of the sudden illness and death of his father René Moreau, who had managed investments for his son and the Bernoullis.[13] Apparently guilt-stricken at not seeing his father earlier, Maupertuis immediately departed, arriving in Paris in early July and proceeded to Saint Malo, where he remained for four months. Maupertuis designated as acting vice president the royal physician Johann Theodor Eller, the director of the experimental philosophy department, which was first in rank at the academy. Maupertuis returned to Berlin and Potsdam in October and remained there to the winter of 1748; his primary residence was with his wife in Berlin.

In 1746 the academy gained more financial support from the king when a royal edict forbade the importation of foreign almanacs. The next year excise officials were given supervision over the sale of all maps from the academy, and they were responsible for making sure that the maps had all the requisite tax stamps on them. Nonetheless, even with this assistance the collection of revenues was insufficient to provide complete funding for the institution.

The Monadic Dispute, Court Relations, and Accolades

At the academy's first meeting in June 1746, Formey announced the topic for the prize competition for 1747: the monadic doctrine. As one of the first official acts of the new academy, the naming of this topic produced a major fissure, for it placed Euler against the Wolffian philosophers who dominated in German universities.

Among natural philosophers, who were widely investigating the theory of matter, most Western scholars accepted Isaac Newton's hard, indivisible, passive atoms as the smallest component of matter while the monadic doctrine, the core of the Leibnizian philosophy, postulated animate, elastic, immaterial, and metaphysical points of energy. The word *monad* seems to have come from the term for point in Euclid's *Elements* and may reflect John Dee's *Mona hieroglyphica* of 1564, which describes a progression from points to monads to matter. In the physical world Leibniz had metaphysical monads be geometric points, infinitely divisible. His article "Système nouveau de la nature et de la communication des substances" in the *Journal des Sçavans* for June 1695 had rejected the concept of passive, perfectly hard atoms, calling the idea contradictory to the law of continuity, the principle that "in nature everything goes by degrees and nothing by jumps or leaps" (*natura non fecit saltum*). Monads, he posited, were endowed with two forces: active *vis motrix* and passive inertia; they had a tendency, *conatus*, to be in motion. The changeability of monads accounted for continuous motion in dynamics, and their elasticity allowed for the conservation of *vis viva* in collisions. Drawing upon this thought and his own differential calculus, Leibniz resolved the seeming paradox between his infinitely divisible monads and continuity in nature. In the labyrinth of the continuum, he believed, reason went awry. Dynamics was the gateway to Leibniz's metaphysics; aggregates of monads generated extended matter.

Though generally agreeing with Leibniz, Wolff and his followers gave monads lesser importance, redefining them as either immaterial souls or generally indivisible—but not passive—atoms, and the Wolffians appealed to metaphysics for justification. Their special metaphysics, which included cosmology, defined connections in the cosmos as rational and mechanical. The literati generally did not accept astrological fatalism, but Wolff's religious critics charged that his physics, together with Leibniz's preestablished harmony, led to determinism and possibly fatalism. To counter the charge, Wolff embraced a physicotheology portraying the universe as an organic clock or natural automaton, having a soul in its teleology.

In Prussia opponents of monads faced hurdles beyond the sciences; they had to confront a widespread perception of metaphysics as—without contradiction—the mother of the other sciences. Formey's statement in the preface to the *Mémoires* (dated 1746 but actually published late in 1745) that metaphysics provides the theoretical basis for the most general laws, the source of evidence, and the foundation for certitude in our knowledge had wide support among Prussian scholars. Few Parisian savants shared such convictions, however. Another problem for critics of monads was that half the Royal Prussian Academy's members were in belles-lettres and not the sciences, and were therefore receptive to the element of mystery and poetry in the monadic doctrine.

Questions about the properties of the least parts of matter were not new to Euler, who like his teacher Johann I Bernoulli and later his friend Maupertuis admired the "great Leibniz" and was most at odds with the Wolffian rendering of his ideas. The subject had appeared in Euler's essay "Recherches physiques sur la nature des moindres parties de la matiére" (Physical research on the nature of the least parts of water), read to the academy in 1744.[14] The topic continued in his writings until at least 1750.

Emulating the Paris Academy, Maupertuis in cosmopolitan fashion drew into the prize competition scholars from the larger republic of letters in Berlin, the Prussian provinces, and abroad; any savant outside the academy could submit an essay. The names of authors were not placed on the papers, and judgment was supposed to be blind. A sealed envelope accompanying each paper had the same motto or pseudonym on the exterior as the paper and held a key identifying the author. The winner would receive a gold medal worth fifty ducats, and the academy was to publish his essay.[15] The academy instructed contestants to examine the monadic doctrine in a clear and exact manner and either refute it with solid arguments that couldn't be countered or deduce from it the major physical phenomena of the universe, giving particular attention to the origin and motion of bodies.[16]

Maupertuis wanted the academy to be an impartial arbiter and not fixed to a particular philosophical dogma. But that was not to be. Since the writings in the early monadic dispute were in German, the academy president did not keep abreast of the dispute until he hired translators and spoke with informed German scholars. Led by Formey, who was aided by Heinius and Philippe Joseph de Jariges, the Wolffians and their partisans at the academy enjoyed superior numbers. Euler was a powerful enemy of this doctrine, and it was Frederick's hope that he would follow Maupertuis's neutrality and moderation; conversely it was Maupertuis, unread in the writings of Leibniz and Wolff, who was shortly won over to agree with Euler and the Newtonians.

Recognizing that literate circles not only in Berlin but across the cultured European world were closely following the debate, Maupertuis and the academy quickly concluded that the competition was too important to have the judgment made by the speculative department alone. Acknowledging the consequences of the prize topic in the sciences and religion, Maupertuis at Euler's prompting named a prize commission including members from all four departments. Formey and Heinius remained on the prize commission, while Maupertuis added members from the other three departments, appointing Euler, and at Euler's urging the physician Johann Theodor Eller, the pastor August Friedrich Wilhelm Sack, and the astronomer Johann Kies. At its head he placed Count Albrecht Christoph zu Dohna, the grand chamberlain to the queen and the private secretary to the king.[17] Like Euler, Dohna was a critic of Wolff. So while this expansion was intended to protect the competition from being accused of bias, throughout this debate Wolff viewed Euler as a puppet master.

Soon after the announcement calling for submissions for the prize, Euler endeavored to show Wolff wrong, and in so doing he ignored an academy regulation that prohibited its directors from participating in its own prize competitions. He thus broke not only that rule but well-established etiquette and social behavior in the sciences when under a thin cloak of anonymity he struck the first blow with his brochure *Gedancken von den Elementen der Cörper, in welchen das Lehr-Gebäude von den einfachen Dingen und Monaden geprüfet, und das wahre Wesen der Cörper entdeckt wird* (Thoughts on the elements of bodies, in which the doctrine of simple things and monads is examined and the true essence of bodies is discovered, E81). The title of Euler's pamphlet recalls Wolff's *Vernünftige Gedancken von Gott, der Welt, und der Seele des Menschen, auch allen Dingen überhaupt* (Rational thoughts on God, the world, and the soul of man, and all things whatsoever, 1720), whose chapter 6 he had rejected. The printer placed on the title page the Wolffian motto *Sapere aude!* (Dare to know!), which later became—through Immanuel Kant—the motto of the Enlightenment.

Employing the term *molecules* for the smallest units of matter, Euler drew on Newton to define as their basic and inseparable property not the extension René Descartes had assigned them but instead impenetrability.[18] (Impenetrability connotes that two bodies cannot occupy the same space at the same time; lacking inpenetrability, a vacuum is not a body.) Noting the infinite divisibility of monads in chapter 6 of Leibniz's *Monadology*, Euler shortly before writing the brochure departed from his earlier position and accepted infinite divisibility as the other essential property for molecules. He set out his thought in a syllogism: "A geometric body is the abstraction of the material, connected to this as the general concept is to the special case. According to Euclid, a geometric body is infinitely divisible. Therefore, it must also be so for the material body." Euler then synthesized the properties of impenetrability and passivity from Newton's atoms, and infinite divisibility from Euclidean geometry and Leibniz's monads, to form his hard and punctual molecules, which offered a distinctive alternative to both. His main criticism was directed at the indivisible and immaterial monads Wolff described. The most difficult point in Euler's syllogistic proof, his attempt to reconcile impenetrability with infinite divisibility, was to establish the stipulation that abstract geometric concepts, in which the two characteristics coexisted, also applied to his physical molecules. Wolffians protested that infinity was a characteristic of God and not of matter, which Euler was attributing to it. This charge—a serious religious criticism in earlier times—he rejected as frivolous.

Establishing the properties of the most elementary bodies, Euler argued, could be accomplished only in the sciences. In demonstrating this he defined *mass* by Newton's law of inertia and discussed uniform density or specific gravity. Inertia, Euler believed, followed the principle of sufficient reason, which Johann I Bernoulli had made the foundational principle of classical mechanics. In being animate and possessing an inner force—the *conatus*, which made them inclined to change their state—monads were said to violate inertia. As already noted, Leibniz had introduced the *conatus* to resolve Zeno's paradoxes of motion by providing for continuous motion as opposed to dividing spatial distances into an infinite number of parts. The paradoxes suggested that this was impossible, as was locomotion. In the case of the hare chasing the tortoise, the hare would never catch up because it would have to pass through an infinite number of points. The mistaken implication was that the sums of infinite series were infinite. But the animate monads do not conform to the laws of mechanics, which made it impossible to verify them empirically and quantify their role in motion. How, Euler asked of monads, could spatially extended material bodies arise from the immaterial? He thought the doctrine to be a delusion of reason.

The anonymity of Euler, who used the academy's printers Ambrose Haude and Johann Karl Spener for the *Gedancken von den Elementen der Cörper*, fooled no one; surely he knew that German learned journals and universities with a strong Wolffian presence would respond. Supporters of Wolff were offended, and the competition provoked public controversy in German letters, pamphlets, journals, and the press. The quarrel brought out the extent of the intellectual separation of universities and journalists from the Prussian academy. Number 125 of Euler's *Lettres à une princesse d'Allemagne* (Letters to a German princess), which were written from 1760 to 1762, describes the breadth and intensity of debate over monads in Berlin in 1747: "There was a time when the dispute respecting monads employed such general attention, and was conducted so heatedly, that it forced its way into all social circles, including the *corps-de-garde*. There was scarcely a single lady at court who had not declared herself for or against monads. In a word, all conversation touched on monads—no other subject could find admission."[19] This declaration, hyperbolic as it doubtless is, suggests that the Wolffians had continued Leibniz's practice of corresponding with literate women about monads.

In north German states, an institutional split emerged between the academy and the universities. Among the first of the numerous rebuttals that Euler's pamphlet received was an anonymous essay by C. A. Körber, who contended that neither Leibniz nor Wolff had maintained that monads comprise physical bodies or could induce motion. Körber opined that Euler had not sufficiently read the philosophical texts to resolve this matter.[20] The Wolffian professor Johann Christoph Gottsched of the University of Leipzig wrote in the *Leipzig Review* that for a distinguished academician to deny the existence of monads and give directions for examining the question of elementary matter during a prize competition concerning them violated the impartiality of the institution. Likely Formey encouraged Gottsched to publish this criticism. Both claimed that Euler in entering the discussion violated the nonsubmission rule for regular members and the condition of impartiality required of an evenhanded competition in the contest. The Wolffians accused Euler of intrigues against them.

In November 1746, Wolff wrote to Maupertuis from Halle condemning Euler's "impudent act of audacity"; he sent copies of two claimed refutations from the *Leipzig Review* and noted that other criticisms of Euler had appeared. His letter was in Latin, for he knew that Maupertuis would not understand German. Wolff perhaps assumed that Maupertuis was ignorant of the *Gedancken*; otherwise he would have forbidden its publication. Wolff declared himself duty bound to make sure that Maupertuis knew of the opposition to Euler in German universities. He refused to enter the dispute directly but decried the position of Euler, who "too overpowered by his favorable fortune [in mathematics], affects a certain supremacy in the Republic of letters." Wolff asserted that Euler's pamphlet sought chiefly "to refute Leibniz and me"; he added that he "prefer[red] the insight of Leibniz in metaphysics and philosophy to the profundity of Euler."[21]

Maupertuis asked Formey to translate the *Leipzig Review* articles and other rebuttals. Formey emphasized passages critical of Euler for improperly opposing monads during the prize competition. As conversations, letters, and rumors on the controversy spread among German savants, Formey informed Maupertuis that there was nothing in the furor that he had a hand in. This was untrue, for Formey had begun corresponding with Gottsched in September and he assured fellow partisans of monads that however great Euler might be in the mathematics and the physical sciences his knowledge of metaphysics was not even superficial. In November Formey apprised Maupertuis that French translations of the German texts in this debate were about to be published. The dispute would soon intensify.

As Prussian courtiers saw it, more significant than the Wolffian comments were criticisms of Euler from Frederick II himself. The two were to have complex relations; though impressed with Euler's accomplishments, the king and the court nobility faulted him for lacking in courtly manners, which had been defined in dictionaries since the seventeenth century, and for failing to display the sparkling wit required of the head of the royal academy. Being friendly was not a court grace, but fawning over the nobility was. Against talented commoners rising in intellectual influence, the charge of being deficient in these courtly social manners was not unusual, and Euler was in good company: The aristocracy had leveled the same charge against Galileo early in his career.[22] In a letter of October 1746 Frederick's brother, Crown Prince August Wilhelm, described his encounter with Euler upon being introduced by Maupertuis. The prince found that while Euler through diligence and the application of his powerful logic had made a deserved name in the sciences, his appearance and somber demeanor darkened and hindered the recognition of his impressive qualities.[23] In his reply in October 1746 to August Wilhelm, Frederick disparaged Euler's epigrams emanating from his construction of new curves, some from conic sections and others from astronomical measurements, and placed him among "the powerful calculators, commentators, translators, and compilers . . . who are useful in the republic of science, but are anything but elegant." The king offered Euler a compliment, calling him "a Doric column" in the essential foundation of the entire edifice of the sciences, "unlike the [later and more elaborate] Corinthian columns, which are simply ornaments."[24] His disagreements with Euler would soon extend to matters of religion.

Meanwhile, exploration and land mapping by the Petersburg Academy of Sciences continued to draw Euler's close attention. In 1745 the *Atlas Russicus* (Russian atlas) had been published, and by March 1746 Euler received and evaluated it. That month he quickly wrote to Johann Schumacher from Berlin; he praised the able scientific, computational, and organizational activity always displayed by Joseph-Nicholas Delisle in working with the large number of astronomical observations, without whom their full value would have been lost. Euler believed that the atlas, especially with Gottfried Wilhelm Heinsius's and his own labor in bringing data into correct explanations, put the geography of Russia ahead of that of Germany. While he later acknowledged that the atlas had errors, he would continue to praise it as he pursued more cartographic studies. Euler asked Schumacher to send him more copies of land maps. Delisle, who was not pleased with the atlas, sharply criticized it.

In the summer of 1746 officials at the Petersburg Imperial Academy tried to persuade Euler to rejoin its ranks. The eighteen-year-old count Kirill Gregor'evich Razumovskij, the brother of Alexei Gregor'evich Razumovskij, Empress Elizabeth's favorite at court, became its president in June, ostensibly for his talent and aptitude in the sciences. His tutor and secretary, the twenty-one-year-old botanist Grigorij Nikolaevich Teplov, rose to power as the new assessor of the academic chancellery, thereby controlling its finances. Along with the empress, both invited Euler to return. The departure of Heinsius and Georg Wolfgang Krafft in 1744 and the imminent departure of Delisle had weakened the Petersburg Academy; reacquiring Euler, at the least for a certain time, was intended to start rehabilitating it. Teplov and Count Razumovskij personally knew Euler, having resided with him when they were his students from 1743 to the summer of 1744 in Berlin, but both showed little interest in the academy. Embarking on trips through Europe, they were away from Saint Petersburg for long periods even before Count Razumovskij was appointed chief of Ukraine in 1750, and they left Schumacher in charge of the academy.

Daniel Bernoulli fancifully suggested that Euler should go to Saint Petersburg for only two years, while Bernoulli would replace him in Berlin. But the Eulers had not forgotten the dangers of 1740 and 1741, and

Euler was pleased to be in Berlin, which made it advantageous for him to correspond with scholars throughout German lands and western Europe. In the tradition of Marin Mersenne and Henry Oldenburg, he was becoming an important circulator of information, though in his case for the mathematical sciences. His letters varied in number, but his output was always impressive: 150 letters in 1746, 90 in 1753; and more than 160 in 1755. A letter in June 1746 from Euler to Schumacher congratulated Count Razumovskij on gaining the presidency but gave personal reasons for declining his offer.[25] Among them were the fact that Euler felt lucky to be living in such a pleasant place; the Berlin Academy had just been established; and he enjoyed the friendship and goodwill of academy president Maupertuis. Having just recovered from a severe fever in the autumn, he didn't want to return to a harsher climate that might damage his health, especially since he was responsible for a growing family. His wife's fear of fires in Saint Petersburg was seemingly also influential. Though he turned down the invitation, Euler remained in contact with Teplov and Count Razumovskij to 1763, a year before others took control of the academy, but the correspondence was of little scientific importance.

While Schumacher's authoritarian behavior at the Petersburg Academy of Sciences troubled Euler, he nonetheless agreed to help Schumacher restore the institution. His letter in French in October 1746 to Teplov scolded members, particularly Gerhard Friedrich Müller, for interfering with Schumacher by pursuing idealistic plans for a perfect academy that hindered the development and administration of useful projects. Even with large salary increases, it was unlikely that French geometers and natural philosophers would move to Russia, but Euler urged Teplov to consult anyway with Maupertuis, with whom he existed "in a very perfect . . . harmony." Acquiring scholars from German-speaking lands was more possible. Through correspondence, journals, and various academy prizes, he developed an extensive knowledge of personalities and research in the sciences in German lands; Euler's list of proposed candidates for the Petersburg Academy included Christian Friedrich Oechlitz in Leipzig for mathematics, the Danish-born Christian Gottlieb Kratzenstein in Halle for physics and mathematics, Johann Theophil Waltz in Dresden for astronomy, and Benjamin Brauser in Danzig for calculating almanacs.[26] Euler noted Oechlitz's receiving the prize of the Academy of Bordeaux for an excellent article. He urged the retention of Delisle, but declared that he could think of no one in German lands who could better replace him than Waltz. He also urged that Daniel Bernoulli be offered a salary, and he wanted the academy to have a good foundation in philosophy in addition to a strong base in physics. He proposed inviting to Saint

Petersburg the philosopher and grammarian Joseph Adam Braun on the merit of his knowledge in physics and philosophy. But he insisted that Russia also educate its own candidates for academic positions.

Euler supported the establishment of a new university in Saint Petersburg to benefit the academy and the empire, and he asserted that the foreign scholars he recommended for the academy could be vital to such a university's founding. Euler also offered to send more essays, "to forward as many as are asked of me. For not having anything else to do, I can use my time here working on my studies." Financing of the Russian Imperial Academy was uncertain; in 1746 the institution sent books to Euler in lieu of his salary; that October he thanked Teplov, but added, "in the future cash will be more agreeably accepted."[27]

As his notoriety grew across Europe, Euler desired in 1746 to gain membership in the Royal Society of London, something he felt would add further prestige to his reputation. Daniel Bernoulli and Clairaut had already been chosen as members. But the requirement that candidates submit their own applications troubled him; he felt that nominations should depend solely on achievement. This seems to explain his delay in pursuing membership; in July 1746 he informed Johann Kaspar Wettstein that he wished to become a fellow of the Royal Society but refused to submit his own name, explaining that he did not want to appear self-promoting. The Berlin Academy's election that year of Folkes, Bradley, and Stirling, all of the Royal Society, does not seem coincidental; but Euler was corresponding with Stirling concerning lunar tables, for his own early computations of lunar motion with differential equations based on Newton's inverse-square law did not satisfactorily describe the motion of the lunar apogee. Bradley's increasingly accurate observations added to others made at the Berlin and Saint Petersburg academies provided Euler with data that made it possible for him to improve his tables and to correct computations. Wettstein offered to nominate Euler, who accepted in November.

Exceeding the Pillars of Hercules in the Mathematical Sciences

To represent the pursuit of scientific discoveries in the mid-eighteenth century, Voltaire had employed the imagery of the Argonaut myth of heroes in search of truths and treasure. For Euler and his rivals the foremost intellectual voyage was for discoveries in differential and integral calculus and their applications leading to a new mechanics, astronomy, and optics. Diderot described them as surpassing the columns of Hercules for the sciences. The pillars, the two pyramidal promontories aside the eastern end of the Strait of Gibraltar, signified in ancient and medieval times the limits of exploration and learning. Diderot extended the thought of Francis Bacon, the title page of whose *Instauratio magna* (Great renewal) depicts a ship passing through them and has the motto from the biblical book of Daniel, *Multi pertransibutat et augebitur scientia* (Many will pass through and knowledge will be increased). Bacon's new empirical method sought to replace Aristotle's syllogisms and not be satisfied with knowledge from ancient authorities; he appealed to the analogy between the voyages of exploration of his time and the advancing of learning. As Diderot's metaphor might extend it, the natural philosophers of the Enlightenment labored to settle the sciences on the solid buttresses arising from the more precise data provided by increasingly superior telescopes, clocks, and surveying equipment. Their primary milieus were the meetings, prizes, publications, and correspondence of the royal science academies in Paris, London, Berlin, and Saint Petersburg, which now eclipsed universities in scientific research.

In 1745 Clairaut had embarked on a fruitful seven-year examination of the lunar orbit, and d'Alembert took up the problem independently. Both began their work in secret. By late 1746, in attempting to describe lunar motion, both had devised series of six first-order differential equations based solely on Newton's inverse-square law of attraction, but these provided only half the secular precession of the lunar apogee, that is, twenty degrees rather than the observed forty degrees. The goal of the search was for not an exact answer but for a close approximation accurate to arcseconds. In computing the motions of three celestial point masses interacting according to Newton's attraction, initial efforts involved holding fixed one of the three points in these problems. In addition to continuing to work on his long-term development of lunar theory, Euler urged gathering more precise results particularly from observatories associated with royal science academies and sharpened by telescope lenses that he was helping to improve. The theoretical advances associated with astronomy and the increasing degree of exactness of the observational data together enabled mathematicians and astronomers to develop better equations. Early calculations of lunar motion with differential equations based on Newton's inverse-square law of attraction were to seek to reduce the gap between observation and theory; but the motion of lunar apogee or apsides, the farthest distance of the moon from Earth, seemed to pose an anomaly. Among Euler's correspondents addressing lunar motion at the time were Cramer in Geneva, Marinoni in Vienna and, above all, Delisle in Saint Petersburg.

Figure 8.2. The Pillars of Hercules, as portrayed on the title page of Francis Bacon's *Instauratio Magna*, 1620.

The preface to Euler's lunar tables published in 1746 in volume 1 of his *Opuscula varii argumenti* announces,

> Their nature and basis of construction would require too much space to explain. I therefore only point out that they are derived from the theory of attraction which Newton with such happy success introduced into astronomy. Although it is claimed of several lunar tables that they are based on this theory, I dare assert that the calculations to which this theory leads are so intricate that such tables must be considered to differ greatly from the theory. Nor do I claim that I have included in these tables all the inequalities of motion which the theory implies. But I give all those equations which are detectable in observations and are above 1/2 arc-minute.[28]

Figure 8.3. Jean-Baptiste le Rond d'Alembert.

Here Euler trusted the inverse-square law. Work on lunar theory was a crucial part in confirming Newtonian dynamics on the Continent, but there seemed to be a challenge to it within a year. Using the calculus of trigonometric functions, Euler arrived at the inequalities of the moon. A major problem involved creating and solving differential equations of the second order. Forces that act on the moon must be considered correctly with respect to the Earth-fixed reference frame. Euler made it standard practice in astronomy to correct theoretically derived terms by comparing them with quality observations. Heinsius judged Euler's lunar tables better than the preceding tables by Jacques Cassini, easier to use, and significant in the development of astronomy.[29] Euler also sent astronomical tables by Charles Leadbetter and lunar tables by Nicaise Grammatico. In 1746 Heinsius wrote to Euler from Leipzig, asking for dates of observations of comets made by English astronomers in 1744. Euler alone met the requirements of applications in practice for navigation, better timekeeping, and a variety of calendars.[30] Each theory had an increasing accuracy.

Another of Euler's principal contributions to the sciences in 1746 came in optics. Having rejected Newton's theory of light a decade earlier, he now presented "Nova theoria lucis & colorum" (New theory of light and colors, E88), which gave the clearest and most comprehensive theory in a medium in physical optics during the Enlightenment.[31] The medium is the ether. The treatise appeared in the first volume of Euler's *Opuscula varii argumenti*, which also contained six fundamental works on mechanics, astronomical tables, smallest particles of matter, prime numbers, and longitude.

Earlier in the eighteenth century, the corpuscular theory in Newton's *Opticks* had competed with the pulse theory in Christiaan Huygens's *Traité de la lumière* (1690). For its inability to explain double refraction in Iceland crystals or the heating of bodies by light considered to be a pressure, Newton objected to Huygens's hypothesis involving a medium; he also took issue with the idea that light was an instantaneous motion in a medium. Somewhat surprisingly, Newton's theory of colors had encountered support as well as opposition in the German states; textbooks by Wolff defended and transmitted it, but by the 1740s Huygens's wave theory of light had been repudiated. Efforts by Johann I Bernoulli and Jean-Jacques d'Ortous de Mairan to produce hybrids of the wave and the corpuscular theory in physical optics had failed.[32]

Euler's "Nova theoria lucis & colorum" propounds a pulse theory but begins by supplying a more complete basis for a wave theory than had existed. After identifying points of opposition between Huygens's and Newton's optics, he characteristically and deftly synthesized consistencies from them. For his new theory Euler was sometimes called Huygensian but, like Cramer, he was not content with how Huygens had developed his optics. Adding to this collection of ideas, original insights gained from his powerful intuition in physics, findings of experiments on sound, and Huygens's analogy between the propagation of sound in air and that of light in an elastic ether, Euler offered a new optical theory: only when it held that light was a pulse motion in the ether did he accept that analogy. While Huygens explained the propagation of light as a tiny displacement of very elastic ether particles, Euler proposed a disturbance of equilibrium in the distribution of density within a subtle, elastic ethereal fluid. In explaining reflection and refraction Euler rejected pulse fronts, envelopes of secondary spherical pulses, which were central to Huygens. He also argued that Huygens could not account well for colors. Euler explained these within the medium by connecting color to frequency of vibrations, an approach close to Nicolas Malebranche's description of vibrations of pressure. Euler did not accept Newton's explication of colors in book 2 of the *Opticks* as propagated by a periodicity or fits of particles. Periodicity, he observed, was also a property of waves. By incorporating frequency into the medium tradition, Euler increased its explanatory range.[33] In treating optics Huygens had employed geometry, and in the *Principia Mathematica* Newton's acoustics had geometric wave equations of vibratory motion. Drawing heavily upon data from a growing body of research on sounds and to a degree synthetic in optics, Euler began to transform Newton's wave equations into algebraic language, introducing differential equations for the phenomena of light.

The quality of Euler's theory and computational techniques, beginning in "Nova theoria lucis & colorum," evoked a broad response.[34] Thinking it too complex, Cramer proposed an improvement upon Newton's theory of light and colors, believing the relation between these two to be similar to that between the Ptolemaic and the Copernican models in astronomy. East of the Rhine, Euler's pulse theory prevailed.

In November 1746, Euler read at the Berlin Academy an incomplete proof of his fundamental theorem of algebra; it held that every real polynomial could now be factored into the product of real linear and/or quadratic factors. That these had roots or solutions had been long recognized, but Euler now offered an incomplete proof. His blemished proof had been evolving since at least 1739, when in a letter to Johann I Bernoulli he first expressed his position. He had discussed this in correspondence with Nikolaus Bernoulli, who corrected his computational errors and perfected techniques from Descartes in removing the second term of an equation and the factorization.[35] As these letters show, Euler had the essentials of his proof by 1744, but he still mistakenly assumed that what he had shown to be true for real numbers would apply to all numbers—that is, that multiplying opposites in signs gave negatives. He was not alone in working on this proof: in December 1746, d'Alembert sent his own flawed version to the Berlin Academy for its *Mémoires*; this was to appear first in print in the issue for 1748. Euler's proof in French, "Recherches sur les racines imaginaires des équations" (Research on imaginary roots of equations, E170),[36] was included in the *Mémoires* for 1749, which were not published until 1751.

Solving the vibrating string problem, used partly to calculate musical frequencies, was among the major topics that engaged Euler in competitions beginning in the late 1740s. All of the leading mathematicians in Europe, including Joseph-Louis, Comte de Lagrange and Pierre-Simon Laplace, were later to enter this contest. After Johann I Bernoulli posed the problem in 1727, Daniel Bernoulli and Euler addressed it from the 1730s onward and in 1743 had linear differential equations for loaded strings with clamped ends. By the end of 1746 d'Alembert had completed his "Recherches sur la courbe que forme une corde tendue mise en vibration" (Research on the curve formed by a stretched string set in vibration), his first published paper on these, in which he derived and solved the linear wave equation for small vibrations of a string with fixed ends. Late that year he sent Maupertuis a packet of his writings, and in correspondence with Euler the following January he noted the equation. Although not the first partial differential equation and not adhering closely to observations, it received wide attention. But d'Alembert added unnecessary restrictions that he held to be appropriate for reaching his solution. In a paper quickly written only months after d'Alembert's but not published until 1749, Euler criticized the stipulation that the string at time $t = 0$ be in the equilibrium position. He had the string at rest with an initial position differing from equilibrium, which would set the string in motion when let go.[37] In modern terminology, his alternate solution allowed any piecewise continuous initial shape for the vibrating string, while d'Alembert's more traditional solution permitted only one initial shape of the string. In his solution of the wave equation, Euler included representing strings with curves that could be described graphically as unbroken, a characteristic of a continuous function. But these functions need not be differentiable at every point, as the stronger, modern, smooth functions and their derivatives must be. Earlier in the Enlightenment the vibrating string problem had been considered to be a discontinuous or mechanical function. Euler's perception of the generality of the associated oscillating functions opened the examination of oscillations generated in continuous media.

While writing the paper "Les loix du mouvement et repos, déduites d'un principe metaphysique" (The laws of motion and rest, deduced from a metaphysical principle) for the academy *Mémoires* of 1746, Maupertuis asked for Euler's help. Maupertuis was proposing a metaphysics based on his principle of least action that closely connected physics and mathematics; he sought the greatest generality, believing that his principle underlay the fitness and movement of animals, the growth of plants, and the rotation of planets. After reading a draft of the paper in December 1745, Euler wrote that the principle was "more general than you propose."[38] The principle of least action required all motions and equilibria to involve an extremum condition that minimized "the quantity of action." "Les loix du mouvement et repos" considered final, not efficient, causes found in applying to dynamics the new variational methods in calculus, but Maupertuis admitted that for extremum problems he could provide only a rudimentary mathematics. The support of Euler was thus important, and the alliance between the two men was growing stronger in Berlin. In a series of papers Euler would soon develop sophisticated variational problems that went beyond the limitations that confined Maupertuis. In May 1746 Maupertuis asked Euler "to correct whatever mistakes there might be [in the paper] and write your comments in the margins." He particularly inquired about the imaginary immaterial planes that he introduced to explain changes in velocity. "Tell me," he wrote, "whether it deserves to be used . . . or should be rejected."[39] Euler tactfully called the planes correct but unnecessary.

Maupertuis must have sent a revised draft retaining this idea, for in December Euler noted that the planes did not weaken his argument, and they were to appear in the published version of the paper.

The timing of his writings shows that Euler could have claimed priority for the principle of least action. To confirm that claim for himself, Maupertuis checked with the Bernoullis on exactly when Euler had written his *Methodus inveniendi lineas curvas maximi minimive proprietate gaudentes* (A method of finding curves that show some property of maximum or minimum, E65). But this was unnecessary, for Euler graciously and steadfastly recognized Maupertuis as the inventor of the principle. "Les loix du mouvement et repos" did not appear in print until 1748; soon afterward a major controversy would erupt over priority for the principle and its universality.

In March 1747, Maupertuis had Euler named vice president at the academy. As the senior director, Johann Theodor Eller should have been by protocol the presidential deputy, but his absences were a liability, and Euler's intellectual leadership prevailed. As vice president, Euler had the task of leading meetings of the academy when Maupertuis was not present. Since his health was fragile, the choleric Maupertuis tried to spend as many winters as possible in France, away from Berlin's harsh climate; the increasing frequency and length of his departures after 1748 would produce a quandary for the administration of the academy.

Though lacking in advanced scientific knowledge, Formey moved ahead in 1747 in opposing Euler in the monad controversy, writing an anonymous rebuttal to Euler's *Gedancken*. Titled *Recherches sur les élemens de la matière* (Research into the elements of matter), it was published early that year before the Berlin Academy's prize commission even collected the competition papers. While completing the *Recherches*, Formey had appealed to Gottsched to help him find a publisher and translator; he also corresponded with Wolff, who encouraged him to proceed with plans to publish the book and saw the manuscript before it went to press. The *Recherches* rejects the competence of "gentlemen of mathematics" (*Messieurs les Géomètres*) to judge "the matter of metaphysics."[40] Here Formey was following Wolffian principles. For Wolff and his followers, the monadic doctrine could be neither proved nor rejected by physics and must be examined in the logical form developed by Leibniz and modified by Wolff. Leibniz had in his *Nouveaux essais sur l'entendement humain* (New essays on human understanding) of 1707 rejected the view embraced by many Aristotelians and John Locke that the mind was a blank tablet written upon by the senses. Leibniz stood closer to Plato, believing that the mind contained the necessary truths of logic, metaphysics, mathematics, and ethics. It could awaken itself to these truths and prove them only through rigorous reflection using its innate, internal, and logical principles. Leibniz held that his monadic doctrine, allowing for continuity throughout physical existence, refuted the notion of hard, discrete atoms. But Leibniz began with the principles of sufficient reason and contradiction, while Wolff's modification, emphasizing contradiction alone and giving sufficient reason a lesser status, went against Leibniz and Johann I Bernoulli.

Formey's *Recherches* based the monadic doctrine on a process of reasoning from tenet to tenet:

1. Experience shows us all bodies perpetually changing their state.
2. Whatever is capable of changing the state of bodies is called force.
3. All bodies, therefore, are endowed with a force capable of changing their state.
4. Every body, therefore, is making a continual effort to change.
5. This force belongs to a body only insofar as it contains matter.
6. It is therefore a property of matter to be continually changing its own state.
7. Matter is a compound of a multitude of parts, denominated the elements of matter. Therefore,
8. As a compound can have nothing but what is founded in the nature of its elements, every elementary part must be endowed with the power of changing its own state.[41]

By the summer of 1747 Euler was attacking several of these tenets of the monadic doctrine and the Wolffian philosophy.[42] Holding that the simplest components of matter were passive and could not set themselves in motion, he found tenet 3 "equivocal and altogether false."[43] He argued that the force to change the position of every body comes not from within the body but from another, outside. Every body, moreover, tends to preserve the same state (inertia) and not change it, as tenets 4 and 8 held. If a philosophy leads to laws contradictory to established laws of motion, Euler believed, its metaphysics must be false. And while Wolff, like Leibniz, placed his faith in a strictly logical inquiry into physical reality, Euler here came closer to Leibniz than were the Wolffians, whom he accused of creating a dogmatic rationalism and abusing the principle of sufficient reason by reducing its role in the sciences. Yet although he preferred Leibniz's theory to the Wolffian variant, Euler's command of mechanics combined with his devout and conservative faith led him to reject the whole monadic doctrine. The Wolffian form joined with the principle of a preestablished harmony he especially judged to be deterministic, contradicting the concept of original sin and the importance of good works. That Euler's chief opponent in the monadic dispute was the Calvinist pastor Formey bespeaks the breadth of debate within the Enlightenment.

While the academic department from which it emanated normally chose the Royal Prussian Academy of Sciences annual prize recipient, Maupertuis named for the selection a commission on which Euler's voice was decisive. Declaring that they were written like the edicts of Caligula, Maupertuis found all thirty essays received unworthy of the prize;[44] he begged his friend Johann II Bernoulli to submit a paper, but Bernoulli declined. Several members wanted dual prizes recognizing both the Leibnizian and Wolffian camps, a standard practice in contested competitions at the Royal Academy of Sciences in Paris; these members feared setting Wolff and his followers against them. But Maupertuis decided to name a single winner. Euler dismissed as "feeble" and "chimerical" the arguments in the essays supporting Wolffian monads;[45] he reported that the final vote of the prize committee was five to two for a modest paper titled "Nunquam aliud Natura, aliud Sapienta dicit" (Never does nature say one thing and wisdom say another[46]) that opposed Wolffian monads, but he had informed the commissioners that he would support any other that took a like position. The names were not given for the vote. Doubtless the two in the minority were Formey and Heinsius, Euler having won over the others. From their correspondence, it appears that Euler had aided the author of the prize paper in preparing his arguments against the Wolffians.[47] Only later, after the heat of debate, did Euler admit that one of the monadist papers was also worthy of the prize.

In June 1747 at a grand public assemblage including the king's sister Amelie and three brothers, the curators, several other nobles, and visiting dignitaries, Dohna announced that the winning paper in the monad competition was "Nunquam aliud Natura" and opened the sealed envelope containing the name of its author. The winner was Johann Heinrich Justi, an advocate and administrator from Sangerhausen in Thuringia. The rude expressions against Wolff in Justi's paper upset Maupertuis; to reduce the animosity that the competition had generated, Maupertuis had Justi strike these expressions before he ordered a translation into French, which he put into a collection with six of the other best essays for and against monads that was published in 1748.[48] Maupertuis and Euler had won this battle, but it was not the end of the monad dispute.

In a letter of August 1747 to Euler, Daniel Bernoulli asserted that Wolff himself had written to his father Johann of the affront the prize decision made to partisans of monads. Daniel feared that Euler would provoke numerous antagonists in metaphysics and bring himself much pain.[49] Later d'Alembert communicated to Lagrange "our friend Euler's" incompetence in metaphysical questions, a view in which Wolffians and the French *philosophes* concurred: "It is incredible that such a great genius in geometry and [algebraic] analysis is so inferior in metaphysics to the smallest schoolboy . . . [and] so absurd. . . . Not all the gods give to the same."[50]

The publication of the *Recherches*, including a translation of the *Gedancken* into French with a commentary and interleaved criticisms, allowed Maupertuis to read more of Euler's theory of matter and the responses to it. In September Euler sent Maupertuis a letter in French about the sharp polemic proceeding with the Wolffians in Leipzig and continued with gossip, rumors, proposed refutations, and unfavorable reviews in journals against Justi's prize-winning paper. The next month a French translation of it and the attack, also from critics in German in Leipzig appeared. This chiefly pitted Leipzig University against the academy. Euler thought Formey to be the chief initiator of the criticism.[51] Maupertuis, who acted as a moderator, concluded that the antagonism between Euler and Wolff was too great for any chance of reconciliation; earlier that year, in May, he had chastised Formey for his part in the quarrel.

In April 1747 Daniel Bernoulli, referring to Euler's participation in other metaphysical controversies, had written to him essentially urging restraint.[52] In effect denying the infallibility of his good friend, Bernoulli cautioned him, "Sie sollten sich nicht über dergleichen Materien einlassen, denn von Ihnen erwartet man nichts als sublime Sachen, und es ist nicht möglich, in jenen zu excelliren" (You should not engage in matters of such kind, for of them one expects nothing but sublime cases, and it is not possible to excel in all fields).[53] His advice was to be silent ("Si tacuisses philosophus mannsisses"). Bernoulli's letter is a treasury of scientific information; after he reported on his own computation of eccentricities in the orbits of Saturn and Jupiter, pointed out childish errors in d'Alembert's hydrodynamics, and praised the ingenuity of Euler's theory of light and colors, he urged Euler to propose his discoveries with less assurance. He rightly claimed that Euler had no greater friend and enthusiast.

In his prize-winning paper on magnets for the Paris Academy in 1744, Euler had rejected Roger Cotes's statement in editing Newton's *Principia* that "gravity is [inherently] found in all bodies universally." Two years later Euler protested that this notion was being wrongly attributed to him, and he endorsed the position of Gerhard Andreas Müller of Weimar in the article "Untersuchung der wahren Ursache von Neutons Allgemeinen Schwehre" (Investigation of the true causes of Newton's general gravity) that attraction arises from the pressure of the ether. While his friend Daniel Bernoulli's study of comets and tides supported the views of Newton and Maclaurin on gravity, Euler equated gravitation with a pressure in the hydrodynamic ether that is greater when the velocity of the ether is less. In hydrodynamics Bernoulli influenced him. In the last chapters of his *Anleitung* and numbers 68 and 69 of his *Lettres à une princesse d'Allemagne*, written in October 1760, Euler dismissed the Cartesian criticism of attraction as being an occult quality. Instead, he observed, it was possible to measure its effects. He posited two chief types of matter: gross matter, such as gold, had a high density, while subtle matter possessed a density many thousands fewer in degree than regular matter; the ether rushes in to fill vacuums and cosmic space. The true cause of Newton's mutual attraction, Euler declared, was not action at a distance but the pressure that it exercised upon nearby bodies.

The monad controversy occurred not in isolation but within the context of assaults on traditional Christianity in Europe. In England Matthew Tindal and Henry Saint John Bolingbroke had led in the espousal of deism, which projected a distant clockmaker God provable by reason, while freethinkers in France and the German states insisted that reason without appeal to the Scriptures could deduce the existence of God, the immortal soul, and divine providence. The influx of French savants won these ideas increasing influence at Frederick's court. In an apologetic tract on Christianity, *Rettung der göttlichen Offenbahrung gegen die Einwürfe der Freygeister* (Rescue of divine revelation from the objections of the freethinkers, 1747, E92), consisting of forty-six octavo pages, Euler put himself among the first critics of Wolffian freethinkers in the German states. At its core the *Rettung* is a Calvinist essay, proposing that the doctrines of predestination and human depravity will urge human beings to control their passions. Extending to physicotheology his arguments against monads, Euler contended that Leibniz's monadic doctrine and principle of the preestablished harmony between the faculties of mind (understanding) and will contradicted the traditional concept of original sin.

The treatise opens with a discussion of the fundamental human faculties of will and mind, and the duty of human beings to perfect to a greater degree their understanding that comes from a knowledge of truth. A rough summary of Euler's thought would be: Scripture guides the will; science uncovers God's workings in nature. The physical universe results from the omnipotence and infinite wisdom of God, so there can be no end to human learning. Together, will and mind bring a closer knowledge of God, who is the highest truth and the greatest good. In providing this, the two map out the way to gain happiness. Scripture does not disclose internal consistency, Euler concedes. Its authority rests not on worldly reason but in our recognition that it is divine revelation.

Euler compared the state of knowledge of the Scriptures with that in geometry, the most certain of the sciences. Only practitioners with a deeper grasp of geometry could unravel the paradoxes and contradictions in their field, he posited, and the freethinkers lacked such ability in biblical studies. Where Bishop George Berkeley in *The Analyst* (1734) limited the sciences to describing and quantifying nature, Euler added their construction of governing concepts. He accepted the existence of mysteries in faith and geometry but held that studies by scholars able in these fields would in time resolve them. In response to the reliance the Wolffians placed exclusively in intuitive reason, Euler pointed to the absurdity of the reasoning that brought some scholars to conclude, as did Zeno's rigidly reasoned paradoxes, that there can be no motion.

Euler faulted doubts among the freethinkers that Earth was created in time and would have an end. Observations, he believed, showed that the planets, which had elliptical orbits because they encountered resistance in the ether, were moving slowly—in modest changes—closer to the sun. In investigating astronomical observations made from the time of Claudius Ptolemy and from documents of Ibn Yunus in Leiden, he noticed a seeming reduction in planetary orbits marking the course from the beginning to the imminent end of Earth. These were not against reason but, to the contrary, stemmed from the contemplation of natural causes. Initially, Euler thought, Earth had been too far from the sun for human beings and animals to originate on it through natural causes. Instead, at some time after the starting of the universe, they needed God to create them.

In closing Euler declared unassailable the divine revelation of the Scriptures. He had sought to "restore to the right path many hearts who have not been too much corrupted as they were thoughtlessly enticed by the temptations of these wretched people,"[54] the freethinkers, whom Euler also termed "hardheads" and "rabble."

The *Rettung* ran through three editions in German and constituted—along with sections of his *Lettres*—the whole of Euler's publications on religion. In letters to Maupertuis he also attacked the freethinkers, and later he prepared a new proof of the existence of God. In numbers 116–120 of the *Lettres*, which were sent by mail to the princess in Magdeburg in April 1761, Euler—like the Christian Platonists—set out three classes of truths: of the senses, of the intellect or reason, and of belief or faith.[55] Establishing these known truths, he posited, depends upon taking precautions to prevent or discover errors and following rules characteristic of each of these to assure the certainty of its proofs. The classes of sensation and belief have guides, just as logic does for correct reasoning. Only some visionaries and philosophers—notably the Pyrrhonists, who doubt everything—have questioned the existence of external physical objects in our world. Euler defined as one of the principal laws of nature the reality of their existence, though he observed that we do not know the true reason for that existence. While delusion and error are possible, careful observation can eliminate them and make the truths of sensation as certain as truths of theoretical geometry. Here Euler's argument is akin to Lockean sensationalism. In the rational intellectual realm, the axiomatic-deductive method of Euclidean geometry and the new algebraic analysis retain primacy as models. Truths of faith rely on the ability of the several people advancing them, their knowledge of the Scriptures, and the reliability of their written historical sources. Convictions in faith are the most subject to error. Groundless reports, rumors, and false testimonies do arise, and who, Euler asked, "believes all that gazetteers and historians have written?"[56] Arriving at historical truths as certain as the truths of reason and of the senses—not later secondary reports, but declarations from direct observation by worthy people—is essential, as are comparisons of primary sources to uncover possible differences and contradictions.

Euler's religious views contributed to his unpopularity among freethinking members of Frederick's court. His relations with the monarch on religion were complex; Frederick, who proclaimed himself a freethinker, maintained that his subjects could say what they please but must obey him. And obedience required guidance and discipline, such as religion provided, though no one religion had the superior claim. Probably in an effort to have better relations with the French Calvinists in Berlin, Frederick later asked Euler to serve on their consistory.

In the eighteenth century the sciences and the humanities—the "two cultures," as later defined by the English novelist and physicist Charles Percy Snow—were diverging. Like Galileo before him, Euler fought to raise the modern sciences from a position subordinate to that of the humanities but not to damage that domain. Yet among Euler and Frederick's favorites, the French *philosophes*, encounters could be strained. The French depicted Euler as a stodgy German, and it is a characterization that continues to resonate. Euler, whose free Swiss burgher manners put him closer to our times, had no use for the shallow witticisms of the *philosophes*—especially not any cleverness critical of religion. Enduring many French slights, he was thick-skinned. Among fellow academicians in the sciences in Berlin and Saint Petersburg, the view of Euler differed markedly from the French depiction of him. The commoners found him vivacious; in conversations he could be animated, charming, occasionally playful, and a good storyteller with a sense of humor. Nicholas Fuss's "Lobrede auf Herrn von Leonhard Euler" describes him as astonishingly erudite, well informed on the classics and, like Leibniz, famed for his encyclopedic knowledge. Euler had "read all the . . . [major and minor classical] Roman authors, knew perfectly the ancient history of mathematics, held in his memory the historical events of all times and peoples, and could without hesitation adduce by way of examples the most trifling historical events. He knew more about medicine, [anatomy, physiology,] botany, and chemistry than might be expected of someone who had not worked especially in those sciences."[57] As Euler's reputation grew, travelers were drawn to his door. Many left amazed that outside his specialties in the mathematical sciences he had absorbed such vast general knowledge.

Euler was in 1746 seeking to be elected a fellow of the Royal Society of London, and he offered to present to the society a manuscript copy of his *Scientia navalis*, which the Petersburg Academy had not yet published. Euler wanted it published as quickly as possible, because he was worried that others would make discoveries similar to his and publish them first. He promised to compare *Scientia navalis* with Pierre Bouguer's *Traité du navire* on navigation and naval architecture, which appeared that year. Euler's ship theory and computations agreed with experience, he asserted, while Bouguer had made mistakes in calculations—especially on rowing—and had failed to cover the field. Euler offered to pay for the printing of *Scientia navalis*, but Royal Society president Folkes would not agree to it. Razumovskij had Schumacher write in August 1747 that Euler's delinquent salary would be paid (which it was by December) and that the Russian academic chancellery with support of the Russian admiralty had renewed its rights to have both volumes of *Scientia navalis* printed in Berlin at the expense of the Petersburg Academy. Razumovskij allocated five hundred rubles for the publication, a decision that drew criticism. Possibly because of costs and uncertainty over who would pay it was impossible to find a publisher in Berlin willing to undertake the difficult project. By February 1747, at any rate, Wettstein had already shepherded through Euler's election to the Royal Society.

That January another argument erupted between Euler and d'Alembert, this time in pure mathematics over the logarithms of negative numbers, one of the two main subjects in their correspondence. Both found negative numbers to be filled with paradoxes. The origins of the controversy, which was conducted in letters delivered between the two by Maupertuis and Euler's Basel cousin, the physicist Reinhard Battier, lay in the article "Recherches sur la calcul integral," which d'Alembert had submitted in 1746 for the Berlin Academy's *Mémoires*. In a letter to d'Alembert sent at the end of December, Euler tactfully rejected d'Alembert's claim in the "Recherches" that $\log(-x) = \log(x)$ for positive numbers x. The next month d'Alembert responded in two letters that he found the new information disturbing. In the second letter he sent Euler some improvements for integral calculus and asked him to "cross out of my treatise the portion where . . . [log(−1)] is discussed."[58]

So far the discussion was amiable, even deferential on the part of d'Alembert. Then in March, d'Alembert defined logarithms as the inverse of the exponential and rejected the usability of the exponential curve with its two branches. To that point his claim was compatible with Euler's position that the logarithm of a negative number is imaginary. But d'Alembert's letter concluded with $-1 = 1/-1$, so this gives $\log(-1) = \log(1/-1) = \log(1) - \log(-1)$, so $2\log(-1) = \log(1) = 0$, so $\log(-1) = 0$. In April Euler rejected d'Alembert's contention "that logarithms of negative numbers are real" and argued that in some metaphysical conditions they may be imaginary. He based logarithms on $e^x = 1 + x/1 + x^2/(1 \times 2) + \ldots$, recognized the ambiguity of the logarithms of +1 and −1, offered counterexamples to d'Alembert's position, and resolved some seeming paradoxes. From Johann Bernoulli's formula, for example, he found that logarithms of imaginary numbers could not be real; otherwise the "$\log\sqrt{-1}/\sqrt{-1}$ could not express the quadrature of the circle."[59] (In his letter the symbol π, not yet standardized, represented what is now 2π.) In August Euler reported that he had removed the contentious portion of d'Alembert's paper and reiterated that the logarithm of −1 is imaginary, $i\pi$. He now saw no more problems in explaining its nature.

The next month Euler read to the Berlin Academy his paper "Sur les logarithmes des nombres négatifs et imaginaires" (On the logarithms of negative and imaginary numbers, E807), but did not forward it for publication; it would appear posthumously in 1862. Euler had proven that if x is a positive real number, all of its logarithms save one are imaginary, while x negative has all imaginary logarithms. Nonetheless, the debate continued until December 1747, when Maupertuis advised Euler that d'Alembert wanted to put aside "his work in mathematics for a little while to reestablish [his] health."[60] Euler wished him a successful recovery. This was not the end of the debate but simply a brief respite.

In his three-page essay "De numeris amicabilibus" (On amicable numbers, E100) in *Nova acta eruditorum* for 1747, Euler was developing—but did not yet state—an algorithm in number theory for generating pairs of amicable numbers. It was to be but one of his hundreds of discoveries at the time in mathematics. He had begun his research with the second type of perfect numbers, which the ancient Greeks held are either 10 or the sums of their divisors—for example, $6 = 1 + 2 + 3$. Krafft's article "De numeris perfectis," in volume 7 of the Petersburg Academy's *Commentarii*, demonstrates that perfect numbers must end in 6 or 8 and proves that two numbers from Niccolò Tartaglia's *Arithmetica* of 1613—130,816 and 2,096,128—are not perfect, nor is the number 511 prime; instead it is 73 x 7. For Krafft, Euler—using Euclid's formula $2^{n-1}(2^n - 1)$—had computed perfect numbers for the primes to $n = 47$.[61] For example, $n = 7$ gives 8,128 and $n = 31$ equals 23,058,443,008,139,952,128. The two were correcting computational errors for n into the 40s in the arithmetic texts of Niccolò Tartaglia and Michael Stifel. In his brief essay published in 1747, Euler proceeded to amicable pairs, in which each of two numbers is the sum of the proper divisors of the other. In classical antiquity 220 (1, 2, 4, 5, 10, 11, 20, 22, 44, 55, and 110) and 284 (1, 2, 4, 71, and 142) were known amicables, since the sum of the proper divisors of each except for the number itself is 220, while Thabit ibn Qurra in the ninth century probably added 17,296 and Ibn al-Banna in the thirteenth century found 18,416. In the mid-seventeenth century Pierre de Fermat independently obtained Ibn Qurra's pair along with 2,620 and 2,924, while Descartes added a fourth pair, 9,363,548 and 9,457,506. Fermat gave a rule for computing amicable numbers from prime factors. Possibly two other pairs were discovered before Euler. His "De numeris amicabilibus" cites the computational technique of Fermat and Descartes but does not reveal his own method. Euler gave thirty amicable pairs with twenty-six of these new, nearly a sevenfold increase over all known previously throughout history. His list includes 2,620 and 2,924 decomposed into their prime factors in item 8.[62]

"De numeris amicabilibus" also examines another challenging problem, "whether . . . there are any odd perfect numbers." Blocked in his effort to solve this, Euler called the matter a "most difficult question," and his characterization is accurate: in number theory the existence of odd perfect numbers is to the present day unsolved.

In 1747 Euler completed "Theoremata circa divisores numerorum" (Theorems on divisors of numbers, E134), which appeared in the *Novi Commentarii* three years later. It presented for the first time his potent computational method for amicable numbers, introduced the concept of number theoretical functions, and gave the first definition of the sigma function, which sums the divisors of a given number n. It was later found that Euler's method did not generate every amicable pair. Expanding upon Fermat's method with prime numbers, Euler—after making some unfruitful substitutions—obtained a total of sixty-one, thereby doubling their number;[63] his lists have a few typographical errors in the results.

During 1747 Euler offered the second in a sequence of four improving proofs of what is usually called Fermat's Little Theorem or the

Euler-Fermat Theorem: in modern symbols, if p is a positive prime number, then $(a^{p-1} - 1)$ is divisible by p, if p and a are relatively prime. In modern Gaussian notation, $a^{p-1} \equiv 1(\mathrm{mod}\, p)$. Euler's first proof in 1736 had employed the technique of mathematical induction, and his second differed little. In the second Euler defined his important phi function $\phi(n)$, denoting the number of integers less than n or equal to it which are relatively prime to n. In adding proofs to the conjectures in arithmetic, Euler contributed to the origins of number theory. The second proof deals with how he determined the factor to show that the fifth Fermat number is not a prime.

In mechanics Euler continued making important advances in 1747, pursuing a general method applying to all types of systems, whether continuous or discrete. He had already devised differential equations of motion to compute discrete systems and elastic bodies, but did not yet have them for perfect fluids and solid media. He was the first geometer to delineate a set of differential equations for the first three types of bodies. Although these equations do not appear in Newton's writings, they are today known to physicists as "Newton's equations." Euler introduced them in the paper "Recherches sur les mouvement des corps celestes en general" (Studies on the movement of celestial bodies in general, E112), apparently presented to Berlin's Royal Academy of Sciences in June 1747 but not published in the academy's *Mémoires* until 1749.

It likely did not take long after finding Newton's equations for Euler to finally free from any restrictions to small motion the principle of linear momentum (Newton's second law) and show that it applied to all systems. He described as "absolutely necessary that these principles [of mechanics] be deduced from the first principles or rather the axioms, upon which the whole doctrine of motion has been established. Ordinarily several such principles . . . it appears must be admitted to the rank of axioms of mechanics . . . but I remark that all these principles reduce to a single one, which can be regarded as the unique foundation of mechanics."[64] Exactly how Euler arrived at this finding regarding the principle of linear momentum is not known; perhaps it was through the study of mutual forces. His results appeared in the article "Decouverte d'un nouveau principe de mecanique" (Discovery of a new principle of mechanics, E177), completed in 1750 but not published in the *Mémoires* until 1752.[65]

In 1747 the investigation of celestial motion, especially for lunar tables and planetary theory, was the other major subject in Euler's correspondence with Clairaut.[66] In September Clairaut rejected the universality of the inverse-square force—that is, that by itself it could describe celestial motion with mathematical exactness in physics and at all distances in astronomy.

Clairaut, the first European geometer to take this revolutionary position, asserted that the motion of lunar apogee posed a challenge. To Newton's $1/r^2$ Clairaut added as a correction a small function, a constant divided by the inverse of the fourth power, c/d^4, sensitive only over small distances and essentially null at stellar expanses. Euler agreed that Newtonian attraction did not appear to account precisely for the motion of lunar apogee but held that in his study of Saturn he had already reached that conclusion. He and Clairaut were not the first to propose modifying Newton's law of attraction. For example, in her book *Institutions de physique* of 1740 Émilie, Marquise du Châtelet had done so for chemical processes. The latest results from lunar and planetary orbits obtained from improving observational astronomy with a precision not before achieved now seemed to differ slightly from Newton's law of attraction, and a small correction seemed needed. For accuracy, calculations of the motion of lunar apogee by use of Newtonian gravitation required force laws. These were "a little different from the ones I supposed," Euler wrote. All "these reasons joined together appear . . . to prove invincibly that the centripetal forces one conceives in the Heavens do not follow exactly the law established by Newton." Euler also did not accept Clairaut's modification, for it failed to account for the orbit of Mercury, so near to the sun that the additional term was too important. In the paper on the motion of Saturn and Jupiter submitted to the French Royal Academy in August 1747 for the Prix de Paris competition, Euler attributed gravitation to the presence of fluid ether. Two of the judges on the jury deciding that contest were Clairaut and d'Alembert.

In September Clairaut had informed Euler of his stratagem for attacking lunar motion with a set of four complicated differential equations, with two second degrees and two first degrees. But these provided only an approximate solution. The two men mistakenly depicted the motion of the moon as a rotating ellipse; while that representation applies to the planets, the sun sharply disturbs the lunar orbit. Clairaut dropped his study of Jupiter, Saturn, and the sun and closely followed Euler's. A critical question facing European astronomers and geometers was whether the secular change in the mean motion of planets over time was cyclical or linear. Confirming that the Newtonian law of attraction by itself explained the mechanical operations of the solar system required resolving the application of this law to the orbits of the two large planets, Jupiter and Saturn. On the Prix de Paris review committee, Clairaut read Euler's paper that had won the prize for 1748 on the inequalities that Saturn and Jupiter "appear to cause in each others' motions, especially at the time of their conjunction."[67] This marked Euler's first effort to accurately describe planetary perturbations.

In "Recherches sur la question des inégalités du mouvement de Saturne et de Jupiter" (Research on the question of the inequalities in the motion of Saturn and Jupiter, E120), the second of two papers on astronomy that he completed by mid-1747, Euler attempted but failed to demonstrate that the theory of attraction sufficed to account for the orbital elements from the slowing of the mean rate of Saturn over time, the speeding up of Jupiter, or vice versa. He accepted the elements given by Cassini in his *Elemens d'astronomie* of 1740 that were founded on the laws of Johannes Kepler. Euler required "some correction [in these elements], since the inequalities caused by Jupiter have been there enveloped in the eccentricity and position of the orbit of Saturn." Clairaut was satisfied that Euler's prize-winning paper showed that secular—as opposed to periodic—mutual perturbations in the orbit of Saturn caused by Jupiter seemed not to be derivable from Newton's inverse-square law. Euler's invention of trigonometric series moved the investigation of perturbations beyond tiresome numerical integration; he performed logarithmic differentiations on them and employed a merging of equations. Drawing upon the latest observations made with the best telescopes, for successive terms in his series Euler approximated special coefficients. He reached these partly by assuming that when Jupiter's orbit is elliptical, Saturn's is circular, and vice versa. In accounting for varying motion, Euler put in his differential equations arbitrary constants, which perturbations actually caused to vary extremely slowly. But Euler's computations with arbitrary terms could lead to mistakes; only slightly better than earlier computations, they contained some erroneous figures that he traced to imprecise observations and most of all to a slight inaccuracy in the inverse-square law at interplanetary distances. Nor did Euler discover the long-term inequalities of Saturn and Jupiter; that remained an unsolved problem for him and undermined his confidence in the inverse-square law. Two years after Euler's death, Lagrange would have the solution. Even without that, the innovative mathematical methods that Euler introduced to describe planetary perturbations were adjudged excellent and had earned him the Prix de Paris. By systematically applying procedures similar to Euler's and improving on them, other geometers and astronomers—chiefly Johann Tobias Mayer, in 1750—succeeded in providing precise and elegant solutions that corrected orbital elements and determined certain terms and coefficients needed precisely to describe perturbations that resisted earlier theoretical computations.

On 3 September 1747 Clairaut sent Euler general equations that he employed to attack the problem of one of the many irregularities in the lunar revolution around Earth—in this case, the motion of lunar apsis or

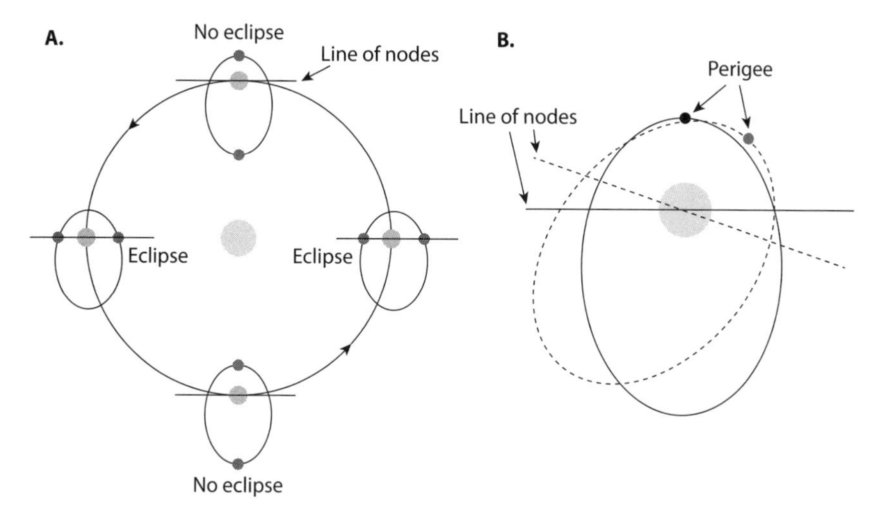

Figure 8.4. The precession of the moon's orbit. A) The Earth-moon system at four points in their orbit around the sun. The line of nodes is the intersection of the moon's orbit with the ecliptic (Earth's orbit). The moon crosses the ecliptic twice a month. An eclipse happens when the moon is in either new or full phase and at the line of nodes. The moon's orbit is slightly elliptical (exaggerated in this picture) and the moon may not be exactly on the line of nodes, so a solar eclipse may or may not be total. B) The moon's orbit precesses so the line of nodes shifts by about 19.3 degrees in one year and the perigee (closest point to Earth) of the moon's orbit shifts by about 40.7 degrees in one year. The line of nodes completes one full precession in 18.61 years. This precession explains why eclipses occur at different dates in successive years. The dotted curves are one year later.

apogee. The obscurity of Newton's treatment of this problem had drawn Clairaut to attempt to resolve it. It was a case of the difficult three-body problem with the moon gravitationally attracted by both Earth and the sun. Yet after many repeated calculations, Clairaut could not explain this phenomenon by the inverse-square law; he kept finding an eighteen-year period of a full revolution of lunar apogee around Earth rather than the observed nine years. On 11 September Clairaut wrote to Euler,

> I was pleased to note that your thinking is the same as mine concerning Newtonian gravity. It appears to be proof to have shown that it is not sufficient to explain such phenomena, but the special role which you assign to the moon does not appear as surprising as my own observations indicated. Instead of attempting to find out what the distance should be, my approach was to find out what the motion of the apogee should be, and in finding it to be only half that observed in nature, I saw the most complete proof of the insufficiency of the inverse-square law. . . .[68]

On 30 September Euler replied,

> I am able to give several proofs that the forces which act on the moon do not exactly follow Newton's law, and the one that you draw from the motion of apogee is the most striking, and in my lunar theory I have clearly pointed this out. . . . Since the errors cannot be attributed to the observations, I do not doubt that a certain derangement of the forces supposed in the theory is the cause. This circumstance makes me think that the vortices or some other material cause of these forces is very probable, since it is then easy to understand that these forces ought to be altered when they are transmitted by some other vortex.[69]

By late 1746 d'Alembert had also devised complex differential equations to describe lunar apogee. In addition to discussing negative logarithms, he communicated with Euler about planetary and lunar theory. In April 1747 d'Alembert expressed to Euler his great pleasure that Euler knew his theory and was applying it well to astronomy. But difficulties remained, and Euler was developing and adding mathematical sophistication to d'Alembert's theory. (D'Alembert was later to accuse Euler of appropriating his original theories without giving him full recognition.) From July to September d'Alembert explained to the Berlin Academy an important paper on his theory for the moon and Earth that remained unpublished until 1749. To prove his priority in lunar theory, in September 1747 Clairaut left a copy of his equations in a confidential *pli chacheté* (sealed envelope) with Paris Academy secretary Jean-Paul Grandjean de Fouchy. Two months later d'Alembert similarly sent his results to Fouchy and the Berlin Academy. As Émilie du Châtelet wrote, "Mr. Clairaut and Mr. d'Alembert are after the system of the world, understandably they do not wish to be forestalled by the essays for the prize."[70]

The thirty-four-year old Clairaut, pleased and perhaps emboldened that Euler's lunar theory and his essay on Jupiter agreed with his own, caused a stir when he pompously announced publicly at the Paris Royal Academy on 15 November that Newton's inverse-square law was insufficient alone to explain all celestial phenomena. He chose this day, the official start of the academic year, because the academy's meeting room would be open to the public and could be full. This promised success for his ideas across the republic of letters. Even though Clairaut's result concerning the movement of lunar apogee was not new, for Newton had discovered the same without differential equations in his *Principia*, the few remaining Cartesians in Paris took heart since the new equations seemed to confirm an error, and this gave greater credence to Newton's critics. Like Euler, Clairaut concluded that higher-order equations could not make up the missing half of the advance of lunar apogee given in Clairaut's six first-order differential equations.

As was typical in "saving the phenomena," Clairaut had as previously noted added a second term, a constant divided by the inverse of the fourth power, to correct the law of gravitation in closer distances and explain perturbations in the lunar orbit. He sent this idea in a letter to the Parisian Newtonian Georges-Louis Leclerc, Comte de Buffon, who attacked the emendation for violating metaphysical simplicity and for suggesting that several forces were involved rather than only one. For decades the French astronomer Pierre Charles Lemonnier had observed lunar motion with the best British instruments, and he also questioned the idea. Both d'Alembert and Euler praised Lemonnier's work, and an argument briefly flared. In December Clairaut, who was working diligently on his lunar theory article for the Paris Academy, suggested to Euler that a medium might account for a very small resistance to the motion of planets and the moon. The next month Euler responded that his studies of lunar motion agreed on the need to correct Newton's inverse-square law, and he promised soon to send Clairaut a packet of papers containing various arguments about it and another through the publisher Marcus-Michael Bousquet that presented exemplary problems from his *Introductio*, which was soon to be released. Busy with the preparation of the *Encyclopédie,* d'Alembert claimed that for the time being fatigue prevented him from participating more in this debate.

In December Clairaut wrote to Maupertuis giving his conclusions about the motion of the lunar apogee. Most likely Maupertuis shared the letter with Euler. Friction was developing among Clairaut, d'Alembert, and Euler. To this point, these three great protagonists of Newtonian celestial mechanics agreed on Newton's law; only Bernoulli disagreed. But now the correspondence was to grow increasingly fierce among the three. Encountering criticism from across Europe, Clairaut worked to justify his position and began to attack Euler on several points, and not always fairly. In January 1748, Euler recognized Clairaut's priority in the new equation with the inverse fourth-power term, but he rejected its universality, for the closeness of Mercury to the sun made the added term overly important and gave mistaken results for its orbit.

During the eighteenth century, the fabrication of dioptric instruments posed many problems. In 1747 Euler embarked on a practical project in physical optics, the construction of better telescopes, which was critical to enhancing research in both observational and theoretical astronomy. This subject was not a temporary diversion for Euler but an interest throughout the rest of his career.[71] Since the refracting telescopes of Galileo Galilei and Thomas Harriot in the early seventeenth century and the reflector telescopes of Newton and David Gregory at its close, astronomers, geometers, and opticians had confronted aberrations from the spherical curvature of lenses, which distorted images, and from refrangibility, which brought chromatic errors in the dispersion of light in reflectors that resulted in blurriness detracting from clarity in telescopic observations. In object glasses various color light rays failed "to undergo the same refraction. For example, red rays form a different focus from the blue, while a mirror reflects them into one and the same focus."[72] Newton's definitive decomposition of white light had made it possible to begin to eliminate chromatic error. He designed a crucial experiment of two prisms that broke white light into its spectrum of component colors and recombined it. In a series of trials Newton failed to reduce colorization by the employment of two optically different media in a two-lens system separated by a space filled with water. The authority of Newton and the empirical support discouraged a search for another solution. Inspired by studies of the eye he'd made when he was a student, Euler attempted to imitate nature in reducing the deviant effects of refrangibility with such colorization; he prepared an apparatus consisting of two convex/concave transparent lenses separated by a shallow space filled by a liquid with an adequately differing index of refraction from the lens glasses. He wanted to reduce the thickness of objective lenses of the telescope and microscope and to increase their power—an attempt still new in physical optics at that time. Apparently it was not until September 1748 that Euler read to the academy his paper "Sur la perfection des verres objectifs des lunettes" (On the perfection of objective lenses of telescopes, E118). The following month he wrote to Goldbach of his efforts to set right a problem in his optical work "in order to create such object glasses which should do the same as the reflector mirrors in the telescopes of Newton and Gregory."[73]

"Sur la perfection des verres objectifs des lunettes," which was published in volume 3 of the *Mémoires* in 1749, holds that Euler had derived a geometric formula that found a balance of one lens on another separated by a liquid that lessened the two aberrations. But this deduction set his work against that of Newton, who had found in a series of experiments that the combination of lenses did not work. Euler's assertion would thus provoke a fierce and lengthy quarrel, primarily with the English optician John Dollond, a staunch partisan of Newton, that seems to have involved national pride. The dispute would appear in the academy's *Mémoires* and the *Philosophical Transactions* of the Royal Society. Euler attributed colors to differing lengths or frequency of waves, but establishing the truth of this lay a century in the future. Certain indices associated with color he took to prove the existence of God.

Academic Clashes in Berlin, and Euler's Correspondence with the Petersburg Academy

The dominance of French among the nobles in Prussia and in the meetings and proceedings of the academy, alongside Maupertuis's ignorance of German, provoked tensions with German academicians not fluent in French. At meetings some spoke French that was next to gibberish; and they considered onerous the need to translate their papers from German or Latin into French. Another potential occasion for resentment came in November 1747 when Maupertuis ordered Formey to tell the four directors that the statutes requiring at least modest productivity in the academy must be strictly followed. Each salaried member had to read a minimum of two papers a year and each associate one. This message must not have had an effect, for in the winter of 1748–49 Maupertuis complained that the senior physicians were not writing articles and not attending meetings of the academy. He believed German academicians to be lacking in *esprit académique*, the self-motivation of members of the Paris Academy, and suspected that they were regularly thwarting his efforts to enhance the position of his institution at court and in society. These circumstances added to his autocratic methods produced increasingly caustic adversarial relations.

In August 1749 the king issued a cabinet order. He decided that salaried members and associates who did not publish any papers would lose their yearly stipends, which would be dispensed to members producing meritorious works. Authors had to submit papers to the president for review fourteen days before they were read to the full academy. Maupertuis informed the king that the department of speculative philosophy and belles-lettres was particularly feeble. Euler, who on average wrote more than ten papers a year in Berlin, sent two letters in mid-November heartily supporting him and asserting that while the mathematics department met all requirements, others were resisting the presidential order.[74] Euler and Formey interceded to attempt to improve relations between German members and Maupertuis, who knew no German until 1750 and never became fluent in the language.

Correspondence especially with the Petersburg Imperial Academy gave Euler an opportunity to touch on a number of important topics. Prominent among them were electricity, exploration, cartography, and economics.

Research on electricity remained popular across Europe. In February Euler recounted the partial cure of a disabled person after fifteen to twenty shocks, and the next month he described the killing of birds with electrical shocks and reported on an article in the *Philosophical Transactions* of the Royal Society of London that examined the effect of streams of electricity on animals. A letter in November to Razumovskij recounted tests by physicians in Venice and at the Academy of Bologna applying electricity in the cure of gout, among other illnesses.[75] Euler had earlier examined reports of electrical research in Paris and Vienna.

In December 1746 Euler wrote to Wettstein about the Russian search for a northwest passage across North America that could provide a navigable water route between Hudson's Bay and the Pacific Ocean,[76] something the English had been searching for since the previous century. Euler called the quest a "glorious undertaking" and surmised that among the large rivers that the Russians discovered emptying from North America into the Pacific, one might be linked to Hudson's Bay. Captain Vitus Jonassen Bering, the Danish leader of the Second Kamchatka Expedition, had believed that no such connection existed, and Russian explorers in the early exploration of these rivers with smaller ships supported his view. From the reported travel of whales, Euler thought that there might be a water connection through the Arctic Ocean, but it would be too long and by the end of summer frozen and too dangerous. The Russian admiralty college had supplied Euler with the data, and he added that he greatly doubted that the Russians would publish the material. Wettstein apparently took this to mean that he should have it printed. Portions of it appeared in London newspapers in 1748 and in two parts in the *Philosophical Transactions* for 1747 and 1748 as "Extract of a letter from Mr. Leonhard Euler" (E107). Now widely distributed, the *Philosophical Transactions* reached a readership well beyond the scientific community. This material on the rivers seemed to jeopardize Russia's interest in keeping secret its strategic military designs and its trade and maritime possibilities. Euler recognized the danger of repercussions against him from Elizabeth's court and the Petersburg Academy. In May 1748, he informed Wettstein that he had drawn from diverse, not well established, and hearsay sources. The release of accurate, well-founded material would have angered the academy and embarrassed him, and he would never have intended to urge the publication of it. This letter to Wettstein alone was clearly insufficient. Euler also directly apologized to the academy for releasing the unproven northwest passage data along with his speculations.

Porcelain was vital to the economies of both Prussia and Russia, and its fabrication methods were a closely guarded secret. In November 1747

Euler warned Razumovskij of an imposter who was making false claims about the techniques in Saint Petersburg; He recommended contacting the German physician and chemist Johann Heinrich Pott, who for a reasonable percentage of the income from a planned factory would share his porcelain-making techniques that produced works of finer quality than those from the procedures in Dresden. He went on to praise the discovery by Andreas Sigismund Marggraf of a way to extract true sugar from beets, a plant grown in Prussia, and to make a great profit from it.[77]

Seeking to restore the prestige of the academy, Teplov in October 1746 had solicited Euler's recommendations for new members. Euler cited excellent candidates in Paris not on Teplov's original list. If the names of Clairaut, d'Alembert, or Alexis Fontaine appeared, he urged, Saint Petersburg should vote for them "without any hesitation."[78] But he added that the French would prefer to stay in Paris with a small salary rather than accept a "large one in Saint Petersburg."[79] Until early December 1747, Schumacher had only begun to consider the candidates Euler had recommended for open positions. To the list for Teplov that included the Germans Braun, Brauser, Kratzenstein, Oechlitz, and Waltz, Euler added the names of Johann Kies in astronomy and Johann Gottlieb Gleditsch in botany and economics. Schumacher sent travel funds for Braun. He cited 660 to 860 rubles as reasonable salaries, while Euler urged more than 1,200 for each man. Schumacher now knew that Bernoulli would not come, so an offer would go to Oechlitz, and if he declined, to Johann Samuel König.[80] Bernoulli had suggested König, but Euler had not read his work and considered Oechlitz the greatest young geometer in German lands. This determined the order for possible invitations.

Euler kept up close contacts with the Petersburg Academy. In November 1747 the academy on Razumovskij's order continued to pay Euler's entire annual salary. That month he submitted an evaluation of papers by Georg Wilhelm Richmann and Lomonosov, finding most of them to be excellent; their publication, he believed, would make the *Commentarii* more important than works of other science academies in Europe. From correspondence Euler learned that a great fire had destroyed the observatory at the Russian Academy and seriously damaged other parts of the Kunstkammer Museum, including its library. The following January Euler wrote to Delisle in Paris that the diligent efforts of cadets (and he might have added officers) had saved almost the entirety of the library's books and journals, along with the museum collection of Peter I that featured technical instruments.

The Euler Family

From 1746 on, Euler and his family were comfortable in Berlin. The next year came Herrmann Friedrich, who would live to be three and a half. Having grown close to Maupertuis, Euler asked him and his wife to be the new son's godparents.

Euler's many children helped give rise to a description of him as possessing a powerful concentration, being undisturbed by noises in his surroundings, and writing his immortal works with "a child on his knee and a cat on his shoulder." While his ability to center his attention on a problem to the exclusion of all else was astounding, the portrayal of him at work with child, cat, and manuscript seems a creation of a later colleague, Dieudonné Thiébault, a grammar professor at the war college in Berlin. Not until 1765 did Thiébault become a regular member of the belles-lettres department of the academy, so he did not know Euler when the children were young, and he had access mostly to secondhand stories, for Euler was to return to Saint Petersburg in 1766. In 1785 Thiébault returned to France, his book with this written portrait of Euler unpublished.[81] Euler's son Johann Albrecht gives a different account of home life, declaring that when he was young, his father had little to do with general household tasks, including the care of children, when he was writing.

Euler actively assisted relatives beyond his immediate family members. In 1746 he arranged for two of his Vermeulen nephews, Georg Wilhelm and Karl Rudolf, to join the Prussian military; both had been students at the gymnasium in Saint Petersburg, where their father was a wine merchant. This did not pass unnoticed by Frederick II. He sent Euler a note on their placement and his gratitude for their enlistment.

The Euler residence on Behrenstrasse was a busy and welcoming place. In addition to students sent by the Petersburg Academy up to 1757 to reside there and study under Euler, scholars traveling to and from Russia visited him. In August 1747 Christian Kratzenstein wrote a note of appreciation for Euler's recommending him to the Russian Academy; the following January he inquired whether Euler could explain the illusions of the Italian magician Peladine, who had performed in Berlin, and announced that he was reading natural magic.[82] At the end of January, Kratzenstein visited Berlin during his move to Saint Petersburg from Halle. Johann Georg Gmelin, who had informed Euler of the debate between Josias Weitbrecht and Richmann over the principle of the conservation of *vis viva*, stopped on his way from Russia to Tübingen, and in October 1747 Delisle thanked Euler for the kindness shown to him and his wife during their visit on their journey from Saint Petersburg to Paris.[83]

One sign of the continuing depth of Euler's affection for his hometown of Basel was his correspondence with Johann Schorndorf, the postmaster there; begun in 1743, it continued for thirteen years. In 1747 Johann Conrad Henninger, the brother-in-law of Schumacher from Saint Petersburg, asked Euler to help clarify the business relationships in a certain Trau family, who were apparently from Basel. Euler requested assistance from Schorndorf and then quickly prepared a Trau genealogy that was sent to Russia. For contributions to this and for sending his initial letter of inquiry to Henninger, Euler thanked the postmaster.

Euler scarcely had time for leisure, though music remained his favorite recreation. He continued to play the clavier and invited composers to his house to present their new works. He also improved his chess game. Letters to Wettstein reveal his preference for a certain English pipe tobacco; many came with a plea for this preferred tobacco and his thanks for an earlier supply. At the zoo, where Katharina and he enjoyed taking the children, he particularly liked to watch the bear cubs playing. They also took the children to marionette shows, a popular entertainment at the time, and Euler was known to laugh robustly at the antics of the marionettes.

Letters by Frederick II reveal that Katharina and Euler were attending the theater. The monarch was upset that Euler would take a pen and paper to plays, occasionally jot notes during the performance, and even leave before the end. On 25 August 1749, Frederick wrote, "During the presentation, one of Euclid's sons realized that [he, Euler] had misplaced his greedy imagination without hearing, without seeing and without speaking. At first he began, while dreaming at first to calculate the effects of the voice on the hall, the theater, optics, and great circle of the oval; having finished and finding no more to do and chilly feelings and boredom, without having finished the act he awoke and left the rest to the devil."[84] Such behavior reflecting Euler's concentration was a major factor contributing to Frederick's prejudice against mathematics in general and his criticism of Euler's personality.

Euler continued to endure sporadic high fevers and a developing loss of eyesight. Knowledge of these fevers he continued to keep within a small circle that included Daniel Bernoulli, and he diverted attention from himself through inquiries about the health of others. (Among his colleagues illness was common.) In 1746 and 1747 Parisian gossip concentrated on Maupertuis's sickness, and d'Alembert wrote to Euler inquiring about it. In December 1747 Euler wrote a letter to d'Alembert, noting that Maupertuis had informed him of a spell of illness for d'Alembert, who had to set aside his research for a time. He wished d'Alembert a speedy recovery, and the ailing *philosophe* responded appreciatively to Euler's concern. While Euler worked quietly through his fevers, his weakening vision was well known.

Formey's correspondence relates that the socially cordial Maupertuis, who resided in Berlin and was also given quarters at Sans Souci in Potsdam, had an active social calendar in the capital, holding dinners for a circle including academicians, men of letters, dignitaries from outside Prussia, and nobles. These were part of what the *philosophes* called the code of sociability, a kind of bond of good citizenship sealed by convivial relations within the growing community of readers. Maupertuis's house was known for his Siberian and Icelandic dogs and his inclusion of a young albino African servant, whom he used in his studies of racial differences, albinism, and theories of generation—subjects of topical curiosity among the sociable worlds of the salons and generally the reading public but not yet investigated within the academy. Euler and his wife attended many dinners at Maupertuis's residence.

Chapter 9

The Apogee Years, II: 1748 to 1750

The year 1748 began sadly with news of the death on 1 January of the patriarch of the Bernoulli family, Leonhard Euler's former teacher Johann I (Jean I), "the old lion" as mathematicians knew him. It came after a short bout of eye pain. Euler's uncle Heinrich Brucker, the pastor of Saint Peter's Church, delivered the funeral oration. Brucker praised Bernoulli for the candor and weight of his arguments, which could move even opponents. The tombstone declares,

> Hoc sub lapide requiescit
> vir quo maiorem ingenio Basilea non tulit
> Saeculi sui Archimedes
> non illis Europae luminibus
> Cartesiis, Newtonis, Leibnitziis
> Mathematum Scientia secundus
> Johannes Bernoulli.
>
> Under this stone rests
> a man as great an intellect as any Basler.
> The Archimedes of his age
> Second to none of the luminaries of Europe
> in the mathematical sciences
> the Descartes, the Newtons, the Leibnizes
> Johannes Bernoulli.[1]

Considering Euler a worthy successor, the city magistrates and the regents of the University of Basel issued a protocol, seeking a written declaration from him in Berlin indicating whether he wished Bernoulli's "professionem matheoseos" and indicating that upon his refusal the position would be offered to Johann II (Jean II) Bernoulli, who was a professor of rhetoric at the university. Euler did not respond; the University of Basel could not compete financially or scientifically with his post in Berlin, and he would not challenge his close friend. The regents in Basel noted simply that circumstances prevented him from coming. In response that March to Euler's letter (now lost) of sympathy to the Bernoullis, Daniel stressed that his father had been steadfast in his admiration of Euler. After a short break, Johann II received the mathematics chair, a position he merited having won the Prix de Paris four times. Daniel Bernoulli wanted the chair in physics, which he received in 1750.

Throughout his career, Euler was a champion of Daniel Bernoulli, though his efforts to once entice him to Berlin had failed. In mid-January 1748 Euler came to Bernoulli's defense against critics at the Petersburg Imperial Academy of Sciences. As a major part of the academy's rehabilitation, Euler had proposed that Bernoulli return to Russia. In April 1747 Bernoulli received the invitation, but by September he declined it. He pointed out that when he asked for the consent of his seriously ill eighty-year-old father he had not gotten it. This was critical to Daniel, who continued to list himself as "the son of Johannes"; his father had asked him to agree not to leave Basel before his imminent death, and Daniel obeyed what was essentially Johannes's final wish. Even if he were in Russia, he would have had to return to Basel for the probating of the will. When his refusal reached Russia, an indignant Kirill Gregor'evich Razumovskij took it to be a slight toward the academy. In light of Razumovskij's reaction, Johann Schumacher issued a warrant canceling Bernoulli's position and future annuities from the academy. Euler held this an injustice, for Daniel Bernoulli had merely obeyed his father and had in no way showed disrespect for the academy. An advocate of justice, Euler did not hesitate to challenge the academic administration, the nobility, or the government, especially on financial issues. His distinguished reputation kept him from the retributions that might have faced other commoners.

Astonished at the warrant against Bernoulli, Count Hermann Karl von Keyserling, who had been the president of the Petersburg Academy in 1733, asked Euler to undertake every possible effort to have it lifted. Since all negotiations within the academy passed through his hands, Euler was able to send to Count Razumovskij's secretary Grigorij Nikolaevich Teplov a detailed report that, so he believed, showed Bernoulli to be blameless for declining the offer. The manner in which a person of Bernoulli's achievements at the academy and distinction across Europe was treated, Keyserling feared, might carry negative consequences for the institution. During the year, Euler continued to urge Schumacher to reconsider the warrant but, it seems, with little effect. Euler also commented that Bernoulli was angry that no one had contacted him about his salary. But payment of salaries for members outside Russia remained uncertain, even for Euler. The actions of the academy, he complained, were making Bernoulli out to be an enemy.

The *Introductio* and Another Paris Prize

While attacking problems in mechanics, Euler continued to concentrate on infinitary analysis. Through the 1720s it had remained a method for solving geometric problems represented through curves with symbols, and appropriately manipulating them, while from the 1730s on geometers not only added new procedures and results but were also fundamentally changing the structure of the field. In his *Mechanica sive motus scientia analytice exposita* (Mechanics of the science of motion set forth analytically) Euler had skillfully applied the distinctive abstract and general methods of analysis to the concrete discipline of mechanics. This was an important part of its widespread application across the sciences. By the 1740s analysis was being well developed, but Euler was the first to stress an explicit distinction between the methods of geometry and the techniques of algebraic and higher analysis as well as to accentuate the advantages of the latter.[2] He sought to transform analysis from methods for solving a restricted number of problems into a conceptual framework with general application in a new and increasingly autonomous field, destined to displace Euclidean geometry from its two thousand years of primacy in mathematics.

By 1748, infinitary analysis still lacked a structure that identified and skillfully organized the fundamental principles of calculus along with a program for the development of the field. Euler's two-volume *Introductio in analysin infinitorum* (E101 and E102), probably the most influential mathematical textbook in modern history, his great treatise on functions was published in 1748. It began to advance the necessary framework, for it masterfully arranged the basic concepts and rules of calculus taken principally from the writings of Isaac Newton, Gottfried Wilhelm Leibniz, John Wallis, the Bernoullis, Brook Taylor, Pierre Varignon, and Guillaume-François-Antoine, Marquis de L'Hôpital and it offered for the most part a methodical and comprehensive theory of algebraic equations as well as three elementary transcendental ones: the exponential, the logarithmic, and the trigonometric.[3] It established the term analysis as generally used today in mathematics.

Some of Euler's contemporaries, together with later scholars, have erroneously taken the word *analysin* in the title as a mistake for *analysis*, but it was not; Euler employed the Greek rather than the Latin accusative. The word *analysis* had come into use in the West after the advent of Neo-Latin, the Ciceronian reforms of Latin during the Renaissance. In classical antiquity and the Middle Ages a Greek word such as *analysis* would have been slightly Latinized, its accusative being *analysim*, with the *m* characteristic of the Latin accusative. From the Renaissance into the eighteenth century, scholars used the proper Greek accusative in this case. Even in the large Renaissance Latin dictionary, *analysis* appeared too late to become a Latinized Greek word. Euler treated it as completely Greek.

As a precalculus text, the *Introductio* does not employ differential and integral calculus. The eighteen chapters in book 1 address elementary functions, continued fractions, infinite products, infinite series, and number partitions. Since the time of Leibniz continental geometers had developed an abundance of elementary algebraic and transcendental functions that they resolved in infinite series. Although of growing importance on the Continent, in 1748 they were not yet central to calculus. In the *Introductio* Euler expanded the foundation of analysis of infinite quantities beyond Newton's kinematics and geometric curves, centering it instead on functions; while L'Hôpital's *L'Analyse des infiniment Petits* (Analysis of the Infinitely Small) of 1699 barely mentions them, Euler's *Introductio* is replete with them. In the preface to the *Introductio* he noted that "since all of analysis is concerned with variable quantities and functions of such variables, I have given a full treatment of functions."[4] Euler's first chapter, "De functionibus in genere" (On functions in general), defines "a function of a variable quantity [as] an analytic expression composed in any way whatsoever of the variable quantity and numbers or constant quantities."[5] In this period before set theory, Euler's variables were general or universal entities derived from abstraction;[6] they were continuous, differentiable, and expandable into Taylor series. Euler's early definition of them pointed out these properties in aid of his search for generality, with few restrictions but also ambiguities, and he classified them as a first step in building a theoretical framework for analysis.

Among the numerous contributions that the *Introductio* made to precalculus was Euler's expansion of Newton's general binomial expansion to irrational functions, which he called a "universal theorem."[7] He put logarithms and exponentials on an equal basis as well. Previously exponentials had been considered simply the inverse of logarithms. Following Henry Briggs and Adriaan Vlacq, by a technique of successive square roots Euler calculated logarithms between 1 and 10 in base 10. He had earlier made the number *e*—today known as Euler's number—the base of hyperbolic logarithms and allowed the variable to take on imaginary values. [8] He was the first to fully bring out the significance of the exponential function in infinitary analysis. In book 2 Euler from $\log e = 1$ computed *e* to 23 decimal places, beginning with 2.718281828459. . . . [9]

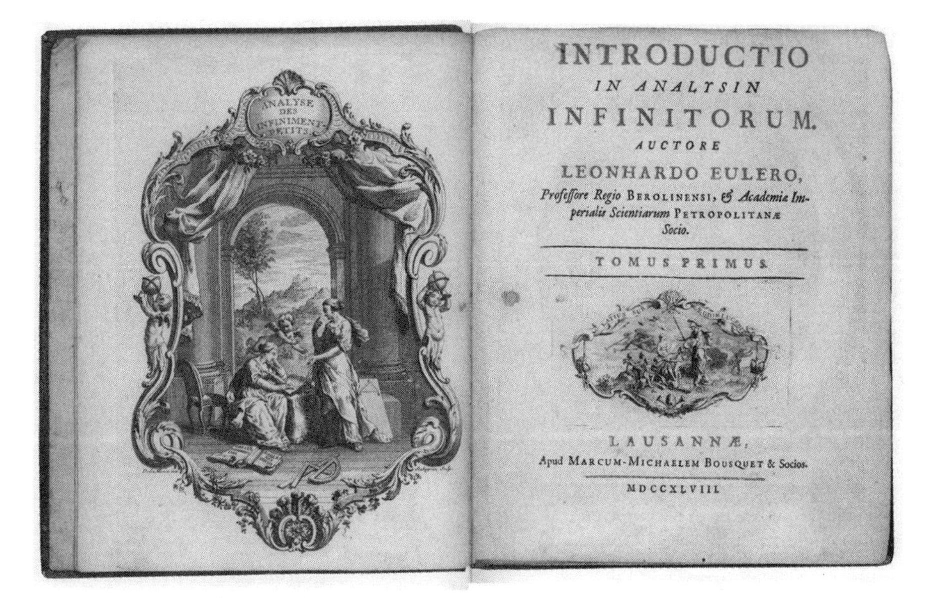
ANALYSE DES INFINIMENT PETITS

INTRODUCTIO
IN ANALYSIN
INFINITORUM.
AUCTORE
LEONHARDO EULERO,
Professore Regio BEROLINENSI, & *Academiæ Imperialis Scientiarum* PETROPOLITANÆ *Socio.*

TOMUS PRIMUS.

LAUSANNÆ,
Apud MARCUM-MICHAELEM BOUSQUET & Socios.
MDCCXLVIII

Figure 9.1. The frontispiece and title page of Euler's *Introductio in analysin infinitorum*, 1748.

The importance of the *Introductio* exceeds its structure. Euler never attacked problems in isolation, and its detail, intersections, cohesion, and extension are impressive. Peerless in computation, Euler relished formulas. Chapters 2 and 3 in book 1 contain rules for formally combining, manipulating, and transmuting functions "into other forms." Their definition was evolving. Chapters 6–11 compute trigonometric functions and their logarithms "more easily . . . than . . . in previous times." "For the sake of brevity," Euler explained, he was representing half the circumference of a circle "by the symbol π."[10] William James had given this symbol in his *New Introduction to Mathematics* (1706). By expressing forty-five degrees as the sum of two arc lengths a and b, deriving the relations from $\tan(a + b) = 1$, and letting $\tan a = 1/2$, which means that $\tan b = 1/3$, Euler improved upon John Wallis's value for $\pi/2 = 2 \cdot 2 \cdot 4 \cdot 4 \cdot 6 \cdot 6 \ldots/1 \cdot 3 \cdot 3 \cdot 5 \cdot 5 \cdot 7 \cdot 7 \ldots$, obtaining a rational series more quickly converging than Leibniz's hardly converging $\pi/4 = 1 - 1/3 + 1/5 - 1/7 + \ldots$.[11] But even after his various, intense computations approximating π, in chapter 8 of book I Euler took the value to 127 decimal places from the master computer Thomas Fantet de Lagny, whose result from 1716 was published five years later in "Mémoire sur la quadrature du cercle"—that is, 3.14159365. . . .[12] The error in the 113th decimal place was not found until 1794. During his last years,

Fantet de Lagny had become a friend of Pierre Maupertuis.[13] Chapters 10, 11, and 15 of book 1 compute more complex expressions. Euler deftly manipulated the infinite series expansions of trigonometric functions and infinite products and playfully made substitutions in them, studying a form of $e^{ix} = \cos x + i \sin x$ and $(e^{x} + e^{-x})/2 = \cosh x$, and in computing the even number zeta functions to twenty-six places he applied products of prime numbers.[14]

In inventing and making standard many modern symbols through the influence of his books and articles, Euler demonstrated the talent of a master notation builder. Beyond adopting the symbol π probably from Wallis and e from Johann I Bernoulli, the *Introductio* introduces the trigonometric functions as *cos.*, *sin.*, *tang.*, *cot.*, *sec.*, and *cosec.*, along with *arcsin.*, *arcos.*, and *arctan*. It differs slightly from modern symbols by having *cosec.* rather than *csc.* and for logarithms to base e, lx for our $\ln(x)$. Like almost everyone in his century, Euler used xx instead of x^2, which aided typesetters. Euler had earlier introduced $f(x)$, Σ, and i and j for infinitely large numbers. He had not yet settled on i for $\sqrt{-1}$,[15] nor had he accepted for infinity the lazy-eight: ∞.[16]

Book 2, consisting of twenty-two chapters, unifies Cartesian or analytic geometry in the plane and is the earliest work to give that field in its modern sense, separate from synthetic geometry. To express its topics in a clear, general way, it employs equations. Euler argued that this was most advantageous for conic sections previously treated geometrically or in analysis with awkward equations. The *Introductio* has a six-chapter appendix on surfaces in space.

In the *Introductio* Euler demonstrated a solid historical knowledge of techniques of his predecessors, but he did not give the history of each problem, for that would make the size of his book unreasonable.[17] Biblical inspiration is also evident. Using logarithms, for example, Euler calculated the rate of population growth; Enlightenment officials and savants considered this growth a characteristic of a good kingdom. Following a literal interpretation of the Bible, Euler had the entire human population as six after Noah's flood and then assumed a population two hundred years later of approximately a million. This gave an annual rate of growth of 6.25 percent. This was long before the English clergyman and political economist Thomas Malthus predicted that population grows geometrically, outstripping food production, which increases arithmetically. The growth would later be checked by poverty, famine, and disease, along with war and Darwinian natural selection. Euler rejected the possibility that the same growth rate could continue for the next four hundred years, for it would have the population reach 166,666,666,666. Earth, he observed, could not sustain that number.[18] This analysis somewhat anticipates Malthus. Assuming instead an annual growth rate of 1 percent, Euler computed a hundredfold population increase over 463 years.

Euler's *Introductio*, his *Institutiones calculi differentialis* (Foundations of differential calculus, E212) and his three-volume *Institutionum calculi integralis* (Foundations of integral calculus, E342, E366, and E385) were the first books to neatly provide the conceptual framework and program for calculus. This trilogy would do for calculus what Euclid's *Elements* had done for theoretical geometry and Al-Khwārizmī's *Algebra* had done for its field.[19] When Euler sent Christian Goldbach a copy of the *Introductio* in June 1748, he remarked that he had nearly completed the manuscript for his *Institutiones calculi differentialis*; it was not published until 1755, however. Many consider the magisterial *Introductio* Euler's most beautiful publication. It surpassed the mathematical sections of Christian Wolff's *Elementa matheseos universae*, which had been published in five volumes from 1732 to 1742.[20] The *Elementa*, which treats arithmetic, geometry, plane trigonometry, finite and infinite analysis, geography, mechanics and optics, retained the older primacy of geometry, such as basing numbers on lines and commensurability.

In April 1748 Alexis Claude Clairaut informed Euler that he had won the Prix de Paris for that year for his essay, "Recherches sur la question des inégalités du mouvement de Saturne et de Jupiter" (Research on the question of the inequalities in the motion of Saturn and Jupiter, E120).[21] The academy required that the authors of essays be anonymous. On each essay they had to put a catchphrase for use later to pair with a name and thus identify the author. Euler's was

> Ponderibus librata suis per inane profundum
> Sidera, quo vis alma trahit retrahitque sequunter.
>
> The stars balanced by their own masses follow through
> the empty abyss, where the fostering force pulls back and forth.

Pierre Charles Lemonnier, having laboriously investigated the perturbations in the orbit of Saturn caused by Jupiter, had set the competition problem: to formulate a theory that can explain "the inequalities [of motion] that these two planets appear to cause mutually, principally near the time of conjunction." Essentially, when Jupiter is catching up with Saturn, it pulls that planet back, causing it to go into a closer orbit, where Saturn unexpectedly speeds up. Jupiter's passing of Saturn causes the ringed planet to go to a more distant orbit, thus slowing it down.

The challenge was to determine whether these perceived anomalies could be accounted for solely from Newton's inverse-square law of attraction or whether there was a discrepancy; the astronomers sought not another force but a modification of that law itself. The complicated interactions of these planets and their elliptical orbits made for quite difficult math, using calculus to describe them, but it was possible.[22] Employing results of observations by Jacques Cassini and especially by Lemonnier at the Paris Observatory, Euler improved the methods of computing these secular inequalities, meaning those that develop over a long period of time. The new trigonometric series that he invented moved the study of perturbing motions beyond tiresome numerical integrations. Having gathered records and data from the latest observations made with the best telescopes, Euler calculated coefficients for successive terms of these series. For varying motion, his new differential equations had arbitrary constants, but comparison of the variance found that the perturbation actually caused was extremely slow. His computations confirmed Lemonnier's results. Euler attributed the mistakes in Lemonnier's figures to the imprecision of observations and most of all to a slight insufficiency in the inverse-square law.

The possibility of insufficiency posed a serious challenge to the dominant Newtonian celestial dynamics; over the next three years, studies by Clairaut, Jean-Baptiste le Rond d'Alembert, and Euler, primarily of the motion of lunar apogee, would remove that challenge. Clairaut's letter in April 1748 suggested that the question on Saturn merited further research and should be the topic for another prize. Here some friction appeared; the implication in the letter was that Euler's essay was mediocre, winning the Prix de Paris only because the Paris Academy had received nothing better.

Competitions and Disputes

In 1747 and 1748, Clairaut and d'Alembert sought general methods to determine perturbations arising from mutual actions among celestial bodies. Clairaut was pleased when Euler agreed with him, doubting the level of exactness that could be provided in these cases through the inverse-square law alone. A great number of phenomena—such as the general orbits of planets and their satellites (according to Kepler's three laws), the movement of lunar nodes, and the theory of tides expressed by Daniel Bernoulli and Colin Maclaurin in the 1740 Prix de Paris competition—were consistent with the inverse-square law of attraction, but the law was coming up against a few apparent celestial discrepancies. This made Clairaut, like Euler, stress the considerable labor and great care that making his case had demanded of him. In the motion of lunar apogee Clairaut thought—to his surprise—that he had found an anomaly. Attempting to crack what is now called the three-body problem, both Clairaut and d'Alembert were personally interested in the subject of the contest for 1748. "Messieurs Clairaut and d'Alembert are after the system of the world, [because] they understandably do not wish to be anticipated by the prize [contest] pieces," remarked Émilie, Marquise du Châtelet.[23]

Other astronomical problems for Euler in 1748 were the solar and lunar eclipses observed from Berlin. Two years earlier for his *Opuscula varii argumenti* (E80) Euler, using calculus and calibrating the latest astronomical observations, had prepared tables good to thirty arcseconds, and they were the most accurate to that time. In July 1748 he eagerly awaited the solar eclipse which, as he had predicted, occurred on 25 July. He had computed its starting time to be 10:17:45 and its ending time to be 13:24:00. The solar eclipse actually began at 10:18:00 and ended at 13:24:30. But Euler's computations of the length of the annular phase, when the sun forms a crescent with horns around the moon, were far off: his prediction was five hours, ten minutes for observations, but it lasted only one hour, twenty minutes.

Although he had been trained in astronomy by Joseph-Nicholas Delisle and was capable of making his own observations, Euler still participated only rarely in the observation process, likely because of his poor vision. Possibly assisted by Christine Kirch, who kept a weather diary for years, and her sister Margarethe, he placed a telescope in a darkened room, constructing a camera obscura, which projected the sun's image onto a white screen. The enterprise used Euler's and Johann Kies's separate computations of when the eclipse image would be at its maximum. If precise, that would coincide exactly with the circle of the sun; but as the eclipse reached its maximum point, the horns of the annulus moved outside the solar disc, and Euler concluded that refraction effects from a lunar atmosphere were magnifying the solar image. To produce that refraction, the moon would need to have an atmosphere about 1/200 the density of Earth's atmosphere. But Euler was not confident that he had discovered a lunar atmosphere. "This celebrated question," he wrote, "has agitated astronomers for a long time: whether the moon has an atmosphere has not yet been decided."[24] His prediction for the lunar eclipse on the night of 8–9 August was also off by five minutes. The eclipse computations, especially for the moon, could have drawn attention to the inverse-square law but did not.

To give more precise figures for both predictions, Euler wrote the essay "Sur l'accord des deux dernieres eclipses du soleil et de la lune avec mes tables" (On the agreement with my tables of the last two eclipses of the sun and the moon, E141), which was published in the *Mémoires* of the Berlin Academy in 1750.[25] He concluded that his positioning of Berlin on the map accounted for the mistakes. "Yet, toward correcting the error in the duration of the annulus," he asserted, "I must note that in my calculations I had assumed the latitude of Berlin was 52°, 36'; now in fact the last observations that Mr. Kies made with the excellent quadrant that Mr. de Maupertuis gave to the Academy only gave its elevation to be 52°, 31', 30", so I had placed Berlin too far north by 4', 30". [Upon making this correction,] the duration of the annulus would have considerably lengthened and would have been very close to my calculation."[26] But there remained a problem with calculating lunar motion; Euler apparently did not relate his work to other work on the moon, Saturn, and Jupiter.

In England James Bradley, who had been the royal astronomer since 1742, was investigating the moon's nodes over a period of 18.6 years. He had discovered the aberration of stellar light, the displacement of starlight in the direction of Earth's movement around the sun, and reported it in the London's Royal Society's *Philosophical Transactions* for 1729. Now he was on the verge of a second major finding, the nutation or apparent motion of fixed stars caused by the slight wobble or gyration in the precessing Earth's axis. In March 1747 Euler tried unsuccessfully to persuade Johann Kaspar Wettstein to forward records of a number of Bradley's observations of the passage of the moon's nodes through the Greenwich meridian. Bradley completed those observations a year later, and he waited until then to announce his discovery of Earth's nutation. Newton had already identified the precession as the result of the attraction from the sun and moon on Earth's equatorial bulge, but he could not compute this three-body problem. To detect nutation Bradley, who needed telescopes with calibrations of fewer than nine seconds, had installed in the Greenwich Observatory better telescopes and filar micrometers capable of calibrations as small as seven seconds.[27] Among the results obtained from employing these devices were variances: plus or minus nine seconds in the plane of Earth's orbit and plus or minus seventeen seconds in the pace of the precession of the equinoxes. Skillful employment of Bradley's discoveries and data from improved telescopes made possible the goal of observational accuracy in arcseconds and laid the foundation for modern astronomy.

In 1748, the correspondence between Euler and Clairaut on the motion of the moon, Jupiter, and Saturn blossomed. Euler's correspondence with d'Alembert continued to cover a range of mathematical topics from the fundamental theorem of algebra to properties of curves, but he now wanted to stop corresponding on a long-standing mathematical matter between them, the logarithms of negative numbers. Yet d'Alembert continued to debate him in letters.[28] Finding little in these to convey anything about the logarithms of negative numbers, Euler took them to be principally argumentative. The debate diminished and for Euler ended in September, when he wrote to d'Alembert that "the matter of imaginary logarithms is no longer so familiar to me that I may rigorously respond to your remarks."[29] By October d'Alembert abandoned the argument. Gabriel Cramer, whom Euler admired as a model historian of mathematics, had edited the correspondence between Leibniz and Johann I Bernoulli on these logarithms, and it provided a crucial source on the origins of the dispute. After reading it, Euler would submit in 1749 for the *Mémoires* a lengthy article, "De la controverse entre M[ss]rs. Leibnitz & Bernoulli sur les logarithmes des nombres négatifs et imaginaires" (E168), supporting his position. Its appearance in print two years later would deepen the division between Euler and d'Alembert.

Another scientific dispute that Euler entered in 1748 was over the nature of space and time. In his *Principia Mathematica* Newton had defined the two as absolute and real, eternal and uniform and "without relation to anything external." Query 28 of his *Opticks* added that space was the "Sensorium of God." In correspondence with Samuel Clarke in 1715–16, Leibniz had attacked Newton's theology and asserted that space and time were relative, which was a position that was difficult to defend before the inventions of non-Euclidean geometries. Voltaire's *Éléments de la philosophie de Newton*, Maclaurin's *An Account of Sir Isaac Newton's Philosophical Discoveries* (published posthumously in 1750), and Euler's "Réflexions sur l'espace et le tem[p]s" (Reflections on space and time, E149, completed and delivered at the Berlin Academy in February 1748 but not published in the *Mémoires* until 1750) were among the works responding to this specific criticism of Newtonian science on the Continent.[30] In his paper, which outlined the relationship between metaphysics and rational mechanics, Euler called the truths of mechanics "indubitably constant," founded in the nature of bodies themselves. Metaphysics, he argued, which studies the same natures, must agree entirely with the laws of mechanics. Space and time must be absolute and real for they are assumed by the incontestable laws of mechanics, which they envelop. Euler rejected the counterclaim of metaphysicians (in essence Leibniz and the Wolffians) that they are relative, imaginary, and destitute of all reality. Arguments with conclusions about them differing from the truths of mechanics, Euler maintained, must contain some hidden logical fallacy. Like Newton, he required absolute space and time to provide coordinates to define inertia and the motion of solid and fluid bodies; these, he insisted, contradicted the view of time as being simply a succession of events.

Through 1748 the monadic debate sustained its force in print, at the academy, and within Berlin's literate classes. The Wolffians considered the prize decision over monads to be an outrage. Both sides issued polemical essays. Breaking with academic protocol, Johann Heinrich Justi published a paper with caustic remarks about Wolff immediately after the announcement of the prize. Intrigues persisted. In the *Göttingische Zeitung von gelehrten Sachen* (Gottingen newspaper of learned things) Johann Georg Sulzer, the Swiss mathematician and a friend of Wolff, asserted that Euler's authorship of the *Gedancken von den Elementen der Cörper, in welchen das Lehr-Gebäude von den einfachen Dingen und Monaden geprüfet, und das wahre Wesen der Cörper entdeckt wird* (Thoughts on the elements of bodies, in which the doctrine of simple things and monads is examined and the true essence of bodies is discovered, E81) was common knowledge, that many academicians disagreed with the selection of the prize-winning paper, and that monads continued to be a dominant topic in Berlin.[31] The collection of papers from the competition commissioned by Maupertuis—containing Justi's essay in French and German, as well as six other papers—now appeared;[32] all but two authors were anonymous. One essay advanced point atomism, which was close to Euler's view; another sought to separate metaphysics from mathematics; and two papers proposed a revised monadology but opposed the Wolffian model.

In a letter to Maupertuis in September Euler complained of the accusation that he had attempted to suppress the truth and did not judge impartially; he expressed the fear that the malicious refutations directed against Justi, in which Jean-Henri-Samuel Formey had participated, had harmed the academy and would spawn "more impudent works."[33] Maupertuis tried to appease Wolff by inviting him to submit a tract to be included in the *Mémoires*, but Wolff declined to do so. In May he wrote to Schumacher in Saint Petersburg about Euler, whom he believed was overreaching his area of competence:

> Herr Euler, der seinen wohl verdienten Ruhm in höheren Mathematik genieszen könnte, will nun mit Macht in allen Wiszenschafften dominieren, darauf er sich doch niemalen gelegt, und da es ihm so wohl an den ersten Gründen, als an Belesenheit fehlet, die zu einer historischen Erkäntniss erfordert wird: wodurch er so wohl seinem eigenen Ruhme sehr schadet . . . als auch die Akademie der Wiszenschafften zu Berlin in viele Schande bringet, wovon der durch ihn erregte Monad-Streit eine klare Probe ableget.

> Mr. Euler, who could enjoy his own well-deserved fame in higher mathematics, wants now with authority to dominate in all the sciences, to which he was though never inclined, and for first reasons as without wide erudition required for historical insights, whereby he does great harm to his own reputation . . . and through the excited monad strife that lacks a clear examination also brings much disgrace to the Berlin Academy of Sciences.[34]

Wolff held that Euler was not at all a philosopher but an "immodest or unashamed pettifogger." But he thought it still possible in 1748 to debate with Maupertuis, whom he considered more clever and more polite than Euler. And he believed that if academy president Maupertuis could read German writings, he would be more understanding about monads.

Although Euler had triumphed in the monadic competition, he failed in June 1748 to stop the election of the French physician, philosopher, hedonist, and materialist Julien Offray de La Mettrie as an ordinary member of the belles-lettres department of the academy.[35] In a letter to Schumacher, Euler indicated that Wolff throughout his defense of monads had attempted unsuccessfully to force him into the circle in Prussia promoting the French Enlightenment. (Euler remained antagonistic to its more radical members.) Open to a breadth of religious views, the academy also elected by unanimous acclamation the prefect of the Vatican Library, Cardinal Angelo Maria Quirini, whom Voltaire admired.[36] La Mettrie's influential *L'homme machine* (Man, a machine), which appeared in 1748, treated men as marionettes to a blind power, declared sensory experience alone to be trustworthy, and presented freethinking and atheism as a way to happiness, a liberation from the dominance of theologians.[37] Euler's opposition to La Mettrie put him directly at odds with Frederick II, who wanted to be known for the wisdom, wit, and learning of his court; the addition of a leading radical French *philosophe* would help build his reputation and agree with his preference for freethinking. Laboring to centralize the Prussian government and make his kingdom a European power, Frederick was a leader in the third phase of absolutism in eighteenth-century Europe, with rulers now known as enlightened despots or reform absolutists, and expected to advance the cause of learning along with the progress of the mechanical arts. In Potsdam La Mettrie found a place not only of refuge but also of advancement. Frederick permitted him to practice medicine and made him a court reader.

Frederick's action reflected tolerance only within court and professional circles, and brought out his view of the academy as a *Freistätte* (free institution), a bulwark against the churches' intolerance. While many French *philosophes*—among them Voltaire—were denouncing religious superstition, *l'infame*, very few espoused atheism. Far more attractive to them and to Frederick was deism, which projected a distant clockmaker God. La Mettrie's book had aroused such a howl of protest even in Paris that he was forced to seek refuge in relatively tolerant Leiden, which in turn compelled him to leave for Berlin. Even Frederick believed that La Mettrie, whose deterministic views undercut the monarch's concept of freedom, had gone too far.[38]

Decrial, Tasks, and Printing *Scientia navalis*

The election of La Mettrie and a barrage of criticisms against Euler in 1748 distressed him; he feared that in Berlin the taste for the arts and belles-lettres was supplanting mathematics, which would make his position in the acdemy useless, and that he might lose his job. Euler's critics charged that his arguments about monads were an attack on culture itself. With extreme discretion, he began to inquire about a post in England. In March he had written to Wettstein that "there is no other country where I would rather establish myself than England."[39] In May he praised the British parliament's plan to naturalize foreign Protestants, which Basel had not done, and he proclaimed his disenchantment with Berlin. Only England was now a suitable location for him; yet no English offer was forthcoming. Later Euler's Dutch relatives attempted to draw him to a Dutch university, but he rejected the idea; Dutch universities did not offer a stage comparable to Berlin in increasingly powerful Prussia, nor would they provide him a reduced teaching assignment to free most of his time for research. And in 1748 he he dismissed the idea of a return to Saint Petersburg; there the prestige of the fine arts was supplanting that of mathematics, eliminating the likelihood of a permanent post for him.[40]

Maupertuis did not see the situation in Berlin so direly. In 1748 he wrote to the king about the merits of the academy. "Our chemists surpass all other chemists in Europe," he asserted, "[and] our mathematicians can rival those at all other academies."[41] By "our mathematicians," Maupertuis meant Euler.

Had Euler known of Frederick's disparaging reference to him in a letter in late November 1748 to Voltaire, his search for a position elsewhere might have intensified. During the year Frederick had been attempting to draw Voltaire to Berlin. Urging Voltaire to finish *Le Siècle de Louis le Grand*, the king likened him to Virgil and called him the greatest man in Europe. Probably to impress Voltaire, Frederick—an admirer of authors from ancient Athens and Rome (their names are sprinkled throughout his letters)—sought an example from classical mythology that suggested Euler's eyesight problem and his intellectual stature. In reference to Voltaire's mistress and intellectual companion Émilie du Châtelet, who kept him from coming to Berlin, Frederick wrote that if "Mme. du Châtelet is a woman of substance, I propose to pay her for borrowing her Voltaire. We have here a great Cyclops of geometry, whom we will offer in exchange for that *bel esprit*. . . . If she agrees to the deal, there is no time to lose: our man has only one eye left and the new curve that he is calculating presently might very well cause him to go entirely blind before our deal is concluded."[42] It is surprising that the letter seems to suggest that Euler is the equal of Voltaire. The king must have suspected that the exchange would intrigue Madame du Châtelet, who was deeply engaged in scientific studies; on several occasions he reiterated his graceless offer.

Throughout 1748 Euler continued to prepare and distribute calendars, almanacs, and geographical maps, the academy's main sources of funding. Euler's calendars employed the new Gregorian style. Since the nobility still had a fondness for astrological predictions, his almanacs included them. In May commissioner David Köhler negotiated as a new privilege for the academy a 5 percent excise tax on the calendars, almanacs, and maps, which could add eight thousand Reichsthaler to academy revenues. But even with royal backing it was difficult to have booksellers in Berlin collect and pay this tax. Determined to bring a more effective administration for the academy, Euler needed the cooperation of Formey, whose election in 1748 to the influential position of standing secretary at the academy was intended to strengthen his role in opposing Euler in its philosophical battles. For two years the virtual successor to Philippe Joseph de Jariges, Formey would serve as a trustee on the editorial board for the *Mémoires*. He was to last as secretary so long that Voltaire would call him "eternal" in that position. As historiographer, Formey needed to proofread works for academic publications, including the almanacs, but he wanted to pass that task to the new historiographer. Yet Maupertuis, committed like Euler to maintaining the highest quality in publications, insisted that Formey make everything ready for printing. This involved gathering legible copies of manuscripts, mainly from their authors; having translations made of Latin and German papers; and drawing contributions from foreign members to offset mediocrity in any volume of the *Mémoires*. Formey wanted the royal printer Ambrose Haude to hire a proofreader and handle this task entirely, but Maupertuis required the royal printers to make the first corrections of the printer's proof sheets, after which Formey would make the final corrections; without this process, he felt, mistakes could be made that would detract from the reputation of the academy. Maupertuis wanted to establish the same impeccable standards that Bernard le Bovier de Fontenelle had set in Paris, and Euler concurred. Maupertuis urged Formey to take Fontenelle as his model.

Euler's correspondence—primarily with Schumacher, but also with Razumovskij and Teplov (the three were the most important officers of the Imperial Academy of Sciences and Arts)—continued in 1748 and dealt almost entirely with administrative matters. The departure of Johann Georg Gmelin and Delisle was drawing sharp criticism. In returning to Tübingen, Gmelin had broken his word and contract; Razumovskij agreed with Schumacher that Gmelin should have to honor his pact, but Euler believed that Gmelin's decision to leave was not malicious; he attributed it partly to the wishes of the parents and wife. The Duke of Württemberg named Gmelin a court botanist and forbade him to return to Russia. Euler counseled the selection of one of two options: strip Gmelin of any advantages from his contract with the Petersburg Academy or assume good intentions and forgive him; the tone of his letter suggests a preference for the latter. Euler saw the benefit in retaining connections with Gmelin, who had just sent him most of the second part of his *Flora sibirica* and promised to complete the rest. Forwarding this material to Saint Petersburg in December, Euler pointed out that neither Württemberg nor Prussian law would force a return.

Euler was also involved in the later part of the Delisle affair. Academy officials were angered at the astronomer's criticism of its atlas and his carrying off of valuable research materials when he left in early 1747. After Delisle returned to Paris, Euler informed him that a warrant had been issued against him in Saint Petersburg, and that Razumovskij was cutting all contacts with him. It took Euler a while to ease the volatile situation; this he would do in 1753 by arranging the printing in volume 12 of the *Nouvelle Bibliothèque Gèrmanique* of an essay by a Russian officer on the geographic discoveries that Delisle had published regarding the north and south of Russia. In 1748 Razumovskij's labors to improve the institutional base for the sciences won him an honorary diploma and foreign membership from the Berlin Academy. Euler was delighted. Maupertuis asked him to let Razumovskij know that the Berlin members had approved the honors in the extraordinary manner of unanimous acclamation, not by the common procedure of reviews. Razumovskij, in return, asked Euler to express his thanks.

Among other topics in letters to Russia was Euler's invitation to Mikhail Lomonosov to submit a dissertation on saltpeter for the 1749 Berlin Academy annual prize competition. Lomonosov's explanation involved his theory of the elasticity of the air. Euler was convinced that there would be no better submission, and that Lomonosov's winning would be a credit to both the Berlin and Petersburg academies, whereas Razumovskij had no opinion on the value of Lomonosov's participation in the competition. It took some encouragement to get Lomonosov to send his article.

Schumacher and Euler wanted to upgrade the Russian Academy's chemical laboratory; the two also discussed the hurried publishing of Euler's *Scientia navalis* (E110 and E111) and the first volume of the academy's *Novi Commentarii*. Euler reported on a new comet and the observation of the solar eclipse made from Berlin. He was studying the lunar atmosphere and sent to Russia copies of astronomical calendars. The Imperial Academy of Sciences and Arts now had vacant positions. Euler wanted these filled, above all, with Russians in the near future; for the moment it was not possible. Among candidates he suggested were astronomer Jean-Philippe Loys de Chésaux from Lausanne and mathematicians Christian Friedrich Oechlitz at the University of Leipzig, (Gian Francesco) Jean de Castillon in Berlin, and G. L. Schmid in Berne, but none were accepted. Of Isaac Bruckner (Brucker) from Saint Petersburg, who was a master of mathematical instruments, Euler noted that he had come to Berlin as a geographer. Bruckner had lived with Euler for a time and must have been part of the Euler table society of scholars in Saint Petersburg. In December 1748 Euler asked the Imperial Academy to set the topic for its first prize competition scheduled for 1751.

After more than a decade's delay, Euler's two-volume *Scientia navalis* on the construction and propulsion of ships was finally published in 1749.[43] That it appeared even then was due to the initiative of Razumovskij the previous year; when he and the Russian Admiralty approved funds for printing both volumes, Razumovskij had ordered that it be issued as quickly as possible. The academy was supposed to pay a Berlin printer, which would allow Euler directly to make all corrections, but there was opposition in Saint Petersburg to this payment. This, the size of the allotment, and difficulties in printing with the poor paper available in Berlin made it impossible to start the project, and this meant canceling the contract. The printing could not be done without Euler's examination. He checked the paper for printing in Saint Petersburg and was pleased to find that its quality was quite good. Euler dated the prologue of the *Scientia navalis* 25 January 1749. For the title page he asked that in addition to "Prof. Honorario Academiae Imper. Scient." after his name, the position "et Directore Acad. Reg. Scient. Borussicae." be inserted, and he requested copies of the work. Both wishes were granted.

Figure 9.2. Portrait of Mikhail L. Lomonosov. MAE (Kunstkamera) RAS (ML-00041).

Scientia navalis, Euler's third landmark book during his tenure in the Prussian Academy, provided optimal ship designs, a precise general definition of stability, and positions of the equilibrium of ships. A general theory of navigation was insufficient to attain these. The second volume of *Scientia navalis* particularly treats the maximum stability, handling, and speed of ships, features that often oppose one another in practice. Some advantages come at the expense of others: gaining the greatest speed and maintaining exact direction, for example, may impede each other; the water also produces such effects as swaying and rolling. Euler endeavored to perfect naval theory by reconciling with ship construction the art of piloting, knowing to what extent to use one method to enhance a variety of properties. The previous lack of mathematical methods and computations in naval construction and naval science meant that progress had been piecemeal, misapprehended, and intermittent. Euler not only presented a completely new theory for the field but was also to introduce novelties in eleven papers in naval theory, including in 1749 a "Memoi[r]ee [*sic*] sur la Force des Rames" (Memoir on the force of oars, E116).[44]

Exceptional in both theoretical and applied mathematics, the *Scientia navalis* continued Euler's program for founding rational mechanics. Soon after its release, he was concerned that the text was too difficult for navigators and quickly began a revision to simplify it. The long delay in

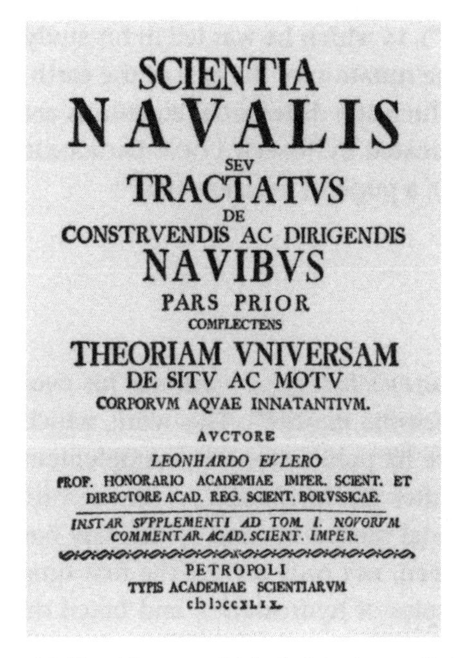
SCIENTIA
NAVALIS
SEV
TRACTATVS
DE
CONSTRVENDIS AC DIRIGENDIS
NAVIBVS
PARS PRIOR
COMPLECTENS
THEORIAM VNIVERSAM
DE SITV AC MOTV
CORPORVM AQVAE INNATANTIVM.
AVCTORE
LEONHARDO EVLERO
PROF. HONORARIO ACADEMIAE IMPER. SCIENT. ET
DIRECTORE ACAD. REG. SCIENT. BORVSSICAE.
INSTAR SVPPLEMENTI AD TOM. I. NOVORVM
COMMENTAR. ACAD. SCIENT. IMPER.
PETROPOLI
TYPIS ACADEMIAE SCIENTIARVM
cIↃIↃccXLIX.

Figure 9.3. The title page of Euler's *Scientia navalis*, 1749.

publishing the two-volume text allowed Euler's competitor Pierre Bouguer, the winner of the Prix de Paris in 1727, to precede him in print, getting out his magnum opus *Traité du navire, de sa construction et de ses mouvemens* in 1746 and thereby to claim priority for central concepts in naval science, such as the metacenter that Euler had earlier discovered. Probably to avoid a dispute over priority, Euler recognized the contributions of Bouguer twice near the close of his prologue. But Euler was the first to clearly establish the principles of elementary hydrostatics, providing variational solutions using differential equations. His *Scientia navalis* and his articles on elasticity initiated continuum mechanics.

A Sensational Retraction and Discord

In 1749, when the rivalry of Euler with Clairaut and more notably with d'Alembert was expanding and becoming heated in argument but not yet antagonistic, the issue remained of devising a more accurate approximation of the three-body problem that would confirm the exactness of or add a small correction to Newtonian celestial dynamics. The motion of lunar apogee continued to be the central problem.

Clairaut discovered in 1749 that while second-order approximations are negligible in two-body problems, they are crucial in his computations of three bodies—in this case the motion of lunar apogee. Thus he had to reverse his conclusion from the previous year that a small modification for Newtonian dynamics was needed. The second-order equations give a recession of the apogee of 3°, 2', 6" per lunar cycle, or about 40° per solar year. The actual value is 3°, 4', 11". At a May meeting of the Royal Academy of Sciences in Paris Clairaut, acting as a master strategist, reversed himself by retracting his statement that lunar motion was contrary to Newton's inverse-square law of attraction and announced that he had found a way to show them in agreement—a *volte face* that d'Alembert reservedly endorsed. Clairaut dropped the inverse fourth power term that he had briefly added to account for gravitational attraction. His letter in June to Cramer declared that he had grasped this correction only six months after proclaiming the reverse in letters sent to the Bernoullis, to England, and to Italian cities. Clairaut sent a copy of his new result in a sealed packet to Martin Folkes of the Royal Society of London, not to be opened until he gave his approval. He did this to forestall others from correcting him and to provide time to complete more calculations.

When Euler learned of the retraction, he redid his own computations in July and found not the slightest error. Mistakenly treating the lunar orbit as a rotating ellipse, he believed that he needed only first-order approximations with differential equations. He considered beyond doubt his calculations of the motion of the moon and Saturn, which would put Clairaut in error. While Daniel Bernoulli had pointed Clairaut in the right direction and supported his findings, neither he, Euler, nor d'Alembert knew the new procedure. Clairaut still kept his method secret so that others would not quickly challenge him in lunar theory. He apprised Cramer that Euler had written twice that year and the next describing "his fruitless efforts to find the same theory as I, and he begged me to tell him how I arrived at [it]."[45] One letter in June refers to young Augustin Nathanael Grischow's visit to Paris with questions for Clairaut, especially the errors in his first lunar computations. Clairaut gave his new results to Grischow, who forwarded them to Kies in Berlin. Euler informed Maupertuis, but was obstinate on resolving the question of lunar inequalities. In July 1749 Clairaut wrote of their clashing results, "I suspect that it is because you neglected the terms which are derived from the square of the perturbing forces within your calculation that when you integrated your second order differential equations you did not attain the correct result. It was after using these terms that I overcame and found as closely as possible the true motion of the apogee."[46]

This explanation still did not completely satisfy Euler, who wanted to see the entire method of calculation that had led to Clairaut's latest results and how he had arrived—perhaps accidentally—at a critical infinite series. Nor did Euler lessen his effort upon receiving a letter in August from Daniel Bernoulli, who was preparing tables of Saturn's orbit and had already exchanged ideas with Clairaut on his latest work on lunar apogee. Busy preparing the first volume of the *Encyclopédie: Dictionnaire raisonné des sciences, des arts et des métiers* (Encyclopedia: Systematic dictionary of the sciences, arts, and crafts), d'Alembert wrote to Cramer in September. Agreeing that Clairaut had been the first of the three to resolve the apogee problem, he claimed to have a method that would be shown to be better. But Euler did not wish to wait; from mid-1749 onward he undertook an initiative to have the entire successful method released so that Clairaut could be mathematically proven correct or—more likely, he thought—wrong.

Clairaut's results heightened the search for a better lunar theory. Euler believed that the computation of the lunar apsides would be decisive in deciding whether the Newtonian law of attraction needed a small correction. This was a third critical test for Newtonian dynamics. The partisans of Newton had defeated the Cartesians over the shape of Earth in a first major test and the flow of the tides in a second, while a fourth, computing the return of Halley's Comet, lay in the future. Meanwhile an occasion emerged for disclosing Clairaut's method for computing lunar apsides. In its regulations written in 1747, the Russian Imperial Academy had projected scientific contests. It would follow the model of the Paris Academy in accepting only scientific debates and none in the humanities. Schumacher and the Saint Petersburg academicians wished to avoid the invective that had followed Berlin's decision against monads, so he counseled Razumovskij to take caution and wait before holding a prize competition. Another matter concerned Schumacher: Razumovskij wanted the academy to regain the prestige it had lost after the departure of Euler, Gmelin, Gottfried Wilhelm Heinsius, and Delisle in the early and mid-1740s, but the essays of its resident members Lomonosov, Georg Wilhelm Richmann, and Christian Nikolaus Winsheim had thus far gone unknown in learned circles in western and central Europe; they would appear in 1750 in the first volume of the *Novi Commentarii*. Schumacher wanted that volume published before the contest; he sought significant prize questions, well-founded evaluations, expert judges respected across the learned world, and the design of a medal for the winner.

Euler saw his chance. In 1749 he sent Teplov four possible topics for the annual prize competition. On 15 July he informed Schumacher that he had sent these to Teplov, and Teplov then instructed Schumacher to rewrite them in "another handwriting" so that the academicians, not knowing their source, would discuss them and possibly propose other problems. At a special meeting in August, the members of the Imperial Academy immediately recognized the topics as Euler's and proposed no others. For their initial prize, they selected the first problem: "To demonstrate whether all the inequalities observed in lunar motion are in accordance with Newtonian theory—and if they are not, to demonstrate the true theory behind all these inequalities, so that the exact position of the moon at any time can be computed with the aid of this theory." They passed over the other three: investigating the duration of the solar year, showing whether Newtonian gravitation could explain all planetary inequalities, and examining whether Earth had a perturbation similar to Saturn's as Saturn approached Jupiter. They chose the first as the most significant scientific topic of the day, predictions of the moon's motion among the stars being crucial for determining geographical coordinates in mapmaking and for obtaining navigational positions at sea.

A few days later, after Schumacher informed him of the topic selected, Euler wrote to Teplov, calling it the best question that the academy could have chosen and noting that "at present this topic holds the interest of most of the chief savants [those in the sciences] of Europe and opinions are divided on it." Euler's own computations seemed to him too obvious to permit any doubt about whether Newtonian attraction explained the motion of lunar apogee. The question, he noted,

> is not only the most profound and the most worthy of all those the Academy had to select from, but that from which the most complete answer may be expected as well. Until now Clairaut has kept his new method a secret, but as soon as he published it, in as much as calculations are involved, one of two alternatives must necessarily follow: either the rest of us must be convinced of his correctness, or else we will find some error in his work. Whatever the result, astronomy, physics, and analysis will enjoy a promising and important profit too great to be ignored.[47]

The choice agreed with Euler's strategy to entice Clairaut to reveal how he had made his revised calculation of lunar apsides. While Teplov counseled urgency in starting the contest, in order to avoid acrimony similar to the monad debate in Berlin Razumovskij accepted Schumacher's proposal to delay it to 1750–51.

In July 1749 d'Alembert's *Recherches sur la précession des equinoxes* was published, which posited that while the moon orbited in an ellipse, the equinoctial points precessed because of the effect of gravity on the sun and moon on Earth's equitorial bulge when the celestial equator crossed the ecliptic. The phenomenon of their advancing through the zodiac had been known since antiquity. Newton was the first to attempt to account quantitatively for it by seeking the ratio of lunar and solar gravitational forces acting upon the equatorial bulge of Earth; the first edition of his *Principia Mathematica* gave the ratio from tidal data as 6.33:1, and the second edition gave it as 4.48:1. The actual value is 2.17:1; thus Newton's results were terribly wrong. Bradley's discovery of the nutation of Earth's axis and a wish to show that Newton's gravitation could explain precession motivated d'Alembert to make his computation of it. Following Bradley, d'Alembert found the ratio to be 2.35:1. While d'Alembert claimed that he had carefully handled the problem and sought the utmost clarity, Cramer correctly pointed out that the book was disordered and contained confusing diagrams, clumsy equations, and typographical errors; to deter competitors, d'Alembert had rushed it to print. On 20 July he sent a copy of his book to Euler through Grischow, but he received no acknowledgment until 3 January 1750.

Young Grischow was a reliable contact who had been at the Paris Observatory since 1747 to participate in astronomical observations. Having become friends with Lemonnier, Clairaut, and d'Alembert, he helped forward to Euler their correspondence. In March 1749 Grischow reported to Euler his observations of a solar eclipse and comets, along with his travels with Lemonnier to England and Edinburgh. He wrote that he had purchased a stop clock for Euler and two telescopes for Maupertuis and the academy, and that he had been made a corresponding member of the Paris Academy. Euler asked d'Alembert to tutor Grischow in mathematics. Grischow inquired about the position in mathematics and physics at the Berlin Academy that had opened with the departure of Euler's cousin Reinhard Battier, a post that briefly went to Grischow.

In December Euler reviewed d'Alembert's *Recherches* at the Berlin Academy. Euler had earlier studied the problems of nutation and precession but without success in resolving them; these presented a further test of Newtonian dynamics. D'Alembert's solutions caused Euler to once again take up studying them. Neither Euler nor anyone else could understand d'Alembert's awkward method. It did not offer a general mechanics of rigid bodies, which Newton had lacked, a project that Euler had contemplated in his *Mechanica* and the first volume of *Scientia navalis*,[48] but d'Alembert correctly gave for the first time an essential procedure for computing the precession of equinoxes that followed Newton's inverse-square law. It took Euler until March 1750 to overcome the obstacles to finding the technique needed to describe precession exactly.

Chiefly leading to the break between Euler and d'Alembert was competition over not the three problems in pure mathematics and theoretical astronomy but a fourth in fluid mechanics. In 1748 the Berlin Academy had announced for its next competition, to be held two years later, a prize question regarding the resistance of fluids. The next year d'Alembert submitted for the prize a long and clumsy paper, "Essai d'une nouvelle théorie de la résistance des fluides," which mentioned the work of Clairaut and Euler. Rather than attempting to trace a portion of a fluid through a conduit and calculate accelerating force, as was typical, d'Alembert introduced the concept of fluid pressure and from it introduced two correct partial differential equations almost entirely for axially symmetric flow. This allowed him to describe at "any point" in a field the state of the fluid. These were the first such equations in hydrodynamics for the field. Despite its defects the *Essai* marks a turning point in physics, for it applies partial differential equations to represent fluid mechanics. But d'Alembert painfully knitted his equations into a fabric of conjecture and error that obscured his contribution. Its rejection for the prize the next year by the Berlin Academy was to sour relations between him and Euler.

There were further disagreements for Euler. Bernoulli informed him that another edition of Robins's *New Principles of Gunnery* had appeared in 1749. Robins, who had begun his criticism of Euler in his review of Euler's *Mechanica*, extended his comments to artillery. To check on Robins's latest differing views, Euler asked Wettstein to send him a copy. As was noted in chapter 8, Euler's highly significant essay of 1749, "Sur la perfection des verres objectifs des lunettes" (E118), contradicts Newton by asserting that a refracting telescope can be constructed with lenses free of color distortion. It was to set off a long dispute with John Dollond, a staunch partisan of Newton in London. Apparently neither man knew that Chester Moor Hall had in 1733 built several refracting telescopes with crown and flint glass without a color distortion. Euler studied related color indices and their implications.

State Projects and the "Vanity of Mathematics"

From its origins in the Electoral Brandenburg Society of Sciences, the Berlin Academy had a practical purpose in the spirit of Leibniz's motto "theoria cum praxis." This did not mean that the academy was to be a technical school. Rather, under Frederick II it was supposed to be a center for excellent speculative research in the sciences that would result in applications for *le bien publique* and for the power of the state. In the *Mémoires* for 1745 Frederick repeated a statement from Formey:

> On regarde aujourd'hui un grand mathématicien, un habile physicien, un homme de lettres qui excelle dans quelque genre que ce soit, on les regarde, dis-je comme ils méritent de l'être, c'est-à-dire, non-seulement comme des gens qui font honneur à leur patrie par la sublimité de leurs connaissances, mais comme des citoyens utiles, sous le pas desquels naissent, ou du moins peuvent naître les découvertes les plus intéressantes pour le bien publique.

> Today a great mathematician, a talented physicist, a man of letters who succeeds in whatever he undertakes is considered as such scholars deserve to be, not only as men of letters who do honor to their countries with their superior knowledge, but as useful citizens under whose very footprints sprout or at least can sprout the discoveries most relevant for the public good.[49]

While turning the agricultural realm to projects that would increase its commerce and output, Frederick retained the betterment of agriculture as the foremost state enterprise of his academy's botanical and chemical departments, as well as to a degree the physical department. Among these were the quality, variety, and uses of plants, together with the richness of soil and an increase in the extent of lands under tillage. Projects that Frederick assigned to Euler before the end of the first two Silesian Wars included grading the brine in the Schönebeck salt mines, aiding in the building of dams and bridges in eastern Friesland, and setting measures for draining the Oder marshland. Later on, practical assignments increased in number.

While among the nobility Euler was not popular and Frederick II did not warm to his personality, the monarch valued his contributions to applied mathematics and the prizes that he won abroad, most of them in Paris. In *Histoire de mon temps* (History of my times, 1745), the king had praised Euler as "an ornament of the court." After the end of the War of the Austrian Succession in 1748, Frederick II assigned him several more state projects.

The first among three notable tasks for 1749 was given in April and required the leveling of the Finow Canal to improve navigation on it. About fifty kilometers in length and having seventeen sluices or locks at the time, the Finow Canal, joining the Oder and Havel Rivers, had first been completed in 1620 but suffered decay from neglect and flooding. Reopened in 1746, it was important to the Prussian king and state for it made possible a river system that traversed Prussia's interior. By connecting Magdeburg and Berlin with Stettin by water, it also made Stettin a second Prussian maritime port; the only other was Hamburg. Along the canal in Brandenburg, industry was to develop rapidly. On a detailed trip gathering many details affecting the canal with colonel Johann Friedrich Balbi of the Prussian Army Corps of Engineers, Euler took his fifteen-year-old son Johann Albrecht to assist him. In May 1749, Euler sent a report to Frederick. Although it benefited from the contributions of his team, Euler was the sole author; he concentrated on manually regulating the pressure in the many successive locks.

On 15 September 1749, well before Euler completed his studies of hydraulics that year, Frederick posed him a second problem, one that belonged to what today would be termed *recreational mathematics*: a proposed lottery. Euler was to examine use of a kind of lottery employed in most large Italian cities and the Low Countries. Frederick wanted to know as soon as possible what hazards such a lottery might incur in Berlin and what profits could be made; in the wake of the two Silesian wars, he needed a way to fund pensions for widows. In two days Euler figured— using the calculus of probabilities— how much each player should pay to have a fair price for a ticket and under what circumstances the state and banks could increase profit margins.

The lottery proposal had ninety tickets with numbers marked on them consecutively, and a player would draw five tickets at random. Euler gave three ways to proceed. The player might choose a number ahead of time and see whether the number was on one of the five drawn tickets. The probability of gain was 5/90 or 1/18. The player would set the prize, eighteen times the amount bet. In another method, called *ambes*, the player was to select two numbers and the amount paid if the two numbers were on the five tickets chosen. The probability of winning was $(5 \times 4)/(90 \times 89) = 2/801$. Operating from a set minimum wager of only the 2/801 portion of 100 ecus, the bank would gain more than 511 percent. The other way to play, by *terns*, had the player give three numbers. The chance of winning here was $(5 \times 4 \times 3)/(90 \times 89 \times 88) = 1/11{,}748$. Euler computed methods for the banks to gain greater or lesser profits, the lesser to encourage greater participation. He identified sizable bank profits: 44 percent for the first method, 94 percent for *ambes*, and 240 percent for *terns*.

The official Prussian lottery did not begin until 1763. Fourteen years earlier, before Frederick's request, Euler had experienced success with a different lottery. In April 1749 he wrote to Goldbach about winning a prize: "I have won today in a lottery 600 Reichsthaler, which was just as good as if I had won the Paris prize this year."[50] But over the course of Euler's career, the calculus of probability was never significant in his mathematical studies; he gave it little attention. Games of chance were subsequently to inspire him to compose only eight papers on probability, most of which later dealt with the Genoese lottery.[51]

On 27 September, less than two weeks after Euler submitted his report on the lottery, the king asked him for his expert advice regarding a machine—a water pump operated by a windmill—intended to power the lavish water fountains at the royal residence Sanssouci in Potsdam.[52] The project occasioned Frederick's first invitation to Euler to the palace for a personal meeting.

After construction of the palace was completed a year earlier, Frederick had hired the Dutch architect Johann Baumann to design a system of water fountains for its park and a Dutch garden technician from Amsterdam to oversee the hydraulics for their operation. The jet of the major fountain was supposed to reach a height of at least a hundred feet, eclipsing the jets at Versailles. A canal from the nearby, slowly flowing Havel River, a series of barrel-like wooden pipes comprised of eight hundred drilled-out tree trunks (each twenty-four feet long), and water pump stations driven by windmills were intended to feed water to an elevated reservoir on the Höneberg behind the castle. A difference in height from it to the park and its fountains below of approximately 150 feet was considered sufficient to provide the high pressure required for the principal fountain and four smaller ones. But from the first trials since March 1749 the wooden pipes burst at the lower end of the canal long before the water jets reached the desired level. Euler had mistakenly assumed that the engineers would subsequently use metal pipes, but they replaced the barrel-like tubes with entirely drilled out tree trunks; these too burst. To Euler it was evident in October that wooden tubes could not handle the high pressure. Their replacements, girded with metal rings and protected against sudden pressure changes by five large copper tanks along the pipeline did not burst but produced an inefficient rate of flow to the reservoir, since the inner diameters of the metal ringed tubes were too small.

Although Euler began with exhausting calculations, his initial attempts to obtain from these the most efficient operation of such pumps by regulating the pressure on them and on fountain pipes did not succeed. Ignorant of the laws of hydraulics and lacking the knowledge of that subject required to complete a project on a scale this large, the Dutch fountain makers were unsuccessful, and the king expressed contempt for them. After early difficulties, Euler was able to determine the efficiency of the pumps by making what turned out to be correct calculations. He demanded thicker walls and larger tubes, advice that was ignored.

In October Euler submitted computations for the tubes, for constructing a windmill, and for ways to improve both. Late that month Frederick thanked him for this work.

The effort to pump the water up a hill behind Sanssouci into another reservoir to feed cascading waterfalls lasted another decade; it too suffered from flawed pipeline constructions and was abandoned. The project was a fiasco. Since Euler worked on this project for only a little over a month, his name does not appear on the list of its participants. Still, the king placed some blame on him and mathematics for its failure. In January 1778 he wrote to Voltaire, "I wanted to make a fountain in my Garden. . . . Euler calculated the effort of the wheels for raising the water to a basin, from where it should fall down through canals, in order to form a fountain jet at Sanssouci. My mill was constructed mathematically, and it could not raise one drop of water to a distance of fifty feet from the basin. Vanity of Vanities! Vanity of mathematics."[53]

The study of hydraulics, essential to this fountain problem, was not new to Euler. Hydraulics belongs to the applied science and engineering that studies the practical application of fluids in motion; its theoretical foundations lie in fluid mechanics, also known as hydrodynamics. When Euler resided with Daniel Bernoulli in Saint Petersburg, he had assisted Bernoulli in preparing his *Hydrodynamica* and had read Johann I Bernoulli's *Hydraulica*. Having completed his *Scientia navalis*, Euler understood well how far hydraulics was from hydrodynamics. From 1746 on, moreover, he corresponded with d'Alembert on hydraulic pressures as Daniel Bernoulli examined them. But when Euler briefly took over from Baumann the problem of the Sanssouci pumps, he omitted the effects of friction, and with embarrassing results. Through the late eighteenth century and beyond, his unsuccessful effort with the Sanssouci fountains was taken to symbolize the breach between theory and practice.

Some German scholars have attributed the failure of the fountains to the pure mathematician Euler for lacking in physics and not drawing upon practical experience.[54] Faulting mathematics for the water-art project's failure, Frederick believed that mathematics was no longer so productive as in the previous century. The record of Euler's insightful suppositions concerning the pipes contradicts both of these assertions. His equations presaging those of motion for ideal fluids had no way of being confirmed quickly, for he lacked successors to extend his ideas in hydraulics, which required reaching the level of achievement in the field that would be attained in the next century by Simeon-David Poisson, Claude-Louis-Marie Henri Navier, and First Baronet George Gabriel Stokes, whose equations covered friction.[55]

By the end of 1749 Euler designed for use at Sanssouci a horse-powered hydraulic machine capable of raising thirty-six tons of water to a height of five feet in twelve seconds. He discussed it at length in correspondence with Maupertuis. This offers another example of the intimate and mutually beneficial interrelations that he saw between science and technology; Euler always considered science a productive force and crossed from theory to application in a sure, well-organized, and thorough way.

The König Visit and Daily Correspondence

In September 1749 Maupertuis had the Berlin Royal Academy elect his friend Johann Samuel König as a foreign member. König had a modest career as a tutor, mathematician, philosopher, librarian, and translator, moving around various cities and countries in northern Europe. He was drawn to Leibnizian philosophy, beginning his study of it in 1731 under Jakob Hermann and continuing in 1735 under Wolff. Shortly after the return of the Lapland Expedition in 1737, the Bernoullis introduced him in Basel to its leader, Maupertuis. A year and a half later Maupertuis acquainted him with Voltaire and Émilie du Châtelet, whom he tutored in mathematics and Leibnizian thought; Voltaire would blame König for converting her to Leibniz's views. Voltaire and Émilie du Châtelet introduced him to the eminent French entomologist René Antoine Ferchault de Réamur, and the two men's discussions prompted König to write a paper on the structure of honeycombs that was so highly regarded that in 1739 Maupertuis and Clairaut had him elected a corresponding member of the Royal Academy in Paris. Two years later he translated into German Maupertuis's anonymously penned *Sur la figure de la terre*, a work praised for its elegance and romance. After accepting a position at Franeker University in 1744, König refined his 135-page *Oratio inauguralis, de optimis Wolfiana et Newtoniana* (Inaugural oration on the optimism of the Wolffian and Newtonian philosophies), which was published in 1749. That year the election of König to the Berlin Academy unwittingly opened the angriest scientific controversy of the century.

König, a zealot for Leibnizian science, traveled to Prussia to visit Wolff in Halle and see relics of the great Leibniz. In September he proceeded to Berlin to express appreciation for his election and in an act of self-destructive honesty to show Maupertuis a manuscript in Latin titled "De universali principio aequilibri et motus, in vi viva reperto deque nexu inter vim vivam et actionem, utriusque minimo" that challenged his host's priority claim for the principle of least possible action and the extent of its application to the laws of mechanics. He wanted the manuscript published, but Maupertuis did not read it. A few days after his arrival, König was at Maupertuis's house, along with Francesco Algarotti and Formey, to discuss the priority strife between Leibniz and Newton over the invention of calculus. Maupertuis sided completely with the English and proclaimed Leibniz a plagiarist. When König interjected "my poor friend" and vigorously defended Leibniz, an argument ensued. The egotistical and autocratic Maupertuis was infuriated with his subordinate König's unfortunate choice of words and in effect the undercutting of his status as academy president; his visitor was clearly not his social equal. After Maupertuis replied with that same phrase for his guest, König quickly left. Shortly thereafter he apologized and traveled home to the Hague in Holland, where he had just become the court librarian to Prince William IV of Orange. König, in Maupertuis's later evaluation, was more than a fanatic, which would have been bad enough; he was also *un frippon*, a scoundrel.[56]

In addition to Euler's research, about one-third of the more than 130 extant pieces of correspondence exchanged with others in 1749 address or touch upon the administration of the Berlin Academy, his wide interests, and health or deaths among scholars and practitioners in the sciences and their patrons. To Giovanni Marchese Poleni in Padua Euler suggested utilizing astronomy to confirm or correct historical chronologies. Sixteen letters to Maupertuis discussed the status of the building of the Berlin Observatory, the data from Charles Marie de la Condamine on the shape of Earth, hydraulic machines, problems with the pumps at Sanssouci, the construction of a telescope with plans from Jan Andrej (Johann Andreas) von Segner, Euler's own textbook on arithmetic and geometry, and the wave theory of light in an elastic medium. Martin Knutzen requested Euler's recommendation for promotion to the rank of professor at the University of Königsberg. One of Knutzen's students, Immanuel Kant, wrote to Euler in August 1749 to request an appraisal of his first article "Gedanken von der wahren Schatzung der lebendigen Kräfte" (Thoughts on the true estimation of living forces, 1747) that selectively examined related basic Cartesian, Newtonian, and Wolffian ideas. Although seeking to improve upon Leibnizian thought and written from a Wolffian standpoint, Kant's article arbitrated at times between Newtonian and Wolffian positions.[57] This is the first known extant letter from Kant.[58] But Euler considered the issue of the measure of these forces settled; there is neither a record of a response nor of Kant later sending a promised appendix. From Saint Petersburg Schumacher asked Euler in February to find a copy of *Gegenwärtig Zustand* (Present situation), a history of the Russian monarchy published in Frankfurt; he reported that Lomonosov had agreed to translate Euler's *Scientia navalis* into Russian and noted that Euler would be sent fifty free copies of the book to distribute to his colleagues, friends, and scholars in the field; he also mentioned that Lomonosov had agreed to submit his own paper on saltpeter to the Berlin Academy.

In March 1749 Euler and Schumacher discussed Nikita Ivanovich Popov's method of observing eclipses. In a July letter, also to Schumacher, Euler recommended that the Russian Imperial Academy purchase a fourteen-foot telescope of the deceased Berlin mechanist Bümmler and nominated Christlob Mylius for its vacant astronomy position.

In his letters to Wettstein Euler expressed the impatience among subscribers to the overdue *Uranographia Britannica*, a celestial atlas with fifty-one star charts by the British physician and amateur astronomer John Bevis. However, only a few copies were ever published, in 1750, and the printer shortly thereafter went bankrupt. Euler also cited astronomical documents dating back to Ptolemy, the seeming shortening of planetary years, the number of oscillations that a pendulum of given length makes over time, and the return of the Second Kamchatka Expedition explorers with results that did not support the existence of a northwest passage in North America. On almanacs he gave their prices, explained how they presented dates (Julian or Gregorian), and observed that Köhler benefited from their sales. Euler regretted that he could not remove astrological sections, for Köhler thought that this would harm sales. Euler also reported

Figure 9.4. Martin Folkes, in an etching by William Hogarth, 1741.

that the Dutch wished Isaac Bruckner to provide them his secret method for finding longitude to two-thirds or even one-half-degree accuracy to give them an advantage over the British in navigation. He asked Wettstein to pay his respects on behalf of his family to the antiquarian Martin Folkes, the president of the Royal Society, and to all of its members. In December Euler expressed infinite gratitude to Wettstein for supplying him with a crate of books and his favorite English tobacco. Tobacco, like almanacs, was a recurring theme in his letters with Wettstein.

Family Affairs

From 1748 to 1750 Euler and his family continued to live happily in Berlin. A genial man, he was not withdrawn from society there. While five of his and Katharina's first eight children reached adulthood, none of the four children delivered after 1747 lived beyond infancy. In April 1749 the twins Ertmuth Louise and Hélène Eleonora were born, but they died in August at seventeen weeks of age. In a letter of September 1749, Euler expressed his grief in commiseration with Wettstein, who had also just lost a child.[59]

In a letter to Goldbach in April 1750, Euler made one of his few mentions of Katharina. He referred to the premature birth the previous month of a son, August Friedrich (who would survive only five months) and

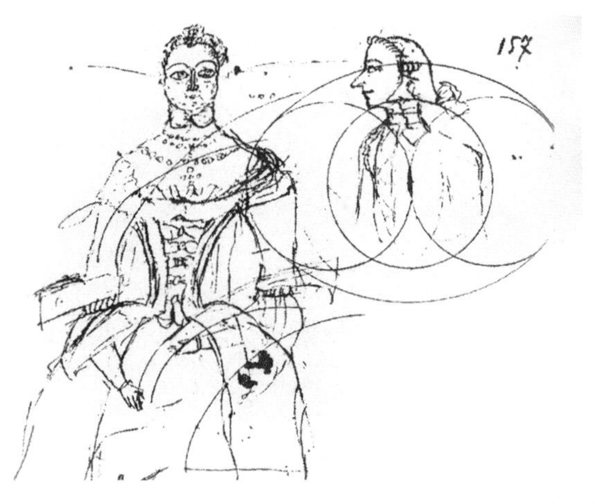

Figure 9.5. Sketches presumed to be of Katharina and Leonhard Euler made by their son Johann Albrecht between 1766 and 1770.

the serious illness of Katharina. No official portrait of her survives, but a sketch made between 1766 and 1770 by Johann Albrecht Euler apparently includes his parents at an earlier time. If so, we have a rough likeness of Katharina, and an indication that Leonhard Euler was by more modern standards a snappy dresser.[60]

Chapter 10

The Apogee Years, III: 1750 to 1753

From 1750 to 1753 Leonhard Euler continued at the peak of his career in Berlin. He dominated the Royal Prussian Academy of Sciences and Belles-Lettres, where he directed the mathematics department, and he helped to administer, judge competitions at, and nominate new members for the Russian Imperial Academy of Sciences in Saint Petersburg. From the Paris Academy he captured a double Prix de Paris in 1752. Throughout the decade Euler maintained his prolific correspondence and writings, particularly in astronomy, differential calculus, analytic geometry, mapmaking, rational mechanics, fluid mechanics, number theory, optics, and topology. In applied mathematics, his articles appeared mainly in the Berlin Academy's *Mémoires*; in number theory and other theoretical mathematics his work was published mostly in the Saint Petersburg *Novi Commentarii* or the Leipzig *Acta eruditorum*. Euler also investigated electricity, magnetism, and music theory. In part seeking to establish higher research standards in the sciences, he engaged in four ongoing major scientific disputes, regarding hydrodynamics, the motion of lunar apsides, the precession of equinoxes, and the vibrating string problem, meanwhile being drawn into a bitter quarrel between Pierre Maupertuis and Johann Samuel König. Euler's labors and style reflected and contributed substantially to a growing scientific collaboration and its attendant rivalries across Europe, mainly at the royal academies. On the Continent, Paris and Basel defined two points of a triangular flow of research. Whether Berlin or Saint Petersburg was the third point depended upon Euler's presence.

Euler and his family remained in Berlin. From about 79,000 inhabitants in 1735, the city was now approaching more than 100,000, and by 1757 would meet that figure. Under Frederick II, a king attentive to culture and science, the city had its opera, theater, and renovated science academy, along with many salons, clubs, and reading circles. Near to Berlin were the art, music, plays, and stunning gardens of his summer palace of Sanssouci in Potsdam. Johann Joachim Winckelmann, a student of Homeric and classical Greek art, sculpture, and literature, introduced his generation to a full and free humanity of a kind fundamental to Enlightenment classical idealism. In Berlin Winckelmann influenced the dramatist and philosopher Gotthold Ephraim Lessing, who with his close associate Moses Mendelssohn, known as the Jewish Martin Luther, combated religious intolerance. In 1750 the baroque composer Johann Sebastian Bach died in Leipzig, and in Königsberg the philosopher Immanuel Kant was beginning his career.

Frederick II was one of the three foremost reform absolutists of late eighteenth-century Europe; the other two were Joseph II of imperial Austria and Catherine the Great of imperial Russia. In the final years of Europe's *ancien régime*, Euler was to endorse ideas central to the king's goal, the transforming of Prussia into a more efficient and prosperous state and a power in Europe. He would be affected at least to a degree by the movement toward the implementation of these concepts. Frederick proposed a concept of the ruler as "the first servant of the state" (*le premier serviteur de l'état*). The codification of Prussian law and an end to censorship of speech and the press opened the state to ideas cultivated in western Europe, and freedom of the press would be raised as an issue in the controversy pitting Maupertuis against König. The Codex Fridericianus, issued in 1747, established the Prussian judicial body. A greater monument to Frederick's interest in legal reform was the Allgemeines preussisches Landrecht (Prussian Civil Code), completed on the basis of the Project des Corporis Juris Fridericiani from 1749 to 1751 by the eminent jurist and baron Samuel von Cocceji. In 1794, after Frederick's death, this work of vast labor and erudition would come into force, combining German and Roman law and supplementing them with the law of nature.

Within the German *Aufklärung* (Age of Enlightenment, a name given by Kant), Berlin was beginning to undergo a *Sattelzeit*, a saddle or bridge era, subjecting old estates to reason and clearing a way for an expansion of freedom. But more important as seen by contemporaries is that Frederician Berlin remained an armed camp. Amid ongoing antagonism with imperial Austria, Frederick closely allied with the nobility, restoring manpower and equipment expended in the first two Silesian Wars, ordering muskets, artillery, and supplies of ammunition and building a financial base to fund three future military campaigns. During the 1750s Euler found Berlin a pleasant city, but it was a view that later he would call a mistake; he would be drawn into the Third Silesian or Seven Years' War, in part as a translator. Until after its end in 1763, only the small French community that had earlier fled religious persecution and more recently a few French intellectuals and other scholars evading religious and political censure accepted Berlin's spartan character and came to reside there.

Competitions in Saint Petersburg, Paris, and Berlin

In December 1749, the Imperial Academy of Sciences in Saint Petersburg sent to press the program for its first prize contest, which was on the motion of lunar apogee. Early in January, after Johann Schumacher informed Euler of the selection of that problem, Euler wrote to Grigorij Nikolaevich Teplov commending the choice, noting that "at present it holds the interest of most of the top men of science of Europe, and opinions are divided on the question [of the cause of this motion]."[1] Although Euler employed only Isaac Newton's inverse-square law for the problem of the lunar apse, that of the great inequality, or any other problem in celestial mechanics, he declared that he differed entirely with Alexis Claude Clairaut's reversal dictating that Newtonian attraction was sufficient to account entirely for it. The results of his many computations of the motions of the moon and Saturn seemed to question the notion that Newton's inverse-square law could be universal. Euler suggested that a small correction was needed to account for irregularities in motion from turbulence. Accurately computing lunar inequalities arising from eccentricities required extensive differential equations, possibly with uncertain coefficients and constants. The calculations were tortuous; after Euler failed in his attempt in 1749 to learn Clairaut's new method, he had proposed the Russian prize topic as a means of obtaining it. His letter to Teplov (cited earlier) adds that

> not only is this question the most profound and the most worthy of all those the Academy had to select from, but the most complete answer may be expected from it as well. Until now Clairaut has kept his new method a secret, but as soon as he publishes it, inasmuch as calculations are involved, one of two alternatives must necessarily follow: either the rest of us must be convinced of his correctness, or else we will find some error in his work. Whatever the result, astronomy, physics, and analysis will enjoy a promising and important profit too great to be ignored. All those minor inequalities that we occasionally encounter not only in lunar motion but in the motion of all planets depend upon the result.[2]

Conference secretary Christian Nikolaus Winsheim sent out the prize program in a letter circulated to academies and universities, and foreign journals announced the contest in 1750. The prize was to be awarded the next year. The competition gave a new impulse to the advance of Russian astronomy after setbacks from the departures of Gottfried Wilhelm Heinsius and Joseph-Nicholas Delisle; it helped restore the stature and authority of the Petersburg Academy and increased its foreign contacts.

In January 1750 Clairaut wrote to Euler that while he found "very interesting" the topic of the Russian Imperial Academy's prize competition for the next year, he was concerned that none of its members were competent to choose the winner, and that Euler, not his Russian colleagues, must have proposed the topic.[3] He remained coy on it. In February Jean-Baptiste le Rond d'Alembert, who was actively engaged in research on the matter, wrote to Euler that he doubted that he would compete. He explained that the prize was too small, he lacked an adequate number of lunar observations on which to base his paper, and one year seemed too short a time to write it, especially while he was composing his "Preliminary Discourse" for the *Encyclopédie*, editing the science parts of its first volume, and several articles. The "Preliminary Discourse" publicly presented the intellectual and emotional spirit of the French Enlightenment, and moved Frederick II to exclaim, "Many men have won battles and conquered provinces, but few have written a work as perfect as the preface to the *Encyclopédie*."[4] In his study of lunar apogee, d'Alembert was examining lunar tables, the limits of errors of observation, and approximations from calculus, and now, for the first time, he expressed support for Clairaut's reversal. Almost a year later, in January 1751, d'Alembert told Euler that it was not convenient for him to participate in the Saint Petersburg prize contest, although he had finished his lunar theory the previous autumn with results, including his method, that he thought were superior to Clairaut's work.

As the contest began, the chancellery of the Imperial Academy had two tasks to complete. The first was to prepare a medal for the winner, but in March its officials decided that the time to have a medal struck was inadequate. The second was to complement the internal review panel with a select panel of distinguished external judges whose decision could not be disputed; this was not a problem. In August 1750 Euler expressed to Kirill Girgor'evich Razumovskij his readiness to serve as a judge; the papers would contain much algebraic analysis that needed to be addressed. Heinsius, Georg Wolfgang Krafft, and Christian Goldbach joined him.

While reexamining lunar motion in June, Euler grew impatient waiting for the publication of Clairaut's new method. In March he had received a copy of Clairaut's article "Systeme du monde dans les principes gravitation universelle" from 1745, which cited discrepancies in computing the motion of the lunar apogee. In the summer of 1750 some points still remained unclear. In July Euler sent Clairaut a detailed analysis of the article; he also sent a copy of his own *Scientia navalis*. Clairaut's new computations, he was convinced, must have errors, for some points were completely unclear; Euler sought to identify and correct them. Upon receiving this letter Clairaut decided to enter the contest. "I cannot flatter myself beforehand with hopes for the success of my paper in the sense of being awarded the prize," he wrote back, "for I have no way of knowing whether someone else will approach the problem in a manner superior to mine. But I am entirely convinced that in view of the contributions that I have made the prize can no longer be deferred." In the first part of his statement Clairaut was probably referring to d'Alembert. He now planned to submit for the contest his paper "Théorie de la lune déduite d'un seul principe de l'attraction" (Theory of the moon deduced only from the principle of attraction). In a letter of July 1750 he stated that it was nearly complete.

In December the Russian Imperial Academy received its first two contest papers, from the Copenhagen geometer Dietrich Siegvert Brumundt and the Freiburg theologian Gottfridus Magnus Maria Stapff; both were in Latin. In February two in French were circulated; one was by Clairaut. The papers had been duplicated within two weeks of receipt and sent to the review panel members. Euler expressed the general evaluation of the external committee. The two articles in Latin were far from answering the question, and the first in French was worthless. After making comparisons of hundreds of lunar observations with positions given in his tables and completing arduous computations, based only on Newtonian attraction, Clairaut wrote at the end of December that he had arrived at no error more than five minutes. He was certain that further computations would find that observation and theory agreed even more closely. The prize was not yet awarded, for the academy decided to extend the competition to June 1751. Euler did not anticipate any benefit from this and Clairaut was annoyed, since had he known of the longer period up front he would have improved his computations. Nonetheless, Clairaut in the meantime recommended that the Paris Academy elect Euler an associate foreign member, but Louis XV chose instead a royal physician. Clairaut urged Euler to meet Voltaire, who in July 1750 had arrived in Potsdam;[5] from that point on Voltaire was to be constantly occupied with the king in either Berlin, Potsdam, or Charlottenburg.

In 1750 Euler's competition with the Paris academicians intensified. Whether it was possible to quantitatively deduce the exact precession of the celestial equinoxes remained for astronomers one of the most significant questions, as was that of whether the nutation of the Earth's axis could be similarly determined. In March Euler answered these by obtaining the astronomical equations for both. He cleared up d'Alembert's confusing mathematics and on March 7 wrote to him declaring what was still an absolute necessity for treating precession, a theory of gyroscopic motion, in which he had far outdistanced his Parisian colleague:

> I applied myself repeatedly and for a long time to the problem of precession, but always encountered an obstacle . . . and above all this problem: given a body turning about any axis freely, and acted upon by an oblique force, to find the change caused both on the axis of rotation and in the movement. . . . I must confess that I could not follow you in the preliminary propositions you employed, for your way of carrying out the calculation was not yet familiar to me. . . . But now that I have succeeded better in the investigation of this same subject, having been assisted by some insights in your work, by which I was gradually enlightened, I have become able to judge your excellent conclusions.[6]

Submitted for the Berlin *Mémoires* dated for 1749, Euler's article "Recherches sur la précession de équinoxes et sur la nutation de l'axe de la terre" (Research on the precession of the equinoxes and the nutation of Earth's axis, E171);[7] it used elaborate trigonometric equations and published a novelty. That novelty, which appeared in a result to be published in "Découverte d'un nouveau principe de mecanique" (Discovery of a new principle in mechanics, E177), was responsible for his breakthrough to the principle of angular momentum. Completed and delivered at the Berlin Academy in March 1750, it was not printed in the *Mémoires* until the next year. Following a method only a little different from that of d'Alembert's, its final version solves these problems without second-order differentials, which proved negligible. Euler helped perfect the work of others by developing needed differential equations. He found d'Alembert's *Recherches sur la précession des équinoxes* (his foremost book on the subject from 1749) to be muddled and difficult. But a series of Euler's articles, one beginning with the same title as d'Alembert's book, do not recognize the priority of that author or mention his name; an omission that astounded and angered d'Alembert. He was particularly upset that the dating of Euler's article as 1749 made it difficult for readers to know who was first.

Since Euler was writing not only on the precession of equinoxes but on the topics of lunar theory and fluid mechanics, both of which d'Alembert had introduced in their correspondence, the French *philosophe* feared that Euler was preempting his ideas. During this period, as d'Alembert became known for his editing of the *Encyclopédie* and his wit in salons, he believed that Euler was essentially blocking his progress in scholarship. Not known for his tact or discretion, in 1751 he sent a lengthy, bellicose complaint to the Berlin Academy noting all the items he claimed Euler had derived from him. The letter did not attack Euler, for whom d'Alembert had the highest regard, but asked for justice regarding points on the originality of Euler's work.[8]

To late June the correspondence between the two men had been friendly, but d'Alembert had now sent an angry letter, probably to Jean-Henri-Samuel Formey, that the academy declined to publish; only a rough draft of it survives, and there is no record of Euler's immediate response. From these complaints Euler felt obliged to insert in the Berlin Academy *Mémoires* for 1750, published in 1752, his "Avertissement au suject des Recherches sur la Précession des Equinoxes" (E180), in which he denied priority for a solution for precession, attributing it instead to d'Alembert.[9] Euler, writing in the third person, declared that "he had not written his article until he had read the excellent work of Mssr. d'Alembert on this material; and that he makes no pretense to the glory due to him, who first resolved this important question."[10] He added that "we are indebted solely to Mssr. d'Alembert for this resolution, where he gave the first account of the nature of the curves involved," known as the bird's beak because of their shape. Not satisfied with his own initial solution, the next year Euler devised what are known as Euler angles (or, more simply, Eulers)—that is, the three separate angles into which general three-dimensional rotations may be decomposed. In this case, the two axes of the spinning, precessing Earth form these.

Fluid mechanics offered another controversy. At the Berlin Academy Euler headed the three-member review jury for its annual prize in 1750 on papers dealing with the resistance of fluids. It found d'Alembert's "Essai d'une nouvelle théorie de la résistance des fluids" to be the best entry, introducing the initial partial differential equations of fluid mechanics set in a field. But Euler believed that the paper's awkward analysis left it far from meriting the prize; having derived some necessary equations in chapter 2 of his own *Neue Grundsätze der Artillerie* (E77), Euler judged the article a clumsy pioneering effort. He had the jury reject d'Alembert's paper, along with all others on fluid mechanics; this was consonant with his search for higher standards. This prize decision, and not their debate over negative logarithms, most soured relations between the two men. The academy returned the articles to the authors and urged them to compare their predictions with results of measurements from experiments. No evaluation of d'Alembert's paper by Euler is known. The instruction to check whether quantitative results matched predictions may have been a pretext to avoid detailing the tortuous reasoning of the "Essai d'une nouvelle théorie" and its failure to account for a single flow through the partial differential equations it gave; it also failed to mention the research of Clairaut and Euler on hydrodynamics. Knowing that his paper was significant, d'Alembert thought that only Euler could have brought about this negative decision. The junior astronomer Augustin Nathanael Grischow, a member of the review jury and a friend of d'Alembert, inappropriately confirmed his suspicion. The academy postponed awarding the prize to 1752, and d'Alembert chose not to reenter; instead he requested that his paper be returned and published it independently with sarcastic notes on the judges.

Shortly before the announcement of the prize decision in fluid mechanics in 1750, d'Alembert wrote a letter to the Berlin Academy that called Euler's solution for the tedious problem of the vibrating string of uniform thickness incorrect. After d'Alembert first wrote on vibrating strings in 1746, he had discussed the topic in letters with Euler, who published his own "De vibratione chordarum exercitatio" (On vibrations of excited strings, E119), *in Nova acta eruditorum*[11]; and for 1751 Euler presented a first draft of a paper "Sur la vibration des cordes" (On the vibration of strings; later translated from Latin as E140).[12] D'Alembert was the first to solve the wave equation with a partial differential equation, which undergirds the study of oscillations generated in a continuous medium. To 1750 Euler added little except a criticism of d'Alembert's restrictions—for example, resolving the wave equation for given end conditions by limiting the shape of strings to the sine function. The debate was just beginning.

Maupertuis's *Cosmologie* and Selected Research

Immediately preceding the second act of the Maupertuis-König affair was the publication of Maupertuis's *Essai de cosmologie* in 1750, which expanded upon his Berlin article from 1746 on least action and the general laws of mechanics, and he added what he deemed proof of God's existence. This eclectic collection examines differing ideas on motion and force from René Descartes, Gottfried Wilhelm Leibniz, Nicolas Malebranche, and Newton. Maupertuis prepared the book for only a small circle of readers that included leading mathematicians, natural philosophers, literate nobles, and men of letters known to him, and he worked with Johann II (Jean II) Bernoulli in having a hundred elegant copies printed at his own expense.[13] The circulation of this first edition restricted to his small group of readers must have lessened the response to it.

Maupertuis's *Essai* shifts the emphasis in mechanics from force to action, which could be better measured. He found confusing the Leibnizian association of force with monads. In culminating in the principle of least possible action, Newtonian dynamics received its crowning glory when Maupertuis restated, "Here then is this principle, so wise, so worthy of the Supreme Being. Whenever any change occurs in nature, the quantity of action is always the smallest possible." It was, Maupertuis argued, the final law of mechanics that Descartes and Newton had pursued, but he claimed to have discovered it, and insisted that it had universal application in physics.

The title *Essai de cosmologie* alone must have troubled the Wolffians, for it seemed that Maupertuis was offering a dangerous alternative to Christian Wolff's *Cosmologia generalis*, the second edition of which had appeared in 1737. Following upon his ontology, Wolff's cosmology consisted of two parts, the scientific and the experimental. The rational scientific component included the general principles of the visible world, and the experimental had experience confirming them. Wolff was convinced of the utility and certainty that his new route in the study of nature provided. To make the real world intelligible, Maupertuis required a concordance between mechanics and metaphysics. Joining principles of physics with the metaphysical concepts of infinite power and wisdom rather than that of design, he gave what he deemed was a new proof of God's existence. Maupertuis admitted, however, that he had only a sketch and not a full explanation of the system of the world.

The origins of Maupertuis's discovery of the principle of least action can be traced through three articles. The first, "Le principe universel du repos et de l'équilibre des corps" (The universal principle of rest and equilibrium of bodies), published in the Paris Academy of Science's *Mémoires de l'Académie* in 1740, was an inquiry into the center of gravity and into elastic curves describing equilibrium in dynamics and hydrodynamics that gave a preliminary formulation.[14] The second, "Accord des différentes lois de la natur qui avoient jusqu'ici paru incompatibles" (Accord among different laws of nature previously appearing incompatible), was a speech given in April 1744 at the Paris Academy that offered an early statement of the law of least possible action. It held that equilibrium in rigid, flexible, elastic, or fluid bodies exists when the sum of all quantities of action is the least possible, and examined the Newtonian conception of the connection between optical density and the speed of light. It presented results from Maupertuis's studies of the propagation of light, together with reflection and refraction in homogeneous and nonhomogeneous media. Its description of refraction employs the sine law of Willebrord Snell and the research of Descartes, and it draws upon Pierre de Fermat's techniques of maxima and minima along with Leibniz's research on optics in a resistant medium. But as Euler later argued, Leibniz explained refraction using a principle completely different from that of least action, one that was "entirely unknown" to him,[15] and in a letter in the spring of 1746, Euler had indicated that he still did not have a copy of the text of the article. That year Maupertuis published his third article, "*Les loix du mouvement et du repos*" (The laws of movement and of rest) in the Berlin *Mémoires*; in it he called the law of least possible action the "general principle" of mechanics. This meant that he had reduced to the same principle the laws of motion and equilibrium.

The first section of Maupertuis's *Essai de cosmologie* emphasizes geometric arguments and criticizes natural theology but agrees with Newton on God's occasional intervention to inject forces into the universe. A second section founds the fundamental truths of mechanics in the metaphysics of final causes. Maupertuis belonged to an older generation in natural philosophy that had not yet broken with metaphysics, while Euler and d'Alembert looked to geometric proofs.

In 1750 Euler finished his *Institutiones calculi differentialis* (Foundations of Differential calculus) and wrote fourteen articles, all but one published during the decade. He was working to remove paradoxes from calculus. At the academy he gave twenty-four presentations. In number theory, he attempted to prove Fermat's conjecture that every integer is a sum of two, three, or four squares. (E242) But this is not true for two squares. While he proved that every integer of the form 4n + 1 is the sum of four integer squares and demonstrated that every rational number is the sum of four rational squares, a final proof eluded him for decades. More significant than this attempted proof is the article's foreshadowing of group theory.[16] Later Euler was to prove Fermat's Little Theorem twice, basing his reasoning not on binomial expansions, as his first two proofs had, but on properties of group theory. He also submitted to the Berlin Academy that year a paper titled "Emendatio laternae magicae ac microscopii solaris" (Improvement of the magic lantern and solar microscope, E196).[17] He sent a copy to Schumacher that appeared in the 1750–51 volume of the Petersburg Academy's *Novi Commentarii*. The paper gave methods to better magic lanterns; these predecessors of slide projectors were important to public optical spectacles during the Enlightenment. They projected a still picture contained in a transparent slide; illumination depended on an oil lamp. Euler was also the first to develop a solar or projection microscope.

Occasionally Euler introduced a subject that would not be developed until the next century. In "Elementa doctrinae solidorum" (Elements of the doctrine of solids, E230), read at the academy in November 1750,[18] and its shorter sequel "Demonstratio nonnullarum insignium proprietatum, quibus solida hedris planis inclusa sunt praedita" (Proof of some not insignificant properties of solid bodies enclosed by planes, E231) in September 1751, he explored questions of *analysis situs* in attempting to found a branch of mathematics, *stereometria*—that is, solid geometry.[19] These two articles were for the *Novi commentarii* for the years 1752–53, which appeared in print in 1758.[20] "It astonishes me," Euler wrote to Goldbach in November 1750, "that these general properties of stereometry have still, as far as I know, been noticed by no one else."[21] Although many theorems existed about surface area, volume, and angle measures for polygons, no one before Euler, except for Descartes, had investigated the combinatorial properties of polyhedra. Euler's "Elementa" begins by stating that just as figures of plane geometry require studies of lines and angles, stereometry must consider the inclination of planes and solid angles; the analogy for classifying polygons does not suffice for polyhedra. He provided the crucial items needed and a nomenclature for them: points or *anguli solidi* (S); faces, which he represented with the Latin term *hedra* (H), and edges, defined as "the junctures where two faces come together along their sides and are sometimes called *acies* (A)," Latin for the edges on a weapon. He notably brought out the significance of edges. Proposition 4 of the "Demonstratio" gives his famous theorem $S + H = A + 2$ (in English notation, S = vertices, A = edges, and H = faces).

Euler's first proof in the "Elementa" is flawed and incomplete; he gave a more convincing but still incorrect proof the next year. His beautiful formula was correct, but he overreached, not giving certain conditions for this to be so and many counterexamples. Euler thought there were two ways to decompose a polyhedron; from the outside, where he could remove pyramids with each apex being a vertex of the polyhedron, or from inside by removing pyramids having a common apex and a polyhedron face as the individual base; he chose the first method. By taking away pyramids until a tetrahedron was reached, he found that $S + H - A = 2$. To prove his theorem required recognizing that in the decomposing process the quantity $S + H - A$ was unchanged. His formula must meet certain conditions, and he recognized that it did not apply to all polyhedra, limiting

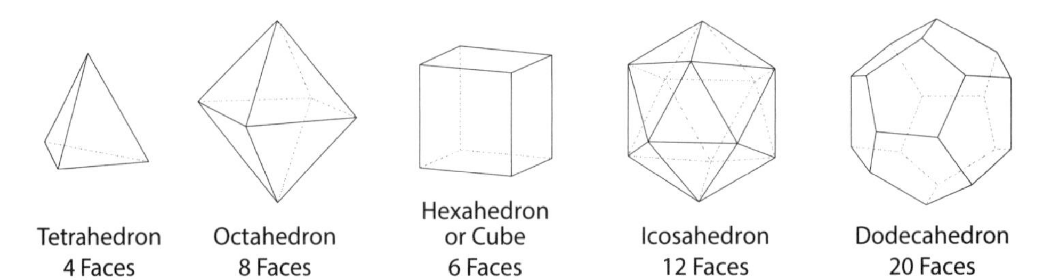

Figure 10.1. Five regular polyhedra or Platonic solids. Euler's formula: F (Faces) – E (Edges) + V (Vertices) = 2.

it to solid bodies bounded by planes, but he did not give many counter-examples. Euler's work together with that of successors who differed by preferring to examine the second decomposition opened the path to the modern definition of a polyhedron. The left side of the modern form of his marvelous relation V − E + F = 2 defines what became known as the Euler Characteristic. For polyhedra, those characteristics are topologically invariant.

Academic Administration

In 1750 the Berlin Academy was building a leading reputation, but Maupertuis's desire for greater accomplishments was met with uneven results. Unsatisfied with the dogmatic German Wolffians as well as with the French savants who rejected metaphysics, he looked to the Swiss to offer a middle path. Producing more men of science than were needed in the homeland, the Swiss were chosen in greater numbers than the French for the academy;[22] Berlin was one stream in the Swiss exodus that also included Amsterdam, London, Munich, Paris, and Saint Petersburg.

In a letter of 1746, Maupertuis had pleaded with Johann II and Daniel Bernoulli to come to Berlin, saying that they alone could "put our Academy above all the Academies of Europe."[23] The Bernoullis declined but were named foreign members and agreed to recommend candidates. Still, the budget was limited, and was mainly directed at current members. In 1750 the academy expelled two Swiss candidates based on poor performance. During two prior years in Berlin, the young Basel mathematician Daniel Passavant had turned out to be among the "laziest and most dissipated of men." Having no research projects of his own, he was assigned tasks with translations that he accepted reluctantly. Maupertuis judged him completely unfit, reporting to the king that this Basler had done almost nothing beyond offering instruction in the house where he resided. Passavant was expelled and his position left vacant; in April 1750 he returned to Basel. (From 1755 to mid-1759, Euler would correspond regularly with him and his cousin Franz Passavant, a Basel official.)

Having accomplished little in Berlin, the physicist and mathematician Reinhard Battier, Euler's cousin, departed in 1750 to become tutor to the son of the Duke of Gotha. Grischow, whom the academy had sent to the Paris Observatory for two years, briefly followed him.

"Will we ever have any but mediocre people?" grumbled Maupertuis. Only two Bernoulli candidates for 1750, the young rhetorician Johann Bernhard Merian, who was, like Euler, the son of a Basel pastor, and the mathematician Johann Georg Sulzer, a teacher from Joachminsthal's gymnasium, succeeded in being named ordinary members. Even though not known for his research, Merian was warmly endorsed by the Bernoullis as capable of doing the institution the greatest honor. Arriving by April 1750, he was to be tireless in academy projects, forming a trio with Maupertuis and Euler in its disputes. Merian was to be one of the most active members and the most reliable ally of Euler at the academy.

The restless Euler had an array of administrative tasks. He watched over the library, observatory, botanical garden, publication of scientific papers, and preparation and sale of the almanacs and maps. The library that he oversaw was perhaps the oldest institution in the academy. The Franco-German historian and antiquarian Simon Pelloutier was the librarian; he and Euler followed the rules that Leibniz had set out in 1700 for a scientific research library. Although not large, the academy library contained most of the best mathematical and physics books and journals, along with scriptural texts and periodic literary writings. It grouped these acquisitions according to their use in the active research of the members. Its comprehensive exchange program of texts and journals with other leading science academies and learned societies provided most publications. Euler would personally correspond to remove gaps in runs of leading journals such as the *Philosophical Transactions* of the Royal Society of London and Leipzig's *Acta eruditorum*.

A painting by Adolph Friedrich Erdmann von Menzel titled *Tafelrunde* (Roundtable) from 1850 has conveyed a misimpression about Frederick II's relations with the academy, particularly whether he attended meetings or met with its regular members. In the mid-eighteenth century, roundtable or discussion groups were a means of bringing scholars together. In his equivalent table society, Euler fostered research. The *Tafelrunde* painting pictures the principal entrance area of Frederick's palace Sanssouci, which consists of two halls, the rectangular entrance hall and the oval marble hall under the central dome. Guests entered through the marble hall, which was the chief reception room; ten pairs of Corinthian columns subdivided it. Menzel imagined Frederick hosting a discussion at the king's new *Tafelrunde* in the marble hall with Voltaire on the left and the nine male muses of the German Parnassus, most of them French intellectual refugees under the king's protection, all of whom Voltaire outshone. As the residence of the muses in ancient Greek mythology, Mount Parnassus was known as the home of learning, music, and poetry. Frederick praised Voltaire as a genius who surpassed Homer and Virgil, and thus should be an ornament of his Parnassus. Frederick also recognized Voltaire as a scoundrel. In time Voltaire would criticize the Frederician muses, for example calling Jean Baptiste de Boyer, Marquis d'Argens less than mediocre. The guests at the table are sometimes mistaken as all being from the Prussian Academy of Sciences. While Francesco Algarotti and Maupertuis belonged to the *Tafelrunde*, Voltaire seldom attended academy meetings but met occasionally with other academicians, mostly his friend Joseph de Francheville, and the king never interacted in 1750 or later with academicians as a whole—at any meeting or with Euler. He did not associate separately with any regular academician in Berlin except for Merian and his father-in-law Charles Etienne Jordan.[24] Jordan, who became the vice president of the academy, had been a childhood friend of Frederick. Except for Voltaire, the *Tafelrunde* members were apparently nobles.

As Euler's close engagement with the Russian Imperial Academy continued in 1750, going far beyond the prize competition and making Saint Petersburg his de facto second home, the Petersburg Academy concentrated on its newer publication, the *Novi Commentarii*, which was divided into four classes: mathematics; astronomy; physicomathematics (or mechanics) and experimental physics; and the physical sciences, consisting of anatomy, botany, and chemistry. In November 1750 Schumacher asked Euler to find a good mathematician to be its editor and prepare registers for the old *Commentarii*, and the next year he asked Euler to undertake those tasks himself. Cartography remained a strong mutual interest. Euler wanted clarification on a report that Delisle had published in Paris a map of all the Russian discoveries from the Second Kamchatka Expedition. He pointed out that Delisle did not have a complete record and had promised to provide better maps as soon as possible. Delisle's article for the Paris Academy *Mémoires* of 1750 and his maps of some of the Russian discoveries appeared two years later.

The nasty affair surrounding Johann Georg Gmelin's decision to accept a position at the University of Tübingen caused a stir throughout 1750. It continued to anger the Petersburg Imperial Academy officials, who held that he remained under contract with them. In April Euler wrote to Teplov that he disapproved of Gmelin's actions. In fact, Gmelin had actually wished to complete his time with the academy and did not want to insult it, but Euler explained that Gmelin feared losing a favorable retirement from his own country; a solid retirement was absolutely necessary, Euler wrote, and being deprived of one drives a man to desperation. He asked Gmelin to send a letter declaring his mistake in not honoring his Russian contract and a contrite confession for his behavior. This pleased Razumovskij. At the end of April, Gmelin thanked Euler for helping to have the academy officials accept his move to Tübingen; in another letter in August, Gmelin asked for Razumovskij's forgiveness. This was only one

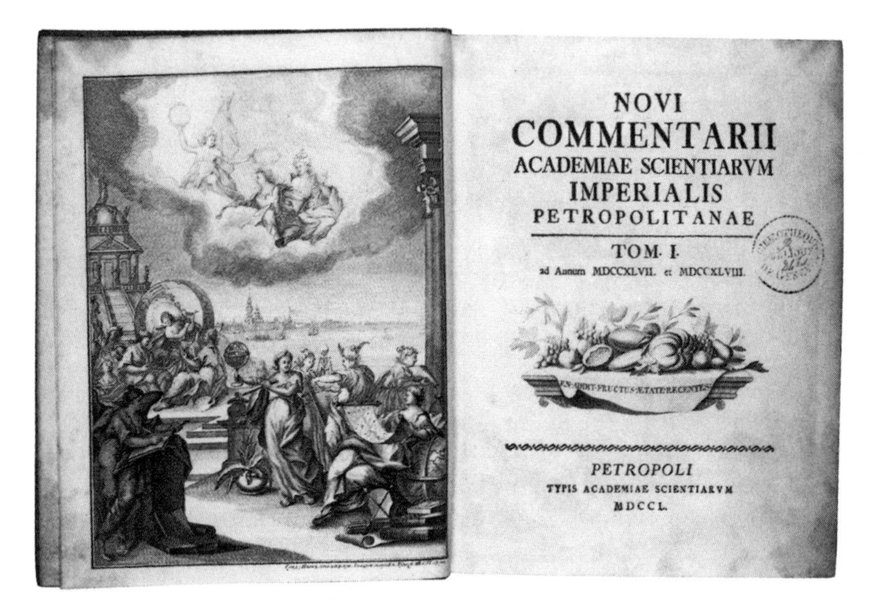

Figure 10.2. The frontispiece and title page of the *Novi Commentarii Academiae scientiarum Imperialis Petropolitanae* for 1750.

component of the affair; Euler also negotiated a new contract for Gmelin to deliver six volumes of his work on Siberian flora to the academy, which would pay two hundred rubles for each. Gmelin was pleased. There would be no payment for the first volume, since when he completed it he was still under contract to the academy; separate compensation, Gmelin agreed, would be unfair. For its preface, Euler obtained a fifteen ruble recompense for him. The new contract was not completed until after Euler returned from his Frankfurt visit in July; he was upset with Schumacher for not forwarding it to him there.[25]

The correspondence between Euler and Teplov in 1750 also covered reviews, descriptions of research, and a rift between Euler and Schumacher. In April Euler evaluated two papers sent to him: one on chronometers and another on oars. He praised the first, which proposed making a better self-winding watch. Euler had earlier considered this idea by applying a pyrometer to a pendulum. Changes in temperature would make alterations to the instrument that appeared to result in a self-winding mechanism. Euler cautioned that this was not a perpetuum mobile. The second paper on oars he found pedestrian; it described a new way to construct oars but was of no practical advantage. In April Euler criticized Schumacher for failing to transmit to the academy a proof concerning astronomy that he had sent and for the delay in purchasing the excellent Blumler telescopes that the academy needed, adding that Blumler's widow was destitute. In August Euler received fifty free copies of his *Scientia navalis*. As thanks for these he submitted two articles on analysis for the *Novi Commentarii* along with a note that reiterated his desire to be named a commissioner to examine with perfect impartiality the prize papers for 1751.

In November 1750 Euler received the second of three invitations from Razumovskij to return to Russia, but again diplomatically declined. He said that his health was deteriorating by the day and that the great geniuses of mathematics—Newton, Leibniz, Jakob Hermann, Wolff, and the Bernoullis—had made their most important discoveries before their fortieth year; what they did afterward did not compare with their work in the earlier period. Euler, who was rapidly approaching fifty, claimed that his capacity was substantially diminished. In addition, he thought that neither his health nor that of some of his children would permit so lengthy a trip and the harsh climate of Saint Petersburg, nor could his seventy-two-year-old mother endure it.

Family Life and Philidor

In 1750, two of three major events for the Euler family involved the deaths of children. In March the twelfth child, August Friedrich, was born. Carried for only seven months, his was a risky delivery, but Euler's wife Katharina recovered rather easily. Extremely weak, August Friedrich survived only nineteen weeks. In December Euler's three-and-a-half year-old son Hermann passed away; having grown quite close to their youngest son, Euler and his wife were deeply distraught.

The third major event came in June: Leonhard and Katharina, along with Johann Albrecht, who was sixteen, traveled to Frankfurt am Main to the home of Katharina's cousin, the Dutch historian and theologian Johannes Michael van Loen, to meet his mother, newly arrived from Basel, and bring her to Berlin. The travels did Katharina much good, helping to restore her health after the last birth experience. The entire trip went from Potsdam to Magdeburg, Halberstadt, Duderstadt, Kassel, Marburg, Giessen, and Frankfurt am Main and back. The round trip took twenty days. Euler, who had earlier met with his brother in Frankfurt, was there from 28 June to 3 July. Widowed since 1745, his mother had remained in Basel with the older of Euler's two sisters and her son Johann Heinrich and his second wife. When Johann Heinrich died, she did not move to live with her daughters but accepted Leonhard Euler's invitation to be with his family. On the return trip the family went to Hanau, where Euler's younger sister and brother-in-law welcomed them, and on to Fulda, Wach, Eisenach, Gotha, Erfurt, Naumberg, Merseburg, and Wittenberg, finally reaching Berlin on 10 July. Curiously, Euler went only as far as Frankfurt on this trip rather than proceeding up the Rhine to Basel, where he could have seen old friends and relatives. Perhaps he lacked the time for an extended trip, or perhaps a strain in his relations with Daniel Bernoulli was the reason. Since 1741 Bernoulli had regularly sent Euler three or four letters a year. Suddenly in January 1750 Bernoulli's correspondence to Euler stopped. Though he did not yet know of Clairaut's latest work on lunar theory, Bernoulli proclaimed Newtonian dynamics correct and criticized d'Alembert on hydrodynamics, the theory of winds, and the vibrating string problem. The next year he calculated the proper frequencies for transverse vibrations. At this time, Euler questioned Clairaut's conclusions on lunar motion. Bernoulli's next letter to Euler came in October 1753, while Euler apparently sent none for that three-year period.

In 1750 Euler, who acted as an advocate and protector for his extended family and often asked Frederick for positions for relatives, tried to assist Johannes Michael van Loen. His mother-in-law had been born a van Loen. This was not his Dutch relative's first attempt to obtain a post in Prussia. In 1746 he had declined Frederick's offer of the presidency of the consistory of the Reformed Church in Berlin and asked for Euler's help in obtaining another post. Euler felt responsible for having van Loen move to Frankfurt am Main. In July 1750, after his visit there, Euler wrote to the king and Maupertuis that it would be useful to have van Loen in Berlin, and he prepared a résumé. In December the king responded that neither a law court position nor the associated salary could be arranged. Van Loen persisted, and in 1755 Frederick proffered the presidency of the Prussian counties of Lingen and Tecklenburg. After writing again to Euler and hesitating, van Loen accepted that post.

Chess remained one of Euler's few recreations. In June 1750 Goldbach wrote from Moscow about it, having read in the newspaper that the twenty-five-year-old French chess champion and composer François André Danican-Philidor was creating a furor in Berlin and Potsdam by defeating its best chess players. To earn extra income, Philidor simultaneously played and won three games blindfolded. Goldbach guessed that Philidor would know of Euler's chess skill. Since Philidor stayed most of the time in Potsdam teaching some noble military officers, Euler responded on 3 July that he had not seen him.[26] But Euler would have been no match for Philidor, who was probably the greatest chess player of the century. Since the young French chess master kept a mistress, some of the Potsdam nobility were upset and unexpectedly forced him to leave early; otherwise, Euler thought, he would have probably found the opportunity to play the Frenchman.

In 1748 Philidor had written in French a book on chess; Euler owned its English translation, *Chess Analyzed*, which appeared two years later. He found quite beautiful the ways it presented for playing chess; its approach was unique, stressing principles over moves and making pawns the "soul of the game." "Seine grösste Stärke" (His greatest strength), Euler wrote to Goldbach, "besteht in Verteidigung und guter Führung seiner Bauern, um dieselben zu Königinnen zu machen, da er dann, wenn die Anstalten dazu gemacht, pièce für pièce wegnimmt, um seine Absicht zu erreichen und dadurch das Spiel zu gewinnen" (is in the defense and in the skillful moves of his pawns in order to transform them into queens. Then, after proper preparations are made, capturing piece after piece in order to achieve his intention and win the game).[27]

Rivalries: Euler, d'Alembert, and Clairaut

In 1751, relations between Euler and d'Alembert deteriorated, and research in fluid mechanics appears to be the chief reason. In "Recherches sur le mouvement des rivieres" (Researches on the motion of rivers, E332), an early draft of which was written in 1750–51, Euler fulsomely praised d'Alembert: "To the profound speculations of Messrs. [Daniel and Johann I] Bernoulli and d'Alembert we owe all that has been discovered in this science up to this time."[28] But apparently this recognition was not in the early draft, which in any case was unpublished. Euler revised this paper to answer d'Alembert's general criticisms. Presented at the academy in 1760, the revision was not printed until 1767. In the calculus of variations it contains the Euler-Lagrange Equation for the first necessary condition and others to assure for definite integrals with arc lengths along curves an extremum—either a maximum or a minimum. Their solutions are functions with a stationary functional or mapping of a vector space to its underlying field. When these functions reach a maximum or minimum, the derivative is zero. Most likely not before the revision was the statement on d'Alembert along with Euler's contributions included in the article. In 1750 a reading of papers by d'Alembert likely prompted Euler quickly to perfect his colleague's concepts and equations in hydrodynamics. His first attempt to combine internal pressure with the momentum principle nearly succeeded; that took until 1752.

From Clairaut—with whom Euler had far better interactions—the Russian Imperial Academy received in February 1751 the paper "Théorie de la lune." A copy went to chief judge Euler, who wrote to its author a month later, "I have with an infinite satisfaction read your paper, which I have waited for with much impatience. It is a magnificent piece of legerdemain, by which you have reduced all the angles entering the calculation to multiples of your angle v, which renders all the terms at once integrable. . . . I must confess that in this respect your method is far preferable to what I have used."[29] Clairaut had finally given accurate computations showing that the motion of lunar apogee could be reconciled with Newton's attraction alone; it had taken arduous labor to accomplish this. The first two dissertations in the competition, from Copenhagen and Freiberg, were in Latin. These and a third in French Euler called worthless, and he rejected them; the decision was indisputable. The four foreign judges and the panel of resident academicians (Grischow, Christian Gottlieb Kratzenstein, Nikita Ivanovich Popov, and Georg Wilhelm Richmann, who were those academicians with competency on the subject), together with Mikhail Lomonosov and Winsheim,[30] voted unanimously for Clairaut.

"The splendid contest paper in French [by Clairaut] is more and more to my liking," Euler wrote to Schumacher, "and even though I did not at first share the view that the Newtonian theory is sufficient to account for all the lunar inequalities, as asserted in this paper; now after having investigated this problem using my own methods, I have arrived at a different conclusion and have changed my opinion."[31] Not entirely satisfied with the computations, Euler had perfected them. After devising his new method to account for apsidal motion that avoids variables and derives an equation between the longitude and the true anomaly of the moon, besides extending his complicated approximations of it far enough and studying more observations of lunar motion, he had "finally discovered the source of insufficiency in my earlier methods" of computation and his errors. Since his method gave the same results as Clairaut's, Euler wrote to Clairaut in early April with excitement:

> I am now altogether clear concerning the motion of the lunar apogee, and I find, as you do, that it entirely conforms to Newton's theory. . . . [S]ince now two completely different methods lead to the same conclusion, no one will refuse to recognize the correctness of your research. For myself, knowing whereof I write, I salute you on this happy discovery, and I even dare to say that I regard this discovery as the most important and the most profound that has ever been made in mathematics. I ask your pardon a thousand times for having doubted the rightness of your retraction.[32]

Clairaut's prize paper, Euler asserted, was exemplary in showing that gravitational attraction alone is completely satisfactory "to explain the motion of lunar apogee." Along with new lunar studies, Clairaut's paper imparted "quite a new luster to the [gravitational attraction] theory of the great Newton."[33] But Euler still harbored a belief, which would be reinforced later by his study of magnetism, that Newton's inverse-square law needed a small corrective factor negligible in this case. He called Clairaut's computations on lunar motion of the highest degree of difficulty, and he stated that prior efforts of other astronomers to show the accord between Newtonian attraction and the motion of lunar apogee had been inadequate. For Euler this was an important stage in developing his own second lunar theory.

In response to contestants who claimed that they had not received enough time to complete their papers the academy extended the competition to June 1751. This upset Clairaut; had he known of the extra time, he could have improved his calculations. At the academy's public meeting on 6 September 1751, Clairaut was named the contest winner with a prize of 100 ducats. Euler lavishly praised the outstanding clarity of Clairaut's calculations in the appendix and how little they differed from observations made at universities and observatories.

The Maupertuis-König Affair: The Early Second Phase

When König's article "De universali principio aequilibri et motus" (The universal principle of equilibrium and motion) appeared in March 1751 in Leipzig's *Nova acta eruditorum*, the second act of the Maupertuis-König affair began. It startled Maupertuis; König had submitted his critique without informing him of plans to publish it.

König's article addressed the two central issues of the debate. It denied the usefulness and generality of the principle of least action, replacing it as a universal principle with Leibniz's conservation of *vis viva* that was applied to equilibrium conditions, and it examined who deserved priority for that principle (which is today known as the Euler-Maupertuis Principle): should it be named for the great Leibniz or for Maupertuis? König asserted that the origins of the principle lay in antiquity with Aristotle's axiom that nature in all its operations does nothing useless and seeks the simplest solution. He sought to establish Leibniz's precedence for the principle but was unable to do so, for he based his claim on a copy of a portion of an unpublished letter supposedly from Leibniz to Hermann in Basel in October 1707. The letter covers a series of writings with signs, unclear marginal notes, and some faint lines that indicate how best to place the interjections. König possessed a fragment; he did not yet make it clear that the famous Bernese bookseller Samuel Henzi, the clerk of the saltworks, had supposedly purchased it for fifteen crowns and copied it. But Henzi had been charged with sedition and was beheaded in Bern in 1749 for taking part in an unsuccessful uprising in the city against its patrician families; his large collection of papers had been destroyed or scattered. This all occurred five years after König's exile from that city for a decade for signing a petition deemed too liberal. The desired letter was not listed among the seventy-eight published ones between Leibniz and Hermann; it apparently supports the claim that Leibniz had much earlier defined the quantity of action in the same way Maupertuis had and projected a maximum or minimum. König represented the principle of least action as a poor chimera and Leibniz's principle of the conservation of *vis viva* as fundamental; he was not successful in making either case. In place of what was to be called the Euler-Maupertuis Principle, he projected a Leibniz-Euler-Maupertuis law, denying the originality of the last two. König further rejected Maupertuis's hard atoms as contradictory to physics and Leibniz's conservation laws. In its exposition, his article was neither clear nor reinforced by persuasive mathematical demonstrations.

König was no stranger to controversy in Berlin. He had submitted one of the six papers published by the academy on the monadic doctrine. Not winning the academy prize, he was apparently bitter.[34] It is possible that König thought his article would not harm his friendship with Maupertuis; in the sciences it should be possible to question a friend's writings. But König, who knew directly of Maupertuis's vanity, suspicions, and insecurity, declared that he had learned of the principle in 1749 but delayed his response for two years out of deference to Maupertuis.[35]

The full scope of the quarrel in the Maupertuis-König affair can be understood only within the intellectual discourse at the academy. Its disagreement fell at the heart of the division within the academy that ranged from the conservative Euler and the Newtonian Maupertuis against the Wolffians and freethinkers over mathematics, science, and religion. Sharpened by the lingering anger among the Wolffians over the monad prize decision, it bore characteristics of the dispute between Newton and Leibniz over priority for calculus as well as the rifts between Cartesians and Newtonians at the Paris Academy—rifts that Maupertuis knew well. The whole was driven by the character of König and the personality of the academy's autocratic president, who was at once arrogant and fragile in self-confidence. Beyond a personal insult, the quarrel involved the

Figure 10.3. Johann Samuel König.

competition between the renovated science academy and the Prussian universities, along with questions of the literary style of the sciences, the principal source of their authority, and their status in higher education, which presaged Kant's work on the strife among faculties. For Euler it was the validity of Maupertuis's mechanics, not religion, that remained the most critical element in the controversy. In light of the flood of books, anonymous journal articles, pamphlets, and reports in gazettes and newspapers in the Dutch, French, and German-language presses published about the affair, the struggle for freedom of the press and of the republic of letters from censorship does not appear to be a major factor. Neither the principle of least action nor its author challenged a religious dogma.

With its counterclaims that rejected the generality of the principle of least action and of Maupertuis's priority for it, König's "De universali principio aequilibri et motus" came as a blow to his pride and personal integrity. The paper found the antecedents of the principle of least action in Aristotle's axiom that nature in its operations seeks the simplest solution. It considered associated work with minima in antiquity by Euclid and Hero in optics and Ptolemy in geography, together with the development of the Aristotelian principle in modern times by Descartes, Nicholas Engelhard, Fermat, Malebranche, Willem 'sGravesande, Leibniz, and

Wolff in mechanics and optics. Citing the work of these men brought the accusation that Maupertuis had plagiarized from them and other sources. For a new member of the academy to attack the president's research so blatantly was, as Euler wrote to young Merian, audacious and provocative.[36]

Since König had been his friend, Euler initially remained out of the debate; at first he was seen as an impartial judge. Shortly after, Euler and Merian joined with Maupertuis and planned a rebuttal. Even though Euler himself deserved priority for the principle, he supported Maupertuis. A theorem on trajectories of bodies attracted by central forces in Euler's *Methodus inveniendi lineas curvas maximi minimive proprietate gaudentes* (A method of finding curves that show some property of maximum or minimum, E65) beautifully applied Maupertuis's principle: an element of the curve multiplied by the speed always gives a minimum. The principle had already appeared in the second appendix on elastic curves of the *Methodus inveniendi*, stating, "Since all natural effects follow some law of maximum or minimum, no doubt in curves described by projectiles attracted by forces, there exists some property of maximum or minimum. . . . [N]othing at all takes place in the universe in which some rule of maximum or minimum does not appear."[37] Still, Euler's defense of Maupertuis was unfaltering. Lacking the mathematical skills to formulate the differential equations for his law, Maupertuis asked the Bernoullis to prepare the scientific arguments for its generality and the differential equations for its applications, but they declined. In a series of articles in the academy *Mémoires*, Euler did both himself.

Gaining the assistance of Frederick II in the king's role as protector of the academy, Maupertuis turned the complete force of the institution against König. Considering the dispute to be about power and prestige as much as science, Maupertuis found the article an assault not only on his achievements and honor but on the rising reputation of the royal academy in Western Europe. He wanted to indict König as a liar and a fanatic and "to bury [him] in the mud as he deserves."[38] The academy split over Maupertuis's action, while many in the larger republic of letters would consider the procedures and judgment of the institution an abuse of power and an injustice. Yet by 1751 Maupertuis had displayed the good along with the bad at the academy. He had assembled as foreign members more than eighty of Europe's most distinguished scholars. A free institution, the Berlin Academy also opposed intolerance on the part of local churches at a time when Frederick's Potsdam court was a haven for French refugee scholars.[39] In Paris Denis Diderot, the coeditor of the *Encyclopédie*, praised Maupertuis, calling him "the rival of Voltaire."

Maupertuis was writing a treatise, *Lettres sur le progrés des sciences* (which would be published shortly after his *Cosmologie* in 1752), not unlike Francis Bacon's *Advancement of Learning*, that urged navigational exploration, geographic measurements, a new round of experiments using the latest concepts, the development of new equipment, the collection of curiosa in anthropology, and the formation of a global college of savants to pool and disseminate knowledge. His labors and his associations defined him as a friend to intellectual freedom.

In the years since the monad prize decision, attacks on Maupertuis had subsided. Now in 1751 many Wolffian philosophers in the universities, along with German and Dutch journalists, who had renewed their criticism of Maupertuis after the publication of his *Essai de cosmologie*, began buffeting him; they challenged the originality of his new law and often condemned his autocratic management of the Berlin Academy. In two replies König rejected what he termed the tyranny of the academy; the public alone, he deemed, was "his natural judge." Besides many in the German press, university faculty members—especially those in the influential circle of Johann Christoph Gottsched, a professor of logic and metaphysics at the University of Leipzig—supported König. Those in the Gottsched circle treated Maupertuis with hostility, particularly for his haughty supervision of the Berlin Academy, and they questioned the universality of the principle of least action. Modeled on the French Academy, they spread Enlightenment thought in the German states. Gottsched, who had translated Pierre Bayle's *Dictionnaire historique et critique* (Historical and critical dictionary) into German, attributed the principle to Maupertuis's misunderstanding of Leibniz's dynamics. The circle disseminated ideas in its monthly journal *Das Neueste aus der anmutigen Gelehrsamkeit* (The latest from engaging scholarship).

Since Gottsched wrote in German, Maupertuis asked his ally Abraham Gotthelf Kästner to translate the articles of Gottsched and others on the topic and send them to him; prior to this he had not known the extent of the dispute. In 1749 Kästner had praised his work on the principle of least action, and Maupertuis had unsuccessfully sought to bring him to Berlin. In June 1751, he advised Maupertuis not to respond to König, for it would be as if "a general of an army wanted to fight a duel with some vagabond volunteer."[40] Like Gottsched, Kästner was in Leipzig, and he was a good source of information from that city, and he was one of the few German geometers and philosophers to remain on good terms with Maupertuis.

Outside of Prussia the response was not so negative. In 1751 in the entry "Action" in the *Encyclopédie*, d'Alembert recognized Maupertuis's priority and endorsed his claim to the universality of the principle of least action.[41] He traced the law to the reading of a paper by Maupertuis at the Paris Academy in April 1744, which later appeared in print as "Accord du différentes loix de nature," and a second article, "Le principe universal du repos et de l'équilibre." D'Alembert saw Euler's *Methodus inveniendi* as a possible challenge to Maupertuis's priority, but he defended Maupertuis's claim on the grounds that his Paris Academy speech had preceded his article's publication. Maupertuis's studies of collisions, equilibrium, and refraction, d'Alembert pointed out, were essential in reaching his principle. D'Alembert declared that on refraction Maupertuis agreed with Newton, whose explanation differed from that of Fermat and Leibniz in having the rays approach the perpendicular. In defining the quantity of action, rather than multiplying time by speed to obtain space covered (as was normally done), d'Alembert showed that Maupertuis presented an inverse relationship. Where E is space traversed in the initial setting of speed V, and e the space traversed in the second with speed v, $E/V + e/v$ = a minimum—that is, $dE/V + de/v = 0$. Supposing $VdE + vde = 0$ gives $E\,V + e\,v$ = a minimum, which is Maupertuis's principle. It agreed with Newton's mechanics. Maupertuis did not stop here, d'Alembert observed, but sought to reconcile the fundamental truths of mechanics with final causes in metaphysics.

In October 1751, Maupertuis brought König's article to the attention of the full Berlin Academy and took measures to bring the affair to a close. Euler had already, in a letter in December 1745, accepted Maupertuis's priority. Even though he could not yet rigorously show the general application of the principle of least action, he was convinced of its truth and importance; Euler's own work on elastic curves in the *Methodus inveniendi* flowed naturally from it. After this letter, he consistently backed Maupertuis. He agreed with the Wolffians that the fundamental principles of mathematics and physics were founded in metaphysics, but "from which . . . that of Leibniz and Wolff is still far off."[42] Presented to the Berlin Academy in 1751 and published in its *Mémoires* for 1751, Euler elegantly established in "Sur le principe de la moindre action" (On the principle of least action, E198) the universality of Maupertuis's principle by demonstrating its application to refraction, reflection, the orbits of planets, and many other natural phenomena. He showed that it facilitated solving other problems of natural phenomena besides mechanics.

With Euler having addressed and resolved the content of the related science, Maupertuis still saw little chance of success in disproving Gottsched and journalistic critics. Recognizing no great advantage in admonishing the often anonymous gazetteers, another task that Euler would undertake, Maupertuis decided to concentrate on the authenticity of

Leibniz's letter. On 7 October, he and the academy had Formey write to König requiring him to produce the original letter by Leibniz within four weeks; correspondence, he knew, had been crucial in the calculus priority dispute between Newton and Leibniz. In the mid-eighteenth century, the standard for resolving priority in the sciences was publications or the testimony from reliable letters. In the spring of 1745 König had obtained several missives written by Leibniz, but it was not known whether he possessed the one in question. On 10 December, König responded from the Hague explaining his search for the full letter, his infinite respect for Maupertuis and the academy, and his chagrin at the controversy. He good-naturedly volunteered to send his copy of the fragment. After Henzi's execution in July 1749 his estate, including his collection of papers, had been dispersed, so König could not locate the original. The date 1707 for the letter posed a problem, for it did not match the time of the correspondence between Leibniz and Hermann. A meeting of the academy on 9 December 1751 discussed the affair; at the next, on 23 December, the beginning of its Christmas break, Maupertuis again demanded from König proof of the authenticity of at least a portion of the letter written by Leibniz's hand.

To increase the search for it, the sullen Maupertuis had the academy investigate the letter's existence. Since König had impugned the honor of the royal academy, the president asked Frederick II, who strongly supported him, to have his judiciary request the regent of Bern, the Marquis d'Argenson, and the magistrates of Basel, who had the Hermann papers, to seek the alleged letter by Leibniz; that Bern housed the remainder of Henzi's collection made the city the most likely location for it. Through early 1752 Maupertuis corresponded with Johann II Bernoulli, who headed the commission set up by the city magistrates that sought extensively to locate in Basel any document that might destroy König's position; Hermann's brother also belonged to this commission. Initially Maupertuis asked that the search be kept secret "until we are in a position to close off all his escape routes";[43] he still wanted Bernoulli to lead in championing his cause. To d'Alembert he complained that he also opposed König's article, for it contradicted his *Essai de cosmologie*.[44] But Euler's correspondence with Maupertuis from late 1751 to the winter of 1752, when the subject was most important to the president, mentions the affair only twice, in March and April 1752. At the same time that Maupertuis and Johann II Bernoulli wrote to each other, Maupertuis likely attempted to get a larger involvement from Euler in the affair, for Euler had attended all the academy sessions and knew the oral arguments and written evidence submitted. Some gaps in the extant letters suggest that the two men may have exchanged other letters before March that have since been lost. The debate would continue in 1752 even more intensely.

Two Camps, Problems, and Inventions

At the academy, other skirmishes between Euler and the Wolffians added to the debate over monads and the contentious Maupertuis-König affair. In Prussia the traditional metaphysics advocated by followers of Wolff was collapsing; in the 1750s the subject was giving way to the sciences in academy prize competitions. In particular, Wolff's special metaphysics consisting of psychology, cosmology, and rational or natural theology had been largely displaced. As this part of Wolff's thought was losing favor in Berlin, the philosophy department in 1749 selected determinism as the prize topic for 1751. This went against Leibniz and Wolff insofar as religious critics held that the preestablished harmony underlay determinism. For years Wolff had strenuously argued against this interpretation; the preestablished harmony did provide a connection between mind and matter, he believed, but it did not present all of the interactions. As criticism of Leibnizian philosophy grew in Berlin, even many who wanted its demise opposed a confrontation. Throughout the 1750s the Berlin Academy remained divided into two camps: Formey, assisted by Johann Philipp Heinius, Philippe Joseph de Jariges, and Sulzer, supported Wolffian metaphysics; Euler, along with Maupertuis, Merian, and André P. de Premontval, another Basler, opposed them vocally and in writing. Tensions persisted between the two parties. The concept of a preestablished harmony remained essential to the Leibnizian and Wolffian philosophies; Euler again rejected that concept, perceiving it as a determinist contradiction dangerous to the fundamental Christian tenet bearing on original sin. D'Alembert added that the topic raised the question of whether humans possessed freedom. Kästner, Maupertuis's friend and an opponent of Formey, had since 1745 regularly sent letters to Euler on such topics as Newton's dynamics, the pulse theory of optics, and the theory of colors. He won the academy's annual prize for 1751 for his anti-Wolffian article "Les evénemens de la bonne et de la mauvaise fortune dependent uniquement de la volonté, ou du la permission de Dieu" (Outcomes of good and bad fortune depending uniquely on the will or the permission of God). This received a sharp rebuke from Formey.

In 1751 Euler completed the manuscript for what has been referred to in this book as his second lunar theory, and he produced eighteen articles, including a proof of Fermat's Little Theorem of integral numbers as the sums of squares, which appeared in 1760, as well as another paper on the inequalities of Jupiter and Saturn. He continued to pursue the partition of numbers, introducing that subject in two articles in volume 13 of the Saint Petersburg *Commentarii* for 1751.

The first, "Observationes analyticae variae de combinationibus" (Various analytical observations about combinations, E158), answered the question that Philipp Naudé had posed in 1741: How many ways can 50 be summed from seven different integers?[45] It is possible that Euler rediscovered the tool of generating functions, formal power series in one indeterminate and coefficients encoding information about sequences of numbers, and began to develop comparisons with them to apply to the partition of numbers. He claimed that "a most perfect solution" for the Naudé problem is the number 522. He did not yet have rigorous proofs for this work.

Euler's other related article for 1751,"Découverte d'une loi tout extraordinaire des nombres, par rapport à la somme de leurs diviseurs" (Discovery of an extraordinary law of numbers in relation to the sum of their divisors, E175), still pursued but did not yet achieve proofs for his partition of numbers, and it ended with one of his chief discoveries in number theory, the pentagonal number theorem. Rather than having a random treatment of products and series, Euler found the law's simple recursion formula for a partition that, he commented, "is quite certain, although I cannot prove it."[46] Extending the concept of triangular and square numbers, pentagonal numbers are figurate numbers of which the points form pentagonal arrays. This article gave a version employing divisor theorems relating to quadratic reciprocity (QR). The later QR theorem is often called the most beautiful in mathematics, permitting one to determine when a number n is a perfect square modulo p. The number n is a perfect square modulo p if there is some perfect square that, when divided by p, leaves a remainder of n. An example for modulo in Carl Friedrich Gauss's notation is $5^2 = 25 \equiv 1(12)$—that is, 1 modulo 12, or $25 - 1 = 24$ and $24/12 = 2$. Euler's article gave conjectures developed in his correspondence with Goldbach from 1742 onward but did not explicitly state or prove the quadratic reciprocity theorem. He later proved that if p is a prime, exactly half the numbers smaller than p will be perfect squares modulo p or quadratic residues; otherwise half are nonresidues.

In 1750 Euler completed a paper titled "De partitione numerorum" (On the partition of numbers, E191) that appeared in volume 3 of the *Novi Commentarii* for 1753. In it he introduced superscripts for various partitions and strengthened his command of generating functions. This article was not his last on the topic.

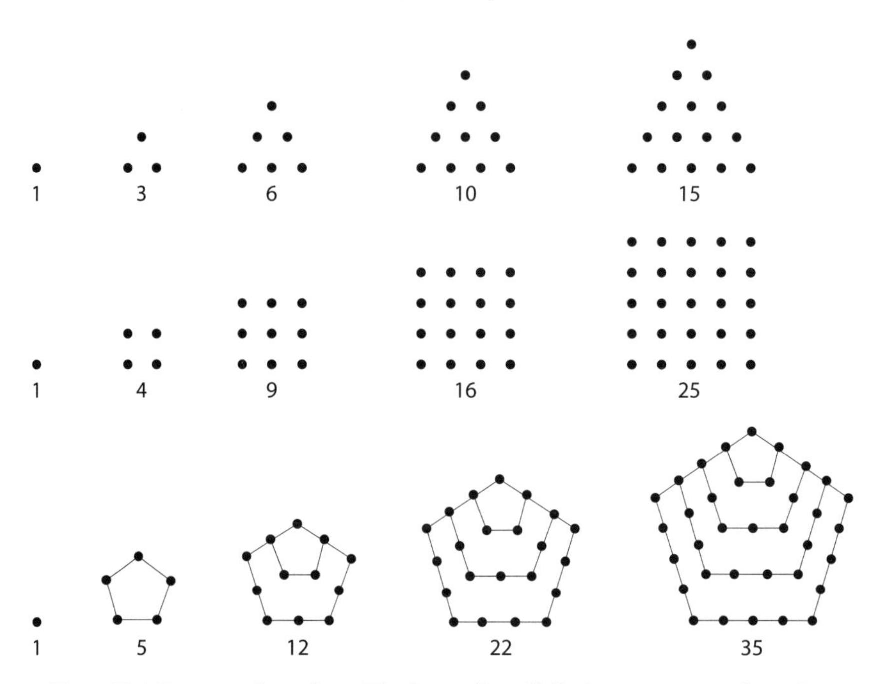

Figure 10.4. Pentagonal numbers. The form $n(3n - 1)/2$ gives a pentagonal number; these begin with 1, 5, 12, 22, 35, 51, ... Each is one-third of a triangular number.

Expanding the approach of his contemporaries and, not unlike David Hilbert later, Euler set out in August 1751, in his first draft of "Questiones physicae," a list of physical problems for mathematicians and physicists at the Petersburg Academy to solve. He wanted these questions to encourage and direct research for several years, and he advised that some be selected as prize problems. This first draft, in a letter to Schumacher, gave seventeen problems; in November Euler added one more. This was his second list of problems; his first, containing seven problems, he had read in Berlin in 1742. In November 1755, Euler would send a second draft of the eighteen problems, but they were not read to the Petersburg Academy until July 1757.[47] Among their topics were the cause of the fluidity of water and a physical explanation for why metal dissolves in certain materials and from others is precipitated. Other questions inquired about the shape of snow, the physical cause of the aurora borealis, and the cause of the variation of the magnetic needle in response to location.

In the intricate debate over a general solution for the vibrating string problem, the relationship between Euler and d'Alembert worsened. The circulation of Euler's results in 1751 provoked a verbose controversy. In solving that problem with arbitrary functions, the two differed in their points of departure. D'Alembert required one initial shape in equilibrium and continuity or smoothness among all functions in calculus. This allowed him to provide with his differential equations classical solutions to the wave equation. A shifting of the debate from the physical problem to the nature of function gave a greater advantage to Euler, whose approach to the physical problem was powerful. Boldly for his time, he proposed discontinuous functions with piecewise continuity or what in the twentieth century would be called the problem's weak solution. Most leading mathematicians, including Daniel Bernoulli and Joseph-Louis, Comte de Lagrange, soon entered this contest with its new class of functions. Even after most mathematicians accepted Euler's alternate solution, d'Alembert tenaciously continued the debate over the vibrating string problem. It lasted over twenty years; not until shortly before his death would d'Alembert admit defeat.[48]

On 23 December 1751 Euler received for review a copy of the collected papers of Count Giulio Carlo de' Toschi di Fagnano, who had made the initial analysis of elliptic integrals. Fagnano's best work came in three papers on the topic reprinted the previous year in the three-volume *Produzioni matematiche*, the first of which dated to 1718. Proceeding in a modern style of theorem and proof with novel methods, notations, algorithms, and analytic transformations, Fagnano gave many new results on elliptic integrals that included what is now called the addition formula for arc lengths of ellipses and "trisected" the lemniscates, curves in the form of the figure 8, with the two parts symmetrical. The symbol for infinity, ∞, is an example. Since Fagnano had applied for membership in the Berlin Academy, which he would later win, it fell to Euler to be his reviewer. Both men recognized that these elliptic integrals could express the arc lengths of the ellipse, cycloid, hyperbola, and lemniscate. Carl Gustav Jacob Jacobi called the day of its delivery the birthday of elliptic functions. For over forty years, beginning in 1727, Euler worked on the theory and the application of these integrals to diverse subjects, including the oscillation of pendulums of large amplitude and his initiating of the theory of mathematical elasticity, this latter particularly in a letter in December 1738 to Johann I (Jean I) Bernoulli. Among other problems that he now also addressed were the measurement of Earth and the three-body problem in astronomy.

At the time that Euler was developing his classical theory of fluid mechanics, he was pleased to find practical applications—in this case in studies of hydraulic machines by Jan Andrej (Johann Andreas) von Segner, who was then at the University of Göttingen. Building upon Daniel Bernoulli's reaction effect, Segner produced a simple waterwheel. In it water flowed through a cylinder, having at its low point several horizontal paddles bent in one direction; as a result, counterpressure forced a turning of the cylinder in the opposite direction. Euler's articles built upon this base to describe crude water turbines theoretically and to accurately compute their efficiency; he showed that for good efficiency very high speeds were necessary. Euler was the first to explain clearly how material disturbance caused a weakening in streams, which was crucial in the construction of water turbines, and he would provide the foundation for the modern theory of turbines.

Botany and Maps

Among Euler's many tasks at the academy in 1751 were overseeing the progress of the observatory, which was nearing completion; seeking the publication for subscribers in Berlin of *Uranographia Britannica*, an astronomical atlas by the English physician and amateur astronomer John Bevis;[49] requesting copies of Edmond Halley's publications on trade winds and magnetic needles; ordering plants; and preparing and distributing for sale the academy's almanacs, calendars, and geographical maps, still the main source of its funding. In these and many other scientific tasks, his correspondence with Johann Kaspar Wettstein—who though not a geometer or natural philosopher displayed a wide knowledge of the sciences—was important. Information in the correspondence flowed both ways. Euler was able to provide the British scientific world with firsthand information on the Continental academies. The *Uranographia* was finished by 1750, but the publisher went bankrupt; it was not printed until 1786.

In a speech Euler branded botany as an unimportant trifle. He gave it only limited support, considering it last in his department, and proclaimed that for solving problems only mathematics in principle deserved financial support. He had little to do with Johann Gottlieb Gleditsch, the guiding spirit and director of the botanical gardens and the small mulberry plantation outside Berlin. With the academy still short on funds, Euler found ten thousand thaler to pay for having the plantation walls repaired and through Wettstein ordered mulberry seeds and 220 bushes from Virginia and Canada; he prized the Virginia bushes for having the largest leaves. In June Euler recognized that these plants were not available in London but might be found in France. He needed them early in the planting season, and they would be grafted the next year with older, local plants. The garden and plantation became popular sites for visitors. To line the avenues around the academy, Euler ordered a dozen kinds of trees, including two types of tulip trees, an ash, a cedar, and a Virginia cypress. He later requested exotic plants, some of which perished, mainly from exposure during transportation. But two magnolias from Virginia and most of the fruit trees survived. In July Euler's comments to Wettstein on the start of the planting season and in October on moving mulberry bushes near the end of the season suggest an interest in plant science. In carrying out his many tasks at the academy, Euler had to work with accountants, builders, carpenters, gardeners, printers, tax collectors, and technicians.

His correspondence with Wettstein deals often with Euler's desire for tobacco, his preference being for the Virginia leaf. In April 1751 his supply ran out, and in June he was reduced to smoking inferior types. Euler had earlier loaned to Grischow, then at the Imperial Academy in Saint Petersburg, some tobacco and books intended for his own use. Once Grischow received his salary, he quickly repaid the loan. In July Euler requested twelve pounds of tobacco. Since the price was uncertain, he recommended that it be paid out of income from books sent to Wettstein for sale. In October he expressed gratitude that his supply of tobacco had arrived.

From Wettstein in England Euler requested the new maps of England, Scotland, and North America, especially of discoveries around Hudson Bay for Maupertuis and for the preface that Euler himself was writing for the *Atlas Geographicus: omnes orbis terrarum regiones in XLI tabulas exhibens* (Geographic atlas: Representing in forty-one maps all regions of the Earth, E205), a school atlas that would ultimately be quite successful. (In 1752 Euler would convince Johann Christoph Rhode from Magdeburg to visit the academy and had his guest named its geographer. Among Rhode's earliest projects was helping to prepare the school atlas, the initial edition of which did not appear until 1753.) In Berlin the run of the *Philosophical Transactions* of Britain's Royal Society was incomplete, so Euler asked for a missing issue, number 433 from 1734, and Halley's astronomical tables. By November 1751 the plants, maps, and *Philosophical Transactions* issue had arrived in good condition. These items were so inexpensive that Euler expressed gratitude to Wettstein, who was paid twenty-eight pounds sterling; he also sent Wettstein sixty copies of the Berlin Academy's genealogical almanac in French and two in German for sale. Copies of the astronomical almanac in Latin, which included the names of all academicians, were not available.

The correspondence with Wettstein for 1752 addressed astronomy, astrology, alchemy, sale of the almanac, the Bible, cosmology, the ether, longitude, magnetism, medicine, and the Royal Society of London. That year Euler observed that John Dollond had objected to his theory of refraction. Dollond's criticism of Euler's theory arose from its opposition to Newton's; Dollond did not address the question of the eye's construction. In addition, it was typical of Euler to reject arguments based only on one past authority, even a revision of Newton. He recognized that astrology was recently in greater fashion than astronomy in England, and he considered specious the alchemical art of making gold. Euler thought that Newton was independently a great chemist. For his transmission of significant information, Euler happily informed Wettstein in February 1752 that the Berlin Academy wanted to name him a foreign associate, and in April, on behalf of academy president Maupertuis, Euler conveyed to Wettstein the academy's awarding him that post.[50] The Wettstein letters and his scientific correspondence made Euler a modern Marin Mersenne, a center for scientific correspondence.

The Maupertuis-König Affair: The Late Second and Early Third Phases

By late February 1752, investigations regarding the first formulation of the principle of least action were ending in the fruitless search in Bern and Basel for the missing letter by Leibniz that Frederick's judiciary had requested to help settle the dispute between Maupertuis and König. While initially considered impartial and best able to make a scientific judgment, Euler did not wish to enter another controversy at the same time as his disputes with d'Alembert over the problem of vibrating strings and fluid mechanics and with Clairaut over lunar motion. Maupertuis still most wanted Johann II Bernoulli to be his chief defender. Had the letter in question been sent to Hermann in Basel, as some thought, its information surely would have been forwarded to his close friend Johann I Bernoulli. After a thorough search, Johann II Bernoulli failed to uncover it, but found three other authentic letters. One on action and force by Leibniz, Bernoulli pointed out, did not mention the principle. The younger Bernoulli rejected the idea that Hermann had received news of this principle earlier and failed to mention it. He discounted the possibility that Hermann would not have informed Johann I Bernoulli of such an important discovery. Asked to decipher these letters, Euler admitted having problems. They, like Leibniz's letters in Hanover, had several words whose meanings Euler did not know, signs within the text indicating placement of marginal notes, and hastily drawn lines through a line of text either to indicate where the materials belonged or to indicate potential deletions, but often neither was clear. The lines drawn completely through a line of the text Euler took to be attempts to scratch them out. In 1707 the word *actio* might refer to the influence of one body upon another, Leibniz's a priori measurement of force, or the product of quantity of matter, speed, and path. By March 1752 Euler, it appears, had still not decided against König's claim.

The desire to locate the missing letter by Leibniz remained intense. Probably at Euler's request, Maupertuis asked Kästner to search through the manuscripts in the collection at the University of Leipzig library. On 31 March Euler's ideas about the alleged letter were evolving, and he wrote to Maupertuis about deciphering difficulties. His reference to the Leibniz letter suggests that he thought that it might be handwritten by that author. But from the confusing interjections Euler deduced a second interpretation: that the fragment was counterfeit. The silence of Hermann on the principle in subsequent letters to Johann I Bernoulli reinforced that argument. Euler and Maupertuis may have had preceding exchanges on the subject that are now lost. In May 1752 Maupertuis gave a far different account of Euler's analysis of the letter fragment. They had, he wrote, "been able to decipher very well Leibniz's original letter despite his extremely ridiculous style of writing and had managed to find some new evidence against König. The letter is shown to depend entirely on action and force, and it does not mention the principle that König attributes to Leibniz."[51] Johann II Bernoulli made the latter point. On 5 April Kästner had reported failing to discover in Leipzig even one letter from Leibniz to Hermann, much less the desired letter. At this time König increased his attacks on Maupertuis.

With the academy investigations now completed, Maupertuis called an extraordinary meeting for 13 April 1752. Two noble curators—the Pomeranian Peter Christoph Karl von Keith and Count Sigismund Ehrenreich von Redern—led it, as happened only on the most important occasions. Present were three directors, thirteen ordinary members, one foreign associate, three guests, and the perpetual secretary Formey, together with two honorary members, the king's sister, and three brothers. Maupertuis intended not to be present. Keith read a brief letter from the president, giving as reason for his absence that he was ill as well as being the subject for the gathering. For this meeting, Maupertuis wanted Euler to be the decisive authority. The academy report on the possible authenticity of the Leibniz letter included an exhibition of results of all official searches initiated by Frederick and also by Formey, recapitulating König's responses to Formey in February and to Maupertuis in March. Euler read "Exposé concernant l'examen de la lettre M. de Leibnitz, allegué par M. le Prof. Koenig, dans le mois de mars 1751 des Actes de Leipzig, à l'occasion du principe de la moindre action" (Exposé concerning the examination of the letter of Leibnitz, as alleged by Prof. Koenig in the March, 1751 issue of the Acts of Leipzig on the occasion of his publication of the Principle of Least Action, E176, later reprinted in the *Jugement de l'académie royale des Sciences de Prusse sur une lettre prétendue de M. de Leibnitz*).[52] To be sure that the German-speaking members not fluent in French understood him, he spoke in Latin. The *Jugement* would be published in June in Berlin and in the *Journal des Sçavans* in Paris.

Since March Euler's position in the debate may have hardened. The *Jugement* gives more attention to the fragment, and the charges reflect more the conceit of Maupertuis. König had weakened his case by providing several improbable explanations to account for the law and the loss of the letter. Absent proof of the letter in Leibniz's hand in the surviving estate of Henzi and in Basel and Leipzig, the *Jugement* declared that

> il est assurément manifeste que sa cause est des plus mauvaise, et que ce fragment été forgé, ou pour faire tort à M. de Maupertuis, ou pour exagérer, comme par un fraude pieuse, les louanges du grand Leibniz.

> there are manifest assurances that its cause [or source] is more evil and that this fragment has been forged, either to malign Maupertuis or by a pious fraud to exaggerate praise for the great Leibniz.[53]

But the reputation of Leibniz surpassed any need for such inflation. Though the *Jugement* refers to Henzi, it intimates that König was attempting to damage the reputation of the president. Its severity also suggests that another—Maupertuis—may have been involved in writing it. Two internal matters damaged support for this institutional condemnation. Sulzer asked to speak in defense of König but was silenced; at the academy Maupertuis held all authority, and members were not allowed to criticize him. This created a strong, largely secret bitterness that did great damage to the academy. For the April meeting, thirteen ordinary academicians opposed to the autocratic management of Maupertuis and possibly upset at being pressured to condemn König did not attend. Those members present voted unanimously for the judgment, voicing their support and signing the document. Absence was dangerous, for members depended entirely upon Maupertuis for their salaries. In addition, they feared the disfavor of the king.

The academy victory was not to be sweet for Maupertuis and Euler. Largely through his harshness and stubbornness together with hostility from the press, Wolffian philosophers, and Voltaire, Maupertuis was to lose in the public arena what he had won in the scientific. The second edition of Maupertuis's *Cosmologie* and the academy's condemnation unleashed a wave of criticism. Opponents within the academy and the presses in Prussia, Holland, and France denounced the judgment. Had the president misused his power? What had principally been an internal dispute in the academy quickly shifted to the public sphere.

König protested the judgment of forgery and asked the academy to reverse it. At its meeting on 8 June the body reviewed the matter and reaffirmed its decision. König now saw little chance of overturning the judgment.

During the final quarter of 1752 the lives of three eminent men—Euler, Frederick I, and Voltaire—would intersect in one of the most celebrated intellectual incidents in history.

On 13 September 1752 the third act of the affair commenced when König angrily returned to the Berlin Academy his diploma of foreign membership and issued his "Appel au Public du jugement de l'Académie Royale de Berlin sur un fragment de lettre M. de Leibnitz, cité par M. König," holding that the fragment—while not written in Leibniz's hand—contained the content and form with markings indicating the authenticity of the original. He cited in full four letters from Leibniz that he had purchased from Henzi, rejected the related articles of Euler on the principle of least action in general, and denounced the tyranny of the Royal Prussian Academy. In his "Appel au Public" König wrote in the third person,

> Il ne reconnait aucun Superieur, aucun juge particulier. Le Public seul est son juge naturel.

> He recognized no Superior or particular private judge. The public alone is his natural judge.[54]

The reading public was largely the republic of scholars. Mainly abandoning the Leibniz letter, König repeated that the genesis of the principle of least action stretched from Aristotle to Wolff, and he claimed to have found errors in Euler's determination of equilibrium. In the scientific community, the "Appel au Public" gained him much sympathy. Voltaire, who was about to enter the fray, wrote to his niece that the persecutor of König in Berlin was already considered an "absurd tyrant," and public opinion would quickly shift even more in that direction.

Responding principally to König's attempts to rebut the judgment of the academy and press criticisms, Euler and Merian accepted it as an obligation to make a strong defense of Maupertuis. In September Euler read copies that Merian had given him of the Hamburg *Freye Urtheile* and a Leipzig literary gazette. Both supported König and were against the academy's position. They found the principle of least action in predecessors of

Maupertuis, including Wolff. In his detailed history of the affair, *Mémoires pour server à l'histoire du jugement de l'Académie* (1752), Merian declared that the academy, which had adopted "the discovery of Maupertuis and the geometrical applications of Euler," had the authority to defend the "reputations of its great men . . . and publicly repulse injuries done to them."[55] He disagreed that this quarrel represented a setback to freedom of the press, as critics claimed. His position appears to reflect not elitism but the question of competency. Qualified scholars, not the general reading public or university philosophers, had to set high standards in the sciences. Merian viewed the polemic as a break from the rational and civil discussion that was sought in the sciences.

Even after it was long apparent that no final decision was possible, Euler continued his effort to resolve the König-Maupertuis affair scientifically. In November Merian read to the academy a long letter translated from Latin that Euler had sent in September, "Lettre de M. Euler à M. Merian" (E182) also published as the *Dissertation sur le principe de la moindre action* (Dissertation on the principle of least action, E186). Following upon König, the letter investigated related findings from the ancients (among them Aristotle, along with Euclid and Hero in catoptrics and optics and Ptolemy in geography) through the moderns, including how Descartes, Fermat, 'sGravesande, Engelhard, Leibniz, and Wolff had responded to Aristotle's axiom. In the letter Euler held that the conservation of *vis viva* was not a form of the ancient principle of equilibrium in a lever; he explained that 'sGravesande's law for elastic bodies was correct and that Descartes and Fermat, who differed on refraction, dealt only with light waves, to which Maupertuis had not limited himself. For curves of elastic bands, or *elastica*, Daniel Bernoulli had in strikingly singular research affirmed the verity of the extremum principle in demonstrating, Euler argued, that "the entire force stored in the curved elastic band may be expressed by a certain formula, which he calls the potential force, and that this expression must be a minimum" for these elastic curves.[56] In equilibrium the academicians found none of Leibniz's living forces. In arriving at his more general law, Maupertuis had not drawn simply upon others. Preceding writings in mechanics, including Leibniz's, showed not the slightest trace of this principle. Neither Leibniz's articles in *Acta eruditorum* in the 1680s on resisting motion and planetary motion nor his two-part *Specimen dynamicum* on planetary motion, the first part of which appeared in 1695, approached the debated principle. Euler denied that Maupertuis's law was a corollary of Leibniz's law of the conservation of *vis viva* and rejected as absurd the notion that Maupertuis's law was derivative from others, which could imply that his president was a plagiarist.[57]

Euler also rejected the notion that it was chimerical or of lesser importance. The principle was not only beautiful, wrote Euler, but fertile in deriving curves and highly important for all of philosophy. It was thus not, as König called it, a poor chimera, while König's law of a nullity of living forces was not vital but sterile and disagreeable. Applying that law, geometers arrived through a number of steps at a definite minimum, and König's principle thus did not require that initial condition.[58]

In establishing in a number of articles the universality of Maupertuis's law in dynamics, Euler brought under one rule curves as small as those describing the motion of light. Extrema are axiomatic in nature, Euler contended, and through their application to elastic and inelastic collisions by the derivation of differential equations natural phenomena could be explained by final causes, as Maupertuis's metaphysics held. In his double method, maxima and minima were equally satisfactory to discover final as well as efficient causes. In this larger quarrel, Euler surmised, only obstinacy made König continue to hold his position.[59] Euler was proceeding to destroy the objections to the significance of the law that König considered invincible.

In the last paragraph of the preface to his *Dissertation* Euler replied to Voltaire's attack on Maupertuis, citing his lack of standing in the sciences and questioning his behavior in what should have been a rational scientific dispute, though Euler did not mention Voltaire by name: "There is another type of person, who wished to insinuate himself into this dispute without having the necessary knowledge; and as such he understood absolutely nothing. He chose to treat this situation by assumptions without science and without decency; it is certainly not for him that we have chosen to write."[60] While Voltaire was not ignorant in the sciences—his satirical *Diatribe du Docteur Akakia, Médecin du Pape* (Story of Doctor Akakia, physician to the pope) refers, for example, to hard body collisions, centrifugal forces, and the polar flattening of Earth—he still lacked the scientific competency essential to resolve the dispute.

In late December Euler lectured to the academy from his "Exposé concernant l'examen de la lettre M. de Leibnitz," probably composed in October, declaring that Maupertuis had persuasively shown his principle to be a general law of nature, applicable not solely to equilibrium but to motion as well.

Even after diligent searches the alleged fragment from Leibniz to Hermann had not been proven authentic; no evidence had been uncovered, a finding with which Merian agreed. Euler's commentary looked to other considerations as well. Would Leibniz, who had made major discoveries, have mentioned this particular discovery only to Hermann? He was, after all, a good friend of Johann I Bernoulli. In any event, Leibniz's living force was not equal to Maupertuis's estimate of action, which was essential for arriving at the principle of least action. The work of König, whose command of mechanics "was worthy only of contempt," did not compare with the discovery of the principle or with the other fruitful accomplishments of Maupertuis, the "illustrious president."[61] Euler found König's grasp of mechanics weak and confused in his work between dynamics and phoronomy or kinematics, which did not consider the causes of motion.[62] Euler had already expanded upon the scientific arguments that came in this "Exposé," but the expanded study would not see publication until the following year.

Writing to Schumacher at the end of December 1752, Euler made use of an epigram from the nearly centenarian Bernard le Bovier de Fontenelle to dismiss Maupertuis's antagonist Voltaire:

> Voltaire à de l'esprit, il est vray: Mais, mais, mais,
> Les mais à son régard ne finissement jamais.
>
> Voltaire has a fine mind, it is true: but, but, but
> The buts regarding him are endless.[63]

Voltaire's widely accepted depiction of Maupertuis and Euler as isolated and having no defenders was wrong. The vote at the academy on the judgment against König showed substantial support for them. In the face of the storm of hostility that the Berlin Academy encountered in the press and at universities, most of Europe's leading men of science, including Daniel Bernoulli, Krafft, Pieter van Musschenbroek, and d'Alembert, supported Maupertuis's priority and the importance of his principle. Bernoulli, who sympathized with Maupertuis after König's "Appel au Public," did not set aside his rejection of Maupertuis's laws of collision for rigid bodies.

In Berlin a strong defense of Leibniz and criticism of Euler remained through 1752. Mathias Roholfs, a Berlin calendar maker, had recourse to a literary form of frequent use among polemicists:

> Gedankloser Zeitvertreiber,
> A – x aequal plus minus – Schreiber
> Mit Recht ist Leibniz Dir verhast.
> Als trägt er schon, was Dir gebühre,
> Macht er die bitterste Satire
> die nur auch Dich vollkommen passt.
> Sein Rechenkusten sollte zeigen,
> Das Denken nicht sey Rechnern eigen

Und das ein mathematicus
Noch mehr also Euler wissen muss.

Thoughtless time killer,
A – x equals plus minus – writer
Rightly you hate Leibniz.
As if he has what you claim
He makes the bitterest satire
that completely matches you alone
His calculations should show
That thought is not only by reckoners
And that a mathematician
Still more thus Euler must know.[64]

Voltaire, a Newtonian, had helped Maupertuis defeat the Cartesians at the Paris Academy and after taking lessons from 'sGravesande in Leiden wrote *Eléments de la Philosophie de Newton* (Elements of Newton's philosophy, 1738). In August 1741 Voltaire thanked Maupertuis in a letter for sending an early draft of his *Essai de cosmologie*, which liberated him from depending upon Wolff's cosmology.[65] In *Elementa matheseos universae* Wolff agreed with Christiaan Huygens and Fontenelle that life existed on other planets. Comparing the intensity of light on Jupiter at different longitudes and latitudes with that of Paris, he calculated that Jovians may reach a height of 13 819/1440 Latin feet. In *Micromegas* (1752), Voltaire derided Wolff for his method and this measurement.

Even as the relations between Voltaire and Frederick II were souring, partly from Voltaire's efforts at being a secret agent for the French, the king again urgently invited him to Berlin after Voltaire's lover Émilie, Marquise du Châtelet, died. Voltaire returned to Berlin and to the royal residence Sanssouci in Potsdam in July 1750 as poet in residence; he negotiated favorable terms that included having the king advance sixteen thousand French pounds to pay for the trip, receiving rooms in the palaces in Berlin and Potsdam, taking supper daily with the king in Sanssouci, obtaining the gold key of a court chamberlain, and having a salary. Voltaire no longer considered Frederick a philosopher king, however, and the monarch called him a scoundrel.[66]

While dominating the intellectual circle around the king, consisting mainly of French learned refugees, Maupertuis lacked the urbanity of Voltaire. Brilliant and biting, Voltaire generally dominated conversation, devoted himself to producing plays of his own and others, and found allies in d'Argens and Algarotti. Although Voltaire spread hyperbole in praise of Frederick, it took only six months for tensions to arise between them.

Maupertuis, who was nearly at the top of the constellation of courtiers surrounding Frederick, would not allow Voltaire into this circle. In May Voltaire wrote to d'Alembert of the unsociability of Maupertuis,[67] who now also refused to follow Voltaire's advice to support the nomination of abbé Guillaume Raynal to the academy. This apparently widened the gulf between the two men; Maupertuis's failure in 1750 to block Raynal's election, which Frederick wanted, did not placate Voltaire. By November the break was beginning with Voltaire, who now wanted the presidency of the academy.[68]

More than Voltaire's acting as a secret agent, his ostentatious style of living at court and illegal attempts to increase his sizable fortune angered the king. When Frederick objected to his extravagances by ending the supply of coffee and sugar to him, Voltaire burned costly candles all day in his salon. His position with the king had most deteriorated when, five months after his arrival, Voltaire became involved in a scandal involving Saxon exchequer stock dealing that led to a nasty court case. In November 1750, Voltaire had the Berlin banker Abraham Hirschel procure in Dresden worthless *billets Saxons* or stocks from the Second Silesian War. On a false rumor that Frederick was planning to pay their full value, Voltaire sold them to Hirschel. Usually Jews could not take Gentiles to court, but Hirschel was a *Schutzjude* who enjoyed legal protection in Prussia. The lawsuit against Voltaire had discreditable charges and countercharges. When Maupertuis refused Voltaire's request to concoct an alibi to justify him in the stock swindle, Voltaire's developing hostility grew more profound. On the basis of possibly forged papers and a technicality, Voltaire won the case, but it damaged his reputation; it was a Pyrrhic victory: he won just two thousand thaler while becoming a laughingstock in Paris and Berlin.

Lessing, who translated the trial documents for the court, wrote that it was easy to see why Voltaire had won: he "war ein grosser Schelm als er" (was the bigger crook of the two). Voltaire demanded from Cocceji a certificate for the arrest of Hirschel but was refused; Hirschel was imprisoned briefly and quickly released.[69]

The furious Frederick wrote an epigram on Voltaire's avarice and briefly considered dismissing him in disgrace, but he could not yet be separated from his *beau genie*. In February the case against Hirschel ended with a final report from Voltaire to Cocceji. As Voltaire sank in Frederick's esteem, Maupertuis rose to greater prominence. Rather than conciliate him, Maupertuis protected Anglivie de La Beaumelle, an enemy of Voltaire.[70] There were, La Beaumelle wrote, greater poets than Voltaire but none better paid.[71] Angered, Voltaire began to seek an effective way to attack Maupertuis.

Planetary Perturbations and Mechanics

In the 1750s the prime astronomical consequence that a planet perturbs the motion of other planets solely in accordance with Newton's law of attraction had not yet affected the planetary tables; European astronomers and geometers had yet to confirm this result.[72] While the perturbation could not be computed for the inner planets since the relations of their masses to that of the sun were unknown, Newton had found the masses of the large planets Jupiter and Saturn by employing the distances and periods of their moons. In his *Principia Mathematica*, book 3, proposition 13, Newton gave the first of three values for the influence of Jupiter on Saturn's irregularities. Another matter was determining whether irregularities in the mean motions of the planets were cyclical or linear, beginning with the large planets Jupiter and Saturn. Since the time of Johannes Kepler and his three planetary laws astronomers had approximated the inequalities, and since the publication of the Rudolphine Tables in 1627, planetary tables had remained without improvement in their accuracy.

In 1752 the Paris Academy awarded its double prize to Euler for the article "Recherches sur les inégalites de Jupiter et de Saturne" (Research on the inequalities of Jupiter and Saturn, E384).[73] A major effect of Newton's theory of universal gravitation was that each planet must perturb the others' motions. In May Euler informed Goldbach of the Paris Academy's receiving his own submission. Euler's winning article examined astronomical observations made in 1598, 1686, 1716, and 1745, and made comparisons of their resulting planetary tables; it was a significant advance over another that he had written on the same subject in 1748. As computations became more complex, he had moved from geometrical manipulation to analysis. Among his chief innovations in the article was treating trigonometric values not as geometric lengths or diagrams but as ratios. He next used the trigonometric series to find approximations of closer values. To explain perturbations in the longitudes of both planets, Euler wrote of mutually induced changes in orbits and expunged the circular arcs that had accounted for these. His innovations included introducing periodic functions, irrational functions, and primitive statistical procedures to derive secular inequalities, eccentricities, and mean motions. He discovered secondary eccentricities. But the omission of crucial factors in his computational formulas based on Kepler's laws led to errors. His flawed algebra indicated that the mean motions of both Jupiter and Saturn were in slow acceleration with a period of 324,000 years; he could not explain that great inequality. Mainly for the methods employed, the Prix de Paris article for 1752 provided a successful pattern for research. Euler returned to the topic four years later, and for Johann Tobias Mayer and others the mathematical methods and computations were to prove fruitful. Euler was the first to determine the perturbations of the Earth's motion and systematically to compute them. The paper on Jupiter and Saturn was not published until 1769, by which time Lagrange had attempted to show that Jupiter was accelerating while Saturn was slowing. The turning point in determining that inequality did not come until the work of Pierre-Simon Laplace in 1785.

Euler's article "Decouverte d'un nouveau principe de Mecanique" (E177) was published in volume 6 of the *Mémoires* in 1752, two years after its completion. Finding insufficient the principles of mechanics then accepted for solving problems, Euler employed differential equations for the general motion of a rigid body, especially that of three dimensions. Using Newton's second law, Euler devised general equations and changed the face of mechanics, as he presented the so-called Newtonian equations in Cartesian rectangular coordinates. The approach fundamentally differed from those of his time; it was the first use of the Newtonian equations in print in a general form. Since he examined these problems in mechanics using volume elements, his means of solution could apply to any type of body. The article transformed and—to the present day—dominated the mechanics of extended bodies; few works have made comparable contributions to the field.

Music, Rameau, and Basel

In 1752 France's greatest eighteenth-century composer, Jean-Philippe Rameau, whose operas held sway over the nation's musical repertory, completed a brochure, the complicated *Nouvelles réflexions sur la démonstration du principe de l'harmonie*, attempting to explain all the fine arts and the *esprit géométrique* with the principle of the *corps sonore* (sonorous body) based on mathematical proportions.[74] The principle involved vibrating systems that emitted harmonic partials. First of all a scientific verity, it was the single unique principle of music underlying all rules governing musical practice. Rameau saw it as fundamental to all the arts and sciences. In April Rameau sent a copy to Euler and asked for comments on that treatise and on his *Démonstration du principe de l'harmonie* (1749–50), for which he claimed to have discovered the universal principle of *corps sonore*. For sources he looked to the ancient Pythagoreans, Descartes, and Malebranche, along with the experiments of Marin Mersenne and especially the acoustics of Joseph Sauveur on overtones, pitch, limits of human audibility, and the velocity of sound. Rameau developed a mathematical model of music that built on them and was neither lifeless nor mechanistic; its theory complemented the judgment of the ear.

Yet Rameau possessed only an elementary knowledge of mathematics, primarily of ratios and numbers, and a faulty method that he applied arbitrarily. He had submitted his *Démonstration* to the Paris Academy, since he considered it a scientific work, but in 1699 Fontenelle, the perpetual secretary of that academy, had argued that for bringing order, clarity, measure, and exactness to knowledge not only of the sciences but in all fields of learning the application of mathematics was essential. "A work on ethics, politics, criticism," he wrote, "will be better, other things being equal, if done by a geometer."[75] During the High Enlightenment, Rameau was one among many scholars adopting the *esprit géométrique*. His belief that mathematical rapports or relations existed in music concurred with efforts of natural philosophers to discover such connections among natural phenomena defined in mathematical terms.

While recognized as a master craftsman, Rameau sought praise as a musical genius and desired the title of *savant*, a higher appellation than *philosophe*, for making seminal contributions to learning. Rameau had already received acclaim from the *philosophes*, and for two of his ballet operas Voltaire had provided libretti.[76] D'Alembert called Rameau the

Figure 10.5. Jean-Philippe Rameau.

greatest musical genius of eighteenth-century France, and in *Rameau's Nephew* Denis Diderot, who had studied composition under him, praised him for delivering the French from the plainchant of Jean-Baptiste Lully that had "been droning out for over a hundred years" but added criticism of Rameau for writing "such reams of incomprehensible visions and apocalyptic truths on the theory of music, of which neither he nor anyone else ever understood a word."[77] Rameau sought to understand the principles and rules underlying the elaborate structure of music with roots embedded in theory; in his attempt to bring out the rational, mathematical, and scientific bases of music, he craved the approval of Johann II Bernoulli and Leonhard Euler more than he did flattery from the *philosophes*.

Not until fifteen years after Euler's *Tentamen novae theoriae musicae* was published would Rameau criticize it in his *Nouvelles reflexions*. During this time Euler had mostly stopped working on problems in musical theory. His belief that tones differing from each other by one or more octaves "are considered alike but are not the same" was ambiguous;[78] they were not harmonically identical. In "Extrait d'une repose à M. Euler sur l'identité des octaves" for 1753, Rameau concentrated on this statement, taking it to mean that these tones were not alike, a notion that he rejected.[79] Women singing in a choral group an octave higher than men, for example, seemed to be singing the same notes. But this was a fact that Euler knew. Rameau also disagreed with Euler's device of *exponens*, giving the degree of agreeableness or pleasantness for notes and chords by ratios of numbers, reducing the harmonic to a melodic relationship of lowest common multiples. The formula is $s - p + 1$, where p is the number of prime factors in the lowest common multiple and s the sum of the prime numbers. A major fifth consists of two notes with a 3:2 ratio, so $s = 3 + 2 = 5$ and $p = 2$. The *exponens* for it would be $s - p + 1 = 5 - 2 + 1 = 4$. Euler made tables of the degrees of agreeableness of chords. In his theory, the lower the degree, the greater the pleasure. Compared to the ancients, Euler was expanding the sphere of consonance. Although mathematically imposing, his musical theory had no place for the law of octaves and was nonfunctional; for the musical performance of its time it made no sense. Since Euler wanted his theoretical research in music to promote the improvement of musical creation and performance, to be understood by musical practitioners his later work would reduce to a minimum the level of mathematics involved.

In September 1752 Euler responded mildly and constructively to both of Rameau's writings that were sent to him. Both men based their work on the principle of harmony, which Euler asserted that Pythagoras had discovered, and he cited the contributions of the ancient Pythagoreans to music. He promised to show Rameau's letter to Frederick II at an opportune moment. A longer response from Euler, sometimes dated 1752, was more likely from 1754, after Rameau sent another letter.

During 1752 Euler is thought to have considered a return to Basel. Many threads still tied him to the city—especially his relations with Johann II and Daniel Bernoulli, and he had paved the way for many of his countrymen to come to Berlin and had helped them make progress there. In 1752 Euler reportedly remarked that if he had ten thousand thaler he would purchase a large Swiss country house near Basel. In June he asked mayor S. Merian of Basel to grant citizenship for his wife, three sons, and two daughters, but this was denied because immigration was essentially closed. In addition, for his assistance in a foreign lawsuit against the city the magistrates had awarded him a gold medal.[80]

Strife with Voltaire and the Academy Presidency

As long as physics was foremost in the dispute over the principle of least action, Voltaire did not enter that fray, for he feared making a mistake in Newtonian dynamics. But when Maupertuis wrote a bizarre, eclectic, and dull collection of writings dealing with scientific miscellany and speculations titled *Lettre sur le progrès des sciences* (Letter on the progress of the sciences), Voltaire in an array of short pamphlets titled the *L'histoire du docteur Akakia et du natif Saint Malo* would at the end of 1752 seize the opportunity to criticize and disparage the ideas Maupertuis's work contained on science and polite society. Maupertuis had intended to show himself a universal scholar in his collection, which parallels the work of Fontenelle and anticipates that of Jules Verne and Carl Sagan in its exploration of Paracelsan alchemy and the search for the philosopher's stone. The academy president offered odd conjectures on vivisecting criminals, dissecting the brains of monkeys and giants to find their souls, and blowing up the Egyptian pyramids to study their interiors. He reported on Patagonians with tails and on drilling a hole through the middle point of Earth.[81] The robust mixture of solid science with the absurd made the whole look laughable.

In mid-September 1752 Voltaire unexpectedly entered the König-Maupertuis affair, turning it from a scientific quarrel into a literary dispute, perhaps the most notorious of the eighteenth century, and escalating it. It went further, becoming a court intrigue directly involving the king. This scene was Sisyphean, with Maupertuis having risen high, only to be tripped and cast down by the agile and cunning Voltaire.[82] A more lethal adversary than König, Voltaire attacked Maupertuis with ridicule and employed his comic ability to make his readers laugh at the faults of others. Voltaire wanted to show this nobleman of the German second empire to be a dreamer and a charlatan. His participation in the argumentation suggests nationalistic competition as well between two Frenchmen for favor of the Prussian monarch; for years the quarrel between Voltaire and Maupertuis amid their partisans was to appear in rival versions in letters, learned journals, literary gazettes, pamphlets, brochures, small books, and anthologies in French, German, Dutch, and Latin. Invective increased as the combatants debated personality, literary style, and open administration. As the harshness grew on both sides, a polite and civil order in literary and scientific dialogues appeared to be disintegrating. Voltaire's denunciation of despotism preceded the coming break between the philosopher king Frederick and the foremost French *philosophe*. Voltaire wanted to besmirch the dinners at Sanssouci with their pomp, flowing wigs and diamonds, merriment, wines, and sarcastically polished conversation as the central location of the Enlightenment in Prussia.[83]

Voltaire's public entrance into the dispute between Maupertuis and König came in the lively, anonymous article "Réponse d'un académicien de Paris à un académicien de Berlin" in the *Journal des sçavans*. Disregarding his own Newtonian views, Voltaire restated König's position and innocence, denounced Maupertuis's handling of the controversy, and asserted that the academy vote against König only reflected fear of the king. The indictment in the "Réponse" enraged Frederick; he saw its charges as libel against the president of his academy and an insult to royal honor. Acting as the academy's protector and worrying about the effect of the quarrel on Maupertuis's health, the king quickly wrote several short missives. These included in mid-September an anonymous rejoinder titled "Lettre d'un académicien de Berlin à un académicien de Paris," which defended the academy's judgment. A bumbling work, it called the "Réponse" an imposition and an effrontery that was defamatory and malicious.[84] In the republic of letters, everyone knew who both authors were. Apparently the king hoped that his reply would restrain the *philosophe*, but he was mistaken. In the world of polemics, Voltaire accepted no master. Even in the face of Frederick's opposition, he refused to stop. When Voltaire's criticisms continued, the king soon published a second edition of his rejoinder with the Prussian arms on the title page. This alienated Voltaire further, and he was determined to inflict as much damage as possible and then leave Prussia. In this conflict he had a pen that was to be a match for the eagle, the scepter, and the crown. Frederick counseled Maupertuis not to "listen to the buzzing of these insects in your ears."[85]

What had begun as a series of reproaches from the king became a rupture when with the full force of his vitriolic pen Voltaire replied to Maupertuis's *Lettre* by issuing clandestinely with a faked permission from the royal printing press his caustic satire *La diatribe du docteur Akakia, médecin du pape* or *L'histoire du docteur Akakia et du natif de Saint Malo*. This was not the first time that Voltaire used such a false method to publish. Written in November 1752, the *Diatribe* ridiculed Maupertuis, who was from the port city of Saint Malo, as a tyrant and buffoon. Voltaire carefully chose the title; in Greek *akakia* means "guileless" or "lighthearted," and indeed Maupertuis's nickname in Berlin was Mr. Lighthearted. In the satire, Doctor Akakia (Voltaire) pillories the young, ignorant, and hapless native from Saint Malo (Maupertuis), trying to cure him of his insufferable haughtiness. His opponent's *Lettre* gave Voltaire the opportunity to deride the various ideas on the sciences or polite society they contained, and his comments were extended to ideas on love and physiology in Maupertuis's *Essai de cosmologie* and *Venus physique* reprinted in 1751, which included paying no doctor whose patients did not recover, proving from a simple algebraic equation that God exists, and studying dreams to change the soul. Voltaire's native believes that humans can live for eight or nine hundred years by plugging all their pores.

Section 10 of the *Diatribe*, titled "Examen des lettres" (Examination of the *Letters*) praises Newton, indicates that Maupertuis should take little credit for knowledge of the polar flattening of Earth, and calls the dispute over the principle of least action insignificant and "miserable," a "war between rats and frogs."[86] Voltaire thought it wrong to impugn with the charge of forgery König's letter written by Leibniz that contradicted Maupertuis's claim to the discovery of the law. The accusation, Voltaire believed, was made only to deny to another the glory of discovering the law. Without proof, he wrote that "it is evident that the Leibniz letters are by Leibniz." So the letter at issue was supposedly not forged, and Maupertuis's claim was prohibiting an honest defense.[87]

In Frederick's Berlin publishing the *Diatribe* was dangerous. Had anyone besides Voltaire written it about the head of the royal academy of sciences, he would have been executed. Voltaire, who was quite ill, volunteered to help review it, and trembled as he began to read the manuscript of his satire to Frederick in his chambers. The two were heard reading it page by page, guffawing, drinking wine, and casting each page into the fire. After laughing heartily until tears rolled down his cheeks, the king ordered Voltaire to destroy all copies and forbade writing any more of it and publishing it; the *philosophe* swore to follow these injunctions but obeyed none of them. As it happened, Frederick's orders were too late anyway: Voltaire had already sent the manuscript to Leiden, and Dutch presses printed thousands of copies; and editions sold rapidly in Paris and the German states. Copies were even shipped to Berlin. Filled with clever wit, the *Diatribe* evoked a burst of laughter across literate Europe from Potsdam to London, Paris, the Hague, Rome, Saint Petersburg, and Vienna.

While the method Doctor Akakia selected to dispose of his adversary through the principle of least action—that is, killing him with a bullet traveling at the square of its velocity—says little for Voltaire's humor, it refers to Leibniz's concept of living force, mv^2, a concept that Voltaire actually opposed; he also rejected the principle of its conservation. In assuming that this quantity of living force is a minimum principle, Voltaire observed, Maupertuis understood only half the law.

Taking Voltaire's "Réponse d'un académicien de Paris" as an affront to his honor, and further angered by Voltaire's tricking the royal printer to publish the *Diatribe*, Frederick had the police seize copies of it and had them privately destroyed in his chambers; its sale was then banned in Berlin. This did not stop Voltaire, and so at 10:00 a.m. on the day before Christmas the king had the public hangman burn copies of the book at the Place de Gendarmen and all official places in Berlin, so that the large crowds leaving churches at midday would see the evidence of royal displeasure. To calm Maupertuis, Frederick sent him a sympathetic letter with a pinch of the ashes of the books. The Berlin populace was fascinated at the fall of a royal favorite. For a time Voltaire remained quietly out of sight at a chateau in a suburb of Berlin; although assured privately of royal protection, he departed three months later, planning enlarged editions of the treatise. Whether the *Diatribe* significantly impaired the health of Maupertuis cannot be determined; he had earlier suffered a series of lung illnesses, and from these his condition was worsening.

Articles 15 and 19 of a sequel to the *Diatribe*'s first edition, which describe Euler as the lieutenant general enlisted to support Maupertuis, praises him as "a very great geometer" belonging to the line of scientific luminaries from Copernicus, Kepler, and Leibniz through Johann I Bernoulli. It affirms that he created the differential equations for the principle of least action, and it declares that scholars who understand his work find it to be that of a genius.[88] Part 1 of article 19 holds that Euler admitted that he had never learned philosophy, and "has been misled into the opinion that one could understand it without learning it. In future he will be content with the fame of being the mathematician who in a given time has filled more sheets of paper with calculations than any other."[89] Part 2 asserts that while Euler was the phoenix of algebra his formulas led in one instance to the notion that a body dropped through a hole to the center of Earth would return to the surface. Voltaire also questioned Euler's trust in resolving scientific paradoxes by extensive and incomprehensible computations when logical analysis would suffice. Voltaire considered empirical research foremost in the sciences. Later editions of the *Diatribe* omitted references to Euler by name but continued to mention "the professor."

As early as 1752 Frederick was considering a successor as president of the Berlin Academy because Maupertuis's health was deteriorating. The presidency, of course, remained a matter of deep interest to Euler. Had the choice depended solely on scientific eminence, he would already be president, for his research and scientific achievements far surpassed the successes of Maupertuis; he was the greatest living mathematician. But the king did not understand the subject; from the beginning of his reign, conversations circulated in court circles about the worth of calculus and other advanced mathematics. This led Jordan to write for Frederick a short account in Latin of their importance, but the king did not read Latin. He did not grasp that Euler's contributions to the mathematical sciences were more significant than the writings of the Paris *philosophes*. For the academy presidency Frederick still wanted more than a leading scholar in any particular field. His ideal person had to be a statesman and an engaging partner at his table: witty, poetic, and freethinking. Neither the religious family man Euler, for whom poetry was not a favorite subject and who was often and most comfortable at his writing desk, nor any other German fit that model. Frederick looked to Paris for the combination ideally of leading French *philosophe*, science editor of the famous *Encyclopédie*, competitor to Euler in the sciences, man of the world, and man of noble ancestry that he found in d'Alembert.[90] Frederick sent Henri Alexandre de Catt, the academy secretary, to offer d'Alembert the presidency with a handsome salary of twelve thousand francs, which was more than had gone to Voltaire, along with free residence in a royal palace and meals at the royal table. Catt was also supposed to bring him to Berlin, but d'Alembert declined the offer. Frederick did not relent. The next May he had Maupertuis meet with d'Alembert in Paris. D'Alembert again refused to leave the rich intellectual and social life of the Enlightenment in Paris. Frederick's pursuit of him would continue.

Chapter 11

Increasing Precision and Generalization in the Mathematical Sciences: 1753 to 1756

In the first two months of 1753, Leonhard Euler and Johann Schumacher exchanged six letters; their main topics were Voltaire's *Diatribe du Docteur Akakia, Médecin du Pape* (Story of Doctor Akakia, physician to the pope) and the final preparation, printing, and distribution of Euler's *Theoria motus Lunae exhibens omnes inaequalitates* (Theory of the motion of the moon which exhibits all of its irregularities, E187). To Schumacher's request for some copies of the *Diatribe* Euler responded that in Berlin sale of the confiscated and burned pamphlet was forbidden. But Euler also wrote to Gottfried Wilhelm Heinsius in Leipzig and had the bookseller Lancke there send copies to Schumacher, who wondered what place would accept the controversial Frenchman if he left Potsdam. Euler noted that a German translation of the *Diatribe* was forthcoming. He and Schumacher concentrated on the *Theoria motus Lunae*, which Euler declared crucial to proving that his computations precisely described physical reality. Euler drew up a list of people to receive a free copy on behalf of Kirill Girgor'evich Razumovskij. In February Euler submitted the final mathematical calculations for the book.

In letters from the first two months of 1753, Schumacher indicated twice that he had made changes to Euler's annual salary, apparently increasing it. Euler wrote that he awaited the third part of Johann Georg Gmelin's *Flora sibirica*, and he praised the successful study of mathematics by the Imperial Academy of Sciences adjunct Semjon Kirillovič Kotel'nikov in Saint Petersburg.

In February Pierre Bouguer wrote twice from Paris. He promised to send his book on practical navigation, *Nouveau traité de navigation et de pilotage*; Alexis Claude Clairaut had earlier alerted Euler that it was about to appear. Bouguer referred to Euler's writings on astronomical refraction, as well as his own, and inquired about Euler's research on the impenetrability and elasticity of bodies along with that of Euler and Isaac Newton on the diffraction of light. Bouguer reported that the Royal Academy of Sciences in Paris had discussed the principle of least action, and that the *Journal des Sçavans* for December would carry his article on the judgment in Berlin concerning the supposed letter from Gottfried Wilhelm Leibniz. Bouguer's second letter closed with the news that the Paris Academy had named Joseph Jêrome le Français de Lalande as an adjunct.

The Dispute over the Principle of Least Action: The Third Phase

At the Royal Academy of Sciences in Berlin—after Euler fiercely criticized Johann Samuel König's "Appel au Public du jugement de l'Académie Royale de Berlin sur un fragment de lettre M. de Leibnitz, cité par M. König," and defended the originality and universality of Pierre Maupertuis's principle, and Johann Bernhard Merian issued a letter of derision—König released in February 1753 his "Défense de l'appel," which still largely attributed Maupertuis's principle of least action to precursors; Voltaire claimed to have seen it circulated the previous month. This was offered as König's final word in the quarrel with Maupertuis, but it did not stop the dispute, which was to last long beyond the deaths of the two men.

In March 1753 Euler spoke to the Berlin Academy again on the principle of least action. That year the academy published his expanded *Exposé concernant l'examen* in a single volume and the article "Examen de la dissertatio de principio minimae actionis una cum examine objectionum Cl. Prof. Koenig, inserée dans les Actes de Leipzig, pour le mois mars 1751" (Examination of the dissertation of Prof. Koenig, inserted in the Leipzig Acta for the month of March 1751, E199).[1] Essentially Euler's third response to König, this account was methodic, masterful, and more scientifically advanced than his first response, which had been historical in nature, and his second response, which had been scientific. Since this volume carried the imprimatur of the curator Christoph Karl von Keith, it officially endorsed the academy's censure. Released separately from the *Mémoires de l'Académie* for 1750, it contains correspondence between Euler and Maupertuis, short writings of Euler on the principle of least action, and a long letter dated September 1752 from Euler to Merian, along with a longer postscript. These repeat the claim that the principle of least action was not derivative and argue that its proven universality confirms its significance.

For Euler this dispute did not require a restraint on discussion with the press. All citizens of the republic of letters, he reasserted, had a right to express doubts about research in journals. But he appealed to sound scientific authority. In contrast to the incisive, learned judgment of the academy, the materials against Maupertuis in periodicals were written, he believed, by slanderous and scientifically illiterate literary gazetteers and journalists. Euler called König's press allies public quibblers and petulant censors speaking from malice who lacked adequate knowledge to make decisions in the sciences. Generally the press distrusted Euler's ability to be impartial and to follow only mathematical arguments in this case. This must have been vexing to a scholar who championed rational scientific methods, performed the most advanced mathematical computations, and advocated the policy of freedom of the academy from royal control. He also set himself against König's claim to speak as the definitive authority, and posed himself as a judge, arbiter, or censor rather than an investigator in this mechanics' controversy.

During the Enlightenment the development of mathematical physics raised questions about causality and prompted a significant debate about the proper realm of the application of geometry, meaning all of mathematics. In his *Discours preliminaire* and in his entry "Application" in the *Encyclopédie* in 1751, Jean-Baptiste le Rond d'Alembert had represented geometry as a touchstone to physics, stemming from it, and warned against the danger of carrying the new *esprit de calcul* too far. In the "Examen de la Dissertation de M. le Professeur Koenig" Euler agreed, declaring that logicians, metaphysicians, and geometers could not provide answers to causality in mechanics.[2] Euler thus placed limits on the application of mathematics in natural philosophy.

In March 1753 Voltaire ended his disastrous visit to Berlin and traveled to Leipzig.[3] Simultaneously a collection of controversial writings, titled *Maupertuisiana*, gathered from the debate (possibly by Voltaire himself), appeared first in Hamburg and soon in a German translation in Leipzig. An engraving in it caricatures Maupertuis as Don Quixote riding on a broken-down nag and tilting with a lance against windmills. Quixotic or impractical, Maupertuis shouts "tremble" as he attacks imaginary enemies in his pursuit of glory. A satyr in the corner of the engraving proclaims "Sic itur ad Astra" (Thus one goes to the stars), but in the end, it is understood, the flawed hero is bound to lose.

In May 1753 Christian Wolff wrote to Schumacher that the strange debate between König and the president of the Berlin Academy and its council had caused a great sensation.[4] Of all places, he found it odd that the controversy had occurred in Berlin—where, Wolff assumed, his philosophy was dominant. He wished that the dispute had never happened. The interference of Euler particularly disturbed him, as did the academy's findings. A self-proclaimed enlightened moralist, Wolff urged a fundamental study of his ethics so that scholars would learn better about making such criticisms. His books imparted enough instruction, he thought; only scholars and practitioners in the sciences were still lacking. Whether the Wolffians concocted the letter in question as a subterfuge to challenge the achievements of Euler and Maupertuis was generally rejected, for there was no known perpetrator of the letter.[5] But as was later shown, a passage near the end of the possible full letter approximates Euler's message in December 1743 to Marcus-Michael Bousquet, the publisher of Euler's *Methodus inveniendi lineas curvas maximi minimive proprietate gaudentes* (A method of finding curves that show some property of maximum or minimum, E65). This suggests that Bousquet participated in the falsification.

Prominent among the scholars now supporting Maupertuis and Euler were d'Alembert and Georg Wolfgang Krafft, the latter of the University of Tübingen. In 1753 Krafft called Maupertuis the sole author of the principle of least action, and d'Alembert praised even his critic, Euler. His entry "Cosmologie" in volume 4 of the *Encyclopédie* the next February attempted to give an impartial account of the origin of that principle.[6]

Writing under the pseudonym O, d'Alembert began the "Cosmologie" entry by defining the principle as part of general physics, which by means of metaphysics brings out analogies and connections to discover general laws of physics and determine which are universal. Nature, he assumed, was simple and unified. D'Alembert credited Jean-Henri-Samuel Formey with establishing that Wolff had introduced the field of cosmology in his *Cosmologia generalis* of 1731, basing it in metaphysics. Yet while the second edition of Maupertuis's *Cosmologie* in 1752 also accepted metaphysics and final causes, it had elicited a hostile response from the Wolffians. D'Alembert showed that Leibniz's living force and Maupertuis's quantity of action were proportional but not the same. For clarification he recommended replacing with "change in velocity" the words "change in the nature" in Maupertuis's abstract. He did not accept Leibniz's principle of the conservation of *vis viva* as a general law of nature and, like Euler, he argued that it produced a minimum only in the case of a nil result. The two laws were not identical: Leibniz's was a law of nullity and not of minimum. D'Alembert praised Maupertuis for being the first to reduce to the same principle both rigid and elastic bodies and reconcile refraction with final causes, which he thought should have pleased the Leibnizians.

D'Alembert represented Maupertuis's law, then, as original and—as Euler and others also argued—not as a consequence of the ancient principle of the lever. But since no copy of the letter in Leibniz's hand could be found, d'Alembert did not know how the public could decide on its authenticity. To confirm the universality of the principle, d'Alembert argued, the reader should consult Euler's *Exposé concernant l'examen* and the adroit defense of Maupertuis from 1753 in the Berlin Academy *Mémoires* that elegantly computed many physical applications. In addition, d'Alembert noted that Maupertuis's algebraic proof of the existence of God was not superior to other proofs and that it rejected charges of atheism against Maupertuis. The last accusation he found simply a fashionable way to criticize the *philosophes*.

The "Cosmologie" entry concludes with strong comments that contrast with the dispassion of its scientific portions; these were probably largely intended for Voltaire. Since Voltaire's support was critical for the *Encyclopédie*, d'Alembert did not criticize him directly. He likened the character the affair had acquired to the intensity of religious disputations that he decried as closed-minded, and he commended Maupertuis's conduct:

> Nous davons ajoûter que M. de Maupertuis n'a jamais rien répondu aux injures qu'on a vomies contre lui à cettte occasion, & dont nous dirons: ne nominetur in vobis, sicut decet philosophos. Cette querelle de l'action, s'il nous est permis de la dire, a ressemblé à certaines disputes de religion, par l'aigreur qu'on y a mise, & par la quantité de gens qui en ont parlé sans y rien etendre.
>
> We should add that Mr. de Maupertuis has never responded to the insults that have been spewed against him on this matter, to which we say: put an end to the charges against you that are unfitting. This debate regarding action, may we say, is similar to certain arguments surrounding religion for the bitterness that has been injected and by the number of people who have spoken without hearing a word. (O)

D'Alembert also praised Maupertuis in his *Discours preliminaire* and defended him in the article "Force" in volume 7 of the *Encyclopédie* for 1757, which largely repeats earlier points from the article "Cosmologie." The notion that Maupertuis's action and Leibniz's *vis viva* were the same was absolutely arbitrary; they were not. In "Force" d'Alembert referred back to part 2, chapter 4 of his own *Traité de dynamique* of 1743, attributing the principle of the conservation of *vis viva* to Christiaan Huygens and Johann I (Jean I) Bernoulli.[7] His "Cosmologie," he declared, had explained the question of least action with great exactitude. He considered it a great error of König's "Appel au public" to have the public judge the authenticity of the alleged Leibniz letter.

Most remembered is not d'Alembert's "Cosmologie" or "Force" articles but Nicolas, Marquis de Condorcet's eulogy of Euler read to the Paris Academy in 1783. After d'Alembert died, Frederick II named Condorcet, another mathematician, to be the secret president of the Berlin Academy, so Condorcet had more than a detached connection with the polemic, and presented the censure of König as a flaw in Euler's character. Critics have found the charge of forgery and the hardness of the language as exceeding appropriate criticism,[8] and there has been a general belief in König's innocence or naïveté. Accepting the first, Condorcet portrayed Euler as too loyal to Maupertuis in a one-time lapse of judgment: "[I]t cannot be hidden that Euler too harshly condemned König; painfully we are obliged to count a great man on the list of enemies of this persecuted scientist. Fortunately, Euler's life places him above more serious suspicions . . . [about] this wise and peaceful geometer whose only fault was an excess of gratitude [toward Maupertuis], which was a selfless sentiment and for which he was wrong [in scientific debates] for the only time in his life."[9]

The claim that König was being persecuted seems overdramatic. He was not naive, having accused Emilie, Marquise du Châtelet of recopying his mathematics for her *Institutions de physique* of 1740 and having engaged in the debates about monads in Berlin and the continuing Newton-Leibniz priority dispute for calculus;[10] nor had he lost his position. Although Maupertuis wrote once to Holland about König's retaining his post, throughout the dispute König did in fact retain it. In mechanics Euler had shown König erroneous in detail about the principle of least action. Condorcet's statement suggests either that he had not read Euler's *Dissertatio de principio minimae actionis* and *Exposé concernant l'examen* or that he did not accept their arguments. Condorcet also did not address what Euler called the tergiversations by König in the affair, his making of almost contradictory statements. What seemed most to bother Euler was König's obstinacy in not accepting the decisive scientific evidence supporting Maupertuis; Euler, as usual, was promoting higher standards in scientific research. The notion that his philosophical opposition to the Wolffians induced the decision is doubtful; he opposed the dominance of philosophy and religion over the sciences. Absence of proof that the fragment in Leibniz's handwriting was authentic, combined with advances in the principle of least action contrary to König, was what guided Euler's thinking the most. Not until 1898 did research indicate that the letter in question may have existed (though not addressed to Jakob Hermann but to the French geometer Pierre Varignon),[11] but this letter has been largely rejected for bolstering König's position.

For Voltaire the Maupertuis-König affair partly represented a final step in the bitter priority battle over calculus pitting Gottfried Wilhelm Leibniz and Johann I Bernoulli on one side against the English on the other. The Newtonian Maupertuis strongly continued to defend the English position. In the sciences Euler and Merian characteristically stood on the English side, against the Wolffians, even though they did not completely agree with the Newtonians. In his opposition to the Wolffian philosophers, Euler lacked adequate means; as the philosophical disputes unfolded, Merian—who translated works of David Hume—would try to act as a mediator.

Administration and Research at the Berlin Academy

In 1753 the state of affairs at the academy frustrated Maupertuis. Even though the Sorbonne and the Paris Parlement were attacking scholars such as George-Louis Leclerc, Comte de Buffon, and the French monarchy threatened to suspend the publication of the remaining volumes of Denis Diderot's *Encyclopédie*, he was unable to recruit leading French savants from Paris. Though Berlin was the capital of Prussia, it was but a small town compared to London, Paris, Naples, or Saint Petersburg, and had only begun to develop the culture, diversity in learning, and science that those cities enjoyed. It had no great libraries, no university, no medical faculty, and only limited contacts with the West. Not even the brilliance of Frederick II's suppers and music at Sanssouci could quickly change this; his royal initiatives and financial support were among the chief elements in a rising country. Still, the failure to lure French thinkers to Berlin was a blow to Maupertuis's vanity. From early 1753 on, as his health worsened, perhaps in part over the König dispute, Maupertuis began to stay away from the city for longer periods. Plagued by his lung disease and considering the climate of Berlin potentially fatal, he spent more time in the healthy "native air" of Saint Malo.

From May 1753 to July 1754, Euler took on an additional role at the academy. In April Maupertuis, believing that he was dying and wishing not to omit anything from his duties, had asked that during his absences Frederick allow him to relinquish presidential duties to Euler, who would provide leadership "by his probity, his brilliance, and his zeal."[12] Though Frederick lacked confidence in Euler's managerial skills, opposed his policy for an independent science academy, and regretted his lack of wit and dash, he accepted the proposal to make Euler the acting president. During this time in his reign the king was at his closest to academy operations.

For adding luster to his court, the king and his noble court urgently wanted to recruit and retain distinguished scholars. Beginning in April 1753, even before the astronomer Johann Kies succeeded the deceased Krafft at the University of Tübingen, Euler pursued Johann Tobias Mayer, who was then at the Georg-August Academy in Göttingen for two years correcting lunar tables and independently examining lunar motion and lunar eclipses; Mayer claimed that the works of Euler in mechanics and calculus had made his advances possible. In a long treatise he maintained that the moon could not have an atmosphere. Under the leadership of the Hungarian Jan Andrej (Johann Andreas) von Segner, he was active in planning the building of the Göttingen Observatory. Euler asked Mayer not to reveal the offer from Berlin for a position at the academy; in the event that Göttingen increased his salary in a counteroffer, Euler asked to be informed so he could work out something better. For its post in astronomy the Berlin Academy treasury had only 550 Reichsthaler, which was slightly less than Mayer's total income of 600 to 800 Reichsthaler in Göttingen, but for preparing a calendar Mayer could earn another 150 Reichsthaler. Frederick was expected shortly to increase academy funds and salaries, the latter by reducing the number of academicians. When Maupertuis returned, Euler assured Mayer, he would provide a higher salary.

In 1753 Euler had good relations with Frederick. To strengthen his position, he had Johann Karl Hedlinger, the Swiss man who prepared medals at the academy, create a medallion for the monarch praising his achievements. In May Euler even sent the king a fruit basket from his garden near Schoneberg. Frederick responded with a note of thanks and soon after gave an assurance that in a pending inheritance dispute he would support Euler's nephew Georg Vermeulen, who was a Prussian officer. There were now more personnel matters to resolve at the academy. In July the chemist Johann Heinrich Pott asked the king to appoint his son-in-law, Dr. Kurella, as his collaborator and the second professor of chemistry. Frederick asked Euler to review the request, but Euler found Kurella clearly unqualified, and Frederick forwarded to Euler a copy of his refusal to make the appointment. In August the king accepted Euler's nomination of Waldemar Christofer Brögger to be the second professor of anatomy. In October Euler proposed Karl Philipp Brandes to be the second professor of chemistry, and the king agreed. But Pott strongly objected. Seeing a difficult situation with Pott, Brandes chose to go to Marburg instead. In having him elected a foreign member of the academy in

Figure 11.1. Tobias Mayer.

1755, Euler showed tenacity, and when Pott retired in 1760, he would have Brandes reappointed as an ordinary member.

While conducting research in the physical sciences, geography, and technology to develop water turbines and improve clocks and telescopes, in 1753 Euler kept central the fields of theoretical astronomy and differential calculus. Almost a year after the 1752 publication of Clairaut's Petersburg Academy prize–winning paper "Theorie de la lune déduite d'unseul principe de l'attraction réciproquement proportionelle aux quarrés de des distances" (Theory of the moon deduced from a sole principle of attraction reciprocally proportional to the square of distances) with the imprimatur of Razumovskij, there continued among Clairaut, d'Alembert, and Euler a fight over priority for the most accurate lunar theory that fully agreed with Newtonian dynamics.[13] Each appealed to his network of support through correspondence and publishers, and all three kept in contact with Gabriel Cramer, who became the central observer in this controversy. Mayer joined the three protagonists in shaping a more exact lunar theory, using differential equations devised by Euler.[14]

During 1753 Euler and Schumacher discussed Euler's eighteen-chapter *Theoria motus Lunae*, which contained the second of his three lunar theories based on calculus.[15] He had completed most of the main body of the text in 1751. Euler's comments on this text and his next important book of the early 1750s, the forthcoming *Institutiones calculi differentialis* (Foundations of differential calculus), indicate problems in printing, including the representation of new curves, and disagreements about costs. In April Euler expressed delight that President Razumovskij thought highly enough of the *Theoria motus Lunae* to offer to send payment for the costs of the printer and bookbinder. That month Schumacher replied that Razumovskij considered the *Theoria* extraordinary, advancing two hundred rubles to pay for it and requesting a French translation; in July Euler gave the actual cost of 633 Reichsthaler, and the Russian Imperial Academy paid for its publication. Translations, especially into German and Russian, could be unreliable, but the French translation of the *Theoria* made by one of his Swiss students was of high quality. Even so, Euler sent six corrections. Two years earlier, Euler had apparently presented an early version of the paper at the Berlin Academy. The foreword stated that the book was not intended for a prize competition and that Clairaut's prize paper had inspired its revision.

In the *Theoria motus Lunae*, one more of his landmark books, Euler was taking the lead in transforming theoretical astronomy into a modern mathematical science. In a letter of 10 April 1751 to Clairaut, Euler wrote, "I have the satisfaction of writing you that I am now altogether clear concerning the motion of lunar apogee and that I find it, as you do, entirely in agreement with the inverse-square law (Newton's theory)."[16] In a letter of 29 June he praised Clairaut's new method profusely: "The more I consider this happy discovery, the more important it seems to me, and in my opinion it is the greatest discovery in the theory of astronomy, without which it would be absolutely impossible ever to succeed in knowing the perturbations that the planets cause in each other."[17]

More than a century after planetary and solar tables had become Keplerian following the Rudolphine Tables of 1627, they were transformed as Newtonian tables, taking into account perturbations of each body caused by others in universal gravitation. Euler, Clairaut, and d'Alembert provided crucial analysis to compute perturbations in the aphelia and nodes of planetary and solar motion produced among them. Since Euler found the classical expansion method of perturbations ineffective, he proposed a variant; he devised a single-step procedure with five ordinary differential equations and applied trigonometric series to describe the motion of planets. All computations used the same mean motion arguments. Euler's procedures gave closer approximations than Clairaut's for the intermediate orbit and the three-body problem.[18] Since some effects were exceedingly small, Euler could work from the series for them to make good approximations. His text and Mayer's writings, which resembled Euler's, were two of the first to confirm that lunar apses follow Newton's inverse-square law of attraction. Having two computational methods achieve the

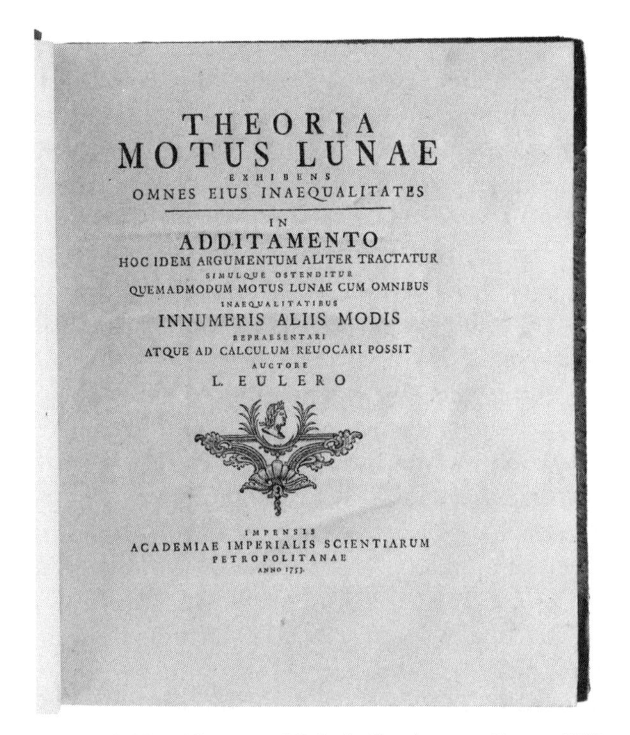

THEORIA
MOTUS LUNAE
EXHIBENS
OMNES EIUS INAEQUALITATES
IN
ADDITAMENTO
HOC IDEM ARGUMENTUM ALITER TRACTATUR
SIMULQUE OSTENDITUR
QUEMADMODUM MOTUS LUNAE CUM OMNIBUS
INAEQUALITATIBUS
INNUMERIS ALIIS MODIS
REPRAESENTARI
ATQUE AD CALCULUM REUOCARI POSSIT
AUCTORE
L. EULERO

IMPENSIS
ACADEMIAE IMPERIALIS SCIENTIARUM
PETROPOLITANAE
ANNO 1753.

Figure 11.2. The title page of Euler's *Theoria motus Lunae*, 1753.

same result, so Euler believed, placed this finding beyond doubt. In his seventy-five-page appendix he added another computational method, an initial form of the calculus of variations, with new formulas to sharpen calculations.

In the spring of 1753 the Göttingen Royal Society of Sciences and Humanities published in volume 2 of its *Commentationes Societati Regiae Scientiarum Gottingense* Mayer's "Novae tabulae motuum solis et lunae" (New tables of solar and lunar motion), which improved equations to account for lunar anomalies contained in Euler's Prix de Paris–winning paper of 1748, "Recherches sur les irregularités du mouvement de Jupiter et de Saturn," developed other formulas with added sine terms (that is, varying like a sine curve or wave), and used superior astronomical equipment in making observations of the distance of the moon from a fixed star, improving significantly upon Clairaut's prize-winning treatise and Euler's *Theoria motus Lunae*. The Berlin Academy had reprinted these writings of Clairaut and Euler in 1752. Euler, Christian Michaelis in Halle, and others urged Mayer to formulate a method to find longitude at sea and better lunar tables.[19] These were vital to the improvement of navigation.

In his pursuit of the British Parliament prize, Mayer's method by 1755 differed in part from Euler's method and was partly inspired by Clairaut. He devised a multistep procedure for his computations. In astronomy a multistep procedure is not iterative but has different arguments within each step. Mayer's approximate solutions for differential equations and higher-order terms and equations were distinct from Euler's; Mayer wanted tables verified by observation. The relations between tables and positions were still complicated; Mayer reduced to thirteen the number of needed differential equations and arranged them in tables, surpassing all of his contemporaries. He informed Euler that, using these tables, "I shall be able to calculate the longitude of the Moon so correctly in observations that the error amounts to not more than 2 minutes."[20] The tables of Clairaut, d'Alembert, and Euler had errors as large as four or five minutes. This quickly brought Mayer an international reputation and set off a competition for his services. Mayer wrote his results in *Theoria lunae iuxta systema Newtonianum*, which was mainly completed in 1754 but not published until 1767, when Nevil Maskelyne prepared it for the press.

In July 1753 Schumacher inquired about the number of copies and the printing costs of Euler's *Institutiones calculi differentialis*. Euler wanted five hundred copies, and gave the printing costs, explaining also the technical problems with the press and the difficulties in distribution and the management of sales. As it had for Euler's *Scientia navalis,* the Petersburg Academy agreed to fund the publication of the *Institutiones*, even with a non-Russian publisher; these efforts on the part of the academy were important to both the German and the French streams of the Enlightenment in fostering discoveries, especially in the mathematical sciences, and in the wider dissemination of scientific knowledge.

In mathematics Euler pursued abstraction equally with the field's empirical roots and applications. He often admonished his fellow geometers that, despite their point of view, number theory was not a waste of time. His letters to Christian Goldbach in 1753 address proofs of Pierre de Fermat's Last Theorem for $n = 3$ and 4, Diophantine analysis, and prime numbers of form $4n + 1$. The next year in "Solution generalis quorundam problematum Diophantaerum quae rulgo nonnisi solutiones speciales admittere videntur" (The general solution of certain Diophantine problems which are ordinarily thought to admit only special solutions, E255), Euler wrote that "the Diophantine method, if further developed, will redound to the benefit of the whole of Analysis, so that [I am] far from feeling sorry for the prolonged efforts . . . [that I] devoted to that branch of mathematics."[21] He was now creating elliptic integrals. Just beneath the surface of that accomplishment lay the close connection between the integrals and Diophantine algebra.

Inspired by Count Giulio Carlo de' Toschi di Fagnano's concepts and extending his earlier work on elliptic integrals, Euler had written in January 1752 one of two important papers on the subject, "Observationes de comparatione arcuum curvarum irrectificabilium" (Observations on the comparison of arcs of nonrectifiable curves, E252). Despite having the Eneström number E251, "De integratione aequationis differentialis" (On the integration of the differential equations) was written in 1753, a year before the "Observationes" (E251); this may explain the appearance of Euler's addition theorem from E251 in the printed form of E252.[22] For a function f, addition theorems give $f(x + y)$ in combinations of $f(x)$ and $f(y)$. These theorems were important in creating the theory of elliptic integrals and functions. Euler's two articles offer theorems on arcs of ellipses, hyperbolas, and lemniscates, coordinates being the sum or difference of an algebraic function or arcs comprising multiples. The addition theorem lies at the heart of the general theory of elliptic integrals and functions. Both of these papers were finally published in the Petersburg Academy's *Novi commentarii* for 1761.

In October 1753 and again probably before late April 1754, Daniel Bernoulli wrote to Euler about the vibrating string problem. The first letter indicates that after reading Euler and d'Alembert, he had written an article on a simpler solution, and he asked whether his works on the vibration and sound of elastic rods were now in print in Saint Petersburg. In the second letter Bernoulli, who had been criticizing d'Alembert in hydrodynamics, questioned d'Alembert's latest article on vibrating strings in the Berlin *Mémoires*. Bernoulli had solved the one-dimensional wave equation in a new way by applying an arbitrary function consisting of a trigonometric series. But d'Alembert asserted that trigonometric series cannot represent discontinuous functions. Euler also rejected using these series for the restrictions

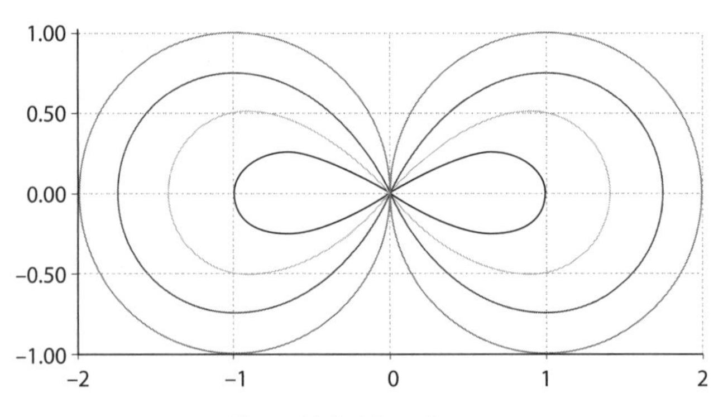

Figure 11.3. A lemniscate.

they imposed prevented a fully general solution. He wanted further studies of discontinuous functions and partial differential equations, which he saw as opening a whole new section of analysis. In "De integratione aequationis differentialis" Euler asserted the central importance of the partial differential equation—with appropriate boundary and initial conditions—for the entire theory of the vibrating string. It contained the first notation for partial derivatives, but because this article was without major new ideas it had few responses except for criticism. D'Alembert declared that Euler's solution did not cover all forms of a string possible initially. At this time the correspondence among the three was testy.

In 1753 Euler entered another Prix de Paris competition. In the yearly alternation of prizes between the theoretical and applied sciences, the topic for that year fell into the latter category: ship propulsion, including the best construction of a ship engine. Daniel Bernoulli won the prize, and Euler shared the *accessit* (honorable mention). Bernoulli's work, which rested on chapter 23 of his *Hydrodynamica*, contained a serious error that the academy missed. Considering forces arising from water pressures, Euler's "Maximes pour arranger le plus avantageusement les machines destinées è élever de l'eau par le moyen des pompes" (Maxims for the most advantageous arrangement of machines employed to raise water via pumps, E208) had earlier given simple rules for mechanics in designing pumps and aqueducts, and it proposed a mechanical principle for building a jet-reactive engine for a ship. This principle was to be similar to that of turbine equations describing uniform motion in a hydroreactive ship. Euler had read a draft of this article at the Berlin Academy in February 1750; it would not be published until 1754.

For the 1753 competition Euler submitted his seventh nautical paper, "Mémoire sur la maniere la plus avantageuse de suppléer à l'action du vent sur les grands vaisseaux" (Essay on the most advantageous manner of moving ships without the force of wind, E413).[23] Its five ways for propelling ships successively increased in sophistication, though the first two seem almost silly. The first had a large vertical surface suspended ahead of the boat; pulled back like a hoe, it was to move the boat forward. The second had two large paddles attached to the sides of the boat and a shaft that turned mechanically. Neither idea was applied to ships, but from this work Euler developed equations of energy and fluid resistance. His third proposal had four paddles, somewhat foreshadowing the paddle-wheel boats of the next century, and the fourth method's mechanically operated fan was not unlike later propellers. The final proposal was futuristic, gaining energy from the different levels of waves to move the ship, which raised a problem of stability.

Prominent in Euler's correspondence and research throughout the 1750s, and especially in 1753, were questions about atmospheric electricity. Of electricity he wanted to know its physical causes and examined reports on its medical applications. Electricity and its applications variously involved experiments with science, medicine, and entertainment. Conducted by men of science and artisans, they were popular at universities, science academies, and public demonstrations, exhibiting such marvels as Stephen Gray's electrified boy, the electrified Venus, and the circulation of electricity through 180 gendarmes. In 1752 Euler had obtained and studied a copy of Benjamin Franklin's *Experiments and Observations on Electricity* in French translation. In a letter in October 1753 to Johann Kaspar Wettstein, Euler reported that he was saddened to learn in July of the death of Georg Wilhelm Richmann, accidentally electrocuted while he was making measurements in Saint Petersburg related to Franklin's kite experiment. Euler credited Franklin with the "unhappy discovery" of the electrical nature of lightning. Attempting likewise to find exactly how lightning is related to electricity, Richmann had put on his roof a bar of iron supported by a mass of pitch; the bar was wired to his chamber. As a thunderstorm approached, he prudently moved clear of the wire, but somehow he brought his chest closer to it and was killed instantly. The Saint Petersburg newspapers reported the tragedy and, dismayed at this news, physicists in France and Germany who were conducting research in the same direction immediately abandoned their projects. In Berlin Johann Nathanael Lieberkühn and the physician Christian Friedrich Ludolf, who had already fixed iron bars on their houses, had them removed.

In Saint Petersburg Richmann had worked with Mikhail Lomonosov, who proposed a prize question on the causes of electricity and a precise theory of electrical forces. Schumacher sent Lomonosov's article "On Phenomena in the Air Caused by Electrical Forces" to Euler. In December 1753 Euler thanked Schumacher for notifying him of the prize question on electricity for 1755; he praised the acuity of Lomonosov and rated his findings higher than the discoveries of purely experimental physicists and scholars that at times involved fantasy. Euler rejected the idea that electricity and magnetism arose from forces in imponderable fluids; he investigated the role of ether and how the air could become electric. He did not set electricity in isolation but looked for connections with magnetism and optics, principally in an underlying ether.

Deeply engaged in the study of geometric optics during his Berlin years and beyond, in the 1750s Euler investigated the rainbow, colors, and especially the dispersion of light. Together with his mastery of the mathematical sciences he displayed a strong interest in related practical scientific technology. He wanted to construct a two-foot telescope with lenses capable of detecting Jupiter's satellites. Developing a compound achromatic refractor to improve telescopes was a difficult scientific challenge; Euler wanted to remove all chromatic aberration, which limited telescopes' abilities, and to reduce the thickness of lenses. Approaching like a physicist the problem of these aberrations, Euler sought mistakenly to find an analogy to the nature of the eye or a mathematical law covering all refractive indices for different colored rays in differing media. Proceeding from incorrect analogies to the nature of the eye and experiments with water-filled lenses, he concluded that a compound achromatic objective consisting of different materials was possible, but problems blocked his efforts.

In 1753 John Dollond, an English optician to the king, sent a letter to the Royal Society of London on the arrangement of lenses in refracting telescopes that appeared as an article in the society's *Philosophical Transactions*. He corrected errors in results from Newton's experiments and maintained that by building upon Euler's mathematics, he had devised largely achromatic lenses. He did not release his optical measurements that indicated that refractive indices do not exist for diverse colored rays in different media and that each must be checked individually instead. For a long time Dollond held a royal monopoly on the manufacture of these telescopes. Since Dollond did not cover Euler's strongest arguments for the analogy to the eye, dismissed other optical arguments simply because a revision of Newton said otherwise, and did not release his measurements, Euler did not yet respond to him in articles in 1753 and 1754 or to mistakes in his mathematics, but a dispute emerged that was to continue through the decade. The Royal Society, of which Dollond was a member, declared that Euler was criticizing Newton. To the contrary, Euler believed that his research on lenses agreed with Newton's and that he had merely applied calculus to Newton's hypotheses.

In 1753, when geography and cartography were important subjects at the Berlin Academy, Euler's work in both fields used information from the Second Kamchatka Expedition and the effort to produce reliable maps of imperial Russia. Refining Joseph-Nicholas Delisle's methods, and with new geodetic measurements and the help of Johann Christoph Rhode, he prepared forty-one maps in color for the preface to the *Atlas Geographicus: omnes orbis terrarum regiones in XLI tabulas exhibens* (Geographic atlas: Representing in forty-one maps all regions of the Earth, E205, 1753), designed principally for use by schoolboys. The preface, printed in Latin and French, consists of nine divided pages, and asserts that the atlas draws all regions of Earth more distinctly than other maps. Observations and the lengths of simple pendulum swings for oscillations of one second made possible many corrections. Research on the oscillations of a second of the simple pendulum in different regions showed the inequality among meridians, which nevertheless produced a spheroid Earth.

In order to represent distinctly all regions of Earth, Euler made corrections—some considerable—using better figures for longitude and latitude. In determining longitude, two methods were in serious competition; both involved a question of technology. One of them, developing a precise clockwork to put on a ship to withstand its strong motions, was not yet possible. Astronomers relied chiefly on the steady perfection of the telescope and in 1731 the invention of the double reflection quadrant in part to measure better angles between celestial bodies. The Royal Society made more sophisticated catalogs of the stars. For his study of longitude Euler could begin with lunar or solar eclipses or the motion of the four moons of Jupiter. The annual Berlin almanacs listed every eclipse visible from the city, but to obtain the most accurate longitude, Euler found that a study of the complex motion of the moon offered the most precise method. It was of utmost importance for navigation. The moon traveled hourly at night the equivalent of its diameter among the background stars; while a faster moving moon would have increased accuracy, there was enough motion to obtain the longitude. Euler traced the lunar nightly path on a celestial globe across the fixed stars, which he considered to be at rest. It was hoped that the motion of the moon might give the time, just as would the hands on a conventional clock.

For obtaining greater accuracy, observational astronomers relied solely on instruments—principally the significant improvement of the telescope. Euler stated that greater precision required these refined instruments and the development of a new six-inch telescope for oceanic observations. He created computational methods with a strong observational base that gave a more precise lunar theory. Improving measurements achieved by Mayer, Euler reduced from six minutes to less than one the errors in computing lunar motion. Consequently, for the moment his results on longitude and latitude surpassed the calculations made by his predecessors. In making these adjustments to arrive at the same projection, Euler claimed that he was able to depict every region of Earth with fidelity.

The Charlottenburg Estate

In October 1753 Euler's mother, accustomed to living in the countryside, had him purchase for her—at the price of six thousand Reichsthaler—an attractive, bucolic estate with a large house in Charlottenburg, then just west of Berlin, where she was to live comfortably until her death in 1761. Charlottenburg was one of three towns, along with Berlin and Potsdam, that had a royal residence, and Frederick II had spent much of his youth at Queen Sophie Charlotte's garden palace.

Euler's mother's lovely house and garden, its rich arable land for raising grains, its meadows, and its plentiful supply of wood for heating and cooking attracted the frugal Euler, and from the estate his family drew all of its needed agricultural goods.[24] These would be especially important during the shortages in 1756, the first year of the Seven Years' War. Euler began with six horses and ten cows. As a schoolchild in Basel he had spent time away from his parents and now, though devoted as a father, he sent all of the children to live with their grandmother, together with a private tutor for the youngsters and a steward.[25] This moving of family members decreased the size of his household by more than half.

When Euler found another position for his children's tutor, he wrote to the Basel mechanist Johann Dietrich, who had studied with Daniel Bernoulli, asking him to recommend that his nephew François Dietrich, who had been teaching in London, accept the position of Euler house tutor; Wettstein was asked in 1755 to offer him the job. The elder Dietrich had been in contact with Euler, and he made magnets that Euler sold in Berlin. He also constructed quite accurate instruments for finding magnetic inclination. Euler's son Johann Albrecht used these in preparing his prize paper for the Petersburg Academy in 1756, one of the seven international prizes that he won. Euler agreed to pay François Dietrich sixty ecus per year, and the tutor was required to live with Euler's mother and take meals at her home. Dietrich arrived safely and found Berlin to his liking; Euler promised to see whether there were opportunities of greater benefit for him.

To earn extra income, Euler rented out the unoccupied rooms in his Berlin house; this permitted him especially to provide convenient lodging there for the young students sent from Russia. He kept up close contact with his children, and always enjoyed taking them to marionette shows and to the zoo. For relaxation, exercise, reflection in solitude, and visits, he often took walks to the Charlottenburg estate, which was about a mile from his Berlin house.

Wolff, Segner, and Mayer

In February 1754 a noisy verbal fight in public between chemists and physicians at the academy over their candidates for the same position upset Frederick II; he complained to Maupertuis that a "certain degree of anarchy reigns in your academy" and, deciding that its secretary Formey was an imbecile, he implored Maupertuis to return as quickly as possible.[26] Though Maupertuis still considered the climate of Berlin potentially fatal, he would return to the city in July 1754 and depart for the last time in the summer of 1756. Until the end of the Seven Years' War in 1763, Euler was the academy's acting president. Uncomfortable with the administrative limits set by Maupertuis, he proved conscientious, financially tightfisted, and occasionally stubborn but always equitable. To his death in 1759, Maupertuis corresponded with Formey and Euler about the business of the academy, particularly its finances. Although d'Alembert had declined to come to Berlin in 1752, he continued to his death in 1783 to advise Frederick through correspondence. In effect, he was the academy's secret president or president in absentia.

Even with Maupertuis absent at the start of 1754, Frederick continued to push for building his realm in learning and the sciences, and this required the recruitment of a larger number of distinguished scholars. The monarch showed confidence in Euler's identification of talent and was aware of his extensive correspondence with German scholars and practitioners in the sciences. Euler was already expanding his influence beyond the Berlin Academy to universities throughout Frederick's domain.

In April 1754 Wolff, an Immortal (the term for members of l'Académie française, the foremost learned body on the French language), died at the age of seventy-five. In June Johann Joachim Lange, a University of Halle professor of mathematics, wrote to Euler about student interest in spherical trigonometry for astronomy and algebra. He realized that the future teaching of physics and mathematics instruction at the university depended considerably upon Euler, whom Frederick had appointed to choose a successor to Wolff as professor of mathematics and departmental chair; the monarch saw this selection as important to Halle's attaining eminence among German universities. Euler replied that no talented Swiss candidate would accept the post. While German universities did little to promote mathematics and physics, two scholars located there would qualify: Segner, a Hungarian from Göttingen, and Georg Friedrich Bärmann from Wittenberg. Since both had good positions, Euler was unsure of their interest. Although certain that Daniel Bernoulli would decline, Euler first offered his friend the Halle post. Upon the king's reception of Bernoulli's refusal in September, Euler was free to negotiate with Segner. This posed a challenge for the generally plainspoken Euler, for Frederick wished him to maneuver deftly in opposing the Wolffians and negotiating the smallest possible salary.

Euler wanted Segner—who was a severe critic of the mathematical and scientific foundations of the Wolffian philosophy though he shunned metaphysics, religious debates, and the old charges of atheism against them—as department director. In 1741 Segner had identified errors in the improved edition of Wolff's *Elementa Matheoseos*, particularly on the algorithms of imaginary numbers. In not correcting these mistakes, Euler believed the Wolffians were damaging their cause. By mid-1754 he mistakenly thought that the Wolffians were a spent force.[27] Since 1742 Segner and Euler had regularly exchanged scientific ideas and shared new knowledge, addressing the principles of mechanics, the nature of mathematical and physical truths, Euler's optics, Pierre de Fermat's conjectures, conic sections, Colin Maclaurin's *Geometrica organica* of 1720, cartography, chemistry, and the nature of the barometer and thermometer. Segner's hydraulic machine and mathematics were major topics. His reputation for being a talented teacher and author of good course handbooks in arithmetic and geometry was growing, and his exchange of ideas with Euler on texts had helped in his teaching. Euler expected him to continue to contribute to mathematics, physics, mechanics, and technology, especially improving hydraulic turbines and telescopes.

In September 1754 Euler invited Segner to Halle, and he quickly agreed to come. In a letter to Frederick II, Euler regretted that Bernoulli had declined an offer but expressed his joy at Segner's acceptance. In October the monarch instructed his minister Karl Ludolf Danckelmann to complete the arrangements. Several times Euler and Segner discussed Johann Albrecht Euler's winning the prize of the Göttingen Royal Society of Sciences in 1754 for his article on using water power to drive grinders and other machines. Euler had Segner elected a foreign member of the Russian Imperial Academy, and in November he asked Segner to send his clever mathematics student Johann Matthias Matseo to Saint Petersburg as a surveyor or architect. Segner observed that Franz Ulrich Theodor Hoch Aepinus, who would soon come to Berlin, was drawn to the research of Euler. In November the king congratulated Euler for obtaining Segner and expressed his satisfaction but complained about the cost.

Euler was not done with recommendations for Halle. He asked Johann Joachim Lange about instruction in mathematics and physics at the university, along with the instruments there for experimental physics. The Wolffians had wanted the relatively weak Lange to get the chair position and advance the Wolffian philosophy. Euler successfully recommended the distribution of courses between Lange and the vigorous Segner; he saw cooperation between them as a duty and urged that introductory lectures in the sciences begin with foundations. In November the king expressed his approval of these curricular plans. Lange informed Euler that Wolff had assembled an impressive collection of instruments, including an apparatus from Pieter van Musschenbroek in Leiden. Euler believed that it was vital to fund equipment for chair positions in the sciences for expanding and adding new details in research. He persuaded Frederick to purchase these instruments from Wolff's son for Segner's research. By January 1755, the king agreed to pay between eight and nine hundred Reichsthaler for them, but he did not sign the appointment until 6 March, declaring that the affair over the hiring and equipment was effectively concluded.

In April 1755, Segner and his family arrived in Halle, which allowed him to begin teaching his classes in the summer semester. In his correspondence with Euler, the two discussed improving mathematics instruction. Euler wanted Lange and Segner to enhance physical and mathematical instruction with themes of interest to students. He suggested lectures in logic, new chemistry courses, and curricular reform; he also proposed a program in the theory of hydraulic turbines. Although Segner faced difficulties from the Wolffian tradition, he was pleased to be in Halle. His appointment put the Wolffians at a disadvantage.

In April 1754 Gerhard Friedrich Müller became the conference secretary of the Petersburg Academy, and Euler recommended to him Mayer, the mathematician Georg Bärmann from Wittenberg, and Abraham Gotthelf Kästner, a mathematician and physicist from Leipzig; Euler lamented the poor condition of German universities and praised the great progress of Kotel'nikov in the study of higher mathematics. Through August of that year Euler made other German nominations, observing that as a geographer Mayer had far more to offer in Russia than in Göttingen. But he clearly wanted to gain Mayer for Berlin, where he offered to also name him director of the academy's geography section. Although eager to work with Euler and the other scholars at the Berlin Academy, Mayer indicated in August that he was hesitant to accept a post with an annual salary under 700 Reichsthaler, but he would accept the Berlin offer of 650 Reichsthaler plus another 100 for moving expenses, which he thought hardly sufficient.[28] Mayer looked forward to finding original observations by Gottfried Kirch and to constructing the calendar in a more popular and more convenient format with lunar ephemerides following his new tables, which were accurate to within thirty seconds. Using the ancient Babylonian number base 60, astronomers had 1 degree = 60 minutes = 3,600 seconds. Euler foresaw one of Mayer's first duties in Berlin as being the preparation of the lunar tables with explanations to justify their precision; the tables, useful for determining longitude, would aid navigation. Euler promised to have the Berlin Observatory turned completely over to Mayer.

In July 1754 Euler encouraged Mayer to enter the competition for the £20,000 sterling prize that the British Parliament had promised in 1714. It sought to perfect to within one minute the method for determining the longitude of the moon at sea and thereby to reduce errors in the positions of ships to less than fifty nautical miles. The surest method for this, Euler said, would be the study of lunar motion. Earlier attempts to find longitude foundered in efforts to establish the exact locations and motion of stars measured from sea; finding from sea observations the true positions of more fixed stars could confirm the longitude of the moon as given in Mayer's tables.[29] Euler had removed the obstacles posed in these general stellar studies. If he could find the lunar longitude at sea with the same exactness to thirty seconds, compared to the efforts of others—including Newton—that were off by at least four minutes, he would deserve the longitude prize. At that time Mayer was working on such a method, expecting observations to confirm his computed results. Assuring total errors of less than one minute would give the precision needed to win the prize. Probably only Euler fully understood Mayer's methods, which also delayed the acceptance of his tables.

On 27 August Maupertuis modestly exceeded Mayer's financial request by authorizing a salary of seven hundred Reichsthaler plus an additional hundred for moving expenses; he also offered free housing. In another letter that same day asking Mayer to come to Berlin no later than October, Euler enclosed the official academy letter with these terms and the academy seal; he pledged to help familiarize the recruit with the new position, and he promised that the academy would quickly recognize Mayer's diligence and research with an increase in salary. In a letter dated two weeks earlier, Müller on behalf of the Russian Imperial Academy had asked Euler to act as its intermediary in extending to Mayer an offer at a higher salary. Not until the end of August did Euler inform Mayer of the Russian offer and recommend its rejection. On 15 September Mayer asked Euler to transmit to Müller his refusal to accept the Russian post.

When Mayer requested that the Georg-August Academy accept his resignation, though, he was met with a surprise: the Hanoverian government asked him what conditions would persuade him to remain. King George III, whose father had founded the University of Göttingen in 1737, wanted to retain his services at whatever cost. Maupertuis recognized that tensions between Hanover and Prussia would influence the decision. To Mayer's surprise, the Hanoverian authorities exceeded his conditions, offering him a salary larger than that of the Berlin Academy and even to name him the sole director of the nearly completed observatory. (Until this time the management of the observatory had been a shared position.) In a skillful letter dated 6 October 1754, Mayer reported to Euler the grounds for his decision to remain in Göttingen, and that he was especially pleased to be named the sole director of the observatory. He added that among Hanoverian authorities circumstances caused him "to be thought of more highly than I deserve."[30] To Euler he said that he would be eternally grateful for the Berlin offer, and feared that "you would now become impatient with me and deprive me of a part of your invaluable favor and written association."[31]

In 1754 d'Alembert, who like Clairaut, Euler, and Mayer continued to investigate lunar motion, published his *Recherches sur differens points importants du systeme de monde*, which had been completed three years earlier. Relying less on observations than on mathematical techniques, he was one of the first to apply to celestial mechanics an iterated sequence of approximations. In improving lunar theory Clairaut, not Euler, was d'Alembert's chief opponent, criticizing his "long and tedious calculations" and finding his work careless and his tables inferior.

In his lunar tables in 1753 and his *Theoria lunae juxta systema Newtoninanum* completed in late 1754, Mayer systematically derived lunar inequalities arising from Newtonian gravitation, providing a better set of lunar tables, accurate enough to find longitude to within a half minute. Upon the opening of the Göttingen Observatory in the summer of 1753 and its completion the next year, which included the installation of a superior six-foot-radius mural quadrant, he began an enhanced series of observations in positional astronomy. His *Theoria lunae* depended on improved practical techniques of observation and removed instrumental errors. From his many observations and exhaustive computations in lunar theory, Mayer created formulas with more accurate terms yielding, unlike Euler's computations, a true rather than approximate lunar anomaly. His results relied on progressively modified corrections. Already in the preface to his lunar tables of 1753 Mayer claimed that his numerical results were more accurate than those of the celebrated Euler, Clairaut, and d'Alembert. In the preface to his *Theoria lunae* Mayer noted that they had agreed that their results had inaccuracies of three minutes or even more than five minutes of longitude. By their own admission, none of these geometers had errors of less than three minutes, which was insufficient to determine longitude at sea within a degree.

In the appendix to his *Théorie de la lune* and *Recherches sur differens points*, d'Alembert expressed doubts about the accuracy and validity of Mayer's work, since the methods that underlay Mayer's formulas were unclear and his tables ignored terms that d'Alembert considered large. For one instance, d'Alembert suggested that Mayer had employed the little-understood equations of condition, a statistical technique for ascertaining simultaneously a large number of unknowns. In responsc, Euler praised Mayer's findings and rebuked d'Alembert for his "unfounded and jealous allegations."[32] To Mayer he wrote in June 1754 with a touch of German nationalism,

> Now, to come to your important discoveries, I first of all congratulate you wholeheartedly on them, and wish that their importance would soon be known to everyone. Your first tables have indeed aroused as much applause as amazement; only jealousy has already let itself be seen more than clearly. You have undoubtedly seen M. d'Alembert's new treatise on the Moon, in which he refers in a contemptuous way to the tables and mocks the previously given exact determination itself; he does not believe that you have made use of the theory and the claim that he and M. Clairaut were the first who investigated the theory of the Moon. Notwithstanding my already long published tables whose quite new construction permits the theory to be easily satisfied, although the errors are still too large, he will deny me almost all parts [of it.]. You would not be behaving badly if you by chance refuted this braggard to the fame of the Germans . . . you brought things so far through your untiring industry that your tables denote the longitude of the Moon accurately to 30 seconds and the latitude even to 10. There is no further improvement to be hoped for from this side [i.e. theoretically], and M. d'Alembert himself stated that his tables are still over 5 seconds in error.[33]

A New Correspondent and Lessons for Students

In *Pensées sur l'interprétation de la Nature* (Thoughts on the interpretation of nature, 1754), a short collection of aphorisms on epistemology, Diderot proclaimed that his contemporaries were "at the dawn of a great revolution in the sciences" and that pure mathematics was approaching its limit. He added, "I would almost go as far as to assert that, within the next hundred years there will hardly be three great geometers in Europe. This branch of science will just cease at the point where the Bernoullis, Euler, Maupertuis, Clairaut, Fontaine, and d'Alembert have left it. They have built the pillars of Hercules and no one will pass beyond. Their works will endure in centuries to come, like the Egyptian pyramids, massive and laden with hieroglyphs, an awesome picture of the might and resources of the men who built them."[34] Recognizing that the abstract mathematical sciences had made great progress, Diderot criticized them and the other sciences for concentrating "on comparing and combining known facts rather than collecting new ones." His Baconian treatise decried the neglect of critical empirical studies and looked to the power of newly developed scientific instruments. Having written articles on calculus and mathematics for the lottery and working closely with d'Alembert, his science editor for the *Encyclopédie*, Diderot was well informed in mathematics. But like the notion that Newton and Leibniz had reached the boundaries of that science, his belief that the growth of pure mathematics was essentially complete would quickly be proven incorrect. Euler was to continue making seminal contributions to calculus and other areas of pure mathematics and to be the principal scholar in transcending their limitations. Early in the next century, Carl Friedrich Gauss and Augustin-Louis Cauchy would lead much as Euler was presently.

In late June 1754 Euler gained a significant new correspondent who would extend the horizons of mathematics. From the Sardinian capital of Turin, the unknown eighteen-year-old Giuseppe Lodovico LaGrangia, later called De la Grange or Joseph-Louis Lagrange, took the initiative of writing in Latin to Euler. This was the first of thirty-seven letters between the two men that continued until 1775; the correspondence was extraordinarily rich in ideas. Lagrange's initial letter examines the formal analogy between the binomial formula $(x + y)^n$ and the successive differentiation $d^n(uv)$ of a product of two functions u and v. He also asked if the famous Wolff was dead. This first letter reveals a novice not yet able to seriously gain Euler's attention; he did not answer it, but he characteristically kept it for his records. Shortly afterward, in reading the correspondence between Leibniz and Johann I Bernoulli, edited in 1745, Lagrange saw that Leibniz deserved priority for the analogy between powers and differentials. Fearing a charge of plagiarism, which did not occur, he wrote quickly to Fagnano. Lagrange's first letter to Euler ends with a promise soon to send some of his ideas on the theory of surfaces and maximal and minimal values, a foretaste of the main subject of their letters over the next decade in what Euler would eventually name the calculus of variations. Lagrange's letter suggests that he had begun to reformulate and probably made some advances in Euler's variational calculus and its applications.

During 1754 and 1755 Euler pursued magnetism, optics, and the education of Russian students working with the Royal Academy of Sciences in Berlin. He agreed with Daniel Bernoulli that Johann Dietrich of Basel was an extraordinary practical mechanic who had markedly improved the magnet needle. Employing Dietrich's methods and instruments, Euler investigated how to make magnets stronger. At Schumacher's request, he asked Dietrich to keep his new instruments secret. Euler observed that English and French journals contained many mistaken articles on the physical causes of magnetism, which was later made a topic for the academy's annual prize. His letters also address the theory and construction

Figure 11.4. Joseph-Louis Lagrange.

of dioptric telescopes and microscopes. (Dioptrics is the theory of the refraction of light, the study of the bending of light at boundaries between two media.) Euler discussed with Müller his teaching of Russian students, deeming the three he currently had to be his most talented. In 1754 the future mathematician and academician Mikhail Ivanovič Sofronov arrived in Berlin and was to reside with Euler for nine months until April 1755. Later Euler would comment that at one point Sofronov had been his most brilliant student, but at this time he expressed concern about Sofronov's problem with alcohol. That year Stepan Rumovskij joined Kotel'nikov in Euler's home; both aspiring astronomers would remain there until the summer of 1756. Kotel'nikov had already been in Berlin since 1752 as a classmate of Johann Albrecht Euler and the Genevan Louis Bertrand.

The daily lessons of Euler's students covered mathematics, mechanics, hydrodynamics, and astronomy. He related to Schumacher and Müller his close attention to the beginning of university instruction of these young scholars before they had any scientific accomplishments. "When the student is not pressed," he wrote to Müller, "he can still bring something better to light. Before he begins to publish, he must labor over a long time to collect a stock of ideas and discoveries. Once he comes in possession of these, he needs afterward only to explain these further according to others and to bring their studies to perfection."[35]

In August 1754, Euler issued a fourth list of problems, sending it to the Petersburg Imperial Academy; he recommended these problems for academy research and some for prize competitions. The list contained fourteen mathematical problems, broadly defined, written on the reverse of the second draft giving the list of physical problems that he had prepared. It was his first list for the academy on mathematics. Among its topics were acoustics, the theory of the Archimedean screw, a Newtonian wave theory, a theory of a moving ship in a turbulent sea, the three-body problem, and the bringing of telescopes and microscopes to their highest perfection. While intended for the next year, the list of problems was not actually read to the Russian Academy until July 1757.[36]

In 1754 Jean-Philippe Rameau wrote directly to Euler about the octave. In Rameau's theory of harmony, it was a fundamental element and essential for resolving the seventh, the first dissonance, into consonance, but it was not a mathematical changing point in music. Euler responded with "Nouvelles reflexions de M. Rameau," directed against the Encyclopedists, especially Jean-Jacques Rousseau, who was in conflict with Rameau. Where Rameau defined harmony as the basis of music and had it depend upon mathematical relationships, providing the basis for its modern theory, Rousseau in his *Lettres sur la musique française* (1753) and *Examen de deux principes avancés par M. Rameau* (ca. 1755) held to the primacy of melody, criticized the use of mathematics, and praised Italian over French music.

Euler tutored his son Johann Albrecht, who made rapid progress in mathematics and the natural sciences; it had by now become apparent that Johann Albrecht's interest lay more in application than in theory. The two had collaborated before, when Johann Albrecht had won the prize of the Göttingen Royal Society in 1754. Both that academy and its prize had been established through the efforts of Albrecht von Haller. Johann Albrecht Euler's paper on the most advantageous way to use water to run mills and machinery had aroused great expectations for his future.[37] His father helped him with the paper but expected him to make further progress by continuing to apply himself to the sciences, and Johann Albrecht benefitted from having Kotel'nikov and Bertrand as classmates. Further to encourage Johann Albrecht, Maupertuis included him among the six people nominated in December 1754 for membership in the Berlin Academy, though with a meager annual salary of two hundred Reichsthaler. The members unanimously approved the nomination, and Euler was joyful at his son's recognition; Johann Albrecht was only twenty.

Having recently learned of the promotion, Goldbach in April 1755 offered heartfelt congratulations to both father and son: "I am certain," he wrote, "that he has already acquired extraordinary knowledge in mathematics. Nevertheless he will still have to endure, in case he is a lover of scholarly disputes, to be contradicted by his antagonist";[38] he stressed that Johann Albrecht would need to respond independently rather than as Leonhard Euler's son. Euler's lengthy response to Goldbach maintains,

> Es ist jetz[t] das mathematische Studium so weitläufig, dass es eine lange Zeit erfordert, ehe man sich in allen Teilen so fest setzen kann, dass man ohne Anstoss etwas namhaftes darin zu leisten imstand kommt; dahero er freilich ohne meine Hülfe noch nichts Sonderliches würde zum Vorschein bringen können. Insonderheit muss er sich ja in keine gelehrte Streitigkeiten mischen. . . .
>
> It is now [true] that the studies of mathematics have become so multifaceted that it takes much time before one can gain solid ground in all parts, so that without incentives he can accomplish something substantial therein. So [Johann Albrecht] independently without my help would still not be capable of bringing anything considerable to light. Especially, he must participate in no scholarly disputes. . . . [39]

Institutiones calculi differentialis and Fluid Mechanics

In 1755 Euler published another masterwork in mathematics, his two-part *Institutiones calculi differentialis* (E212), the second component of his trilogy on calculus.[40] The book was probably begun around 1727, but it was mostly finished in 1748 and completed two years later, when Euler was forty-three. For the previous decade, he had worked on it steadily.[41]

Institutiones calculi differentialis is the first textbook to organize systematically the hundreds of important discoveries made since the time of Leibniz and Newton. Today it is mainly remembered for the definition of the concept of function, which stressed not the role of formulae but the more general idea of a formal correspondence between two sets of numbers. This new definition, which looked forward to the modern concept of mapping between two sets, was probably motivated by the controversy with d'Alembert and Daniel Bernoulli over the vibrating string.[42] The book began with the first didactic presentation of the calculus of finite differences. Euler viewed differential calculus as the limit of the calculus of finite differences when the differences become "infinitely small." This was a sound idea, but Euler did not possess a formal theory of limits, and so his program failed at a fundamental point. He had to resort to the idea that the differential, an infinitely small quantity, is "a true zero" and to formalize differential calculus as a "calculus of zeroes." This led to a series of hazardous statements. For example, corollary 83 in chapter 3 noted that the differential "could certainly not be anything but zero, for if it were not = 0, then a quantity equal to it could be shown, which is against the hypothesis" and that "there are not so many mysteries hidden in this concept as usually believed. The mysteries have rendered suspect to many people the calculus of the infinitely small. Doubts that remain we shall thoroughly remove in the following pages."[43] The exactness of mathematics, Euler argued, required that the differential equal precisely zero. Thereby he refined the classical notion of infinitesimals as quantities tending to zero. He wrote, "[E]very quantity can be diminished until it vanishes completely, and is reduced to nothing. But an infinitely small quantity is simply an evanescent quantity and therefore actually equal to zero."[44] His vague definition of infinitesimals as quantities smaller than any fixed number looked back to the ideas of Johann I Bernoulli, and it would remain the accepted formulation of calculus for several decades. Euler's work had the merit, however, of having removed the indeterminacy of higher-order differentials and making what we now call the derivative of a function more prominent than the differential.[45]

The second part of *Institutiones calculi differentialis* contains an impressive array of important results, many of them found by Euler himself. Chapters 5 and 6 elaborate his summation formula for the Basel Problem and what would later come to be called the Euler-Maclaurin Formula. The results of the Bernoulli numbers were many, starting from their generating formula and going on to their application at the summation of power series and connection with the Riemann zeta function. Euler found several properties of these numbers, which had applications in many fields of mathematics, and their computation provides a challenging problem even today. With them Euler obtained exact sums of power series of even reciprocals. Among the questions that Euler studied in chapter 6 are the partial sums of the harmonic series, the Euler constant gamma, the value of π, and approximated formulae for large factorials.

The book was to be extremely influential, and is now considered one of the most important scientific texts of the eighteenth century. But its first impact on the scientifically educated public was slight and disappointing: after six years, 406 of the 500 copies of the first edition remained unsold.[46] The state of university education in calculus at that time suggested that a wider readership did not yet exist for a book written on that level.

In transforming infinitary analysis and applying it skillfully to astronomy, mechanics, and optics, along with advancing analytic number theory, Euler was bold, nearly tireless, agile, and occasionally bizarre. He sought to bring more coherence, order, and simplicity to each of these fields and to reduce seeming complexities and disjointed areas. Carrying out his massive computations required that he devise fruitful analogies, algorithms, and formulas. In the invention of unorthodox methods for summing infinite series with deft interpolations, approximations, and substitutions, he was unmatched. Like the Bernoullis, Euler had a sharp sense of the problem, looking for other geometers and natural philosophers with worthwhile solutions illuminating the whole. His analytical intuition was peerless. When temporarily blocked from making further computational advances, he turned to pursue other topics and awaited a breakthrough whenever he or others made them. Possessing them, he returned to problems to refine methods and to provide another round of extensive computations.

In these calculations Euler was essentially a formalist. He placed great demands on himself and was almost always correct; to confirm his equations in mechanics, astronomy, and optics, he required the most accurate and thorough observations and experiments. He sharply distinguished between results derived from measurements and those devised from reason; he did not ignore or seek to remove empirical measurements but wanted to lessen the need for them. Although he made few errors and had scarce lapses in rigor, his work has been portrayed as "happy-go-lucky analysis," for his methods in calculus do not conform to the standards of rigor of the late nineteenth century.[47] This evaluation might apply to another with a lesser intuition or without a penetrating sense of problems, but is mistaken for Euler. As a matter of fact, in his day satisfactory proofs of convergence were only beginning to emerge; Euler had supplied one, the integral test. Only with the work of Augustin-Louis Cauchy in the early nineteenth century would come a general theory of convergence and much of the rigorous foundation for calculus.

In his *Novum Organon* of 1620, Francis Bacon had offered a powerful symbolic metaphor in his aphorism 1.95, rooted in antiquity, of ants (mindlessly gathering), spiders (producing works from their own substance, like pure rationalists), and bees (observing, collecting, digesting and analyzing, and transforming) that might be applied to Euler. These terms were commonly used at the time for the study of nature. Bacon argued that good natural philosophers were unlike ants and spiders. In this metaphor, which Bertrand Russell found somewhat unfair to the ant, Euler would be a splendid bee.

Expanding upon d'Alembert's pathbreaking but imperfect work in fluid mechanics, from 1752 to 1755 Euler wrote a three-part series of masterful articles that laid the foundations of modern fluid dynamics. He had read to the Berlin Academy, reportedly in August 1752, a treatise titled

"De motu fluidorum in genera" (On the motion of fluids in general). According to Clifford Truesdell, the article was probably published much later in 1761 under the title *Principia motus fluidorum* (E258).[48] In 1753 Euler read another paper that became the first in his series, completing its final version in 1755: "Principes généraux de l'etat d'equilibre des fluides" (General principles of the state of the equilibrium of fluids, E225). In 1757 all three parts were published in volume 11 (for 1755) of the academy's *Mémoires*.[49] The articles reduce to differential equations the entire theory of hydrostatics. Euler was perfectly conscious of the importance of this step. At the beginning of the article he wrote, "The generality I here take on, far from dazzling our enlightenment, reveals to us the true laws of Nature in all their brilliance, and there we shall find even stronger reasons to admire her beauty and her simplicity."[50]

This article contains the differential equations known today as the Euler equations for hydrostatics. He introduced the pressure (p) in the modern sense: not as a force against the walls of the vessel but instead as a measure of the internal stress of the fluid.[51] For dynamic principles he proved that there is no essential difference between compressible and incompressible fluids. These three articles on fluid motion marked a turning point, with the continuum view of matter advanced as an essential principle; this was a fundamental contribution to rational mechanics.

At the Berlin Academy in September 1755, Euler presented the second section of the treatise, "Principes généraux du mouvement des fluides" (General principles concerning the motion of fluids, E226), which places the entire theory of fluid motion on the same level as his principles of hydrostatics and their equilibrium in the previous article. Here Euler obtained the dynamical equations for ideal fluids, one of the greatest achievements of all time in mathematical physics. It is difficult to summarize the paper, but every single result appears in modern treatises. For example, Euler gave a counterexample to the hypothesis that velocity necessarily admits a potential. He showed that the particles of a fluid have fewer restrictions than applied for solids. He also gave boundary conditions for related partial differential equations. To clarify when velocity potential existed, Euler gave counterexamples of simple vortex flows. The theory of fluid motion he reduced to the solving of various analytic formulas.

At the beginning of the paper Euler derived the partial differential continuity equation $\partial u/\partial x + \partial v/\partial y/ + \partial w/\partial z = 0$, where u, v, and w are the components of the velocity. His derivation is similar to one of d'Alembert's in d'Alembert's paper for the 1750 Berlin Academy prize, but Euler did not mention d'Alembert. Relations between the two were touchy, and the reason for this silence is unclear. It does not seem likely that Euler's failure to mention d'Alembert was simply an oversight, and it deepened the chasm between the two men. Their correspondence in 1750 contains no mention of fluid mechanics.

The third section of Euler's treatise, "Continuation des recherches sur la théorie du mouvement des fluides" (Continuation of the research on the theory of fluid motion, E227) augmented "Principes généraux du mouvement des fluides." It applied the theory developed in the second section to specific examples, offering a detailed theory of flow in tubes, and for the first time dealt with compressible fluids. Euler put his work in historical perspective, noting, "However sublime are the researches on fluids which we owe to Mssrs. [Johann I and Daniel] Bernoulli, Clairaut, and d'Alembert, they flow so naturally from my two general formulas that one cannot sufficiently admire this accord of their profound meditations with the simplicity of the principles from which I have drawn my two equations, and to which I was led immediately by the first axioms of mechanics."[52]

This statement, sometimes interpreted as Euler's disparaging the achievement of his predecessors and inflating his own, followed his pattern of praising significant accomplishments of his predecessors. He took a more general point of view that recognized prior work.

A New Telescope, the Longitude Prize, Haller, and Lagrange

In the four years after 1752, Euler had broken off relations in astronomy with d'Alembert and stopped corresponding with Clairaut. No evidence of a falling out with Clairaut exists; more likely their interests diverged. After Euler failed to be elected a foreign member of the Paris Academy of Sciences in 1753, which d'Alembert called a great injustice, the next vacancy occurred in 1755 when Abraham de Moivre died; but that position was reserved for the president of the Royal Society of London, the astronomer George Parker, Second Earl of Macclesfield. Undeterred, Clairaut probably helped arrange in June to have Euler elected simultaneously as an associate member. The honor, Euler wrote to Wettstein in August, "has given me a great deal of pleasure."[53] That letter also expressed his impatience in waiting for the latest volume of the *Philosophical Transactions* of the Royal Society. Euler requested two copies, one being for Johann Theodor Eller. A letter in November addressed his extensive work in dioptrics to perfect the telescope. His design of an achromatic nine-foot telescope with multiple lenses, the surfaces of each improved to reduce refraction and give greater clarity, differed from the telescope of Dollond that had been examined in the previous volume of the *Philosophical Transactions*. Euler believed that his design was superior, and he urged craftsmen in London to manufacture his new telescope.

In May 1755 Euler informed Mayer that the English, after exhibiting complacency in basing lunar tables on Newtonian theory, would recognize the superiority of Mayer's recently improved lunar tables and should award him the maximum prize of the £20,000 sterling that the British Parliament had promised since 1714.[54] Mayer had perfected to within one degree the method to determine the location of the moon, which allowed for obtaining longitude at sea to a half degree and latitude to ten minutes, giving the precision required to win the prize. The discovery of this method alone deserved part of the longitude award, and at this time no further improvements were likely. So encouraged, Mayer sent to English officials with Hanoverian connections and then to the English Board of Longitude his lunar and solar tables that applied differential equations from Euler, derived the formulas on which Mayer had based his latest tables, and thus claimed he was due the longitude prize. Euler was very curious about this matter and asked to be kept informed of its progress. James Bradley, the royal astronomer, unsuccessfully recommended Mayer's tables for the longitude prize. Euler believed that the English commissioners had done Mayer an injustice by explaining that since they claimed the tables were in print, they were ineligible as an original work for the award. Euler knew that Mayer's latest tables allowed for computing longitude to thirty seconds compared to the previous four-minute differential in others, that they were not the ones in the Göttingen *Commentationes*, and that these new tables were yet unpublished. Since the competition winner had to demonstrate practical applications, Mayer's method had to be tested at sea, but the Seven Years' War, beginning in 1756, precluded this for a time. The English wanted John Harrison's clockmakers to develop an alternative that was more exact.[55]

After the Swiss anatomist Albrecht von Haller expressed interest in working for Frederick II as the curator and a professor at the University of Halle, in May 1755 the king instructed Euler to invite him to those positions. Haller, a distinguished naturalist and physician as well as a previous student of Hermann Boerhaave in Leiden, had retired two years earlier as a professor of medicine, botany, and surgery at the University of Göttingen. Euler was not enthusiastic about the potential appointment of Haller, believing that Segner would contribute more to scholarship. Complying with the king's wishes, he proposed a salary of two thousand Reichsthaler; in mid-August Haller requested three thousand Reichsthaler, stressing that he preferred the leisure and freedom of his native

Bern to joining a prominent scientific institution with a modest salary. He also mentioned that he did not wish to undertake the Prussian post for another decade. Frederick found excessive Haller's requests as regarded salary and benefits. He apparently thought that Haller was negotiating with Prussia only to leave Hanoverian service and gain more benefits from the University of Göttingen. The monarch then left the further negotiations to Euler, who offered 2,400 Reichsthaler, which Haller also rejected. In October and November Euler and Haller exchanged letters, and in December Haller declined the Halle professorship because his terms had not been met. Instead he chose to remain in Bern, where he had moved in 1753, to enjoy its liberty and to participate in public service contributing to improving the school system, the care of orphans, and city economics. Haller wrote to Maupertuis that he was unsuited by nature to court ceremonials and the king's roundtable with its conduct that ran counter to his timidity.[56] Frederick was happy with Euler for attempting to increase the honor of Prussia and his royal self, and he was pleased that Euler had him elected one of the ten foreign members of the Paris Academy of Sciences in 1756. The relations between the two men were at their zenith.

In August 1755 Euler received the celebrated second letter from Lagrange, which was revolutionary in that it proposed a way of eliminating the tedium of Euler's geometric considerations in the *Methodus inveniendi* by reducing variational problems entirely to analytic techniques. In a succinct description, Lagrange introduced a new algorithm, analogous to ordinary differentiation of functions and denoted by the Greek letter delta (Δ), which reduces to a set of algebraic rules the solution of variational problems. In September Euler wrote that Lagrange had bought the theory of maxima and minima to a high point of perfection in its generality and usefulness.[57] Lagrange's letter started a major development in what Euler now renamed the calculus of variations. To be sure to give Lagrange full credit for the discovery, he withheld articles on the subject until Lagrange had perfected his ideas and published them in the 1760–61 volume of the *Miscellanea Taurinensia* (Turin miscellany).

Anleitung zur Nauturlehre and Electricity and Optimism Prizes

Besides making prolific and seminal contributions to mathematics and introducing a succinct notation, through his style Euler advanced and shaped the field perhaps more than any of his contemporaries.[58] To his stress on collaboration, participation in fierce competitions, and open exchange of methods and results through publications and extensive correspondence, Euler added another dimension to his style in the Lagrange case. While he already gave credit to others where he deserved it, such as in the case of Maupertuis and the principle of least action, he now went further by not simply applauding the achievement of a talented junior scholar but by generously allowing that scholar time to prepare and publish his work first; this would resolve any question of priority. In his "Dialectics of Painting" (ca. 1647) the great French artist Nicolas Poussin, who emphasized the importance of clarity, logic, and order in the arts, had defined the term *style*—in a way that may be applied to Euler—as "a particular manner and skill . . . which comes from the particular genius of each individual in his way of applying and using ideas. This style, manner, or taste comes from nature and intelligence."[59] Having Lagrange's powerful tool, Euler proceeded over roughly the next decade and a half to invent the analytical form of the calculus of variations.

Possibly as early as 1755, but more likely a few years later, Euler completed his *Anleitung zur Naturlehre* (Instruction for natural philosophy, E842) in 162 folio pages, his most important work on the nature and laws of the theory of matter. It again rejected Wolffian monads.[60] This book did not affect the continuing debate over the smallest amounts of matter, for the particular manuscript, which was kept among Euler's numerous writings in Berlin (most of which were published) was fated to be forgotten for almost a century until Paul Heinrich von Fuss discovered it in 1844, publishing it in the *Opera postuma* in 1862. Euler's increasing attention to astronomy and geometric optics, his belief that he had already disproved monads, and Wolff's death in 1754 may explain its being set aside; Euler did not want to argue with a recently deceased opponent. Chapters 1–5 examine the basic properties of matter, centering on impenetrability, while chapters 6–11 explain forces. Chapters 12–19 discuss the ether, subtle celestial air, and mutual attraction or gravitation.

Continuing the recent investigations in Berlin and Saint Petersburg of the physical causes of electricity, the Russian Imperial Academy made it the prize topic for 1755. At the time, that institution was at an ebb in its research; according to its *Protokolii* (Protocols), essentially its conference minutes, only three to five members attended most meetings. Johann Albrecht Euler submitted a paper, "Disquisitio de causa physica electricitatis," that won the prize. It explained electricity as stemming from the least and most elastic part of the ether that penetrates the pores of all matter; rubbing together two bodies of different masses produces this electricity. Since the elder Euler had written on the topic and knew Lomonosov's research, it was thought that he had written the paper, but his position with the Imperial Academy in Russia made him ineligible to participate. In September 1755 the conference secretary Müller sent an inquiry to Euler requiring that the author indicate his name on the first and last page of the dissertation. Euler responded that while he had developed the general theory, his son Johann Albrecht deserved the prize, having worked out the particulars. The text and his son's name were in Latin.

For 1755 the philosophy department of the Berlin Academy had chosen as the prize topic an examination of the plausibility of a version of Alexander Pope's optimism from the French translation of the *Essay on Man* that *tout est bien* (all is for the best, or whatever is, is right). During the devastating Lisbon earthquake on the morning of 1 November, All Saint's Day, many died in churches, and it was followed by a tsunami and five days of fire that destroyed most of the city, so the effect was to challenge the notion of optimism. This particular competition popularized a supposed affinity between Pope and Leibniz, who could be trivialized as proposing a theory of the best of all possible worlds. Johann Georg Sulzer, Christoph Martin Wieland, and their Wolffian allies in Zurich still believed that a Leibnizian triumph was possible. Sulzer had opposed the selection of the prize topic, worrying that this might be used as an opportunity to belittle Leibniz, but he thought that in the academy vote Johann Philipp Heinius and Formey were with him and Merian alone could not be effective. This would defeat Maupertuis's *Essai de cosmologie*, which found excessive most final considerations in physicotheology and rejected Pope on the problem of evil. Pope's *Essay*, published in 1733–34, asked how a divine creator who is good could allow evil and projected a distant clockmaker God. French freethinkers embraced its universalist spirit, while Euler opposed a narrow interpretation of Pope,[61] believing the problem of evil to be beyond human understanding. In the final vote Formey turned to back Maupertuis, and A. F. Reinhard won the prize challenging Pope's system. Before the winning article was published in autumn, an anonymous English-language treatise titled "Pope, a Metaphysician" argued against depreciating Leibniz and his theory; its authors, Moses Mendelssohn and Gotthold Ephraim Lessing, had entered the competition after the winner had already been chosen.[62] The academy was widely and bitterly criticized for giving the prize to what was deemed a mediocre article.

Chapter 12

War and Estrangement, 1756 to July 1766

Chief among the matters that Leonhard Euler discussed with the Russian Imperial Academy of Sciences from the end of 1755 through the first half of 1756 was the problem of paying for the printing of his two-volume *Institutiones calculi differentialis* (Foundations of differential calculus), which was under agreement with the publishing and paper trading firm of Haude and Spener in Berlin. Euler wanted to be sure that the best paper be used; after Ambrose Haude's death in 1748, the firm had been expected to use only paper of high quality. Apparently the Petersburg Academy had agreed to fund the publication. In September 1755 Johann Schumacher awaited from Euler two copies of the *Institutiones* and seven volumes of the work of Polybius. As the printing progressed, Euler found his first estimate of 229 rubles to be too low; instead he needed about 335. Since Schumacher was seriously ill, he tried to collect payment from Grigorij Nikolaevich Teplov in November. Euler's letter the next month to Teplov claimed that Kirill Gregor'evich Razumovskij had promised to pay for this project. Another letter from December wished Schumacher good health and repeated the financial request.

As usual in matters of finance, Euler was tenacious. In January 1756 he informed Razumovskij about his problem in obtaining the payment, and the next month he took it up with the academy's conference secretary Gerhard Friedrich Müller. A lengthy letter to Schumacher in February lists the exact cost as 325 rubles plus 2 percent interest. Euler believed that the greater expenditures might be retrieved from the subsequent sales of the book for 1.5 Reichsthaler. But payment at this time was absolutely necessary.

The Antebellum Period

In March, April, and May 1756, Euler repeatedly requested that Müller, his main correspondent in the Petersburg Academy that year, provide funds to cover the printing of his *Institutiones*, but all of his entreaties failed. In a letter in June Euler called it shameful that since that academy was not sending funds, he had to pay the cost from his personal means. He was fortunate to have sufficient income from winning the double Prix de Paris in 1756 for his fundamental article "Investigatio perturbationum quibus planetarum motus ob actionem eorum mutuam afficiuntur" (An investigation of the perturbations by which the motions of the planets are affected by their mutual action upon each other, E414).[1] Responding in June 1756 Müller expressed regret at Euler's using personal funds for the printing of his *Institutiones*, complimented him on winning the Paris prize, praised the research of the Milanese mathematician and physicist Paolo Frisi on electricity, and sent questions for Franz Ulrich Theodor Hoch Aepinus.

Euler continued in his attempts to prepare native scholars to fill academy and university posts in Russia. In December 1755, he had reported to Teplov that under his instruction in Berlin Semjon Kirillovič Kotel'nikov and Stepan Rumovskij had made great progress in mathematics, surpassing other foreigners in it. A letter to Schumacher the next April reviewed Mikhail Ivanovič Sofronov's grades. In June Müller wrote that Kotel'nikov and Rumovskij had returned to Saint Petersburg. For his education of Russian students, his publications in the academy's *Commentarii* and *Novi Commentarii*, and his later research circle in Saint Petersburg, Euler is credited with establishing a Russian school of mathematics. It is clear that he also founded the continuing Russian school of astronomy.

Euler's letters to Müller cover the arrival of the chemist Ulrich Christoph Salchow in Saint Petersburg to join the academy and the appointment of Gerhard Andreas Müller to the chair of physics at the University of Giessen, along with Euler's own research, especially on the path of Venus and on optical technology. In January 1756 Euler inquired about a rumored comet not yet seen in Berlin and announced his invention of a more powerful magic lantern, a machine that could magnify the representation of objects and project images of all sorts on a wall. For at least five years since writing "Emendatio laternae magicae ac microscopii solaris" (Improvement of the magic lantern and solar microscope, E196) in 1750 he had worked to perfect it,[2] addressing the problem of the more that objects were magnified the less distinct they became. The name *magic lantern* Euler traced back to a wish to suggest magic or witchcraft. It differed from the common camera obscura, a device that passed a bright light through a small hole in a thin material so that the rays would not scatter but cast images upside down on a flat surface opposite to the hole; applying only to figures painted on glass, the camera obscura was more limited than the magic lantern. Euler gave his improved version of the lantern to his

Russian pupils and another to Razumovskij. Like Galileo Galilei, he recognized the support that recreational devices might generate for further projects. After turning to dioptrical systems in "Régles générales pour la construction des telescopes et des microscopes" (General rules for the construction of telescopes and microscopes, E239, 1756) Euler did not pursue new theoretical research on magic lanterns.[3] In April 1756 he claimed that his optical research had produced the best construction to date of the telescope and microscope, and he recommended the purchase of magnetic needles from Swiss mechanist and instrument maker Johann Dietrich.

Numbers 196 and 197 of Euler's *Lettres à une princesse d'Allemagne* (Letters to a German princess), sent in 1762, describe the magic lantern in detail. Euler introduced larger wicks on lamps inside the machine and chimneys to let smoke escape and thereby help illuminate objects as much as possible. He observed that rays from the sun strengthened by what is called a burning glass could provide light for the lantern and turn it into a solar microscope. The camera obscura remained a popular topic; Francesco Algarotti would explain it in his *Saggio sopra la pittura* (Essay on painting) in 1764.[4]

During the first four months of 1756, Euler and Christian Goldbach exchanged six letters. Covering applied mathematics, M. Chauchot's winning of the Prix de Paris for 1755 on the rolling and pitching of ships, and the difference between metaphysical and mathematical proofs, the correspondence still stressed number theory. Euler attempted to prove the binary Goldbach conjecture that it is possible to express every even integer greater than 2 as the sum of two primes (for example, $10 = 7 + 3$ and $12 = 5 + 7$), but he was not successful. He gave fourteen theorems on properties of prime numbers and proofs, including by contradiction. These theorems dealt with such issues as when large numbers are prime.

In April Euler shared with Pierre Maupertuis the second letter from Joseph-Louis, Comte de Lagrange that had come the previous August. In a condensed form understandable only to Euler, it presented Lagrange's precise new algorithm for solving isoperimetric problems. Euler would soon coin the term *calculus of variations*, and asserted that "all of dynamics and hydrodynamics can be developed with astonishing ease by the single method of maxima and minima."[5] Euler believed that Lagrange had attained the greatest perfection in the field; his algorithm reduced to analysis Euler's abstruse geometric methods of finding maxima and minima, and it solved some of the problems in Euler's *Scientia navalis*—and others elsewhere—that Euler's method could not. Lagrange sought to demonstrate that by applying his algorithm to ordinary techniques in dynamics and hydrodynamics researchers could solve the most difficult problems. Euler was ecstatic. Although he had corresponded with Alexis Claude Clairaut and Jean-Baptiste le Rond d'Alembert, he had never addressed a genius comparable to himself in the sciences. Although Lagrange was only eighteen Euler treated him as an equal, and Maupertuis was also taken with Lagrange's sagacity. The letter justifies the principle of least action, which Maupertuis called the supreme law of physics. Lagrange's effort to prove its "greatest possible universality" much pleased Maupertuis,[6] and he recommended that Lagrange be elected to the Berlin Academy. Euler invited him to leave Turin and its artillery school and come under the protection of the Prussian king. In May Lagrange declined but asked Euler to thank Maupertuis for the proffered honor.

On 3 June Maupertuis presided for the last time over the Royal Academy of Sciences in Berlin. In November Lagrange was to demonstrate generally how to approach rapidly with a line and solve the brachistochrone curve with one end point free, a problem that Euler had previously solved only for special cases.

In June 1756 Maupertuis, his weakened lungs unfit for the harsh Berlin winter to come, departed from the city for the healthier climate and airs of Paris, Bordeaux, Toulouse, and Saint Malo; he would never return to Prussia. His increasing disaffection with the German intellectual community, his displeasure with the paucity of German achievements in his academy, and the tug of his loyalty between Prussia and France also figured in his decision. During his last two years in Berlin, Maupertuis had attended only sixteen meetings of the academy, but he oversaw finances, membership, and publications. Even after the Seven Years' War commenced in 1756, correspondence continued between the enemy capital cities Paris and Berlin. While in Prussia Maupertuis had retained a connection to France and now, in France, he would in part remain loyal to Frederick II, though this disturbed many of his friends; he and d'Alembert would even congratulate the Prussian monarch on his victories. Maupertuis wanted to return to Berlin via a route passing through Hamburg, but that port was blocked to him, and was it not possible to make a sea journey to Prussia leaving from Bordeaux.[7] Frederick recommended that his ailing president go to Italy to recuperate and lengthened the time allocated for his vacation, but Maupertuis did not go to Italy.

Maupertuis confidently left the academy in the hands of Euler, who resumed the office of acting president, a position that he had held during Maupertuis's earlier absence. While they differed on some membership selections, the two were now close friends; Euler was the dominant figure in the sciences, and Maupertuis in society, and both were religious. That Euler was a Reform Protestant and Maupertuis a Catholic presented no obstacle to their collaboration. The president trusted Euler, which strengthened his influence at the academy. Diametrically opposed to both of them in religion, Frederick II sought to add to the academy such free spirits as Louis Isaac de Beausobre in 1755. For the moment the king, occupied with the war, left the acting president in control of the academy's selection of members and finances; Euler believed both powers to be critical to an independent and professional institution. (Their subsequent loss to Frederick would eventually influence his decision to leave Berlin.)

In absolutist Prussia the old patronage system was shifting. While personal patronage ties still existed in the sciences with maneuverings and brokering among ministers, well-situated aristocrats, generals, and senior academicians were replacing them in the etiquette of the royal court.[8] As president of the Berlin Academy at midcentury, Maupertuis had sought honors such as salaries, titles, and recognition reflecting his noble status. Other leaders benefited from the patronage system, including Clairaut, d'Alembert, and Euler, but they and most of their academic colleagues did not consider such patronage necessary for participation in the life of the sciences. A commoner by birth, Euler at this point saw no need to pursue noble rank. While the academicians were free in selecting their intellectual stance, they publicly justified and gained financing for their scientific research and technical work by demonstrating its utility to the state and the public. Euler had to navigate the interests of the higher nobility at the royal court, including the state ministers, but again mainly met resistance because the role of science was not yet established in the Prussian royal state. While no longer courtiers except on special occasions, the academicians were not yet independent professionals, a status that Euler attempted to build for them. Euler and Maupertuis, who were not competitors, did not face conflicting ambitions. Only after the war would the king reassert total authority over the academy. Through letters exchanged with Euler and with Johann Bernhard Merian, who was named librarian of the academy in 1757 and often replaced Jean-Henri-Samuel Formey as secretary, Maupertuis continued to submit articles and helped to oversee the publications and finances of the academy.

Euler persisted in his effort to draw Lagrange to Berlin. Although in July 1756 Lagrange did not want to leave Turin, the next month Euler included his name on the list for election as a foreign member of the academy. Once he was so named Lagrange changed his mind and wanted to move to Berlin. His opposition to any attempt to disprove the generality of the principle of least action and his praise for the research of Maupertuis strengthened his case for regular membership in the Berlin Academy. "I count myself very fortunate," Lagrange wrote, "to be able, after the learned Euler to contribute in some manner to the universal application of that law." It "will always be esteemed as the most beautiful and important discovery of mechanics."[9] He did not, however, agree with Maupertuis that it was a fundamental metaphysical law; to that point only Euler understood its total application. Lagrange apprised Maupertuis that he would accept an advantageous offer, meaning a larger salary. Maupertuis instructed Euler to act as an intermediary in the recruitment and immediately arrange suitable compensation. In September Euler reported that to recommend Lagrange for the position Maupertuis awaited the return of the king to Berlin. In October Lagrange wrote to Euler that his salary must surpass what he was making in Turin.

In the fall of 1756, the disruption of the Seven Years' War brought a break to postal service between Berlin and Turin that effectively stopped the initial negotiations. In January 1757 Maupertuis recommended that the salaries freed up by the deaths of several members of the physics department, among them Johann Nathanael Lieberkühn, be combined to pay Lagrange. But Euler's efforts to obtain that salary foundered. Maupertuis had left Berlin, and in February Euler wrote to him, "We see little possibility of electing Lagrange. In the present circumstances [the war], no one dares make the proposal to the king."[10] And Lagrange's articles were not yet available. His salaried position in Berlin would not come for nearly a decade.

Just before the war started, Euler's Russian students and adjuncts had returned home, though Euler kept in touch with Müller. He was pleased with the progress of these students but had regularly criticized the Petersburg Academy for not fully funding their living expenses. Euler's correspondence with Müller continued without interruption throughout the war; in December 1756 he requested for a friend a copy of the horoscope of Nostradamus, published in Geneva in 1565, that was difficult to find.

Into the Great War and Beyond

In early 1756 a sharp realignment of alliances, called the diplomatic revolution, replaced the old arrangement in Europe that had pitted France and Prussia against imperial Austria and Britain. In January Frederick II secretly signed the Convention of Westminster with his old enemies the British to avoid an attack from Hanover, but he failed to inform his French allies of the convention, and they viewed it as a betrayal. Apparently Frederick thought unlikely a *rapprochment* between imperial Austria and France, but that reorientation became imperative for the Habsburg monarch Maria Theresa and leaders in her court after the loss of Silesia to Prussia in the first two Silesian Wars ending in 1748. Negotiations headed by the Austrian foreign minister and later state chancellor Prince Wenzel Anton Kaunitz had Austria join France in 1756, and France agreed to end its wars against the Austrian Habsburgs. This shift in partners made for a confessional (meaning religious) match, with Catholics facing Protestants. Orthodox Russia under Empress Elizabeth also joined Austria and France, and these three European powers formed a strong coalition that confronted Prussia. Frederick's unprincipled diplomacy and his taking of Silesia from Austria awakened a lust for *revanche* against him. The other states in coalition also wanted to reduce the sudden growing power of Prussia.

In August 1756 the Seven Years' War began with Frederick's invasion of Saxony. He claimed that this was a preemptive act of self-defense, but no evidence existed that Saxony was involved in preparing the plans against him that were under way in Vienna and Saint Petersburg. War on the part of Prussia against Austria and Russia was imminent. Britain, wanting to avoid military intervention, provided Prussia mostly with financial aid. The war was popular; the states in the west experienced a surge of patriotism, the concept of nation was developing among all classes beyond the concept of royal states. The Seven Years' War was to involve the very survival of Prussia and at the end confirm its new status of European power. While its principal theater was in central Europe, a clash in commercial interests between Britain and France took it to Africa, Canada, the Caribbean, India, and the Philippines; it was the first truly global war.

The conflict opened with Prussian victories in Frederick's conquest of Saxony and appropriation of its finances. A month after the king entered Dresden in September 1756, Euler reported to Johann Kaspar Wettstein that Frederick had ordered *Te deum* (based on Psalms 20:6) sung in all churches. Since there are two numberings of Psalms, it is either the prayer for the king in time of war or, more likely, thanksgiving and glory in victory along with prayers for the king. In mid-1757, Frederick defeated the imperial Austrians at Prague and Kolin. Although the Prussians had the initiative through 1757, these were not the decisive battles that Frederick sought. Instead the Austrians inflicted heavy losses on his forces, and the war was costly; the Prussian economy suffered sharp losses. A triumph late in 1757 against a Franco-Austrian force at Rossbach and Leuthen marked a brief resurgence of the Prussian position. In a letter to Maupertuis that December Euler mentioned the first of these victories. But worse times lay ahead for Prussia.

While early in the war Euler made little of his Russian connections, they did not prevent him from proposing to Müller two scholars for positions in Russia. In 1755 the physician Franz Ulrich Aepinus and his auditor and mathematician Johan Carl Wilcke had come to Berlin from the University of Rostock and studied electricity together. During the next two years, Euler recommended Wilcke to teach physics at Moscow University, but he went instead to the Royal Swedish Academy of Sciences. To succeed Georg Wilhelm Richmann at the Petersburg Imperial Academy Euler guided through the election of Aepinus, whom he found exceptional in physics and astronomy and who had been named director of the Berlin Observatory. A deep friendship developed between the two: Aepinus boarded with Euler and participated in his table society meetings. In October 1756 Euler informed Müller that Aepinus wanted to leave Berlin for Saint Petersburg. The next January Euler gained from Frederick the release of Aepinus, who left for Russia as the war raged. From Aepinus Euler had learned of electrostatic analogies, elastic electric fluids, and the latest on magnetic phenomena. He told Maupertuis that he considered Aepinus's work the foremost at the time in electricity. Aepinus was writing *Tentamen theoriae electricitatis et magnetismi* (An attempt at a theory of electricity and magnetism), which further develops Benjamin Franklin's one fluid theory of electricity and bases electrical phenomena on action at a distance without giving its cause. After learning of this fundamental electrical research, Euler and his son Johann Albrecht turned more to the subject.

In July 1756 Euler entered into a lawsuit against Count Peter Karl von Keith, a neighbor whose land bordered his Charlottenburg estate, over damages relating to the boundary and divisions of a private cemetery. This challenge demonstrated Euler's independence; a commoner, he confronted a landed aristocrat and curator of the Berlin Academy. The dispute took more than a year to resolve, and its appearance often in Euler's correspondence suggests its significance to him. He cited the case in letters to Maupertuis in August and October 1756, and in January 1757 reported Keith's death. In February of that year he commented on d'Alembert's attack on him for continuing the case. In September 1757 Euler won the lawsuit; he was awarded five thaler, but each side had to pay one hundred thaler, the cost of court work and the lawyers.[11] Letters to Maupertuis in January, September, and December 1757 and February 1758 also mention the case.

In 1756 Louis Bertrand of Geneva, who later became a professor of mathematics there, was among Euler's correspondents; Euler called him "one of my best friends."[12] From 1752 to 1756 Bertrand had resided with him in Berlin; possessing a good knowledge of mathematics when he arrived, he received an advanced level of instruction from Euler. Bertrand succeeded so ably that Euler believed that he should be a member of the Berlin Academy.

In Berlin Bertrand's main task was to tutor Frederick II in mathematics. With the beginning of the Seven Years' War, the king granted Bertrand permission to travel to London. In a letter from August 1756 Euler asked Wettstein to make travel arrangements for Bertrand; the same letter expressed Euler's worry that a few verses in the almanac that he had prepared at the academy were being falsely credited to Voltaire, when their correct attribution should have been to Joseph Francheville. Reduced now to smoking plugs,[13] Euler pleaded in the same letter for a dozen pounds of Virginia tobacco.

In a letter of November 1756 Bertrand wrote that Wettstein was still not in London and that John Dollond's telescope following Euler's directions had failed; he also judged as unfounded Johann Tobias Mayer's hope for an astronomy prize from the British Parliament. In December Bertrand wrote that the Berlin Academy would acquire some South American platinum, which Andreas Sigismund Marggraf was studying. He also sent potatoes and, for Johann Albrecht, logarithm tables. Probably later in December he sent minerals including platinum from Peru and tobacco from Virginia.

The Seven Years' War brought many hardships to Berlin and the academy, and Euler's task to keep the academy active was difficult. Individual scientific research proceeded nearly without interruption, but despite a tenfold increase in Saxon appropriations and British subsidies, the economy suffered sharp losses. In the summer of 1757 academic salaries were cut in half. Since few funds existed for repairs of academy buildings and property, these began to deteriorate. From 1757 to 1760 no new members were elected, and this lessened the infusion of new ideas and programs. The war disrupted the posting of letters from Berlin to Turin, Paris, London, and Saint Petersburg; Euler's correspondence contained a mélange of news concerning research, the academy's almanacs, lenses and the telescope, maps, tobacco, military campaigns, austerity, invasions and near invasions in wartime Berlin, administration at the Berlin and Petersburg academies, the publication of the *Novi Commentarii*, and various prize competitions. From 1757 to 1762, Euler's correspondence with Goldbach, who was growing old, almost stopped.

In 1756 and 1757 Euler's research included his improved description of the magnetic field by recognizing both declination and inclination, which was important for finding longitude. A compass did not point exactly north but had an angle of fifteen degrees from Berlin; that angle was the magnetic declination. In 1755 Johann Albrecht had completed his eighty-four-page article on magnetic inclination. Employing inclination compasses made by Dietrich and sent by Daniel Bernoulli, Euler's "Recherches sur la déclinaison de l'aguillle aimantée" (Research on the declination of the magnetized needle, E237) introduced a mechanical frame of reference giving the relative positions of the geographic and magnetic poles of Earth. Using spherical trigonometry, Euler determined from any place on Earth the true positions of the two magnetic poles positioned on two different meridians. His article was not to be published until 1862 in his *Opera postuma*.

Two of Euler's seminal articles, "De integratione aequationis differentalis $mdx/\sqrt{(1-x^4)} = ndy/\sqrt{(1-y^4)}$" (On the integration of the differential equation $mdx/\sqrt{(1-x^4)} = ndy/\sqrt{(1-y^4)}$, E251) and "Observationes de comparatione arcuum curvarum irrectificabilium" (Observations on the comparison of arcs of unrectifiable curves, E252),[14] together giving a unified theory of elliptic integrals, were in the *Novi Commentarii* for 1756–57; the volume was not published until 1761.

In early 1757 Euler and Dollond independently made progress correcting spherical lenses to construct an achromatic telescope.[15] Euler's work involved a convex and a concave lens separated by a small space filled by a liquid with an index of refraction differing adequately from that of the glass; the shape of the lenses was circular or parabolic.

At this time, astronomers disagreed over whether the moon had an atmosphere. A solar ring and missing rays from the fixed stars produced a paradox: either the moon had an atmosphere or it did not. Assuming the invisibility of rays from the fixed stars passing through a thin lunar atmosphere, Euler accepted the concept of an atmosphere on the moon and believed that it had inhabitants.[16] In September Euler responded to Wettstein's request for a copy of his *Introductio in analysin infinitorum* (Introduction to the analysis of the infinite). He did not think that it would fit English tastes with its Newtonian fluxions, but agreed to send a copy to a bookseller if Wettstein paid all shipping fees and a price of 3.5 ecus.

In August news came of the death of Johann Samuel König, who had been unable to complete his book on Leibnizian physics. With Lagrange's support, Maupertuis in France felt assured about the success of his principle of least action; despite his precarious health and the war, he contemplated a return to Berlin, but this remained impossible. In October 1757 a paper comparing the rational psychological principles of Gottfried Wilhelm Leibniz with the countering critical empirical views of the abbé Étienne Bonnot Condillac was read at the academy. Merian, who opposed the ideas of Leibniz, found correct the comparison favoring Condillac. At a meeting in January 1758, Merian referred to a work of Maupertuis on evidence, certainty, and the contingency that the president always thought possible in the sciences. The next month Merian read his own article "On the Proof of Natural Laws."[17] Outside the mathematical sciences Marggraf, known for his precise laboratory techniques and one of the last in chemistry to defend the phlogiston theory of combustion, attempted in 1758 to manufacture soap by deriving soda from common salt. Employing microscopic and flame tests, Marggraf checked the difference between the alkali in potash and that in common salt and determined the difference between them in solubility. In June Euler placed before the academy a paper by Aepinus on an attempt to find from the principle of least action the center of oscillations.

After the Prussians defeated the French in early 1758 at Crefeld, near Düsseldorf, William Pitt proposed sending six thousand British troops to join Frederick. The Prussians were euphoric over the prospect that the war would soon end with a peace favorable to them. But from late August 1758 on the Prussian situation deteriorated. Prussian troops singing Lutheran hymns fought the Russians to a devastating draw at Zorndorf. In October Frederick suffered a major loss against the Austrians under Leopold von Daun near the village of Hochkirch. The next month the monarch held up at Dresden, briefly stabilizing his position. Frederick had visibly changed; his hair grayed, and he was missing half his teeth.

From the beginning of the war, the pensive and reserved Euler displayed an enthusiasm and patriotism for Prussia and its king. He had offered to decipher and translate all letters from Russian officers intercepted by Prussian forces. In 1758 there were so many that they took time away from his regular correspondence. In February he urged the British to support the "just causes" of Frederick—that is, the freedom of the Germans and the defense of the Protestant religions. In October he wrote to Maupertuis, "We have just intercepted the mail that [Field Marshal and Count Wilhelm von] Fermor had sent 25 September to his [Russian] court and I have been asked to examine all the Russian letters which number in the hundreds. . . . Fermor's reports to the empress are singularly interesting, since they still tell her of their continuing victory. One can see that this has no other purpose than to misguide the Tsarina, since all of Europe is convinced that this is nothing but pretense and that victory is outside of Fermor's grasp."[18]

In April 1758, Bertrand wrote from London describing the war and the political situation in England, a meeting of the Royal Society of London, and his pleasure at learning that Johann Albrecht Euler was preparing a book on mechanics. Bertrand judged the British telescopes made by James Short to be superior to those on the Continent and reported that the previous year Dollond had found a process to produce from two types of glass—flint and crown with differing refractive indices—lenses that eliminated differences in refractions, which produce achromatism. The advocates of Newtonian optics had denied the existence of such lenses. In experiments that year Dollond had built his lenses strictly by Euler's formulas, hoping to prove Euler's theory wrong and Isaac Newton's right. The experiments were widely known, and their outcome was eagerly awaited in Europe. The experiments vindicated Euler, giving him a triumph in the long optical debate. The realization of excellent lenses was still to be about five years in coming. In October, Bertrand sent Euler a supply of tobacco from Geneva and invited Johann Albrecht to spend the summer with his family. He found Dollond's new telescope not completely satisfactory but still better than its predecessors.

Losses, Lessons, and Leadership

In 1759 the Russians planned a new campaign against Frederick II. "We have just intercepted the mail from Vienna to Saint Petersburg that contained the war plans for the campaign," Euler wrote in March to Maupertuis, "and the king was extremely pleased. . . . It is perhaps the only news that we know for certain. . . . "[19] Still, in August Frederick suffered at Kunersdorf on the eastern side of the Oder River his worst defeat by the joint Russian and Austrian forces. The Prussians had nineteen thousand killed, wounded, or missing, and the Russians and Austrians nearly fifteen thousand.[20] If the Austrian and Russian forces had cooperated more closely, the Prussian situation would have been worse. In the war's first three years, the Prussian forces were being bled white with the loss of a hundred thousand highly trained men.

In February 1759 Euler wrote to a disheartened and ailing Maupertuis, who was increasingly pessimistic about the war, "That God protect our great King! That Frederick lives, reigns, and triumphs; that is the wish of us all along with Formey."[21] On 6 March Euler expressed his highest Prussian patriotism by writing to Frederick that he hoped that his youngest son Christoph Euler would prove worthy of the honor of the king's inducting him into the Prussian field artillery regiment. At the age of sixteen, Christoph had volunteered, and Frederick lauded this as an act of true patriotism. On 14 March Frederick wrote thanking Euler for "his show of loyalty and his attachment to his monarch." Maupertuis remained allegiant to Frederick, and in France this provoked a charge that he was a spy.

Early in 1759 Euler sent Maupertuis improved telescopes but noted that problems with the lenses remained; he faulted the glasses used in them and wrote to England for the manufacture of better glass. Although not as successful as Maupertuis had been, Euler pursued the king's support for the sciences.

In 1759 Euler announced that he had completed a significant part of his *Institutiones calculi integralis* (Foundations of integral calculus), the last volume of his trilogy on calculus; initially to have been two volumes, the work had expanded to three. Meeting difficulty in locating a publisher, and unable to find a German editor, Euler edited the volumes himself. In 1759 he shared another Prix de Paris for a paper in applied mathematics titled "Dans le roulis & dans le tangage" (Concerning the pitching and yawing of a vessel, E415).[22] For navigators and shipbuilders, its new calculations and reasoning regarding the effects of these actions comprise one of his most notable works on the stability and speed of vessels.

In April 1757 Euler had written to Goldbach that he was investigating a chess problem. In March 1759 he read a paper on the problem, "Solution d'une question curieuse qui ne paroit soumise à aucune analyse" (Solution of a curious question that does not seem to have been subjected to any analysis, E309). "I found myself one day in a company" the article began in French, "where on the occasion of a game of chess, someone proposed the question: To move with a knight through all the [sixty-four] squares of a chess board, without ever moving two times to the same square, and beginning with a given square."[23] Today this problem is called the knight's tour. Euler credited Bertrand with giving him the idea; it was the only mathematical problem in chess that he addressed, and he solved it.

Making the knight's movements a matter of rational skill, Euler took a systematic approach and generalized the problem. He examined an open tour, the case in which the knight may not move directly from the last point of his route to the first, or not be "reentrant upon itself," and he considered closed tours, which are reentrant. Euler constructed types of routes for each. He transformed a partial open tour into a closed tour, giving seventeen similar routes or symmetric invariants. This was computationally inefficient.

Such chess puzzles in Berlin reflected the wide popularity of chess in the city. The automaton, a complex clockwork machine known as the Turkish player or the Turk, was invented to defeat human chess players. When wound, it powered about a dozen flawless moves, designed to re-create

Euler's version of the knight's tour. The Hungarian noble engineer Johann Wolfgang, Ritter von Kempelen invented the machine for Empress Maria Theresa in 1769, and it would also appear in Leipzig in 1784. The Turk consisted of a carved life-size figure dressed in oriental garb that sat behind a cabinet with a chessboard on top; it usually won, which caused a mild sensation.[24] In Prussia the interest in the relations among intelligence, concealment, and mechanism appears to have been greater than anywhere else in Europe. In chess circles the knight's tour problem was pursued into the nineteenth century but has since been largely forgotten. It has little to do with the proper playing of chess.

After deciding not to go to Italy, in 1759 Maupertuis headed from the south of France toward Prussia. He wanted to reach Berlin and Magdeburg, where his wife was residing in the Prussian court, but he only got as far as Switzerland. After a few months in Neuchâtel, he had another bout of illness and turned back to Basel and the home of his friend Johann II (Jean II) Bernoulli, where he had been welcomed the previous fall. In his last hours he was with a Capuchin monk and Johann III, the fourteen-year-old son of his host. On 27 July Maupertuis died; he was buried in the neighboring village of Dornach in a Catholic cemetery.[25] While Euler had held a leadership role in the administration of the Berlin Academy since 1753, he was now officially in charge.

In September 1759 Euler agreed to privately instruct the fourteen-year-old Friederike Charlotte Leopoldine Louise, a second cousin of Frederick II; he did so chiefly because her father, his close friend Friedrich Heinrich, who would become margrave in 1771, had asked him to. She was thus also known later by the variable title Princess of Brandenburg-Schwedt; she, her younger sister Louise Henriette Wilhelmine, and their cousin Sophie Friederike Dorothea were the three German princesses of the time. For the frugal Euler the chore was a way to earn extra money besides gratifying his love of teaching. He had taught noble children from the time of his arrival in Berlin, but what was unusual for the time was the fact that his student was female. Euler had read Bernard le Bovier de Fontenelle's best-selling *Entretiens sur la pluralité des mondes.* Writing *Mondes* with élan, Fontenelle had sought to reach a wider reading public, including women. Euler had read also Algarotti's *Neutonianismo per le dame* (Newtonianism for the lady) and perhaps Voltaire's *Micromegas.*

In noble circles the friendship between Euler and Friedrich Heinrich was unusual, for Euler was not popular among Frederick II's entourage. The two men shared an interest in music; during many visits they would perform duets, with Euler playing the clavier. Friedrich Heinrich raised his two daughters alone in Berlin. His wife, Maria Leopoldine of

Figure 12.1. Princess Friederike Charlotte Leopoldine Louise von Anhalt-Dessau (later Brandenburg-Schwedt) in a pastel by Susan Petry, 2010.

Anhalt-Zerbst, was the daughter of Leopold, known as the Old Dessauer. After the birth of the second daughter, Maria Leopoldine suffered from postpartum depression and became so violent that she was sent to Kolberg Fortress in Silesia. Never recovering from her madness, she remained there for the rest of her life, and Frederick II forbade her husband visitation rights.

Euler's initial lesson plans for Friederike Charlotte Leopoldine Louise, beginning in September 1759 in Berlin, have not survived. This was not Euler's only teaching project at the time; he was recommending that a second edition of his *Einleitung zur Rechen-kunst* (Introduction to arithmetic, E17) in German be prepared for gymnasium students in Russia. For the advance of science he wanted to enlist sound pedagogy, reliable textbooks, and refined popularizations along with acute original works.

In mid-September 1759, on behalf of the academy's office of the president, Euler sent Frederick a pair of field glasses, a telescope, and information on his optical research perfecting both devices. Employed as spyglasses, telescopes allowed Frederick to observe the movement of enemy forces. On 15 September, the king sent Euler a handwritten letter from Waldau: "I wish to thank you for the small field glasses that you sent me. . . . I commend you for the care that you have taken to render useful to men a theory that your studies provide and your application to the sciences. My current duties do not permit me to examine everything that you have sent me with the attention that all that comes to me from you merits; I reserve the right to do that when I will have a little more time to do so. . . . On this I pray to God that He keep you under his holy and worthy protection."[26] Apparently Euler hoped that Frederick, despite the strains between them, would name him president of the academy.

In the first months after his death, academicians repeatedly spoke of Maupertuis's watchful eye on their research. Formey, who delivered a eulogy in January 1760, hoped that if he was not named the president of the academy he would at least be the most trusted adviser to the king regarding its affairs. Instead Frederick turned to Jean Baptiste de Boyer, Marquis d'Argens, a French noble and *philosophe* and the director of the historico-philological department at the Berlin Academy. D'Argens, who entered into academy affairs by September, would write an anecdotal history of it. Among other things he, as well as Dieudonné Thiébault, mentioned an unusual occurrence. Soon after learning of Maupertuis's death, the botanist Johann Gottlieb Gleditsch claimed that one afternoon at about 3:00 p.m., as he entered the meeting room of the academy, he had seen to his left the ghost of Maupertuis standing immobile in a corner next to the great clock. Gleditsch noted the visage but busied himself at his work cabinet. The apparition stayed for about ten minutes, its eyes set on him, and then disappeared. To his colleagues Gleditsch insisted that it was as real as if a live figure were standing there.[27] Formey gave a different version, maintaining that the vision terrified Gleditsch. Not prone to superstition, most of his fellow academicians were amused and ignored the story; others claimed that Maupertuis's restless spirit had not yet found peace and continued to wander. When the king heard of the incident he was angered more than amused, viewing it as a slight to Maupertuis. After Voltaire continued his assault on Maupertuis, the king wrote an ode praising his president as noble, true, and a model for others.[28]

With the death of Maupertuis, d'Alembert in antagonistic France, and Frederick immersed in the Seven Years' War, in 1760 Euler managed the Berlin Academy in close relationship with Merian. No longer having to report to Maupertuis, Euler was now actually the president, though without the official title. In this position he would pursue as always his goal of choosing candidates solely on merit, thereby helping to establish the independence of the academy.

Once Johann Heinrich Pott retired, Euler brought from Marburg as an ordinary member the chemist Karl Philipp Brandes, who had been an associate member for five years. Upon the death of Johann Theodor Eller, the director of the physics or experimental philosophy department and personal physician to the king, a successor was elected. Among the twenty-three academicians present, the vote was thirteen for Marggraf, six for

Johann Gleditsch, and four for the physicist Christian Friedrich Ludolff.[29] D'Argens reported this result to Frederick II in September, and in January 1761 the king accepted the choice of Marggraf, calling it well deserved. Acting solely upon the initiative of the academy and his authority, Euler now nominated nine candidates and presided over their unanimous election. While most of them were foreign, coming from Bologna, the Hague, Paris, or Switzerland, there were also three Germans: Johann Essais Silberschalg was elected in March, followed by the physician and chemist Christian Ludwig Roloff and the dramatist and philosopher Gotthold Ephraim Lessing in October.[30] The botanist and surgeon Johann Jacob Huber from Göttingen and Kassel was by birth Swiss, originally from Basel. Frederick, who did not know the foreign candidates, apparently believed that he could comment only on the Germans Huber, Silberschlag, and Lessing; he hardly knew the first two, though he had enough acquaintance with Lessing, whom Voltaire had slandered. Occupied with the Seven Years' War, the king validated the elections without comment or requiring any proof of accomplishments. With Frederick's mild disdain for German candidates, he was never pleased with the election of Lessing, and he planned to respond to the next election in April 1761, when the popular poet and fabulist Christian Fürchtegott Gellert, one of two coryphaei in Leipzig, and Johann Heinrich Lambert were the German candidates. Election required the approval of the king. Gellert would fail to win election, for Frederick wanted most of German literature excluded from the academy and concluded that with Lessing there were already too many Germans.

On two other matters Euler was not so successful. He argued for an increased salary for Marggraf in recognition of his great service and research, but Frederick replied that he lacked the funds for it; during the war he issued no new salaries and no raises were granted. From 1756 to 1759 the *Mémoires* had been published regularly under Maupertuis. Although the issues for 1760–63 were mainly completed on time and ready for the press, d'Argens advised the king not to print these, for d'Argens needed the paper for state business and the academy did not. D'Argens recommended that their publication be suspended as long as there was no president; thus, a volume for the years 1760–63 appeared only in 1764. Only after the Prussians and Austrians signed the Treaty of Hubertusburg in February 1763 would new life be infused into the academy.

When Maupertuis died, the Swiss comprised between one-fourth and one-third of the membership of the academy, and it appeared that Euler led a disciplined regiment. In an attempt at what might be termed a humorous and contradictory inspection Thiébault found the Swiss to have a dangerous appetite for control of everything. He noted, making an exception for Lambert and Merian, that he had "seen how [Johann Georg] Sulzer, [Nikolaus von] Béguelin, [Jakob] Wegelin, etc., ruled the entire academy."[31] Thiébault declared that the Prussians, who had a harsh political order, did not think of that course of action, that the French, who enjoyed a degree of power in their homeland, were satisfied to enjoy equality; and that the Swiss, who were free at home, wanted to dominate the academy. Thiébault portrayed the Germans as tending "to enjoy natural laws" and the French enjoying "the privilege of liberty," whereas the Swiss pursued "absolute authority."[32]

That the tasks in leading the academy in 1760 did not detract from Euler's prolific science writing is demonstrated by his nine articles published that year. Keeping up his productivity, he would publish a total of fifty-four articles in the following five years. In 1760 the publisher Anton Ferdinand Röse in Rostock offered to issue either Euler's book on integral calculus or another on rigid bodies. Euler had been working assiduously on the latter topic, and by 1758, he also arrived from his study of torque at the Euler equations, in which angular velocity and the diagonalizing vectors and tensors of inertia appeared.[33] By 1762 Euler completed his masterwork *Theoria motus corporum solidorum seu rigidorum* (Theory of the motion of solid or rigid bodies, E289), but it took until 1765 for Röse to get it into print. Euler was thorough in the planning for it, proposing the type, the format, and the attainment of subscriptions. When the *Theoria motus corporum* appeared, it still had only twelve subscribers. (Publishers sought preorders from subscribers to underwrite production costs.) Euler corresponded with Wenzeslaus Johann Gustav Karsten, a professor of mathematics in Rostock, who wrote the preface maintaining that the book was not taken in part from a 1761 article by d'Alembert.

In 1760 the Paris Academy of Sciences awarded two prizes: one on the making of glass and the other on the variation of planetary motion. Euler's son Karl, who won the prize in astronomy, was twenty; a student planning to pursue medical studies, he had little interest in the mathematical sciences. Euler is credited with editing the paper, but doubtless he wrote it.

In September 1757 Euler had reported to Maupertuis that his "correspondence with Saint Petersburg is completely broken,"[34] and his letters to Müller did stop in 1758 and were apparently curtailed to only two in 1759. In May 1759 Müller sent him a copy of the fourth volume of the *Novi commentarii* and reported that the printing of the fifth volume was complete; he also expressed his wish that the correspondence between them would no longer lapse.[35] Their letters resumed and increased from fourteen in 1760, to fifteen in 1761, to nineteen in 1762, and to twenty-three in 1763. Although Prussia and Russia were on opposite sides during the Seven Years' War, the postal service continued between them, albeit with delays. The growing correspondence with Müller has been taken to suggest Euler's wish to return to Russia, but it testifies more to his efforts to improve the Imperial Academy of Sciences and Arts, in part by nominating candidates for positions; in 1760 he nominated Mayer and Lambert, along with the physicist Johann Arnold from Erlangen, the chemist Johann Cartheuser from Frankfurt, the astronomer Georg Moritz Lowitz (Lovits) from Göttingen, and Gerhard Andreas Müller from Giessen. Euler failed to fill a vacant position with Caspar Friedrich Wolf, a field doctor in the Prussian Army with a medical degree from the University of Halle; Wolf's *Theoria generationis* of 1759 refuted the theory of fully preformed embryos of organisms and confirmed a gradual developmental process of epigenesis. Euler also requested abstracts of articles for volumes 5–9 of the *Novi commentarii.* He proudly commented on the Bavarian, Paris, and Saint Petersburg prizes won by Johann Albrecht, requested the payment of his own stipend, and inquired about Mikhail Lomonosov.

Euler's correspondence with Müller concentrated on the magnet, Franklin's theory of electricity, and many discoveries in applied optics, especially to improve the microscope and telescope, including the lenses and research of John Dollond, Clairaut, and Samuel Klingenstierna. In astronomy he studied the paths of comets, Joseph Jêrome le Français de Lalande's expedition in 1761 to observe the transit of Venus, and Stepan Rumovskij's observations.

Of the events that transpired in the fall of 1759, most were obstacles to Lagrange's coming to Berlin. After a three-year interruption, his correspondence with Euler resumed. In October Euler informed Lagrange of the death of Maupertuis and reported that Lagrange's larger written work had not returned to Berlin. He recommended that Lagrange seek a publisher in either Lausanne or Geneva for his two articles on his new method, since Berlin was in financial straits.[36] Euler had received the first volume of the *Miscellanea Taurinensia* (Turin miscellany) and expressed his admiration for Lagrange's excellent work presenting the most difficult differential equations, calling it Lagrange's "true chef d'oeuvre."[37] A second letter in October, which discussed the calculus of variations, discontinuous functions, and sounds, dismissed as pretense Jean-Phillipe Rameau's *Demonstration du principe de l'harmonie* of 1750 regarding the extent of the same chord occurring at different intervals. Euler declared that he avidly awaited the publication of Lagrange's two articles in the next volumes of the Turin *Miscellanea.*

In 1756 Euler had completed two articles, "Elementa calculi variationum" (E296), in which the term *variation* appears, and "Analytica explicatio methodi Maximorum et Minimorum" (An analytical explanation of the method of maxima and minima, E297). He began these by citing the research using geometric methods of Johann I (Jean I) Bernoulli on the brachistochrone and Jacob Bernoulli on isoperimetric problems. Euler reportedly spoke on these articles in Berlin in 1756 and sent them to Saint Petersburg four years later, but they were not put in print until 1766 in the Petersburg *Novi commentarii*.[38] His son Johann Albrecht had also applied Lagrange's new algorithm to solve a problem in 1757, the results of which were published in 1764 in the Bavarian Learned Society's *Abhandlungen* (Proceedings), and Bertrand in Geneva applied the new algorithm with great success. While Euler always recognized Lagrange's priority, Lagrange was upset that it was Euler who publicly announced the new algorithm. In 1769 the Bavarian Learned Society became the Bavarian Academy of Sciences and Humanities.

Although the relations between Euler and Lagrange had cooled, the two continued to correspond. In 1760 and 1761 Lagrange published in the Turin *Miscellanea* two articles on his new method. He did this quickly, since Euler had agreed to withhold his own articles on the subject until these appeared. These articles rigorously derived what has since come to be called the Euler-Lagrange Equation, the first necessary condition for an extremum. Though they are the mark of a genius, Lagrange's articles are difficult to read; in contrast, Euler's future writing on the subject was more accessible. Even though he recognized Lagrange's priority and crucial role in developing the calculus of variations, in which Euler solved with clarity a substantial number of problems, many would come to believe that Euler was its actual inventor.

In the conflict between d'Alembert and Euler, in which the vibrating string problem remained an issue, Lagrange had entered the debate in 1759 by sending d'Alembert his treatise "Nouvelles recherches sur la nature et la propagation du son" (New research on the nature and the propagation of sound). Although d'Alembert did not accept Lagrange's analysis of the vibrating string problem in an elastic medium, he was impressed, writing, "If I am not wrong, you are destined to play a great role in the sciences and I applaud your success in advance."[39] He promised to defend Lagrange and his interests, and the two became good friends. In politics d'Alembert was essentially his foster father. While Lagrange had earlier contact with Euler, their later friendship was never so close and warm. These two strong personalities addressed problems in the mathematical sciences only, including some critical comments from Euler. In research Lagrange tacitly followed Euler.

Responding to the rumors that d'Alembert would become president of the Berlin Academy, and with a large salary, Euler in an expansive letter to Lagrange in early October 1759 vilified the *philosophe*. "D'Alembert," he wrote, "has tried to undermine by various sophisms [Euler's solution of the vibrating string problem], and that for the single reason that he did not get it himself. He announced that he had a crushing refutation in press, but I ignore that claim. He believes that he can deceive the semi-learned by his eloquence. I doubt whether he is serious [about producing a refutation] unless perhaps he is thoroughly blinded by conceit. . . . What disorder he would cause if he were to become president. But I await all with serenity, having decided to abstain from all activities with him."[40] This letter stated that the late Maupertuis "of happy memory" had not wanted Euler to publish in the Berlin *Mémoires* his rigorous demonstration disproving d'Alembert and transform his academy into a playground featuring d'Alembert's rhetoric.

In the spring of 1760, twenty-six-year-old Johann Albrecht Euler wed Anna Charlotte Sophie Hagemeister, who was the same age; the marriage made Formey an uncle to Johann Albrecht. Although the groom's annual salary was supposed to double, from two hundred to four hundred Reichsthaler, the higher salary was not paid until 1763; the couple thus materially depended upon Leonhard Euler and lived with him. Over the next two years Johann Albrecht and his wife had two children. From late in 1760 on, the Euler home must have been crowded, because a military action against the estate in Charlottenburg would force all members of the family, except for Christoph, to return to Berlin. Johann Albrecht was active in research; during his years in Berlin, he published fifteen articles in such journals as the *Philosophical Transactions* of the Royal Society of London and the publications of the Paris and Bavarian Academies.[41]

Before the Russians approached Berlin in the fall of 1760, many fled the city. The royal court moved to Magdeburg, eighty-three miles away. Euler was asked—probably by the future margrave Friedrich Heinrich—about leaving, but he chose to stay. Distressed at the distance that made it impossible to give lessons to Friedrich Heinrich's daughter in person, he agreed to supply private instruction by mail as far as the subjects permitted and to have occasional visits. From April 1760 to May 1762, he sent 234 letters and would spend the following two years revising them for publication. But it was impossible to publish them in a Prussia torn by war and in financial distress.

In 1760 the Seven Years' War went badly for the Prussians. The Austrians and Russians cooperated more closely, and the French in the west along with the Swedes in the north united against Prussia. After the victory at Kunersdorf in August, the Russians planned to move into Berlin but were delayed on a mission to help remedy Austrian disappointments in Saxony, where the Austrians saw that the exploitation of Saxon resources

Figure 12.2. Johann Albrecht Euler, by Emanuel Jakob Handmann, 1756.

was crucial to the Prussian war effort and Prussian strength there was increasing. Not until early October 1760 did the Russians, led by Gottlob Heinrich Totleben and Zhakar Grigorevich Chernyshev, reach and occupy Berlin along with the Austrians.

The plundering of Euler's Charlottenburg estate by the Russians was a low point in his relations with them but did not produce a break. Chernyshev, who had formerly visited Euler, knew that he was a salaried member of the Petersburg Imperial Academy, and thus a military guard, a *salvegarde*, was to be dispatched to protect Euler's house and estate; more generally, Totleben ordered troops not to harm the civilian population. Before Chernyshev's message was sent, however, the Prussians quickly retreated and abandoned Charlottenburg to the Cossacks, who devastated Euler's estate.[42] The protective guard arrived too late. When General Totleben learned of Euler's loss it was reported that he declared that he had not come to make war on the sciences and would reimburse Euler. In a letter to Müller in October 1760, Euler submitted a detailed request for payment from the Russian government to cover his losses, including "four horses, twelve cows, many small animals, a lot of oats and hay."[43] Added to the wreckage of all furniture in the house, the damages by Euler's reckoning cost eleven hundred Reichsthaler, or seven hundred rubles. Taking into account future losses, such as being unable to cultivate his lands for the coming year, he arrived at a total just below twelve hundred rubles. Possibly the amount that Totleben paid was only for the year. From Berlin, Charlottenburg, and surrounding lands, the Russian generals extracted ransom and cash contributions of nearly 2.5 million Reichsthaler, though they remained in Berlin less than a week, until 11 October 1760.

Euler's relations with the Russian officers remained good.[44] He spent time in their company with major general Aleksey Michailovich Maslov and the physician Konstantin Ivanovich Schepin, who were friends.[45] Chernyshev expressed regret over the estate losses from the Cossacks, noted Euler's application for restitution, and denied the rumor that the Russian monarchy had offered Euler a guaranteed three thousand rubles in salary to return to Russia. It is thought, apparently mistakenly, that Chernyshev reported the damages to Empress Elizabeth, who paid a second time, in the amount of four thousand rubles, to cover Euler's losses. Instead, Razumovskij later mentioned this destruction to Empress Catherine II, who came to the throne in 1762; the payment appears to be part of her effort to draw Euler back to Russia. He did not receive these funds until 1763. In the meantime, Euler submitted a detailed bill seeking a full payment from the Prussian government.

In May 1761 Euler took his son Karl to the University of Halle to earn a medical degree. Since Karl was known for being talkative, the conversation must have been lively. After Frederick II spent the winter in Leipzig, the Austrians turned mainly to Silesia and the Russians to the Baltic region; Euler's trip was thus not to a combat zone. In Halle he stayed at the house of his longtime friend Jan Andrej (Johann Andreas) von Segner. From 1755 to 1766 their correspondence covered family and personal matters as well as political events, such as the occupation of Halle in 1759 and 1760. Among the topics that they discussed after 1760 were mechanics, the logarithms of negative numbers, the properties of logarithmic functions, the observation of the transit of Venus from Halle, and the designing of better objectives for telescopes. Instruction at the university, especially with Segner's textbooks such as the *Elementa Arithmeticae, Geometricae* of 1758 and editions that would later run to 1768 and Segner's *Cursus mathematicus*, was a prominent topic. In these books Segner introduced a whole series of Euler's symbols. Through his correspondence with the king, Segner, and Johann Joachim Lange, Euler had a fruitful and substantial influence on the development of curricula in the exact sciences at the University of Halle during the second half of the eighteenth century.[46]

After a brief visit with Segner, Euler proceeded to Magdeburg to spend a few weeks meeting with Friedrich Heinrich and instructing his daughter the princess. The entire trip in May took three weeks.

Rigid-Body Disks, Lambert, and Better Optical Instruments

Over a twenty-year period beginning in 1761 with "De frictione corporum rotatantium" (On the friction of rotating bodies, E257),[47] Euler published four articles concerning the movement of a rigid body circling a moving axis. The other three were "Du mouvement de rotation des corps solides autour d'un axe variable" (On the motion of rotation of solid bodies around a variable axis, E292, 1765); "De mouvement d'un corps solide quelconque lorsq'il tourne autour d'un axe mobile" (On the motion of a rotating solid around a mobile axis, E336, 1767); and "De effectu frictionis in motu volutorio" (On the effect of friction on associated motion, E585, 1781, 1785),[48] the third of which was read at the academy in April 1775. The Euler disk demonstrates such motion. On a clean and smooth surface, a circular and homogeneous metal disk is spun. It starts by rotating around the vertical axis but friction makes it tilt and roll along a circular path. As the tilting of the axis increases, the circular path widens and the pitch from the contact point of the disk and the surface is higher. The rising pitch seems to suggest that the speed of the disk's motion is increasing. But friction dissipates energy and on the surface the disk abruptly halts. To explain this requires a set of differential Euler equations, Euler angles, and other measures.

By February 1762, Karl had completed his medical studies and graduated. He was soon named physician to the French colony in Berlin and earned a good reputation. After the recent death of Empress Elizabeth, Segner hoped for a peace treaty with the Russians. In 1762, after three years of service, nineteen-year-old Christoph Euler became a lieutenant in the Prussian artillery. Frederick had previously limited to the nobility the positions of officers in the military, but under the pressures of war had to add commoners. Euler's letters to Müller indicate that Christoph's promotion did not begin to better his own relations with Frederick.

In 1762 Frederick asked d'Alembert to compare Lambert to Euler as a scholar. Lambert, who possessed a reputation as a mathematician, cosmologist, physicist, and logician,[49] was from Mullhausen in Alsace, which then belonged to the Swiss cantons. Self-educated since he was nine, he may have briefly attended courses at the University of Basel. While a tutor in Switzerland in his twenties, Lambert continued his autodidactic learning, demonstrating by the age of thirty a strong command of calculus and demonstrative geometry along with fluency in French and Latin as well as a good knowledge of Greek and Italian. He was a tutor, referring to himself as the *Hofmeister*. Traveling with his students in 1758, he met Mayer, Pieter van Musschenbroek, and d'Alembert. Lambert was known for originality in his *Photometria* (1760) and *Cosmologische Briefe* (1761) the next year projecting the Milky Way as a system and the universe as populated with other creatures like human beings.[50] In April 1761 Euler nominated Lambert to be a regular member of the Royal Academy of Sciences in Berlin, and he was unanimously elected. But late during the war Frederick, working to reestablish his control over the academy, withheld approval.[51] D'Alembert asserted that Lambert and Euler were equals in the sciences, and in the same proportion as Pierre Bayle to René Descartes and Newton. But in abstruse metaphysics, d'Alembert placed Descartes, Newton, and Leibniz under Bayle and the French philosopher Pierre Gassendi. The king responded that "our century has the rage for curves." Frederick had earlier suggested the unworthiness of mathematics in the fine arts by citing Euler's attempt to compose a minuet, following a simple arrangement within sections of $a + b$. From his field camp the king happily teased d'Alembert with a satirical epistle concluding with a mathematical proof implying an artistic mathematical poetry. D'Alembert responded with a poem based on quotations from the letter titled "Reflexions de la géométrie sur la poésie (Reflections of geometry on poetry).[52] The two had a good rapport.

The importance of royal academy prize competitions in generating significant papers contributing to the development of the sciences in the eighteenth century persisted in 1762, when the Petersburg Imperial Academy divided its annual prize between Clairaut and Johann Albrecht Euler; The assigned problem was to account for perturbations in the motion of comets. Clairaut had submitted the article "Recherches sur la comète des années 1531, 1607, 1689, et 1759" (Research on the comets of 1531, 1607, 1689, and 1759) as an addendum to his paper of 1758 announcing the return of the most recent of these, what would later come to be known as Halley's Comet. Johann Albrecht sent an article titled "Meditationes de perturbation motus cometarum ab attractione planetarium orta" (Meditations on the motion of comets arising from the attraction of planetary perturbations). The academy published the papers in 1762, but the paying of the prize money was delayed. The articles confirmed Clairaut's belief that academic competitions, even when small, could considerably advance the sciences by producing "better papers than heavily endowed awards" and generating "reverberations . . . throughout the entire scientific republic."[53] Since at the time the elder Euler was believed to be directing and outlining his son's research, this competition may be interpreted as a renewal of his correspondence with Clairaut that had suddenly stopped in 1752. Amity and respect marked the correspondence between the two; apparently a divergence of interests and not disagreement had led to the decade-long silence between them.

It was not the theoretical study of lunar or cometary motion but an applied science, the technical development of optical instruments, that now drew Clairaut and Euler together. Neither of them had adequate lenses. Refining a manufacturing process that assembled elementary lenses of crown and flint glass, Dollond had had been the first to succeed in producing good achromatic lenses. In 1762 Euler read the article "Considérations sur les nouvelles lunettes d'Angleterre de Mr. Dollond, et sur le principe qui en est le fondement" (Considerations on the new English glasses of Mr. Dollond and the principle on which that is based, E380), which was not printed until 1769. In September 1763 Clairaut resumed corresponding with Euler; the exchange lasted only a few months. Lalande had reproached Clairaut for not checking on Euler's work. A letter from Clairaut discussing telescopes with double lenses noted the "very grand generality of your formulas" in articles on optical instruments in the Berlin Academy *Mémoires*,[54] but urged Euler to be clearer and more succinct in his research on the manufacture of achromatic lenses. The next January Clairaut sent a letter bridging differences between his method and Euler's and commenting on Dollond's use of flint and crown glass. Euler designed a microscope consisting of three convex achromatic lenses and a concave lens between the ocular and objective lenses.

To 1762 the longitude prize competition posed by the British Parliament remained unresolved. The Seven Years' War had impeded Mayer's efforts; the presence of French troops often disrupted observations at the Göttingen Observatory and made it dangerous to go there after dark. The competition, moreover, could not be completed until observations were made at sea, and this could not be undertaken during wartime. In February 1762 Mayer died; the following year his widow sent his final manuscript of corrected tables of lunar and solar motion to London. The British royal astronomer Nevil Maskelyne carefully edited these and published *The British Mariner's Guide* in 1763. These and Mayer's longitudinal method provided the first practical and reliable analytical means that reduced the complexity of Euler's computations to determine longitude at sea. Mariners around the globe were able to grasp and employ them, but the prize was not awarded for another two years.

As 1762 began, Prussia was on the verge of collapse. Then an event that Frederick II called the "miracle" turned the impending defeat into victory. On 5 January Empress Elizabeth of Russia died of a stroke. Her son and successor, the feebleminded Peter III, an admirer of Frederick, withdrew Russian troops that were opposing him. In July Peter's thirty-three-year-old wife Catherine, a German from Anhalt-Zerbst, led a military coup that removed him. Her ascent to the Russian throne was to influence Euler's decision to return to Russia. While he had steadfastly resisted efforts to draw him back to Saint Petersburg under Elizabeth, he was open to the possibility with Catherine II, patroness of the arts and sciences, on the throne. Her Winter Palace was to become the main building of the Hermitage Museum. In central Europe the fighting now came to a close; In February 1763 the Austrians and Prussians signed the Treaty of Hubertusburg, ending the Seven Years' War for them. The Austrian monarchy recognized the borders of 1756 with Prussia, under which Prussia kept Silesia but withdrew from Saxony. Prussia corroborated its status as a European power and was a growing competitor to imperial Austria as the leading German power.

The Presidency of the Berlin Academy

After Maupertuis died, the rumor had grown that d'Alembert or another Frenchman would come to Berlin as the new president of the academy. The other, whom d'Alembert proposed, was the polyhistor and encyclopedist Chevalier Louis de Jaucourt, but Frederick II rejected him for not being among the first rank of scholars. In a letter to Müller in December 1759, Euler had repeated that d'Alembert was holding out for a larger salary. In addition to the polemical correspondence between the two, in 1761 came d'Alembert's publication of his *Opuscules mathématiques* (Short mathematical treatises), which contained a paper refused for publication by the Berlin Academy that revived his claim to priority over Euler for solving "the argument over a string"—that is, the vibrating string problem. Like Daniel Bernoulli's animosity toward d'Alembert, Euler's did not abate but flared in early 1763. In a letter in June to Müller he described d'Alembert's impertinence and noted that Clairaut considered his French colleague's work a disgrace. In addition to their competition in the sciences was Euler's opposition to the ideas of the French Enlightenment in religion and philosophy.

As 1763 began it was announced that d'Alembert would come to Berlin to visit Frederick. Information was conflicting on whether he was to be president of the Berlin Academy. Prince Vladimir Sergeevič Dolgorukij, the Russian ambassador to Prussia, thought d'Alembert would be so named, while Müller called it highly unlikely; in April 1763 Euler observed that nothing was yet known with certainty about this matter. But once the appointment seemed about to happen, he denounced in June d'Alembert's "intolerable arrogance" and conceit. Euler questioned how this "esprit créateur" and "homme qui embrace tout"—this dilettante—could lead the Royal Prussian Academy and set its future course.[55] To Euler, Bernoulli's past complaints against d'Alembert now seemed justified.

Frederick worked diligently to draw d'Alembert to Berlin. The two met in Cleves that May, and then d'Alembert came to Potsdam, residing for three months in Charlottenburg. He intended to compliment the king on the peace ending the Seven Years' War. Royal courtiers were unsure of how to treat so distinguished a *philosophe*; it was assumed that he would be named president of the Berlin Academy or to a high political post. Frederick flattered d'Alembert and offered a position combining the income and privileges of Maupertuis and Voltaire. He promised a salary of more than twelve thousand francs that compared very favorably with d'Alembert's income in Paris of seventeen hundred, along with lodging at Sanssouci and dinner at the monarch's table. But d'Alembert refused to leave the charms of Parisian society and his mistress Julie-Jeanne de Lespinasse, to whom his attachment was continuing to grow; together with Baron Anne-Robert Jacques Turgot and others, he would in 1764 help her set up her own salon in Paris. A hypochondriac, d'Alembert knew the harsh Berlin climate had damaged Maupertuis's health and might be harmful to his own. Following his refusal, the king was to keep the presidency vacant; Frederick became more authoritarian, while Euler for a time attended to much of the general administration. In Voltaire's absence and with d'Alembert's rejection of the presidency Euler, having oversight of daily operations at the academy, possibly still hoped to become the permanent president. If so, he was sadly mistaken.

Through correspondence the king would remain in close contact with d'Alembert, who was the shadow president of the academy until his death twenty years later. In January 1764 the king declared himself the deputy president and ruled over the academy in that position as long as d'Alembert continued to reject the presidency; in time, Frederick hoped, he would change his mind. Scarcely would the king make any change without consulting d'Alembert; proposals for Potsdam would first have to go through Paris. Ordinary members of the academy were to correspond with d'Alembert through Henri Alexandre de Catt, the king's secretary, who would forward personal letters. The idea of a shadow president angered Euler; over the previous decade it was not principally the French but he and his son Johann Albrecht, at times with an internal membership circle including Gleditsch, Marggraf, Merian, Pott, Sulzer, and Johann Peter Süssmilch, as well as Lagrange in Turin, who had through their research made the Berlin Academy the first authority in Europe on their subjects.[56] To Euler, having a shadow president of a prestigious institution was unacceptable.

Once it was understood that d'Alembert would not accept the presidency, he was more known in Berlin for his sociability; it was Euler he was most eager to meet, but not in Potsdam. Before that visit in 1763, d'Alembert urged Frederick to make Euler the president of the Berlin Academy, but the king declined. D'Alembert proposed a higher salary, which the king granted. At the meeting in late August Euler was surprised at the deference that d'Alembert paid him and was perhaps embarrassed not to have recognized that d'Alembert regarded him so highly. Amid churning hostility the two had not communicated for a decade, but with their Berlin talk that changed. Johann Albrecht Euler greatly appreciated the treatment of his father by d'Alembert. Infinitary analysis has abundant computational formulas used daily that Euler increased and, according to Nicolas, Marquis de Condorcet, kept "at the very forefront on his mind and could, if asked, recite . . . by heart. When d'Alembert saw Euler in Berlin, he was astonished that Euler's mental capacity was absolutely effortless."[57]

The meeting enchanted and gratified Euler. In October he wrote to Goldbach that "our friendship is perfect, and one cannot tell me enough times of the pleasant things that M. d'Alembert has said on my behalf to the king."[58] Euler would be forever grateful for d'Alembert's words, and peace was restored between them. The many compliments that Euler was getting from one of their heroes surprised the Francophiles in Frederick's court. But Euler still found himself in a difficult situation with the king regarding finances and membership at the academy: a major reason for his proposal to leave Berlin had been the possibility of d'Alembert's becoming president; but after d'Alembert's lavish praise for him, it was unlikely that Frederick would let him return to Russia.

Having switched largely from scientific research to literary writings, d'Alembert could not compete with Euler in the sciences. Had d'Alembert remained in the field, competition with him would be serious, but since at least 1754 d'Alembert had been principally a *philosophe* and author and less a working mathematician. The old rivalry between the two had now essentially ended. Euler began to recommend that d'Alembert accept the presidency to facilitate gaining his own release from Berlin.

What Soon Happened, and Denouement

Even before d'Alembert arrived, Euler was earnestly considering leaving Berlin and returning to Russia. The first indication came in letters in March to Müller. Euler stated that if d'Alembert became president of the Royal Prussian Academy, he would leave Berlin immediately. To avoid any financial loss, in May 1763 Euler shrewdly put up for sale the Charlottenburg estate and its house. Once that was completed, he would be free to move. His primary purpose for owning the estate, as a place for his mother to live, had come to an end in 1761 upon her death. By mid-July 1763 he sold it for 8,500 Reichsthaler. During most of that year Euler and Müller corresponded about Euler's return to Russia. In May Euler wrote negatively about d'Alembert's visit to Berlin and offered to reconsider a proposal that Razumovskij had made to him in 1746 of an annual salary of two thousand rubles and five hundred rubles a year for his son Johann Albrecht. Euler remarked that he must find a position elsewhere and could not yet return to Russia. Johann Kaspar Taubert, Schumacher's son-in-law, now ran the academy. Once he learned of Euler's intention, as transmitted to Müller, he and Teplov worked to gain Euler's return. Taubert managed the academy as if he had inherited it. He treated most academicians as minor clerks. Since Euler wanted to return, Taubert urged Razumovskij to offer an annual salary of fifteen hundred rubles and five hundred for Johann Albrecht. He pointed out that Euler's current salary of twelve hundred Reichsthaler in Prussia was only equal to four or five hundred rubles, but he did admit that "the cost of living in Saint Petersburg is higher than in Berlin."[59] Taubert also wanted Euler to replace Müller as secretary.

When Catherine II was informed of Euler's offer, she quickly ordered Teplov to begin negotiations to invite him. Her court was already developing plans for the reestablishment of a strong Imperial Academy of Sciences and Arts. On 7 July she authorized a letter offering Euler the title of ordinary academician along with the positions of director of the mathematical department and conference secretary of the academy with an annual salary of eighteen hundred rubles. The exchange rate varied, but that amount was roughly 50 percent more than two thousand Reichsthaler.[60] The proposal also included housing, light and heat, coverage of incidental expenses, and five hundred rubles for traveling from Berlin to Saint Petersburg. Johann Albrecht was to be an ordinary professor at the academy with an annual salary of six hundred rubles, also with five hundred rubles for moving expenses.[61] Teplov also guaranteed Euler's future son-in-law, Johann Jakob van Delen, a place in the military service to the Russian monarch.[62] Impatient to receive a reply, Teplov spoke of his amity and estimation; Euler was assured that he had the right to change the terms and should not delay his arrival. Nevertheless, Euler declined the offer, partly because he did not want to take Müller's position away from him; but he pointed out that he could hope for nothing better than to pursue his studies under the protection of Russia. He was especially occupied with completing his two-volume great treatise that increased the application of calculus to astronomy. After meeting with d'Alembert and calm was restored between the two, Euler wrote to Teplov in July that he must wait for a more favorable time to depart from Berlin.[63] His wife, he commented that month to Müller, still complained that people had not been allowed to take their money out of Russia, and she retained a fear of the danger of fires in Saint Petersburg; Euler also wondered about the varying value of currency. He now wished that d'Alembert would accept the presidency of the Berlin Academy so that he himself could leave Berlin. In August he assumed new tasks vital to the Berlin Academy's finances, and in October his attitude about leaving seemed to have changed. Müller insisted that the ties with Saint Petersburg must remain unbroken.

Hardly had the peace been arranged between Prussia and Russia when Euler began with new energy to work on the *Novi commentarii*. The seventh volume needed to be completed and the eighth prepared. Euler's distance from Saint Petersburg and the uncertainty of the postal service during the war had made this task most difficult. Euler was now interacting with d'Alembert, having given him the incomplete volumes 7, 8, and 9 of the *Novi commentarii*.

In 1763 Frederick asked the devout Euler to head the venerable and powerful consistory of the French Calvinist Church in Berlin. In Protestant churches the consistory, made up of elected officials—including, in this case, church elders—was the governing body; the monarch's relations with it and thus the French Huguenot colony were strained. Though profoundly differing with the king in matters of religious doctrine, Euler agreed to take the position; he had already served on various church councils. As the consistory head and an elder of the Friedrichstadt congregation, he significantly influenced offers of church positions, issues of orphanages and charities, and measures to increase church attendance.

The consistory appointed Euler to a commission to improve the catechism.[64] Following his father's methods, Euler proposed expanding the teaching of catechism during the regular service at the expense of sermons. For classes children were separated by gender and ability; every Sunday afternoon they would meet and take questions from assigned sections of readings from the pastor. Euler decided that Sunday classes should begin by verbally reciting selections from the reading. The pastor was to instruct the children in greater detail and add more catechism questions that students would be randomly asked to answer. Answers had to be from the catechism, word for word. In preparation for these classes the students were required to attend two weekly sessions with the pastor. To make the catechism more instructive and interesting for the students, the teacher was to include books, especially on stories from the Old and New Testaments. Euler wanted students to be accustomed to speaking in public. He urged that the pastor not be elevated in a pulpit, but should stand among the worshippers or the children so as to better respond to questions. Euler proposed that Bibles be placed on desks so that students could check passages from their catechism and become more familiar with the Scriptures. He wanted chapels in local churches to provide space for instructing children from the orphanage and the charity school, as well as other poor children.

Following the Seven Years' War, governmental funds were short in Berlin. At the time Euler was refining his proposal for five classes of lotteries. In August 1763 Euler communicated a proposal for another source of income, porcelain manufacture. On 14 August, Dietrich Peter Denffer (Janssen), a royal painter in Courland, wrote to Euler about the decline of the Saxon porcelain industry there; he thought it possible to improve its manufacture and offered to send Frederick II the manuscript of the book that he was writing on the subject. On 30 August Frederick responded to Euler that the idea was agreeable and that perfecting the fabrication of porcelain was more important than discovering the secret of its composition; that, he felt, would be a great achievement.

In the summer of 1763 the wealthy Dutch cadet officer and baron Johann Jakob van Delen, who would later marry Euler's daughter Charlotte, joined with Euler's relatives in Holland to try to get him to accept a position of professor with a salary of five thousand gulden (an amount equal to about 2,140 rubles) in a Dutch university.[65] For Euler's wife this would mean a return to her home country, and it meant that she would not be separated from future grandchildren. To check on this post, Euler's son Karl and van Delen traveled to Holland. Euler wavered briefly, but he wanted to be released from all teaching, and the Dutch university would not agree to this. In November Euler informed Goldbach that if the position of the president of the Göttingen Academy remained unoccupied, he might apply. Unsure of the stability of the new Russian government, Euler had not yet settled on a return to Russia.

In December the personal relations between Frederick and Euler soured further. On 2 December Euler asked permission for the marriage of his younger daughter Charlotte to Baron van Delen. The next day the king rebuffed the request, banning the marriage, saying that van Delen's first responsibility was to the military and to make advances through the Prussian officer ranks. He blocked the marriage for three years. Euler was now even more open to making a major change in his location.

From 1763 to 1765 Euler's correspondents included Lalande, Lambert, and Segner, commenting on an array of scientific topics. Lalande discussed the comet of 1759, noting that the motion of Earth had no effect on its course, expressed his doubts about observations of the transit of Venus in 1761, and noted the preparation of his two-volume text *Astronomie*. He reviewed the problem of nutation in the precession or wobbling of Earth, and wrote of Lagrange's arrival in Paris, the publication of Alexis Fontaine's *Le calcul intégral* (Integral calculus), Johann Albrecht Euler's winning of the Petersburg Imperial Academy's prize on the paths of comets, and the optical research of Rudiero Josep Boscovich. Lambert addressed partial differential equations, thermodynamics, and the three-body problem. He covered an article that he was writing for the Petersburg *Novi commentarii* and mentioned that he would possibly present that academy a copy of his *Pyrometrie* of 1760 on fire and heat. Also among Lambert's many topics were infinite series, properties of curves, imaginary logarithms, differential and integral calculus, fluid mechanics, and the improvement of telescopes. From the years 1763–65, twenty-seven of Segner's letters to Euler survive. He reflected on his *Neues Organon* of 1764 and a second edition of his *Einleitung in die Naturlehre* (Introduction to Natural Philosophy) of 1746.

During the years 1763–65 Euler spent much time writing, revising, or editing three books. His yearly production of writings was over eight hundred pages, a substantial body of work by any standard. In 1763, Euler had ready for the press most of the manuscript of 1,700 pages for his three-volume *Institutionum calculi integralis*, but in these unfavorable times it went unpublished for five years. The *Diarium litterarum*, a journal from the town of Griefswald, sought a publisher—perhaps through another person like Karsten. Observing that the Swiss printers annually published a number of poor writings, the *Diarium* declared that it would go counter to the *Aufklärung* of the era that a "work written for the world and all times by the fault of a publisher lies unprinted."[66] But pure mathematics characteristically had a limited readership; to cover printing costs, a wider public was needed.

One day in 1763 Christoph Jetzler, a furrier from Schaffhausen, arrived at Euler's house. At twenty-eight or twenty-nine years of age, Jetzler wanted to pursue an academic career, but family pressures forced him into another profession. After the death of his father, however, he began to follow his interests in the sciences. He traveled to Berlin, knocked on Euler's door, and asked permission to copy sheet by sheet from Euler's textbook on integral calculus. Euler provided a room and agreed to help him overcome difficulties in comprehending the new theory.[67] During Jetzler's stay in Berlin his mother sent him cherries and apple wedges, and for Euler plums, which he dearly loved. After months of writing himself to exhaustion, Jetzler returned home in the late spring of 1764. In gratitude he sent Euler a large round of cheese, a practice that was a Swiss custom at the time. This was not the end of their connection; Euler offered to help explain problems in science, including some in astronomy two years later. Jetzler continued his studies and became professor of physics and mathematics at the University of Schaffhausen.

By 1764 Euler's *Lettres à une princesse d'Allemagne* were revised, now in French for his students and to reach a larger readership. Continuing his research in optics, in 1765 he completed most of the manuscript for his *Dioptricae* (Dioptrics, or general theory of lenses, E367); it would be four more years before it saw publication.

Frederick II had always wanted complete control of the Royal Prussian Academy. Any thoughts that he had only momentarily applied his authority in the 1740s to fashion an effective institution were dispelled when, freed of his many military duties in late 1762, he began to tighten his control over its administration; at home and abroad it was soon understood that everything would lie in his hands. He started with two crucial changes, from which Euler lost out in his efforts to make the academy independent and professional: the monarch reasserted and extended his authority over the naming of new members and in the oversight of finances. When the majority of members of the academy wanted to elect the German art historian and archaeologist Johann Joachim Winckelmann, who was profoundly influenced by classical Greek art and ancient Rome, Frederick followed the advice of Euler and Marggraf, who opposed Winckelmann's election. This was not simply an expression of the king's high regard for Euler; he had allotted a possible salary of one thousand Reichsthaler for Winckelmann, who wanted twice that amount, and he apparently did not realize the man's scholarly distinction. The rejection of Winckelmann and, over the next decades, the rejection of philosopher and historian Johann Gottfried von Herder, who would win the Berlin Academy prize several times, showed Frederick's condescension toward German candidates, manifested also in his unhappiness with Lessing's election and perhaps Winckelmann's conduct. To Frederick the wishes of the majority of academicians meant little or nothing; his negative decisions on membership for Winckelmann and Herder isolated the academy from leading national German scholarship. Reacting against Euler's independence in the 1760 elections, in January 1764 Frederick declared that the academy should choose no new member until he named a president; and he inserted the clause that he alone had the right to name all future members. He did not hold to the provision about a new president, but while accepting some advice kept to the absolute reservation concerning new members, which went a step beyond his earlier veto of names on lists of candidates and the free election under Euler. In the future, during Frederick's reign, the regular members of the academy had no influential voice in selecting candidates.[68]

As of late 1763, Euler had not yet definitely decided to return to Saint Petersburg. In October he wrote to Goldbach that, he feared, "the idea of transforming this [Berlin] academy into a French academy still exists,"[69] by which he meant an academy open to radical ideas of the French Enlightenment. "As much as I am upset at the prospect of leaving, I still must prepare myself for this eventuality." The selection of members from 1764 to 1766 supported the idea of a French-like academy. These included seven Frenchmen, suggested by d'Alembert, and two Swiss, one of them the nineteen-year-old Johann III Bernoulli. After Euler wrote in October 1763 to his friend Wenzeslaus Johann Gustav Karsten, who was seeking a new post in Butzow, that few vacant positions existed in Berlin and these required a perfect knowledge of French, Johann Albrecht informed Euler that the king had ordered the improvement of the academy. In response Euler invited Johann III Bernoulli, who arrived at the end of the year. It was important to be patient and not upset other members or the king, since his selection was not yet approved, and no salary was yet available. In November and December he assisted André Pierre Le Quay de Premontval as a foreign member on psychiatry,[70] and the assembly of the academy found his work brilliant. In 1764 Euler officially recommended Bernoulli to the king's agent, and Frederick agreed to appoint this young member of the famous Basel family. That year the king commanded the academy to greet as new members Bernoulli, the theologian and military historian Quintus Icilius, and Johann Castillon, who taught at the artillery school and had completed a translation and commentary on Newton's *Universal Arithmetick*. Euler regretted that almost all foreign members were French freethinkers whose selection he could not block. These included in 1764 Louis de Jaucourt, a prolific writer for Denis Diderot's *Encyclopédie*; François-Vincent Toussaint, the author of *Les mouers* (On manners) of 1748, a book that the French courts had burned; and in 1765 the wealthy, controversial hedonist Claude Adrien Helvétius, the author of the popular *De l'esprit* (On the mind) that derived morality from self-interest founded in the love of pleasure and sought the greatest happiness for the greatest number. The Sorbonne condemned Helvétius's tract and the public executioner burned it; it was also placed on the Catholic Church's index of banned books. Euler, of course, greatly disliked Helvétius.

Late in 1763, the sale of the Berlin Academy's almanac had become the central issue in the brewing financial dispute with the king; its sale remained a state monopoly and still provided most of the funds to support the academy. Frederick believed that it must earn more income, so he proposed that its administration be changed. But Euler's fellow directors Marggraf and Johann Philipp Heinius continued to grant Euler a free hand, and he gave control to the almanac's chief commissioner David Köhler. It was rumored that Köhler would be dismissed from selling the almanac, for the king correctly agreed with the rumors that he was skimming off perhaps one-fourth of the funds from sales.[71] Discounting these charges, Euler supported Köhler, whom he appreciated for always promptly paying his salary; he worried that the removal of Köhler would weaken his administrative position at the academy. Euler took the selection of the commission as a vote of royal mistrust and a personal insult; prior to this time he had managed everything alone, but now he was one among many. He wanted nothing detrimental to occur to Köhler.[72] Sulzer complained about Köhler and reported to the king. He suggested that many academicians feared Euler, muffling complaints about the almanac's mismanagement, but only when Henri Alexandre de Catt joined Sulzer did he get a response from Frederick.

In February 1765 the king issued an order to compile a new record of the academy's finances and a possible reorganization of them. During the war the academy had spent twenty-five thousand Reichsthaler, but Frederick believed that much could have been saved though tighter controls on spending. By royal command, the budget had to include rebuilding the chemical laboratory, constructing adjacent residences, expanding the enclosure of the botanical gardens, and repairing all buildings used by the academy. But Frederick, discontented with the past finances, held Euler responsible; the king replaced the committee of four noble curators with a commission of inquiry consisting of six salaried members elected by the academy to manage finances and salary. It was not clear whether this was a permanent change in the administration. The committee consisted of Euler, Merian, Sulzer, Beausobre, Castillon, and Lambert, and Euler was elected its chief. They were charged with examining why the revenues from the sale of the almanac and of maps had declined so sharply during the Seven Years' War. Köhler, then the treasurer, had real enemies in the academy—most notably Formey. At this point Sulzer, Lambert, and Catt were not favorably disposed toward him. Unable to agree on a common proposal, the commission submitted three plans for gathering content for the almanac and controlling its sale. Sulzer and Beausobre favored publicly leasing out the almanac; Lambert wanted to administer it himself; and Euler fought to keep it for Köhler, but under new regulations. Most commission members projected that the almanac sales with their changes would increase from thirteen thousand to sixteen thousand ecus.

On 13 June 1765 Euler, risking his prestige and acting as a curator without consulting other commission members, wrote directly to the king in support of Köhler. The letter had the opposite effect to what Euler desired, however; it not only alienated the other commission members but made the king decide to turn over the sale of the almanac to the private sector. Three days later, Frederick responded sharply in French to Euler, "I understand that I cannot reckon curves, but I do know that 16,000 ecus are more than 13,000."[73] He also warned Euler not to be fooled and to accept the majority report of the commission: "The rights accorded to my academy must be used to support our men of science, not to feed a corrupted treasurer who is already paid well enough."[74] Not only did the monarch decide to lease out the calendar, but he refused to answer two further letters from Euler on the subject. When Euler finally agreed in a third letter to drop the subject, Frederick replied in March 1766 that he was pleased to see him desist on this issue.[75] Euler broke with the other members of the committee, arguing that they did not have the right to submit their report, since he had not signed it.

The king and likely all academicians had another reason to be angry with Köhler: after the war, he had exchanged the academy's financial reserves in metal coinage for paper currency that soon lost two-thirds of its value. At that time, and upon the royal decision on the almanac sale, the collaboration between Euler and Köhler ended, yet the commission continued to work with close attention to the mathematics department directed by Euler. That field received the only increased salaries awarded in 1764, bringing them to four hundred Reichsthaler. One recipient was Johann Albrecht Euler. Still, the election and the implementation of financial changes beginning in 1763 eroded the independence of the academy, which gave Euler added reason to depart.[76] By 1765 Euler's decision to leave would be final and irreversible.

By the fall of 1765 the commission members began to call themselves the economic commission, and they had to address another important matter regarding the organization of the archives of the academy. The members discovered that numerous documents that they thought were in the academy's archive were actually still in the homes of Köhler and the academy secretary Formey; this gave them reason to study the formation of a new archive and then to implement a proposal for it. The archives consisted of two departments: one gathered academic records, and the other financial and administrative documents. In September Frederick sanctioned an archivist independent of the permanent secretary and the academic treasurer with a salary of 250 Reichsthaler. The commission wanted no direction from these officials. The archivist would also be registrar and supply the commission with data. On 3 October Euler proposed that each department have its own archivist, with Formey remaining in charge of academic works and Köhler, because of his detailed knowledge of them, to oversee the financial and administrative archives. The commission rejected Euler's recommendation, and in response to the two decisions he resigned from it in November.

In January 1764 Lambert had arrived on his own initiative in Berlin, reaching an academy that would challenge his brilliance and offer funding. The Swiss, including Euler and Sulzer, warmly greeted him as a compatriot. Wanting to meet the scholar considered a world wonder, Frederick granted Lambert an audience that has proven famous. According to Thiébault, for the meeting with the king Lambert dressed oddly in traveling clothes and was described as looking "like a bear."[77] The conversation went like this:

> Frederick: "What science have you especially studied?"
> Lambert : "All, Sire."
> Frederick: "Are you also a good mathematician?"
> Lambert: "Yes, Sire."
> Frederick: "Who has instructed you in mathematics?"
> Lambert: "I, myself."
> Frederick: "Then you are a second Pascal."
> Lambert: "Yes, Sire."

Upset with what he took as conceit the king, upon making off to his cabinet room, laughed and fumed, "He is the biggest imbecile I have ever seen."[78] If the monarch did say this, he did not grasp that Lambert was right in his responses, including that about Pascal. Lambert, a successful autodidact, had by 1764 proved that π and e are irrational, which put him close to the level of Pascal in mathematics. His proof that e is irrational gave formalism to what Euler had shown but still lacked rigor. Until January 1765 the king refused to confirm Lambert as a salaried member.[79] But four months after Lambert's arrival, Euler wrote enthusiastically, "He is in all respects a person possessing enough talent for a whole academy."[80] And Immanuel Kant considered Lambert the greatest philosopher of the century, called his *Photometria* a classic book of physics, and foresaw that his name was to be forever linked with several important mathematical discoveries.

In a later conversation with Thiébault, Lambert ranked himself among living geometers; responding to a query from Thiébault, he stated without fear or hesitation, "The two foremost living geometers are Euler and d'Alembert or d'Alembert and Euler." He did not settle on a final ordering.[81] Euler, he opined, was more naive and d'Alembert had more finesse, subtlety, and elegance. Speaking with no vanity, he said, but following the dicta of complete objectivity, he himself was the third geometer.

In 1764 Euler continued his considerable correspondence with Müller. In March he thanked Müller for sending publications of Aepinus and the works of Johann Ernst Zeiher, who was leaving Russia for Wittenberg as professor of mechanics. Euler suggested Lambert as a replacement. The next month Euler successfully urged the Petersburg Academy to name as foreign members d'Alembert, Lalande, and the Spanish glassmaker Raphael Pacecco, whose glasses improved light perspectives and telescopes. Zeiher and Marggraf independently tested the different glasses, and Euler found Pacecco's research a necessary introduction to the deepest theory of colored light rays. In July Euler enclosed in a package a letter to Aepinus from Karsten, who wanted to move to Saint Petersburg.

In late October Euler and Lambert met in Berlin to discuss with the Russian envoy Dolgorukij the rehabilitation of the Petersburg Imperial Academy, providing Dolgorukij with exhaustive information on the Berlin Academy's framework. On 9 November the visiting Russian chancellor, Count Mikhail I. Voroncov, invited Euler and Lambert to dinner to discuss the reorganization, and Dolgorukij joined them. These conversations were important to the Russian effort to draw Euler back to Saint Petersburg. At Voroncov's dinner they exchanged ideas on what "benefit might accrue to the [Russian] government [from this development] . . . and how members might combine their individual strengths in the general interest of the nation."[82] The next day Euler reported that Lambert had impressed his Russian hosts, who suggested that he enter their service, but he declined. Euler declared that in royal academies individuals were supposed to make most scientific contributions; instead he and Lambert stressed the need for more teamwork, which Euler found lacking in nearly all academies; he felt that institutions fostering collaboration should achieve more general applications and greater profits.[83] It was thought that collaboration would result between Euler and Lambert, but instead sharp differences soon emerged between them on such matters as finances and the Berlin Academy's almanac. Even so, the idea that Lambert worked to drive Euler from Berlin is erroneous. At the end of November, Euler met almost daily with Voroncov, and each time they addressed the reform of the academy. That some of Euler's proposals were implemented before his return pleased him.

At the academy, Euler enjoyed the friendship and assistance of Merian, who had come to Berlin only because he had lost out four times in the lottery to be a professor at the University of Basel.[84] Intellectually talented, with a well-rounded education and great energy, he earned a European reputation for his critical philosophical writings. Aided by his admirable wife, Merian turned his house into a center of intellectual life in Berlin.

By contrast Sulzer, a Wolffian, opposed Euler. Sulzer was important in restoring to a strong position the Wolffians, who had been displaced at the academy in the 1750s. While he had been made a member in 1750, Sulzer had received no salary from Maupertuis. In the intellectual life in Berlin, he had become a robust leader, but by 1763 he wanted to return to Switzerland. In 1764 Castillon joined Sulzer in the Wolffian resurgence. This development must have discouraged Euler. To keep Sulzer in Berlin, Frederick offered him a position in the new *Ritterakademie* (knights' academy) with a sufficient salary.

After the end of the Seven Years' War, this knights' academy was planned and royally founded in 1765 in Berlin with its goal being to teach the sons of nobles and attempt to develop for the king a close connection with the science academy, but those academicians seldom showed interest in it. At the age of twenty-nine and with a distinguished record of publications, Johann Albrecht Euler applied to teach mathematics there, but Frederic generally did not want joint appointments between the science academy and the knights' academy, and they were rare; one such joint appointment was of Thiébault, who personally edited the king's writings in French and would teach French grammar at the knights' academy. Johann Albrecht's application was rejected with the excuse that he was too young, but that this was an evasion was made clear when eighteen-year-old Frédéric Castillon, the son of Jean, received the position. Johann Albrecht wrote in disgust to Karstens that Frédéric's knowledge of mathematics was at best at the level of a gymnasium pupil, and the Eulers took this as another slight in Berlin.

In 1765 Euler's landmark work *Theoria motus corporum solidorum seu rigidorum* (Theory of the motion of solid or rigid bodies, E289) was published,[85] and it reflected Euler's recognition that his earlier *Mechanica sive motus scientia analytice exposita* on the motion of point masses was lacking in several ways. He not only brought it up to date but went beyond. Called his second *Mechanica*, *Theoria motus corporum solidorum* treated the kinematics and dynamics of particles without referring to his predecessors—mainly Galileo, Christiaan Huygens, and Newton. While not completely without geometry, most of it is analytical, providing and calculating differential equations. Euler based dynamics either on d'Alembert's principle, sum $(F) = 0$, which was a variation of Newton's second law, sum $F = ma$, or on the principle of action and reaction.

Theoria motus corporum solidorum applied to a solid the system of equivalent forces. A didactic work, it addressed particular problems, which it solved by use of a general theory. The text began with a six-chapter introduction containing illustrations and necessary additions to the motion of solid bodies, followed by a nineteen-chapter "Tractatus" on the mechanics of rigid bodies. The book gave inertia as its elementary force and employed three permanent axes to describe it. Euler examined moments of inertia in wires, plane figures, and major solids, and he proceeded to the general case in which the moments were unequal. He also noted the importance of rotational motion and studied varieties of rotation under different forces; he had earlier thought that straight line motion was sufficient. In 1758 he made the important discovery of the equations of motion for the rotation of rigid bodies, and he subsequently solved that problem in 1759. The stress on rotational motion was something new, and Euler also investigated gyrations when a heavy pendulum is moved. *Theoria motus corporum solidorum seu rigidorum* closed with a five-chapter supplement dealing principally with friction.

In designing mathematical models, algorithms, and differential equations to predict the motion of rigid bodies, Euler advanced two fundamental laws for that motion that are today known by his name. The first describes translational motion, changing the velocity of the center of mass, and the second describes how the moment of forces and connections controls the change of angular momentum.

Theoria motus corporum solidorum is the last piece in Euler's ongoing study of mechanics, which took over thirty years to complete. He had earlier devised the differential equations for the motion of fluid, elastic, and flexible bodies. *Theoria motus corporum solidorum* presents his clear and detailed analytical revision of the whole theory of solid bodies.

The first section of the book contains a lengthy discussion of the nature of space—whether it is Newton's absolute or Leibniz's relative case. In 1748 Euler had addressed the subject at the academy in his article "Reflexions sur l'espace et le tems" (Reflections on space and time), which was published two years later; he now examined thoroughly the reasons for and against each view, and his account was considered the best treatment of the subject for over a century. Presenting himself less as a philosopher than a mathematician, Euler showed his mastery of the subject.[86]

Also in 1765, Euler received an honor from the British. The desire of the British Parliament not to give foreigners its prize for devising a more practical method for finding longitude aboard a ship when it is not in view of land was a reason for delay in the making of the final test for it, as was British participation in the Seven Years' War; the award was not given until that year. British officials had urged the clockmaker John Harrison to construct a marine chronometer that gave longitude better than Mayer's lunar method and indicated that the prize of ten thousand pounds for Harrison would be doubled when it was possible to make many more of these devices. After his first test of what became his famous clock in 1764, Harrison thus won that major part of the prize. Mayer's widow and his heirs received part of the smallest bounties for the prize, three thousand pounds, for his lunar tables. It was only after Clairaut protested in May 1765 in *The Gentleman's Magazine* that he and Euler had not been recognized for building the foundation of the theory of lunar motion and for Mayer's analytical method that Euler would receive an unsolicited gift of three hundred pounds, the equivalent of more than thirteen hundred Reichsthaler,[87] for his contributions to the theory of lunar motion. It came as a pleasant surprise.

In the next year, Euler would return to Russia. The reasons for that are more complex than the refusal of Frederick II to name him the president of the Berlin Academy as well as to the Russian increase in his salary by 50 percent, though both were important. The position of Frederick on the presidency had not changed: the candidate must be a distinguished scholar, and ideally a *philosophe*, a *héro de salon*, a man of the world, and a fit partner at the royal table. Frederick believed that only among the French could he find such a figure. While Euler had wanted the presidency since his arrival, his failure to obtain it does not seem decisive in his decision to leave Berlin; loss of control over electing academy members and over finances and salaries was most critical. Sulzer and Thiébault saw as the chief cause for the break the royal letters to Euler on these two issues. After the war, Euler had tried without success to have Frederick guarantee the future of his younger sons, but for Christoph no further promotion in the officer ranks was possible in the Prussian military, and there would be no position for Karl; both of these facts disappointed Euler. After the war, Frederick appointed only aristocrats to fill officers' and governmental positions. A dedicated family man, Euler took as the cruelest blow in his personal association with the king this refusal to promote the careers of his sons.

Through 1765 Euler's relations with the Prussian court deteriorated for these reasons and one more: his gaze was now fixed on Saint Petersburg, and negotiations were under way for his return to Russia. Catherine II considered the acquisition of Euler crucial to revitalizing her science academy. In November, after a break of more than fifteen years, Euler wrote to Joseph-Nicholas Delisle that since the death of Maupertuis the selection of new members in Berlin no longer came from the academicians. Euler had thus decided to return to Saint Petersburg, where Catherine promised radical reforms for her academy. These included naming eight additional members for new positions, and Euler was supposed to select these. He asked Delisle for his help in listing possible candidates.

At the end of 1765, Johann Albrecht Euler wrote to Karsten that his father was in a bad situation. In November Euler wrote to Beausobre about the Berlin Academy's arguments over economics and attempts at reconciliation. Some members did not want to debate with Euler and sought ways to resolve differences; others related criticism of him to the king and said that Euler should resign his directorship. His departure, they thought, might follow. Insulted, in response Euler cited his twenty-four years of faithful and able service. He now decided that he and his family should leave; to remain would be disgraceful. He had not yet formally made the request to depart, but at the end of November he left his director's position in Berlin. On 24 December he sent a letter asking to return to Russia with the terms he and his son desired and awaited the response from the Petersburg Academy's chancellor Voroncov. Essentially Catherine had said to give Euler a blank sheet of paper on which to write his requirements, and grant him anything that he wanted. Partly out of fear of totally losing his sight, Euler insisted on terms for his family as well as himself. He requested the positions of vice president and director of the mathematical department of the academy with an annual salary of three thousand rubles, an increase of half over the amount earlier offered him by Taubert. (The official exchange rate at the time was made 125 rubles equal to 400 Reichsthaler.) Euler requested a yearly annuity of one thousand rubles for his wife and a physics professorship for his son Johann Albrecht with a salary of one thousand rubles; his son Karl would have the position of physician for the academy and Christoph a post of army officer in artillery. Euler wanted a staff of talented young mathematicians assembled around him. His eldest daughter was promised to meet appropriate suitors, and the family was to be treated like nobility.

On 6 January 1766 Catherine II replied enthusiastically to Euler's desire to return. Since the position of vice president was new, it would still have to be defined. While the academy lacked the funds for his annual salary, Catherine would allocate money, for she felt that Euler was deserving of the three thousand rubles.[88] Although there was no vacant position in physics, she would add one for Johann Albrecht, and Euler's wife Katharina was to receive an annuity. Positions would be open for Christoph in the artillery and Karl in medicine, either in private practice or working with the academy. Catherine would provide a state-subsidized residence exempt from military quartering, along with heat and lighting. She would cover all moving expenses, as determined by Euler and Dolgorukij, and all of Euler's property could be brought to Russia duty free. For a scholar of Euler's European-wide reputation, Catherine considered these terms an inexpensive bargain. On Christmas eve, Voroncov wrote that Catherine had accepted Euler's request. In January Dolgorukij officially confirmed this decision. In her January letter Catherine wrote to Voroncov, "I am sure that my academy will rise from the dust because of this acquisition, and I congratulate myself in advance with the returning of this great man to Russia."[89] She promised Euler that she would not make any changes in the academy before his arrival.

As he lost control over the administration of the Berlin Academy in 1765, Euler had begun to separate himself from its activities and made no secret of his desire to relocate. In July, after Euler was excluded from public celebrations, the Prussian state minister J. L. Dorville asked him to nominate candidates for the royal librarian, and Euler gave three names: Christoph von Sax of Utrecht, Georg Christof Hamberger from Göttingen, and the fictional Jacob Daniel von Wegelin from Saint Gall, the last being a figure from Euler's sardonic response to Dorville's wish to have a Swiss candidate. On 3 February 1766 Dorville asked Euler to begin negotiations with Wegelin, but the selection committee of Catt and Formey quickly realized its mistake. To avoid embarrassment, the next day Euler wrote that Wegelin had left Berlin and withdrawn his name, leaving the way open for the fittest candidate. One interpretation of the proposal of an imaginary candidate is that it shows Euler seeking release from Prussia.

On 2 February, shortly after Euler received Voroncov's and Dolgorukij's letters, he formally asked to be released from his Berlin post. He and an extended family of fourteen, except for Christoph in the Prussian Army, were ready to leave that month; Euler knew that Frederick would not permit Christoph to go. Tensions grew at the Prussian court; with Maupertuis dead and d'Alembert unwilling to accept a post in Berlin, the king did not now want to lose his most distinguished man of science. Frederick and his advisers chose not to reply to Euler's request. Catt, the king's reader, wrote Euler that he hoped the news of it was only rumor. He suggested that Euler might leave the economic commission but remain as a director of the academy. While Euler mistakenly thought that Catt was against him, actually he—and from Paris, d'Alembert—urged release to avoid a public scandal. Catt described Euler as a great and honest man,[90] and it is likely that this praise helped move Frederick to grant Euler's wish. Johann Albrecht kept a record of his father's appeals. On 4 February Johann Albrecht wrote to Karsten, "My father yesterday applied for leave for himself and his entire family, which consists of eighteen souls." On 15 February he noted that his father "had already twice asked for his release, but to this hour still has received no answer. We live in the most terrible uncertainty and probably will remain for some time in that state. . . . "[91] After Euler wrote to Jacob von Stählin, the academy secretary in Saint Petersburg, that his ties with Berlin were completely broken, Stählin raised the issue with Frederick. On 18 February Frederick's secretary informed Stählin that the king would not let others make his decision, which was his to be made no matter the cost to him. Clearly academics did not have a right to end at any particular time their contracts with the crown. Euler did everything he could to be cashiered; in March he and his eldest son stopped for the most part attending the Berlin Academy's meetings. Euler's thoughts were now focused beyond Prussia.

Gaining release from Frederick proved more difficult than the negotiations with Schumacher in 1741. Regarding Euler's first two letters the king remained silent, a familiar strategem of the monarch. On 15 March Euler again requested permission from the king to leave Berlin. "Sire!" his letter began, "I have always worked hard at the service of Your Majesty, but I cannot be indifferent to the wonderful offer made me and my family by the Russian Court."[92] Two days afterward, Frederick responded, "I would like to let you know that you will please me if you desist from this demand" and commanded Euler to "never again write about this subject."[93] The king's refusal led Euler to ask d'Alembert for assistance, but it was not soon forthcoming. Euler's perseverance bespoke Swiss stubbornness and liberty, but there was no progress with the king. Euler's son did not know whether his father would have to stay or could depart. "Unfortunately," declared Johann Albrecht on 5 April 1766, "the king persists in his utter silence and will neither let my father leave nor have the good grace of offering him satisfaction or service. My father, however, upholds his Swiss freedom, and becomes more and more eager for his departure."[94] On 8 April Catt promised to show Euler's letter to the king, and Catt discussed it a week later. Not until late April did d'Alembert write a letter promising to help improve Euler's relations with the king and urging him to stay in Berlin. He dared not risk Frederick's anger by backing Euler. Apparently the letter was not delivered, since Euler had made his decision.

On 29 April Euler and Johann Albrecht began to say farewell to the members of the Berlin Academy. The next day Euler again politely asked for his release. He pointed out that all academicians who asked to leave gained permission without any problems. He observed that it would not be sensible to turn down the good offer that the Russian court had extended to his entire family. He now requested leave for his eldest son, who was a member of the academy, and his second son, Karl, a physician treating impoverished French Huguenots in Berlin. He realized that it would be difficult to have Christoph, an artillerist, join them.[95] Frederick likely did not wish to antagonize needlessly "my dear sister" Catherine, who was applying pressure. On 2 May the monarch wrote in French a terse two-line statement without a word of thanks: "With reference to your letter of 30 April, I permit you to quit in order to go to Russia. Fréderic."[96] The king seemed anguished, and even so, d'Alembert believed that Frederick did not grasp the severity of the loss. On 8 May, Catt extended best wishes to Euler on his new path.

During their final month in Berlin, three of Euler's children were married—his son Karl and two daughters. Ten days before the family's departure for Russia, Karl Euler wedded Anna Emilie Bell. Little is known of the elder and sixth child, Helene, who married the first major Karl von Bell, a quartermaster on the Prussian general staff and later a Russian officer. He was perhaps a noble relative of Karl Euler's wife. Frederick finally allowed Baron Johann Jakob van Delen to marry the younger daughter, Charlotte. All three couples were to return to Russia with Euler. Only the van Delens would eventually leave the Euler fold; in the summer of 1769 they moved to van Delen's country estate in Hückelhoven in the duchy of Jülich on the Lower Rhine River in Holland.

On 23 May Leonhard and Johann Albrecht Euler officially bade farewell to members of the Berlin Academy; they attended their final meeting on 29 May. On 9 June the Euler party left Berlin. It was an emotional scene attended by Prussians of royal blood, including the future margrave Friedrich Heinrich and his daughters, joined by Euler's colleagues, neighbors, and students. Of the family, only Christoph stayed; Frederick decreed that he must remain in the military, but in a year the king was to relent.

The time from Euler's departure from Berlin to his arrival in Russia was not uneventful. His party left the Prussian capital in a small wagon caravan, going first to Warsaw. In 1762 Prince Adam Kazmierz Czartoryski, from one of the wealthiest Polish noble families, had refused the crown; instead it went to his thirty-one-year-old cousin Stanislaw August Poniatowski of Poland, a favorite of Catherine II. A patron of the arts and sciences, Czartoryski invited Euler to Warsaw and Poniatowski agreed. The Polish king and his court warmly feted him for ten days, and Poniatowski conveyed his regards for Catherine. The Eulers proceeded next to Mitau (Jelgava), the capital of the duchy of Courland, where the duke hosted them for four days, perhaps at the nearly completed Rastrelli Palace. They then made their way to the Baltic and boarded boats for the nearby seaport of Riga. Supposedly a second boat carrying luggage and supplies sank; we only know of this alleged event from a derisive letter written by Frederick II and a note of it in Thiebault's *Souvenirs*. Since neither observed the event or gave any evidence of it, it seems unlikely that it occurred. Recognizing his guest's distinction, the duke in Riga granted Euler free lodging and furniture along with luggage, service, and two grenadier guards, as well as reduced prices for some of the best items. The city meanwhile honored "the great Euler" for his many achievements.

After corresponding with d'Alembert in early June, Lagrange indicated that he had not wanted to work under Euler. On 26 July Frederick wrote to d'Alembert thanking him for having Lagrange agree to accept the Berlin post. To gain Frederick's endorsement, d'Alembert had claimed that Lagrange was a *philosophe* and man of the world (though he was neither) and the foremost geometer. Lagrange's acceptance took part of the sting off Euler's leaving. A letter from Frederick to d'Alembert was tasteless; he remarked that "the one eyed monster . . . has been replaced by another who has both eyes." Of the alleged boat sinking, Frederick continued, "A ship loaded with his x, z, [and] his kk, has wrecked; everything was lost. It is a pity, because there was something that could fill six foliate volumes of papers full of numbers from beginning to end. Apparently now, Europe will not have the fun of reading them."[97] No other record has yet been found of the ship's sinking. For ten years there was no correspondence between the king and Euler.

Immediately after Euler's departure for Russia, the Berlin Academy returned for a final time to the debates over the Leibnizian and Wolffian philosophies by setting a eulogy of Leibniz as its prize topic for 1768. The Wolffians had lost recently, as in 1763 when Moses Mendelssohn won the prize and Immanuel Kant the *accessit* (honorable mention) arguing that the proofs of metaphysics were not so clear and comprehensive as those in Euclidean geometry. But in 1768 the academy awarded the prize to the modest *éloge* by the French astronomer Jean Sylvain Bailly of the Paris Academy. According to the prize committee, the article brought "praise to neither Leibniz nor its author,"[98] but this competition has been called the final defeat of Euler by the Wolffians at the academy, where for a more than a decade an eclectic version of their philosophy dominated.

Chapter 13

Return to Saint Petersburg: Academy Reform and Great Productivity, July 1766 to 1773

On 28 July 1766, after an eight-week journey, Leonhard Euler and his family arrived in Saint Petersburg. He was fifty-nine. It was a joy to return to the country where he had started his career, and he was pleased to have good relations with the new government, but none of his old friends and colleagues from his first period in Russia remained in Saint Petersburg. Christian Goldbach had died in 1764, and in April 1765 Gerhard Friedrich Müller moved to Moscow. Euler wished Müller success but lamented that he would not see his "oldest and best friend" in Russia.[1] The ostentatious Catherine II had Euler greeted in royal fashion and exceeded the terms of his contract. On her initiative, she had sent him 10,800 rubles from the state treasury; of that sum, 8,800 rubles went for the purchase of a large, two-story brick and wood house of her own choosing. Located on the bank of the Greater Neva River, it had been built for Prince Alexander Kurakin around 1720. The remainder was spent on furniture. Euler and his wife, their daughters and spouses, and his staff moved in along with Johann Albrecht Euler and his family, who took rooms on the first floor. Together with Karl Euler, the household numbered eighteen. For a time Catherine also provided one of her cooks. Since the area lacked sufficient permanent bridges across the Neva, buildings along the side of the quay that contained Euler's house were still unfinished. Its location a few blocks from the Kunstkammer, his workplace, was excellent for Euler. In accordance with the order of Peter the Great that buildings in this area be directed toward the Neva, each of the two floors of the house had thirteen windows facing the river. The yard had a coach house that could accommodate sixteen carriages.

Already in the evening of the settlement for the house, Euler's near blindness became apparent; he could not read the documents that he had signed a few hours earlier or even distinguish between a blank page and another with writing. The vision in his previously healthy left eye was failing from a developing cataract, and his hearing was deteriorating.

Soon after Euler's arrival, Catherine granted him and his two oldest sons a lengthy audience. He asked to be named academic counsel, a position that would make him a nobleman. Whether his family had encouraged him to seek the post or he did this on his own initiative is not clear. He saw the advantage in Russia of having *-tschin* added to a family name. But while support existed to raise Euler to the nobility, the empress delicately rebuffed his request. Not yet willing to take the step of making a commoner a noble on the basis of intellectual achievement, she exclaimed to Euler that his fame was "better than any noble title."[2] Catherine also placed the Euler family on the same level socially as the nobility, sharing status and privileges that were not yet available to the middle estate in imperial Russia. But she did not list him among the six categories of nobles there. Euler was pleased that his wife would receive a widow's pension upon his death, and it was important to him that his sons had guaranteed positions in Russia. Johann Albrecht was named the professor of experimental physics at the Imperial Academy of Sciences and Arts with a salary of one thousand rubles along with proposed free housing. At this early audience, Euler asked Catherine to make his son Karl a crown physician to travel with the royal court to Moscow. Catherine declined that request, for she wanted Karl to remain in the capital city, belonging to an academic commission for codifying the laws and compiling a new law book. But physicians were most welcome in Russia, and she named him a crown physician as well as physician to the academy. Christoph, who had not yet arrived in Russia, was designated a colonel in the military. That his sons had good positions and his daughters were married well relieved Euler of "the burden of caring for his children, which had always worried him during his . . . [later years] . . . in Berlin."[3] Freed of that responsibility, he could concentrate more on reorganizing the Petersburg Academy.

Restoring the Academy: First Efforts

Under Empress Elizabeth the Petersburg Imperial Academy of Sciences and Arts had badly deteriorated. In 1765 it suffered a great loss with the death of Mikhail Lomonosov. Catherine II charged Euler with the task of restoring its early splendor; like Peter the Great, the empress viewed the sciences as vital to converting imperial Russia from a backward realm on the fringes of Europe into a powerful modern state on the Continent. Citing Jacob von Stählin and Franz Ulrich Theodor Hoch Aepinus as exemplars, Euler during his first audience with Catherine emphasized that the academy should be concerned not only with speculation, which would make it a burden on the state if it did nothing else, but even more with general applications.[4] Here Euler agreed with the empress and her court, who stressed utility. Catherine pledged more fully to support research in abstract mathematics, mechanics, and optics as well as theoretical, computational, and observational astronomy and their applications. Her academy had the geography section continue to prepare reliable maps and atlases for exploration and trade, to pursue hydraulics in flood-prone Saint Petersburg, and to advance naval science. Its projects also encompassed metallurgy, mineralogy, and natural history. The academy, like that in Berlin, had to address a range of technical problems; it was to evaluate technological contrivances, inventions, and proposals and to improve upon them. This made up a major part of Euler's proposed work. In support of the effort to revitalize the academy, the empress promised to add to its funding. She planned to raise the pensions of ordinary professors to a thousand rubles, provide members housing near the institution, and within a few years add eight new academicians.[5] Crucial to Euler was her pledge to reorganize the academy in a way that would give more autonomy to scientific pursuits and reduce internal strife. In Catherine's later audiences with Euler, many of them lasting several hours, she listened with close attention and made comments.

Immediately Euler and his son Johann Albrecht began rehabilitating the academy, the university, and the gymnasium. As the eldest and most distinguished member of the academy, the doyen, Euler was a vice director, presiding at all the twice-weekly general sessions and calling conferences in the absence of the president. He was also to select new members engaged in high quality research. Even with his declining vision, Euler felt it his duty to be at the academy each weekday morning to organize its operations and direct it. As Euler's eyesight worsened in September, Johann Albrecht began to assist him more. They set out to improve the operations of the academy by better arranging meetings, distributing papers for discussion ahead of sessions, publishing the *Novi commentarii Academiae Imperialis scientiarum Petropolitanae*, ordering supplies and equipment, and seeing to official correspondence. Euler resumed leadership in the preparation of maps, and Müller in Moscow promised to help. Euler welcomed his offer, for he planned to make eight small maps that together would be an atlas of the Volga region. Skilled in several languages, Euler proposed opening a school in Saint Petersburg to instruct translators. In August 1766, Müller wished him good fortune and invited him to visit Moscow, but the decline in Euler's eyesight prevented him from accepting. In October he wrote that the academy appreciated Müller's research in history, and especially on Siberian geography, and had accepted his suggestion to draw and publish a large map of Siberia. Euler reported a sudden and serious deterioration of his vision that left him unable to distinguish among letters of differing sizes on a white page.

There was a brief interruption in Euler's labors. At the end of August and the beginning of September 1766 he suffered from another serious fever. The *Protokolii* (Protocols) of the regular academic meetings reported his illness and absences on 21 August and 1 September. On 15 October he reported that his fever had lasted only several days, while the eye malady was leading to the loss of the remaining vision in his left eye from the cataract on the lens.[6] But Euler refused to let this hinder the performance of the sessions of the academy or his participation in them. The most important publications, papers, and letters now had to be read to him; he determined what was to be written in correspondence and dictated it. He had to make computations mentally—even the most difficult ones—and likewise dictate them. He still attempted to read documents, but had scant success. Johann Albrecht recorded his computations. Another hand wrote the letter of 15 October, and Euler only signed his name. Even in the onset of near blindness, his astonishing memory, rich imagination, steady willpower, insatiable curiosity, and disciplinary intuition continued to serve him well, and his addiction to research and delight in solving the most difficult problems encouraged him confidently to proclaim, "One more distraction removed" in reference to his sight.[7]

In October 1766, Catherine II accepted Euler's proposal to replace with a new commission the chancery that had supervised the academy, and she divested the president, Count Kirill Gregor'evich Razumovskij, of his legal powers but not the office. The president had to be a noble, so there was no discussion of Euler's assuming that position. Though a student of Euler in 1743 and 1744, Razumovskij was little interested in the sciences. Elizabeth had appointed this prominent courtier to the presidency in 1746 when he was only eighteen; he was now the hetman, or chief, of the Don Cossacks and the Ukraine. To improve the management and operation of the academy, Catherine created a directorship as an auxiliary post to the president and transferred presidential powers to the appointee. She added sufficient staff to run the academy departments. The academic commission belonged to the new directorship, and five academicians were among its members: Euler, his son Johann Albrecht, Semjon Kirillovič Kotel'nikov, Johann Lehman, and Stepan Jakolevič Rumovskij. They were to be joined by the conference secretary Stählin and the new director. *Primus inter pares* on the commission, the director would report directly to the empress. That and the supervision that the nobles exercised over the academy's operations ensured Catherine's preponderance.

The academic commission was ordered to draw up detailed plans for radically reorganizing the institution, to manage it until a new director was appointed, and then to assist the new manager. From his initial academic session after returning to Russia in August 1766 and through the following decade, Euler worked to improve the academy's administration. At the end of 1766 he gave the committee a five-page memorandum in French titled "Plan for a Reestablishment of the Imperial Academy of Sciences," which called for enlarging the number of members, simplifying the editorial board, more effectively overseeing the publications of the academy to enlarge profits, and increasing financial payments for foreign scientists invited to Russia. In February 1767, Euler proposed an increase in salaries and an end to the censorship of imported books. He urged reducing teaching assignments and providing more adequate funding for the academy through an expanded sale of books, calendars, and journals. Intrigues and trifling quarrels continued, and Euler wanted to find means to lessen this internal strife. While in Berlin he had supported the pioneering efforts in Saint Petersburg of Lomonosov to educate native Russian scientists, and he appreciated Lomonosov's commitment to Russian poetry, rhetoric, and history along with chemistry and physics. In both Prussia and imperial Russia, Euler attempted to develop a measure of autonomy for the sciences as a profession. Like Lomonosov, whom Razumovskij had named to head the educational branch of the academy from 1760 to 1765, Euler proposed giving academicians a greater role in the control and operation of their institution. Euler differed with Lomonosov in being less willing to accept noble domination of the academy and briefly gained limited independence for it. Opposition from the first two directors and from courtiers attached to the academy was soon to block all of Euler's administrative proposals.

For the first director of the Petersburg Imperial Academy, in 1766 Catherine had named the general and count Vladimir Grigor'evich Orlov, the younger brother of her bearlike libertine lover, the general Grigorij Grigor'evich Orlov; he would hold the post until December 1774. In the summer of 1762, the two brothers had been leaders in Catherine's coup against her husband, Peter III, and she rewarded them with governmental offices. Catherine instructed Vladimir Orlov to bring the academy into order. Occasionally one of the vice directors, Alexey Reshevski of the Junker nobility of eastern Prussia, filled in for him. Euler and his son enjoyed fairly good relations with both. Yet Orlov, who in character was not unlike Johann Schumacher, joined with courtiers attached to the academy in reestablishing the despotic bureaucracy that Lomonosov and Euler opposed, and the policies lessened the institutional authority of Euler and his son Johann Albrecht. Regarding the administration of Orlov, Count Sigismund Ehrenreich von Redern of the Royal Academy of Sciences in Berlin, whose atlas Euler had published in 1762, exclaimed to Euler during a visit to Saint Petersburg, "what an extraordinary kind of person you have for the president [actually director] of the academy who is against all scholars, regards the academy as useless, and believes like [Jean-Jacques] Rousseau that science would make the world more evil."[8]

Initially the empress hoped that Orlov, a graduate of the University of Leipzig, would provide leadership at the academy. But preoccupied with the events leading to the great Pugachev Rebellion against serfdom and soon the rebellion itself, he came to neglect his duties. In 1773 and 1774, the Cossack Yemel'yan Ivanovich Pugachev was claiming to be Peter III, who had died in prison. Appearing on the Siberian frontier Pugachev led millions of serfs who along with Cossacks, Ural miners, and Old Believers in the Orthodox religion sought respect and liberty. Roughly three thousand landowners were killed before the uprising was crushed.

Upon his return to Russia, Euler maintained close ties with the Berlin Academy. In 1766 he sent eleven articles to be published in volumes of its *Mémoires* for different years on such topics as consonance and dissonance in music, improved microscopes, the knight's tour in chess, the rotation of celestial bodies, and the propagation of sound, including the disturbed state of a slice of air—that is, a wave. Jean-Henri-Samuel Formey informed him about the arrival of Joseph-Louis, Comte de Lagrange in Berlin and difficulties in selling Euler's house. Two years later, in November, Euler gave the Berlin Academy (in his son Johann Albrecht's handwriting) a short account of his memoirs after he lost his sight, and urged a tighter collaboration between the two academies.[9]

The Grand Geometer: A More Splendid Oeuvre

During his second Saint Petersburg period, Euler's correspondence fell sharply, but neither administrative duties nor near blindness stopped his extraordinary productivity. Alone, as lead author, or in the supervision of printing, he presided over a stream of articles and books numbering 415, which amounts to 52 percent of the total publications for his career.[10] Of these, more than 150 were to appear after his death. In Russia Euler's articles addressed a range of topics, primarily in pure and less so in applied mathematics, ranging over algebra, theoretical and computational astronomy, integral calculus, dioptrics, hydraulics, rational mechanics, music, and number theory. He published articles on practical questions, including actuarial mathematics with tables on the problems of mortality and reproduction, and he occasionally explored recreational problems, such as magic squares.[11] Many articles dating from the time after he would retire from the academy lack the ingenuity of his earlier papers.

The writing was not simply an upsurge at the start of the second Saint Petersburg period; Euler had already composed many works initially published at that time in Berlin. The exceptional flourish began more precisely in the final half decade of Euler's Berlin period. Sixteen of the eighteen articles printed in 1767 he had completed in Berlin, along with twenty of twenty-two issued in 1768, and fourteen of sixteen for 1769. Those numbers fall to two of nineteen for 1770 and three of ten for 1771. Clearly Euler did not singly write or dictate all parts of the articles and books from the second Saint Petersburg era. Vital support came afterward from the small research circle that had formed over the years around him. When he was the lead author, members from his team assisted him, mainly in making computations. That collaboration had begun with his son Johann Albrecht and the astronomers Kotel'nikov and Rumovskij. In February 1767, Christoph Euler moved from Berlin, entering a Russian artillery regiment as a colonel. That year Euler began to instruct Petr Borisovič Inochodzev and Ivan Judin in astronomy. His research circle was not yet complete.

By 1766 the Petersburg Imperial Academy agreed to print two multivolume writings by Euler and in the next year a third, all of them composed in Berlin. Two were his monumental *Institutionum calculi integralis* (Foundations of integral calculus, E342, E366, and E385) and his *Dioptricae* (Dioptrics, or general theory of lenses, E367). In Saint Petersburg Euler concentrated on having both published. It was owing to a proposal of Stählin, who found a copy among Euler's papers after his return to Saint Petersburg, that the *Lettres à une Princesse d'Allemagne, sur divers sujets de physique et de philosophie* (Letters to a German princess on diverse subjects of physics and philosophy, E343, E344, and E417; "physics" here means the elementary physical sciences) were to go to press.[12] Each of these texts comprised three volumes. Euler's labors on the first two works and his advancing blindness may at least partly account for the brief misplacement of the letters.

On 7 August 1766, at the first conference that he attended upon his return to the Petersburg Imperial Academy, Euler submitted the entire text for his *Institutionum calculi integralis* on ordinary and partial differential equations, most of which were linear. Since 1759 he had searched fruitlessly in German lands for a copy editor and printer for the manuscript; he recognized that for any press printing this book would be difficult. By December 1763, the manuscript of the text had been completed.[13]

In January 1767 Euler wrote to his young disciple Lagrange that he was pleased to have so worthy a successor in Berlin but regretted that Lagrange had not come to Saint Petersburg. The previous October Lagrange had arrived in Berlin through the efforts of Jean-Baptiste le Rond d'Alembert, and Euler and d'Alembert continued to be his two principal teachers and advisers. While his rapport with both was good, he remained closer to d'Alembert and wanted not to work directly under Euler, who pressured him to increase his number and body of publications. In 1765 Euler had noted d'Alembert's rejection of their solution to the vibrating string problem, and he asked whether that action conformed to the law of continuity. He commented on the new edition of Gottfried Wilhelm Leibniz's manuscripts by the publisher R. E. Raspe, and on the first edition of Leibniz's *Nouveaux essais sur l'entendement humain* (New essays on human understanding, 1703) with remarks on John Locke; Euler also observed that the Seven Years' War had delayed the publication of the Berlin Academy's *Mémoires*. In 1766 Euler remarked that Lagrange had been the first person he recommended to Catherine II, and he praised her plans for her academy of science. In the relationship between Euler and Lagrange, a letter in January 1767 began a new phase. Euler recognized that Lagrange would not come to Russia. As the director for the mathematical section in the *Mémoires* of the Berlin Academy, Lagrange made it his primary responsibility to assure that volumes begun by Euler and containing several of his articles would be published.[14] Lagrange would later mention an essay by Euler on the motion of mass points around two fixed centers of attraction and another on tautochrones in resisting media.

Euler's letter in January also reported that the Petersburg Imperial Academy would publish his text on integral calculus. Finding a publisher for his books continued to be a challenge. Although originally projected to require two volumes, the text had contained so many emendations since 1759 that three volumes were now necessary. The title page lists Euler as vice director of the Imperial Academy as well as a member of the Paris Academy and the Royal Society of London. In December 1767 Lagrange expressed delight that the Imperial Academy had agreed to have Euler's *Institutionum calculi integralis* printed and called it a "true service" to scholars in the sciences.

Between 1768 and 1770, the academic press published the three volumes in quarto of the *Institutionum calculi integralis* (E342, E366, and E385).[15] It completes Euler's trilogy on calculus that includes the *Introductio in analysin infinitorum* (Introduction to the analysis of the infinite) and the *Institutiones calculi differentialis* (Foundations of differential calculus). Together they came to more than 2,500 pages, and represented the

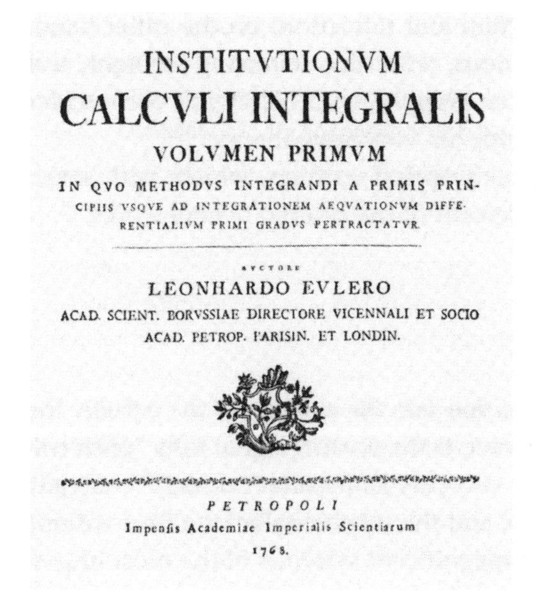

INSTITVTIONVM
CALCVLI INTEGRALIS
VOLVMEN PRIMVM
IN QVO METHODVS INTEGRANDI A PRIMIS PRINCIPIIS VSQVE AD INTEGRATIONEM AEQVATIONVM DIFFERENTIALIVM PRIMI GRADVS PERTRACTATVR.

AVCTORE
LEONHARDO EVLERO
ACAD. SCIENT. BORVSSIAE DIRECTORE VICENNALI ET SOCIO
ACAD. PETROP. PARISIN. ET LONDIN.

PETROPOLI
Impensis Academiae Imperialis Scientiarum
1768.

Figure 13.1. The frontispiece of Euler's *Institutionum calculi integralis,* 1768–1770.

first thorough group of texts on calculus.[16] The claim that they formed the basis of the curriculum of modern calculus is exaggerated, but they contained most elementary solutions of differential equations and all basic cases of integrability. The *Institutionum calculi integralis* was a novel work offering a comprehensive catalog of solutions to partial and ordinary differential equations by an evolving concept of elementary functions and by power series, many of which appear in a course separate from calculus. In his *Introductio* Euler had already offered the general theory of algebraic and elementary transcendental functions; he represented functions by a single analytical expression composed of variables and constants.[17] Before him there had been few methods of solution and few applications. He refined, simplified, and extended each of these and added hundreds of innovations in the theories of those equations that are particularly helpful in their application to mechanics. The text results from Euler's objective over several years of presenting calculus in a didactic book.

Volume 1 of the *Institutionum calculi integralis* consists of an introduction, describing the general nature of integral calculus, followed by three sections that comprise what might be termed ten chapters.[18] Each chapter is the elaboration of several problems with solutions and several corollaries with proofs. The first section integrates elementary functions, especially the transcendental logarithmic, exponential, and trigonometric, and examines an evolution by infinite products. The second section covers the integration of second-order differential equations via multiplication, approximations, and infinite series. The third resolves partial differential equations of third and higher orders. Euler's definition of integration as antidifferentiation was questionable, since he did not prove that antidifferentials existed. He integrated only continuous functions, and in his examples the antidifferential always existed. The resolution of this problem would lie outside the eighteenth century. Since many simple functions could not be integrated by reducing them to the most elementary forms, Euler compared integration to inverse arithmetical operations, such as addition to subtraction or multiplication to division. He did not achieve a complete theory of definite integration.

Chapter 6 in the second section of volume 1 dealt with elliptic integrals, and here Euler was continuing to extend the work of Count Giulio Carlo de' Toschi di Fagnano. Like Fagnano, he emphasized the relations between integrals and algebraic results.[19] Of the so-called integrals, the arc lengths of the ellipse, hyperbola, and lemniscates gave the typical cases. Euler had shown that the method of indeterminate coefficients led to what he called the complete integral and what we would today term the general integral. In this field, these were Euler's horizons for his work. Perhaps he alone saw here the germ of an entirely new branch of analysis.

The second volume was divided into two parts. The first, consisting of twelve chapters, treated ordinary differential equations of second order, and the second part differential equations of third and higher orders, their construction by the quadrature of curves, and elliptic integrals. After receiving the two volumes, the first from the printer and the second from Formey, Lagrange lauded Euler's research as erudite, ingenious, and—as always—fruitful.[20]

The third volume contained two parts on resolving differential equations. While actively pursuing elliptic integrals, Euler had for a time set aside what is now considered the closely allied area of intricate Diophantine equations.[21] He returned to it only when composing the last section of his *Vollständige Anleitung zur Algebra* (Complete guide to algebra, E387 and E388). A seven-chapter appendix of the *Institutionum calculi integralis* refines what Euler called a new branch of integral calculus, the calculus of variations, following the analytic methods of Lagrange, whom he called the creator of this field. For it Euler offered a higher degree of perfection of its methods and computations.

In August 1767, Euler wrote to director Vladimir Orlov about the publication of the *Lettres à une Princesse d'Allemagne*.[22] Probably that month, Euler informed his friend of noble rank, Friedrich Heinrich, that the first two volumes would be published in several months. The *Lettres*, principally Euler's *apologia* rendering his defense of positions that he had taken in scientific disputes and a related history of science, are the most successful high scientific popularization of the Enlightenment. They also convey briefly Euler's religious, moral, and spiritual positions, core elements of which will be noted below. In Russia the *Lettres* were published first in French, the language of the nobility and the second language of Europe. That language together with the popular style and clarity of the *Lettres* allowed the work to reach the largest group of readers. A comment from Denis Diderot in *Rameau's Nephew* (written in the 1760s but not published until 1805) might apply: "One needs a profound knowledge of art or science to have a good grasp of their elements. Works of classic rank can only be produced by those who have grown gray in harness. The middle and end illuminate the obscurity of beginnings."[23] The first two volumes of the *Lettres*, E343 and E344, appeared in 1768 in Saint Petersburg and the last (E417) in 1772. Each is hefty: the first volume has 315 pages, the second 340, and the third 404.

From the final years of Euler's life to the present, the *Lettres* have met with phenomenal success. By 1800 they had gone through thirty editions and were translated into eight other languages: Danish, Dutch, English, German, Italian, Russian, Spanish, and Swedish.[24]

That the 234 sets of lessons sent in French to the future princess Friederike Charlotte Leopoldine Louise from April 1760 to May 1762 remained unpublished in Prussia possibly owed at least partly to tensions from 1772 on between Frederick II and Friedrich Heinrich and between the king and Euler. Throughout his stay in Berlin, Euler likely wanted to have them put into print. He probably prepared in Russia the first German translation, while between 1768 and 1774 Rumovskij translated the *Lettres* from French into Russian, the first of five Russian editions leading up to 1808.[25]

The *Lettres* are a treasure trove from the sciences, a distinctive encyclopedia of knowledge, and they give Euler's mature positions as well as outcomes of earlier research and speculations. He began simply, for his student had little knowledge of natural philosophy or mathematics. But he proceeded quickly to more difficult topics in steps and well-chosen examples lightened with an occasional touch of humor. The text showed his insight, how he worked through problems, and the clarity of his explanations. It contained no mathematics.

The initial three topics—extension, velocity, and sound—were among the seventy-nine dealing with general physical science that comprised the first of three natural divisions within the *Lettres*. For René Descartes,

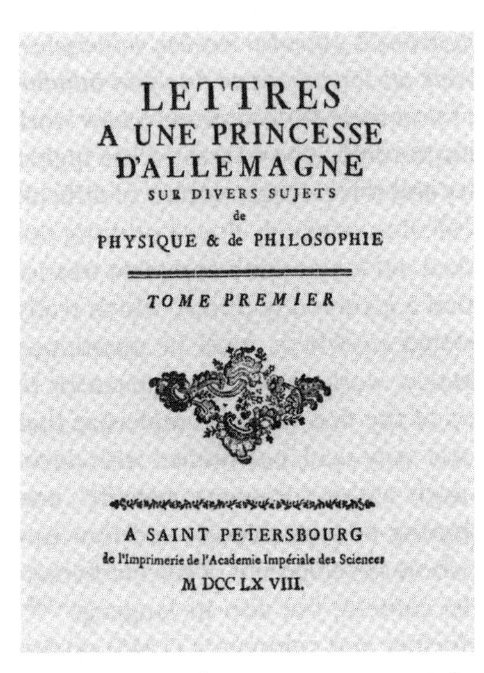
LETTRES
A UNE PRINCESSE
D'ALLEMAGNE
SUR DIVERS SUJETS
de
PHYSIQUE & de PHILOSOPHIE

TOME PREMIER

A SAINT PETERSBOURG
de l'Imprimerie de l'Academie Impériale des Sciences
M DCC LX VIII.

Figure 13.2. The frontispiece of Euler's *Lettres à une Princesse d'Allemagne,* 1768–1770.

extension was the essential property of matter. But Euler assigned two other fundamental properties from the work of Isaac Newton: impenetrability and inertia. He explained these in letters 69, 70 and 74, which make impenetrability the most general property. After giving Earth's circumference as 25,020 miles early in the *Lettres* Euler—like Newton—computed the velocity of sound. For the first edition of his *Principia Mathematica* Newton had multiplied frequency by wavelength and found the velocity of sound to be 968 English feet per second. Basing his calculation upon new frequency measures of organ pipes, in the second edition in 1713 he increased the figure to 1,020 feet. Euler gave a closer approximation of 1,142 feet per second; the actual value is 1,107. Euler treated light as analogous to sound, as a vibration in the ether. The figure for the speed of sound was minuscule measured against the greatest velocity known, that of light which since the time of Galileo Galilei had not been thought to be instantaneous. Euler calculated that it moved 12,000,000 English miles per minute. That this is not much higher than the actual value of 11,176,943.8 miles per minute could mean that Euler rounded the figure to make its dimension clearer to his student. Letters 4 to 17 proceeded to music, the air, and the atmosphere. Letters 17 to 45 explained light, starting with the systems of Descartes and Newton on the subject, both of which Euler rejected, and proceeded to his pulse theory, optics, the theory of colors, dioptrics, reflection, vision, and the structure of the eye. His opposition to Newton's corpuscular optics meant not that Euler was anti-Newtonian but that he found fault with Newton's optics on questions of reflection and refraction.

The ether, seen as an extremely tenuous and elastic form of matter filling in nature all of otherwise empty space, remained the fundamental concept of Euler's physics.[26] With it he continued to explain most physical phenomena—mechanical, optical, magnetic, and electrical. Ether first appeared in letter 19. Euler's application of ether to explain celestial motion has been interpreted as Cartesian,[27] but it was not Descartes's ether. Two considerations regarding ether guided Euler. He sought to remove the neo-Cartesian charge that mutual attraction was a secret or occult property, and he rejected Newton's action at a distance in explaining it.

Letters 45 to 79 addressed gravity and its effects, along with mechanics, cosmology, the tides, and the theory of matter—especially impenetrability. Euler praised Newton for the great discovery of universal gravitation and gave the falling apple example that Voltaire had used. Newton, he wrote, had looked to how the force acting upon the descending apple would be affected if the height of the tree were the distance of the moon. Euler's account helped this example make its way into folklore. A dozen letters discussed attraction, a property with which all celestial bodies are endowed; its effects depended upon mass and proximity. The enormous distances of the stars from us prevent them from affecting the planets in our solar system. Euler accepted attraction over impulsion, for it would not lead to false consequences on such questions as lunar motion and the shape of Earth. Cartesian impulsion required action by contact and rejected action at a distance. Following Bernard le Bovier de Fontenelle on the plurality of worlds, letter 60 found it highly probable that there were inhabitants on other solar planets and what are now called exoplanets around fixed stars. Euler projected an infinite number of the last. Continuing studies of the motion of comets and the moon, he held, had confirmed the exactness of the inverse-square law of mutual gravitational attraction. Letter 61 gave priority to Johann Tobias Mayer for achieving the high degree of precision needed for exactly determining the moon's motion. Euler credited Descartes with identifying the influence of the moon on the tides, a position that he believed the ancients had held. But he rejected Descartes's idea that the way the moon exercised that influence was to press on the ether as it moves, causing the tides; Euler instead attributed the effect to the moon's attraction.

After rejecting in letter 76 Christian Wolff's animate monads endowed with force, in letters 125 to 133 Euler returned to make another extensive critique of them. His historical and philosophical account reviewed the monad debate from the Royal Prussian Academy's prize competition in 1747 as well as the response of the literate community in Berlin to the quarrel over the principle of least action between Pierre Maupertuis and Johann Samuel König. The main defense the Wolffians made of monads came from their use of the principle of sufficient reason; they assumed that to compose matter these ultimate components must exist. Still, the existence of the monads remained unexplained. Believing that the Wolffians begged the question, Euler portrayed them as viciously attacking Maupertuis. In letters 49 and 78 he noted that Voltaire had chuckled at him for his idea about a stone tossed down into a hole drilled through the center of Earth. He declared that while he now found untenable his position about its return, he saw no harm in supposing it and calculating results. Truth suffered nothing, he insisted, from the jesting pleasantries, and he confessed to having made jokes about monads.

The second section of the *Lettres*, 80 through 133, inquired into philosophy, religion, logic, and the Euler-Venn diagrams; it covered topics as diverse as liberty, ethics, language, forms of syllogisms, divisibility, monads, and the certainty of scientific, moral, and historical truths. In religion Euler adopted ontological and epistemological arguments against three groups: the Wolffians, the mechanistic materialists, and the idealists. He considered as well the predecessors of idealism, the solipsists represented by Nicolas Malebranche, for whom a person's own mind alone exists and knowledge from outside is unsure. He criticized skeptical freethinkers and the French Encyclopedists, and argued that the world without matter as proposed by the idealists is incomplete, and that God must create Leibniz's best of all possible worlds. But letter 85 opposes Leibniz's notion of a preestablished harmony between mind and matter as a fiction leading to a materialistic determinism destructive to liberty. For Euler the connection between mind and matter remained a mystery. Letters 92 and 93 assert that the preestablished harmony would detract from spirits, a vital part of Euler's world that Wolffian metaphysics threatened.[28] He concentrated in religion on the perplexing topics of the origin of evil, the permission of sin, and the existence of calamities. The Eulerian dualism underlay these, and was more complex than the Cartesian version: Euler held that materials and spirituality, or body and spirit, comprise the universe. The two are totally different. Material bodies have the three general properties, while the chief qualities of spirit are intelligence, will, and liberty. Spirits are not limited by space or time, and God operates through them; a spirit without liberty would be like a body without extension. The importance of liberty he deemed equivalent to that of impenetrability in the physical realm.

Even God cannot restrict that freedom for spirits and intelligent beings. Evil does exist and troubles the spirit. Euler maintained that virtue makes for happiness, a central goal for spirits and human beings.

The most exhaustive and authoritative science popularization written during the eighteenth century, the *Lettres* critically examined in greater depth than did other works the complex and changing major Enlightenment natural philosophies, the Cartesian, Newtonian, Leibnizian, and Wolffian, and presented each in terms understandable to the educated European reading public. As one of the few leading men of science, well grounded in all four schools of thought and with a sure command of them, Euler could comment adeptly on each. In neither method nor content was Euler a Cartesian, as has been sometimes thought; instead he brought together consistent ideas in the sciences from different natural philosophers and new thought in mathematics and physics introduced by some members of the Bernoulli family, and he combined these with his original insights. This was the Eulerian system. Alexandre Koyré included the *Lettres* among the prominent Newtonian popularizations—Henry Pemberton's *View of Sir Isaac Newton's Philosophy*, Voltaire's *Philosophical Letters* and *Elements of Newton's Philosophy*, Count Francesco Algarotti's *Newtonianism for the Ladies*, Colin Maclaurin's *Account of Sir Isaac Newton's Philosophical Discoveries*, and Pierre-Simon Laplace's subsequent *System of the World*[29]—but Euler's work was far greater.

The letters in the second section demonstrate that Euler's methods in the sciences and learning in general were not purely theoretical. In letter 115 he, like the Christian Platonists, divided all human knowledge and proofs of truth into three classes: sensory or physical, based on evidence from the senses; intellectual, based upon reason; and historical or moral, based on faith through the testimony of authoritative figures. Proof of a claim varied according to the class in which it lay. In the sciences Euler began his research by modifying his theoretical constructs with empirical data. Letter 99 on the highest perfection of method followed the analytical procedures employed by the Bernoullis, combined with critical empiricism practiced and taught by Joseph-Nicholas Delisle. Letters 105 and 115 added Lockean sensationalism to faith and reason. Here Euler's methodology shared ground with that of the Wolffian Alexander Gottlieb Baumgarten, whose *Metaphysica* of 1739 and *Aesthetica* of 1750 held that the senses had their own rules and perfection. Each class, Euler explained, was liable to error and may mislead, but with sound precautions the same degree of conviction could be produced in each. Euler eschewed the unlimited use of reason the Wolffians thought necessary for obtaining scientific and philosophic truths, and he did not hold theology to be the queen of the sciences; nor could authoritative figures remove the confusions among moral convictions. And while physics could offer an external test for affirming the validity of a philosophy, metaphysics could not confirm physical postulates, laws, and operations.

The third section, letters 134 to 234, was devoted entirely to physical questions. Euler composed this after he visited the future princess in Magdeburg in May 1761, at which time she said she was no longer able to understand his letters completely and asked him to restrict himself to physical questions. Prominent topics here were the nature and causes of electricity and its visible manifestations in sparks from discharges, thunder, and lightning. Euler devoted at least seventeen letters to electricity, 138–154, and nineteen to magnetism, 169–187. The disequilibrium and elasticity of the ether were crucial to Euler's explanation of both.

The division here of the *Lettres* differs from that in the original three volumes. The second of those volumes ended with electricity, and the third commenced with the celebrated problem of longitude. The last section in each also covered concave and convex lenses, the telescope, the microscope, stellar distances, and the search for generalization in the sciences.

Letters 155 through 169 began with two of six methods for determining the longitude of a ship,[30] and the narrative here could be seen as a brief history of science. Euler recognized that all six methods had defects and would require corrections. The first depended upon carefully measuring the direction and length of a journey and placing the result on a map to obtain, as he commented, a rough approximation of a position at sea. Euler noted that the currents in the Atlantic Ocean made a voyage from Africa to America take less time than the reverse. His second method for obtaining longitude, the most classical, depended upon a proposed precise timepiece, a recording of the time passed between a single event—such as the sun at noon at a reference point—and the given location. From these Euler computed the angle of rotation of Earth, knowing that it turned fifteen degrees in an hour. The difference between the two points gave the angle of rotation, which provided the difference in longitude. It was only with the accurate timepieces invented by John Harrison in the 1760s that his method became useful. Euler referred to the British Parliament prize competition on the subject of longitude but did not mention Harrison; he wrote these letters in 1761, four years before he was awarded a small portion of that prize.

Theoretically Euler's next three methods for finding longitude resembled that making use of the clock, while the last employed a magnetic needle. In place of the sun's zenith as the reference point, all but the last appealed to some astronomical phenomenon. A third method used eclipses of the moon and eight simultaneous equations. Each eclipse had to be measured at two different places and the findings compared with known longitudes. For the least possible error, the observations had to be combined.[31] Still another method relied on eclipses of the satellites of Jupiter, and a fifth observed the daily motion of the moon and determined its velocity. Each procedure required making tables and comparing the result at the departure point with that for the final location. The lunar method using eclipses carried in practice the highest degree of precision, but it was not possible to ascertain without an error the moon's true place for every moment. Guided by prize competitions of the Paris Academy of Sciences, Euler explained a sixth method in which sailors employed a magnetic needle and its declination to chart their way at sea.[32]

Letters 101 to 108, written in February and March 1761, introduce what are today known as Venn diagrams, though that is a misnomer.[33] Diagrammatic representations in logic were not original with Euler; they appeared in some eighteenth-century treatises on the subject and it is possible that Johann Heinrich Lambert employed them shortly before Euler's *Lettres*. In letters 101 and 102 Euler stressed the need for a disciplined language in representing general ideas and expanding upon them; he employed circles in diagrams to explain different forms of syllogisms and hypothetical propositions. The rules of reasoning held that if two propositions in a syllogism were accepted, the third that followed necessarily from them must also be true. Euler explained joint and independent areas of affirmative positive universal, negative universal, affirmative particular, negative particular, and more with two or three intersecting circles. An example is "for every A is B; but no C is B, and no B is C; therefore no C is A."

In the article "On the Diagrammatic and Mechanical Representation of Propositions and Reasoning" more than a century later, in 1880, the Cambridge mathematician John Venn accepted only the diagrams in logic that he called "Eulerian circles." He added ovals for representations and found that the same diagrams could be utilized to analyze different lists of propositions by closely following which compartments were empty. Today the representations may be called Euler-Venn diagrams. Euler further proposed a design for a logic machine, but no record exists that the machine was ever constructed.

Another multivolume masterwork by Euler, the *Lettres* are a principal document of the Enlightenment. Their passion for learning reflects that period's faith in education, including support of female learning. In the mid-nineteenth century, some scholars mistakenly believed that the title referred not to an actual person but to a technique in writing. The *Lettres*, with their insightful explanation of the sciences and his core religious

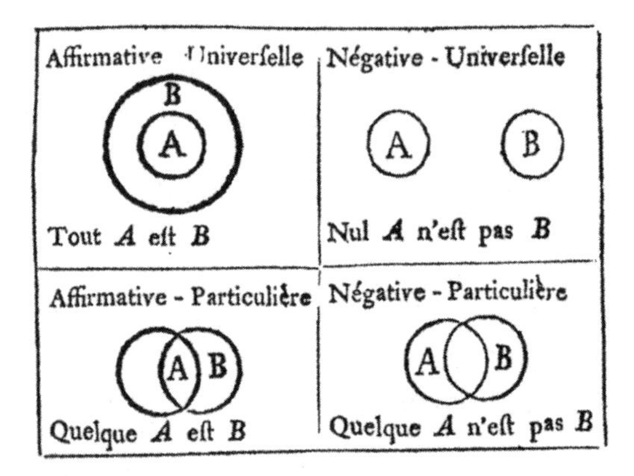

Figure 13.3. Two circles indicating the amount of rational overlap from Euler's *Lettres à une Princesse d'Allemagne.*

and spiritual positions, offer probably the best rounded view of Euler's character.

During the time of Catherine the Great, Russia was growing receptive to new thought in philosophy and the sciences. In Saint Petersburg the Ukrainian *philosophe* and Voltairean Jakov Pavlovič Kozel'sky was prominent in adjusting to Russia the new critical rationalism from the West. To the Russian reading public he helped introduce d'Alembert and Diderot. After serving in the Russian Army, he published in 1764 two pedagogical texts, one on arithmetic and another on mechanics, that circulated widely. In 1768 his *Filosofskiya predlozheniya* (Philosophical propositions) appeared in Russian. While he knew the writings of Descartes, Christiaan Huygens, and Newton in the sciences, Kozel'sky looked especially to Euler. He followed Claude Adrien Helvetius and Euler in viewing space as an abstraction and went further; devoid of matter it was "nothing, but . . . considered in connection with bodies is extension."[34] But for Euler absolute space was a physical reality determining inertial motion.

One notable negative comment associated with the *Lettres* generally endures to the present. Aimed at the author, it came from d'Alembert. In writing to Lagrange, d'Alembert expressed his eagerness to read through Euler's forthcoming *Institutionum calculi integralis* (1768–72); Lagrange, he posited, might examine the *de Physique et de Philosophie* part of the original title of the *Lettres*, which revealed that "our friend is a great analyst but quite a poor philosopher."[35] This statement of displeasure with Euler reveals that d'Alembert had not read closely the *Lettres*. It was perhaps thought that these were similar to Newton's study of the apocalypse or revelation and the prophecies of Daniel that continued to his later years,[36] but neither appears in the *Letters*. Spirituality, religious faith, and dreams do, but they are not analyzed. The original title of the *Lettres* may have contributed to d'Alembert's final remark, which long resonated among Euler's critics.

The influence of the *Lettres* in German culture went beyond the sciences. Among others, Immanuel Kant, Johann Wolfgang von Goethe, and Arthur Schopenhauer praised them. Although Euler was comparatively weak in philosophy, Kant read the *Lettres* before criticizing Wolffian dogmatic rationalism. Most philosophers agree that Kant's transcendental idealism, based upon the view that space and time were not abstractions from the physical world, was directly indebted to Euler, as apparently was Kant's treatment of the impenetrability of atoms.[37] In *Materialen zur Geschichte der Farbenlehre* (Data on the history of the theory of colors, 1790–1808) and *Zur Farbenlehre* (On the theory of colors, 1810) Johann Wolfgang von Goethe, an early leader of the romantic *Sturm und Drang* movement in the German states, proposed a sensory approach to color as a counterbalance to Newton's mathematical optics. He criticized Newton for his trust of mathematics over the sensation of the eye but did not, however, offer a complete theory. Goethe regarded Euler highly as an original thinker, who with his concept of the achromatic eye had disproved part of Newton's corpuscular optics. In his magnum opus, *Die Welt als Wille und Vorstellung* (The world as will and representation), which was published as one volume in the first edition (1818) and as two volumes in the second (1844), Schopenhauer—via the application from Kant of the principle of sufficient reason—envisioned an endlessly violent and ultimately irrational world. Still, in chapter 15 of volume 2 of the second edition, he lauded Euler's *Lettres* for their insight, precision, and clarity of concepts, presented with a new charm. Schopenhauer likened Euler's achievement "als hätte man ein schlechtes Fernrohr gegen ein gutes vertauscht" (as that of a man who had exchanged a poor telescope for a good one). The *Lettres*, he wrote, revealed cogently the fundamental truths of mechanics and optics.[38]

From 1766 to 1777 Euler concluded roughly forty years of intermittent studies on geometrical optics, and was the first natural philosopher to construct a mathematical pulse theory of light. Elevated by their experimental component, his optical studies included probes of reflection, the dispersion of light, the phenomena of colors, the course of light rays in the atmosphere, and refraction in different fluids. Upon returning to Saint Petersburg, Euler kept dioptrics a major topic in his studies. Dioptrics, the branch of optics that examines the refraction of light through different surfaces and their systems, is often connected with studies of the eye;

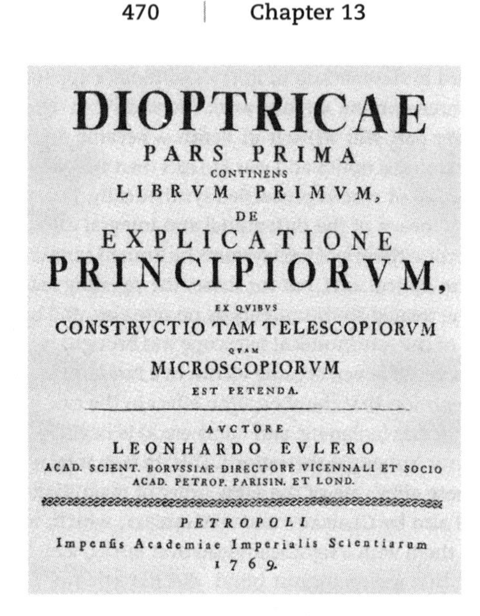
DIOPTRICAE
PARS PRIMA
CONTINENS
LIBRVM PRIMVM,
DE
EXPLICATIONE
PRINCIPIORVM,
EX QVIBVS
CONSTRVCTIO TAM TELESCOPIORVM
QVAM
MICROSCOPIORVM
EST PETENDA.

AVCTORE
LEONHARDO EVLERO
ACAD. SCIENT. BORVSSIAE DIRECTORE VICENNALI ET SOCIO
ACAD. PETROP. PARISIN. ET LOND.

PETROPOLI
Impenſis Academiae Imperialis Scientiarum
1769.

Figure 13.4. The frontispiece of Euler's *Dioptricae,* 1769–1771.

it investigates the properties of the eye as an optical instrument. Among Euler's predecessors in this subject were Johannes Kepler and Descartes. In 1757 Euler had submitted two articles to the Berlin Academy on dioptrics, particularly the construction of composite lenses. In April 1765 he sent the Paris Academy the treatise "Précis d'une théorie générale de la dioptrique" (E363). Following this in October 1768, Euler presented to the Petersburg Academy his grand work on the subject, *Dioptricae*, and the academy published its first volume (E367) in 1769, its second (E386) the next year, and its third (E404) in 1771.[39] Seven of the *Dioptricae*'s chapters reissued Euler's previous papers from his systematic study of optics with results, many of which the French, notably Alexis Claude Clairaut and d'Alembert, had since surpassed. Euler wrote to Daniel Bernoulli that Wolfgang Ludwig Krafft had helped him redact the entire work for publication. The *Dioptricae* was another didactic work that for a long time was a successful textbook in central Europe and Russia.

The *Dioptricae* presented a universal optics with practical applications. Volume 1 explained the general theory of optics and covered the construction of dioptric instruments in general, including visual confusions in working with them. In the mathematical description of optics, Euler noted, the geometrical side dominated. In elementary geometric optics, Euler's formulas described properties of lenses and the passage of light through a system of them; his was the first such complete theory. Volumes 2 and 3 presented his research on perfecting optical instruments and explained rules that applied to them. The second volume concentrated on building the telescope, including scopes with convex ocular lenses; the third dealt with the construction of simple and composite microscopes. Euler's research had made possible a considerable improvement of these instruments. Under Euler's direction, Anders Johan Lexell had helped to edit the *Dioptricae*. Throughout his research on dioptrics, Euler never departed from his belief in the absence within the eye of chromatic aberration; he even took it as an indication of the existence of God.

A Further Research Corpus: Relentless Ingenuity

As 1766 closed, Euler wanted to have the cataract on his left eye removed, and in Paris the royal physician Antonio Nuñes Ribeiro Sanchez had developed a procedure that greatly encouraged him; in October he sent Sanches a letter inquiring about the surgery. He knew Sanchez, who in a letter in 1740 had inquired about mathematical methods applied to morality, to which Euler responded with comments on the application of mathematical analysis to moral philosophy. In 1749 Sanchez reported rumors in Paris about new maps based on discoveries made by the Second Kamchatka Expedition, expressed his honor at being associated with the Petersburg Imperial Academy, and cited his connections with the academy's work on electricity. But in 1767 apparently no surgeon in Saint Petersburg could perform Sanchez's operation, so it would have to wait.

In February Euler was pleased by his son Christoph's arrival from Berlin. Catherine II had again intervened with Frederick II, this time to allow Christoph to leave his Prussian military service early.

Euler's continuing productivity in research after 1765 has a good example in "Solutio facilis problematum quorumdam geometricorum difficillimorum" (Facile solutions to some difficult geometric problems, E325), which is contained in volume 11 of the *Novi commentarii.*[40] In December 1763 Euler had sent a draft for reading at the Petersburg Imperial Academy and it appeared in print in 1767. The "Solutio" concentrates on centers in triangles and provides formulas for the center of gravity, the center of the inscribed circle, and the center of the circumscribed circle as defined by the side lengths of the triangle. Knowledge of these intersections made it possible to find the triangle from which they came. Placing the triangle ABC with point A at the origin and with the aid of elementary geometric considerations, Euler obtained formulas for reaching and finding the three important points of intersection of any triangle that is not equilateral, that is, the *orthocenter*, the intersection of the triangles

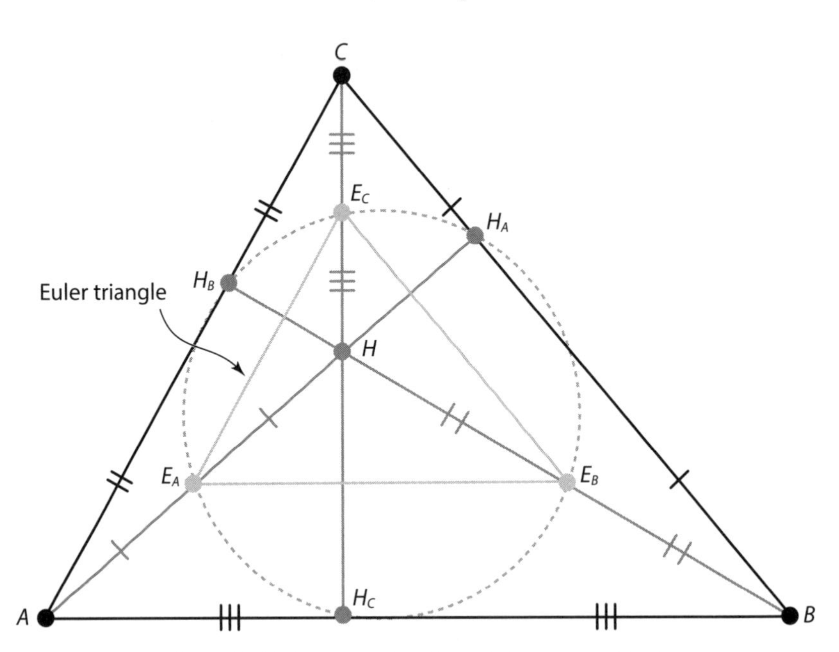

Figure 13.5. The Euler triangle and the triangle ABC.
The Euler triangle is the triangle $E_A - E_B - E_C$.

three altitudes; the *centroid*, of the three medians from the triangle's vertex to the opposite side; and the *circumcenter*, of the triangles three perpendicular bisectors, must lie on a straight line. This is called the Euler line, which is not shown above. It is considered the most famous line in triangle geometry. For allowing him to discover a beautiful geometric result after making difficult calculations, the case in which the three intersection points are collinear—that is, that always lie on the same line—Euler credited powerful techniques from elementary geometry and the methods of analytic geometry, and he worked out an example.[41] In its simplicity, this property of triangles seems an extension from Euclid himself. Euler's techniques suggested a probable, nonanalytic proof of it.

In volume 12 of the *Novi commentarii*, issued in 1767, Euler published separately from the *Dioptricae* the article "De novo microscopiorum genere ex sex lentibus compositio" (On a new type of microscope composed of six lenses, E350),[42] a draft of which had been read at the Petersburg Imperial Academy of Sciences in May 1763. The new microscope improved upon Euler's earlier models, which had only four lenses, but neither in France nor in England could artisans yet fabricate an instrument of this level of sophistication; in Saint Petersburg Euler and his assistants were to continue their efforts to develop such lenses. Vincent Chevalier, who built a six-lens

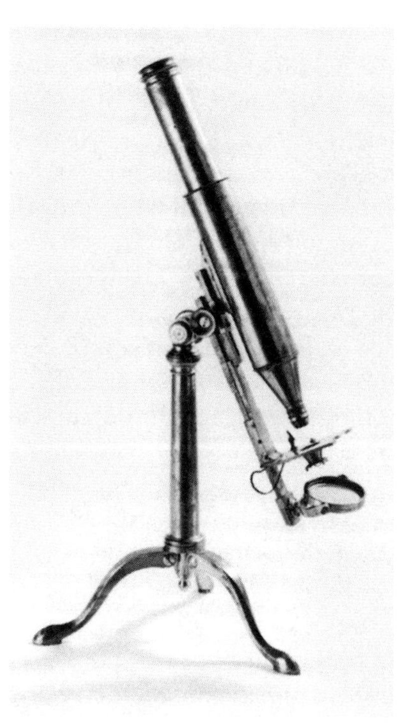

Figure 13.6. The Euler microscope.

version of the microscope for the Paris Academy in 1824, would call Euler "the illustrious and true inventor of the achromatic microscope."[43]

Volume 16 of the *Mémoires* of the Berlin Academy for 1767 contained two articles by Euler that related to his part in the origins of demography, especially the vital statistics of births and deaths, and thus the fundamentals of life insurance.

The first article, "Recherches générales sur la mortalité et la multiplication du genre humain" (General researches on the mortality and multiplication of the human race, E334), written probably seven years earlier, addressed the two independent principles of mortality and reproduction.[44] Since available tables of life and death varied considerably and Euler wished to examine both in a general form, he proceeded without referring immediately to actual registers of them, giving several ratios for computing survival yearly from infancy to old age and introducing the concept of the probable duration of life. The number of infants born in a given year was N, and he sought to know how many would survive after a given number of years, n, or $(n)N$. Given M men of the same age, he asked how many would be alive after n years; the number sought was $(m + n)$

$M/(m)$. To find the probability of the survival of men after n years, Euler employed $(m + n) \geq m/2$. The indication "greater than" gives the probability that those individuals would survive after the psychological dread of dying. In this paper and "Sur les rentes viageres" (On life annuities, E335), Euler computed life annuities. The second part of "Recherches générales sur la mortalité" considers population growth; Euler did not take account of wars, plagues, and emigration. He conjectured that population was a closed entity and that the laws of the birth rate and mortality were invariant. Population growth, then, would be a geometric progression. Given a census and list of deaths the next year, Euler showed how to construct a mortality table. Knowing the number of people in a given year, the number of newborns that year, and the duration of life over a hundred years, he could compute the number of people after a century.

The Berlin pastor and academician Johann Peter Süssmilch pioneered demography by collecting related vital data and in his major work *Die göttliche Ordnung in den Veränderung des menschlichen Geschlects aus der Geburt, dem Tode und der Fortpflanzung* (The divine order in the changes in the human sex from birth, death, and reproduction of the same, 1741, 1761–65) proposed a beautiful and harmonious divine order in the universe rather than the chaos apparent in nature, and Euler cited this study in his own "Recherches générales sur la mortalité." Süssmilch avoided algebraic calculations, leaving innovations in them to others. His chapter 8 (On the rate of increase of population and the time in which it doubles) views fluctuations in the number of people as part of a divine plan and systematically treats the expansion of population. Either Euler or another extensively reworked and revised it.

In a letter in an unknown hand to Vladimir Orlov in August 1767 about his serious illness, failing eyesight, and the project to reform the Petersburg Imperial Academy, Euler expressed his commitment to fulfilling his duties at the academy and teaching new adjuncts, for leaving those positions vacant would damage the academy. As adjuncts Euler recommended the younger members of his research circle. In 1767 Wolfgang Ludwig Krafft arrived; he was the son of Georg Wolfgang Krafft, who had collaborated with Euler in astronomy during his first Saint Petersburg period. Either later that year or early the next, Euler sent Orlov a letter thanking him for the concern about his eyesight and reported that a difficult illness had left him inactive for several weeks. It was a high fever caused, he said, by overwork. He now dictated much of his correspondence to his son Johann Albrecht. In 1768 the younger Krafft and the twenty-six-year-old astronomer Petr Borisovič Inochodzev were made adjuncts. Euler's research circle was still incomplete.

In 1768 Euler won another Prix de Paris prize for his paper "Meditationes in quaestionem utrum motus medius planetarum semper maneat aeque velox, an successu temporis quampiam mutationem patiatur? & quaesit ejus causa? " (Reflections on the question, Does the mean motion of planets always remain at a constant speed, or does it over time undergo some change? And what might be the cause? E416). The essay, first read in Berlin in 1760, was published in volume 8 of the Paris Academy's *Recueil des Pièces qui a Remporté les Prix de l'Académie Royale des Sciences* in 1771. The optimistic Euler sent a circular to the entire academy mentioning that he thought that he had finally obtained a general solution to the three-body problem. But he was mistaken, and Lagrange and d'Alembert criticized his assertion.

From 1768 on, Euler continued to follow closely the research of Lagrange. Their correspondence reveals a friendly competition on different subjects, but chiefly on number theory. In February Euler praised Lagrange's research on tautochrone curves in resistant media, which was important in mechanics, calling his new method for solving them more ingenious than those of his predecessors, except for Alexis Fontaine, and declared his pleasure at Lagrange's paper on this subject in volume 10 of the *Commentaires.* Euler and Joseph Jerôme le Français de Lalande highly esteemed the mathematical talent of Fontaine and thought Lagrange's method a rediscovery, but they did not cite its origins.[45] From mid-1768 to May 1769 Lagrange wrote three articles on number theory, the domain so dear to Euler that his contemporaries had practically left it for him alone. The third concerns the solution of a quadratic Diophantine equation, $x^2 - dy^2 = \pm 1$, where x and y are positive integers and d is a fixed integer that is not a square, which Lagrange identified with the work of Pierre de Fermat. In 1732 Euler had given it the name Pell-Fermat equation; it is thought that Euler was mistaken in believing that John Pell had worked on the equation, but it does appear in a book that Pell at least helped to write. Having shown how to reduce certain Diophantine equations to the Pell equation and demonstrated its connection with continued fractions,[46] in 1765 Euler had an algorithm to solve certain of these equations, including the case when $d = 61$. This is imposing, since the smallest x is a ten-digit number and y is one of nine digits. In 1769 Lagrange could solve a special case, but he did not yet have a general solution, a topic that Euler revisited the next year.

Upon Euler's return to Saint Petersburg, the correspondence between him and d'Alembert had ceased, while that between d'Alembert and Lagrange had continued. In June 1769, d'Alembert wrote to Lagrange about the excellence of Euler's *Institutionum calculi integralis*, but when anticipated by the assiduous Euler in a new mathematical result that he too had made but not yet published, described him to Lagrange perhaps in comic disgust as "*Ce diable d'homme*."[47] Euler responded that "d'Alembert usually rushes to express his doubts about the results obtained by other scientists while he himself never allows anyone to declare any doubts about his research."[48] The letters of the two men to Lagrange show occasional traces of mutual displeasure. In the eight volumes of *Opuscules mathématiques*, published in Paris, d'Alembert transmitted his latest findings in mathematics, mechanics, lunar theory, the precession of equinoxes, the shape of Earth, and other topics germane to celestial mechanics. In December Lagrange thanked Euler for sending him through Formey and the Haude and Spener Library the first two volumes of the *Institutionum calculi integralis*, praising its ingenuity and fruitfulness.

During the early part of his second period in Saint Petersburg, Euler steadily continued his important research in rational mechanics, hydrodynamics, and acoustics. From 1760 on, and possibly as early as 1752 according to Carl Gustav Jacob Jacobi, he had worked to complete a four-section essay on these. In 1761 he had published, while still in Berlin, *Principia motus fluidorum* (Principles of the motion of fluids, E258),[49] which generalized his earlier results in the theory of fluid motion and included setting forth the differential equations for hydrostatics. The next three sections of this essay on fluid motion appeared in print from 1769 to 1771.[50] "Sectio prima de statu aequilibrii fluidorum" (First section on the state of equilibrium of fluids, E375), presented at the Petersburg Academy in 1766, was printed three years later.[51] The second section, "Sectio secunda de principiis motus fluidorum" (E396), deals with the principles of fluid motion, and the third, "Sectio tertia de motu fluidorum lineari potissimum aquae" (E409), with the linear motion of fluids—particularly water.[52] The fourth section, "Sectio quarta de motu aeris in tubis" (E424), on the motion in air jets from a minimum of agitation and the acoustics of them, appeared in print in 1771.[53]

Masterful in the didactic skills for teaching algebra and resolving equations, Euler now completed what was to be his most popular and best-selling mathematical text, his influential two-volume *Vollständige Anleitung zur Algebra* (Complete guide to algebra, E387 and E388). Favoring here the vernacular over his usual preference for Latin, the language of learning, Euler had the Petersburg Imperial Academy publish it in Russian in 1768–69 and the next year in German. A few passages from the first volume suggest a starting date of composition of 1765–66, but Euler stated in a preliminary editorial report that in 1767, soon after the loss of most of his sight, he had dictated it in Saint Petersburg to his assistant,

Vollſtändige
Anleitung
zur
Algebra
von
Hrn. Leonhard Euler.

Erſter Theil.
Von den verſchiedenen Rechnungs = Arten,
Verhältnißen und Proportionen.

St. Petersburg.
gedruckt bey der Kayſ. Acad. der Wiſſenſchaften 1770.

Figure 13.7. Title page of part 1 of Euler's *Vollständige Anleitung zur Algebra,* 1770.

an unemployed tailor with no knowledge of algebra. In a letter from September 1769 to the Göttingen mathematician Abraham Gotthelf Kästner, Johann Albrecht Euler wrote, "I am presently caught up in a work on algebra that my father is dictating to his tailor since he went blind."[54] Johann Albrecht noted that the tailor, who lacked mathematical training and was of modest academic ability, could by the end of the process apprehend the difficulties in algebra and became capable of solving complicated problems in it; this letter seems testimony to Euler's superb teaching skills.

Another rumor has it that, to establish self-discipline, Euler dictated the *Algebra* after his cataract operation. But that operation did not come until 1771, after several editions of the *Algebra* had appeared, and the highly disciplined Euler did not need such self-control.

Translated into French by Johann III Bernoulli in 1773 and later into English, Dutch, and Italian, the *Algebra* introduces a beginner to the subject, from natural numbers employing algebraic and arithmetic principles and practices moving in steps through the basic theory of equations and selected series expansions and elementary functions—for example, the logarithmic and number theoretic, including quadratic forms. Bernoulli had agreed to translate this work only to show a wider readership how far Euler and Lagrange had perfected studies on the subject. Since Euler had no revisions in his dictation, Bernoulli in a few places found it necessary to suppress and revise parts of calculations and to insert illustrations.[55] The entire second section of volume 2 of Euler's *Algebra* was devoted to Diophantine analysis; it gave his proof of Fermat's Last Theorem for the case $n = 3$.

Most of the *Algebra* is dependable, though Euler did err in handling complex numbers by misusing the product rule for square roots. The text became an important model for teaching its subject, competing with books by Newton, Clairaut, and Maclaurin. In July 1773 Lagrange, who had carefully studied the German edition, sent Euler the French translation by Johann III Bernoulli, to which he added a three-hundred-page set of valuable notes, primarily on continued fractions and Diophantine analysis. The French edition appeared the next year. In mathematics, only Euclid's *Elements* has been more widely printed in German lands than Euler's *Algebra*, and the latter had comparable sales, principally through its appearance in Reclam's Universal Library series from 1883 until 1942.[56]

Since his arrival Euler, as the head of the academic conferences and vice director of the Petersburg Academy, participated in planning and equipping the expedition to Siberia to observe and measure the 1769 transit of Venus, when that planet passed between Earth and the sun. Since Copernicus the mean distance between Earth and the sun had been the basic astronomical unit in our solar system. From Kepler's third law of planetary motion, stated in his *Harmonice mundi* in 1619,[57] it was possible to give precisely the distances of the planets from the sun in astronomical units; for any two planets that law had the squares of times of revolution around the sun proportional to the cubes of their mean distance from it. But it was almost impossible to express these distances in terrestrial units. By 1700 telescopic observations had led to a consensus figure of the Earth lying at eighty million miles distance from the sun. The transits of Venus provided a way to measure accurately the mean horizontal solar parallax and by use of similar triangles and observations from different latitudes to find precisely the average and the perihelion and aphelion distances of Earth from the sun measured in Earth radii. *Solar parallax* when obtained by triangulation refers to the difference in the sun's position viewed from the center of Earth and a point one Earth radius distant viewed from the center of the sun. Observations of the sun's position relative to the stars from widely separated observatories were employed to determine it. The difference between the two positions gave the sun's angle of parallax. With the solar parallax known and the

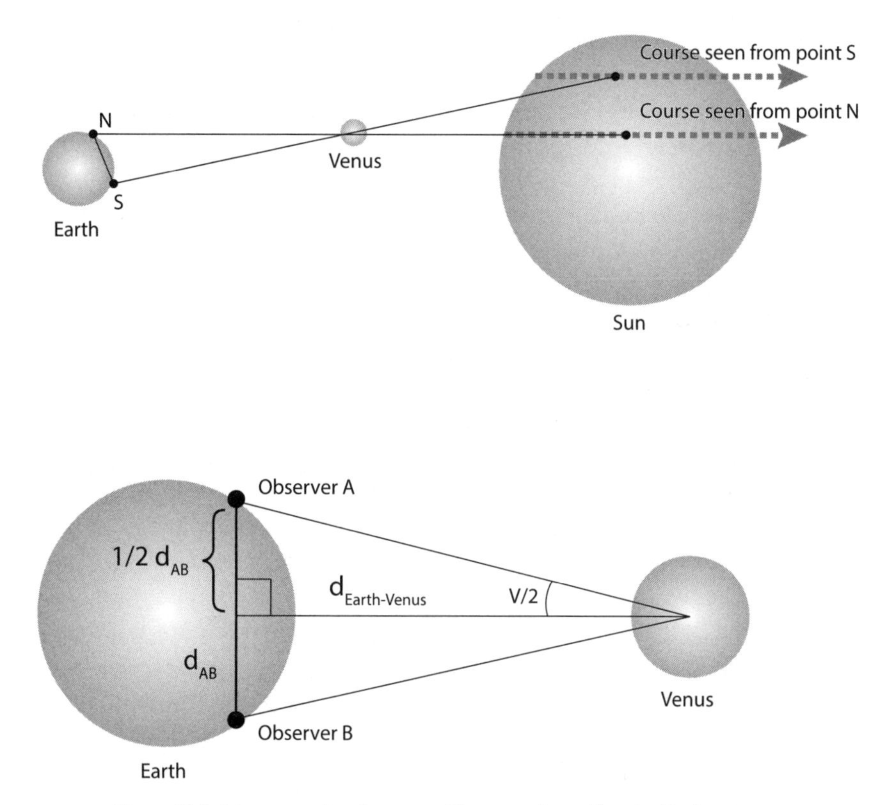

Figure 13.8. Lines crossing between Venus and parallactic displacement. With Kepler's third law and basic trigonometry it is possible to calculate the distance from the earth to the sun as 93 million miles. The equation is $\tan(V/2) = (\frac{1}{2}\, d_{A-B}) / d_{Earth-Venus}$. The illustrations are not to scale.

mean Earth radius, it was possible to calculate an astronomical unit giving the mean distance from Earth to the sun.

After about 1725 international cooperation on scientific expeditions had been significant. Even in 1761 during the Seven Years' War, the Paris and the Petersburg Imperial Academies—and, on the other side of the war, the Royal Society of London—worked together to measure the transit of Venus.[58] The transits occur in pairs eight years apart, with these doubles separated from each other by more than a century.[59] During the eighteenth century the two years for transits were 1761 and 1769. Long before 1761 Euler had studied this phenomenon and designed improved telescopes for more precise measurements.

In 1753 observations of the transit of Mercury failed to determine solar parallax accurately, but the growing press of newspapers and journals in France called for more astronomical expeditions. In Paris Delisle, heading the research on the transits of Venus, corrected past data; in 1760 he determined that India and the East Indies were the best locations for observing the approaching transit and determining solar parallax. The shortest duration for observing the transit would be in Tobolsk, Siberia. Delisle's colleague Lalande issued a call to action: "The occasion presented to us by this celebrated phenomenon is one of those precious moments of which the benefit, if we let it escape, will not be compensated later—neither by the efforts of genius nor by dint of hard work, nor by the munificence of the greatest kings. It is a moment that the past century envies us. . . ."[60] In 1761 Rumovskij led a team to Selenginsk and Tobolsk to make astronomical observations and record meteorological data; his was one of six European astronomical expeditions. At the invitation of the Petersburg Imperial Academy (in the Seven Years' War, France was an ally of imperial Russia), the French astronomer Jean–Baptiste Chappe d'Auteroche of the Paris Academy participated, but bad weather prevented gathering satisfactory data. In 1761 Euler corresponded with Jan Andrej (Johann Andreas) von Segner about the transit of Venus, and from then until 1768 Lalande sent Euler letters about observations of it as well as of lunar and solar parallaxes, along with Clairaut's descriptions of telescopic lenses in Paris and London.[61] After Delisle died in 1768, Lalande in Paris directed the study of the Venus transits.

In 1767 Euler wrote to Vladimir Orlov of astronomers who were preparing to observe the approaching Venus transit. That year and the next, most of Euler's articles dealt with perfecting telescopes and microscopes. In 1768 the Petersburg Imperial Academy of Sciences ordered from Peter Dollond in London a number of telescopes and other precision optical instruments, many with achromatic lenses. In October Euler corresponded with Daniel Bernoulli on his own research in dioptrics and his developing work on a new lunar theory. The next month Bernoulli sent Euler a letter on Dollond's telescopes and errors in lunar tables. In 1769 Euler described Dollond's application of his mathematics to making achromatic lenses. Better equipped and prepared than previously and now the academy's chief astronomer, Rumovskij undertook in 1769 a second trek that led him to Kola, Siberia, to measure the transit. His second expedition had thirteen members, including Johann Albrecht Euler. Although its results were more accurate than those compiled in 1761, poor environmental conditions again hampered it and the data included some guesses.

Among the topics in celestial mechanics that Leonhard Euler investigated was that of planetary perturbations. Of his articles on inequalities in which planetary motions are perturbed, "Annotatio quarundam cautelarum in investigatione inaequalitatum quibus corpora coelestia in motu perturbantur observandarum" (Note of certain precautions to be taken in the investigation of the inequalities by which the heavenly bodies are perturbed in their motion, E372) appeared in 1769.[62] Irregularities in the mean motion of Saturn caused by Jupiter raised a problem, for they did not seem to conform to a repeating cycle, as Newton's theories of gravitation suggested they would; this gravitation was viewed as a necessary but possibly not a sufficient condition. Euler thought that a celestial ether might cause the observed deceleration; this had theological overtones, for it suggested that the solar system might collapse into the sun. Lagrange briefly considered the influence of comets, and throughout the 1770s the Paris Academy kept it among its prize topics. In 1770, after determining solar parallax from the transit of Venus and longitude from solar eclipses in "Exposito methodorum, cum pro determinanda parallaxi solis ex observato transitu Veneris per solem" (Exposition of methods, from an observed transit of Venus across the sun, E397), Euler investigated the perturbation of the motion of Earth arising from the action of Venus in "De perturbatione motus terrae ab actione Veneris oriunda" (On the perturbation of the motion of Earth due to an action arising from Venus, E425).[63] He used rotating coordinates and worked on the restricted three-body problem for the sun, Earth, and Venus. Lexell would compute the tables of perturbations that disagreed substantially with the calculations by Mayer and Nicolas Louis de Lacaille. Urged by Lalande, Laplace wrote to Euler. Discovering that he had made a major error, Lexell in a 1779 essay reported and corrected his mistake.

In 1769 Lexell had come from Uppsala to Saint Petersburg as an adjunct; he knew of the return of Euler and the academy's preparations for its observations of the transit of Venus, and seeking to gain acceptance in the academy and to be able to participate in that project, he sent the academy a paper on a new integration method in calculus. When Euler judged the paper very favorably, Orlov suggested that Lexell probably had the help of an experienced mathematician; Euler responded that only he himself or d'Alembert could have written the paper. Lexell was to be an important addition to Euler's research circle; under Euler's direction, he studied the solar parallax and the comet of 1769, and two years later he investigated irregularities in lunar motion. Lexell reported his findings to the academy in March 1771 and was made a professor of astronomy.

The turning point in astronomy in the study of the great inequality of Jupiter and Saturn did not come in Euler's lifetime, but was solved in 1785 by Laplace, who was able to explain the 929-year period of acceleration and deceleration of the mean motion of Saturn and Jupiter with gravitation alone by making perturbations powers and products of second and

Figure 13.9. Anders Johan Lexell.

higher dimensions. This advance produced far more accurate planetary tables and was seen as "extricat[ing] Newtonian theory from one of its greatest perils."[64]

Throughout his career Euler followed various studies in the sciences and medicine. In 1769 the Swiss naturalist Charles Bonnet sent him his *La palinogénésis philosophique*, which advanced preformation, the theory that in reproduction the complete animal exists in miniature in the male germ cell or, according to the ovists, the female germ cell. Corresponding with nearly all of the leading men of science of his time, Bonnet expressed enthusiasm for Euler's works. His earlier discovery of parthenogenesis—virginal reproduction—seemed to support the ovist position on preformation; Bonnet's book also addressed final causes, holding that human beings would become angels. Perhaps in February 1770, Euler responded with a note of thanks to Bonnet for the book and declared that in light of the experiments, observations, and studies by Caspar Friedrich Wolff, a Saint Petersburg academician since 1767, he accepted the theory of epigenesis. In a letter in March, Euler commented on epigenesis, the freedom of the will, and the immortality of the soul, along with Wolff's research on the anatomy of the eye.[65]

Around 1770 Euler began to study the construction of lunes (two intersecting circular arcs on the same side of common chords), a problem that in the 1720s Nikolaus II and Daniel Bernoulli, along with Christian Goldbach, had investigated. Work on the origins of lunes lay in classical Greek antiquity; according to Aristotle's Analytica posteriora (Posterior analytics) and the work of Proclus, Hippocrates of Chios failed in the fifth century B.C. to square the circle but succeeded in solving the related problem of squaring a lune. Hippocrates discovered three cases, beginning

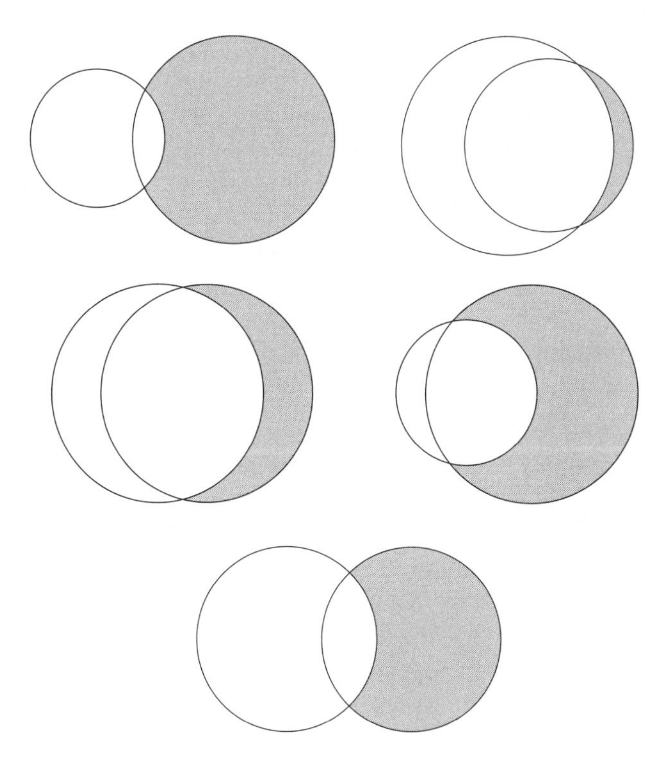

Figure 13.10. Lunes.

with two right isosceles triangles circumscribed by a circle and a semicircle constructed on a hypotenuse.[66] The ratio of the inner sides (p_1:p_2) had to be commensurable. Hippocrates's three cases had angle ratios of 1:2, 1:3, and 2:3. In the late sixteenth century, the French mathematician François Viète had rediscovered the 1:3 case and added another lune for 1:4. In "Considerationes cyclometricae" (E423), presented to the Petersburg Academy in July 1771, Euler gave two additional lunes with angle ratios of 1:5 and 3:5.[67] He was unaware that the Finnish mathematician Daniel Wijnquist had found these five years earlier.[68] Euler argued that his essay was comprehensive, and it was not possible to find other lunes; this work was not, he knew, a prelude to the quadrature of the circle.

In differential geometry Euler had begun with curve theory but in the 1740s turned to the theory of surfaces, becoming the first mathematician to do so. While he did not achieve a complete theory of surfaces, he would arrive at significant results. At the Petersburg Imperial Academy of Sciences in 1770 his essay "De solidis quorum superficiem in planum explicarre lice" (About solids, whose surfaces can be developed on the plane, E419), was read; stretching his theory of surfaces, it introduced developable surfaces, an entirely new concept. Euler pointed out that while elementary geometry held that cones and cylinders could be flattened out or developed into a plane, the sphere lacked this property. He inquired what other types of surfaces have it, calling this question characteristic and important.[69]

In 1770 and 1772 Euler received two more Prix de Paris awards for two articles, "Réponse à la question proposée par l'Académie Royale des Sciences de Paris pour l'année 1770" (E485, cowritten with Johann Albrecht Euler[70]) and "Réponse à la question proposée par l'académie royale des sciences de Paris pour l'anné 1772" (E486). The papers for both answered a question posed by that academy on bettering methods for lunar theory: participants were to provide equations for irregular motions of a satellite and explain why the new theory gave the secular equation for lunar motion.[71] After allowing for inequalities or irregularities covering a short period of time, the secular equation expressed algebraically the magnitude of the remaining inequalities in a planet's motion. Both articles appeared in 1777 in volume 9 of the *Recueil* of the Paris prizes. The introductory words for the second essay state that completing it took "extreme labor"; Euler shared the 1772 prize with Lagrange, who refined Euler's approximations for the three-body problem. This research showed that Euler's effort in celestial mechanics continued.

As the *Protokolii* of the academy indicate, until his surgery in September Euler attended almost all academy conferences in 1771, directing the two that were held weekly. Usually seven to eleven members attended, including Euler, his son Johann Albrecht, Johann Karl Fischer, Kotel'nikov, Krafft, Erik G. Laxmann, Lexell, Aleksey Protassov, Rumovskij, Stählin, and Caspar Friedrich Wolff; they covered all the major sciences, together with medicine. Astronomy and geography continued to be prominent. In January Euler's son Christoph made astronomical observations in Perevoloschena and Gluchov that were applied to the improvement of Russian maps; academicians also kept daily meteorological data. On 14 January, Euler submitted six articles for the *Novi commentarii* on topics including trajectories, the brachistochrone problem, and the motion of elastic bodies. That month Johann Georg Gmelin commented on the Linnean genera and system, and he reported finding unknown plants and curiosities in Siberia. In February the academy acquired Kästner's *Dissertationes mathematicae et physicae* (1756–66) from the University of Göttingen as well as articles from its library. On 13 April Euler submitted three articles on the motion of fluids and flexible bodies. He regularly had reports read on the travels of Peter Simon Pallas, a professor of natural history, and particularly his Siberian journeys, including a brief excursion to Tobolsk and a visit to Beijing. Müller, who met with Pallas, called for trips to the Kirghiz steppes, and in April they compared Russian and German calendars. The next month Johann III Bernoulli sent a book on astronomy for the library. In June Lagrange's new analytic theory was reviewed and the secretary thanked the Royal Swedish Academy for books and for a globe and instruments sent from Stockholm. On 4 July Euler submitted seven articles for the *Novi commentarii* on topics including cyclometry, Diophantine equations, infinite series, and determining longitude, the last of which was discussed at the next meeting. In August the conference considered the purchase of the supplement to Diderot's *Encyclopédie*. After missing sessions on 17 and 19 August, Euler was absent from conferences from September 1771 to May 1772.[72]

The Kulibin Bridge, the Great Fire, and One Fewer Distraction

Even though nearly blind, Euler was consulted on technological projects, especially for creating underlying theories and providing fundamental designs, just as he had been for Segner's waterwheel, and aided by his assistants he oversaw the evaluation of technological plans and inventions. He often enjoyed an advantageous position, for he not only understood the latest developments in mathematics and physics but had begun to apply them.

During the late eighteenth century in fast-growing Saint Petersburg and across Europe, the construction of permanent bridges over wide and swiftly flowing rivers posed a significant problem, including those for the Greater Neva River; it was critical for Saint Petersburg, situated on the islands in its delta, to replace its old pontoon bridges.

In 1771 the talented, self-educated mechanic and inventor Ivan Petrovic Kulibin submitted a first draft of a proposal to build a wooden, single-arch bridge over the Greater Neva with a span of 298 meters (the typical bridge at this time spanned only 50 to 60 meters); this was the first contact between Euler and Kulibin. Kulibin addressed the strength and load-bearing capacity of the bridge, and after several failed attempts provided a one-fortieth-scale model. Placing on it a weight fifteen times that of the model, the academicians tested it, and it passed. While Euler found the computations correct, the "gentlemen academicians" reviewing it called it doubtful, since Kulibin could not go on from this small model to ascertain the weight of the actual bridge. Ice was also an obstacle to

Figure 13.11. Ivan Petrovic Kulibin.

its construction and success. The proposed bridge went unconstructed. In December 1772 Kulibin wrote to vice director Reshevski that he had deduced from a series of experiments a rule allowing him to reach from his small model the load-bearing capacity of the actual bridge, but it took until 1775 to present his conclusions. The bridge was never built.

Kulibin, whose father was a trader in Nizhny Novgorod, had been drawn early to building mechanical tools and clock mechanisms, and had constructed an egg-shaped ornamental clock with a complex automatic mechanism that he had sent to Catherine II in 1769. As a result, she had invited him to Saint Petersburg to join the mechanical workshop of the academy, and there the imaginative Kulibin studied mechanics, physics, and chemistry and found new ways to cut glass for better microscopes, telescopes, and other optical instruments. He became the workshop's foreman and chief technician. From 1771 on, Euler kept in regular contact with Kulibin, often visiting along with assistants Kulibin's three-story workshop on Vasilievsky Island, which was not far from Euler's house. Their interactions included discussions and efforts concerning improving telescopes and microscopes.

About noon on 22 May 1771, a great fire broke out in the lane behind the academy of sciences on Vasilievsky Island. Euler's wife Katharina had long feared such an event. For three hours the fire spread in the land between the seventh and twenty-seventh blocks from the center of the city, destroying about 550 wooden and even some brick houses near the Neva, among them Euler's. Helpless, almost blind, and in his bedclothes on the second floor of his house, Euler might have died in the general confusion had not his Basler handyman Peter Grimm quickly recognized the danger, found a ladder, and climbed into the house to rescue him. Euler reportedly resisted the efforts to remove him, wanting to stay to try to save his writings; Grimm had to put him over his shoulder and carry him down the ladder. After securing his safety, Grimm probably saved some documents. The fire destroyed Euler's library and furniture, but the quick action of Vladimir Orlov, who was likely at the nearby government buildings, retrieved most of the manuscripts. One loss was the worksheet for the prize paper on lunar theory to be submitted for the Royal Academy of Sciences in Paris in 1772; Johann Albrecht Euler had to rewrite its lunar theory and remake all of its calculations.

In reporting news of the fire in a letter in September to his friend the astronomer Jacques Mallet-Favre in Geneva, Nicolas, Marquis de Condorcet noted that Catherine had promptly resolved to make up the loss, giving Euler six thousand rubles to reconstruct the house, which was rapidly done, and to purchase furniture in the meantime. An appropriate house was found to rent for the essentially blind geometer. The refurbished house to which Euler soon returned was impressive: it had four large columns supporting its front porch and double wooden doors for the main entrance. A large marble foyer separated the house into four large apartments and at the back of the foyer two facing marble stairways led from the first floor to the second. Each apartment was essentially independent. A drinking house next to Euler's had also burned; responding to Euler's request, in June 1771 Catherine agreed "not to build a new drinking house in place of the old one but instead to give the plot to Professor Euler."[73]

In September 1771 the nearly blind Euler hoped to have his vision restored through an operation to remove the cataract in his left eye. He had followed such operations in Paris and perhaps knew that the composer Johann Sebastian Bach had undergone two of them. As Euler aged, the lens of his eye slowly clouded, which made his vision fuzzy, making it progressively more difficult to see. Since late 1766 he had been almost blind. Baron Michael Johann Baptist von Wenzel, a famous eye surgeon who had recently developed one of the earliest cataract operations in Europe, traveled to Saint Petersburg. To the mid-eighteenth century, physicians and surgeons had mainly neglected treating eye diseases; the field was filled with quacks, and there was controversy over whether to use the old couching treatment or extraction. Wenzel had a rising reputation, and in 1772 would become oculist to King George III of Great Britain. Euler's son Johann Albrecht and nine physicians gathered around him on the surgery table to observe the operation. Wenzel made a large incision in the cornea and removed the cataract with little pain, cleaning the cloudy lens. The cataract was described as appearing like a small flattened pea the color of beeswax. The operation took only three minutes, there was little or no bleeding, and Euler had been deemed a good patient. For a time he had bandages over his eye.

The result of the operation was the restoration of sight to Euler's left eye, though he was instructed to avoid bright light and not to overstrain his eyes with reading and writing. But the incision was large and the eye weak after the operation. Moreover, Euler was not told to remain as still as possible with bandages over his eyes for a few months or to confine himself to his bed with sandbags placed around his head to immobilize it. Nor were steroid drugs available to strengthen the eye. As a result, in October Euler suffered a complication, possibly an infection, that left him almost blind once again and occasionally in pain. But even without the "distraction" (as he called it) that his sight had provided, he could still perform the most difficult computations in his head. Nicholas Fuss is reported to have related that once, when two of Euler's students were summing a converging infinite series to the seventeenth term and differed by one unit in the fiftieth place, he made the correct calculation in his head.[74] That extent of his computational skills is difficult to believe, but Euler could still neatly write his name.

Persistent Objectives: To Perfect, to Create, and to Order

From 1764 on, Euler had prepared his 775-page landmark work *Theoria motuum lunae, nova methodo petractata* (The theory of lunar motion, treated by means of a new method, E418),[75] which is generally taken to constitute his second lunar theory but is here designated his third; his research and computations on lunar theory in the late 1740s, resulting in his "Tabulae astronomicae solis & lunae" (E87), was published originally in his *Opuscula varii argumenti* in 1746. This was his first, and his *Theoria motus lunae exhibens omnes eius inaequalitates* (Theory of the moon that exhibits all its irregularities, E187) his second. There were also the influences of several papers on astronomy, including the Prix de Paris winners for 1748 and 1750 on the great inequality of or irregularities in the orbits of Jupiter and Saturn and the papers from 1770 and 1772 on lunar theory.[76] Toward the end of his time in Berlin and since his return to Russia Euler had considered what would contribute the most distinction to his legacy,[77] and had decided that it would be a new lunar theory, one of the most complex problems in celestial mechanics. Displaying characteristic stubbornness, he was dissatisfied with his two previous theories. Not even eyesight problems slowed him; in 1767 he worked steadily on methods, algorithms, and computations, and the next year he had a preliminary draft that he presented to the academy, but much remained to be done. In May 1771 he wrote to Lagrange that he had spent almost an entire year intensely preparing the new lunar theory.[78] Euler sought to produce lunar tables that were both better than the semiemperical ones of Mayer and Clairaut and a great deal clearer and easier to use. While his tables were not to be as exact as Mayer's, none of the roughly twenty other lunar theories of the time compared to his in their fit with practical considerations from observations or in accuracy. Three of Euler's assistants—his son Johann Albrecht, along with Krafft and Lexell—assisted him, mainly with able calculations, and they are named on the title page of the *Theoria motuum lunae*. Euler particularly praised Lexell. Guiding the complex and abstruse computations of the members of his circle must have been a difficult challenge for the nearly sightless Euler.

In Saint Petersburg, astronomy continued to be a central topic for Euler. During his career he wrote more than a hundred papers and books on mostly theoretical, astronomical investigations; these fill ten volumes of his *Opera Omnia*. This work was part of the mid- and late-eighteenth-century flourishing of astronomy in Europe, responding principally to the needs of navigation, commerce, exploration, and academy prize competitions. Among Euler's spectrum of many contributions to astronomy, his third lunar theory is the most important.

The monumental work comprised three lengthy volumes. It replaced the single-step method of his earlier *Theoria motus lunae* (E187) for computations on lunar motion with a multistep procedure. The first volume investigates differential and numerical equations of condition required for his brilliant choice of coordinates; he employed a two-coordinate system: one fixed and the other moving. Euler integrated differential equations, since what began as small errors of differentials would grow over time. Examination of their integrability greatly increased the accuracy and lessened computational labors. He tested perturbations of the lunar orbital elements, including eccentricity and mean longitude, for more exactness and to refine his secular equations. Euler had acute astronomical as well as mathematical insights: since gravitation stemming from the sun on the moon was double that from Earth, the main irregularities or perturbations of the moon derived from the influence of the sun. For the theory of perturbed motion in celestial mechanics, Euler was the first and chief

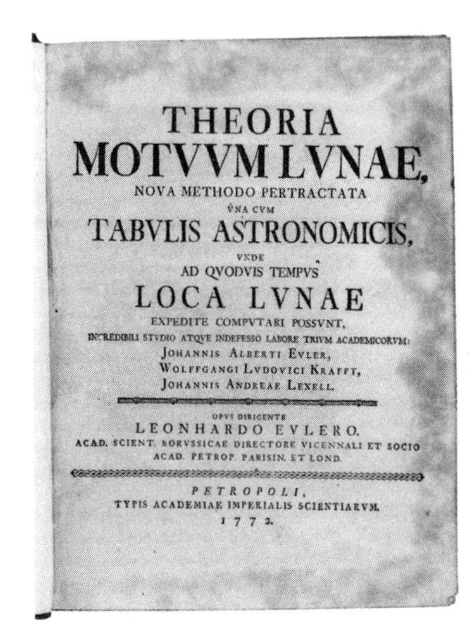

THEORIA
MOTVVM LVNAE,
NOVA METHODO PERTRACTATA
VNA CVM
TABVLIS ASTRONOMICIS,
VNDE
AD QVODVIS TEMPVS
LOCA LVNAE
EXPEDITE COMPVTARI POSSVNT,
INCREDIBILI STVDIO ATQVE INDEFESSO LABORE TRIVM ACADEMICORVM:
JOHANNIS ALBERTI EVLER,
WOLFFGANGI LVDOVICI KRAFFT,
JOHANNIS ANDREAE LEXELL.

OPVS DIRIGENTE
LEONHARDO EVLERO.
ACAD. SCIENT. BORVSSICAE DIRECTORE VICENNALI ET SOCIO
ACAD. PETROP. PARISIN. ET LOND.

PETROPOLI,
TYPIS ACADEMIAE IMPERIALIS SCIENTIARVM.
1772.

Figure 13.12. The frontispiece of Euler's *Theoria motuum lunae,* 1772.

provider in depth of its mathematical methods. His assistants had a large store of better observations that he had been collecting from at least his time in Berlin and from systematic ones made at the Petersburg Observatory in the Kunstkammer tower.

The second volume made comparisons of Euler's equations with Clairaut's and constructed improved astronomical tables using the new equations, formulas, and observations. Among the problems in Euler's lunar theory was describing two simultaneously acting centripetal forces. From his article on the elliptical orbit of the comet of 1769 and his collaboration with Lexell, Euler dictated article 86 of the *Theoria motuum lunae*, which was essentially the first place to give the formula—to be rediscovered by Carl Friedrich Gauss in 1801—for computing the path of Ceres. Rather than pursue a solution for the general case of the difficult Newtonian three-body problem, which he now held out no hope for achieving, Euler in *Theoria motuum lunae* examined special, restricted cases and found a more effective method for approximating them to have more exactness with actual positions. More than 450 of the book's pages integrate as many as thirty-one differential equations, most of which he had invented.[79] As in 1753, he employed multiple observations to refine differential coefficients. While Newton had created the geometrical form of celestial mechanics, Euler founded through his research in astronomy its analytical form.[80] Euler's lunar theory was the most advanced of the time, and even today part of the manuscript record dealing with this work remains to be examined.[81] It was not to be surpassed, or perhaps better perfected, for a century.

The shape of Saturn's rings was another topic that Euler now examined, and it speaks to his approach and belief in the simplicity of the most basic laws of nature and the ultimate methods to reaching them. In 1777 he wrote in Latin, "The solution of this problem is highly significant, in that via a path involving complicated calculations we finally arrived at very simple formulae. There can be no doubt, therefore, that another, much simpler, path to the solution exists that in fact might have been anticipated; however there is no shame in expounding the present solution, since it involves noteworthy devices that may be of use in other researches."[82]

During the eighteenth century, Euler was the principal creator of the theory of elasticity; he had investigated it at least as early as 1732 in his article "Solutio problematis de invenienda curva, quam format lamina utcunque elastica in singulis punctis a potantiis quibuscunque sollicitata" (Solution to the problem of finding curves which are formed by an elastic strip when a force is applied to a single point, E08). Along with others, he sought to organize and unify theories on elasticity and to provide a general mathematical theory for it, but that lay fifty years in the future with the work of Augustin-Louis Cauchy. Euler's outstanding essay "Genuina principia doctrinae de statu aequilibrii et motu corporum tam perfecte flexibilium quam elasticorum" (The natural principles of the equilibrium and the motion of perfectly flexible and elastic bodies, E410) represented a pinnacle in that search;[83] read at the Petersburg Academy in 1771, it was published that year in volume 15 of the *Novi commentarii*. Here Euler introduced the concept of shear force, but also accomplished far more. Applying the first principles of mechanics to provide a simple, direct way, he recast the entire corpus of known results on elastic and flexible bodies. Part of his originality in this matter was knowing that the eighteenth-century models of elastic lines and other physical structures were too restrictive, idealized, and lacking internal stress; existing solutions were developed from principles either unclear or too specialized. At the start of the 1771 essay, Euler noted that "we are still far distant from a complete theory sufficient to determine the figure of elastic surfaces and bodies."[84] His challenge was to offer a "more precise" and lucid study of "simple threads, whether perfectly flexible and elastic, as they have been treated up to now by the geometers."[85] He took pains to achieve this. In "De gemina methodo tam aequilibrium quam motum corporum flexibilium determinandi, et utriusque egregio consensu" (The equilibrium of the twin method and determining the motion of flexible bodies, and each in excellent agreement, E481, published in 1776) he unified the elastica and the catenary in a single conceptual framework.

The 1771 essay also reinstated and extended the distinction made by Jakob (Jacques) Bernoulli between laws of mechanics and constitutive laws that defined certain kinds of continuous bodies. Without any special hypothesis about composing materials, Euler was able to derive methodically the general differential equations of equilibrium and motion for a deformable line in a plane.

From his early studies under Johann I (Jean I) Bernoulli on number theory Euler had worked mostly in fits and starts. From the release of the *Algebra* to his death, it was his favorite pastime; Diophantine equations and the search for prime numbers particularly interested him. In the years after 1769, nine of his papers on the subject appeared in print. While his earlier papers had broken ground, created powerful tools, transformed the subject for mathematical development, and been influential in converting Lagrange to the subject, most of the papers during this period, filled with intricate calculations, seem a random collection of difficult exercises; for the nearly blind Euler they were essentially transcripts of conversations with his assistants. With the exception of Lagrange, he recognized that among his contemporaries higher arithmetic was not a popular subject. Notably, to a lesser degree, d'Alembert influenced the younger Lagrange to move into this field of research, and in number theory Lagrange was to base most of his work on Euler. With Euler at this point the only eighteenth-century mathematician seriously working in number theory, Lagrange had no other mentor.

In 1768 and early 1769 Lagrange had written three articles on number theory. As he informed d'Alembert, he was seeking a direct and general solution for indeterminate equations of second degree and cited an essay by Euler in volume 11 of the *Novi commentarii* as prompting his interest.[86] Lagrange stressed the importance of the problem; he greatly desired to draw Euler's attention to it again and to solicit his most recent publications on it. Probably in late 1769, Formey had sent to Euler Lagrange's great essay on indeterminate problems, published in the latest volume of the *Nouveaux Mémoires* of the Berlin Academy. During this time, mathematicians were attempting to solve particular cases of the equation, which in its origins went back to Archimedes of Syracuse's cattle problem,[87] and earlier Euler had discussed it with Goldbach. Lagrange found the problem charming and was pleased that his research had merited Euler's attention. In 1770 Euler wrote a continuation of Lagrange's articles in which he used finite differences to provide a general but not a strict proof of a particular case, the Pell-Fermat equation, $x^2 - ay^2 = b$. Filled with admiration for Euler and indebted to him, Lagrange wrote in February, "I am most happy that my work on indeterminate problems should have been found worthy of your attention; the favorable opinion of a scientific savant of your rank is most flattering for me, particularly in a field where, to my knowledge, you are the one and only competent judge. . . . [I]t seems to me that only Fermat and you have been successful so far in dealing with such matters, and if I have been so lucky as to add to your discoveries, I owe it to nothing else than the study I have made of your excellent works."[88] Employing a modification of Euler's argument concerning finite differences, Lagrange now presented the first rigorous general solution for all equations of the second degree with two unknowns.[89] Today Euler's results are expressed in the language of congruences, which were introduced later by Gauss.[90] The set of integers with addition and multiplication modulus or divisible by a prime number formed a field, which in substance Euler knew but had missed as it bore on this problem. Perhaps since Lagrange did not use the proof in his paper, Euler did not later quote it. Euler's tract on these proofs went unpublished in his lifetime.

In 1772, hardly a year after receiving Lagrange's papers, Euler submitted three papers on what have been known since then as power residues; these deal with divisibility.[91] If for an integer α in the congruence $x^n \equiv \alpha (\text{mod } m)$ is solvable for an integer n greater than 1, the number α is a residue of degree n modulo m. For $n = 2$, the power residues are quadratic. Euler also investigated the cubic and biquadratic (fourth power) cases. His essay on quadratic residues, "Observationes circa divisionem quadratorum per numeros primos" (Observations about the division of squares by prime numbers, E552), separated his earlier conjectures from what was now basically the law of quadratic reciprocity.[92] He pursued prime divisors of different forms of numbers, such as $X^2 + Y^2$, surpassing Fermat. "I derive as much pleasure from investigations of this kind," Euler confessed, "as from the deepest speculations of higher mathematics. I have indeed devoted most of my efforts to questions of greater moment, but such a change of subject always brings me welcome relaxation. Higher analysis, moreover, owes so much to the Diophantine method that no one has the right to dismiss it altogether."[93]

After Euler's production of articles briefly ebbed in 1770 and 1771 during his emphasis on the *Theoria motuum lunae* and following his cataract operation, it resurged in 1772, and in May he resumed directing the sessions of the academy. On 18 May 1772 he submitted thirteen articles saved from the fire on such topics as elliptic integrals, prime numbers, and map projections. The subjects that the academy addressed later that year included volume 2 of Gmelin's *Flora sibirica,* Lexell on the comet of 1769, an essay on the transit of Venus, Lalande's contributions in Paris to integral calculus, Daniel Bernoulli's solutions in fluid mechanics, and vibrating chords. On 3 December Euler submitted eight more articles on Diophantine problems and differential equations.

In 1772 Euler's correspondence fell sharply, consisting of seven letters with only two by him. His health, the fire, and the absence of a personal secretary, together with the uncertainty that mail to and from Russia would be delivered, seem to account for this. In February Lagrange expressed his wish that Euler's eye operation had been successful, reported that the French edition of the *Algebra* was in press, and said that he had received volume 14 of the *Novi commentarii* with great satisfaction regarding Euler's research on orthogonal trajectories and mean proportions. In August Euler sent Stanislaw August Poniatowksi of Poland geographical coordinates for Barnaul and some points in Moldavia. The next month Stanislaw thanked him for these and earlier results for the mouths of the Danube and Dniester Rivers. After a six-year break, in October Lambert sent Euler his additions to logarithmic tables and discussed selected infinite series and Lambert's achromatic glasses for the nearsighted. Mathematicians were now integrating rational functions via logarithms. A letter from Euler with only a date of 1772 on it comments on a number theory article by Johann III Bernoulli submitted the previous year for the *Nouveaux Mémoires* of the Berlin Academy; this was only his second—and the last—letter that he sent that year.

Chapter 14

Vigorous Autumnal Years: 1773 to 1782

At the start of 1773 Leonhard Euler was sixty-five years old. For the library of the Imperial Academy of Sciences and Arts in Saint Petersburg he now faced two matters. In January the Nuremberg historian and bibliographer Christoph Gottlieb Murr attempted to have him purchase for Russia some handwritten tables of Johannes Kepler that were in Frankfurt, noting that their publication could bring further fame to Russia. In February Jean-Baptiste le Rond d'Alembert sent the sixth volume of his *Opuscules mathématiques*. Euler did not respond directly to either d'Alembert or Murr; his correspondence with d'Alembert had ended by late 1766. But until at least 1774, d'Alembert kept in contact by continuing to write to Euler's son Johann Albrecht.[1] Murr sent five more entreaties, which Euler gave to the royal court and the academy. Empress Catherine II was interested in "the celebrated Kepler's" manuscripts. The academy's *Protokolii* (Protocols) in November reported academy director Vladimir Grigor'evich Orlov's learning that the empress had acquired all of these for the academy, and that a catalog had been placed in the archive. Murr would participate as well in later library acquisitions for the academy. In 1773 Euler dictated only two letters. The first, transcribed by Johann Albrecht in May, was to Euler's good friend, the margrave Friedrich Heinrich. Although his vision was so poor that he could neither read nor write, Euler said that his general health had recovered and that he was happy. He mentioned that the reconstruction of his house had so enlarged it that he could live there with his sons and their families, and also noted having sent to the margrave through Jean-Henri-Samuel Formey six copies of volume 3 of his *Lettres à une Princesse d'Allemagne* (Letters to a German princess). Throughout the year, life remained calm in Saint Petersburg. Begun in the autumn, the rebellion of the Ural Cossacks led by Yemel'yan Ivanovich Pugachev, a pretender to the throne who claimed to be the deceased emperor Peter III, was distant from Saint Petersburg but caused turmoil and was a serious danger to Catherine. The *Pugachevschina* threatened the noble social order in Russia.

After completing his *Theoria motuum Lunae* (Theory of lunar motion, E418) in 1772, Euler had turned to his articles; the number of articles rose from eight in 1772 to nineteen the next year. The subjects for 1773 were a Diophantine problem, a differential equation, the vibrating chord (a topic that took five essays), the tautochrone in fluid and rare media, the collision of bodies, and a new proof that every integer is the sum of four squares. The persistent vibrating chord dispute with d'Alembert, Daniel Bernoulli, and Joseph-Louis Lagrange was the leading documented mathematical controversy of the century. None of the competitors moved far from his original position, but Lagrange moved closer to d'Alembert; many if not most geometers of the time accepted Euler's resolution.

Euler's near total blindness required him to seek additional assistance. Since the young people he considered in Saint Petersburg could not dedicate the time that he desired, early in 1773 he turned to Daniel Bernoulli to recommend from Basel a capable assistant to help with his research and to be his personal secretary. Proud never to have lost his Swiss accent, Euler liked to employ the Basler dialect and challenge others to remember expressions.

The Euler Circle

In July Bernoulli complied by sending his most talented student, Nicholas Fuss, who was from a humble Basel family, to live with Euler and be his research assistant and secretary. It was a superb match; the sympathetic young man who came to the Russian capital possessed an excellent combination of talent and adaptability. Applying these, along with a great drive, he helped the aging geometer to continue to produce an impressive flow of articles. For these last years of Euler's life Fuss was his closest associate, reading to him every morning the collected correspondence and the daily paper, and placing on the table in the study whatever mathematical literature related to the problem for the day. Euler would develop a general plan of attack for the problem being addressed, and on a later morning, after help from members of the circle in refining ideas and making computations, he led the preparation of the paper set out in Latin for printing. In 1784 Fuss was to marry his mentor's granddaughter Albertina,[2] Johann Albrecht's second daughter. The arrival of Fuss in 1773 and the inclusion a year later of Mikhail Evseevič Golovin, a relative of Mikhail Lomonosov, increased Euler's research circle. In November Euler dictated a letter written in an unknown hand to Orlov observing that he wanted to have these two students of his, Fuss and Golovin, apply their talent and zeal to scientific activities—meaning that he wanted them admitted to the academy. Among the other members of the circle were Johann Albrecht and Christoph Euler, Semjon Kirillovič Kotel'nikov, Wolfgang Ludwig Krafft, Anders Johan Lexell, and Stepan Jakolevič Rumovskij; all had been Euler's pupils and were now his disciples.

Euler set the topics and initial problems for each of his theoretical and computational papers, most of them brief, and interacted directly with every member. At the study in his residence his research team worked around a large table with a chalkboard in the middle. Although Euler could write distinctly on it in large symbols even after 1771, apparently his assistants used it most. When alone, the nearly blind Euler would often walk around the table for exercise and left a sheen from running his hand along its edges. During working sessions participants made calculations that Euler mentally reviewed and reworked to help eliminate errors. After the sessions all papers on the table were put into a large portfolio. Working with individual members of the circle that by 1783 would comprise eight, half the regular members of the academy, Euler completed on average another article every twelve days. For Euler's papers published after 1773, Fuss made the computations for more than 160 and Golovin 70. In the last decade of Euler's life, Fuss and Golovin did most of the recording from Euler's dictation, though some papers were written down and computations were made by his sons. More than three hundred of the papers completed or written in collaboration with members of his research circle during his second Saint Petersburg period appeared posthumously. These comprised more than one-third of his total writings and were published for more than thirty years after his death.

Elements of Number Theory and Second Ship Theory

Euler's profound contributions to both theory and application in the sciences in Berlin and Saint Petersburg had made him famous. Apparently his blindness served to increase his recognition, which was now perhaps second only to that of Voltaire. Stories of the power of Euler's computational genius expanded; one had him, on a sleepless night, calculate the first six powers of the twenty smallest integers. Nicolas, Marquis de Condorcet, considered it to be as many as the first hundred integers.

"It appears," wrote the French mathematician Adrien-Marie Legendre in the preface to his *Essai sur la théorie des nombres* of 1798, "that Euler had a special inclination towards . . . investigations [of properties of numbers] and that he took them up with a kind of passionate addiction, as happens to nearly all those who concern themselves with them."[3] This was indeed the case when, by 1773, Euler had turned more to examining number theory, starting with Diophantine equations and primes, and now concentrated on it. By trial and error throughout the previous decade he had explored the question of whether large numbers satisfied a primary criterion; this work resulted in two papers, "De numeris primis valde magnis" (On very large prime numbers, E283) and "Quomodo numeri praemagni sint explorandi, utrum sint primi, nec ne" (How very large numbers are to be tested for whether or not they are prime, E369). In 1773 he directed Lagrange's attention to this new research; "I am sure," he dictated, "that this will lead to very important discoveries."[4] Euler was preparing a textbook on the elements of number theory, *Tractatus de numerorum doctrina capita sedecim, quae supersunt* (Treatise on the theory of numbers, consisting of sixteen chapters, E792). Although he had actually begun work on it in the years 1748–50, worked on it again around 1756, and readdressed it shortly before his death, this unfinished treatise was not published until 1849 in volume 2 of his *Commentationes arithmeticae*. The textbook defines and classifies prime numbers and multiples; it examines factorizations, including the frequency of prime numbers in them, along with quadratic and cubic residues, divisors in general, and divisors of numbers of the form $x^2 + y^2$ and $x^2 + 2y^2$. Euler may have realized that all quadratic residues were biquadratic, and he did search for conditions in which given numbers had the biquadratic character. His book, which contains some incomplete proofs, is nonetheless filled with clear definitions and brilliant isolated remarks, and hidden within these is the crown jewel of number theory, the law of biquadratic reciprocity. The book appeared in print only after Carl Friedrich Gauss, Peter Gustav Lejeune Dirichlet, Carl Gustav Jacobi, and Ferdinand Gotthold Eisenstein proved all of Euler's guesses and far more. It was reprinted in his *Opera postuma* in 1862.

In April 1773 Euler's last book, *Théorie complete de la construction et de la manoeuvre des vaisseaux, mise à la portée des (!) ceux, qui s'appliquant à la navigation* (Complete theory of the construction and maneuver of ships brought into the reach of everyone involved in navigation, E426), his second ship theory, was presented at the Petersburg Academy, and it was published in French later that year in Russia. It presents the laws that Euler had discovered on the force that wind exerts upon flexible and rigid bodies and applies these to calculating the disposition, form, and size of sails, equipment especially important to ships prior to the invention of the steam engine. Attentive to the needs of the Russian Navy, Euler was well aware that his previous *Scientia navalis* was too abstract and profound to be understood and applied by ordinary seamen. He took several chapters from the earlier work and removed all of the theory of fundamental mechanics and most of his proper ship theory. As the title of his *Théorie complete de la construction et de la manoeuvre des vaisseaux* suggests, it offers an entire ship theory comprehensible to most regular navy personnel, navigators, mast makers, and shipbuilders.[5] Most of it comes from Euler's Prix de Paris–winning papers of 1759 and 1761, the first on pitch and roll and the second on the lading of ships.[6] The book comprises three parts: on the equilibrium of ships, on the resistance to the motion of ships in their courses, and on and their masts and steering. It concludes with a supplement of two brief notes; the first computes the resistance encountered by the prow of a ship, and the second proposes and analyzes a device consisting of a sail attached to a roller to pull a ship upstream.

The Diderot Story and Katharina's Death

In September 1773 the French *philosophe* Denis Diderot, the editor of the famous and controversial multivolume *Encyclopédie*, accepted the invitation of Catherine II to come to Saint Petersburg. He was very ill upon arrival, seemingly almost near death. From 1759 on Catherine had read the first volumes of his great work, and from the beginning of her reign, she had been his patron. Princess Ekaterina Romanova Dashkova visited Paris in November 1770, dining and speaking extensively with Diderot on four evenings from roughly five o'clock until midnight about culture, art, politics, science, and literature; she had participated in Catherine's rise to power and had asserted the empress's innocence in the death of her husband.

Diderot stayed in Saint Petersburg for five months, and was given the privilege of being able to contact the monarch at any time. She was delighted to speak with him rather than with her many courtiers, whom she found coarse and boring. Catherine found Diderot's mind to be extraordinary and requested a memorandum on founding a university that he recommended be open to students without distinction by social order,[7] but she thought impractical his recommendations on judicial reform and agricultural servitude. It was reported that to block his annoying habit of gesticulating and hitting her knee during conversations, she had a table placed in front of her. To help relieve Diderot's strapped financial situation, Catherine purchased his library, while permitting him as long as he lived to keep it and providing him an annual salary as its librarian.

Although Diderot was closely associated with materialism and criticisms of religion, both of which Euler strongly opposed, on 25 October

Figure 14.1. Denis Diderot.

1773 Euler set aside his misgivings and proposed that Diderot and his closest friend Frederick Melchior Grimm, a minor *philosophe*, be named corresponding members of the academy, and charged the secretary, his son Johann Albrecht, to notify them in writing of their election *honoris causa* and to expedite the academic diplomas for them. Three days later Diderot sent a grateful response to the "gentlemen of the academy," declaring that "everything that I created I would gladly trade for a page of the writings of Monsieur Euler, or for work of my new colleagues."[8] Diderot submitted diverse questions on the natural history of Siberia and its minerals, and the academy promised to deliberate on these and respond. On 4 November Euler presided at the session that Diderot and Grimm attended for their induction into the Imperial Academy of Sciences and Arts. While there was a negative disposition toward Diderot among members, they maintained a cordial encounter. Johann Albrecht Euler introduced them to each member, but his father had warned him to avoid arguments in general. Diderot spoke briefly on the natural history of Siberia, metallic ores, mountains, salt lakes, and a variety of vodka, and the academicians discussed these matters. Erik G. Laxmann, a specialist on Siberian fauna, was given the questions that Diderot had submitted. The empress attended the meeting and reportedly was amused slightly by Diderot's negative comments on religion in answering younger members' questions. In December Laxmann presented responses to queries at the academy; these were forwarded to Orlov for his approval before sending them to Diderot.

A story circulated widely in Berlin a few years later about Diderot's visit to Saint Petersburg and his encounter with Euler that has no confirmation in the Russian records;[9] it is an anecdote that apparently was initially only a joke. Dieudonné Thiébault had come to Berlin in March 1765. Frederick II, who had made him a professor of literature, enjoyed conversing with him about European men of letters. The monarch disliked Diderot, who along with Jean-Jacques Rousseau had criticized the treatment of wounded soldiers in Prussia, and the king liked to invent tales. Thiébault's book, *Mes souvenirs de vingt ans de séjour à Berlin*, published in 1801, includes the original tale about Diderot, put out possibly by a member of Frederick's court to belittle the encyclopedist. The story goes that a learned philosopher met Diderot in Saint Petersburg at the imperial court and with the secret acquiescence of the empress proposed to prove algebraically to him that God existed. "Sir," the philosopher solemnly exclaimed in French, "$(a + b^n)/n = x$, therefore God exists. Respond!" This was not entirely out of character for the time; Pierre Maupertuis had previously attempted to provide an algebraic proof of the existence of God. Still, when the atheist Diderot momentarily failed to recognize the sham and searched for an effective rejoinder to the pretended and meaningless proof, he sensed supposedly from the peals of laughter by courtiers that the court was playing a joke on him.[10] Dieudonné's account does not give Diderot's response, but it holds that this misadventure deeply embarrassed Diderot and was possibly a reason he soon requested permission to return to France. Since Diderot had begun his career as a mathematics teacher, he would have seen the hollowness of this supposed proof; his *Mémoires sur différens sujects de mathématiques,* a collection of five articles on subjects pertaining to higher mathematics, was well regarded. In Paris the censor commented that Diderot presented the articles "with great sagacity," and the reviewer, Abbé Guillaume Thomas François Raynal, found them "filled with erudition and genius."[11] Moreover, it is unlikely that the Russian nobility at court would have understood what was happening. Through the years the story was to be twisted; in 1872, in *A Budget of Paradoxes*, Augustus de Morgan modified it by identifying the philosopher or mathematician as Euler, perhaps to show Euler's strong religious convictions.[12]

What actually happened? Correspondence between Lexell and Johann Albrecht Euler indicates that no debate occurred at the academy.[13] After the conference the group went to Orlov's house. Franz Ulrich Theodor Hoch Aepinus, who did not attend the conferences, was present and engaged Diderot in an argument. Lexell wrote that Aepinus clearly won.

In November 1773 Katharina, Euler's wife of almost forty years, died at the age of sixty-six. On 17 November, Leonhard Day, she had said that she did not feel well and had gone to bed, where she remained for four days. On the morning of 22 November she called her family together to bid them farewell and gave well-intentioned cautions about the future; she passed away at 11:30 a.m. The marriage had been happy, and the loss was a severe shock to Euler. Almost a hundred mourners and thirty carriages were at the funeral three days later; Johann Albrecht Euler recorded the cost of the funeral as more than five hundred rubles.[14] The loss greatly complicated Euler's domestic life, for Katharina had managed the entire household while he did little or nothing with it.

The Imperial Academy: Projects and Library

Recoiling from the restrictions on the freedom that Lomonosov and later he himself had attempted to win for the academy, Euler wanted in 1774 to withdraw from the academic commission on which the director had the principal authority. On 1 February Euler dictated to Johann Albrecht a letter to Orlov requesting that he and his son be released from collaborating with the commission, citing his poor vision and failing hearing; the next day he asked to be relieved of the activities of the geography department, again citing his poor vision. On 3 February he announced that neither he nor Johann Albrecht would attend further meetings of the academic commission, but both would remain active at the academy. In a September letter to Orlov concerning Fuss and Golovin, Euler said that the best way he could now serve was through the education of students.

Yet on only a few occasions, and these for reasons of poor health, did Euler miss academy conferences in early 1774. In January Krafft gave rules to be followed in the construction of good barometers and thermometers. Peter Simon Pallas had sent several reports from Siberia, along with plants for the botanical garden, and Johann Georg Gmelin had sent materials on White Russia and Lepechkin on the coast of the Baltic Sea. Books for the academy library came gratis from the Royal Swedish Academy of Sciences, including the volume of Charles de Geer's *Mémoires pour servir à l'histoire des insectes* for 1773, as well as from the Göttingen Academy of Sciences. Volumes of Diderot and d'Alembert's *Encyclopédie* enriched the library. In February the academy studied selected Chinese writings; Euler had his paper read on the principles of harmony in music; Lexell compared Euler's lunar tables with those of Johann Tobias Mayer; and Johann Albrecht read a report on meteorological observations for 1773. In March Lexell presented a paper by Euler on plane geometry and another by Krafft on square numbers. In April the conference received from Pallas a box of materials on his Siberian travels, among them a report on insects, and an account of his attempt to have the Dutch publishers Brouwer and Bragge print his findings. Orlov demanded that the academy comment on Pallas's papers; he did not want a foreign press to reveal Russian discoveries and secrets on possible trade routes or products. Euler and Kotel'nikov argued that the academy had the right to publish these materials on its own, while Lexell opposed publication without the consent of Pallas.

For the remainder of 1774 the academy was active in part with library acquisitions and its journal. On behalf of d'Alembert, it received in June from the Paris Academy of Sciences the second edition of his *Traité de dynamique* and five volumes of his *Opuscules mathématiques*, together with French writings covering music, history, literature, philosophy, the theory of fluid resistance, and destroying the Jesuits. Murr arranged an exchange of Euler's publications for source works by Christian Friedrich Ludolff, Cuthbert Tonstall, Willibrord Snell, Peter Apian, and Georg Peuerbach; Johannes Kepler on comets as well as the birth of Jesus; and a letter of Nicolas Fatio de Duillieur to Jacques Cassini. On 22 August the academy received from the American Philosophical Society volume 1 of its *Transactions* for 1769–71. For much of October, illness forced Euler to be absent. On 17 November Fuss submitted seven articles by Euler, including one on elasticity, for volume 19 of the *Novi commentarii* that Fuss had transcribed. As the *Protokolii* minutes note, three days later two Latin articles of Daniel Bernoulli on sums of trigonometric series were received, and Euler submitted for reading a paper on the sums of progressions of some sine and cosine functions.

In 1774 Euler had additional papers read and published at the academy. Among them was his "Nova series infinita maxime convergens perimetrum ellipsis exprimens" (A new infinite series that expresses the perimeter of an ellipse and converges very rapidly, E448). For over forty years he had worked on elliptic integrals, and this article offered his most advanced method for finding the arc length of an ellipse.[15] His "Demonstrationes circa residua ex divisione potestatum per numerous primos resultantia" (Demonstrations concerning the remainders resulting from the division of powers by prime numbers, E449), read at the academy in 1772, was published in 1774 in volume 18 of the *Novi commentarii*. It is significant partly for its terminology: Euler had used the word *imaginary* in relation to numbers with two different meanings, but not in the sense of an imaginary number; in the eighteenth century, the adjective *imaginary* could mean what is now called a complex number or, more vaguely, an element of something that does not exist. To the end of the eighteenth century, the status of negative and complex numbers was unclear and caused a scandal in mathematics, but Euler's discoveries were to be decisive in the acceptance of complex numbers as a legitimate entity in analysis.

Since 1770 Euler had been studying developable surfaces in geometry. The title of his article "De solidis quorum superficiem in planum explicare licet" (On solids whose [entire] surfaces can be flattened to a plane, E419), published in 1772, stated the goal of his search. There he proved that the line element of a surface was the same as that of a plane. Euler scheduled a reading of "De curvis triangularibus" (On triangular curves, E513), which examined orbiforms, curves of constant breadth, a property shared by the circle, and triangular curves—that is, closed curves having three cusps;[16] the article drew upon analytical and geometrical principles, and resolved a fundamental question about the property of surfaces developable into a plane, demonstrating that the evolvents of the triangular curves—curved lines formed as they evolve—were also of constant breadth. Euler subsequently encouraged research on developable surfaces. These articles were outstanding mathematical achievements, a formative contribution in the founding of differential geometry, and clear examples of Euler's expanding the horizons of mathematics. Other writings dealt with perfecting reading glasses and with the oscillation of a pendulum of a given weight.

Among the topics of Euler's papers presented in 1774 and published later were the motion of comets, vibrating musical chords, the values of a family of given integral formulas, a table of primes up to one million and beyond, Isaac Newton's expansion of powers of a binomial case, and a proof of Newton's binomial theorem.

The Russian Navy, Turgot's Request, and a Successor

After the death of Peter the Great, the Russian Navy had declined, and to revive the fleet, Catherine II turned partly to foreign officers and technicians. She recruited a contingent from England, including the admiral Charles Knowles, to advise on building warships and dockyards. During Russia's war with Turkey from 1770 to 1774, Knowles was in or near Saint Petersburg, engaged mainly in administrative activities, and Euler met often with him. In 1775, a year after returning to England, Knowles published his translation of the treatise *Extrait du mechanism des mouvements des corps flottans* (Extract on the mechanism of the motions of floating bodies), originally published in 1735 in Paris by Cesar Marie de la Croix, the minister in charge of the French fleet. Two articles by Euler on the *Extrait* are from the time that he was writing his *Scientia navalis*. Knowles observed that during his stay in Russia he himself had tested de la Croix's principles on the motion and the stability of battleships and frigates built there and found them to be correct.

In August 1774, a day before he became the comptroller general of France, Baron Anne-Robert Jacques Turgot wrote to Louis XVI, with the backing of d'Alembert, to request that the monarch order the publication of two writings by "the famous Leonard Euler, one of the greatest mathematicians of Europe"[17] that would be useful as textbooks to be introduced into all French naval and artillery schools. Before this Turgot, one of a few statesmen *philosophes*, had corresponded about Euler with Condorcet, who claimed to have studied the work of Euler for fifteen years. One of the proposed books was Euler's *Théorie complete de la construction et de la manoeuvre des vaisseaux*, which had become available the preceding year. Asking for an improved second edition, Turgot invited Euler to make corrections to the French language. The other work that Turgot wanted was a translation into French of Euler's German translation *Neue Grundsätze der Artillerie* of Benjamin Robins's *The New Principles of Gunnery* with the extensive commentary by Euler. Turgot recognized that Euler had ownership of these and proposed a remuneration. The printing would cost about five thousand francs; Euler was to be sent one thousand rubles "as a token of the esteem [Louis XVI] has for your works which you so justly deserve." These two amounts were to be "paid from the secret accounts of the navy."[18] It would take roughly a year to gain the final royal approval.

Pleased to have so worthy a successor in mathematics, in March 1775 Euler passed to Lagrange the mantle of foremost mathematician in Europe, as Johann I (Jean I) Bernoulli had done in his letter of 1745 that addressed Euler as "incomparable" and "the prince of mathematics." "Il est bien glorieux pour moi," Euler wrote, "d'avoir pour successeur à Berlin le plus sublime géomètre de ce siècle. . . . J'ai parcouru avec la plus grande avidité les excellens mémoires dont vous avés enrichi les derniers volumes de Berlin et de Turin. . . . (It is most flattering to me to have as my successor in Berlin the outstanding geometer of this century. . . . I have gone most avidly through the excellent works with which you have enriched the latest volumes in Berlin and Turin.)[19]

Euler wrote that response just after receiving Lagrange's *Recherches d'arithmétique* of 1775; Lagrange had wondered whether it would meet with Euler's approval, and Euler was more enthusiastic than anticipated. But at no point did Euler's letter in March specifically mention the *Recherches*; he did not comment on Lagrange's brilliant discovery that "infinitely many forms" in number theory with the discriminant δ—a numerical invariate—can be reduced to "a small number" for each value of δ.

Euler's research had two outstanding characteristics that were exhibited in his response to Lagrange. First, he was receptive to thought from whatever source and extremely quick in mastering it; he would employ the new ideas as points of departure for a series of research efforts and swiftly refine, introduce, and develop better equations. Second, he would never abandon a topic or finding that had roused his curiosity; he liked to return relentlessly to conjectures after methods that he or others would develop enabled him to deduce them better, as well as to reconsider theorems. Even after finding a proof, Euler pursued another that was simpler or more direct.[20] Upon considering Lagrangian reduction, he wrote an article titled "De insigni promotione scientiae numerorum" (On an important progress of the science of numbers, E598) that began by citing, and then proceeded to refine, Lagrange's *Recherches*, which clearly demonstrated results from Euler's past conjectures.[21] Euler's paper was read to the Petersburg Academy in October 1775; in it Euler stated, "All the theorems that I formulated long ago in vol. XIV in the old [Petersburg] *Commentarii* [for 1773] have acquired a much higher degree of certainty . . . and there seems no doubt that whatever in them is still to be desired will soon receive a perfect proof."[22] He, most of all—but also likely his circle of assistants—took great delight in obtaining proofs beyond those of Lagrange. The article was published only posthumously, in 1785; still beyond the range of Lagrange's new methods was the law of quadratic reciprocity, which Euler had formulated in works on quadratic residues dating from 1772. To the present day, scholars disagree on whether Euler actually reached that law.[23] In "De insigni promotione scientiae numerorum" he confidently declared that he expected the proof of it and of other conjectures in number theory to be "quam . . . mox expectare licebit" (likely to come soon).[24] It would take another twenty-five years fully to prove that fundamental law, and it required Gauss to do so.

At the Academy: Technical Matters and a New Director

In September 1775 at an academy conference, a report by Euler was read on a technical matter, Ivan Petrovic Kulibin's third proposal to build a bridge across the Greater Neva River. As Fuss wrote to Daniel Bernoulli in 1777, "Over the last little while so many projects for building bridges over the Neva have been submitted here that this undertaking has almost become a joke."[25] Most builders assumed that a scale model able to support an adequate proportion of weight would be of sufficient strength when at the full size needed for the bridge, but such judgments left out factors such as the actual length required (1,057 English feet) and the bridge's potential bending. Although ignorant of mathematics, Kulibin was able to derive his rule in a third attempt with a model of a chain of uniform elements successively decreasing in size toward the center, comprising a bridge that met the load-bearing capacity required. Euler's report stated that in 1772 by experimental means Kulibin's rule for load-bearing capacity was confirmed. But in the modeling of bridges, Euler wanted more. In September 1775 he set up "a precise investigation of the question of inferring from the strength of a model to that of an actual bridge of whatever size. . . . "[26] Working with his assistants for about a year, Euler developed a precise theory for modeling bridges of any size, reaching conclusions similar to those that Kulibin had found independently and experimentally. Euler's article, published the next year under the title "Regula facilis pro diiudicanda firmitate pontis aliusue corporis similis ex cognita firmitate moduli" (A simple rule for determining the strength of a bridge or other similar body from the strength of a model, E480), gave a brief and accessible account of some of his calculations.[27] As regards mechanics, it examined three-dimensional regions in general. To reach a wider readership, it was published in the independent journal *Ezhemesiachnye sochineniya, k pol'ze I uveseleniyu sluzhaschie* (Monthly writings serving purpose and enjoyment) in an English translation. This journal, edited by Gerhard Müller, was influenced by the European Enlightenment.

In the transformation of the exact sciences and mathematics, Euler continued to make significant contributions to mathematical physics. In "Nova methodus motum corporum rigidorum determinandi" (A new method of determining the motion of rigid bodies, E479), read at the academy in October 1775, he gave a new derivation of the motion of a rigid body.[28] The paper was the first to give the integral form in orthogonal Cartesian coordinates of the principles and the moment of momentum. These laws, expressed by six nonlinear partial differential equations, represented "the fundamental, general, and independent laws of mechanics for solving all kinds of motions of all kinds of bodies,"[29] and they have been called Euler's two laws of mechanics.[30] It is not clear, however, whether Euler considered these laws as fundamental and independent axioms of mechanics. He had probably arrived at the law of rotational momentum much earlier, starting from Newton's second law of motion applied to mass points; but this was not stated explicitly in the article. Whatever the case, the work was an important advance in the development of mechanics.[31]

Also in October 1775, Turgot informed Euler that Louis XVI had agreed to fund the printing of the naval and artillery books. The support of the king—along with that of a great court minister who was respected across Europe—humbled Euler. These affirmed his prestige and confirmed that the esteem for him extended to the highest social and political circles on the Continent. Turgot invited Euler to remove any errors in the originals, and the improved second edition of the *Théorie complete de la construction et de la manoeuvre des vaisseaux* that added supplements by Lexell, appearing in 1776. The study received acclaim rarely achieved in the applied sciences; its Parisian editors praised "the great good that Euler's many discoveries had done for the French and other enlightened nations."[32] In 1776 Henry Watson published an English and Simone Stratico an Italian translation. The next year a Dutch rendering (of which the English title would be *Extract of*. . .) appeared in The Hague, and in 1780 a second Italian translation was brought out by Gaetano Carcani.

Through the end of 1775 Euler submitted articles for the academy's *Novi commentarii.* On 4 September he was ill and absent from academy conferences, but Fuss delivered six of his articles, which included studies of geographic projections, continued fractions, and elastic curves; these would be kept in the archives.[33] Three days later Lexell read one of the papers, which was on the subject of pentagonal numbers, and the academy planned the mathematics section of the twentieth volume of the *Novi commentarii*; it would include two articles by Daniel Bernoulli, four by Euler, and one by Lexell. On 28 September the academy reviewed the fourteen journals it had received. They included the *Mercure historique et politique de France*, *Journal encyclopédique*, *Journal des Sçavans*, *Gazette litteraire de Berlin*, *Gazette de Leyde*, and the *Hamburger Correspondent* newspaper. On 9 October three more articles by Euler were read, and the academy discussed obtaining more books for the academic library that might be of great use to its research. Three days later Sergei Gerassimowitsch Domaschnev, the new director of the Petersburg Academy since Orlov had left the position in December 1774, wrote that Catherine wanted to be instructed in the content of Kepler's manuscripts; Krafft and Lexell were to respond to him. On 19 October Lexell presented six more of Euler's articles on mathematics and three on geographical projections. Lexell read papers by Euler on 23 and 26 October, when the topic was rigid bodies, and Fuss presented the ninety-eighth article that he had transcribed from Euler. The academy considered publishing a work of Pallas on the Kalmuks and a catalog of plants. Euler turned briefly to agriculture, publishing a paper written eight years earlier, "Nachricht von einem neuen Mittel zur Vermehrung des Getreides" (Report on a new means of increasing the production of grains, E341A), but he did not continue with this subject. In November Lexell, Krafft, and Fuss read more papers by Euler on mathematics and physics.

Neither Domaschnev nor Orlov before him sought to enhance the Petersburg Academy; both had been appointed for being major supporters of Catherine's coup. Domaschnev, a talented graduate of Moscow University, seemed at first a better choice for director, but he was actually worse; he was arrogant and arbitrary toward members and conspicuously violated their institutional rights. Despite protests from Euler, he left vacancies unfilled, pilfered funds from the academy's treasury, and was known to take books ordered for the library for his personal collection. Domaschnev rarely came to the academy conferences; his first visit was on 11 December 1775, when he spoke to the whole assembly. Calling himself an admirer of the enterprise, he praised the contributions made to the national well-being of Russia from the application of reason and from progress in technological inventions and the sciences.[34] The academy, he declared, was fortunate to have the support of its protector, the genius Catherine. Academy secretary Johann Albrecht Euler, now known by the French variant of his name, Jean Albert, replied that the academy was grateful for these assurances of Domaschnev's support for its research. Three days later Domaschnev checked the academic protocols cited for making Golovin and Fuss adjuncts. To meet high scholarly standards he thought it important to have the publication of articles in Latin.

A Second Marriage and Rapprochement with Frederick II

After his wife Katharina died in 1773, Euler was determined not to follow the custom of the time that had elderly parents living in the care of their children; and neither did he wish to live without marital companionship. Although he realized that his sons would oppose a remarriage, the aging scholar quietly began without discussion to search for a modest middle-aged woman to be his wife. Since after losing most of his eyesight he rarely traveled outside his home or participated in social activities, he searched among the small group of women who visited his house.

About 1771 a Frau Müller, a client of Johann Albrecht in financial management, had introduced Frau Metzen into the group. In the following years, Frau Metzen often took meals with the Euler family. Several times widowed, and formerly a housekeeper for the empress's lieutenant general Fedor Bauer, a hero in the war against the Tatars and Ottoman Turks in 1770–71, Frau Metzen was of minor social status. Euler proposed to her in late 1775, and she initially accepted.

Returning home one day from church services during the Christmas holidays of 1775, Euler suddenly announced on the street to his grown sons his decision to marry Frau Metzen; his family responded angrily and tried to have the weeping fiancée renounce the proposal on the spot. The sons raised an array of arguments against the marriage; Johann Albrecht, who held Frau Metzen to be of "low descent," argued that Saint Petersburg's noble society and the empress would not understand his father's choice.[35] Frau Metzen was thus forced to reject the proposal. More than social status appears to have motivated the sons: they knew the event would affect their impending inheritance. When Euler remained committed to the marriage, the family split, but then he suffered another serious fever and was under the care of his son Karl for two weeks. At the end of that period the sons had effectively blocked the marriage and familial harmony was restored. But Euler was resolute in working on his own to select a mate and arrange a future wedding ceremony.

On 15 January 1776 Golovin presented to the academy his translation into Russian of Euler's *Théorie complete de la construction et de la manoeuvre des vaisseaux*, which added many comments on calculations that Euler had made and explained the results. Initially Euler had prepared this book for instructional use in the Russian naval school, but after reading the first half of the translation, Rumovskij urged publication of the total work and the translation was sent to the academic commission. Benjamin Wilson of the Royal Society of London sent a letter on his experiences with light from phosphorus, and Euler explained Wilson's results. On 25 January the academy's conference rejected a proposal to reduce the size of volume 20 of the *Novi commentarii*. On 1 February Fuss read an article on divergent series by Euler, who presented a paper in French on establishing salaries that based its calculations on probabilities. Johann Albrecht announced that he had gotten a letter on meteorology from their former colleague, the Swiss mathematician Nicolas Béguilen, in Berlin. Domaschnev proposed new due dates for collecting articles for the next *Novi commentarii*. They would be received during the six months beginning in September 1776, and the papers were to be distributed to the appropriate department of the academy for review; the conference reserved the right to include important discoveries reported after the deadline. The title of the volume would be changed to the *Acta* (Transactions) of the Imperial Russian Imperial Academy. The conference accepted these terms unanimously. Domaschnev further proposed that articles be accepted in languages other than Latin, and especially in French and German; this prompted debate, and a decision was put off. On 22 February the academy was assigned an important technical project: Catherine ordered the academicians to examine a proposal by José (Osip) de Ribas, a captain in the Imperial Corps of Cadets, to build a bridge across the Greater Neva that differed from Kulibin's plan. Domaschnev selected Euler, his son Johann Albrecht, Kotel'nikov, and Rumovskij to oversee the project; at the same time he added the adjuncts Fuss and Golovin to this committee to review the description and calculations for another bridge by Kulibin.

On 4 March, with Domaschnev absent, the academicians prepared a report for him stating that beyond its scale model the single-arch bridge designed by Ribas had failed to hold up in traversing the distance for crossing the Greater Neva. They referred to the five-book series *La science des ingenieurs dans la conduit des travaux de fortification et d'architecture civile* (The science of engineers in the conduct of works of fortification and civil architecture) of 1729 by Bernard Forest de Belidor.[36] The academy awaited a new version of Kulibin's proposal. On 7 March Fuss read an article by Euler on an "illustrious" theorem by Lagrange; it was apparently the theorem involving δ from volume 24 of the *Mémoires* of the Royal Prussian Academy of Sciences and Belles-Lettres in Berlin. On 11 March after Fuss read a paper by Euler on integrating a given formula, Lexell described a catalog of books that had belonged to Nils Rosén von Rosenstein of Uppsala, deceased since 1773, from which to consider acquisitions for the library. Daniel Bernoulli sent a letter of gratitude for publications sent to Basel in January. On 14 March 1776 Domaschnev returned for a meeting, and four days later Fuss read a paper by Euler on integrating an irrational formula, while Euler presented five analytical papers for volume 20 of the *Novi commentarii*; early in April the academicians arranged the astronomical section of this volume.

On 11 April Pallas presented the beginning of his first volume on the political, physical, and moral history of the Kalmuks, promising to supply the rest together with letters from related travels of academicians. On 15 April Euler submitted two more papers for publication, and three days later Golovin provided a German translation of an article by Kulibin on the bridge. The next week Domaschnev returned, a series of experiments on electricity were conducted, and Benjamin Wilson of the Royal Society of London was proposed as a foreign member of the Petersburg Academy. The conference on 29 April included Krafft's research on electrical conduction.

An important matter outside the academy now required Euler's attention. For a decade after their strained separation, Frederick II and Euler had not reestablished a fairly good relationship. Catherine II was the mutual benefactor in ultimately encouraging reconciliation; working to gain the favor of the Russian empress, Frederick had nominated her in 1766 to be an honorary member of the Berlin Academy of Sciences. After she refused that offer, he nominated her in September 1766 to be a foreign member, and she accepted. Now, a decade later, she wished to nominate Frederick for a position *honoris causa* in the Petersburg Academy; this was a step toward strengthening the bonds between Prussia and imperial Russia as new powers in Europe. That encouraged Euler to renew his contacts with Frederick; he was also pleased to discover the king's shift away from the radical French Enlightenment.

The Berlin gazettes published some calculations related to the monarch's plans to establish widows' pensions, and this provided a reason for Euler to contact the king directly. This was not a new topic for Euler; in April 1769 he had provided the Petersburg Imperial Academy with the beginnings of a new type of computation for establishing a fund for annuities to widows in Russia. This masterwork on insurance mathematics was presented to the academy in February and May 1776 and published that same year. Its full title was *Eclarcissemens sur les établissemens publics en faveur tant des veuves que des morts, avec la déscription d'une nouvelle espèce Tontine aussi favorable au public qu'utile à l'etat calculés sous la direction de Monsieur Léonard Euler* (Clarifications on the public establishment of funds for the benefit of widows and deceased persons, with the description of a new kind of tontine favorable to the public and useful to the state, evaluated under the guidance of Mr. Leonhard Euler, E473).[37] The work was clear and would be intelligible even to readers who were not insurance specialists, and its third chapter was inventive. Euler's tontine was a special annuity insurance provided through the interest on a government loan, but it had drawbacks: it depended upon the number of participants allowed, required equal deposits at the beginning, and varied with the interest projected. His insurance plan rested "on the [correct] principles of probability," Euler declared, "so that no one has the right to complain."[38] In a letter to Frederick dated early April 1776 he cautioned the monarch that the Berlin newspaper figures were generally too small to provide the required payments and that circumstances could double or triple the costs, which might ruin a company. After declaring that the Berlin computations rested on false grounds and that instances involving second marriages were not covered, he promised to send two tables with corrections developed for the same purpose in Saint Petersburg. On 16 April Frederick sent a letter thanking Euler for his intent to send the detailed information but noting that the enclosures were missing. The king added that in establishing widows' pensions he had acted only from "the purity of his motives." Frederick was taken with the interest that Euler showed in "his old compatriots," and Euler was impressed with the monarch's warmth. In September he sent Frederick his recently printed brochure on widows' pensions, the *Eclarcissemens*, and J. A. Kritter made a German translation. Working with constant and with variable installments, Euler resolved two problems that actuaries had treated separately. In October Frederick sent another note of thanks.

On 18 April 1776 Golovin presented at the academy conference a German translation of an article by Kulibin about the project of constructing a new large wooden, single-span model of the bridge over the Greater Neva. Kulibin was expected imminently to submit a model, the third version, for a static load test. Euler had his response to Kulibin's proposal entered into the conference minutes. "Although I very much doubt that Kulibin's model could support 12 times greater than its own weight," he observed, "I do not know whether it is appropriate to express such an opinion without verifying it."[39] Kulibin's model was not completed until almost the end of the year. In December the academy formed an evaluation committee that included Lexell, Krafft, and the adjunct Petr Borisovič Inochodzev; perhaps Euler chose each. The committee tested the model, which was one-tenth the size of the bridge, for a maximum load of 3,500 *poods*.[40] From a final series of trials over two months, the committee in January 1777 found the arguments for the model to be persuasive and decided that the project was worthy of approval. Euler had actually worked for a year on the enterprise via a priori arguments. In correspondence with Daniel Bernoulli, Fuss also praised the work of Kulibin. But the immediate construction would take four to five months, and there was not enough time to obtain sufficient choice wood for the scaffolding before winter. A wooden bridge, moreover, would not last long, and constructing a stone bridge entailed problems with cutters and others preparing parts. The committee report in March therefore "rejected the design as impractical though theoretically sound."[41]

In July 1776 Euler had apprised his children without consulting them that he intended to marry his departed wife Katharina's half sister Salome Abigail Gsell, who was fifty-three;[42] he was sixty-nine. As the younger daughter of Georg Gsell and his third wife Marie née Graff, Salome Abigail was the granddaughter of Maria-Sybilla Merian; born and educated in Russia, she spoke only broken German. She was a small, humble woman devoted to her home, and apparently she already lived at the Euler house. The sons could not dismiss their mother's sister as being of lower class, and her age meant that there would be no other heirs and competitors for their inheritance. The wedding was formally announced in church on 24 July. Only the day before had Euler revealed to his sons the terms of his will, which again provoked sharp emotions. Johann Albrecht was to receive one-twelfth of the house and a portion of the remaining silverware and library. All else, including furniture, china, linens, drapery, carriages, clocks, and mirrors, would go to Euler's widow. Salaries were rearranged: while Euler gave each child two hundred rubles annually and free housing, they were to pay his widow two hundred rubles rent yearly. Johann Albrecht believed that as the co-owner of the house and assistant to his father he deserved more. The other sons, who had grown up in the meantime, also found the terms unjust.

On Thursday, 28 July 1776, around five o'clock, some thirty guests began to arrive at the Euler house. The grandchildren were dressed in their best clothes for the celebration, and Fuss was the master of ceremonies. The pastor of the Reform Church arrived after six o'clock. Euler, his bride, the pastor, and an artillery captain named Haecks proceeded to the study to sign Euler's last will and testament. Haecks was the witness. Johann Albrecht received the document and silverware inventory for safekeeping. Euler gave his bride a copy of these that he had dictated. After the signing, the document was to be sent to the empress. The thousand-ruble pension for Euler's first wife would be transferred to his second. According to Johann Albrecht, tension over the will persisted during the wedding reception. It was to lessen over the years, but probably never completely went away. The evening concluded with a delightful celebratory supper. Salome Abigail received generally warm respect from the family, except for the pouting over Euler's will. She was to manage the entire household and was devoted to caring for her husband until his death.

According to Fuss, Euler's second marriage created a minor sensation. "Fools laugh at him," he wrote. People whom Fuss considered sensible regretted "that a wise man in his final years would do something foolish."[43] But others recognized the importance of the nearly blind Euler's not having to depend upon his children.

In Saint Petersburg, the Euler family attended a Reform—essentially Calvinist—Church, which was probably near the academy and home. Since their return to Russia Euler and his son Johann Albrecht had tried unsuccessfully to end a rift among the French, Swiss, and German members in the church, and later Fuss joined in these efforts.

During the last half of 1776, Euler attended almost every academy conference. On 12 August the institution celebrated the empress's birthday. Moscow University sent a Latin dissertation and the Paris Academy its *Histoire* for 1772. The academy received a volume on astronomy from 1776 by Jean III Bernoulli and praise from Formey in Berlin. The secretary gave a brief life of Daniel Bernoulli and requested a portrait; the academy began a journal of medicine. On 19 August Fuss read papers of Euler on the orbits of comets, hypergeometric series, and integration; Golovin followed with another six. In September the academy considered problems in chemistry and meteorology, and Domaschnev urged the preparation of a complete geography of the Russian empire. The members discussed the academy's annual prize, and Pallas agreed to prepare an extract of the winning article. On 30 September the secretary reported on publications sent to the academy, and the Royal Society of London provided volume 66 of its *Philosophical Transactions* for 1776 and the Greenwich Observatory's *Observations, 1765–1776*. J. H. de Magellan sent his *Description of Octants and Sextants*, published in 1775. In October Magellan proposed that the Royal Academy correspond with the large London libraries and requested a copy of Euler's *Lettres à une Princesse d'Allemagne*. Of two articles by Euler presented on 16 December, one dealt with Saturn's rings and the other with the strength of columns. A week later the academy voted on foreign members; thirty had been nominated and twenty approved, including Johann (Jean) III Bernoulli; Georges-Louis Leclerc, Comte de Buffon; the Marquis de Condorcet; Johann Gleditsch; Albrecht von Haller; Lagrange; Andreas Sigismund Marggraf; Nevil Maskelyne; and August Ludwig Schloezer. From religion, the military, and the nobility the director added a dozen names for honorary members. The prize committee commended the investigation of Jan Baptist van Helmont's *Chymia praepart intellectum* (Chemical philosophy arranges understanding [in medicine]). On 16 January 1777 Euler attended his last meeting at the academy, except for that of the inauguration of Princess Dashkova.

End of Correspondence and Exit from the Academy

Frederick II's efforts to establish guidelines for salaries in 1776 prepared the way for him to be named a member *honoris causa* of the Petersburg Imperial Academy of Sciences on 9 January 1777. As its dean and doyen, Euler was charged to write to Frederick; on 17 January he sent his gratitude for Frederick's accepting the membership and declared his profound respect for the monarch. Membership for anyone outside the sciences was the first of its kind at the Petersburg Academy. Euler spoke of it as a "brilliant prerogative" to redouble its efforts for the advancement of science.[44] He continued to be a contact between the Petersburg and Berlin Academies. On 1 February Frederick wrote thanking him and congratulating the academy on having a dean of Euler's talent and merit, noting that under the management and direction of Euler it had achieved distinction reached by few academies. On 3 February the letter arrived from Potsdam for the absent Euler.

After 1776 Euler's correspondence substantially diminished and soon ceased. In March 1775 his letters with Lagrange had already ended. His last letter treated the topics of elliptic integrals, paradoxes in integration, and the lost proofs of Pierre de Fermat.[45] In the sending of these letters, Formey in Berlin had acted along with Johann Albrecht in Saint Petersburg as an intermediary. In 1777 Euler dictated only a letter to Frederick and another to Stanislaw August Poniatowski, the king of Poland, and in May of the next year the last letter with a short three pages giving his observations on Nicolas de Béguelin's article on prime numbers in the Berlin Academy *Mémoires* for 1775; the letter would later be published as "Extrait d'une lettre de M. Euler à M. Béguelin, en Mai 1778" (E498). Béguelin, he began, had based his research on the beautiful property that all numbers "which can only be represented as $x^2 + y^2$ in one way are either prime or twice a prime,"[46] when the numbers x and y are relatively prime. For similar formulas of the form $nx^2 + y^2$, Euler presented a rule for finding what he called a *valeur convenable* and later a *numerus ideonus*, a suitable number n that gives primes. His rule covers all numbers $nx^2 + y^2$ that are less than $4n$. This allowed him to find all suitable n "with ease." He listed sixty-five of these numbers from 1, 2, and 3, . . . to 1320, 1365, . . . and to 1848. These are not random but follow a progression. Among the primes that Euler found are 1,702,009 and 11,866,009. He pointed out an error in his table of primes in volume 19 of the *Novi commentarii*. The number 1,000,009 had to be removed, for it equaled 293 × 3,413. Euler had gotten that result in 1749.

During his final years, Euler separated from the academy, partly because of his worsening hearing but also likely out of disgust with the administration under Domaschnev. Euler had been withdrawing from business operations since 1774, and as the minutes of the *Protokolii* show,[47] on 16 January 1777 he stopped attending sessions altogether. But he did not lose contact: he continued to work with his circle and submitted a stream of articles, many read at conferences by Fuss, Golovin, or Krafft on topics ranging across the mathematical sciences, including his research in astronomy, fluids, mechanics, optics, and pure mathematics. Among the subjects in pure mathematics were continued fractions, the integration of irrational formulas, and paradoxes in the calculus of variations. For the 1777 volume of the *Acta* two articles examined the rings of Saturn, the first addressing their apparent figure with reference to tables by Cassini and Halley. Articles for that year also dealt with Benjamin Wilson's optics, a more highly perfected lunar theory, and—from his study of vibrating strings and astronomy—a quick way of determining coefficients in trigonometric series, what today are usually referred to as Fourier series; this last article was not published until 1798.[48]

Mapmaking and Prime Numbers

In 1775 Euler had returned to mapmaking, a project he had already pursued for a half century. During the eighteenth century, terrestrial cartography was an important subject. Perhaps Euler was influenced by Johann Heinrich Lambert's *Anmerkungen und Zusätze zur Entwerfung der Land- und Himmelscharten* (Notes and comments on the composition of terrestrial and celestial maps, 1772). Lambert's book provided a major foundation for the applied science of modern mathematical cartography. Stereographic projections or the planisphere had been known since antiquity, and because in them all meridians and parallels were circles they were easy to use. In a series of three papers for the 1777 volume of the *Acta*, Euler contributed to the foundation of the field.[49] The first paper studied analytically the mapping of the sphere on the plane, the second proved geometric results for great circles and circles of latitude under both Mercator and stereographic projection, and the third described the equidistant conic sections that Joseph-Nicholas Delisle had rediscovered and along with Euler employed in the Russian atlas in 1745. It was helpful that these conic sections preserved scale along constant latitude paths. Lagrange's article "Sur la construction des cartes géographiques" of 1778 notes that in constructing maps most modern cartographers applied the stereographic function. They needed only three points to trace the entire curve of meridians or parallels. In antiquity these projections had been only for star charts and astrolabes.

From 1777 to 1783 Euler dictated to his assistants another twenty-four articles on number theory that would be published posthumously; these are separate from the nine papers he wrote on the subject after 1770 that appeared in print during his life. The posthumously published articles for 1778 include seven on prime numbers;[50] in these Euler generalized Fermat's criterion that a nonsquare integer, which can be written as $4n + 1$, is prime only when it can be represented as the sum of mutually prime squares. As noted earlier, he found "suitable" or "idoneal" numbers with which to determine primacy. These are the odd integers N for which any integer that can be written in only one way as the quadratic form $x^2 + Ny^2$, with x and y rational, is a prime. Since it is easy to test whether a number is even or a perfect power, this gives a fast way to test whether a number is prime. After excluding trivial cases, Euler gave sixty-five such integers from 1 to 1,848 suitable for testing what large numbers are prime. Applying his new method he discovered a few previously unknown primes larger than 10^7. Using $N = 1{,}848$, for example, he established the primality of $18{,}518{,}809 = 197^2 + 18{,}480{,}000$.[51] Throughout his research on primes Euler sought to find counterexamples. When not occurring only in obvious cases, these exceptions were used in proving a rule.

In June 1778 Fuss related Euler's results to Béguelin. Earlier he had reported them to his teacher Daniel Bernoulli, who in March answered, with little regard for number theory,

> What you tell me, on your behalf and that of Mr. Euler, is infinitely more exalted, no doubt; I mean Mr. Euler's beautiful theorem on prime numbers and his new method of deciding whether any given number however large, is prime or not. What you have taken the trouble to tell me about this question seems to me very ingenious and worthy of our great master. But, pray, is it not doing almost too much honor to prime numbers to spread such riches over them, and does one owe no deference at all to the refined taste of our time? I hold in due esteem whatever comes from your pen and admire your high ability to overcome the thorniest difficulties; but my admiration is doubled when the topic leads to some useful piece of knowledge. That includes, in my view, your deep researches into the strength of beams.[52]

In 1778, Golovin's translation into Russian of Euler's *Théorie complete de la construction et de la manoeuvre des vaisseaux* was published. Not to be outdone by the largesse of Louis XVI, Catherine gave Euler two thousand rubles. Euler's article on iterating the exponential function, read the previous year, also appeared.[53]

A Notable Visit and Portrait

In the evening of a day at the end of July 1778, Johann III Bernoulli arrived in Saint Petersburg and immediately sought to find Euler's house. The grandson of Euler's teacher Johann I Bernoulli, a son of Johann II (Jean II), and a child prodigy, he had come in 1763 to the Berlin Academy as a mathematician and astronomer, and there he had been Euler's colleague. In January 1765 he had cooperated with Johann Albrecht Euler on meteorological observations. The visit to Saint Petersburg was joyful for the entire family and especially for the Euler patriarch, evoking many fond memories. On his trip from Berlin to Russia, young Bernoulli had journeyed through Brandenburg, Pomerania, and Königsberg, where he met Count Hermann Karl von Keyserling and Immanuel Kant, who was a good and lively man who reminded him of d'Alembert.[54] In Saint Petersburg he stayed at the Euler house. He went through Euler's eleventh notebook and watched him posing for his portrait by Joseph Frédéric Auguste Darbès.[55]

Figure 14.2. Johann III Bernoulli.

Today Johann III Bernoulli is known almost entirely through his travel reports. One of them notes that Euler's general health remained good because he pursued a moderate and regular style of life. Most of Euler's vision, Bernoulli remarked, "had been long lost and for some time entirely . . . he cannot recognize people by their faces, nor read black on white, nor write with pen on paper; yet with chalk he writes his mathematical calculations on a blackboard clearly and in rather normal size; these are immediately copied by his adjuncts, either Mr. Fuss or Golovin, usually the former, into a large book, and from these materials are later composed memoirs under his direction."[56] In December Bernoulli was elected a foreign member of the academy.

In 1778 Darbès completed his portrait of Euler. From 1773 to 1785 Darbès had resided in Russia and worked in Courland and Saint Petersburg; he would later become known for his portraits of illustrious men, including Friedrich Wilhelm II and the young Johann Wolfgang von Goethe. From the portrait of the elderly Euler, the engraver Samuel G. Kütner and others made copies; it is of great value, for it was painted from life—Euler had sat for it. It shows a keen and strong-willed intellectual wearing a light-brown dressing gown with a fur collar and a velvet beret in dark green tones; around his neck is a pouch. The forehead is high and the facial features strong, with perhaps a slight smile. Contemporaries thought the portrait one of the best of Euler, if not the best. According to Paul Heinrich von Fuss, his father Nicholas preferred this picture to all others of Euler and

Figure 14.3. Portrait of Euler by Joseph Frédéric Auguste Darbès, 1778.

believed it most resembled its modest, conscientious, and benevolent figure.[57] The identity of the "old man in a hat" that the picture represented was soon lost in Russia and not rediscovered until the signed portrait was given in 1928 to the Tret'akovskii Gallery in Moscow. When a catalog was being prepared for the gallery, the question arose as to who was the subject.[58]

Loss of vision and worsening hearing increasingly kept Euler at home, but Johann III Bernoulli reported that he "gladly receives visitors daily at any time." As Fuss remarked, Euler was known for his encyclopedic erudition; he had read the major and minor writers of ancient Rome, along with histories from antiquity to his time and of peoples across Asia, Europe, and the Mediterranean. He knew medical and herbal medicines and chemical science. Since early in his career his learning had been well rounded; visitors were astonished at the breadth of his knowledge, much of it seemingly unrelated to his long concentration on advancing mathematics and the natural sciences. Princess Dashkova, the future director of the academy, claimed to have visited him often, and once Grigorij Aleksanseovich Potemkin came. According to legend, Catherine II stopped her coach near to Euler's house, wondering whether to greet him or to enter. He did not like to travel but did have a dacha at Duderhof outside Saint Petersburg where he went to relax.

Magic Squares and Another Honor

Euler now turned to magic squares, a subject in recreational mathematics. Magic squares of order n arrange n^2 numbers in a square with sides n so that the sums of all rows, columns, and big diagonals are equal. For millennia, magic squares had intrigued mathematicians and especially puzzlers, number lovers, and amateur mathematicians. The artist Albrecht Dürer's master engraving of 1514 of the winged personification Melancholy, the product of an age that associated creative genius with melancholy, has the figure amid the tools of geometry and a magic square of order 4. In his knight's tour paper, Euler listed squares of order 8 in which the knight visits all cells once and only once. Two articles—"De quadratis magicis" (On magic squares, E795), read at the academy in October 1776,[59] and "Recherches sur une nouvelle espèce de quarrés magiques" (Investigations on a new type of magic square, E530), presented in March 1779 and published three years later—give the germs of ideas developed from a systematic study of these problems. The latter is the only article that Euler published in a Dutch journal; in 1775 he and Johann Albrecht had been named members of the Academy of Vlissingen in Holland. In the issue containing "Recherches sur une nouvelle espèce de quarrés magiques", the journal notes their membership was perhaps a gift of thanks for their work and the article.

Figure 14.4. Albrecht Dürer's engraving *Melencolia I*, 1514.

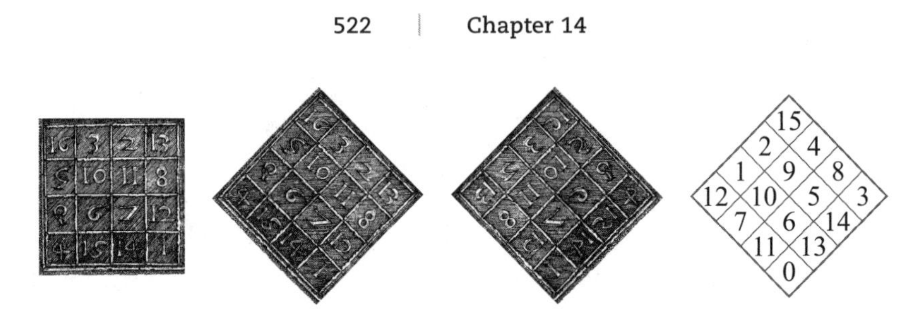

Figure 14.5. Euler's magic squares.

"Recherches sur une nouvelle espèce de quarrés magiques," (E530) a long paper on pandiagonal magic squares with hundreds of examples, generalizes the concept of Latin squares—$n \times n$ squares, having numbers 1 through n—each appearing exactly once in a line, a column, and both diagonals in a Greco-Latin square. An $N \times N$ Greco-Latin square has two sets of n elements. Traditionally Latin letters represent the first elements and Greek letters the second—hence the name. The goal is to place in each box or cell one Latin and one Greek letter such that no two pairs are the same—that is, that all $2N$ elements differ. As a formula to construct these magic squares, he represented numbers by $mx + n$ generally employing the Latin letters $a, b, c, d, \ldots$ to represent the value 0, and the Greek letters, $\alpha, \beta, \gamma, \delta$, to take the values $1, 2, 3, 4, \ldots, x$ that resulted in all numbers $1, 2, 3, \ldots, x^2$ with no repeats. In E795 for the case $x = 3$, Euler found it impossible to have two diagonals of the three letters but it was possible to make the sum of the diagonals equal to that of the lines. His method is shown below.

a	b	c
b	c	a
c	a	b

It is useful for deriving magic squares of order 4. Euler found sixteen variations of four letters that meet the required conditions and noted that there are twenty four variations of four letters with altogether 576 different figures. Many of these differ in structure. When $x = 4$, the method succeeds simply with Latin letters and an appropriate combination of them with Greek letters. An example is

aα	bβ	cγ	dδ	1	4	14	15
bγ	aδ	dα	cβ	13	16	2	3
cδ	dγ	aβ	bα	8	5	11	10
dβ	cα	bδ	aγ	12	9	7	6

Proceeding to the case $x = 5$, Euler symmetrically arranged the Latin letters in rapport with the Greek in the columns that will reach 14,400 permutations. For $a = 0$, $b = n = 5$, $c = 2n = 10$, $d = 3n = 15$, $e = 4n = 20$, $\alpha = 1$, $\beta = 2$, $\gamma = 3$, $\delta = 4$, and $\varepsilon = 5$, he arrived at: 8 20 2 21 14. This is known as the Greek-Latin quadrature.

In the eighteenth century these were popular problems. Those pursuing them could look to a solution for a card game following a quadrature schema that involved a donkey, a king, women, and jacks; each figure, each color, each row, and each column could appear only once. Euler had an analogous problem that involves arranging thirty-six officers of six different ranks taken from six different regiments.

Throughout "Recherches sur une nouvelle espèce de quarré magique" the chief question was to find what sizes of Greco-Latin squares could be constructed. Believing that a 6×6 magic square did not exist and from his efforts to construct other magic squares, Euler did not have a rigorous proof for order 6 squares but simply an experimental heuristic. Still he came to the general conclusion that it is impossible to construct any Greco-Latin square of sizes $n = 4k + 2$ and not until 1970 was this proven generally wrong.[60] Magic squares are possible for all sizes except 2 and 6.

On a secret mission to Saint Petersburg in 1780 over problems from the first partition of Poland that had occurred in 1772 among Prussia, Russia, and imperial Austria, his old friend the Prussian margrave Friedrich Heinrich visited Euler, who was now confined to bed, his health having deteriorated. The margrave spent several hours holding Euler's hand in conversation at his bedside. Euler knew that he could not pursue mathematics with the margrave; instead they spoke of laws, and of the histories of times and peoples.

Regarding his family, Euler was to receive sad news: his two adult daughters would die before him. In February 1780, his adult daughter Charlotte passed away on her husband's estate near Aachen and was buried in the small church in Hückelhoven; she had been the only one of Euler's adult children to leave Russia. In May 1781 Katharina Helene died in Wiborg along the Russian border with Finland. In 1777 Katharina had married Karl-Josef von Bell, a Prussian immigrant to Russia who had become a colonel and a member of the Russian nobility.

In 1780 at the age of seventy-three, Euler obtained what is probably his last major result in mechanics. (The notion that scientists make their discoveries before the age of thirty clearly does not apply to him.) He demonstrated that the moments of forces are represented by directed segments, and that they can be added by the well-known parallelogram law; in modern terms, Euler had discovered that moments of forces are vectors. Euler's breakthrough is associated with his research on the theory of elasticity and on rigid bodies, which substantially involved moments of forces, and with his general law of the moment of momentum.[61] His discovery appeared in two articles, "De momentis virium respectu axis cuiuscunque inveniendis" (On finding the moments of forces about any axis, E658), and the short "Methodus facilis omnium virium momenta respectu axis cuiuscunque determinandi" (A facile method for finding all moments of forces about any axis, E659). Both appeared posthumously in volume 7 of the *Nova Acta* of the Petersburg Imperial Academy of Sciences intended for 1789 but not printed until 1793. These papers also contained the first appearance of the now familiar formula for the distance between two straight lines in space. Euler's discovery encouraged mathematicians and physicists of the next generation to investigate the vectorial properties of angular velocity and moments of vector quantities.

Euler continued to have articles published, most of them in the *Acta* of the Petersburg Imperial Academy; for 1780 there were thirteen. These include "Considerationes circa Brachystochronas" (E501), "Sur l'effet de la refraction dans les observations terrestres" (On the effects of refraction during terrestrial observations, E502), "Determinatio onerum quae columnae gestare valent" (Determining the load that columns can bear, E508—that is, finding the bending moment that made use of the so-called Young modulus of linear elasticity), and "Investigatio perturbationum quae in motu terrae ab actione veneris producuntur" (Examining perturbations in the motion of Earth affected by the action of Venus, E512). In 1780 Euler wrote a series of articles on secant and tangent methods that went unpublished until 1830. Among the articles printed in 1781 were "De curvis triangularibus" (E513), on curves with three points of inflection, and "Extraits de différentes Lettres de M. Euler à M. le Marquis de Condorcet" (Extracts of different letters from Mr. Euler to the Marquis de Condorcet, E521) on analytical theorems about a few integrals resolved by clever manipulations and the sum of a binomial series converging to $4/\pi$. Among seven more articles, "Trigonometria sphaerica universa, ex primis principiis breviter et dilucide derivata," an introduction to the first principles of spherical trigonometry, and "Investigatio motuum, quibus laminae et virgae elasticae contremiscunt," on a precise theory of vibrating rods, were published in the *Acta* in 1782; "Trigonometria sphaerica universa" (E524) was presented to the academy in March 1782, while "Problème de géométrie résolu par l'analyse de diophante," (E754) on the resolution by Diophantine analysis of problems inspired by rational trigonometry, appeared in 1820 in the *Mémoires de l'Académie Impériale des Sciences de St. Pétersbourg*.[62]

In May 1780 the Massachusetts legislature, prompted by John Adams, John Hancock, and James Bowdoin established the American Academy of Arts and Sciences in Boston to recognize excellence in these fields. The rival learned body to the American Philosophical Society in Philadelphia, the US academy in 1781 began to select distinguished foreign honorary members, including the French naturalist Buffon, author of the monumental thirty-six volume *Histoire naturelle* (1749–88); another member chosen that year as an American fellow was Benjamin Franklin. In January 1782 the academy elected Euler, its first member from Russia.[63] That August, Johann Albrecht sent his father's letter of acceptance, which has since been lost.

Chapter 15

Toward "a More Perfect State of Dreaming": 1782 to October 1783

In April 1782, Sergei Gerassimowitsch Domaschnev—without consulting the members of the academy—arbitrarily ordered the removal of Semjon Kirillovič Kotel'nikov, who had given eleven years of distinguished service. Domaschnev moved to transfer Kotel'nikov's position of superintendent of the Imperial Library and his cabinet of curiosities of natural history to Peter Simon Pallas. On 9 and 11 April, the academicians and adjuncts held extraordinary meetings at the home of Leonhard Euler, where they unanimously opposed Domaschnev; later that month they again denounced the director's action. Through the summer, Kotel'nikov continued to attend academy conferences. In August the academicians declared that the dismissal of Kotel'nikov violated the academy's 1747 charter and asked Domaschnev to withdraw it. The letter was met with silence, a common practice among absolute rulers. In November 1782 the academicians dispatched another letter of protest, criticizing Domaschnev's breach of the institutional rules of academic protocol on such matters as budget.

When this too received no response, in December 1782 the members sent an open letter of protest to the academic commission that called for the dismissal of Domaschnev and informed Catherine II of the dispute; Euler's name was among the signatures. Catherine II did not respond immediately; before making a decision, she held an inquest of nearly two months, but its findings eventually brought Domaschnev's release. Under Catherine the Russian government would no longer ignore the few rights of the academy's men of science; the academicians were safe from political dismissal.

The Inauguration of Princess Dashkova

On 24 January 1783, with a stroke of the pen, Catherine II dissolved the academic commission and appointed as the director of the Russian Imperial Academy of Sciences the princess Ekaterina Romanova Dashkova, another favorite of hers. After a somewhat rocky relationship, they had reconciled a year earlier. The princess had supported Catherine in the coup; during this upheaval, Dashkova dressed in an officer's uniform, and a salon she held during her travels to Edinburgh from 1776 to 1779 had among its guests the historian William Robertson, the chemist Joseph Black, the natural philosopher Adam Ferguson, and the economist Adam Smith. "It was always an immense pleasure," she said, "to see [the Edinburgh scholars] whose conversation never failed to be instructive."[1] In Paris in 1781 the princess met Benjamin Franklin, whose erudition and manner impressed her. She also conversed with Denis Diderot and Jean-Baptiste Le Rond d'Alembert and corresponded with Voltaire, and this won her a place among the Enlightenment luminaries. Her appointment as director started the academy on a new epoch. When on 23 January Catherine offered her the position, the princess had been "struck dumb with astonishment," and quickly insisted, "I cannot accept any office which is beyond my capacities."[2] Catherine II dismissed the objection and declared the princess abler than her two predecessors. The princess knew well that the empress possessed the tact, shrewdness, and tenacity to obtain a positive answer, so resistance was pointless. When at the staff meeting with the empress on the morning of 24 January Domaschnev attempted to describe to the princess the duties she would have as academy director, she politely rebuffed him, saying that she "would treat its members with perfect impartiality."[3]

The princess, known as Little Catherine, was thirty-eight; she was Russia's first modern stateswoman, a notable female aristocratic manager, and a woman of letters. Not a scientist, she faced a daunting task. Assiduously reading over the many records of the operations and business of the academy, she conducted a critical assessment of her potential to be director. She saw that in the debt-mired academy, salaries of members were in arrears, academy buildings were in disrepair, and laboratories and libraries were depleted. Important scientific expeditions had yet to be planned. The princess's evaluation convinced her that with strong organizational and administrative skills, both of which she possessed, together with her intellectual ability, she could end the disarray in academy management and make the institution more intellectually distinguished. In being able to report directly to Catherine II, she had greater authority than her two predecessors. Recognizing the importance of the prerogatives of the academicians, she would not be installed unless the academy elected her. Her respect for them pleased the members, who met on Saturday, 28 January, and voted for her unanimously.

Early on the morning of 29 January Princess Dashkova received at her residence, almost directly across the Neva from the new academy

Figure 15.1. Portrait of Princess Ekaterina Romanova Dashkova.
Photo by Ed Owen. Courtesy of Hillwood Estate, Museum, and Gardens (51.66).

building, a delegation of all the professors, administrators, and support staff of the academy. She informed them that she would enter the academy to begin her directorship the next day, and that if they had any business to discuss with her they should come to her room at any hour. After their departure, Princess Dashkova read all evening the reports that she had received about the academy. She was convinced that her slightest mistake would be criticized. She was determined not to forget the names of governmental officials or any officers of the academy.

On 30 January, Princess Dashkova was to be formally presented as director of the Imperial Academy. All professors and others of the academy assembled in the conference hall of the old academy building. Formerly the palace of Princess Praskovea, the sister-in-law of Peter the Great, it had been since 1728 the main building of the institution. On the way to her inauguration, Princess Dashkova stopped to visit at the home of the man she called Euler the Great. She shrewdly begged Euler, who felt honored, to let her enter the academy on his arm and to introduce her. His introduction would be a strong indication of her respect for the sciences and might help reduce her dread of public speaking. She invited Euler, his son Johann Albrecht, and Nicholas Fuss to ride in her carriage. Fuss had to guide Euler, who was almost completely blind. The academicians had long put merit ahead of social rank, and Princess Dashkova allied herself with merit when she snubbed Jacob von Stählin. As the incident was related in her memoirs,

> I noticed that Mr. Stählin, the Professor of Allegory, but with the rank of State Councillor, equivalent to a major-general, had taken his place [the seat of honor] next to the Director's chair and thus wanted, according to the rank he was given God knows why, to play the first personage after me. [Mr. Stählin had received the title and rank during the reign of Peter III, and one could say that his science and he were all as allegorical as his title.] I therefore turned to Mr. Euler and told him to sit down wherever he thought fit, for any seat he occupied would always be the first among all. His son and grandson were not alone in showing appreciation . . . at my remark, and the professors, who all had the highest respect for the venerable old man, had tears in their eyes.[4]

This was the last session of the academy that Euler attended. With support from Catherine the Great, the Petersburg Imperial Academy began to offer tribute to him during his lifetime by bestowing on him alone the grand medal of the academy in February and by commissioning for the assembly hall a mural depicting the allegory of the wisdom of geometry that includes a board filled with formulas and calculations from Euler's third lunar theory.

1783 Articles

The publications of Euler for 1783 include volume 1 of his *Opuscula analytica* (Short analytical works, E531) consisting of thirteen articles in 363 pages. Corollary 8 in chapter 3 gives a formulation very near to the law of quadratic reciprocity; it appears in the conclusion of Euler's paper "Observationes circa divisionem quadratorum per numerus primos (Observations about the division of squares by prime numbers, E552), where he succeeded in separating it from his early conjectures. This essay, read at the academy in 1772, deals partly with Lagrange's reduction theory for proofs. But proof of the law of quadratic reciprocity remained above Euler's and Lagrange's methods. Among other articles in the *Opuscula analytica*, E553 expands functions into continued fractions of functions, E555 examines interpolation methods in what would come to be called Fourier series and pathological functions, E560 proves what Euler knew as Waring's theorem but is now called Wilson's theorem which in its modern form is: if p is a prime, then $(p - 1)!$ is congruent to -1 modulo p. In his time, Euler would have recognized this theorem as: if p is prime, then p divides evenly into $1 + (p - 1) \times (p - 2) \times \ldots 3 \times 2 \times 1$. E562 gives the sines and cosines of multiple angles as an infinite product. Euler wrote E560 after receiving two proofs for Wilson's theorem by Lagrange. His own proof involved a primitive root modulo, the prime n.[5] E570, the last essay for the year, determines longitude by observing the distance between the moon and a fixed star; it appears on in the first volume of the *Opuscula*.[6] (The second volume of fifteen articles in 346 pages would not be printed for two more years.) In 1783 the translation from German into French of Euler's *Neue Grundsätze der Artillerie* was published and introduced into French artillery schools.

The Petersburg Academy's *Acta* and *Opuscula analytica* contain for 1783 a total of twenty-eight Euler essays. Among the topics in the *Acta* are E532, on a series by Johann Heinrich Lambert describing roots of a trinomial equation; E536, on moments of inertia of triangles; E537, on the figure of elastic curves intended to counter objections of d'Alembert; E538, discussing cautions in determining the motion of planets; and E542, on pentagonal numbers.[7]

During 1783 Euler continued to do research at his house with his circle, especially Fuss, that included complex computations, which he had always loved and was still able to complete mentally. His strength and energy never flagged. Domestic life was pleasant, and his wife Salome Abigail had a busy house to manage: his sons Johann Albrecht and Karl and their families resided there, as did Fuss. Eight of Johann Albrecht's children, four sons and four daughters, were present, though only one of Karl's eleven children had survived. Euler enjoyed interacting with the youngsters and assisting in their education.

Final Days

Johann Albrecht, Fuss, Anders Johan Lexell, and the Lutheran minister Abel Burja, who were with Euler—along with Nicolas, Marquis de Condorcet at a distance—provided a record of his final two days. On 17 September his friend Burja, a published author on logarithms in the Berlin Academy *Mémoires*, visited him to discuss mathematics. He found Euler "happy and affectionate as always,"[8] and Euler complained only briefly of dizziness.

To his death Euler remained enthusiastic and continued to have essentially the full force of his extraordinary genius in research computation and teaching. The attacks of vertigo that he had suffered for a year did not slow him, but by early September his dizziness was increasing. The morning of 18 September he experienced another spell of it; at the time he had been teaching four of his grandchildren elementary mathematics, and on this last morning he instructed a grandson, the brightest of them in the sciences. Afterward his guests Fuss and Lexell arrived. He informed Lexell that he had now lost all of his eyesight, and the two discussed the diurnal motion of Earth, the topic for the academy prize in 1783.[9] At lunch Euler turned to the improvement of (Frederick) William Herschel's telescope by combining eye glasses in it and the news of the success that June by the Montgolfier brothers in launching balloon ascents in Paris, the topic of the moment within the republic of letters; news of it had just arrived in Saint Petersburg and was announced in public journals. That morning Euler made difficult mental calculations about the upward motion of hot-air balloons and how high they could rise, and he outlined its differential equation. He took lunch with his guests. With Lexell, who was to be his successor, he discussed the orbit of the planet Uranus, which Herschel had just discovered in March 1781—another subject that was the rage within the European reading public. He asked whether anyone had constructed tables of its motion. Euler enjoyed dictating—mostly to Fuss—difficult computations for his assistants to put on two large writing slates in his study; in this case the topics were aerodynamics and the orbit of Herschel's planet. Later on the day of his father's death Johann Albrecht found these reckonings made by Euler on two of his large blackboard slates; he quickly fleshed out the calculations on balloon aerostatics, and without delay dispatched the four-page essay "Calculs sur les Ballons aérostatiques faits par feu M. Léonard Euler" (Calculation on aerostatic balloons made by the late Mr. Euler, E579), in French to the Paris Academy *Mémoires,* where it was to appear as an *avertissement* in the volume for 1784.

After lunch, he fell faint and went to lie down. He slept for a few hours. He then woke and returned to his family and friends. At tea time, about four o'clock, Euler started joking and playing with one of his grandsons who had just entered the room. Lexell was with them, and he and Euler were sitting on a couch. Euler was smoking a pipe.[10] He had finished a cup of tea and asked his wife whether he had already completed two cups. She said only one, so he asked her for a second. Suddenly, within two minutes, the pipe fell from his hand. "Meine Pfeife!" he exclaimed, and he bent over to reclaim it but was unsuccessful and stood up. For a year Euler had endured vertigo and weakening health, but now he suffered a stroke. Clasping both hands to his forehead, he said, "Ich sterbe" (I am dying) and lost consciousness, which he never regained. All attempts to revive him by cutting his veins, enemas, and Spanish flies failed.[11] He passed away about eleven o'clock that night. As Fuss noted, Euler was seventy-six years, five months, and three days of age. Twenty years earlier, in reflecting on the state of the soul after death, Euler had presumed that there would be a suspension of the union between the body and the soul, in which the senses influence even our dreams. In death, he believed, "we will find ourselves in a more perfect state of dreaming."[12]

When Euler died, his five children had given him twenty-six grandchildren who survived him. The total number of grandchildren was to be forty-five.

Across all of Europe, Euler's death was considered a notable public loss; perhaps only Voltaire was better known than he. The scientific world recognized that it had been deprived of one of its great colleagues: the four major royal science academies in London, Paris, Berlin, and Saint Petersburg, along with societies in Basel, Lisbon, Munich, Stockholm, and Turin, all of which Euler had belonged to, announced their profound loss. Frederick II of Prussia, the king of Poland, and the margrave of Brandenburg-Schwedt sent condolences to Johann Albrecht.[13] Euler was buried on Vasilievsky Island in the Evangelical Lutheran or Protestant section of the Smolensk Cemetery, which was mainly for members of the Russian Orthodox Church.

Major Eulogies and an Epilogue

On 22 September 1783, Fuss formally announced at the academy the death of Euler from apoplexy. Noting the length and brilliance of his mentor's career and the immortality of his name throughout Europe, Fuss declared that he had been "for fifty-six years the glory and adornment of the academy" and proposed a monument for him.[14] Princess Dashkova agreed and had the academy commence planning for a memorial service, which took over a month to complete. She expressed her regrets to the family. The Russian Academy mourned Euler's passing, even as Princess Dashkova kept it fully functioning; it was engaged in studying crystallography, planning a cabinet of fossils, and continuing Wolfgang Ludwig Krafft's research on projectiles. At the time, Black had sent a letter of thanks from Edinburgh for being named a foreign associate. The academicians wanted Euler recognized equally for his genius and for his virtues; on 24 October Fuss completed his "Éloge de Monsieur Léonard Euler," which was to be read at an extraordinary public conference. Three days later Princess Dashkova invited all honorary members and friends of academicians. On 3 November 1783, the Imperial Academy of Arts and Sciences held its memorial meeting for Euler. Princess Dashkova presided, and an archbishop, two counts, a baron, and the Euler family attended. In the eulogy Fuss depicted Euler as an exemplar in the effort of his century "in enlightening the world."[15]

Fuss discussed Euler's prodigious ability, his mental fecundity, and his contributions to fields ranging across mathematics and the mathematical sciences. In those Fuss placed Euler on the same plane with Galileo Galilei, Gottfried Wilhelm Leibniz, and Isaac Newton.[16] Fuss's

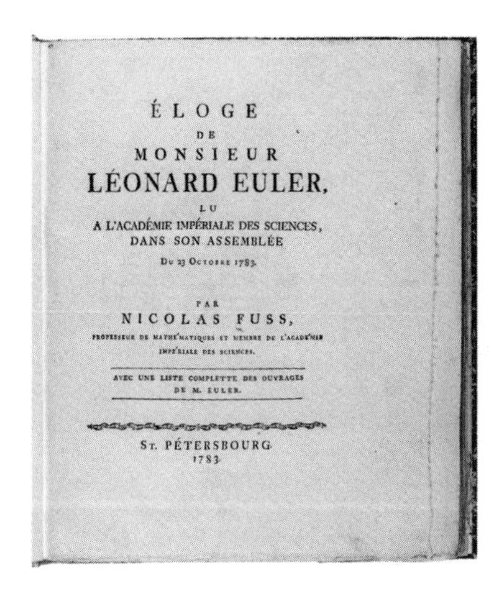
ÉLOGE
DE
MONSIEUR
LÉONARD EULER,
LU
A L'ACADÉMIE IMPÉRIALE DES SCIENCES,
DANS SON ASSEMBLÉE
DU 23 OCTOBRE 1783.

PAR
NICOLAS FUSS,
PROFESSEUR DE MATHÉMATIQUES ET MEMBRE DE L'ACADÉMIE
IMPÉRIALE DES SCIENCES.

AVEC UNE LISTE COMPLETTE DES OUVRAGES
DE M. EULER.

ST. PÉTERSBOURG
1783.

Figure 15.2. The title page of Nicholas Fuss's "Éloge de Monsieur Léonard Euler."

account illumined the personality of Euler and the impression that he made on his contemporaries: ascetic; always ready to engage in conversations, with the rare ability to shift immediately from deep discussion to a casual level and back; gentle and cheerful, enjoying harmless jokes; good-hearted and fair, refusing to hold grudges. Euler, Fuss avowed, had been an incomparable teacher; the student recalled a scholar devoutly religious, unmarked by the skepticism and the crisis of conscience that often went with the mentality of the Enlightenment. He was also hailed as a good husband, father, and friend.

Condorcet, as the secretary of the Paris Academy of Sciences, was to deliver the other principal eulogy, and Jean-Henri-Samuel Formey, his counterpart in Berlin, briefly lauded Euler. Condorcet described him as distinctive in embracing in their universality all of the mathematical sciences. Condorcet lauded Euler's work in algebraic analysis for effecting a revolution, and he faulted Euler only for his part in the dispute with Johann Samuel König. The eulogy closed with a political message in praise of imperial Russia, referring to the honors accorded to Euler—especially the mural of the allegory of geometry and the planned marble bust at the expense of the academy. It was "a country, which at the beginning of the century, we considered as scarcely emerged out of barbarism, [and now] is become the instructor of the most civilized nations of Europe in doing honor to the life of great men and in preserving their memory; it is setting these nations an example, which some of them may blush to reflect, that they have had the virtue neither to propose nor to imitate."[17] In January 1784 at the Berlin Academy, Formey held "the Great Euler" to

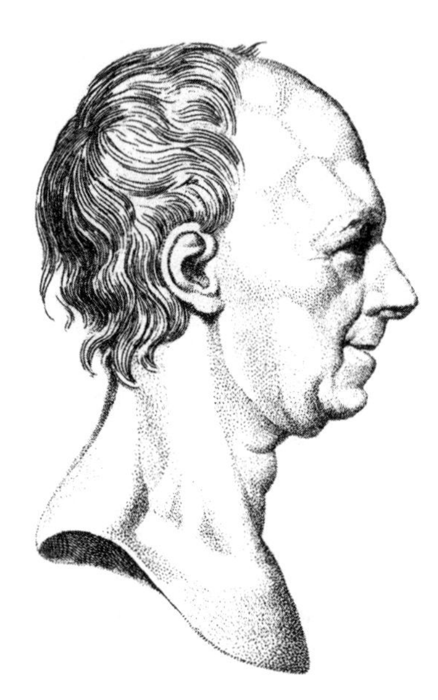

Figure 15.3. Bust of Euler by Jean-Dominique Rachette.

have replaced Newton and Leibniz, noting that he was superhuman in examining, cleaning up, and expanding upon their work and in establishing many new fields.[18] Formey made a flowery parallel of Euler to Tiresias, proclaiming that "this genius never slowed and that relieved of the light of days, as Tiresias, was able to see more clearly than ever and plumb the depths that the immensity of nature . . . [offered through] the laws of mathematics."[19]

At the demand of Princess Dashkova, in January 1785 the officers of the Petersburg Imperial Academy carried out the project for a statue, installing a half-length marble bust by Jean-Dominique Rachette in the library hall of the Kunstkammer.[20]

Euler had promised Vladimir Grigor'evich Orlov to prepare enough articles to appear in the academy's *Nova acta* for the next twenty years but left enough for well over forty, transcribed by his assistants.[21] By the early nineteenth century the modest gravestone of Euler could not be located in the Smolensk Cemetery—not even by Fuss, a participant in the funeral. This recalls the loss of the tombstone of Archimedes by the citizens of Syracuse, rediscovered 150 later by the foreigner Cicero. After years passed, Fuss often went with his son to Smolensk Cemetery, upset over the loss of Euler's stone plaque marking his grave. They searched for it, but it was not found until 1830, overgrown and nearly unreadable at the grave for one of Euler's

Figure 15.4. Euler's tomb. In 1957, it was moved into the Lazarus Cemetery at the Alexandre Nevski Monastery in what is now Saint Petersburg.

daughters-in-law, Emily, the wife of his son Karl. In 1837 the academy replaced it with a simple lasting monument constructed out of pink Finnish granite with the inscription *Leonhardo Eulero, Academia Petropolitana*.[22]

It is not these material tributes, which Euler richly deserved, but his massive and profound writings, including the more than 866 books and essays, along with his extensive correspondence, that are his full monument. Critical studies of these works are enriching and deepening our knowledge of the course of mathematics and the exact sciences in Europe during the mid- and late eighteenth century and beyond. Euler had reached across borders and boundaries in a search for collaboration and competition. The examination of his exchange of letters with the Paris Academy of Sciences and the Royal Society of London, along with the leading geometers, natural philosophers, physicists, and astronomers of his time—d'Alembert, Alexis Claude Clairaut, Joseph-Nicholas Delisle, Lagrange, Johann I (Jean I) and Daniel Bernoulli, Lambert, Pierre Maupertuis, Johann Tobias Mayer, and Jan Andrej (Johann Andreas) von Segner—as well as Christian Goldbach and Mikhail Lomonosov is mostly at an early stage.[23] The jubilee volumes from 1957, 1983, and 2007 attest to the better but incomplete recognition of Euler's legacy in the mathematical sciences of his time.[24] Typically publications and letters in mathematics are unusual in their ability, centuries after the death of the principals, to give impetus to new developments and applications in solid mathematics.[25] Euler's influence upon serious research to the present is extraordinary. For 2010, the *Science Citation Index* lists 1,724 articles examining selected ideas, conjectures, theories, metaphors, and methods of Euler, 298 of which have his name in the title. For men of science before 1900, only essays regarding Newton appear in a comparable number and slightly exceed the writings drawing upon Euler.

Notes

Several endnotes conclude with the Eneström number of an original writing given in the Euler archive.

Preface

1 In recent years, all of Euler's manuscripts have been indicated as well by their description in Mikhaĭlov, 1962. The *Manuscripta Euleriana* was reviewed in *Scripta mathematica* 28 (1967), pp. 210–211. See also Truesdell, 1972.
2 The Euler Archive can be found at http://eulerarchive.maa.org/.
3 Mikhaĭlov, 2007.
4 For the chief sources in Russian archives and collections for Euler's work in his two periods in Saint Petersburg, see Mikhaĭlov, 1962.
5 See Mann, 2011.
6 Daston, 1991; Dauben and Scriba, 2002, pp. xxiv–xxv.
7 Fuss, [1968] 1843, vol. 1, p. xl.
8 Dauben and Scriba, 2002, pp. xxiv–xxv. See also Robson and Stedall, 2009; and Pocock, 2005.
9 Chandrasekhar, 1987; Mac Lane, 1986.
10 Caparrini and Fraser, 2013.
11 See the largely hagiographic Ignatieff and Willig, 1999.
12 Ball and James, 2002; Giusti, 2003; Speiser, 2003.
13 Blanning, 2007, pp. 475–513.

Introduction

1 Blanning, 2000, p. 1.
2 Kant, 1781 and 1787.
3 Similarly, Lagrange thought that with the spectacular Enlightenment contributions the field of mathematics had been mined to the "limit of human accessibility," and a generation later Sylvestre François Lacroix gloomily found analysis "practically exhausted."
4 Blanton, 1988.
5 Wilson, 2007, pp. 140–145.
6 For the changing fields included in the *scientiae mathematicae* during the sixteenth and seventeenth centuries and their new status of primacy in natural philosophy, see Remmert, 2009. For a listing of these subjects, see d'Alembert, 1995, pp. 144–145.
7 Blanning, 2000, pp. 1–11.
8 Euler was engaged in improving telescopes and studying the possibility of life on the moon and the planets. Were he alive today he would be intensely interested in the building of powerful new telescopes, such as the Kepler and the Spektre R, and the search for habitable exoplanets approximately the size of Earth.
9 Habermas, 1989.

Chapter 1. The Swiss Years: 1707 to April 1727

1 Eighteenth-century Europe generally had two calendars, the Julian or "old style" calendar that started in antiquity and the Gregorian or "new style" calendar, a reform named after Pope Gregory XIII, who introduced it in 1582. Catholic lands rapidly accepted the Gregorian calendar; but the people of Protestant and Orthodox Europe, including imperial Russia, did not accept it before 1700. Basel would adopt the new style in 1701, and Euler used it in his autobiography. Since 1700 was observed as a leap year, the two calendars differed after February by eleven days—for example 5 March versus 16 March. Since the new year in the Julian calendar commenced on 25 March, the years for the first three months were one year apart. The dates given in this volume follow the Gregorian calendar.

2 The city much later had the sobriquet *Das fromme Basel* (pious Basel). See Heusler, 1957.

3 Gossman, 1994, p. 67. Robert Merton (1938 and 1973) and Charles Webster (2008) further developed this thesis.

4 Originally named Gerrit Gerritszoon, this brilliant natural son of a priest later adopted the Latinized tautological double name Desiderius Erasmus. Desiderius comes from the Latin noun for "longing" and Erasmus from the Greek for "beloved." His most famous book, translated as *The Praise of Folly*, was written in 1509. See Bonjour, 1988, pp. 235–275.

5 The pseudonym Paracelsus was given to him later, meaning "equal to Celsus"; Aulus Cornelius Celsus was an ancient Roman encyclopedist best known for writing a tract on medicine.

6 See Principe, 2013 and Webster, 2008.

7 See Euler, 1955, especially pp. 34–35; Gekker and Euler, 2007, p. 399; Burmeister, 2009, pp. 99–104.

8 Today Lake Constance is at the intersection of the borders of Switzerland, Austria, and Germany.

9 Hans-Georg lived from 1573 to 1663. He married his first wife, Ursula Ringsgewandt, in June 1594. The daughter of a brushmaker from Nuremberg, she died in 1611, and in 1611 or 1612 he married Eva Reck.

10 Some descendants moved and founded Euler lines mainly in what became Germany, Russia, and Sweden.

11 Bonjour, 1960, pp. 109–115; Stähelin, 1957.

12 Heefer, 2007, pp. 949–952.

13 See Raith, 1972, esp. "Pietismus und Aufklärung," pp. 180–185; and Raith, 1982, pp. 6–31.

14 See Lehmann, 1972, esp. p. 292. Church records indicate that the number of households in Riehen and Bettingen in 1693 was around two hundred and the number of residents fourteen hundred. There were thus a comparatively large number of children per family.

15 Paul's predecessor Burckhardt had lived in one room with his nine children.

16 Anna Marie was to marry the Münster organist and preceptor Christoph Gengenbach, while Maria Magdalene married Pastor Johann Jacob Nörbel.

17 Fellmann, [1995] 2007, pp. 22–23.

18 See Herzog, 1780, esp. pp. 32–34.

19 Fuss, [1783] 2007. Scans of the original publications of Euler are contained in the digital library of the Euler Archive.

20 See Obolensky, 1986.

21 Ibid., p. lii.

22 Fellmann, 1995, p. 14.

23 Burton, 2003, pp. 160–161.

24 Gossman, 1994, pp. 50, 69–71.

25 Spiess, 1929, p. 33.
26 Daniel Bernoulli, letter that included an announcement of Burckhardt's death, 4 September 1743, in Fellmann, 2007, p. 14.
27 See Nagel, 2008.
28 Huber, 1959, pp. 22–24.
29 Truesdell 1984b, pp. 342–343.
30 See the bibliography in Euler, *OO* IVA.2.
31 During Euler's years in Berlin he would clash with König in the controversy over the principle of least action.
32 Fellmann, 2007, pp. 17–18.
33 Johann Jakob Ritter, quoted in Fellman, 2007, p. 18.
34 Darwin, Einstein, and Russell considered some lectures dull, and would occasionally skip them. For Darwin, see Desmond and Moore, 1991.
35 Garber, 2000.
36 Whiteside, 1970.
37 Leonhard Euler, *Meines Vaters Lebens-Lauf* (autobiography), quoted in Fellmann, 1995, p. 12.
38 In book 2, *On the Motion of Bodies*, of the first edition of the *Principia Mathematica*, proposition 10, problem 3, Bernoulli had discovered a mistake. Proposition 10 examines the motion of a projectile operating under the uniform force of gravitational attraction and passing through the resistance of a medium calculated as the product of its density and the square of the velocity. It computes with the density of the projectile at each point to have the body move in a curved line. But for motion lost, Bernoulli found the numerical value to be half that of Newton's calculation. (Both were wrong.) Bernoulli sent his eldest son, Nikolaus II, to London in 1711 with word of his discovery. Personally taking up the problem at nearly seventy years of age, Newton recognized that this finding gave the appearance of a weakness in his lunar theory. The second edition of the *Principia*, issued in 1713, showed that Bernoulli was also mistaken. Increasing the value of resistance noted above to the force of gravity by three halves, Newton had found the correct answer, three-fourths the amount cited in the first edition.
39 It particularly contains what is called L'Hôpital's Rule for calculating the value of $f'(x)/g'(x)$ if for some value of $x, f(x)/g(x)$ is indeterminant as 0/0.
40 Fuss, 1783, p. 52.
41 Ibid.
42 Ibid.
43 The genealogical table of mathematicians in the Bernoulli family, including Daniel I, whose chief work lies in other fields. It follows the table in Fellmann 2007, p. 17.

Genealogy of the Bernoulli Mathematicians

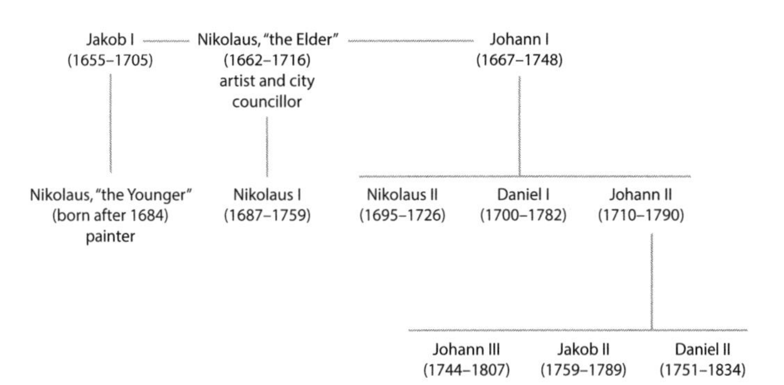

44 Bernoulli 1743, p. 616.

45 Mikhaĭlov, 1999.
46 Wolff, 1860, letter 22, p. 47.
47 Schröder, 1959, p. 275.
48 From 1732 on its title was *Nova acta eruditorum*.
49 Euler, 1726 (E01).
50 Johann I Bernoulli, quoted in Calinger 1999, p. 633.
51 Fellmann, 1995, p. 23.
52 Euler, 1735 (E13).
53 Euler, 1750 (E128).
54 Euler, 1727 (E03). Ian Bruce has made a translation and transcription of E03 that is available at his website, Some Mathematical Works of the 17th and 18th Centuries, at http://www.17centurymaths.com/contents/euler/e003tr.pdf.
55 Euler, 1727 (E03), quoted in Sandifer, 2007a, p. 7.
56 Euler, 1729 (E05); 1738 (E23); 1751 (E173).
57 See Fuss, 1783, p. 52; and Bernoulli, 1743, foreword dated 1 March 1743, p. 389.
58 Maindron, 1881, pp. 15–16.
59 Ibid., pp. 14–17.
60 Gillispie, [1980] 2008, p. 85.
61 The academy transferred to publishers the right to print prize-winning essays, first in 1721 to Claude Jambert and later to Gabriel Martin, Jean-Baptiste III Coignard, and Louis Guérin. See Hahn, 1971, p. 61.
62 Euler, 1727 (E04).
63 Maindron, 1881, pp. 15–16.
64 Jahnke, 2003, pp. 330–339.
65 Euler, 1727 (E04).
66 See Fellmann, 1983a, p. 81.
67 See Mikhaĭlov and Sedov, 1959, esp. 258–259.
68 Euler, 1727 (E04), 35–36.
69 Ibid., p. 35.
70 Euler, 1727 (E02).
71 Daniel Bernoulli, quoted in Euler II.11, p. 143.
72 Voltaire, [1752] 1879; see, in particular, the sections "A Memorable Session," "Peace Treaty," and "Letter to Dr. Akakia."
73 For an account of the *Lettres*, see chapter 13.
74 Laudan, 1968.
75 In 1713 Senator Stähelin, a prominent member of the family, had suffered a spectacular bankruptcy of over 100,000 gulden, the greatest financial loss of the time in Basel. Forced to flee the city, he broke his ankle while jumping from the tower wall; he sought sanctuary in the Capuchin monastery at nearby Dornach. It does not appear that this financial failure had any influence on Benedict Stähelin's call to the faculty, however. Gossman, 1994, pp. 43–44.
76 The subject matter and reliability of Euler's notebooks are only beginning to be critically examined in detail.
77 Ferdinand Rudio, "Vorwort," in Euler, *OO* III.1, pp. x–xi; Maltese, 2003.
78 Maltese, 1992.
79 In a letter of November 1734 to Daniel Bernoulli, Euler still expressed an interest in becoming a medical doctor but apparently thought that the price remained too high. *Bibliotheca Mathematica* 1906–7 (2012), pp. 137–141.

Chapter 2. "Into the Paradise of Scholars": April 1727 to 1730

1 At the archive, see f. 136, op. 1, nos. 129–140. See also Mikhaĭlov and Sedov, 1957, and Mikhaĭlov, 1959, pp. 269–279.

2 Schröder, 1959, pp. 256–280.
3 As the need for administrative expertise grew in European states during the eighteenth century, Prussian university faculty taught finance, administration, and social welfare to bureaucrats who wished to control the treasuries of princes. Wolff, however, was not one of these sorts of teachers.
4 Wolff, 1860, p. x.
5 Christian Wolff, quoted in Fellmann, 2007, p. 30. See also Schröder, 1959, p. 276, n. 12.
6 Daniel Bernoulli to Leonhard Euler, 1 March 1727, in Mikhaĭlov and Sedov, 1957, pp. 10–37 esp. p. 31.
7 Leonhard Euler to Daniel Bernoulli, 18 January 1727, in Mikhaĭlov and Sedov, 1957, p. 30.
8 George, 2003, pp. 33–39.
9 The subject of the large mural at the top of the stairs of the Petersburg Academy of Sciences building is the Battle of Poltava.
10 Lincoln, 2000, p. 25.
11 See Winter, 1958b, esp. p. 3.
12 Boss, 1972, pp. 10–12.
13 Many dozens of unchartered learned societies were formed on the regional, provincial, local, and urban levels and beyond the boundaries of Europe. See Emerson, 1990, pp. 960–979 and McClellan, 1993, pp. 153–165.
14 Aiton, 1985, pp. 309–311.
15 Vucinich, 1963, p. 102.
16 Cracraft, 2004, p. 241; Gouzévitch and Gouzévitch, 2009, p. 367.
17 Winter, 1958b, p. 4.
18 Fontenelle, [1790] 2010, p. 226.
19 Ibid., p. 74.
20 Kopelevich, 1977, pp. 45–46.
21 Gouzévitch and Gouzévitch, 2009, pp. 367–368.
22 Hans, 1963, p. 9.
23 Boss, 1972, p. 73; Vucinich, 1963, p. 58.
24 Vucinich, 1963, p. 71.
25 Saine, 1987.
26 Wolff, 1860, letter 97, pp. 164–166, and letter 100, pp. 168–170.
27 Ibid., letter 11, pp. 18–21.
28 Ibid., letter 8, pp. 12–14.
29 Count Aleksandr Gavrilovich Golovkin, quoted in Vucinich, 1963, p. 71.
30 Tetzner, 1958, p. 137.
31 Daniels, 1973, pp. 17–18.
32 Cracraft, 2004, p. 250.
33 Lundbaek, 1995.
34 Hofmann, 1959, pp. 143–145.
35 Kopelevich, 1983, p. 374.
36 Voltaire, [1732] 2007, p. 47.
37 That style of writing in the sciences originated in the *Dialogo* of Galileo in 1632.
38 Crowe, 1986, pp. 9–37.
39 Spiess, 1929, pp. 63–68.
40 Gibbon, 1952, p. 634.
41 Mikhaĭlov, 2002, pp. 9–10.
42 Leibniz, 2011, contains all of the mathematical papers of Leibniz from different journals.
43 Maclaurin, 1724.
44 Bernoulli, 1727; Mazière, 1727.
45 Bernoulli, 1738, p. 12.

46 Hermann, 1728, pp. 4–18.
47 Ibid., pp. 1–43; Bilfinger, 1728, "*De veribus corpori moto insitis, et illarum mensura,*" pp. 43–121; Wolff, 1728, "Principia dynamica," pp. 217–236.
48 See Gleb K. Mikhaĭlov, "Introduction," in Euler, IVA.2, esp. pp. 62–63.
49 Bilfinger, 1729.
50 Boss, 1972, pp. 110–111.
51 Fontenelle, 1727, pp. 154–159.
52 Hankins, 1965, p. 282.
53 See Hans, 1963.
54 Vucinich, 1963, pp. 79–80.
55 Grasshof, 1966, p. 30.
56 Leonhard Euler, *Autobiography*, in Fellmann, [1995] 2007, pp. 12–13.
57 Ibid., p. 13.
58 About fifteen years later, Euler wrote an article modeling the action of a saw for a sawmill run by human power; he projected the energy for a cut to be proportional to the square of its area.
59 Daniel Bernoulli to Giovanni Marchese Poleni, August 1727, in Mikhaĭlov, 1983, p. 232. See also Truesdell 1984a, p. 315.
60 This paper was lost; it was only found and published in the mid-twentieth century. See Leonhard Euler, "De effluxu aquae ex tubis cylindricis utcunque incilinatis et inflexis," in Mikhaĭlov, 1965, pp. 253–280; see also Mikhaĭlov, 1999.
61 The Dolgorukis were a Russian princely family that included statesmen, military officers, and men of letters. They claimed to have descended from Rurik, the semi-legendary individual who founded the first Russian state.
62 See Fuss [1843] 1968, vol. 1.
63 Vucinich, 1963, p. 81.
64 Euler, 1732 (E08).
65 Boss, 1972, p. 52.
66 Gouzévitch and Gouzévitch, 2009, p. 364.
67 Marty, [1970] 1977, p. 130.
68 See Euler, *OO* IVA.2. See also Fuss, [1843] 1968, vol. 2, pp. 1–94.
69 Leonhard Euler to Johann I Bernoulli, 16 November 1727, letter 190 in Euler, *OO* IVA.2, pp. 77–81.
70 From his study of the differentials of this function and related circle sectors, Bernoulli knew that $dy/y = dx \log(-n)$, but resisted his own findings dating back to 1702 that differed with $\log(n) = \log(-n)$.
71 Euler derived this equation from determining the area of the first quadrant of a unit circle to be $(a^2/4i)(\log - 1)$, which he computed from a formula given by Bernoulli.
72 Yushkevich, 1983.
73 Ibid., p. 161.
74 See Thiele, 2007c, pp. 433–440.
75 Euler, 1740 (E44).
76 David Hilbert, quoted in Thiele, 2007a, p. 237.
77 Euler, 1732 (E09). See also Sandifer, 2007a, p. 44.
78 See Euler, 1957 (I.27).
79 Constantin Carathéodory, "Introduction," in Euler 1952 (I.25), pp. i–xxvi.
80 Euler, 1732 (E09), p. 12.
81 Euler, 1954 (I.27), pp. 1–29.
82 Artin, 1964; Davis, 2007; Sandifer, 2007f, 2007g, 2007h.
83 Leonhard Euler to Christian Goldbach, October 1729, in Fuss, [1843] 1968, vol. 1, p. 3, as translated in Sandifer, 2007a, p. 33.
84 The name of the function comes from Adrien-Marie Legendre, who denoted it as gamma in his *Exercices de calcul integral* (3 vols., 1811–17).

85 Euler, 1738 (E19).

86 James Gregory invented the infinite series for arctan $x = x - x^3/3 + x^5/5 - x^7/7 + \ldots$. After Huygens summed the reciprocals of triangular numbers, that is, $S = 1 + 1/3 + 1/6 + 1/10 + 1/15 + \ldots$, Leibniz cleverly transformed the series, dividing it by 2 into $S/2 = (1 - 1/2) + (1/2 - 1/3) + (1/3 - 1/4) + \ldots$.

87 Yushkevich and Winter, 1959–76, vol. 1, p. 10; Weil, 1984, p. 172.

88 Euler had the equivalent in earlier symbolism of the improper integral $\int_0 - (\log (1 - t)/t)dt$.

Substituting for log $(1 - t)$ its series expansion of $-t - t^2/2 - t^3/3 - \ldots$, Euler integrated the expansion termwise to equal $1/2 + 1/2^2/4 + 1/2^4/16 + \ldots$. Substituting z for $1 - t$ and for $1/(1 - z)$, which is the sum of the geometric series, the expansion $1 + z + z^2 + z^3 + \ldots$, and intuitively applying L'Hôpital's rule, he computed the integral in a second way by parts to equal $-(\log 2)^2 - (1/2 + (1/2^2)/4 + (1/2^3)/9 + (1/2^4)/16 + \ldots) + \sum_{n=1}^{\infty} 1/n^2$. Setting the two expressions equal, Euler determined that $\sum_{n=1}^{\infty} 1/n^2 = 2(1/2 + (1/2^2)/4 + (1/2^3)/9 + (1/2^4)/16 + \ldots) + (\log 2)^2$.

89 Christian Goldbach to Leonhard Euler, December 1729, in Fuss, [1843] 1968, vol. 1, p. 10.

90 Suzuki, 2007, p. 367.

91 See Sandifer, 2007d, esp. p. 109.

92 Nevskaya, 2007.

93 Ibid., p. 279.

Chapter 3. Departures, and Euler in Love: 1730 to 1734

1 Another vigorous opponent of Bilfinger had been James Jurin, the secretary of the Royal Society of London, who after he left continued criticizing academy papers defending the principle of *vis viva*. This led Wolff to write in December 1735, "Er ist noch einer von den blinden Anhängern des Newtons und mit diesen ist nicht viel auszurichten" (He is still one of the blind followers of Newton and with these not much is accomplished); Yushkevich and Winter, 1961, vol. 2, p. 32; see also pp. 83–84.

2 Leonhard Euler, *Autobiography*, cited in Fellmann, [1995] 2007, p. 13.

3 Leonhard Euler, quoted in Kopelevič, 2007, p. 377.

4 Nevskaya, 2007, p. 278.

5 Lake Ladoga did not have such lights when its ice fractured every spring.

6 Truesdell, 1984a, p. 304.

7 Yushkevich and Winter, 1959–76; letters 1–5 are in vol. 1, pp. 45–51; the quotation herein is on p. 49.

8 Fuss, [1843] 1968, vol. 2, p. 412.

9 Ibid.

10 Nevskaya, 1983.

11 Maria Sibylla Merian died in January 1717.

12 Spiess, 1929, p. 72.

13 Müller had brought Juncker to the academy from Saint Gallen in 1731.

14 1 sazhen = 2.123 meters. This gives dimensions of 63.7 meters by 21.2 meters.

15 This description of the house comes from architect Domenico Trezzini, who made an inspection of it in 1744 after the Eulers left.

16 Winter, 1957, pp. 10–11.

17 In 1740 Korff was named to serve in the diplomatic corps in Copenhagen; soon after the death of his patron Biron he was dismissed.

18 See Kopelevič, 1983, esp. pp. 374–376.

19 The letter sent on 18 November, number 2198, is given in Yushkevich and Winter, 1959–76, vol. 2, p. 182.

20 The *Leipziger Magazin für reine und angewandte Mathematik* (Leipzig magazine for pure and applied mathematics) existed only briefly late in the eighteenth century.
21 His paper on ballistics, "Meditatio in experimenta explosione tormentorum nuper instituta" (Meditation upon experiments made recently on the firing of a cannon) was written in 1728 or late 1727 but not published until 1862. See Euler, 1862 (E853).
22 Matvievskaya, 2007, pp. 124–125.
23 Knobloch, 1989; Matvievskaya, 2007.
24 Euler, 1739 (E33).
25 Fuss, [1843] 1968, vol. 2, pp. 8–11.
26 Fitzgerald and James, 2007, pp. 52–53.
27 Euler, 1862 (E853).
28 Vlacq gave the log of 10 as 2.3025851.
29 See Euler, *OO* II.14.
30 Maor, 1994, p. 151.
31 Euler, 1738 (E20).
32 Yushkevich, 1983, p. 162.
33 See Ferraro, 2000 and 2004.
34 Euler, *OO* I.14, p. 35.
35 Pengelley, 2007.
36 The formula that Euler and Maclaurin used was $\sum_{k=1}^{n} f(k) = \int_0{}^{n} f(t)dt + (\sum_{k=1}^{\infty} B_k/k!)$ $(f^{(k-1)}(n) - f^{(k-1)}(0)$, which usually gives a divergent infinite series. But when f is a polynomial, the series is finite.
37 Sandifer, 2007i.
38 Euler, 1862 (E811); translated by Richard Pulskamp as "The True Valuation of the Risk in Games," http://cerebro.xu.edu/math/Sources/Euler/E811.pdf.
39 See D'Antonio, 2007b, pp. 239–261.
40 Fraser, 1991.
41 Euler, 1732 (E10).
42 See Newton, 1999.
43 See González Redondo, 2007, pp. 205–215.
44 The history of the writing and the printing of Daniel Bernoulli's *Hydrodynamica* is given in the "Introduction" to its republishing in volume 5 of his *Werke*. For an English translation, see Carmody and Kobus, 1968.
45 Mikhaĭlov and Sedov, 2007, pp. 176–177.
46 See Fontenelle, [1686] 1990; and Gillispie, 1956.
47 Euler, 1738 (E32).
48 Euler, *OO* I.2, p. x.
49 Fuss, [1843] 1968, vol. 1, p. 24; Weil, 1984, p. 173.
50 Fuss, [1843] 1968, vol. 1, pp. 30–31.
51 Ibid., vol. 1, pp. 602–625; Weil, 1984, pp. 169–176.
52 Euler, 1741 (E54). David Zhao has made a parallel translation into English, and Ian Bruce translated E54 and E26 into English and annotated them.
53 In E54, section 3, Euler wrote, "Signficante p numerum primum formula ap–1 – 1, semper per p dividi poterit, nisi a per p dividi queat"(With a signified prime number p, then formula $a^{p-1} - 1$ can be divided by p always, unless a is able to be divided by p).
54 A generalization, now known as the Euler-Fermat Theorem, holds that if $\phi(n)$ gives the number of positive integers less than n and relatively prime to it, and if a and n are relatively prime, then $a^{\varphi(n)} \equiv 1(\text{mod } n)$. Exactly when $\phi(n) = n - 1$, n is prime. See Fletcher, 1989.
55 Euler, 1738 (E26). See also Edwards, 1977, p. 39.
56 In the paper "Theoremata circa divisores numerorum in hac forma *paa* ± *qbb* contentorum" (Theorems about divisors of numbers of the form *paa* ± *qbb*, E164), *OO* I.2, pp. 194–222, Euler formally proved that divisors of composite numbers $a^2 + b^2$ must be of the form $4k + 1$ for some integer k, of $a^8 + b^8$, $16k + 1$, and of $a^{2n} + b^{2n}$, $2^{n+1}(k) + 1$.

57 For that time, C. Edward Sandifer finds that "exact sciences" is a better translation of *matheseos* than is "mathematics" because of the latter's more limited scope. See Sandifer, 2007a, p. 75.
58 Sandifer, 2007a, p. 75.
59 My thanks to Jukka Pikho of the University of Helsinki for clarifying this.
60 Wolff, 1730, vol. 1, pp. 383–384.
61 Sandifer, 2007a, p. 76.
62 Leonhard Euler, quoted with translation in Weil, 1984, p. 174.
63 For several decades Euler worked at proving Fermat's little theorem at roughly ten-year intervals. In 1747 (in E134) he offered a somewhat different induction proof, and his proof two years later (in E262) depends upon geometric progressions. See Sandifer, 2007a, pp. 203–206.
64 The next perfect numbers are 28, 996, and 8,128.
65 Dunham, 1999, pp. 10–12.
66 These involve an exponent, a power of x, and are solvable by some powers.
67 Yushkevich and Winter, 1965, pt. 2, p. 10.
68 Euler, 1738 (E29).
69 Thiele, 2007a, p. 242.
70 Euler, 1738 (E27).
71 Euler, 1738 (E25). The quotation herein is from Sandifer, 2007a, p. 67.
72 Varadarajan, 2006, chap. 4.
73 Havil, 2003.
74 Euler, 1740 (E43).
75 Ibid., pp. 156–157.
76 Dunham, 1999, pp. 31–35.
77 For a history of the most famous of the three, see Beckmann, 1971; and Posamentier, 2004.
78 For $-1 \leq x \leq 1$. Euler found further that $1 + 1/2 + 1/3 + \ldots + 1/x - \ln x = (1/2)x = B_1/2x^2 + B_2/4x^4 + B_3/6x^6 + \ldots$, providing the Bernoulli numbers.
79 Euler, 1740 (E36).
80 Euler, *OO* IVA.5, pp. 122–124, 206.
81 Mikhaĭlov, 2002, pp. 3–17 and 479–480.
82 Lauridsen, 1889, pp. 61–65.
83 Golder, 1941, p. 170.
84 Waxell, 1952, p. 44.
85 See Figurovskij, 1958, pp. 49–63, esp. pp. 52–52; and Lauridsen, 1889, p. 61.
86 Waxell, 1952, p. 69.
87 Figurovskij, 1958, pp. 52–53.
88 Lauridsen, 1889, p. 69.
89 Lipski, 1959.
90 Fisher, 1977.
91 Letter of Leonhard Euler to Gerhard Müller, end of 1734. The date given here is taken from a note in the upper left corner of the letter.
92 Yushkevich and Winter, 1959–76, vol. 1, p. 37. For Bering's dramatic death, see Sandifer, 2006c.

Chapter 4. Reaching the "Inmost Heart of Mathematics": 1734 to 1740

1 Emil A. Fellmann counts eighty-five works (or 14 percent) of Euler's ultimate total by 1744; see Fellmann, 1983a, p. 31.
2 Later Euler's correspondence with astronomers Clairaut, Heinsius, and Friedrich Christoph Mayer suggested such observations—for example, of comets and lunar motion.

3 Condorcet, 1786, Glaus translation, p. 3.
4 Minutes of the academic conference, 27 January 1735, quoted in Nevskaya, 2007, p. 276.
5 Schmidt, 2008.
6 Ibid., p. 42.
7 Euler, 1741 (E50).
8 Bloch, 1973.
9 Nevskaya, 1983, p. 368.
10 Johann Albrecht von Korff, quoted in Nevskaya, 2007, p. 282.
11 Leonhard Euler to Johann Albrecht von Korff, 10 December 1735, in Yushkevich and Winter, 1959–76, vol. 1, pp. 130–131.
12 Euler, 1738 (E18).
13 Plofker, 2007, p. 151.
14 Euler, *OO* III.2, pp. 73–74, as translated in Weil, 1984, pp. 261–262.
15 Weil, 1984, p. 184.
16 Euler, 1740 (E41).
17 Ibid., p. 79.
18 Dunham, 1999, pp. 46–48.
19 Another of Euler's three solutions was his application of the generalized binomial theorem to expand $[1/(\sqrt{1} - x^2)] \cdot (1 - x^2)^{-1/2}$ into an infinite series—that is, $(1 + a)^n$ assuming $a = - x^2$ and $n = -1/2$. He integrated each term of the series for $x < 1$—that is, when it converges—and arrived at $\pi^2/8 = 1 + 1/3^2 + 1/5^2 + 1/7^2 + \ldots$, the sum of the reciprocals of odd squares. Since any number is the product of a power of 2 and an odd number and because $4/3 = 1 + 1/4 + 1/16 + 1/64 + \ldots$, Euler found that $\pi^2/6 = 1 + 1/4 + 1/9 + \ldots$ by multiplying the two series together. The two series are absolutely convergent, so they can be multiplied. Euler's computations were brilliant and bold, his approach dauntless, and his intuition sure.
20 Calinger, 1996, pp. 135–136.
21 Jakob Bernoulli had introduced the formula $\Sigma_{k=1}^{\infty} k^c = 1/(c + 1) + (1/2)(n^c) - (c/2) B_2 n^{c-1} + c(c - 1)(c - 2)/2 \times 3 \times 4)(B_4 n^{c-3}) + \ldots$, where k and c are integers, and gave what Archimedes had known the value for $c = 2$ and Bachet for $c = 3$; Bernoulli added the case $c = 4$. See Sandifer 2007f.
22 Euler, *OO* I.15, p. 96.
23 Ferraro, 1998.
24 $B_6 = 1/42$, $B_8 = - 1/30$. $B_{10} = 5/66$, and $B_{12} = 691/2730$. See Agoh and Dilcher, 2008, p. 237.
25 Euler, 1750 (E130).
26 Among his correspondents at this time was Giovanni Jacopo Marinoni, the imperial astronomer in Vienna. They exchanged letters on astronomy, geography, the quadrature of the circle, and mistaken efforts to prove the irrationality of π.
27 See Buchwald and Cohen, 2001.
28 Maltese, 2003, pp. 202–208.
29 In the land of the *Ancients' Inferno*, Sisyphus of Corinth, the cleverest of all human beings, can even fool the gods, including Zeus. After a long life Sisyphus goes to Hades, where to keep him from thinking up new tricks and misdeeds the gods condemn him eternally to roll a boulder up a steep hill only to have it slip from his hands when he nears the top and again go to the bottom. Others are assigned to attempt to fill bottomless hogsheads or lift weights that always fall again.
30 Mikhaĭlov, 1965; Truesdell, 1984a, pp. 314–315.
31 Dugas, 1955, p. 232.
32 Euler, *OO* II.1, p. 8.
33 Johann I Bernoulli to Leonhard Euler, 2 April 1737, in Euler, *OO* IVA.2, pp. 152 (with German translation, pp. 155–156) and p. 162 (with German translation, p. 168).

34 Euler, *Mechanica sive motus scientia analytice exposita*, quoted in Polyakhov, 2007, p. 237.
35 Romero, 2007, pp. 232–234.
36 Cannon and Dostrovsky, 1981, p. 1; Suisky, 2009, pp. 117–187.
37 Euler, *OO* II.1, p. 8.
38 Euler, 1752 (E180).
39 Maltese, 2003, pp. 202–208.
40 Lagrange, 1788, preface.
41 Ibid., pp. 38–39.
42 Ibid., p. 13.
43 Robins, 1739, pp. 2–3.
44 Ibid., p. 8.
45 Ibid., pp. 3–4.
46 Ibid., pp. 7–8.
47 Ibid., p. 4.
48 See Smith, 1993.
49 Paul Stäckel, "Editor's Foreword," in Euler, *OO* II.1, p. x.
50 Robins, 1739, p. 1.
51 On 24 March 1736, Euler wrote to Giovanni Jacopo Marinoni in Vienna that "no one had demonstrated the possibility of . . . [completing the Königsberg transit] or showing that it is impossible."
52 Euler wrote "Solutio problematis ad geometriam situs pertinensis" in 1735, but it was not published until 1741; Euler, 1741 (E53). For the English translation, see Struik, 1969, pp. 184–187.
53 Weil, 1984, p. 257.
54 Euler, 1738 (E19).
55 Euler, 1741 (E54), section 3: "Significante p numerum primum formula $a^{p-1} - 1$ semper per p dividi poterit, nisi a per p dividi queat" (If p denotes a prime number, then the number given by the formula $a^{p-1} -1$ is always divisible to p, unless p divides a).
56 Euler, 1741 (E54).
57 Euler, 1744 (E72). See also Sandifer, 2007a, pp. 249–261.
58 Sandifer, 2007a, p. 249.
59 The terms may be expressed as $1/(m^n - 1)$, where m and n denote integers greater than 1.
60 Leonhard Euler to James Stirling, 8 June 1736, in Tweedie, 1922, pp. 178–191.
61 Euler, 1738 (E25).
62 Mills, 1985, pp. 1–13.
63 Mills, 1982, p. 305.
64 Weil, 1984, p. 181.
65 Euler, 1747 (E99). See also Weil, 1984, pp. 164–180.
66 Euler may have also contributed to a serial history of mathematics, consisting of about two hundred pages in four issues, and signed by Peter Bagdonovic, one of the academy's translators.
67 Euler, 1747 (E95).
68 Concerning this, Renau d'Eliçagaray wrote *La théorie de la manoeuvre des vaisseauz*. Renau wrote an article that spans 116 pages in octavo; he was a member of the Royal Academy of Sciences in Paris.
69 The correspondence with Weggersløff appears in Smirnov et al., 1963.
70 Brunet, 1931.
71 Newton, 1687, book III, prop. 20.
72 Todhunter, [1873] 1962.
73 Iliffe, 1993, p. 335.

74 Ibid., p. 368.
75 Voltaire, quoted in Cassirer, 1951, p. 74.
76 Euler, 1738 (E32).
77 Euler, 1755 (E212).
78 See Liebing, 1961, esp. pp. 5–9.
79 Euler, 1738 (E34).
80 Aldridge, 1975, p. 109.
81 Snow, 1964.
82 Nevskaya, 2007, p. 281.
83 Marin Mersenne (1588–1648) was known for transmitting scientific thought before the advent of science journals by acting as a center for the exchange of letters.
84 Grasshoff, 1966, pp. 122–125, 147.
85 Bernoulli was the true author of the book. L'Hôpital was his student, and took notes from Bernoulli, then publishing them.
86 Euler, 1750 (E129).
87 Euler, 1738 (E17).
88 In the 1740s Gottsched was to translate Pierre Bayle's *Dictionnaire historique et critique* from French into German.
89 Euler, 1738 (E17).
90 Ibid., pp. 216, 263–264.
91 See Euler, 1739 (E33), esp. pp. x–xiv. See also Tserlyuk-Askadskaya, 2007, pp. 354–355.
92 Leonhard Euler to Johann I Bernoulli, December 1738, in Euler, *OO* IVA.2, pp. 263–264 (with German translation, pp. 268–269). See also Fellmann, [1995] 2007, English translation, p. 50.
93 Johann I Bernoulli to Leonhard Euler, March 1739, in Euler, *OO* IVA.2, pp. 277 (with German translation, p. 282).
94 In declining health, Johann I had to have Daniel read to him the letters from Euler.
95 Fuss, [1843] 1968, vol. 2, p. 456.
96 Ibid., 464–465.
97 Later it provoked a debate with Jean-Jacques Rousseau, who advocated simplicity in music in contrast to the late baroque grandeur. In comparative obscurity at Saint Thomas's church in Leipzig, Johann Sebastian Bach was composing impressive instrumental music that would include in the future a series of preludes known as *The Well-Tempered Clavier*.
98 Christensen, 2002, p. 759.
99 The velocity of sound in air is C = 1087.93 feet per second + 0.6(t), where t = temperature in centigrade. See Cannon and Dostrovsky, 1981, p. 44.
100 Leonhard Euler to Johann I Bernoulli, August 1737, in Euler, *OO* IVA.2, pp. 161–175.
101 Euler, 1768 (E343–E344), vol. 1, letter 2 (22 April 1760).
102 The source of pleasing harmonies was the "discovery and contemplation of perfection" imparted by order from rules and laws that convey pleasant feelings to the soul. Euler, 1739. (E33), p. 3. See also pp. 19–20, and chapter 2, pp. 26–44.
103 Euler studied stretched strings. When a string vibrates, the weaker tones are always present; in music the relation of tones to overtones comprises a constant interval and forms a progression from the lowest to the highest: 1, 2, 3, 4, 5, ...
104 Euler extended the beginning tones of accord to the proportions 32:36:40:45:48:54:60, but indicated that the ear can scarcely distinguish 36:45:54:64.
105 Although Euler surmised (based on extensive calculations) in chapter 10 of the *Tentamen* that a musical 7 was impossible, he thought that a different schematic would show there were other possible examples of its consonance. He left these questions open, as part of what he called *speculum musicum*.

106 Daniel Bernoulli to Leonhard Euler, January 1741, in Fellmann, [1995] 2007, English translation, p. 51.
107 Mizler, 1752, p. 62.
108 Birke, 1966, pp. 70, 72, 75.
109 Marpurg's "Der Critisches Musicus und der Spree" (Critical music and the Spree, 1749) discusses emotions in music.
110 Leonhard Euler to Daniel Bernoulli, 16 May 1739, in Euler, *OO* IVA.1, p. 25.
111 Truesdell, 1984a, p. 310.
112 Thiele, 1982, p. 34.
113 Bernoulli's letter is scientifically important, inquiring about Euler's summation of the infinite series of the inverse squares of the natural numbers, $\zeta(2)$, and asserting that the Lapland Expedition led by Maupertuis had conclusively proved that the Newtonian, not the Cartesian, shape of Earth is correct.
114 See Bernoulli, 1983.

Chapter 5. Life Becomes Rather Dangerous: 1740 to August 1741

1 *Protokoly zasedanij konferentsii imperatorskoj*, vol. 1, pp. 555–599.
2 Leonhard Euler to Kantemir Antiokh, March 1740, in Yushkevich and Winter, 1959–76, vol. 3, pp. 125–126.
3 Kopelevich, 2007, pp. 46–47.
4 Euler, 1750 (E131).
5 Euler, 1740 (E35), p. 208.
6 Ibid., p. 197.
7 Euler, 1744 (E65). See also Fraser, 2005, pp. 168–180.
8 Carathéodory, 1952.
9 Ibid.
10 Harnack, 1900, vol. 1, pt. 1, p. 17.
11 Vucinich, 1963, p. 95.
12 Preuss, 1846–57, vol. 16, pp. 391 and 394.
13 Spiess, 1929, p. 105.
14 Ostermann, chancellor for Peter the Great, had helped place Anna on the throne.
15 Yushkevich and Winter, 1959–76, vol. 1, p. 45.
16 Daniel Bernoulli to Leonhard Euler, 28 January and 20 September 1740, in Fuss, [1843] 1968, vol. 2, pp. 59–63 and 466–467.
17 Daniel Bernoulli to Leonhard Euler, 20 September 1740, in Fuss, [1843] 1968, vol. 2, p. 474.
18 Daniel Bernoulli to Leonhard Euler, 23 April 1743, in Fuss, [1843] 1968, vol. 2, pp. 522–529; see esp. p. 523.
19 Johann I Bernoulli to Leonhard Euler, 18 February 1740/1 September 1741, in Fuss, [1843] 1968, vol. 2, p. 58.
20 Winter, 1957, p. 27.
21 Leonhard Euler to Gerhard Friedrich Müller, March 1740, in Yushkevich and Winter, 1959–1976, vol. 1, p. 45.
22 Leonhard Euler to Christian Goldbach, 1 August 1741, in Yushkevich and Winter, 1959–76, vol. 2, p. 83.
23 Leonhard Euler to Christian Goldbach, September and October 1740, in Yushkevich and Winter, 1959–76, vol. 2, p. 55n2; and Leonhard Euler to Christian Goldbach, 1 August 1741, in Yushkevich and Winter, 1959–76, vol. 2, p. 83.

Chapter 6. A Call to Berlin: August 1741 to 1744

1 Koser and Droysen, 1965–68, vol. 1, pp. 67–73.
2 Ibid., pp. 71–176; Bartholmess, [1850] 2009; Hartkopf and Dunken, 1967, pp. 11–15.
3 After the Thirty Years' War (1618–48) and the Peace of Westphalia the Holy Roman Empire remained Catholic. But in 1653 it was officially divided into two imperial associations by confessions for resolving religious and other disputes: the Corpus Evangelicorum, or Protestant league, and the Corpus Catholicorum.
4 Harnack, 1900, vol. 2, no. 142, p. 214.
5 Ibid., no. 146, p. 244.
6 Laitko, 1987, p. 74.
7 Harnack, 1900, vol. 1, pt. 1, pp. 247–254. Their correspondence hearkened, Voltaire thought, to the relation between Constantine and the Neoplatonic philosopher Julian.
8 An English translation was appended to John Keill, *An Examination of Dr. Burnet's Theory of the Earth* (1734).
9 Harnack, 1900, vol. 1, pt. 1, pp. 247–254.
10 Ibid., vol. 2, no. 148, p. 248.
11 Ibid., vol. 1, pt. 1, pp. 254–255.
12 Ibid., vol. 2, p. 259.
13 Ibid., vol. 2, chap. 2, pp. 250–256.
14 Beeson, 1992, pp. 136–138.
15 Preuss, 1846–57, vol. 21, pp. 369 ff. and vol. 22, p. 12; Harnack, 1900, vol. 1, pt. 1, pp. 248–249; McClellan, 1985.
16 Preuss, 1846–57, vol. 22, pp. 12ff.
17 Leonhard Euler to Joseph-Nicholas Delisle, 21–24 July 1742, in Euler, *OO* IVA.1, letter 499, p. 99.
18 Frederick II to Pierre Maupertuis, June 1740, in Terrall, 2002, p. 181.
19 Ibid., p. 183.
20 Some accounts have Kaspar as her father, but that is unlikely.
21 Duffy, 1985, p. 30.
22 Beeson, 1992, p. 140.
23 Delisle wrote that he had reviewed a catalog of the books by the astronomer Christfried Kirch covering decades of stellar observations that he and his wife Christina had made. Delisle lacked a few of these books, which he wanted Euler to try to acquire for him.
24 Before the First Silesian War, this was roughly equal to 20,000 francs.
25 Preuss, 1846–57, vol. 20, p. 199.
26 Leonhard Euler to Christian Goldbach, September 1741, in Fuss, [1843] 1968, p. x.
27 Koser and Droysen, 1965–68, vol. 2, p. 9.
28 Radelet-de-Grave, 2007, p. 173.
29 Pierre Maupertuis to Johann II Bernoulli, 1741 in Spiess 1929. For a similar quote, see also Jean Le Rond d'Alembert to Joseph Louis Lagrange, Spiess, 1929, p. 119.
30 Condorcet, 1786, p. 41. "*Madame, parce que je viens d'un pays ou quand on parle, on est perdu.*" Louis-Gustave du Pasquier, [1928] 2008, held that the verb is *pendu* (hanged), not *perdu* (lost).
31 In 1748, Baudan was named general quartermaster of the Russian Army.
32 Euler's letter of 15 December 1742 to Goldbach relates this order. See Fuss, 1843, vol. 2, pp. 169–171.
33 From 1866 to 1914 a bank would occupy the enlarged house, the builders taking care to preserve its impressive facade. On the bicentennial of Euler's birth in 1907, a plaque was placed on it. From 1995 to 1997, 21 Behrenstrasse was modified again and expanded to serve as the residence of the Bavarian representative. Today it sits across from the Comic Opera.

34 Preuss, 1846–57, vol. 20, p. 199.
35 Harnack, 1900, vol. 1, pt. 1, pp. 263–266.
36 Vucinich, 1963, pp. 98–99.
37 Leonhard Euler to Pierre Maupertuis, 18 June 1746, in Euler, *OO*, IVA.6, letter 12.
38 Grigorij Nikolaevich Teplov to Count Kirill Razumovskij, April 1744, in Yushkevich and Winter, 1959–76, vol. 1, p. 65, n. 1.
39 Pekarsky, 1864, p. 60.
40 See Harnack, 1900, vol. 1, pt. 1, p. 237; and Winter, 1957, p. 23.
41 Harnack, 1900, vol. 1, pt. 1, p. 261.
42 Yushkevich and Winter, 1959–76, vol. 1, p. 48.
43 Biermann, 2007, p. 91.
44 Winter, 1957, p. 154.
45 Harnack, 1900, vol. 1, pt. 1, p. 262.
46 Winter, 1957, pp. 23 and 25.
47 Leonhard Euler to Frederick II, 24 January 1743, in Euler, *OO* IVA.6, p. 304.
48 "De observatione inclinationis magneticae dissertatio" (A dissertation on an observed tendency of magnets, E108).
49 Radelet-de-Grave, 2007, pp. 173–174.
50 Normally the monetary amount of the prize was 2,500 livres.
51 Anderson, 1995, p. 103.
52 Winter, 1957, pp. 25–26.
53 See Euler, 1847 (E790).
54 Ibid. See also Euler, *OO*, III.2, pp. 392–415, esp. 405.
55 Other sources that Euler cited were Galileo Galilei's *Discorsi*, Pierre Bouguer's *Essai d'optique*, Johann Bernoulli's *Hydraulica*, Daniel Bernoulli's *Hydrodynamica*, Benjamin Robins's *New Principles of Gunnery*, Euler's own forthcoming *Scientia navalis*, and Giovanni Alfonso Borelli's *De motu animalium*.
56 Leonhard Euler to Christian Goldbach, 15 December 1742, in Fuss, [1843] 1968, vol. 1, p. 171.
57 Their titles are "Determinatio orbitae cometae qui mense Martio huius anni 1742" (Determination of the motion of a comet which can be observed in March of this year, 1742, E58); "Theoremata circa reductionem formularum integralium ad quadraturam circuli" (Theorems concerning the reduction of integral formulas to the quadrature of the circle, E59); "De inventione integralium, si post integrationem variabili quantitati determinatus valor tribuator" (On the resolution of an integral, if after integration the value for the determined variable quantity is assigned, E60); "De summis serierum reciprocarum ex potestatibus numerorum naturalium ortarum dissertatio altera, in qua eaedem summationes ex fonte maxime diverso derivantur" (On sums of series of reciprocals from powers of natural numbers from another discussion, in which the sums are derived principally from another source); and "De integratione aequationum differentialium altiorum graduum" (On the integration of differential equations of various degrees).
58 Daniel Bernoulli to Leonhard Euler, 4 September 1743, in Fuss, [1843] 1968, vol. 1, pp. 529–530.
59 Ibid.
60 See Harnack, 1900, vol. 1, pt. 1, p. 267.
61 Aldridge, 1975, p. 140.
62 Humbert was also the assessor of the French Calvinist consistory. The consistories were Calvinist ecclesiastical courts that included all officials, especially pastors and elders. They dealt with religious and political issues, including challenges from theoretical opposition.
63 Winter, "Introduction," in Euler, *OO* IVA.6, private translation by John S. D. Glaus, p. 6.

64 Preuss, 1846–57, vol. 14, p. 10.
65 Euler, *OO* IVA.5, p. 122. Daniel Bernoulli was made a foreign associate in 1746.
66 Winter, 1957, p. 36.
67 Ibid., p. 7.
68 Euler, *OO* IVA.6, letter 9a, pp. 306–309.
69 Harnack, 1900, vol. 2, p. 264. The physics class included general and experimental physics, natural history, chemistry, botany, and anatomy; the mathematics class included geometry, astronomy, mechanics, hydraulics, meteorology, civil and military architecture, and all branches of theoretical and practical mathematics; the philosophy class included metaphysics, morals, natural law, and the history and criticism of philosophy; and the philology class included literature, ancient and modern history, German history, languages, antiquities, inscriptions, and medallions.
70 "Monsieur," Euler continued, "d'avoir achevé un ouvrage, qui éclaircira aussi bien la question si importante sur la figure de la terre, qu'il augmentera l'admiration que tout le monde a conçue pour vous" (Sir, to have completed a work, which will also shed light on the question so important on the figure of Earth, will increase the admiration that the whole world has for you). See Euler, *OO* IVA.5, pp. 110–111.
71 In November 1747 Clairaut read to the assembled Paris Academy of Sciences his paper "Du systeme du monde dans les principes de la gravitation universelle," which appeared in the academy's *Histoire* for 1749. See Clairaut, 1749, p. 329.
72 Truesdell, 1984a, p. 333. According to Clairaut, the second phase of "l'époque d'une grande révolution dans la Physique," which had commenced with Newton's *Principia*, was well under way.
73 Ten volumes of series 2 of Euler's *Opera Omnia*, nos. 22 through 31, are on celestial mechanics and astronomy.
74 Fuss, 1843, vol. 1, p. 632.
75 Fraser, 1994, pp. 122–124.
76 See Weil, 1984, esp. pp. 132–133.
77 Weil, 1984, p. 261.
78 Euler, *OO* I.14, pp. 177–187.
79 Ibid., p. 203.
80 Euler found that $\log_a(1 + x) = 1/k\ (x/1 - x^2/2 - x^3/3 - x^4/4 - \ldots)$. See McKinzie, 2007b, pp. 143–144.
81 Richard Feynman called this "the most remarkable formula in math." See Nahin, 1998, p. 67. Nearly 125 years later, the Euler identity was essential to the rigorous recognition that not every real number is the root of an algebraic equation, that transcendental numbers must exist.
82 Euler, 1748 (I.8), p. viii.
83 His publications on number theory fill four large volumes in series I of his *Opera omnia.*
84 Philipp Naudé, quoted in Sandifer, 2005e.
85 Euler, *OO* I.2, pp. 163–193.
86 Ibid., pp. 194–390.
87 Leonhard Euler to Alexis Claude Clairaut, April 1742, in Euler, *OO* IVA.5, p. 124 of 120–126.
88 Weil 1984, p. 187.
89 Euler, *OO* IVA.5, p. 5.
90 Weil, 1984, pp. 187, 287–292.
91 Euler, 1746 (E86).
92 Yushkevich and Winter, 1959–76, vol. 2, p. 83.
93 Ibid.
94 Beeson, 1992, p. 150.
95 Winter, 1957, p. 29.
96 Leonhard Euler letter to Christian Goldbach, 4 July 1744, in Fuss, [1843] 1968, vol. 1, pp. 276–278.

97 Winter, 1957, p. 35.
98 Influenced by the writings of Tartaglia, Blondel wrote *L'art de jeter les bombes* (The art of launching bombs, 1683).
99 Euler, 1746 (E77). See also Euler, *OO* II.14.
100 Robins, 1739, p. 11.
101 Fuss, [1783] 2007, p. 378.
102 Euler, 1745 (E77).
103 Volumes 13–14 of the *Commentarii* (for 1741–46) appeared in 1751, and the eleven volumes of the *Novi Commentarii* (for 1747–65) appeared from 1750 to 1767.
104 Yushkevich and Winter, 1959–76, vol. 1, p. 22 and vol. 2, p. 102.
105 Ibid., vol. 2, pp. 72–74.
106 Truesdell, 1984a, p. 298.
107 Yushkevich and Winter, 1959–76, vol. 2, pp. 52–53.
108 Ibid., pp. 56 and 58. Schumacher used *analysis* instead of Euler's *analysin*.
109 Ibid., p. 8.
110 Ibid., p. 59. To a letter of inquiry (now lost) from Euler Schumacher responded that the young Israel Benjamin Gsell, the son of Georg Gsell and his third wife Maria Dorothea Gsell, had not been promised a position at the academy; the mother thought he had.
111 See Zubov, 1958, esp. pp. 28–29.

Chapter 7. "The Happiest Man in the World": 1744 to 1746

1 Bruyn, 2002, p. 56; Taylor, 1997, p. 69. The royal opera house had opened in December 1742; Georg Wenzelaus, Baron von Knobelsdorff, was the master architect.
2 Harnack, 1900, vol. 1, pt. 1, pp. 292–293.
3 Weil, 1984, chap. 3.
4 Leonhard Euler to Christian Goldbach, 4 July 1744, in Yushkevich and Winter, 1959–76, vol. 2, p. 199; and Fuss, [1843] 1968, vol. 1, pp. 292.
5 Yushkevich and Winter, 1959–76, vol. 2, no. 156, p. 34. The results were published in 1749 in *La figure de la terre, déterminée par les observations de Messieurs De la Condamine et Bouguer* (The observation of the earth, as determined by the observations of Messrs. De la Condamine and Bouger).
6 Yushkevich and Winter, 1959–76, vol. 2, p. 70.
7 The list of academy prizewinners begins in Harnack, 1900, vol. 2, p. 305.
8 Harnack, 1900, vol. 2, p. 29.
9 See Fuss, [1783] 2007.
10 Euler, *OO* III.1, pp. 3–15.
11 Harnack, 1900, vol. 1, pt. 1, p. 293.
12 Winter, 1957, pp. 40–41.
13 Ibid., p. 37.
14 Daniel Bernoulli to Leonhard Euler, 7 July 1745, in Fuss, [1843] 1968, vol. 2, p. 576.
15 Terrall, 2002, p. 240; Harnack, 1900, vol. 1, pt. 1, pp. 289–320.
16 Raith, 1983, p. 462.
17 Fuss, [1843] 1968, vol. 1, p. 321.
18 Carlyle, 1969, pp. 254–256.
19 Harnack, 1900, vol. 1, pt. 1, p. 313.
20 Harnack, 1900, vol. 1, pt. 1, p. 295; vol. 2, pp. 270–271. For his position in the philosophy and philology classes, Formey received a 200 Reichsthaler supplement to his salary of 400 Reichsthaler.
21 Fuss, [1843] 1968, vol. 1, p. 292.
22 Oldfather, Ellis, and Brown, 1993, p. 72.
23 Euler, *OO* IVA.6, letter 8.

24 Thiele, 1982, pp. 90–91.
25 Euler, *OO* I.24, p. xi.
26 Carathéodory, 1952, pp. viii–li.
27 Goldstine, 1980, p. 73.
28 Euler, *OO* I.24, p. 255.
29 Ibid., p. 308.
30 Euler, *OO* II.28, pp. 105–251.
31 Euler, 1744 (E67) and 1744 (E68).
32 Bradley, 2007c.
33 In Steele, 1994, p. 348, Thomas P. Hughes refers to Robins as "a founder of modern gunnery."
34 See Truesdell, 1954, esp. p. xxxviii.
35 Edwards, 2007.
36 See Fellmann, 1983b, esp. pp. 303–307.
37 Fellmann, 1983a, p. 73.
38 J. F. Weidler to Leonhard Euler, after 23 June 1744, in Euler, *OO* IVA.1, p. 450.
39 See Winter, 1958b, esp. p. 8.
40 Stepan Jakolevič Rumovskij, quoted in Thiele, 1982, p. 60.
41 Leonhard Euler to Christian Goldbach, 15 August 1750, in Fuss, [1843] 1968, vol. 1, pp. 527–529.
42 Winter, 1958a, p. 33.
43 Yushkevich and Winter, 1959–76, vol. 2, p. 31.
44 Ibid., pp. 79–84; see esp. pp. 81–82.
45 Leonhard Euler to Johann Kaspar Wettstein, January 1746, Yushkevich and Winter, 1959–76, vol. 3, p. 256.

Chapter 8. The Apogee Years, I: 1746 to 1748

1 For the impact of the *Encyclopédie*, which was published in 28 volumes between 1751 and 1772, see Darnton, 1979 and 1982; and Goodman, 1994, pp. 23–25.
2 The *Encyclopédie*'s three thousand subscribers at its start and four thousand by volume 5 made it prosperous. It has been called the grandchild of Pierre Bayle's *Historical and Critical Dictionary* of 1697.
3 See Darnton, 1997; and Popkin, 1998.
4 Truesdell, 1984b.
5 Since Katharina Euler's family included naturalists, there has been conjecture that she may have helped him in this botanical work, but there is no evidence to support it.
6 René Moreau to Daniel Bernoulli, June 1746, in Terrall, 2002, p. 240.
7 Harnack, 1900, vol. 1, pt. 1, pp. 310–311.
8 Leonhard Euler to Pierre Maupertuis, 18 June 1746, in Euler, *OO* IVA.6, pp. 68–69.
9 Hankins, 1970, p. 46.
10 Fuss, [1843] 1968, vol. 2, p. 598.
11 Terrall, 2002, p. 244.
12 Winter, 1957, p. 100.
13 Terrall, 2002, p. 242.
14 Euler, *OO* III.1, pp. 3–15. See chapter 7 of this book for a discussion of the paper.
15 Euler, 1768 (E343–E344), vol. 2, p. 40.
16 Harnack, 1900, vol. 2, no. 175, p. 305.
17 Harnack, 1900, vol. 1, pt. 1, pp. 402–403; Euler, 1840 English edition of the *Letters*, vol. 1, p. 328.
18 Euler, *OO* III.2, pp. 351–366.
19 Euler, 1768 [1872, an English ed. of the *Letters*], vol. 2, letter 125, pp. 39–40. In either a lapse of memory or a failure in proofreading, Euler gave the date of the prize competition as 1748 instead of 1747.

20 Clark, 1999, p. 440.
21 Ibid.; see also Terrall, 2002, pp. 259–260.
22 Biagioli, 1993, chap. 1.
23 Crown Prince August William to Frederick II, October 1746, in Yushkevich and Winter, 1959–76, vol. 1, p. 3.
24 Frederick II to Crown Prince August William, October 1746, in Yushkevich and Winter, 1959–76, vol. 1, p. 3.
25 Leonhard Euler to Johann Schumacher, June 1746, in Yushkevich and Winter, 1959–76, vol. 2, pp. 89–90.
26 Yushkevich and Winter, 1959–76, vol. 2, pp. 93–95.
27 Leonhard Euler to Grigorij Nikolaevich Teplov, October 1946, in Yushkevich and Winter, 1959–76, vol. 2, pp. 93–95. The Teplov letter reported that Kies had married Naudé's eldest daughter.
28 Translation in Wilson, 2007, pp. 123–124.
29 Gottfried Wilhelm Heinsius to Leonhard Euler, January 1746, in Euler, *OO* IVA.1, letter 1004, p. 179.
30 Abalakin and Grebnikov, 2007, pp. 251–254.
31 Euler, 1746 (E88), pp. 72 ff.
32 Hakfoort, 1995, p. 76.
33 Euler, 1746 (E88), p. 39.
34 Hakfoort, 1995, pp. 72, 187–189.
35 Baltus, 2007a, pp. 45–48.
36 Struik, 1969, pp. 99–102.
37 Euler 1750 (E140); and Clifford Truesdell, "The Rational Mechanics of Flexible or Elastic Bodies," in Euler, *OO* II.11, part 2, pp. 237–300.
38 Euler, *OO* IVA.5, p. 56.
39 Terrall, 2002, p. 271.
40 Euler, *OO* II.11, part 2, pp. 260–261.
41 Euler, 1768 (E343–E344), vol. 1, pp. 256–257.
42 Leonhard Euler to Christian Goldbach, 4 June 1747, in Yushkevich and Winter, 1965, pt. 2, pp. 274–275.
43 Euler, 1768 (E343–E344), vol. 1, p. 257.
44 Terrall, 2002, p. 258.
45 Euler, 1768 (E343–E344), vol. 1, p. 40.
46 The expression was taken from Juvenal; another translation is "Nature and wisdom are never at strife."
47 Yushkevich and Winter, 1965, pt. 2, p. 277, n. 3.
48 Terrall, 2002, p. 263.
49 Fuss, [1843] 1968, vol. 2, pp. 624–625.
50 Lagrange, 1882, pp. 147–148.
51 Euler, *OO* IVA.6; Euler, *OO* IVA.1, letter 1518, p. 262.
52 A certain Herr Ramspeck had informed his father of these disputes. In 1748 Ramspeck became professor of mathematics in Basel and the same year switched this position for elocution.
53 Winter, 1957, p. 47.
54 Euler, 1746 (E92).
55 Terrall, 2002, pp. 384–388.
56 Euler, 1768 (E343–E344), vol. 1, p. 18
57 Fuss, [1783] 2007, p. 394.
58 Euler, *OO* IVA.5, p. 257.
59 Ibid., p. 264.
60 Ibid., p. 273.
61 These rapidly produce large numbers. When $n = 19$, the number is 137,438,691,328, and $n = 47$ gives 9,903,520,314,282,971,830,448,816,128.

62 The fourth amicable pair is 5,020 and 5,564; the fifth 6,232 and 6,368; the sixth, 10,774 and 10,856; the seventh 12,285 and 14,595; the eighth 17,296 and 18,416; the ninth 63,020 and 76,084; and the tenth 66,928 and 66,992.
63 For an explanation of his method see Sandifer, 2005a.
64 Euler, 1752 (E177).
65 Ibid.
66 Bodenmann, 2009.
67 See Euler, 1749 (E120), esp. p. 119.
68 Alexis Claude Clairaut to Leonhard Euler, 3 September 1747 in Euler, *OO* IVA. 5, p. 173.
69 Ibid., p. 175.
70 Ibid., p. 336.
71 Habicht, 1983a; Fellman, 1983b.
72 Fellmann, [1995] 2007, English translation, p. 95.
73 Ibid., p. 31.
74 Euler, *OO* IVA.6, pp. 90–92.
75 Yushkevich and Winter, 1959–76, vol. 2, pp. 17, 105–107.
76 Leonhard Euler to Johann Kaspar Wettstein, 10 December 1746, private translation by John S. D. Glaus, EA; in Yushkevich and Winter, 1959–76, vol. 3, pp. 263–266.
77 Leonhard Euler to Kirill Gregor'evich Razumovskij, November 1747, in Yushkevich and Winter, 1959–76, vol. 2, pp. 106–107.
78 Leonhard Euler to Grigorij Nikolaevich Teplov, 29 October 1746, Yushkevich and Winter, 1959–76, vol. 1, Letter 40, pp. 93–95.
79 Hankins, 1970, p. 46.
80 Yushkevich and Winter, 1959–76, vol. 1, pp. 103–115.
81 Thiébault, [1805–13] 2008.
82 Euler, *OO* IVA.1, p. 222.
83 Ibid., p. 104.
84 Preuss, 1846–57, vol. 10, quoted in Pasquier, [1928] 2008, Glaus translation, p. 106.

Chapter 9. The Apogee Years, II: 1748 to 1750

1 Spiess, 1929, pp. 100–101.
2 Ferraro, 2007b, pp. 47–78.
3 Euler, 1748 (E101); see also Euler, 1738 (E19).
4 Euler, 1748 (E101), Blanton translation (1998–99), vol. 1, p. vi.
5 Ibid., p. 3.
6 Ferraro, 2000, p. 112.
7 Ibid.
8 Maor, 1994, p. 11.
9 Euler, 1748 (E101), Blanton translation (1988–89), vol. 2, pp. 318–335.
10 Ibid, vol. 1, p. 101.
11 Ibid., vol. 1, pp. 114–115, 149, 155.
12 Fantet de Lagny, 1692.
13 Legend has it that Maupertuis was at his bedside when he was dying. When it appeared that de Lagny had expired, Maupertuis leaned over his bed and whispered "What is the square of 12?" The answer quickly came: "144." Supposedly these were Fantet de Lagny's last words.
14 Euler, 1748 (E101), Blanton translation (1988–89), vol. 1, p. 140.
15 Nahin, 1998, p. 117.
16 In the second half of the seventeenth century, the English mathematician John Wallis had introduced this symbol.

17 Euler, 1748 (E101), Blanton translation (1998–89), vol. 1, p. xi.
18 Ibid., vol. 1, p. 88.
19 Bos, 1974.
20 Shabel, 2003.
21 Euler, 1749 (E120).
22 Wilson, 2007.
23 Chapin, 1995, p. 37.
24 Euler mistakenly believed that the moon had an atmosphere. "Sur l'atmosphere de la Lune prouvée par la dernier eclipse annulaire du Soleil" (On the atmosphere of the moon, proved by the last ring-shaped eclipse of the sun, E142) described observations made in Berlin of the eclipse of 25 July 1748. The translation herein comes from Sandifer 2008.
25 Euler, 1750 (E141).
26 Euler, 1750 (E141), Fabian and Nguyen translation, Euler Archive.
27 The filar micrometer is an instrument used with a telescope for precise visual measurements—for example, of the distance between double stars or of the diameters of asteroids. The word *filar* comes from the Latin *filum*, "thread."
28 Bradley, 2007a.
29 Leonhard Euler to Jean-Baptiste le Rond d'Alembert, 28 September 1748, in Euler, *OO* IVA.5, p. 294.
30 Euler, 1750 [E149].
31 Thiele, 1982, p. 68.
32 Clark, 1999, p. 444.
33 Leonhard Euler to Pierre Maupertuis, September 1748, in Euler, *OO* IVA.6, p. 262.
34 Harnack, 1900, vol. 2, no. 176, pp. 310–311.
35 La Mettrie lived only three more years; Frederick II wrote a eulogy for him in 1751.
36 Koser and Droysen, 1965–68, vol. 2, p. 284.
37 Winter, 1957, p. 48.
38 Maupertuis, favorite of the monarch and a supporter of La Mettrie, was similarly denounced in Paris for subverting morality in his anonymous book *Vénus physique* (1746) on the process of generation and heredity.
39 Leonhard Euler to Johann Kaspar Wettstein, 5 March 1748; Yushkevich and Winter, 1959–76, vol. 3, p. 275.
40 Ibid.
41 Harnack, 1900, vol. 1, pt. 1, p. 325.
42 Koser and Droysen, 1965–68, p. 237.
43 Euler, 1748 (E101), Blanton translation (1988–89), vol. 1, p. viii.
44 Euler, 1749 (E116).
45 Alexis Claude Clairaut to Gabriel Cramer, 3 June 1749, in Speziali, 1955, p. 227.
46 Alexis Claude Clairaut to Leonhard Euler, 21 July 1749, in Euler, *OO* IVA.6, p. 189.
47 Yushkevich and Winter, 1959–76. vol. 2, p. 20.
48 Wilson, 1987.
49 Harnack, 1900, vol. 1, pt. 1, p. 307.
50 Leonhard Euler to Christian Goldbach, April 1749, in Fuss, [1843] 1968, vol. 1, p. 497.
51 These essays on probability appear in Euler, *OO* I.7.
52 Eckert, 2002 and 2008.
53 Frederick II to Voltaire, January 1778, in Eckert, 2002, p. 1. "Vanity of vanities!" comes from Ecclesiastes 1:2.
54 Fuss, [1843] 1968, vol. 2, p. 8.
55 Euler, *OO* II.2, p. xlvii.
56 Terrall, 2002, p. 295.
57 Calinger, 1979, p. 350.
58 Kant, 1999, pp. 45–46.

59 Leonhard Euler to Johann Kaspar Wettstein, 27 September 1749, Yushkevich and Winter, 1959–76, vol. 3, pp. 283–285 and in a letter in Yushkevich and Winter, vol. 3, pp. 283–285.
60 Fellmann, [1995] 2007, p. 118.

Chapter 10. The Apogee Years, III: 1750 to 1753

1 Leonhard Euler to Grigorij Nikolaevich Teplov, 3 January 1750, in Yushkevich and Winter, 1959–76, vol. 2, letter 109, pp. 186–188. Some letters by Euler are in Russian. See, for example, vol. 3, pp. 123–125.
2 Leonhard Euler to Grigorij Nikolaevich Teplov, January 1750, in Yushkevich and Winter, 1959–76, vol. 2, pp. 286–287.
3 Alexis Claude Clairaut to Leonhard Euler, 19 January 1750, in Euler, *OO* IVA. 5, pp. 191–193, esp. p. 192.
4 Frederick II, quoted in Schwab, 1995, p. x.
5 See Magnan, 1986.
6 Leonhard Euler to Jean-Baptiste le Rond d'Alembert, 7 March 1749 in Euler, *OO* IVA.5, p. 306.
7 Euler, *OO* II.29, pp. 92–123.
8 Hankins, 1970, p. 50.
9 Euler, 1752 (E180).
10 Euler, 1752 (E180), translated by Robert Bradley, at the Euler Archive, http://euler-archive.maa.org/.
11 Euler, 1749 (E119).
12 Euler, 1750 (E140). See also Clifford Truesdell's comments in Euler, *OO* II.11, part 2.
13 A special copy with gilded edges was prepared for Frederick II and the crown princes.
14 Callot, 1964, pp. 107–108.
15 Euler, 1753 (E186), pp. 49 and 53.
16 Suzuki, 2007, p. 375.
17 Euler, 1753 (E196).
18 Euler, 1758 (E230).
19 Euler. 1758 (E231).
20 See Richeson, 2007 and 2008; and Sandifer, 2004c and 2004d.
21 Fuss, [1843] 1968, vol. 1, p. 538, and Yushkevich and Winter, 1959–76, vol. 3, p. 291.
22 At midcentury, nine of the regular Berlin academicians were Swiss: Nicolas de Beguelin, Jean (Giovanni) Castillon, Henri Alexandre de Catt, Leonhard Euler, Johann Heinrich Lambert, Johann Bernard Merian, Daniel Passavant, Johann Georg Sulzer, and Jakob Wegelin.
23 Pierre Maupertuis to Johann II and Daniel Bernoulli, 1746, in Terrall, 2002, p. 266.
24 After the armistice ending World War II in 1945, the Menzel painting and other artworks in the Kaiser Friedrich Museum burned.
25 Although the matter proved difficult, Euler found a tutor for the grandson of Vasilii Nikitich Tatishchev.
26 Fuss, [1843] 1968, vol. 1, p. 545.
27 Ibid.
28 Euler, *OO* II.12, p. lviii.
29 Leonhard Euler to Alexis Claude Clairaut, March 1751, in Euler, *OO* IVA.5, pp. 203–204.
30 Winsheim died in March 1751.
31 Leonhard Euler to Johann Schumacher, 3 April 1751, in Yushkevich and Winter, 1959–76, vol. 2, p. 239.
32 Leonhard Euler to Alexis Claude Clairaut, April 1751, in Euler, *OO* IVA.5, p. 206.

33 Euler, *OO* II.24, p. 1.
34 Immanuel, 1898, p. 422.
35 See Bachelard, 1961; Brunet, 1929, vol. 1, pp. 155–156; and Pulte, 1989.
36 Leonhard Euler to Pierre Maupertuis, 10 December, 1745, in Euler, *OO* IVA.6, pp. 306–309.
37 Oldfather, Ellis, and Brown, 1933, pp. 77–78.
38 Pierre Maupertuis, quoted in Terrall, 2002, p. 266.
39 Harnack, 1900, vol. 1, pt. 1, pp. 320–321.
40 Abraham Gotthelf Kästner to Pierre Maupertuis, June 1751, in Terrall, 2002, p. 298.
41 D'Alembert, 1751.
42 Leonhard Euler to Pierre Maupertuis, 10 December 1745, in Euler, *OO* IVA.6, pp. 56–57.
43 Pierre Maupertuis, quoted in Terrall, 2002, p. 297.
44 Ibid., p. 296.
45 Euler, 1751 (E158). See also Sandifer, 2005b; and Jordan Bell's translation of E158 at the Euler Archive, http://eulerarchive.maa.org/.
46 Leonhard Euler, quoted in Andrews, 2007, p. 225.
47 See Sandifer, 2007c.
48 Hankins, 1970, p. 48.
49 Bevis discovered what is today known as the Crab Nebula and observed the return of Halley's Comet in 1759. He died at the age of seventy-six, falling from his telescope.
50 See Ilic, 2008.
51 Leonhard Euler to Pierre Maupertuis, 31 March 1752, and Maupertuis to Johann II Bernoulli, 20 May 1752, in Euler, *OO* IVA.2, pp. xx.
52 The Berlin Academy's *Mémoires* for 1750 appeared in 1752. See Harnack, 1900, vol. 2, no. 171, pp. 296–303.
53 Euler, 1753 (E199), pp. 69–71.
54 Magnan, 1986, p. 45.
55 Johann Bernhard Merian, *Mémoires pour server à l'histoire du jugement de l'Académie*, quoted in Terrall, 2002, p. 301.
56 Oldfather, Ellis, and Brown, 1993, p. 78.
57 Ibid., pp. 13, 61–63, 89.
58 Ibid., pp. 133–135, 161.
59 Euler, 1753 (199).
60 Euler, E186, p. 144.
61 Ibid., pp. 101–105.
62 This letter (E182) filled pages 520–532 of volume 6, which appeared in print in 1754. See Euler, *OO* II.5, pp. 132–141.
63 Leonhard Euler to Johann Schumacher, December 1752, in Yushkevich and Winter, 1959–76, vol. 2, p. 39.
64 Euler, *OO* II.5, p. xvii.
65 Callot, 1964. p. 107. See also Voltaire, [1732] 2007, pp. 47–63.
66 See Strachey, 1922; and Strachey 1997, p. 7.
67 Brunet, 1929. vol. 1, p. 130.
68 According to Frederick, it started in a conversation at the royal table at Sanssouci. Voltaire was animated in the discussion and Maupertuis was silent. When the two went for a walk in the palace gardens with two officers who tried to engage Maupertuis in the conversation, he had little to say. The conflict would grow.
69 In January 1751, when the king returned to Potsdam, "Brother Voltaire" was secluded in a Berlin suburban chateau; it was presumed that Voltaire was hiding in penitence, but he was also ill.
70 Magnan, 1986, p. 41.
71 Voltaire opposed La Beaumelle, who wanted to succeed Julien Offray de La Mettrie as the king's reader; these circumstances stoked hostilities.

72 See Wilson, 1985, especially p. 132.
73 If no prize was won in a given year, the following year there was a double prize; the topic remained the same.
74 Christensen, 1987, p. 23.
75 Bernard le Bovier de Fontenelle, quoted in Heilbron, 1990, p. 1.
76 Paul, 1970, p. 140.
77 Diderot, [1821] 2006, p. 5.
78 Jean-Philippe Rameau, quoted in Paul, 1970, p. 150.
79 This article appeared in Paris in the journal *Mercure de France* in 1753. See also Gertsman, 2007, p. 345.
80 Spiess, 1929, p. 106.
81 Harnack, 1900, vol. 1, pt.1, p. 342.
82 In Greek mythology Sisyphus, the founder of Corinth, was punished for chronic deceit by being compelled to push a large boulder up a hill and watch it roll back to the bottom, only to begin the uphill push again.
83 See Goldenbaum, 2008.
84 See Mervaud, 1985, pp. 220–223.
85 Ibid.
86 Voltaire, [1752] 2005, p. 14.
87 Ibid.
88 Spiess, 1929, p. 135.
89 Truesdell, 1984b, p. 368.
90 See Terrall, 1990.

Chapter 11. Increasing Precision and Generalization in the Mathematical Sciences: 1753 to 1756

1 Euler, 1753 (E199).
2 Euler, 1753 (E199), independent printing, pp. 151, 171, and 193.
3 Voltaire took with him the chamberlain's key and Frederick's poems.
4 Harnack, 1900, vol. 2, pt. 172, pp. 302–303.
5 Breger, 1999, p. 380.
6 See D'Alembert, 1754.
7 D'Alembert, 1743, pp. 169–186.
8 Terrall, 2002, p. 297
9 Condorcet, 1786, pp. 59–60.
10 Gandt, 2001, pp. 188–190.
11 Brunet, 1929, vol. 1, pp. 155–156; Tuffet, 1967, p. lxx.
12 Brunet, 1929, vol. 1, pp. 155–156.
13 Bodenmann, 2010, p. 31.
14 See Wepster, 2009.
15 Euler, 1753 (E187).
16 Leonhard Euler to Alexis Claude Clairaut, 10 April 1751, in Euler *OO* IVA.5, p. 206.
17 Euler, *OO* IVA.5, pp. 211–213.
18 Wilson, 1980a, p. 146.
19 Wepster, 2009, p. 206.
20 Johann Tobias Mayer to Leonhard Euler, 23 February 1755, in Forbes 1971, pp. 96–97.
21 Weil, 1984, p. 121.
22 Euler, 1761 (E252) and 1761 (E251). See also D'Antonio, 2007a; and Sandifer, 2005c.
23 Sandifer, 2004b.
24 Fellmann, [1995] 2007, pp. 93–94.

25 Yushkevich and Winter, 1959–76, vol. 2, p. 318.
26 Beeson, 1992, p. 261; Terrall, 2002, p. 349.
27 Euler, 1768 (E343); Euler *OO* III.12, letters 78–79, pp. 169–173.
28 Forbes, 1971a, p. 89.
29 See Wepster, 2009.
30 Ibid., p. 93.
31 Ibid.
32 Ibid., p. 88.
33 Forbes, 1971b, p. 86.
34 Diderot, 1999, p. 37.
35 Yushkevich and Winter, 1959–76, vol. 1, p. 27.
36 Sandifer, 2007c, pp. 25–28.
37 Yushkevich and Winter, 1959–76, vol. 1, p. 9.
38 Christian Goldbach to Leonhard Euler, 26 April 1755, in Yushkevich and Winter, 1959–76, vol. 1, p. 376.
39 Fellmann, 2007, pp. 103–104.
40 Euler, 1755 (E212).
41 See Yushkevich, 1983.
42 See Volkert, 1987.
43 Euler, [1755] 2000 (I.10), pp. 69–79, from the Blanton translation.
44 Leonhard Euler, quoted in Ferraro, 2007b, p. 58.
45 See Bos, 1974.
46 Truesdell, 1984a, p. 296.
47 Weyl, 1968a, p. 124.
48 Clifford Truesdell, "Introduction," in Euler, *OO* II.12.
49 Euler, 1757 (E226).
50 Euler, *OO* II.12, p. xx.
51 Rouse, 1978, pp. 45–47.
52 Euler, 1757 (E226) *OO* II.12, p. 316.
53 Leonhard Euler to Johann Kaspar Wettstein, 5 August 1755, in Yushkevich and Winter, 1959–76, vol. 3, letter 288.
54 Forbes, 1971a, p. 97.
55 Harrison was the inventor of the first successful marine chronometer.
56 Spiess, 1929, p. 89.
57 Euler, *OO* IVA.5, p. 38.
58 See Mattmüller, 2008.
59 Blunt, 1967; for the translated passages, see pp. 361–366.
60 Euler, 1862 (E842).
61 See Knapp, 1971; and Harnack, 1900, vol. 1, pt. 1, pp. 403–405.
62 Usually the prize competition drew about a dozen articles. Contributions of leading German scholars—for example, Immanuel Kant wrote his *Optimism* in 1759—and the influence of prize papers caused these competitions to grow in importance.

Chapter 12. War and Estrangement, 1756 to July 1766

1 Euler, 1771 (E414).
2 See Yushkevich and Winter, 1959–76, vol. 1, p. 100, and vol. 2, pp. 191, 405, and 409–411.
3 See Speiser, "Introduction," in Euler, *OO* III.6, pp. vii–xxviii.
4 See Algarotti, 1764, pp. 60–66.
5 There is a similar quotation in Euler's letter of 24 April 1756 to Lagrange which can be found in the *OO* IVA.5, pp. 386–387.

6 Terrall, 2002, p. 355.
7 Harnack, 1900, vol.1, pt. 1, p. 347.
8 Biagioli, 1992.
9 Terrall, 2002, p. 355.
10 Ibid., p. 393.
11 Spiess, 1929, p. 141.
12 Leonhard Euler to Johann Kaspar Wettstein, 9 October 1756, Yushkevich and Winter, 1959, vol. 3, p. 344.
13 The plug is a form of pipe tobacco containing large leaves layered atop each other and pressed to form a block (or "plug").
14 Euler, 1761 (E251). Stacy Langton has translated this paper at the Euler Archive, http://eulerarchive.maa.org/. See also Struik, 1969, pp. 378–383.
15 See Fuss, [1783] 2007, especially p. 385 and Fuss, [1783] 2009, Sheynin trans., pp. 8–48, especially p. 23.
16 Leonhard Euler to Johann Kaspar Wettstein, 26 March 1757, Yushkevich and Winter, 1959–1976, vol. 3, p. 347.
17 Winter, 1957, p. 61.
18 Leonhard Euler to Pierre Maupertuis, October 1758, in Pasquier, [1928] 2008, as translated by Glaus, pp. 90–91.
19 Ibid., p. 91.
20 Marston, 2001, p. 66.
21 Leonhard Euler to Pierre Maupertuis, February 1759, in Euler, *OO* IVA.6, pp. 17–18.
22 Euler, 1771 (E415).
23 Euler, 1766, (E309), p. 310; see also Sandifer, 2006b.
24 There were cases when it was not a machine but had inside the Turk figure a full-size chess master.
25 Spiess, 1929, p. 143.
26 Frederick II to Leonhard Euler, 15 September 1759, in Euler, *OO* IVA.6, p. 377.
27 Thiébault, [1804–13] 2008; Harnack, 1900, vol. 1, pt. 1, p. 364.
28 Pasquier, 1927, as translated by Glaus, pp. 143–144.
29 Winter, 1957, p. 64.
30 Harnack, 1900, vol. 1, pt. 1, pp. 347–348.
31 Spiess, 1929, p. 156.
32 Ibid.
33 Maltese, 2000; Langton, 2007a.
34 Leonhard Euler to Pierre Maupertuis, September 1757, in Euler, *OO* IVA.6, pp. 423–426.
35 Yushkevich and Winter, 1959–76, vol. 1, p. 145.
36 Leonhard Euler to Joseph-Louis Lagrange, October 1759, in Euler, *OO* IVA.5, pp. 423–426.
37 Ibid., p. 425.
38 Euler, 1766 (E296) and 1766 (E297). See also Fraser, 1985.
39 Jean-Baptiste le Rond d'Alembert to Joseph-Louis Lagrange, 1759, as translated by Clifford Truesdell, in Hankins, 1970, p. 59.
40 Leonhard Euler to Joseph-Louis Lagrange, October 1759, in Euler, *OO* IVA.5, p. 420.
41 Karl Euler, 1955, p. 269.
42 Fuss, [1843] 1968, vol. 1, p. 657.
43 Leonhard Euler to Gerhard Friedrich Müller, 18 October 1760, in Yushkevich and Winter, 1959–76, vol. 1, p. 161.
44 Later Totleben was found to be a traitor; he sought to be a spy for Prussia with an assured income after the war.
45 Leonhard Euler to Gerhard Friedrich Müller, 18 October 1760 in Euler, *OO* IVA.1, No. 1806, p. 305.

46 Yushkevich, 1963.
47 Euler, 1761 (E257).
48 Euler, 1765 (E292).
49 Jaki, 1977.
50 Lambert's description of the Milky Way was mostly considered fictional until telescopic observations by William Herschel found numerous nebulae.
51 The next year Lambert, a peripatetic and shy but argumentative scholar, resigned from the young Bavarian Academy of Science, which he had founded. A Protestant, he could not work with the Bavarian Jesuits. He went to Switzerland as a geometer and surveyor. The Petersburg Imperial Academy of Sciences invited him to join, but he waited for a call to Berlin.
52 Spiess, 1929, p. 158.
53 Euler, *OO* IVA.5, p. 11.
54 Euler, *OO*. IVA.5, p. 223.
55 Pekarsky, 1864, pp. 79– 80.
56 Harnack, 1900, vol. 1, pt. 1, p. 357.
57 Condorcet, 1786.
58 Fuss, [1843] 1968, vol. 1, p. 668.
59 Pekarsky, 1864, p. 61.
60 Grigorij Nikolaevich Teplov to Leonhard Euler, 28 March 1763, in Pekarsky, 1864, p. 79, and Yushkevich and Winter, 1959–76, vol. 2, pp. 214–215.
61 Pekarsky, 1864, pp. 78–79.
62 Yushkevich and Winter, 1959–76, vol. 2, p. 433.
63 Leonhard Euler to Grigorij Nikolaevich Teplov, 26 July 1763, in Yushkevich and Winter, 1959–76, vol. 2, p. 435.
64 Unpublished letter of Euler to the consistory, translated by John S. D. Glaus.
65 Yushkevich and Winter, 1959–76, vol. 2, p. 435.
66 Spiess, 1929, p. 175.
67 Fellmann, 2003; Mattmüller, 2008, pp. 45–46.
68 Harnack, 1900, vol. 1, pt. 1, pp. 351 and 358.
69 Fuss, [1843] 1968, vol. 1, p. 667.
70 Jaquel, 1983, p. 442.
71 Harnack, 1900, vol. 1, pt. 1, p. 363.
72 Forbes 1971a.
73 Frederick II to Leonhard Euler, 16 June 1765, in Euler, *OO* IVA.6, p. 390 and Preuss, 1846–57, vol. 20, p. 209.
74 Frederick II to Leonhard Euler, 16 June 1765, in Euler, *OO* IVA.6, p. 390.
75 Frederick II to Leonhard Euler, 17 March 1766, in Euler, *OO*, IVA.6, p. 392 and Preuss, 1846–57, vol. 20, p 210.
76 Goldenbaum, 2008.
77 Harnack, 1900, vol. 1, pt. 1, p. 366.
78 Spiess, 1929, p. 150.
79 Lambert remained calm. If Frederick did not appoint him, it would be a stain on the king's record.
80 Biermann, 2007, p. 89.
81 Ibid.
82 Ibid., p. 87.
83 Yushkevich and Winter, 1959–76, vol. 1, p. 31.
84 Spiess, 1929, p. 155.
85 Euler, 1765 (E289).
86 Spiess, 1929, p. 157.
87 Forbes, 1971a, p. 13.
88 Ibid., appendix X, pp. 87–88.

89 Ibid., pp. 59–92; see esp. p. 63. The translation from Russian is by Alexander Levin.
90 Spiess, 1929. pp. 181–182.
91 Johann Albrecht Euler to Wenzeslaus Johann Gustav Karsten, 15 February 1766, in Fellmann, [1995] 2007, p. 111.
92 Leonhard Euler to Frederick II, 15 March 1766, in Euler, *OO* IVA.6, p. 391.
93 Frederick II to Leonhard Euler, 17 March 1766, in Euler, *OO* IVA.6, p. 392.
94 As quoted in Fellmann [1995] 2007, p. 112.
95 Euler, *OO* IVA.6, p. 391.
96 Ibid., p. 393; and Preuss, 1846–57, vol. 13, p. 47.
97 Fellmann, 2007, pp. 112–113; Euler, *OO* IVA.5, p. 391.
98 Harnack, 1900, vol. 1, pt. 1, p. 411.

Chapter 13. Return to Saint Petersburg: Academy Reform and Great Productivity, July 1766 to 1773

1 Yushkevich and Winter, 1959–76, vol. 1, p. 263.
2 Spiess, 1929, p. 184.
3 Fellmann, [1995] 2007, p. 116.
4 Yushkevich and Winter, 1959–76, vol. 1, p. 263.
5 Ibid., p. 32; Euler, *OO* IVA.1, p. 105.
6 Yushkevich and Winter, 1959–76, vol. 1, p. 32.
7 Leonhard Euler to Antonio Nuñes Ribeiro Sanches, 17 October 1766, in Pasquier, [1928] 2008, p. 116; Euler, *OO* IVA.1, p. 351.
8 Count Sigismund Ehrenreich Redern, quoted in Vucinich, 1963, p. 141.
9 Euler, *OO* IVA.1, p. 16.
10 Fellmann, [1995] 2007, p. 135.
11 In recreational mathematics, a magic square is an arrangement of distinct numbers in an $n \times n$ matrix, in which each row, column, and diagonal has as its sum the same number.
12 Euler, *OO* IVA.3.
13 Ferraro, 2007b, p. 76.
14 Taton, 1983, pp. 414–415.
15 Euler, 1768 (E342), 1769 (E366), and 1770 (E385).
16 See Sandifer, 2004a.
17 Fraser, 1997, pp. 325–327; Lützen, 2007; Thiele, 2007c.
18 Ferraro, 2007b.
19 Weil, 1984, pp. 250–252.
20 Euler, *OO* IVA.5, p. 464.
21 See Euler, 1770 (E385), vol. 3, pp. 195–223.
22 This is summarized in Euler, *OO* IVA.1, letter 1919, p. 325.
23 Kemp, 1963, p. 260.
24 By 1983 the number of editions of the *Lettres* had reached 111. See Fellmann, 1983a, p. 71.
25 This was the first of five editions. The second appeared in 1785, the third in 1790-91, the fourth in 1796, and the fifth with only the first two books in 1808. All editions were published in Saint Petersburg.
26 See Grigor'ian and Kirsanov, 2007.
27 See Mincenko, 1957.
28 Calinger, 1975, p. 230.
29 Koyré, 1968, p. 18.
30 See Euler, *OO* III.12; and Radelet-de-Grave, 2007, pp. 168–169.

31 Wilson, 2007, pp. 132–133.
32 Ibid.
33 See Sandifer, 2004e.
34 Grigor'ian and Kirsanov, 1983, p. 388.
35 Pekarsky, 1864, p. 60.
36 Manuel, 1963.
37 Timerding, 1919.
38 The *Lettres* also inspired the English physicist Michael Faraday. See Caparrini and Williams, 2008.
39 Euler, 1769 (E367), 1770 (E386), and 1771 (E404).
40 Euler, 1767 (E325).
41 Caparrini and Williams, 2008.
42 Ibid.
43 Henry, 2007, p. 188.
44 Euler, 1767 (E334). See the translations by Richard J. Pulskamp and Todd Doucet available at the Euler Archive; see also Ineichen, 2007, pp. 103–104.
45 Montucla, [1758] 1960, p. 77.
46 Sandifer, 2005b.
47 Weil, 1984, p. 169.
48 Lagrange, 1882, vol. 14, pp. 135–245.
49 Euler, 1761 (E258).
50 Mikhaĭlov and Sedov, 2007.
51 Euler, 1769 (E375).
52 Euler, 1770 (E396) and 1771 (E409).
53 Euler, 1772 (E424).
54 Johann Albrecht Euler to Abraham Gotthelf Kästner, September 1769, in Pasquier, 1927, Glaus translation, p. 158.
55 Euler, 1984, p. liv.
56 Fellmann, [1995] 2007, p. 121.
57 Caspar, 1993, p. 290.
58 See Woolf, 1959.
59 The pairings after 1600 are 1631 and 1639, 1761 and 1769, 1874 and 1882, and 2004 and 2012.
60 Joseph Jerôme le Français de Lalande, quoted in Taton and Wilson, 1995, p. 158.
61 They also discussed the path of the comet of 1759, the publication of Euler's *Integral Calculus,* and the reconciliation between d'Alembert and Euler, over which Lalande expressed his joy.
62 Euler, 1769 (E372).
63 Euler, 1772 (E425)
64 Grant, 1842, p. 60.
65 In 1772, Bonnet sent Euler the second volume of *La palinogénésis philosophique*.
66 For Hippocrates of Chios's proof regarding lunes, see Calinger, 1999, pp. 87–89.
67 Euler, 1772 (E423).
68 See 2007b, p. 60.
69 Reich, 2007.
70 Golland and Golland, 1993.
71 See Wilson, 1985.
72 *Protokolii*, September 1771 to May 1772.
73 "Leonhard Euler in St. Petersburg," a report from School No. 27, E. A. Agavelov, pp. 1–30, 1999.
74 Francis Horner, "Euler," in Euler, 1770 (E387), Horner translation, p. xlvii.
75 Euler, 1772 (E418).

76 Euler, *OO* II.24; see also Abalakin and Grebnikov, 2007, pp. 253–255.
77 Leonhard Euler to Joseph-Louis, Comte de Lagrange, 31 May 1771, in Euler, *OO* IVA.5, p. 566.
78 Ibid.
79 Marsden, 1995, p. 186.
80 Kholchevnikov, 2008.
81 Verdun, 2012, pp. 235–239.
82 Leonhard Euler, quoted in Abalakin and Grebnikov, 2007, p. 249.
83 Euler, 1771 (E410).
84 Clifford Ambrose Truesdell, "Part IV. Researches Subsequent to Euler's 'First Principles of Mechanics, 1752–1788,'" in *OO* II.11, pt. 2, p. 391. See also Fraser, 1997b, sect. 12, p. 2.
85 Truesdell, "Part IV," p. 391.
86 Euler, *OO* IVA.5, p. 55.
87 Archimedes's cattle problem is today considered a problem in Diophantine analysis; it examines simultaneous polynomial equations with integer solutions. The problem requires computing the total number of cattle in a herd that meets a series of conditions. The herd is to consist of bulls (in larger numbers) and cows, with each in given proportions according to four colors; it thus has four variables and three equations. The Greek epigram stating it was first translated by Gotthold Ephraim Lessing in 1773. Since it involves huge numbers, both for the number of bulls by color and the total, it was not often pursued. Lessing computed the total number to be the equivalent of 7.76 times 10^{206544}.
88 Joseph-Louis, Comte de Lagrange to Leonhard Euler, February 1770, in Weil, 1984, pp. 167–168; Euler, *OO* IVA.5, pp. 471–477.
89 Euler, *OO* IVA.5, p. 470, n. 7.
90 If two numbers b and c have the property that their difference $b - c$ is integrally divisible by a number m (i.e., $(b - c)/m$ is an integer), then b and c are said to be congruent modulo m.
91 Power residue modulo m. An integer a for which the congruence $x^n \equiv a(\mathrm{mod}\, m)$ is solvable for a given integer $n > 1$. The number a is called a residue of degree n modulo m.
92 Euler, 1783 (E552).
93 Leonhard Euler, as translated in Weil, 1984, p. 253.

Chapter 14. Vigorous Autumnal Years: 1773 to 1782

1 See d'Alembert, 2009.
2 Their eldest son, Paul Heinrich von Fuss, was the permanent secretary of the academy from 1826 to 1855.
3 Adrien-Marie Legendre, quoted in Weil, 1984, p. 325.
4 Ibid., p. 219.
5 Ferreiro, 2007.
6 Euler, 1771 (E415).
7 Vucinich, 1963, pp. 192–193.
8 Denis Diderot to Johann Albrecht Euler, 28 October 1773, in *Protocols*, Zhelnina, 1999, pp. 22–23.
9 See Struik, 1939; Brown, 2007; and Gillings, 1954.
10 Thiébault, [1804–13] 2008, vol. 2, pp. 305–306.
11 Wilson, 1972, p. 90.
12 In a letter of 31 December 1867 to the literary journal *Athenaeum*. De Morgan had first given his account of this tale (The *Athenaeum* Projects: Centre for Interactive Systems Research, City University, London).

13 Stén, 2014, p. 132.
14 Fellmann, [1995] 2007, p. 125.
15 D'Antonio, 2007a, pp. 124–125.
16 Reich, 2007, pp. 484–485.
17 Baron Anne-Robert Jacques Turgot to Louis XVI, August 1774, in Truesdell, 1984a, p. 337.
18 Ibid.
19 Euler, *OO* IVA.5, p. 504, as translated in Weil, 1984, pp. 215–216.
20 See Gautschi, 2008, esp. p. 31; and Weil, 1984, pp. 132–133.
21 Euler, 1785 (E598).
22 Ibid., quoted in Weil, 1984, p. 219.
23 Edwards, 2007, argues that Euler did reach the law, but Sandifer, 2005d, disagrees.
24 Euler, 1785 (E598), quoted in Weil, 1984, p. 285.
25 See Raskin, 2007, esp. p. 319.
26 Ibid.
27 Euler, 1776 (E480).
28 This followed his memoir "Formulae generales pro translatione quacunque corporum rigidorum" (General formulas for the translation of arbitrary rigid bodies, E478), which replaced older treatments in geometry and previous mechanics with analytical formulae for a single point.
29 Truesdell, 1968, p. 260. See also Mikhaĭlov and Sedov, 2007, esp. pp. 175–176.
30 See Langton, 2007a; and Truesdell, 1968, p. 171.
31 See Caparrini and Fraser, 2013.
32 Pasquier, [1927] 2008, p. 147.
33 *Protokolii*, September 1775.
34 *Protokolii*, 4 March 1776.
35 Fellmann, [1995] 2007, p. 126.
36 They particularly considered book 3, chapter 6 and book 4, chapter 3.
37 Ineichen, 2007, p. 111.
38 Leonhard Euler, quoted in Ineichen, 2007, p. 111.
39 Raskin, 2007, p. 321.
40 A *pood*, a unit of weight employed in Imperial Russia, equaled forty Russian pounds, or roughly 16.38 kilograms. Today's equivalent is approximately thirty-six pounds.
41 Vucinich, 1963, p. 173.
42 Ibid., p. 128.
43 Fellmann, [1995] 2007, p. 129.
44 Leonhard Euler to Frederick II, 17 January 1777, in Euler, *OO* IVA.6, pp. 395–396.
45 Leonhard Euler to Joseph-Louis, Comte de Lagrange, 1775, in Euler, *OO* IVA.5, pp. 504–509.
46 Euler, 1779 (E498), Linowitz translation.
47 *Protokolii*, 1774.
48 Euler, 1777 (E487), 1778 (E496), 1778 (E497), 1778 (E504), and 1798 (E704); see Ferraro, 2008c, p. 282.
49 Euler, 1778 (E490), 1778 (E491), and 1778 (E492); see also Heine, 2007.
50 Euler, 1779 (E498), 1801 (E708), 1802 (E715), 1805 (E718), 1805 (E719), and 1806 (E725).
51 Weil, 1984, p. 188.
52 As translated in Weil, 1984, p. 224.
53 Euler, 1778 (E489)
54 Fellmann, [1995] 2007, p. 117.
55 Bernoulli, 1779, p. 245.
56 Fellmann, [1995] 2007, p 119.

57 Henry, 2007, p. 21.
58 Darbès made a copy that is now in the Museum of Art and History in Geneva.
59 Euler, 1849 (E795).
60 See Klyve and Stemkowski, 2007.
61 Caparrini, 2002, pp. 154–156.
62 Euler, 1782 (E524), 1782 (E526), and 1820 (E754).
63 Dvoichenko-Markov, 1965, pp. 53–54.

Chapter 15. Toward "a More Perfect State of Dreaming": 1782 to October 1783

1 Woronzoff-Dashkoff, 2008, p. 129.
2 Dashkova, 1995, p. 200.
3 Ibid.
4 Ibid., p. 334. See also Hoogenboom, 2007, p. 3. The princess probably mistakenly called Fuss his grandson, because before she wrote her memoirs Fuss married one of Euler's granddaughters, Albertina.
5 Weil, 1984, p. 201.
6 Euler, 1783 (E553); 1783 (E555); 1783 (E560); 1783 (E570).
7 Euler, 1783 (E532); 1783 (E536); 1783 (E537); 1783 (E538); 1783 (E542).
8 Pavlova, 1958, pp. 605–608.
9 Letters of Lexell to Jean-Hyacinthe Magellan on 30 September 1783 and to Pehr Wilhelm Wargentin on 10 October 1783 in Stén, 2014, pp. 233–234.
10 Pavlova, 1958, p. 606.
11 Letter of Lexell to Pehr Wilhelm Wargentin on 10 October 1783 in Stén, 2014, p. 234.
12 Euler, 1768 (E343–E344), vol. 1, p. 310.
13 Johann Albrecht transcribed at least fifty-two, and possibly as many as seventy, of Leonhard Euler's articles in Saint Petersburg. From 1766 to 1790 Johann Albrecht had an active correspondence of over five hundred letters exchanged with Formey in Berlin. Written in French in the form of a diary, these letters are often more accurate than the official protocols.
14 Fuss, [1783], 2007, p. 394.
15 Ibid.
16 Ibid., p. 393.
17 Condorcet, 1786, p. 49.
18 See Formey, 1786, especially pp. 8–9.
19 Pasquier, 1927, Glaus translation, p. 131. The blind Theban seer Tiresias was the most famous soothsayer in archaic and classical Greek mythology; an adviser to Oedipus, he was well known for telling the truth. When Tiresias was blinded by a god, neither Zeus nor Athena could undo this but as compensation granted him the inner vision of wisdom and the gift of prophecy. He is referred to in the works of Homer, Euripides, Aeschylus, and Sophocles.
20 In 2007 this sculpted bust was installed at the International Mathematical Euler Institute in Saint Petersburg.
21 From his searches in the Petersburg Academy archives and private holdings of Euler's descendants, Euler's great-grandson Paul Heinrich Fuss found sixty handwritten manuscripts that were published as the *Opera postuma* (E805) in 1862, two years after Nicholas Fuss died. This was the last Euler publication in the nineteenth century. See Ozhigova, 2007.
22 Its inscription includes the year of installation, MDCCCXXXVII, and gives below the place and date of birth and death according to the Julian and Gregorian

calendars. In 1957, when celebrating the 250th anniversary of the birth of Euler, the academy moved his remains and the monument to the Lazarian Cemetery of the Alexander-Nevskii Monastery in Leningrad (now Saint Petersburg). It is near the grave of Mikhail Lomonosov.

23 Matvieskaya, 2008. The author discusses the deciphering of Euler's notebooks from Gustaf J. Eneström in 1913, to Vladimir Ivanovitch Smirnov's study begun in 1954 of those mainly on number theory, through Gleb K. Mikhaĭlov's examination of them in 1957. Their investigation is still unfinished.

24 His stature is supported in the picture from 1957 positioning Euler between Luther and Marx. An example of a useful application today is the Viaduc de Millau, a bridge spanning the valley of the River Tarn between Clermont-Ferrand and Montpellier in France. Opened in 2004, it is the highest motorway bridge in Europe. Its shape is an elegant curve that for stability, vibrations, and wind currents depends on Euler's formulas. It is dedicated to Leonhard Euler.

25 Truesdell, 1984b, p. 340.

General Bibliography of Works Consulted

The *Opera omnia* of Leonhard Euler

Begun in 1911 by the Swiss Society of Natural Scientists, which is today the Swiss Academy of Sciences, the publication of Euler's *Opera omnia* (Collected works) has been a large and formidable task; it consists of four series and more than eighty volumes. The first three series contain seventy-two volumes in quarto of three hundred to six hundred pages each with almost no repetition. The first series covers mathematics; its twenty-nine volumes are printed in thirty-two volume parts, and all are currently in print. The second series concerns mechanics and astronomy; its thirty-one volumes are in thirty-two volume parts. The subdivided volumes provide two equally large works. The twelve volumes of the third series on miscellany are currently in print. The *Opera omnia* mostly republishes works that Euler prepared for printing in all of the branches of the mathematical sciences known during his lifetime. They appear in the original language—chiefly Latin or French. These first three series contain a dozen multivolume textbooks, twelve short works on popular science, a series of prize essays submitted annually to the Paris Academy of Sciences (almost all of them successful at winning), and hundreds of research articles, most of which appeared in the leading scientific journals of Europe. The exact number of Euler's books and articles contained in these series is 819. The total number of pages for these series is approximately 35,000. The Euler Commission in Basel still has two volumes yet to publish in the second series volumes 26 and 27 on astronomy and perturbation theory; these are projected for completion in 2015–16.

The last series of the *Opera omnia*, the fourth, will have ten volumes in twelve volume parts: IVA begins with a vade mecum summarizing the over 3,300 extant letters to and from Euler out of an estimated total of 4,500 to 5,000 letters with over 275 correspondents across Europe in French, Latin, German, Russian, and English, and it is ordered chronologically from 1726 to 1782. To 1998 four volumes—1, 2, 5, and 6—had been printed in series IVA. Volume 4 is scheduled for release early in 2015. That leaves four correspondence volumes—IVA.3 (with Daniel Bernoulli), IVA.4 (with

Swiss correspondents), IVA.8 (with Halle correspondents) and IVA.9 (with Knutzen associates). These last four volumes will probably appear in 2016 and 2017. The planned mutual Russian-Swiss project provisionally titled Euler Heritage (presently still in its early stages) would supply online Euler's originally handwritten literary record as it can be found in the Saint Petersburg Archive of the Russian Academy of Sciences—that is, his previously unpublished diaries, manuscripts, and twelve scientific notebooks. It will also contribute to a critical analysis of these online volumes.

The *Opera omnia* of Leonhard Euler: Volumes Cited

Note: The numbers in the left column below represent the series numbers (in roman numerals), followed by the volume numbers (in arabic numerals).

I.2 *Commentationes arithmeticae*, part 1. Edited by Ferdinand Rudio. Leipzig: B. G. Teubner, 1915.

I.4 *Commentationes arithmeticae*, part 3. Edited by Rudolf Fueter. Leipzig: B. G. Teubner, 1941.

I.7 *Commentationes algebraicae ad theoriam aequationum pertinentes.* Edited by Louis-Gustave du Pasquier. Berlin: B. G. Teubner, 1923.

I.8 *Introductio in analysin infinitorum*, part 1. Edited by Adolf Krazer and Ferdinand Rudio. Berlin: B. G. Teubner, 1922.

I.9 *Introductio in analysin infinitorum*, part 2. Edited by Andreas Speiser. Zurich: Orell Füssli, 1945.

I.10 *Institutiones calculi differentialis.* Edited by Gerhard Kowalewski. Leipzig: B. G. Teubner, 1913. Translated by John D. Blanton as *Foundations of Differential Calculus* (New York: Springer, 2000). Originally published 1755.

I.11 *Institutiones calculi integralis*, vol. 1. Edited by Friedrich Engel and Ludwig Schlesinger. Basel: Birkhäuser Verlag, 1912. Originally published 1768.

I.14 *Commentationes analyticae ad theoriam serierum infinitarum pertinentes*, vol. 1. Edited by Carl Boehm and Georg Faber. Leipzig: B. G. Teubner, 1925.

I.15 *Commentationes analytice ad theoriam serierum infinitarum pertinentes*, vol. 2. Edited by Georg Faber. Berlin: B. G. Teubner, 1927.

I.17 *Commentationes analyticae ad theoriam integralium pertinentes*, part 1. Edited by August Gutzmer. Leipzig: B. G. Teubner, 1914.

I.19 *Commentationes analyticae ad theoriam integralium pertinentes*, part 3. Edited by Alexander Liapounoff, Adolf Krazer, and Georg Faber. Leipzig: B. G. Teubner, 1932.

I.24 *Methodus inveniendi lineas curvas. . . .* Edited by Constantin Carathéodory. Zurich: Orell Füssli, 1952.

I.25 *Commentationes analyticae ad calculum variationum pertinentes.* Edited by Constantin Carathéodory. Zurich: Orell Füssli, 1952.

I.26 *Commentationes geometricae*, part 1. Edited by Andreas Speiser. Zurich: Orell Füssli, 1953.

I.27 *Commentationes geometricae*, part 2. Edited by Andreas Speiser. Zurich: Orell Füssli, 1954.

II.1 *Mechanica sive motus scientia analytice exposita*, part 1. Edited by Paul Stäckel. Berlin: B. G. Teubner, 1912.

II.2 *Mechanica sive motus scientia analytice exposita*, part 2. Edited by Paul Stäckel. Leipzig: B. G. Teubner, 1912.

II.5 *Commentationes mechanicae, Principia mechanica.* Edited by Joachim Otto Fleckenstein. Zurich: Orell Füssli, 1957.

II. 6 *Commentationes mechanicae ad theoriam motus punctorum pertinentes.* Edited by Charles Blanc. Zurich: Orell Füssli, 1957.

II.11 *The Rational Mechanics of Flexible or Elastic Bodies 1638–1788.* Edited by Clifford Ambrose Truesdell. Zurich: Orell Füssli, 1960.

II.12 *Commentationes mechanicae ad theoriam corporum fluidorum pertinentes*, part 1. Edited by Clifford Ambrose Truesdell. Zurich: Orell Füssli, 1954.

II.13 *Commentationes mechanicae ad theoriam corporum fluidorum pertinentes*, part 2. Edited by Clifford Ambrose Truesdell. Zurich: Orell Füssli, 1955.

II.14 *Neue Grundsätze der Artillerie, translated from the English of Mr. Benjamin Robins with many annotations.* Edited by Friedrich Robert Scherrer. Berlin: B. G. Teubner, 1922.

II.17 *Commentationes mechanicae ad theoriam machinarum pertinentes.* Edited by Charles Blanc and Pierre de Haller. Zurich: Orell Füssli, 1982.

II.18 *Scientia navalis*, part 1. Edited by Clifford Ambrose Truesdell. Zurich: Orell Füssli, 1967.

II.19 *Scientia navalis*, part 2. Edited by Clifford Ambrose Truesdell. Zurich: Orell Füssli, 1972.

II.20 *Commentationes mechanicae et astronomicae ad Scientam navalem pertinentes.* Edited by Walter Habicht. Zurich: Orell Füssli, 1974.

II.21 *Commentationes mechanicae et astronomicae (ad scientiam navalem pertinentes) volumen posterius.* Edited by Walter Habicht. Zurich: Orell Füssli, 1978. In the preface to this volume, Habicht gives an excellent analysis of vols. II.18 and II.19.

II.22 *Theoria motuum lunae. Nova methodo petractata.* Edited by Leo Courvoisier. Zurich: Orell Füssli, 1957.

II.23 *Sol et Luna I.* Edited by Leo Courvoisier and Joachim Otto Fleckenstein. Zurich: Orell Füssli, 1969.

II.24 *Sol et Luna II.* Edited by Charles Blanc. Basel: Birkhäuser Verlag, 1991.

II.25 *Commentationes astronomicae ad theoriam perturbationum pertinentes*, part 1. Edited by Max Schürer. Zurich: Orell Füssli, 1960.

II.26 *Leonhardi Euleri Opera omnia. Commentationes astronomicae ad theoriam perturbationum pertinentes*, part 2. Edited by Andreas Verdun. Basel: Springer Verlag, in press, projected for 2016.

II.27 *Leonhardi Euleri Opera omnia. Commentationes astronomicae ad theoriam perturbationum pertinentes*, part 3. Edited by Andreas Verdun. Basel: Springer Verlag, in press, projected for 2016.

II.28 *Commentationes astronomicae.* Edited by Leo Courvoisier. Zurich: Orell Füssli, 1959.

II.29 *Commentationes astronomicae.* Edited by Leo Courvoisier. Zurich: Orell Füssli, 1961.

II.30 *Sphärische Astronomie und Parallaxe.* Edited by Leo Courvoisier. Zurich: Orell Füssli, 1964.

II.31 *Commentationes mechanicae et astronomicae ad physicam pertinentes.* Edited by Eric J. Aiton, with appendixes by Andreas Kleinert. Basel: Birkhäuser Verlag, 1996.

III.1 *Commentationes physicae.* Edited by Eduard Bernoulli, Rudolf Bernoulli, Ferdinand Rudio, and Andreas Speiser. Berlin: B. G. Teubner, 1926.

III.2 *Rechenkunst.* Edited by Edmund Hoppe, Karl Matter, and Johann Jakob Burckhardt. Zurich: Orell Füssli, 1942.

III.5 *Opticae Miscellanea.* Edited by David Speiser. Zurich: Orell Füssli, 1962.

III.6 *Commentationes opticae miscellanea.* Edited by Emil Cherbuliez and Andreas Speiser. Zurich: Orell Füssli, 1962.

III.9 *Commentationes opticae.* Edited by Walter Habicht and Emil Alfred Fellmann. Zurich: Orell Füssli, 1973.

III.10 *Commentationes physicae ad theoriam caloris, electricitatis et magnetismi pertinentes.* Edited by Patricia Radelet-de-Grave, David Speiser, and Karine Chemla. Basel: Birkhäuser Verlag, 2004. This volume is the most recent completed in series III.

III.11 *Lettres à une Princesse d'Allemagne*, part 1. Edited by Andreas Speiser. Zurich: Orell Füssli, 1960.

III.12 *Lettres à une Princesse d'Allemagne*, part 2. Edited by Andreas Speiser. Zurich: Orell Füssli, 1960.

IVA.1 *Briefwechsel.* Edited by Adolf P. Yushkevich, Vladimir I. Smirnov, and Walter Habicht. Basel: Birkhäuser Verlag, 1975.

IVA.2 *Briefwechsel Euler–Johann I Bernoulli und Niklaus I Bernoulli*, vol 1. Edited by Emil A. Fellmann and Gleb K. Mikhaĭlov. Basel: Birkhäuser Verlag, 1998.

IVA.3 *Briefwechsel von Leonhard Euler mit Daniel and Johann III Bernoulli.* 2 vols. Edited by Emil A. Fellmann and Gleb K. Mikhaĭlov. Basel: Birkhäuser Verlag, in preparation.

IVA.4A.4–4.2 *Correspondence between Euler and Christian Goldbach.* 2 vols. Edited by Franz Lemmermeyer and Martin Mattmüller. Basel: Springer Verlag Basel, 2015.

IVA.4A.7 *Commercium Epistolicum.* Edited by Vanja Hug, Andreas Kleinert, Martin Mattmüller, Gleb K. Mikhaĭlov, Fritz Nagel, Norbert Schapper, and Thomas Steiner. Basel: Birkhäuser Verlag, in press, projected for 2015.

IVA.5 *Correspondance Euler avec A. C. Clairaut, J. d'Alembert et J. L. Lagrange.* Edited by Adolf P. Yushkevich and René Taton. Basel: Birkhäuser Verlag, 1980.

IVA.6 *Briefwechsel Euler-Maupertuis und Friedrich II.* Edited by Pierre Costabel, Ašot T. Grigorijan, and Adolf P. Yushkevich in collaboration with Emil A. Fellmann. Basel: Birkhäuser Verlag, 1986.

IVA.7 *Correspondance de Leonhard Euler mit L. Bertrand, Ch. Bonnet, J. Castillon, G. Cramer, Ph. Cramer, G. Cuenz, G. L. Lesage, J. M. von Loen et J. K. Wettstein.* Edited by Siegfried Bodenmann and Andreas Kleinert. Basel: Springer Verlag, in press, projected for 2015.

IVA.8 *Briefwechsel von Leonhard Euler mit Johann Andreas von Segner und anderen Hallenser Gelehrten.* Edited by Andreas Kleinert and Thomas Steiner. Basel: Birkhäuser Verlag, in preparation.

Facsimiles and Recent Editions of the Publications of Leonhard Euler

Euler, Leonhard. 2013. *Commentatione Astronomicae ad Theoriam Perturbationum Pertinentes*, part 2. Basel: Birkhäuser Verlag. Electronic Text.

———. 2012a. *Leonhardi Euleri Commentationes Arithmeticae Collectae.* Edited by Paul Heinrich Fuss and Nicholas Fuss. Charleston, S.C.: BiblioBazaar.

———. 2012b. *Leonhard Euler's Mechanik oder Analytische Darstellung der Wissenschaft von der Bewegung mit Anmerkungen und Erläuterungen.* Charleston, S.C.: BiblioBazaar.

———. 2012c. *Letters of Euler to a German Princess, on Different Subjects in Physics and Philosophy.* Memphis, Tenn.: General Books.

———. 2012d. *Nouveaux Principes d'Artillerie, Comments par Leonard Euler, Benjamin Robins et al.* Charleston, S.C.: BiblioBazaar.

———. 2012e. *Opuscula Varii Argument.* Charleston, S.C.: BiblioBazaar.

———. 2012f. *Scientia Navalis Seu Tractatus de Construendis Ac Dirigendis Navibus.* Charleston, S.C.: BiblioBazaar.

———. 2012g. *Tentamen Novae Theoriae Musicae.* Charleston, S.C.: BiblioBazaar.

———. 2012h. *Theoria Motuum Lunae Nova Methodo Pertractate.* Charleston, S.C.: BiblioBazaar.

———. 2011a. *Dioptric Continens Librum Primum, de Explicatione Principiorum, ex Quibus Constructio Tam. . . .* Charleston, S.C.: BiblioBazaar.

———. 2011b. *Dioptricae*, parts 1–3. Charleston, S.C.: BiblioBazaar.

———. 2011c. *Drei Abhandlungen Über Kartenprojection.* Charleston, S.C.: BiblioBazaar.

———. 2011d. *Elements of Algebra.* Charleston, S.C.: BiblioBazaar.

———. 2011e. *Fonction Zéta de Riemann, Fonction Gamma, Formule d'Euler, Identité d'Euler. . . .* Memphis, Tenn.: General Books.

———. 2011f. *Institutiones Calculi Integralis.* Charleston, S.C.: BiblioBazaar.

———. 2011h. *Methodus Inveniendi Lineas Curvas.* Charleston, S.C.: BiblioBazaar.

———. 2011i. *Theoria Motus Corporum Solidorum.* Charleston, S.C.: BiblioBazaar.

———. 2011j. *Theoria Motuum Planetarum et Cometarum.* Charleston, S.C.: BiblioBazaar.

———. 2011k. *Theorie der Planeten und Cometen.* Translated by Johann von Paccassi. Charleston, S.C.: Nabu.

———. 2011l. *Vollständige Anleitung zur Differenzial-Rechnung Aus dem Lateinischen Übersetzt und mit Anmerkungen und Zusätzen Begleitet*, vol. 1. Charleston, S.C.: BiblioBazaar.

———. 2010a. *Cours D'Arithmetique Raisonée: Theorique et Pratique Sans le Secours D'Aucun Maitre, 1865.* Whitefish, Mont: Kessinger.

———. 2010b. *Drei Abhandlungen Über Kartenprojection.* Whitefish, Mont.: Kessinger.

———. 2010c. *Introductio in Analysin Infinitorum.* Whitefish, Mont.: Kessinger.

——. 2010d. *Lettres de L. Euler à une Princesse d'Allemagne sur Divers Sujets de Physique und de Philosophie.* 2 vols. Charleston, S.C.: BiblioBazaar.
——. 2010e. *Opuscula Analytica*, vols. 1–2. Whitefish, Mont.: Kessinger.
——. 2010f. *Theoria Motus Lunae Exhibens Omnes Eius Inaequalitates.* Whitefish, Mont.: Kessinger.
——. 2009a. *Briefe Über Verschiedene Gegenstande Aus der Naturlehre.* Whitefish, Mont.: Kessinger.
——. 2009b. *Letters of Euler on Different Subjects in Natural Philosophy.* Memphis, Tenn.: General Books.
——. 2009c. *Nouveaux Principes d'Artillerie de M. Benjamin Robins.* Whitefish, Mont.: Kessinger.
——. 2009d. *Opuscula Analytica*, vols. 1–2. Whitefish, Mont.: Kessinger.
——. 2008. *Abhandlungen Über Variations-Rechnung.* Charleston, S.C.: BiblioBazaar.
——. 2007. *Letters of Euler on Different Subjects in Natural Philosophy Addressed to a German Princess*, vol. 1. Whitefish, Mont.: Kessinger.
——. 2007. "General Principles of the Motion of Fluids." *Physica D: Nonlinear Phenomena* 237, pp. 1825–1839.
——. 2007. "Principles of the Motion of Fluids." *Physica D: Nonlinear Phenomena* 237, pp. 1840–1854.
——. 2004. *Leonhard Euler's Mechanik Oder Analytische Darstellung der Wissenschaft von der Bewegung mit Anmerkungen und Erläuterungen.* Ann Arbor: Michigan Publishing.
——. 2003. *Leonhardt Euler, Lettres à une Princesse d'Allemagne, sur divers sujets de physique et de philosophie.* Edited by Srishti D. Chatterji. Paris: Presses polytechniques et universitaires romandes.
——. 1999. *Élémens d'algebre par Léonard Euler*, vol. 1. Elibron Classics Series. Port Chester, N.Y.: Adegi Graphics.
——. 1997. *Letters to a German Princess.* Translated by Henry Hunter in 1795. Bristol, England: Thoemmes.
——. 1984. *Elements of Algebra.* Translated by John Hewlett. New York: Springer.

Books and Articles by Leonhard Euler

The citations for works by Euler are distinctive. Euler's articles, books, and published letters are listed by their year of publication and by Eneström numbers. Articles were often written well before their actual date of publication; when Euler completed each is still being investigated. In some entries in this section, other dates given refer to the time the articles were read at the Berlin or the Petersburg Academy of Sciences or when a first portion was published. For the most part, the Eneström numbers follow a chronological order.

Most of these writings are available in Euler's *Opera omnia*, and almost all of the originals are on the Internet at the Euler Archive. The *Opera omnia* (indicated by *OO*) series, volume, and page numbers follow the title citation and are in parentheses; series and volume numbers are indicated in roman and arabic numerals, respectively.

Dominic Klyve of Central Washington University, Lee Stemkoski of Adelphi University, and Erik Tou of Pacific Lutheran College direct the Euler Archive, which the Mathematical Association of America hosts on the Internet. In this section, "EA" follows the *Opera omnia* citations (within the same parentheses) to indicate that a work can be found in the Euler Archive. To consult the archive, simply go to http://eulerarchive.maa.org/ and click on an Eneström number; then click again on the Eneström number near the bottom of an individual page for photocopies of the original publication and, in some cases, links to English translations.

This section of the bibliography employs the following abbreviations:

Acta Ac. Petrop.	*Acta Academiae Scientiarum Imperialis Petropolitanae* for 1777–82; Saint Petersburg, 1778–86.
Acta erud.	*Acta eruditorum*; Leipzig, 1682–1731.
AHES	*Archive for History of Exact Sciences.*
Comm. Ac. Petrop.	*Commentarii Academiae Scientiarum Imperialis Petropolitanae* for 1726–46; Saint Petersburg, 1728–51.
HM	*Historia mathematica.*
Mém. de Berlin	*Mémoires de l'Académie Royale des sciences et des belles-lettres de Berlin*; 1746–71.
Mém. de Paris	*Histoire et Mémoires de l'Académie Royale des sciences de Paris*; 1699–1790.
N. Acta Ac. Petrop.	*Nova Acta Academiae Scientiarum Imperialis Petropolitanae*; Saint Petersburg, 1783–1802.
N. Acta erud.	*Nova Acta eruditorum*; Leipzig, 1732–70.
N. Comm. Ac. Petrop.	*Novi Commentarii Academiae Scientiarum Imperialis Petropolitanae*; Saint Petersburg, 1750–76.
N. Mém. de Berlin	*Nouveaux Mémoires de l'Académie Royale des sciences et des belles-lettres de Berlin* for 1770–86; Berlin, 1772–88.
Pièces prix Paris	*Pièces qui ont Remporté les Prix de l'Académie Royale des Sciences de Paris.*
Recueil prix Paris	*Recueil des Pièces qui ont Remporté les Prix de l'Académie Royale des Sciences de Paris.*

1726 E01. "Constructio linearum isochronarum in medio quocunque resistente." *Acta erud.* 45 (1726), pp. 361–363. Reprinted in *N. Acta erud.* 6 (1746), pp. 579–581 [E01A] (*OO* II.6, pp. 1–3, and EA).

1727 E02. *Dissertatio physica de sono, quam. . . .* Basel: E. and J. R. Turneisen. Published as a sixteen-page quarto containing two chapters, "De natura et propagatione soni" and "De productione soni." The last page has six *annexa*—essentially, appendixes. Reprinted in *Disputationes anatomicae selectae editit A. Haller* 7, no. 2 (Göttingen, 1751), pp. 207–226 [E02A] (*OO* III.1, pp. 181–196, and EA).

1727 E03. "Methodus inveniendi trajectorias reciprocas algebraicas." *Acta erud.* 46 (1727), pp. 408–412 (*OO* I.27, pp. 1–5, and EA). Ian Bruce has made an English translation of E03 that is available at his website, Some Mathematical Works of the 17th and 18th Centuries, at http://www.17centurymaths.com.

1728 E04. "Meditationes super problemate nautico de implantatione malorum, quae proxime accessere." Originally published anonymously in Paris in *Pièces prix Paris.* Reprinted by Claude Jombert in *Recueil prix Paris* 2, no. 1

(1732); forty-eight pages with two diagrams [E04A] (*OO* II.20, pp. 1–35, and EA).

1729 E05. "Problematis traiectoriarum reciprocarum solutio." *Comm. Ac. Petrop.* 2 (for the year 1727), pp. 90–111, (*OO* I.27, in preparation).

1729 E06. "De novo quodam curvarum tautochronarum genere." *Comm. Ac. Petrop.* 2 (for the year 1727), pp. 126–137 (*OO* II.6, pp. 4–15, and EA).

1729 E07. "Tentamen explicationis phaenomenorum aeris." *Comm. Ac. Petrop.* 2 (for the year 1727), pp. 347–368 (*OO* II.31, pp. 1–17, and EA).

1732 E08. "Solutio problematis de invenienda curva, quam format lamina utcunque elastica. . . . " *Ac. Petrop.* 3 (for the year 1728), pp. 70–84 (*OO* II.10, pp. 1–16, and EA).

1732 E09. "De linea brevissima in superficie quacunque duo quaelibet puncta jungente." *Comm. Ac. Petrop.* 3, pp. 110–124 (*OO* I.25, pp. 1–12, and EA).

1732 E10. "Nova methodus innumerabiles aequationes differentiales secundi gradus reducendi ad aequationes differentiales primi gradus." *Comm. Ac. Petrop.* 3, pp. 124–137 (*OO* I.22, pp. 1–14, and EA).

1735 E13. "Curva tautochrona in fluido resistentiam faciente secundum quadrata celeritatem." *Comm. Ac. Petrop.* 4 (for the year 1729), pp. 67–89 (*OO* II.6, pp. 32–50, and EA).

1736 E15. *Mechanica, sive motus scientia analytice exposita*, vol. 1. Originally published by the Petersburg Academy of Sciences (*OO* II.1, and EA).

1736 E16. *Mechanica, sive motus scientia analytice exposita*, vol. 2. Originally published by the Petersburg Academy of Sciences (*OO* II.2, and EA).

1738 E17. *Einleitung zur Rechen-kunst: Commentationes ad physicam generalem pertinentes et Miscellanea.* Originally published as two books in 1738 and 1740 by the Petersburg Academy of Sciences for use in the academy's gymnasium (*OO* III. 2, pp. 1–162, and EA).

1738 E18. "De Indorum anno solari astronomico." Originally published as an appendix to Euler's friend and academy colleague T. S. Bayer's *Historia regni Graecorum Bactriani*, Saint Petersburg, pp. 201–213 (*OO* II. 30, pp. 5–12, and EA).

1738 E19. "De progressionibus transcendentibus, seu quarum termini generales algebraice dari nequeunt." *Comm. Ac. Petrop.* 5 (for the years 1730–31), pp. 36–57 (*OO* I.14, pp. 1–24, and EA).

1738 E20. "De summatione innumerabilium progressionum." *Comm. Ac. Petrop.* 5, pp. 91–105 (*OO* I.14, pp. 25–41, and EA).

1738 E21. "Quomodo dats quacunque curva invenire oporteat aliam, quae cum data quodammodo juncta ad tautochronismum producendum sit idonea." *Comm. Ac. Petrop.* 5, pp. 143–159 (*OO* II.6, pp. 51–64, and EA).

1738 E23. "De curvis rectificabilibus algebraicis atque traiectoriis reciprocis algebraicis." *Comm. Ac. Petrop.* 5, pp. 169–174 (*OO* I. 27, pp. 24–28, and EA).

1738 E25. "Methodus generalis summandi progressiones." *Comm. Ac. Petrop.* 6 (for the years 1732–33), pp. 68–97 (*OO* I.14, pp. 42–72, and EA).

1738 E26. "Observationes de theoremate quodam Fermatiano, aliisque ad numeros primos spectantibus." *Comm. Ac. Petrop.* 6, pp. 103–107 (*OO* I.1, pp. 1–5, and EA).

1738 E27. "Problematis isoperimetrici in latissimo sensu accepti solutio generalis." *Comm. Ac. Petrop.* 6, pp. 123–155 (*OO* I.25, pp. 13–40, and EA).

1738 E29. "De solutione problematum diophanteorum per numeros integros." *Comm. Ac. Petrop.* 6, pp. 175–188 (*OO* I.2, pp. 6–17, and EA).

1738 E32. "Von der Gestalt der Erden." Originally published as a series of articles in *Anmerckungen über die Zeitungen* by the Petersburg Academy of Sciences from 3 April 1738 to 25 December 1738 (*OO* III.2, pp. 325–346, and EA).

1738 E34. "Dissertatio de igne, in qua ejus natura et proprietates explicantur . . ." *Recueil prix Paris* 3, pp. 5–21 (*OO* III.10, pp. 2–13, and EA).

1739 E33. *Tentamen novae theoriae musicae ex certissismis harmoniae principiis dilucide expositae.* Originally published as a book by the Petersburg Academy of Sciences (*OO* III.1, pp. 197–427, and EA).

1740 E35. *Einleitung zur Rechen-Kunst*, vol. 2. Originally published by the Petersburg Academy of Sciences (*OO* III.2, pp. 163–303, and EA).

1740 E36. "Solutio problematis arithmetici de inveniendo numero qui per datos numeros divisus, relinquat data residua." *Comm. Ac. Petrop.* 7 (for the years 1734–35), pp. 46–66 (*OO* I.2, pp. 18–32, and EA).

1740 E40. "De minimis oscillationibus corporum tam rigidorum quam flexibilium. Methodus nova et facilis." *Comm. Ac. Petrop.* 7 (for the years 1734–35), pp. 99–122 (*OO* II.10, pp. 17–34, and EA).

1740 E41. "De summis serierum reciprocarum." *Comm. Ac. Petrop.* 7 (for the years 1734–35), pp. 123–134 (*OO* I.14, pp. 73–86, and EA).

1740 E42. "De linea celerrimi descensus in medio quocunque resistente." *Comm. Ac. Petrop.* 7 (for the years 1734–35), pp. 135–149 (*OO* I.25, pp. 41–53, and EA).

1740 E43. "De progressionibus harmonicis observationes." *Comm. Ac. Petrop.* 7 (for the years 1734–35), pp. 150–161 (*OO* I.14, pp. 87–100, and EA).

1740 E44. "De infinitis curvis eiusdem generis seu methodus inveniendi aequationes pro infinitis curvis eiusdem generis." *Comm. Ac. Petrop.* 7 (for the years 1734–35), pp. 174–189 (*OO* I.22, pp. 36–56, and EA).

1741 E47. "Inventio summae cujusque seriei ex dato termino generali." *Comm. Ac. Petrop.* 8 (for the year 1736), pp. 9–22 (*OO* I.14, pp. 108–123, and EA).

1741 E50. "Methodus computandi aequationem meridiei." *Comm. Ac. Petrop.* 8 (for the year 1736), pp. 48–65 (*OO* II.30, pp. 13–25, and EA).

1741 E53. "Solutio problematis ad geometriam situs pertinentis." *Comm. Ac. Petrop.* 8 (for the year 1736), pp. 128–140 (*OO* I.7, pp. 1–10, and EA).

1741 E54. "Theorematum quorundam ad numeros primos spectantium demonstratio." *Comm. Ac. Petrop.* 8 (for the year 1736), pp. 141–146 (*OO* I.2, pp. 33–37, and EA).

1741 E57. "Inquisitio physica in causam fluxus ac refluxus maris." Originally published in *Pièces prix de Paris en 1740* by G. Martin, pp. 235–350. See *Recueil prix Paris* 4, no. 9 (*OO* II.31, pp. 19–123, and EA).

1743 E60. "De inventione integralium, si post integrationem variabili quantitati determinatus valor tribuator." *Miscellanea Berolinensia* 7, pp. 129–171 (*OO* I.17, pp. 129–171, and EA).

1744 E65. *Methodus inveniendi lineas curvas maximi minimive proprietate gaudentes, sive solutio problematis isoperimetrici latissimo sensu accepti.* Originally published as a book in Lausanne and Geneva by Marcus-Michael Bousquet (*OO* I.24, and EA).

1744 E66. *Theoria motuum planetarum et cometarum. . . .* Originally published in Berlin by Ambrose Haude (*OO* II.28, pp. 105–251, and EA).

1744 E67. "Beantwortung verschiedener Fragen über die Beschaffenheit, Bewegung und Würckung der Cometen." Originally published anonymously by Euler in 1744 (*OO* II.31, pp. 125–150, and EA).

1744 E68. "Fortgesetzte Beantwortung der Fragen über die Beschaffenheit, Bewegung und Würckung der Cometen" (*OO* II.31, pp. 151–194).

1744 E72. "Variae observationes circa series infinitas." *Comm. Ac. Petrop.* 9 (for the year 1737), pp. 160–188 (*OO* I.14, pp. 216–244, and EA).

1746 E76. "Novae et correctae tabulae ad loca lunae computanda." Originally part of an extract from "Tabulae astronomicae solis & lunae," E87, *Opuscula varii argumenti* 1, pp. 137–168 (*OO* II.23, pp. 1–10, and EA).

1745 E77. *Neue Grundsätze der Artillerie. . . .* Originally published as a book in Berlin by Ambrose Haude (*OO* II.14, pp. 1–409, and EA).

1745 E78. *Dissertation sur la meilleure construction du cabestan.* Originally published in *Pièces prix Paris en 1741* in Paris by G. Martin, pp. 29–87. See *Recueil prix Paris* 5, no. 2 (*OO* II.20, pp. 36–82, and EA).

1746 E80. *Opuscula varii argumenti.* One of three volumes, pp. 1–136 (*OO* II.6, pp. 75–174, and EA).

1746 E81. *Gedancken von den Elementen der Cörper, in welchen das Lehr-Gebäude von den einfachen Dingen und Monaden geprüfet, und das wahre Wesen der Cörper entdeckt wird.* Originally published in Berlin by Ambrose Haude and Johann Carl Spener (*OO* III.2, pp. 347–366, and EA).

1746 E82. "De la force de percussion et de sa véritable mesure." *Mém. de Berlin* 1 (for the year 1745), pp. 21–53 (*OO* II.8, pp. 27–53, and EA).

1746 E86. "De motu corporum in superficiebus mobilibus." *Opuscula varii argumenti* 1, pp. 1–136 (*OO* II.6, pp. 75–174, and EA).

1746 E87. "Tabulae astronomicae solis & lunae." *Opuscula varii argumenti* 1, pp. 137–168 (*OO* II.23, pp. 1–10, and EA).

1746 E88. "Nova theoria lucis & colorum." *Opuscula varii argumenti* 1, pp. 169–244 (*OO* III.5, pp. 1–45, and EA).

1746 E91. "Recherches physiques sur la nature des moindres parties de la matiere." *Opuscula varii argumenti* 1, pp. 287–300 (*OO* III.1, pp. 3–15, and EA).

1746 E92. *Rettung der göttlichen Offenbahrung gegen die Einwürfe der Freygeister.* Originally printed as a pamphlet in Berlin by Ambrose Haude and Johann Carl Spener (*OO* III.12, pp. 1–265, and EA).

1747 E93. "Disquisitio de bilancibus." *Comm. Ac. Petrop.* 10, pp. 3–18 (*OO* II.17, pp. 1–15, and EA).

1747 E94. "De motu cymbarum remis propulsarum in fluuiis." *Comm. Ac. Petrop.* 10, pp. 22–37 (*OO* II.20, pp. 83–100).

1747 E97. "De attractione corporum sphaeroidico-ellipticorum." *Comm. Ac. Petrop.* 10, pp. 102–115 (*OO* II.6, pp. 175–188, and EA).

1747 E99. "Solutio problematis cuiusdam a celeb. Dan. Bernoullio propositi." *Comm. Ac. Petrop.* 10 (an essay from 1738), pp. 164–180 (*OO* I.25, pp. 84–97, and EA).

1747 E100. "De numeris amicabilibus." *N. Acta erud.*, pp. 267–269 (*OO* I.2, pp. 59–61, and EA).

1748 E101. *Introductio in analysin infinitorum.* Originally published in Lausanne by Marcus-Michael Bousquet as two volumes (*OO* I.8 for vol. 1, and EA).

Translated as *Introduction to Analysis of the Infinite*, 2 vols., by John D. Blanton (Berlin: Springer Verlag, 1988–89).

1748 E102. *Introductio in analysin infinitorum*. Originally published in Lausanne by Marcus-Michael Bousquet as two volumes (*OO* I.9 for vol. 2, and EA).

1748 E107. "Extract of a letter from Mr. Leonhard Euler. . . . " *Philosophical Transactions* 44 (1748), pp. 421–423 (*OO* III.2, pp. 373–375, and EA).

1748 E108. "De observatione inclinationis magneticae dissertatio." Originally published in *Pièces prix Paris*, pp. 1–47 (*OO* III.10, and EA).

1748 E109. "Dissertatio de magnete." Originally published in *Pièces prix Paris*, pp. 63–97 (*OO* III.10, and EA). Awarded the Prix de Paris for 1744.

1749 E110. *Scientia navalis, seu tractatus de construendis ac dirigendis navibus*, vol. 1. Originally published by the Petersburg Academy of Sciences (*OO* II.18, and EA).

1749 E111. *Scientia navalis, seu tractatus de construendis ac dirigendis navibus*, vol. 2. Originally published by the Petersburg Academy of Sciences (*OO* II.19, and EA).

1749 E112. "Recherches sur le mouvement des corps celestes en generale." *Mém. de Berlin* 3, pp. 93–143 (*OO* II.25, pp. 1–44, and EA).

1749 E116. "Memoi[r]e sur la Force des Rames." *Mém. de Berlin* 3, pp. 180–213 (*OO* II.20, pp. 101–129, and EA).

1749 E118. "Sur la perfection des verres objectifs des lunettes." *Mém. de Berlin* 3, pp. 274–296 (*OO* III.6, pp. 1–21, and EA).

1749 E119. "De vibratione chordarum exercitatio." Originally published in *N. Acta erud.*, pp. 512–527, perhaps after being written in 1748 (*OO* II.10, pp. 50–62, and EA).

1749 E120. "Recherches sur la question des inégalités du mouvement de Saturne et de Jupiter." *Pièces prix Paris.* Paris: Martin, Coignard et Guerin, pp. 1–123 (*OO* II.25, pp. 45–157). Awarded the Prix de Paris for 1748.

1750 E121. "Conjectura physica circa propagationem soni ac luminis una cum aliis dissertationibus analyticis: de numeris amicalibus, de natura aequationum ac de rectificatione ellipsis." Originally published in Berlin by Ambrose Haude and Johann Carl Spener as part of *Opuscula varii argumenti* 2, pp. 1–22 (*OO* III.5, 113–129, and EA).

1750 E125. "Consideratio progressionis cuiusdam ad circuli quadraturam inveniendam idoneae." Originally published in *Comm. Ac. Petrop.* 11 (for the year 1739), pp. 16–127 (*OO* I.14, pp. 350–363, and EA).

1750 E126. "De novo genere oscillationum." *Comm. Ac. Petrop.* 11 (for the year 1739), pp. 128–149 (*OO* II.10, pp. 78–97, and EA). It was presented to the Petersburg Academy of Sciences in December 1738, and addresses a mathematical rediscovery of resonance.

1750 E128. "Methodus facilis computandi angulorum sinus ac tangentes tam naturales quam artificiales." *Comm. Ac. Petrop.* 11 (for the year 1739), pp. 194–230 (*OO* I.14, pp. 364–406, and EA).

1750 E129. "Investigatio curvarum quae evolutae sui similes producunt." *Comm. Ac. Petrop.* 12 (for the years 1739–40), pp. 2–52 (*OO* I.27, pp. 130–180, and EA).

1750 E130. "De seriebus quibusdam considerationes." *Comm. Ac. Petrop.* 12 (for the year 1740), pp. 53–96 (*OO* I.14, pp. 407–462, and EA).

1750 E131. "Emendatio tabularum astronomicarum per loca planetarum geocentrica." *Comm. Ac. Petrop.* 12 (for the year 1740), pp. 109–221 (*OO* II.29, pp. 1–91 and 109–221, and in EA).

1750 E134 "Theoremata circa divisores numerorum." *N. Comm. Ac. Petrop.* 1 (for the years 1747–48), pp. 20–48 (*OO* I.2, pp. 62–85, and EA).

1750 E140. "Sur la vibration des cordes" (translated from Latin; written 1748). *Mém. de Berlin* 4, pp. 69–85 (*OO* II.10, part 2, pp. 63–77, and EA).

1750 E141. "Sur l'accord des deux dernieres eclipses du soleil et de la lune avec mes tables pour trouver les vrais momens des pleni-lunes et novi-lunes." *Mém. de Berlin* 4 (for the year 1748), pp. 86–98 (*OO* II.30, pp. 89–100, and EA). Andrew Fabian and Hieu Nguyen have made an English translation that is available at the Euler Archive.

1750 E142. "Sur l'atmosphere de la Lune prouveé par la derniere eclipse annulaire du Soleil." *Mém. de Berlin* 4, pp. 103–121 (*OO* II.31, pp. 239–256, and EA).

1750 E149. "Réflexions sur l'espace et le tem[p]s." *Mém. de Berlin* 4, pp. 324–333 (*OO* III.2, pp. 376–383, and EA).

1750 E153. "Demonstratio gemina theorematis Neutoniani quo traditur relatio inter coefficientes cujus vis aequationis algebraicae & summae potestatum radicum ejusdam." Originally published in *Opuscula varii argumenti* 2, pp. 108–120 (*OO* I.6, pp. 20–30, and EA).

1751 E156. *L. Euleri Opusculorum tomus III. continens novam theoriam magnetis . . . Una cum nonnullis aliis dissertationibus analytico-mechanicis.* Originally published as a book in Berlin by Ambrose Haude and Johann Carl Spener. It contains E109, E173, and E174, and is one of three volumes.

1751 E158. "Observationes analyticae variae de combinationibus." *Comm. Ac. Petrop.* 13 (for the years 1741–43), pp. 64–93 (*OO* I.2, pp. 163–193, and EA). Jordan Bell has made an English translation that is available at the Euler Archive.

1751 E164. "Theoremata circa divisores numerorum in hac forma paa ± qbb contentorum." *Comm. Ac. Petrop.* 14 (for the years 1744–46), pp. 151–181 (*OO* I.2, pp. 194–222, and EA). Jordan Bell has made an English translation that is available at the Euler Archive.

1751 E168. "De la controverse entre M[ss]rs. Leibnitz & Bernoulli sur les logarithmes des nombres négatifs et imaginaires." *Mém. de Berlin* 5 (for the year 1749), pp. 139–179 (*OO* I.17, pp. 195–232, and EA). Stacy Langton has made an English translation that is available at the Euler Archive.

1751 E170. "Recherches sur les racines imaginaires des equations." *Mém. de Berlin* 5, pp. 222–288 (*OO* I.6, pp. 78–150, and EA). Todd Doucet has made an English translation that is available at the Euler Archive.

1751 E171."Recherches sur la précession de équinoxes et sur la nutation de l'axe de la terre." *Mém. de Berlin* 5, pp. 289–325 (*OO* II.29, pp. 92–123, and EA). Steven Jones and Robert Bradley have made an English translation that is available at the Euler Archive.

1751 E173. "Nova methodus inveniendi trajectorias reciprocas algebraicas." *Opuscula varii argumenti* 3, pp. 54–87 (*OO* I.27, pp. 253–276, and EA).

1751 E174. "De motu corporum flexibilium." *Opuscula* 3, pp. 88–165 (*OO* II.10, pp. 177–232, and EA).

1751 E175. "Découverte d'une loi tout extraordinaire des nombres, par rapport à la somme de leurs diviseurs." Originally published in *Bibliotheque impartiale*

3, pp. 10–31. Reprinted in *Commentatio arithmeticae* 2 (1849), and *Opera postuma* 1 (1862) (*OO* I.2, pp. 241–253, and EA).

1752 E176. "Exposé concernant l'examen de la lettre de M de Leibnitz alléguée par M. le Professeur Koenig, dans le mois de mars, 1751. Actes eruditorum à l'occasion du principe de la moindre action." Originally published in *Histoire de l'académie des sciences de Berlin* 6, pp. 52–62. It was reprinted in *Jugement de l'Académie Royale des Sciences sur une lettre prétendue de Mr. de Leibnitz*, pp. i–lxx. Berlin: Haude und Spener (*OO* II.5, pp. 64–73, and EA).

1752 E177. "Découverte d'un nouveau principe de mecanique." *Mém. de Berlin* 6, (for the year 1750), pp. 185–217(*OO* II.5, pp. 81–108, and EA).

1752 E180. "Avertissement au suject des Recherches sur la Précession des Equinoxes." *Mém. de Berlin* 6, p. 412 (*OO* II.29, pp. 124, and EA). Robert Bradley has made an English translation that is available at the Euler Archive.

1752 E181. "Recherches sur l'origine des forces." *Mém. de Berlin* 6, pp. 419–447 (*OO* II.5, pp. 109–131, and EA).

1752 E182. "Lettre de M. Euler à M. Merian." *Mém. de Berlin* 6, pp. 520–532 (*OO* II.5, pp. 132–141, and EA).

1753 E186. *Dissertation sur le principe de la moindre action avec l'examen des objections de M. le Prof. Koenig faites contre ce principe.* Berlin: Michaelis (*OO* II.5, pp. 214–250, and EA).

1753 E187. *Theoria motus lunae exhibens omnes eius inaequalitates.* Originally published as a book by the Petersburg Academy of Sciences (*OO* II.23, pp. 64–336, and EA).

1753 E191. "De partitione numerorum." *N. Comm. Ac. Petrop.* 3, pp. 125–169 (*OO* I.2, pp. 254–294, and EA).

1753 E196. "Emendatio laternae magicae ac microscopii solaris." *N. Comm. Ac. Petrop.* 3, pp. 363–380 (*OO* III.6, pp. 22–37, and EA).

1753 E198. "Sur le principe de la moindre action." *Mém de Berlin* 7, pp. 199–218 (*OO* II.5, pp. 179–193, and EA). It was originally read in 1750 at the Royal Prussian Academy of Sciences.

1753 E199. "Examen de la dissertation de M. le Professeur Koenig, inserée dans les Actes de Leipzig pour le mois mars 1751" (translated from Latin). *Mém. de Berlin* 7, pp. 219–245 (*OO* II.5, pp. 194–213, and EA).

1753 E205. *Atlas geographicus omnes orbis terrarum regiones in XLI tabulis exhibens. . . .* Berlin: Michaelis (*OO* III.2, pp. 305–317, and EA).

1755 E212. *Institutiones calculi differentialis cum eius usu in analysi finitorum ac doctrina serierum.* Euler divided this volume on differential calculus into two books; it was published by the Petersburg Academy of Sciences (*OO* I.10, and EA). Chapters 1–9 translated as *Foundations of Differential Calculus* by John D. Blanton (New York: Springer Verlag, 2000.

1755 E213. "Remarques sur les mémoires précedens de M. Bernoulli." *Mém. de Berlin* 10, pp. 233–254 (*OO* II.10, pp. 233–254, and EA). The work considers treatises by Daniel Bernoulli on the vibration of chords.

1755 E217. "Recherches sur la veritable courbeque décrivent les corps jettés dans l'air ou dans un autre fluide quelconque." *Mém. de Berlin* 9 (for the year 1753), pp. 321–352 (*OO* II.14, pp. 413–447, and EA).

1757 E225. "Principes généraux de l'état d'équilibre des fluides." *Mém. de Berlin* 11, pp. 217–273 (*OO* II.12, pp. 2–53, and EA).

1757 E226. "Principes généraux du mouvement des fluides." *Mém. de Berlin* 11, pp. 274–315 (*OO* II.12, pp. 54–91, and EA).

1757 E227. "Continuation des recherches sur la théorie du mouvement des fluides." *Mém. de Berlin* 11, pp. 316–361 (*OO* II.12, pp. 92–132, and EA).

1758 E230. "Elementa doctrinae solidorum." *N. Comm. Ac. Petrop.* 4, pp. 109–140 (*OO* I.26, pp. 71–93, and EA).

1758 E231. "Demonstratio nonnullarum insignium proprietatum, quibus solida hedris planis inclusa sunt praedita." *N. Comm. Ac. Petrop.* 4, pp. 140–160 (*OO* I.26, pp. 94–108).

1759 E237. "Recherches sur la déclinaison de l'alguille aimantée." *Mém. de Berlin* 13, pp. 175–251 (*OO* III.10, and EA).

1759 E239. "Régles générales pour la construction des telescopes et des microscopes, de quelque nombre de verres qu'ils soient composés." *Mém. de Berlin* 13, pp. 283–322 (*OO* III.6, pp. 44–73, and EA).

1760 E241. "Demonstratio theorematis Fermatiani: omnem numerum primum formae 4n + 1 esse summam duorum quadratorum." *N. Comm. Ac. Petrop.* 5 (for the years 1754–55), pp. 3–13 (*OO* I.2, pp. 328–337, and EA).

1760 E245. "De methodo Diophanteae analoga in analysi infinitorum." *N. Comm. Ac. Petrop.* 5 (for the years 1754–55), pp. 84–144 (*OO* I.22, pp. 237–295, and EA).

1760 E247. "De seriebus divergentibus." *N. Comm. Ac. Petrop.* 5 (for the years 1754–55), pp. 205–237 (*OO* I.14, pp. 585–617, and EA).

1761 E251. "De integratione aequationis differerentialis. . . . " *N. Comm. Ac. Petrop.* 6. pp. 37–57 (*OO* I.20, pp. 58–79, and EA).

1761 E252. "Observationes de comparatione arcuum curvarum irrectificabilium." *N. Comm. Ac. Petrop.* 6, pp. 58–84 (*OO* I.20, pp. 80–107, and EA).

1761 E257. "De frictione corporum rotantium." *N. Comm. Ac. Petrop.* 6 (for the years 1756–57), pp. 233–270 (*OO* II.8, pp. 146–177, and EA).

1761 E258. "Principia motus fluidorum." *N. Comm. Ac. Petrop.* 6 (for the years 1756–57), pp. 271–311 (*OO* II.13, pp. 133–168, and EA).

1764 E283. "De numeris primis valde magnis." *N. Comm. Ac. Petrop.* 9, pp. 99–153 (*OO* I.3, pp. 1–45, and EA).

1765 E289. *Theoria motus corporum solidorum seu rigidorum* (*OO* II.3, and EA).

1765 E292. "Du mouvement de rotation des corps solides autour d'un axe variable." *Mém. de Berlin* 14, pp. 154–193 (*OO* II.8, pp. 200–235, and EA).

1766 E296. "Elementa Calculi Variationum." *N. Comm. Ac. Petrop.* 10, pp. 51–94 (*OO* I.25, pp. 141–176, and EA).

1766 E297. "Analytica explicatio methodi Maximorum et Minimorum." *N. Comm. Ac. Petrop.* 10 (for the year 1760), pp. 94–135 (*OO* I.25, pp. 177–207, and EA).

1766 E305. "De la propagation du son." *Mém. de Berlin* 15, pp. 185–209 (*OO* III.1, pp. 428–451, and EA). It was originally read in 1759 at the Royal Prussian Academy of Sciences.

1766 E306. "Supplement aux recherches sur la propagation du son." *Mém. de Berlin* 15, pp. 210–240 (*OO* III.1, pp. 452–483, and EA).

1766 E307. "Continuation des recherches sur la propagation du son." *Mém. de Berlin* 15, pp. 241–264 (*OO* III.1, pp. 484–507, and EA).

1766 E309. "Solution d'une question curieuse qui ne paroit soumise à aucune analyse." *Mém. de Berlin* 15, pp. 310–337 (*OO* I.7, pp. 26–56, and EA).

1766 E317. "Eclaircissemens sur le mouvement des cordes vibrantes." *Mélanges de philosophie et de la mathematique de la société royale de Turin* 3, pp. 1–26 (*OO* II.10, pp. 377–396, and EA).

1766 E318. "Recherches sur le mouvement des cordes inégalement grosses." *Mélanges de philosophie et de la mathématiques de la société royale de Turin* 3, pp. 27–59 (*OO* II.10, pp. 397–425, and EA). This study examines the motion of unequally thick strings.

1767 E325. "Solutio facilis problematum quorumdam geometricorum difficillimorum." *N. Comm. Ac. Petrop.* 11, pp. 103–123 (*OO* I.26, pp. 139–157, and EA).

1767 E332. "Recherches sur le mouvement des rivieres." *Mém. de Berlin* 16, pp. 101–118 (*OO* II.12, pp. 272–288, and EA).

1767 E334. "Recherches générales sur la mortalité et la multiplication du genre humain." *Mém. de Berlin* 16, pp. 144–164 (*OO* I.7, pp. 79–100, and EA). Richard Pulskamp and Todd Deucet have made English translations that are available at the Euler Archive.

1767 E335. "Sur les rentes viageres." *Mém. de Berlin* 16, pp. 165–175 (*OO* I.7, pp. 101–112, and EA).

1767 E336. "Du mouvement d'un corps solide quelconque lorsq'il tourne autour d'un axe mobile." *Mém. de Berlin* 16 (for the year 1760), pp. 176–227 (*OO* II.8, pp. 313–356, and EA).

1775 E341A. "Nachricht von einem neuen Mittel zur Vermehrung des Getreides." *Anmerckungen über die Zeitungen* 6, pp. 109–113 (*OO* III.2, pp. 387–390).

1768 E342. *Institutionum calculi integralis volumen primum, in quo methodus integrandi a primis principiis usque ad integrationem aequationum differentialium primi gradus pertractatur.* This was the first of three volumes published by the Petersburg Academy of Sciences (*OO* I.11, and EA).

1768 E343–E344. *Lettres à une princesse d'Allemagne sur divers sujets de physique & de philosophie.* Volumes 1 and 2 were originally published as separate books (*OO* III.11 and *OO* III.12); the third volume was not published until 1772. The complete books E343, E344, and E417 were edited and translated by David Brewster in two volumes as *Letters of Euler on Different Subjects in Natural Philosophy Addressed to a German Princess* (New York: Harper and Brothers, 1840). Russian edition: *Pis'ma k nemet͡skoĭ print͡sesse o raznykh fizicheskikh i filosofskikh materii͡akh / Leonard Ėĭler; izdanie podgotovili.* Edited by M. A. Bobovich. St. Petersburg: Nauka. The Russian edition has extensive commentary.

1768 E350. "De novo microscopiorum genere ex sex lentibus compositio." *N. Comm. Ac. Petrop.* 12 (for the years 1766–67), pp. 195–223 (*OO* III.7, pp. 104–144, and EA).

1768 E363. "Précis d'une théorie générali de la dioptrique." *Mém. de Paris* (for the year 1765), pp. 555–575 (*OO* III.7, pp. 178–198, and EA).

1769 E366. *Institutionum calculi integralis volumen secundum, in quo methodus inveniendi functiones unius variabilis ex data relatione differentialium secundi altiorisve gradus pertractatur.* This second of three volumes was published by the Petersburg Academy of Sciences (*OO* I.12, and EA).

1769 E367. *Dioptricae pars prima continens lubrum primum de explicatione principiorum, ex quibus constructio tam telescopiorum quam microscopiorum est pretenda.* Originally published as a book in three volumes (1769–71), edited by W. L. Krafft, by the Petersburg Academy of Sciences (*OO* III.3, and EA). The first of three volumes on optics, this study collects earlier treatises.

1769 E369. "Quomodo numeri praemagni sint explorandi, utrum sint primi, nec ne." *N. Comm. Ac. Petrop.* 13, pp. 67–88. Reprinted in *Comm. arithm.* 1 (1849), pp. 379–390 (*OO* I.3, pp. 112–130, and EA).

1769 E372. "Annotatio quarundam cautelarum in investigatione inaequalitatum quibus corpora coelestia in motu perturbantur observandarum." *N. Comm. Ac. Petrop.* 13, pp. 159–201 (*OO* II.26, in preparation, and EA).

1769 E375. "Sectio prima de statu aequilibrii fluidorum." *N. Comm. Ac. Petrop.* 12, pp. 305–416 (*OO* II.13, pp. 1–72, and EA).

1769 E380. "Considérations sur les nouvelles lunettes d'Angleterre de Mr. Dollond, et sur le principe qui en est le fondement." *Mém. de Berlin* 18, pp. 226–248 (*OO* III.8, pp. 102–121, and EA).

1770 E385. *Institutionum calculi integralis volumen tertium, in quo methodus inveniendi functiones duarum et plurium variabilium, ex data relatione differentialium cujusvis gradus pertractur.* Originally published as a book by the Petersburg Academy of Sciences (*OO* I.13, and EA).

1770 E386. *Dioptricae pars secunda continens lubrum secundum de constructione telescopiorum dippticorum cum appendice de constructione telescopiorum cotapticodioptricum.* Originally published as a book by the Petersburg Academy of Sciences (*OO* III.4, and EA). The second of three volumes on optics.

1770 E387. *Vollständige Anleitung zur Algebra*, vol. 1: *Von den verschiedenen Rechnung-Arten, Verhältnissen und Proportionen.* Originally published as a book in German; Leipzig: Philipp Reclam Jr. (*OO* I.1, and EA from a 1920 edition). Translated as *Complete Elements of Algebra* by Francis Horner (London: Longman, Hurst, Rees, Orme, 1822).

1770 E388. *Vollständige Anleitung zur Algebra*, vol. 2: *Von Auflösung algebraischer Gleichungen und der unbestimmten Analytic.* Originally published as a book in German; Leipzig: Philipp Reclam Jr. (*OO* I.1, and EA from a 1920 edition).

1770 E396. "Sectio secunda de principiis motus fluidorum." *N. Comm. Ac. Petrop.* 14, pp. 270–386 (*OO* II.13, pp. 73–153, and EA).

1770 E397. "Exposito methodorum, cum pro determinanda parallaxi solis ex observato transitu Veneris per solem, tum pro inveniendis longitudinibus locorum super terra, ex observationibus eclipsium solis, una cum calculis et conclusionibus inde deductis." *N. Comm. Ac. Petrop.* 14, pp. 322–554 (*OO* II.30, pp. 153–231, and EA). This study collected four treatises.

1771 E404. *Dioptricae pars tertia.* Originally published as a book by the Petersburg Academy of Sciences (*OO* III.4, and EA). The third volume of three on optics.

1771 E409. "Sectio tertia de motu fluidorum lineari potissimum aquae." *N. Comm. Ac. Petrop.* 15, pp. 219–360 (*OO* II.13, pp. 154–261, and EA).

1771 E410. "Genuina principia doctrinae de statu aequilibrii et motu corporum tam perfecte flexibilium quam elasticorum." *N. Comm. Ac. Petrop.* 15, pp. 381–413 (*OO* II.11, pp. 37–61).

1771 E413. "Mémoire sur la maniere la plus avantageuse de suppléer à l'action du vent sur les grands vaisseaux." *Recueil prix Paris* (for the year 1753), pp. 1–47 (*OO* II.20, pp. 190–228, and EA). Awarded a Prix de Paris *accessit*, 1753.

1771 E414. "Investigatio perturbationum quibus planetarum motus ob actionem eorum mutuam afficiunur." *Recueil prix Paris* 8, pp. 3–50 (*OO* II.26, in preparation, and EA). Awarded the Prix de Paris, 1756.

1771 E415. "Examen des efforts qu'ont à soutenir toutes les parties d'un vaisseau dans les roulis & dans le tangage ou Recherches sur la diminution de ces mouvemens dans le roulis & dans le tangage." *Pièce prix Paris* (for the year 1759), pp. 1–47 (*OO* II.21, pp. 1–30, and EA). Awarded the Prix de Paris, 1759.

1771 E416. "Meditationes in quaestionem utrum motus medius planetarum semper maneat aeque velox, an successu temporis quampiam mutationem patiatur? & quaesit ejus causa?" *Recueil prix Paris* 8, pp. 3–47 (*OO* II.12, pp. 92–132, and EA).

1772 E417. *Lettres à une princesse d'Allemagne sur divers sujets de physique & de philosophie*, vol. 3. Originally published by the Petersburg Academy of Sciences (*OO* III.11, pp. 1–312, and EA; letters 155–234 in English).

1772 E418. *Theoria motuum lunae, nova methodo petractata. . . .* Originally published as a book by the Petersburg Academy of Sciences (*OO* II.22).

1772 E419. "De solidis quorum superficiem in planum explicare licet." *N. Comm. Ac. Petrop.* 16, pp. 3–34 (*OO* I.28, pp. 161–186, and EA).

1772 E423. "Considerationes cyclometricae." *N. Comm. Ac. Petrop.* 16, pp. 160–170 (*OO* I.28, pp. 205–214, and EA).

1772 E424. "Sectio quarta de motu aeris in tubis." *N. Comm. Ac. Petrop.* 16, pp. 281–425 (*OO* II.13, pp. 262–369, and EA).

1772 E425. "De perturbatione motus terrae ab actione Veneris oriunda." *N. Comm. Ac. Petrop.* 16, pp. 426–467 (*OO* II.26, in preparation).

1776 E426. *Théorie complete de la construction et de la manoeuvre des vaisseaux, mise à la portée des (!) ceux, qui s'appliquant à la navigation.* Originally published in 1773 as a book by the Petersburg Academy of Sciences. A new edition, corrected and augmented, was published in 1776; Paris: Chez Claude-Antoine Jombert (*OO* II.21, pp. 80–222, and EA).

1774 E448. "Nova series infinita maxime convergens perimetrum ellipsis exprimens." *N. Comm. Ac. Petrop.* 18, pp. 71–84 (*OO* I.20, pp. 357–370, and EA).

1774 E449. "Demonstrationes circa residua ex divisione potestatum per numeros primos resultantia." *N. Comm. Ac. Petrop.* 18, pp. 85–135 (*OO* I.3, pp. 240–281, and EA).

1776 E473. *Éclarcissemens sur les établissemens publics en faveur tant des veuves que des morts, avec la description d'une nouvelle espèce Tontine aussi favorable au public qu'utile à l'état calculés sous la direction de Monsieur Léonard Euler.* Originally published by the Petersburg Academy of Sciences (*OO* I.7, pp. 181–245). Presented to the academy in February and May 1776.

1776 E478. "Formulae generales pro translatione quacunque corporum rigidorum." *N. Comm. Ac. Petrop.* 20, pp. 189–267 (*OO* II. 9, pp. 84–98, and EA).

1776 E479. "Nova methodus motum corporum rigidorum determinandi." *N. Comm. Ac. Petrop.* 20, pp. 208–238 (*OO* II.9, pp. 99–125, and EA).

1776 E480. "Regula facilis pro diiudicanda firmitate pontis aliusive corporis similis ex cognita firmitate moduli." *N. Comm. Ac. Petrop.* 20, pp. 271–285 (*OO* II.17, pp. 220–231, and EA).

1776 E481. "De gemina methodo tam aequilibrium quam motum corporum flexibilium determinandi, et utriusque egregio consensu." *N. Comm. Ac. Petrop.* 20, pp. 286–303 (*OO* II.11, pp. 180–193, and EA).

1777 E485. "Réponse à la question proposée par l'Académie Royale des Sciences de Paris pour l'année 1770. . . . " *Recueil prix Paris* 9, ninety-four pages (*OO* II.24, pp. 101–106). On lunar motion.

1777 E486. "Réponse à la Question Proposée par l'Académie Royale des Sciences de Paris pour l'anné 1772. . . . " *Recueil prix Paris* 9, thirty-eight pages (*OO* II.24, pp. 167–190). On the perfection of methods and equations for secular motion.

1777 E487. "Réflexions de M. L. Euler sur quelques nouvelles expériences optiques, communiquées à l'Académie des Sciences, par Mr. Wilson." *Acta Ac. Petrop.* (for the years 1777–78), pp. 71–77 (*OO* III.5, pp. 351–355, and EA).

1778 E489. "De formulis exponentialibus replicatis." *Acta Ac. Petrop.* (for the years 1777–78), pp. 38–60 (*OO* I.15, pp. 268–297, and EA).

1778 E490. "De repraesentione superficiei sphaerica super plano." *Acta Ac. Petrop.* (for the years 1777–78), pp. 107–132 (*OO* I.28, pp. 248–275).

1778 E491. "De proiectione geographica superficiei sphaericae." *Acta Ac. Petrop.* (for the years 1777–78), pp. 133–142 (*OO* I.28, pp. 276–287, and EA).

1778 E492. "De proiectione geographica de lisliana in mappa generali imperii Russici Usitata." *Acta Ac. Petrop.* (for the years 1777–78), pp. 143–153 (*OO* I.28, pp. 287–297, and EA).

1778 E496. "De figura apparente annuli Saturni pro eius loco quocunque respectu." *Acta Ac. Petrop.* (for the years 1777–78), pp. 276–287 (*OO* II.30, pp. 269–280, and EA).

1778 E497. "De apparitione et disparitione annuli Saturni." *Acta Ac. Petrop.* (for the years 1777–78), pp. 288–316 (*OO* II.30, pp. 281–305, and EA).

1779 E498. "Extrait d'une lettre de M. Euler à M. Beguelin, en Mai 1778." *N. Mém. de Berlin* (for the year 1776), pp. 337–339 (*OO* I.3, pp. 418–420, and EA). Benjamin Linowitz has made an English translation that is available at the Euler Archive.

1780 E501. "Considerationes circa brachystochronas." *Acta Ac. Petrop.* (for the year 1777), pp. 70–88 (*OO* I.25, p. 250–268, and EA).

1780 E502. "Sur l'effet de la refraction dans les observations terrestres." *Acta Ac. Petrop.* (for the year 1777), pp. 129–158 (*OO* III.5, pp. 370–395, and EA).

1778 E504. "De theoria lunae ad maiorem perfectionis gradum evehenda." *Acta. Ac. Petrop.* (for the year 1777), pp. 281–327 (*OO* II.24, pp. 191–234, and EA).

1780 E508. "Determinatio onerum, quae columnae gestare valent." *Acta Ac. Petrop.* (for the year 1778), pp. 121–145 (*OO* II.17, pp. 232–251, and EA).

1781 E512. "Investigatio perturbationum quae in motu terrae ab actione veneris producuntur." *Acta Ac. Petrop.* (for the year 1778), pp. 308–316 (*OO* II.27, and EA).

1781 E513. "De Curvis Triangularibus." *Acta Ac. Petrop.* (for the year 1778), pp. 3–30 (*OO* I. 28, pp. 298–321, and EA).

1781 E520. "Essai d'une théorie de la résistance qu'éprouve la proue d'un vaisseau dans son mouvement." *Mém de Paris* (for the year 1778), pp. 597–602 (*OO* II.21, 223–229, and EA).

1781 E521. "Extraits de différentes Lettres de M. Euler à M. le Marquis de Condorcet." *Mém. de Paris* (for the year 1778), pp. 603–614 (*OO* I.18, pp. 69–82, and EA).

1782 E524. "Trigonometria sphaerica universa, ex primis principiis breviter et dilucide derivata." *Acta Ac. Petrop.* 3, pp. 72–86 (*OO* I.26, pp. 224–236, and EA).

1782 E526. "Investigatio motuum, quibus laminae et virgae elasticae contremiscunt." *Acta Ac. Petrop.* (for the year 1779), pp. 103–161 (*OO* II.11, pp. 223–268, and EA).

1782 E530. "Recherches sur une nouvelle espèce de quarrés magiques." *Verhandelingen uitgegeneven door het zeeuwsch Genootschap der Wetenschappen de Vlissingen* 9, pp. 85–239 (*OO* I.7, pp. 291–392, and EA).

1783 E532. "De serie Lambertina, plurimisque eius insignibus proprietatibus." *Acta Ac. Petrop.* (for the year 1779), pp. 29–51 (*OO* I.6, pp. 350–369, and EA).

1783 E536. "De proprietatibus triangulorum mechanicis." *Acta Ac. Petrop.* (for the year 1779), pp. 126–155 (*OO* II.9, pp. 138–162, and EA).

1783 E537. "De figura curvae elasticae contra obiectiones quasdam Illustris d'Alembert." *Acta Ac. Petrop.* (for the year 1779), pp. 188–192 (*OO* II.11, pp. 276–279, and EA).

1783 E538. "Cautiones necessariae in determinatione motus planetarum observandae." *Acta Ac. Petrop.* (for the year 1780), pp. 295–334 (*OO* II.29, pp. 360–391, and EA).

1783 E542. "De mirabilibus proprietatibus numerorum pentagonalium." *Acta Ac. Petrop.* (for the year 1780), pp. 56–75 (*OO* I.3, pp. 480–496, and EA).

1783 E552. "Observationes circa divisionem quadratorum per numeros primos." *Opuscula varii argumenti* 1, pp. 64–84 (*OO* I.3, pp. 497–512, and EA). For a selection of English translations of portions, see Struik, 1969.

1783 E553. "Observationes analyticae." *Opuscula analytica* 1, pp. 85–120 (*OO* I.15, pp. 400–434, and EA).

1783 E555. "De eximio usu methodi interpolationum in serierum doctrina." *Opuscula analytica* 1, pp. 157–210 (*OO* I.15, pp. 435–497, and EA).

1783 E560. "Miscellanea analytica." *Opuscula analytica* 1, pp. 329–344 (*OO* I.4, pp. 91–104, and EA). Presented to the Petersburg Academy of Sciences in 1773.

1783 E562. "Quomodo sinus et cosinus angulorum multiplorum per producta exprimi queant." *Opuscula analytica* 1, pp. 353–363 (*OO* I.15, pp. 509–521, and EA).

1783 E570. "De inventione longitudinis locorum ex observata lunae distantia a quadam stella fixa cognita." *Acta Ac. Petrop.* (for the year 1780), pp. 353–363 (*OO* II.30, pp. 334–338, and EA).

1784 E579. "Calculs sur les ballons aérostatiques faits par feu M. Léonard Euler, tels qu'on les a trouvés sur son ardoise, après sa mort arrivée le 7 Septembre 1783." *Mém. de Paris*, pp. 264-268 (*OO* II.16, pp. 165–169, and EA).

1785 E580. *Opuscula analytica* 2. Originally published by the Petersburg Academy of Sciences. It contains fifteen articles that are scattered through the Euler *Opera omnia*.

1785 E584. "De insignibus proprietatibus unciarum binomii ad uncias quorumuis polynomiorum extensis." *Acta Ac. Petrop.* 5, pp. 76–89 (*OO* I.15, pp. 604–620, and EA).

1785 E585. "De effectu frictionis in motu volutorio." *Acta Ac. Petrop.* 5, pp. 131–175 (*OO* II.9, pp. 197–238, and EA).

1785 E592. "De relatione fractionum transcendentium in infinitas fractiones simplices." *Opuscula analytica* 2, pp. 102–137 (*OO* I.15, pp. 621–660, and EA).

1785 E598. "De insigni promotione scientiae numerorum." *Opuscula analytica* 2, pp. 275–314 (*OO* I.4, pp. 163–196, and EA).

1793 E658. "De momentis virium respectu axis cuiuscunque inveniendis. . . . " *N. Acta Ac. Petrop.* 7, pp. 191–204 (*OO* II.9, pp. 387–398, and EA).

1793 E659. "Methodus facilis omnium virium momenta respectu axis cuiuscunque determinandi." *N. Acta Ac. Petrop.* 7, pp. 205–214 (*OO* II. 9, pp. 399–406).

1845 E660. *Institutionum calculi integralis volumen quartum.* Its third edition was published posthumously as a book by the Petersburg Imperial Academy of Sciences (*OO* I.13, and EA). It contains 11 supplements.

1798 E704. "Disquisitio ulterior super seriebus secundum multipla cuiusdam anguli progredientibus." *N. Acta. Ac. Petrop.* 11, pp. 114–132 (*OO* I.16, pp. 333–353, and EA). Includes an article from 1777.

1801 E708. "De formulis speciei mxx + nyy ad numeros explorandos idoneis, earumque mirabilibus proprietatibus." *N. Acta Ac. Petrop.* 12, pp. 22–46 (*OO* I.4, pp. 269–289, and EA). Paper read at the academy in March 1778.

1802 E715. "De variis modis numeros praegrandes examinandi, utrum sint primi nec ne?" *N. Acta Ac. Petrop.* 13, pp. 14–44 (*OO* I.4, pp. 303–328, and EA).

1805 E718. "Facillima methodus plurimos numeros primos praemagnos inveniendi." *N. Acta Ac. Petrop.* 14, pp. 3–10 (*OO* I.4, pp. 352–359, and EA).

1805 E719. "Methodus generalior numeros quosvis satis grandes perscrutandi utrum sint primi nec ne." *N. Acta Ac. Petrop.* 14, pp. 11–51 (*OO* I.4, pp. 360–394, and EA).

1806 E725. "Illustrio paradoxi circa progressionem numerorum idoneorum sive congruorum." *N. Acta Ac. Petrop.* 15, pp. 29–32 (*OO* I. 4, pp. 395–398, and EA).

1820 E754. "Problème de géométrie résolu par analyse de diophante." *Mémoires de l'académie impériale des sciences de St. Petersbourg* 7, pp. 3–9 (*OO* I.5, pp. 28–34, and EA). Presented in 1782 to the Petersburg Academy of Sciences.

1847 E790. "Commentatio de matheseos sublimioris utilitate." *Journal für die reine und angewandte Mathematik* 35, 1847, pp. 109–116 (*OO* 3.2, pp. 392–399, and EA).

1849 E792. "Tractatus de numerorum doctrina capita sedecim, quae supersunt." *Commentationes arithmeticae* 2, pp. 503–575 (*OO* I.5, pp. 182–283, and EA).

1849 E795. "De quadratis magicis." *Commentationes arithmeticae* 2, pp. 593–602. Reprinted in *Opera postuma* 1, pp. 140–151 [E795A] (*OO* I.7, pp. 441–457, and EA). Jordan Bell has made an English translation that is available at the Euler Archive.

1849 E799. "Fragmenta commentationes cuiusdam maioris. . . . " *Commentationes arithmeticae* 2, pp. 648–651 (*OO* I. 5, pp. 366–370, and EA).

1862 E805. *Opera postuma mathematica et physica.* Originally published in two volumes. It contains fifty-nine of Euler's previously unpublished papers, E792–E799 and E806–E856.

1862 E806. "Fragmenta arithmetica ex Adversariis mathematicis (*) depromta." *Opera postuma* 1, pp. 157–266. Its articles appear throughout Euler's *Opera omnia* (also at EA).

1862 E807. "Sur les logarithmes des nombres négatifs et imaginaires." *Opera postuma* 1, pp. 269–281 (*OO* I:19, pp. 418–438, and EA with English translation).

1862 E811. "Vera aestimatio sortis in ludis." *Opera postuma* 1, pp. 315–318 (*OO* I.7, pp. 458–465, and EA). Richard Pulskamp has made an English translation that is available at http:///www.cs.xu.edu/math/Sources/EulerE811.pdf.

1862 E814. "Institutionum calculi differentialis Sectio III." *Opera postuma* 1, pp. 342–402 (*OO* I.29, pp. 334–429, and EA).

1862 E837. "De emendatione tabularum llunarium per observationes eclipsium Lunae." *Opera postuma* 2, pp. 354–364 (*OO* II.24, pp. 271–285, and EA).

1862 E842. "Anleitung zur Naturlehre, worin die Gründe zu Erklärung aller in der Natur sich ereignenden Begebenheiten und Veränderungen festgesetzt werden." *Opera postuma* 2, pp. 449–560 (*OO* III.1, pp. 16–178, and EA).

1862 E844. "Théorie générale de la Dioptrique." *Opera postuma* 2, pp. 567–604 (*OO* III.9, pp. 1–48, and EA).

1862 E853. "Meditatio in experimenta explosione tormentorum nuper instituta." *Opera postuma* 2, pp. 800–804 (*OO* II.14, pp. 468–478, and EA). An English translation of selections has been made by Florian Cajori and appears in Smith, 1929, pp. 95–99.

1862 E856. "Fragmentum ex Adversariis mathematicis depromtum (Mechanica. III. N. Fuss.)." *Opera postuma* 2, pp. 824–826 (*OO* I.23, pp. 450–455, and EA).

Correspondence, Eulogies, and Archival Sources for Leonhard Euler

Opera omnia (*OO*) series and volume numbers are indicated herein in roman and arabic numerals, respectively.

Condorcet, Marie Jean Antoine Nicolas de Caritat, Marquis de. 1786. "Éloge de M. Euler." In *Histoire de l'Académie royale des sciences pour l'année 1783*, pp. 37–68. Paris: Imprimerie royale.

Euler, Leonhard, 1840, 1872. *Letters of Euler on Different Subjects in Natural Philosophy: Addressed to a German Princess.* 2 vols. New York: Harper & Brothers, Publishers.

Euler, Leonhard. 1735. "Letter of 10 December 1735 to Korff." *OO* III.2.

Euler, Leonhard, and Samuel König. 1976. *The Letters of Two Notable Swiss Scientists, Leonhard Euler and Samuel König, to Pierre Louis Moreau de Maupertuis, President of the Academy of Science, Berlin, 1737–1759, Relating to the "Least Action" Controversy, The Affairs of the Berlin Academy, Frederick the Great, Voltaire. . . .* Zurich: Hellmut Schumann.

Forbes, Eric G., ed. 1971a. *The Euler-Mayer Correspondence (1751–1755): A New Perspective on Eighteenth-Century Advances in Lunar Theory.* New York: American Elsevier.

——, trans. 1971b. *Tobias Mayer's Opera Inedita: The First Translation of the Lichtenberg Edition of 1775.* New York: American Elsevier.

Forbes, Eric G., and Curtis Wilson. 1995. "The Solar Tables of Lacaille and the Lunar Tables of Mayer." In *The General History of Astronomy*, ed. René Taton and Curtis Wilson, vol. 2, part B, pp. 55–69. Cambridge: Cambridge University Press.

Fuss, Nicholas. 2010. *Lobrede auf Leonhard Euler.* Whitefish, Mont.: Kessinger.

——. 2009. "Eulogy in Memory of Leonhard Euler." In *Portraits: Leonhard Euler, Daniel Bernoulli, Johann-Heinrich Lambert*, compiled and translated by Oscar B. Sheynin, pp. 8–48. Berlin: NG Verlag.

——. 2007. "Eulogy in Memory of Leonhard Euler." In *Euler and Modern Science*, eds. N. N. Bogolyubov, G. K. Mikhaĭlov, and A. P. Yushkevich; translated from the Russian by Robert Burns, pp. 369–397. Washington, D.C.: Mathematical Association of America.

Fuss, Paul Heinrich, ed. [1843] 1968. *Correspondance Mathématique et Physique de Quelques Célèbres Géométres du XVIIIème Siècle*, 1st ed., Saint Petersburg. Reprint, 2 vols., New York: Johnson Reprint.

Glaus, John S., trans. 2003. "Euler's Correspondence with Johann Kaspar Wettstein," Euler Archive, http://eulerarchive.maa.org.

Materialy dlya istorii Imperatorskoj Akademii nauk, vols. 1–10. Saint Petersburg: Akademie Science, 1885–1900.

Mikhaĭlov, Gleb K., ed. 1962. *Manuscripta Euleriana Archivi Academiae scientiarum URSS*, vol. 2. Moscow: Nauka. Reprint 1965.

——. 1965. *Ryukopisnye materialy L.Euler'a v Arkhive Academii Nauk*. Perevod s Latinskogo I. A. Perelmana. Tom 2, Chast 1. Trudy po Mekhanike. Moskva-Leningrad. Nauka. 1965. 574 str. (Trudy Arkhiva Akademii Nauk. Nomer 20).

Protokoly zasedanij konferentsii imperatorskoj Akademii nauk s 1725 po 1803 goda/Procés-verbaux des séances de l'Académie imperial des sciences depuis sa foundation jusqu'á 1803. 1897–1911. 4 vols. Saint Petersburg: Academy of Sciences.

Smirnov, Vladimir Ivanovich. Sostavitely: T. N. Klado, Yu. Kh. Kopelevich, T.A. Lukina i drugie. 1963. *L. Euler. Pisma k Uchenym.* Redaktor i Avtor Predislovia akademik Moskva-Leningrad. Izdatelstvo Akademii Nauk SSSR. 397 c. Imennoi Ukazatel: p. 388–396.

Wettstein, Johann Kaspar. 1976. "Correspondence Euler-Wettstein." In *Die Berliner und die Petersburg Akademie der Wissenschaften im Briefwechsel Leonhard Eulers*, vol. 3, edited by Adolph Pavlovich Yushkevich and Eduard Winter, pp. 256–366. Berlin: Akademie Verlag.

Yushkevich, Adolph Pavlovich, and Eduard Winter, eds. 1959–76. *Die Berliner und die Petersburger Akademie der Wissenschaften im Briefwechsel Leonhard Eulers.* 3 vols. Berlin: Akademie Verlag.

——. 1965. *Leonhard Euler und Christian Goldbach: Briefwechsel 1729–1764*. Berlin: Akademie Verlag.

Bibliographies for Leonhard Euler

Burckhardt, Johann Jakob. 1983. "Euleriana—Verzeichnis des Schrifttums über Leonhard Euler." In *Leonhard Euler, 1707–1783: Beiträge zu Leben und Werk*, ed. J. J. Burckhardt, E. A. Fellmann, and W. Habicht, pp. 511–552. Basel: Birkhäuser Verlag.

Eneström, Gustaf. 1913. "Verzeichnis der Schriften Eulers chronologisch nach den Jahren geordnet, in denen sie verfasst worden sind." In *Jahresbericht der Deutschen Mathematiker-Vereinigung*, vol. 2, pp. 1–388. Leipzig. Eneström enumerates 866 distinct works. The numbers E1 to E866 are now referred to as the Eneström numbers.

Mikhaĭlov, Gleb Konstantinovich. 2007. "Euleriana: A Short Bibliographical Note." *Physica D: Nonlinear Phenomena* 237, pp. xvii–xviii.

Jubilee Anniversary Volumes and Other Critical Studies of Leonhard Euler

Note: Jubilee anniversary volumes will be cited as LEJA (for Leonhard Euler jubilee anniversary), followed by a number indicating a volume's place in the sequence from the date of publication after the first such anniversary in 1907.

Baker, Roger, ed. 2007. *Euler Reconsidered.* Heber City, Utah: Kendrick.
Bogolyubov, N. N., G. K. Mikhaĭlov, and A. P. Yushkevich, eds. [1988] 2007. *Razvitie ideĭ Leonharda Eulera i sovremennaya nauka.* 1st ed. Leningrad: Izdat'svo Nauka. Translated as *Euler and Modern Science* by Robert Burns (Washington, D.C.: Mathematical Association of America). A collection of twenty-seven papers presented at the 1983 conference "The Development of Euler's Ideas in the Modern Era" in Moscow and Saint Petersburg. LEJA 7.
Bradley, Robert E., Lawrence A. D'Antonio, and C. Edward Sandifer, eds. 2007. *Euler at 300: An Appreciation.* Washington, D.C: Mathematical Association of America.
Bradley, Robert E., and C. Edward Sandifer, eds. 2007. *Leonhard Euler: Life, Work and Legacy.* Elsevier Studies in the History and Philosophy of Mathematics 5. New York: Elsevier.
Burckhardt, J. J., E. A. Fellmann, and W. Habicht, eds. 1983. *Leonhard Euler, 1707–1783: Beiträge zu Leben und Werk.* Basel: Birkhäuser Verlag. LEJA 4.
Crépel, Pierre, and Luigi Pepe, eds. 2008. *Bollettino di Storia delle Scienze Matematiche* 28.
Engel, Wolfgang, ed. 1985. *Festakt und Wissenschaftliche Konferenz aus Anlass des 200. Todestages von Leonhard Euler.* Abhandlungen der Akademie der Wissenschaften der DDR. Abteilung Mathematik-Naturwissenschaften-Technik Jg. 1985, No. 1 N. Berlin: Akademie Verlag. LEJA 6. This volume contains thirteen papers.
Eyink, Gregory, Uriel Frisch, René Moreau, and Andreĭ Sobolevskiĭ, eds. 2007. *Euler Equations: 250 Years' Proceedings of an International Conference, Physica D: Nonlinear Phenomena* 237, pp. 1825–2250.
Knobloch, E., I. S. Louhivaara, and J. Winkler, eds. 1984. *Zum Werk Leonhard Eulers: Vorträge des Euler-Kolloquiums im Mai 1983 in Berlin.* Basel: Birkhäuser Verlag. LEJA 5. This volume contains seven papers.
Lavrentiev, Mikhail Alekseevich, Adolph P. Yushkevich, and Ašot T. Grigor'yan, eds. 1959. *Sammelband der zu Ehren des 250. Geburtstages Leonhard Eulers der Akademie der Wissenschaften der UdSSR vorgelegten Abhandlungen.* Moscow: Izdatel'stvo Akademii nauk. LEJA 2.
Schröder, Kurt, ed. 1959. *Sammelband der zu Ehren des 250. Geburtstages Leonhard Eulers der Deutschen Akademie der Wissenschaften zu Berlin vorgelegten Abhandlungen.* Berlin: Akademie Verlag. LEJA 3.
Vasilyev, Vladimir N., ed. 2008. *Leonard Eĭler: K 300-letiiu so dnia rozhdenii / Leonhard Euler: 300th Anniversary.* Saint Petersburg: Nestor-Historia. LEJA 8. In Russian and English.

Sources by Authors Other Than Euler

Note: Jubilee anniversary volumes will be cited as LEJA (for Leonhard Euler jubilee anniversary), followed by a number indicating a volume's place in the sequence from the date of publication after the first such anniversary in 1907.

Abalakin, V. K., and E. A. Grebnikov. 2007. "Euler and the Development of Astronomy in Russia." Translated by Robert Burns. In *Euler and Modern Science*, ed. N. N. Bogolyubov, G. K. Mikhaĭlov, and A. P. Yushkevich, pp. 245–263. Washington, D.C.: Mathematical Association of America. LEJA 7.

Agoh, Takashi, and Karl Dilcher. 2009. "Shortened Recurrence Relations for Bernoulli Numbers." *Discrete Mathematics* 309, no. 4, pp. 887–898.

——. 2008. "Reciprocity Relations for Bernoulli Numbers." *American Mathematical Monthly* 115, no. 3, pp. 237–244.

Aiton, E. J. 1985. *Leibniz: A Biography*. Boston: Adam Hilger.

Aldoshin, Gennadii Tychonovich. 2008. "Leonhard Euler and Aerohydroelasticity" [title translated from Russian]. In *Leonard Eĭler: K 300-letiiu so dnia rozhdenii / Leonhard Euler: 300th Anniversary*, ed. Vladimir N. Vasilyev, pp. 159–170. Saint Petersburg: Nestor-Historia. LEJA 8.

Aldridge, A. Owen. 1975. *Voltaire and the Century of Light*. Princeton, N.J.: Princeton University Press.

Aleksandrov, A. P. 2007. "Opening Speech of the Symposium 'Modern Development of Euler's Ideas.'" Translated by Robert Burns. In *Euler and Modern Science*, ed. N. N. Bogolyubov, G. K. Mikhaĭlov, and A. P. Yushkevich, pp. 3–7. Washington, D.C.: Mathematical Association of America. LEJA 7.

Alexanderson, G. L. 2007. "*Ars Expositioni*: Euler as Writer and Teacher." In *The Genius of Euler: Reflections on His Life and Work*, ed. William Dunham, pp. 61–69. Washington, D.C.: Mathematical Association of America.

Algarotti, Francesco. 1764. *Essai sur la peinture et sur l'Académie de France*. Translated from the Italian (Saggi sopra la pittura). London: L. Davis and C. Reymers.

Anderson, Matthew Smith. 1995. *The War of the Austrian Succession 1740–1748*. New York: Longman.

Andrews, George E. 2007. "Euler's Pentagonal Number Theorem." In *The Genius of Euler: Reflections on His Life and Work*, ed. William Dunham, pp. 225–232. Washington, D.C.: Mathematical Association of America.

Ardema, Mark D. 2005. *Newton-Euler Dynamics*. Berlin: Springer Verlag.

Ariga, Nobumichi. 2006. "Euler's Variational Mechanics" [title translated from Japanese]. *Kagakushi Kenyu* 45, pp. 220–228.

Artin, Emil. 1964. *The Gamma Function*. Translated by Michael Butler. New York: Holt, Rinehart and Winston.

Ayoub, Raymond. 2007. "Euler and the Zeta Function." In *The Genius of Euler: Reflections on His Life and Work*, ed. William Dunham, pp. 113–133. Washington, D.C.: Mathematical Association of America.

Bachelard, Suzanne. 1961. *Les polémique concernant le principe de moindre action au XVIIIe siècle*. Paris: Palais de la Découverte.

Baig i Aleu, Marià. 2008. "Teoría matemática y práctica naval en la Ilustración. Salvador Jiménez Coronado, traductor de la obra de Euler sobre la construcción y la maniobra de los navios." *Quaderns d'Història de l'Enginyeria* 9, pp. 249–277.

Bailhache, Patrice. 1995. "Deux mathématiciens musiciens: Euler et d'Alembert." *Physis Rivista Internazionale di Storia*, new series, 32, pp. 1–35.

Baker, Roger. 2007. "Introduction." In *Euler Reconsidered*, ed. Roger Baker, pp. 1–11. Heber City, Utah: Kendrick.

Ball, J. M., and R. D. James. 2002. "The Scientific Life and Influence of Clifford Ambrose Truesdell III." *Archive for Rational Mechanics and Analysis* 161, pp. 1–26.

Ball, W.W.R. 2007. "Euler's Output, A Historical Note." In *The Genius of Euler: Reflections on His Life and Work*, ed. William Dunham, pp. 89–91. Washington, D.C.: Mathematical Association of America.

Baltus, Christopher. 2007a. "The Euler-Bernoulli Proof of the Fundamental Theorem of Algebra." In *Euler at 300: An Appreciation*, ed. Robert E. Bradley, Lawrence A. D'Antonio, and C. Edward Sandifer, pp. 41–53. Washington, D.C: Mathematical Association of America.

Banichuk, N. V., and A. Yu. Ishlinskiĭ. 2007. "Leonhard Euler and the Mechanics of Elastic Systems." Translated by Robert Burns. In *Euler and Modern Science*, ed.

N. N. Bogolyubov, G. K. Mikhaĭlov, and A. P. Yushkevich, pp. 213–237. Washington, D.C.: Mathematical Association of America. LEJA 7.

Barbeau, E. J. 2007. "Euler Subdues a Very Obstreperous Series (Abridged)." In *The Genius of Euler: Reflections on His Life and Work*, ed. William Dunham, pp. 135–147. Washington, D.C.: Mathematical Association of America.

Barnett, Janet Heine. 2009. "Mathematics Goes Ballistic: Benjamin Robins, Leonhard Euler, and the Mathematical Education of Military Engineers." *British Society for the History of Mathematics Bulletin* 24, pp. 92–104.

——. 2007. "Enter, Stage Center: The Early Drama of the Hyperbolic Functions." In *Euler at 300: An Appreciation*, ed. Robert E. Bradley, Lawrence A. D'Antonio, and C. Edward Sandifer, pp. 85–105. Washington, D.C: Mathematical Association of America.

Barrow-Green, June. 2010. "Euler as an Educator." *British Society for the History of Mathematics Bulletin* 25, pp. 10–22.

Bartholmess, Christian. [1855] 2009. *Histoire philosophique de l'Académie de Prusse depuis Leibniz jusqu'à Schelling*. Paris: Franck. Reprint, University of Michigan Press.

Bashmakova, I. G. 2007. "Euler's Contribution to Algebra." Translated by Robert Burns. In *Euler and Modern Science*, ed. N. N. Bogolyubov, G. K. Mikhaĭlov, and A. P. Yushkevich, pp. 137–153. Washington, D.C.: Mathematical Association of America. LEJA 7.

Becchi, Antonio, Massimo Corradi, Federico Foce, and Orietta Pedemonto, eds. 2003. *Essays on the History of Mechanics: In Memory of Clifford Ambrose Truesdell and Edoardo Benvenuto*. Basel: Birkäuser Verlag.

Beckmann, Petr. 1971. *A History of Pi*. New York: Saint Martin's.

Beeson, David. 1992. *Maupertuis: An Intellectual Biography*. Studies on Voltaire and the Eighteenth Century, vol. 299. Oxford: Voltaire Foundation.

Bell, Jordan. 2010. "A Summary of Euler's Work on the Pentagonal Number Theorem." *AHES* 64, no. 3, pp. 301–373.

Bellhouse, D. R. 2007. "Euler and Lotteries." In *Leonhard Euler: Life, Work and Legacy*, ed. Robert E. Bradley and C. Edward Sandifer, pp. 385–395. Elsevier Studies in the History and Philosophy of Mathematics 5. New York: Elsevier.

Bernoulli, Daniel. 1738. *Hydrodynamica*. Strasbourg, France: Johann Reinhold Dulsecker.

Bernoulli, Johann, I. 1742. *Opera omnia*, vol. 2. Translated by Otto Spiess. Lausanne, Switzerland: Marci-Michaelis Bousquet/Sociorum.

——. 1727. "Discours sur les loix de la communication du mouvement." *Recueil prix Paris* 2, pp. 58–108.

Bernoulli, Johann, III. 1779. *Reisen durch Brandenburg, Pommern, Preussen, Curland, Russland und Pohlen in den Jahren 1777 und 1778*, vol. 3. Leipzig: Kaspar Fritsch.

Bernoulli, René. 1983. "*Leonhard Eulers Augenkrankheiten*." In *Leonhard Euler, 1707–1783: Beiträge zu Leben und Werk*, ed. J. J. Burckhardt, E. A. Fellmann, and W. Habicht, pp. 471–489. Basel: Birkhäuser Verlag. LEJA 4.

Beutler, Gerhard, Leos Mervart, and Andreas Verdun. 2004. *Methods of Celestial Mechanics*, vol. 1. Berlin: Springer Verlag.

Biagioli, Mario. 1993. *Galileo Courtier: The Practice of Science in the Culture of Absolutism*. Chicago: University of Chicago Press.

——. 1992. "Scientific Revolutions, Social Bricolage, and Etiquette." In *The Scientific Revolution in National Context*, ed. Roy Porter and Mikulaś Teich, pp. 11–54. Cambridge: Cambridge University Press.

Bibliotheca Mathematica [1906–7] 2012. *Bibliotheca Mathematica*, vol. 7. Charleston, SC: Nabu.

Biermann, K.-R. 2007. "Was Leonhard Euler Driven from Berlin by J. H. Lambert?" Translated by Robert Burns. In *Euler and Modern Science*, ed. N. N. Bogolyubov,

G. K. Mikhaĭlov, and A. P. Yushkevich, pp. 87–97. Washington, D.C.: Mathematical Association of America. LEJA 7.

——. 1983. "Aus der Vorgeschichte der Euler Ausgabe 1783–1907." In *Leonhard Euler, 1707–1783: Beiträge zu Leben und Werk*, ed. J. J. Burckhardt, E. A. Fellmann, and W. Habicht, pp. 489–501. Basel: Birkhäuser Verlag. LEJA 4.

Bilfinger, Georg. 1729. "De tubis capillaribus dissertatio experimentalis prima." *Comm. Ac Petrop.*, vol. 2 (for 1727), pp. 233–287.

——. 1728. "De viribus corpori moto insitis, et illarum mensura," *Comm. Ac. Pb.* 1 (for 1726), pp. 43–121.

Birke, Joachim. 1966. *Christian Wolffs Metaphysik und die Zeitgenössische Literatur- und Musiktheorie: Gottsched, Scheibe, Mizler*. Berlin: Walter de Gruyter.

Blanning, Timothy Charles William. 2007. *The Pursuit of Glory: Europe 1648–1815*. London: Viking.

——, ed. 2000. *The Eighteenth Century: Europe 1688–1815*. Oxford: Oxford University Press.

Bloch, Marc. 1973. *The Royal Touch: Sacred Monarchy and Scrofula in England and France*. London: Routledge and Kegan Paul.

Blunt, Anthony. 1967. *Nicolas Poussin*. London: Phaidon.

Bodenmann, Siegfried. 2010, January. "The 18th-Century Battle over Lunar Motion." *Physics Today*, pp. 27–32.

——. 2009. "Newton's and Selene's Reconciliation: Clairaut, d'Alembert, and Euler on Lunar Theory." Paper presented at the American Association for the Advancement of Science Annual Convention, Chicago.

Bogdanov, V. I., and T. I. Malova. 2008. "Flooding, Ebbing, and Flowing" [title translated from Russian]. In *Leonard Eĭler: K 300-letiiu so dnia rozhdenii / Leonhard Euler: 300th Anniversary*, ed. Vladimir N. Vasilyev, pp. 221–233. Saint Petersburg: Nestor-Historia. LEJA 8.

Bonjour, Edgar. 1988. *Die Schweiz und Europa. Ausgewählte Reden und Aufsätze*, vol. 8. Basel: Helbing und Lichtenhahn.

——. 1960. *Die Universität Basel. Von den Anfangen bis zur Gegenwart, 1460–1960*. Basel: Helbing und Lichtenhahn.

Borgato, Maria Teresa. 2008. "Euler, Lagrange, and Life Insurance." In *Leonard Eĭler: K 300-letiiu so dnia rozhdenii / Leonhard Euler: 300th Anniversary*, ed. Vladimir N. Vasilyev, pp. 115–127. Saint Petersburg: Nestor-Historia. LEJA 8.

Bos, Henk J. M. 1974. "Differentials, Higher-Order Differentials, and the Derivative in the Leibnizian Calculus." *AHES* 14, no. 1, pp. 1–90.

Boss, Valentin. 1972. *Newton and Russia: The Early Influence, 1698–1796*. Cambridge, Mass.: Harvard University Press.

Bouvier, Nicholas, Gordon A. Craig, and Lionel Gossman. 1994. *Geneva, Zurich, Basel: History, Culture, and National Identity*. Princeton, N.J.: Princeton University Press.

Boyer, Carl B. 2007. "The Foremost Textbook of Modern Times." In *The Genius of Euler: Reflections on His Life and Work*, ed. William Dunham, pp. 69–75. Washington, D.C.: Mathematical Association of America.

Bradley, Robert E. 2007a. "Euler, d'Alembert, and the Logarithm Function." In *Leonhard Euler: Life, Work and Legacy*, ed. Robert E. Bradley and C. Edward Sandifer, pp. 255–279. Elsevier Studies in the History and Philosophy of Mathematics 5. New York: Elsevier.

——. 2007b. "The Genoese Lottery and the Partition Function." In *Euler at 300: An Appreciation*, ed. Robert E. Bradley, Lawrence A. D'Antonio, and C. Edward Sandifer, pp. 203–217. Washington, D.C: Mathematical Association of America.

——. 2007c. "Three Bodies? Why Not Four? The Motion of Lunar Apsides." In *Euler at 300: An Appreciation*, ed. Robert E. Bradley, Lawrence A. D'Antonio, and C. Edward Sandifer, pp. 227–239. Washington, D.C: Mathematical Association of America.

Breger, Herbert. 1999. "Über den von Samuel König veröffentlichten Brief zum Prinzip der kleinsten Wirkung." In *Pierre Louis Moreau de Maupertuis: Eine Bilanz nach 300 Jahren*, ed. Hartmut Hecht, pp. 363–381. Berlin: Verlag Arno Spitz.

Breidert, Wolfgang. 2007. "Leonhard Euler and Philosophy." In *Leonhard Euler: Life, Work and Legacy*, ed. Robert E. Bradley and C. Edward Sandifer, pp. 97–109. Elsevier Studies in the History and Philosophy of Mathematics 5. New York: Elsevier.

——. 1983. "Leonhard Euler und die Philosophie." In *Leonhard Euler, 1707–1783: Beiträge zu Leben und Werk*, ed. J. J. Burckhardt, E. A. Fellmann, and W. Habicht, pp. 447–459. Basel: Birkhäuser Verlag. LEJA 4.

Broman, Thomas. 2012. "Metaphysics for an Enlightened Public: The Controversy over Monads in Germany, 1746–1748." *Isis* 103, no. 4, pp. 1–23.

——. 1998. "The Habermasian Public Sphere and 'Science in the Enlightenment.'" *History of Science* 36, pp. 123–149.

Brown, Bancroft H. 2007. "The Euler-Diderot Anecdote." In *The Genius of Euler: Reflections on His Life and Work*, ed. William Dunham, pp. 57–61. Washington, D.C.: Mathematical Association of America.

Bruckner, Albert, ed. 1972. *Riehen: Geschichte eines Dorfes.* Riehen, Switzerland: Verlag A. Schudel.

Brunet, Pierre. 1931. *L'introduction des thèories de Newton en France au XVIIIe siècle: Avant 1738.* Paris: Albert Blanchard.

——. 1929. *Maupertuis.* 2 vols. Paris: Albert Blanchard.

Bruno, Giuseppe, Andrea Genovese, and Gennaro Improta. 2011. "Routing Problems: A Historical Perspective." *British Society for the History of Mathematics Bulletin* 26, pp. 118–127.

Bruyn, Günter de. 2002. *Unter den Linden.* Berlin: Siedler Verlag.

Bryant, Robert, Phillip Griffiths, and Daniel Grossman. 2003. *Exterior Differential Systems and Euler-Lagrange Partial Differential Equations.* Chicago: University of Chicago Press.

Brylevskaya, Lasissa Ivanova. 2008. "Euler's Jubilee Celebration" [title translated from Russian]. In *Leonard Eĭler: K 300-letiiu so dnia rozhdenii / Leonhard Euler: 300th Anniversary*, ed. Vladimir N. Vasilyev, pp. 316–322. Saint Petersburg: Nestor-Historia. LEJA 8.

Buchwald, Jed Z., and I. Bernard Cohen, eds. 2001. *Isaac Newton's Natural Philosophy.* Cambridge, Mass.: MIT Press.

Buchwald, Jed Z., and Robert Fox, eds. 2013. *Oxford Handbook of the History of Physics.* Oxford: Oxford University Press, 2013.

Bullynck, Maarten. 2010. "Factor Tables 1657–1817, with Notes on the Birth of Number Theory." *Revue d'histoire des mathématiques* 16, pp. 133–216.

——. 2009. "Modular Arithmetic before C. F. Gauss: Systematizations and Discussions on Remainder Problems in 18th-Century Germany." *HM* 36, pp. 48–72.

Burckhardt, Johann Jakob. 2007. "Leonhard Euler, 1707–1783." In *The Genius of Euler: Reflections on His Life and Work*, ed. William Dunham, pp. 75–89. Washington, D.C.: Mathematical Association of America.

——. 1983a. "Euleriana–Verzeichnis des Schrifttums über Leonhard Euler." In *Leonhard Euler, 1707–1783: Beiträge zu Leben und Werk*, ed. J. J. Burckhardt, E. A. Fellmann, and W. Habicht, pp. 511–552. Basel: Birkhäuser Verlag. LEJA 4.

——. 1983b. "Die Eulerkommission der Schweizerischen Naturforschenden Gesellschaft—Ein Beitrag zur Editionsgeschichte." In *Leonhard Euler, 1707–1783: Beiträge zu Leben und Werk*, ed. J. J. Burckhardt, E. A. Fellmann, and W. Habicht, pp. 501–510. Basel: Birkhäuser Verlag. LEJA 4.

Burmeister, Karl Heinz. 2009. "Bodolz als Heimat, berühmter Gelehrter: Leonhard Euler (1707–1783). . . ." In *Jahrbuch des Landkreises Lindau*, pp. 99–104. Bergatreute, Germany: Eppse.

Burton, David M. 2003. *The History of Mathematics: An Introduction*, 5th ed. Boston: McGraw Hill.

Busev, V. M. 2007. "Euler's Handbook on Arithmetic for Use in the Gymnasium" [title translated from Russian]. In *Leonard Eĭler: K 300-letiiu so dnia rozhdenii / Leonhard Euler: 300th Anniversary*, ed. Vladimir N. Vasilyev, pp. 255–264. Saint Petersburg: Nestor-Historia. LEJA 8.

Cajori, Florian. 2007. "Frederick the Great on Mathematics and Mathematicians." In *The Genius of Euler: Reflections on His Life and Work*, ed. William Dunham, pp. 51–57. Washington, D.C.: Mathematical Association of America.

Calinger, Ronald. 2009. Book review of Emil A. Fellmann, *Leonhard Euler*, 2007. *Isis* 100, no. 2, pp. 401–402.

——. 2007. "Leonhard Euler: Life and Thought." In *Leonhard Euler: Life, Work and Legacy*, ed. Robert E. Bradley and C. Edward Sandifer, pp. 5–61. Elsevier Studies in the History and Philosophy of Mathematics 5. New York: Elsevier.

——. 1999. *A Contextual History of Mathematics*. Upper Saddle River, N.J.: Prentice-Hall.

——. 1996. "Leonhard Euler: The First St. Petersburg Years (1727–1741)." *HM* 23, pp. 121–166.

——. 1979. "Kant and Newtonian Science: The Pre-Critical Period." *Isis* 70, no. 3, pp. 349–363.

——. 1975. "Euler's *Letters to a Princess of Germany* as an Expression of His Mature Scientific Outlook." *AHES* 15, no. 3. pp. 211–233.

——. 1968a. "Frederick the Great and the Berlin Academy of Sciences (1740–1766)." *Annals of Science* 24, pp. 239–249.

——. 1968b. "The Newtonian-Wolffian Confrontation in the St. Petersburg Academy of Sciences." *Cahiers d'histoire mondiale* 11, pp. 417–436.

Calinger, Ronald, and Elena N. Polyakhova. 2007. "Princess Dashkova, Euler, and the Russian Academy of Sciences." In *Leonhard Euler: Life, Work and Legacy*, ed. Robert E. Bradley and C. Edward Sandifer, pp. 75–97. Elsevier Studies in the History and Philosophy of Mathematics 5. New York: Elsevier.

Callot, Émile. 1964. *Maupertuis: Le Savant et le Philosophe*. Paris: Riviere.

Cannon, John T., and Sigalia Dostrovsky. 1981. *The Evolution of Dynamics: Vibration Theory from 1687 to1742*. New York: Springer.

Caparrini, Sandro. 2007. "Euler's Influence on the Birth of Vector Mechanics." In *Leonhard Euler: Life, Work and Legacy*, ed. Robert E. Bradley and C. Edward Sandifer, pp. 459–479. Elsevier Studies in the History and Philosophy of Mathematics 5. New York: Elsevier.

——. 2002. "The Discovery of the Vector Representation of Moments and Angular Velocity," *AHES* 56, no. 2, pp. 151–181.

Caparrini, Sandro, and Craig Fraser. 2013. "Mechanics in the Eighteenth Century." In *Oxford Handbook of the History of Physics*, ed. Jed Buchwald and Robert Fox, vol. 2, no. 2, pp. 358–406. (Oxford: Oxford University Press, 2013).

Caparrini, Sandro, and Kim Williams, eds. 2008. *Discovering the Principles of Mechanics 1600–1800: Essays by David Speiser*. Basel: Birkhäuser Verlag.

Carathéodory, Constantin. 1952. "Einführung in Eulers Arbeiten über Variationsrechnung." In *Leonhardi Euleri Opera omnia* I:24, pp. viii–lv. Zurich: Orell Füssli.

Carlyle, Thomas. 1969. *History of Frederick the Great*. Edited by John Clive. Chicago: University of Chicago Press.

Carmody, Thomas, and Helmut Kobus. 1968. *Hydrodynamics by Daniel Bernoulli and Hydraulics by Johann Bernoulli*. New York: Dover.

Caspar, Max. 1993. *Johannes Kepler*. New York: Dover.

Cassirer, Ernst. 1951. *The Philosophy of the Enlightenment*. Translated by Fritz Oelin and James Pettegrave. Princeton, N.J.: Princeton University Press.

Cesiano, Jacques. 2008. "The History of Euler's Theorem about Polyhedrons (1750–1811)" [title translated from Russian]. In *Leonard Eĭler: K 300-letiiu so dnia rozhdenii /*

Leonhard Euler: 300th Anniversary, ed. Vladimir N. Vasilyev, pp. 79–88. Saint Petersburg: Nestor-Historia. LEJA 8.

Chandrasekhar, Subrahmanyan. 1987. *Truth and Beauty: Aesthetics and Motivations in Science*. Chicago: University of Chicago Press.

Chapin, Seymour L. 1995. "The Shape of the Earth." In *The General History of Astronomy*, ed. René Taton and Curtis Wilson, vol. 2, pp. 22–35. Cambridge: Cambridge University Press.

Christensen, Thomas. 2002. *The Cambridge History of Western Music Theory*. Cambridge: Cambridge University Press.

———. 1987. "Eighteenth-Century Science and the *Corps Sonore*: The Scientific Background to Rameau's Principle of Harmony." Unpublished manuscript, Yale University Department of Music.

Clairaut, Alexis-Claude. 1749. "Recherches sur différents points importante du système du monde dans les principes de la gravitation universelle." *Mém. de Paris*, vol. 33, pp. 329–364.

Clark, William. 1999. "The Death of Metaphysics in Enlightened Prussia." In *The Sciences in Enlightened Europe*, ed. William Clark, Jan Golinski, and Simon Schaffer, pp. 423–474. Chicago: University of Chicago Press.

Clark, William, Jan Golinski, and Simon Schaffer, eds. 1999. *The Sciences in Enlightened Europe*. Chicago: University of Chicago Press.

Clemens, H. 1902. *Die älteren Ephemerideausgaben der Berliner Akademie und die Begründung des Astronomischen Jahbuches 20*. Berlin: Akademie Verlag.

Coates, John. 2008. "Euler's Work on Zeta and L-Functions and Their Special Values." *British Society for the History of Mathematics Bulletin* 23, pp. 37–41.

Cohen, I. Bernard, and George E. Smith, eds. 2002. *The Cambridge Companion to Newton*. Cambridge: Cambridge University Press.

Cracraft, James. 2004. *The Petrine Revolution in Russian Culture*. Cambridge, Mass.: Belknap Press.

Cross, Jack L. 2010. *Members of the Russian Academy of Sciences (RANS)*. Memphis, Tenn.: General Books.

Cross, Jim. 1983. "Euler's Contributions to Potential Theory 1730–1755." In *Leonhard Euler, 1707–1783: Beiträge zu Leben und Werk*, ed. J. J. Burckhardt, E. A. Fellmann, and W. Habicht, pp. 331–345. Basel: Birkhäuser Verlag. LEJA 4.

Crowe, Michael J. 1986. *The Extraterrestrial Life Debate 1750–1900: The Idea of a Plurality of Worlds from Kant to Lowell*. Cambridge: Cambridge University Press.

D'Alembert, Jean-Baptiste le Rond. 2009. *Inventaire analytique de la correspondence: 1741–1783*. Oeuvres completes de D'Alembert, series 5, vol. 1, ed. Iréne Passeron with Anne-Marie Chouillet and Jean-Daniel Candaux. Paris: CNRS Éditions.

———. 1995. *Preliminary Discourse to the Encyclopedia of Diderot*. Translated by Richard N. Schwab. Chicago: University of Chicago Press. Original work published 1751.

———. 1754. "Cosmologie." In *Encyclopédie ou Dictionnaire Raisonné des Sciences, des Arts et des Métiers*, vol. 4, edited by Denis Diderot, pp. 295–297. Paris: Briasson, David, Le Breton, and Durand.

———. 1751. "Action." In *Encyclopédie ou Dictionnaire Raisonné des Sciences, des Arts et des Métiers*, vol. 1, edited by Denis Diderot, pp. 119–120. Paris: Briasson, David, Le Breton, and Durand.

———. [1747] 2009. *Réflexions sur la cause générale des vents*. Paris: Chez David. Reprint, Whitefish, Mont.: Kessinger. Awarded the Berlin Academy Prize, 1746.

———. 1743. *Traité de dynamique*. Paris: Chez David.

Daniels, Rudolph L. 1973. *V. N. Tatishchev: Guardian of the Petrine Revolution*. Philadelphia: Franklin.

D'Antonio, Lawrence A. 2007a. "Euler and Elliptic Integrals." In *Euler at 300: An Appreciation*, ed. Robert E. Bradley, Lawrence A. D'Antonio, and C. Edward Sandifer, pp. 119–131. Washington, D.C: Mathematical Association of America.

——. 2007b. "'The Fabric of the Universe Is Most Perfect': Euler's Research on Elastic Curves." In *Euler at 300: An Appreciation*, ed. Robert E. Bradley, Lawrence A. D'Antonio, and C. Edward Sandifer, pp. 239–261. Washington, D.C: Mathematical Association of America.

Darnton, Robert. 1997, 27 March. "George Washington's False Teeth." *New York Review of Books*, pp. 34–38.

——. 1982. *L'aventure de l'Encyclopédie*. Paris: Perrin.

——. 1979. *The Business of Enlightenment: A Publishing History of the Encyclopédie 1775–1800*. Cambridge, Mass.: Harvard University Press.

Dashkova, Ekaterina Romanovna. 1995. *The Memoirs of Princess Dashkova: Russia in the Time of Catherine the Great*. Translated and edited by Kyril Fitzlyon. Durham, N.C.: Duke University Press.

Daston, Lorraine. 1991. "The Ideal and Reality of the Republic of Letters in the Enlightenment." *Science in Context* 4, no. 2, pp. 367–386.

Dauben, Joseph W., and Christoph J. Scriba, eds. 2002. *Writing the History of Mathematics: Its Historical Development*. Basel: Birkhäuser Verlag.

Davis, Philip P. 2007. "Leonhard Euler's Integral: A Historical Profile of the Gamma Function." In *The Genius of Euler: Reflections on His Life and Work*, ed. William Dunham, pp. 167–185. Washington, D.C.: Mathematical Association of America.

Debnath, Lokenath. 2009. *The Legacy of Leonhard Euler*. London: Imperial College Press.

Delshams, Amadeu, and Maria Rosa Massa Esteve. "Consideracions al voltant de la Funció Beta a l'obra de Leonhard Euler (1707–1783)." *Quaderns d'Història de l'Enginyeria* 9, pp. 59–82.

Demidov, Sergei Sergeevich. 2008a. "D'Alembert et la notion de solution des équations différentielles aux dérivées partielles." *Bollitino di Storia delle Scienze Matematiche* 28, pp. 155–166.

——. 2008b. "Leonhard Euler and the Vibrating String Problem" (in Russian with English abstract). In *Leonard Eĭler: K 300-letiiu so dnia rozhdenii / Leonhard Euler: 300th Anniversary*, ed. Vladimir N. Vasilyev, pp. 50–56. Saint Petersburg: Nestor-Historia. LEJA 8. In Russian with abstract in English.

Desmond, Adrian, and James Moore. 1991. *Darwin: The Life of a Tormented Evolutionist*. New York: W. W. Norton.

Dick, Stephen J. 1982. *Plurality of Worlds*. Cambridge: Cambridge University Press.

Diderot, Denis. 1999. *Thoughts on the Interpretation of Nature*. Translated by David Adams. Wiltshire England: Clinamen.

——. [1821] 2006. *Rameau's Nephew*. Translated by Margaret Mauldon. Oxford: Oxford University Press.

——. 1967. *Dennis Diderot's The Encyclopedia; Selections*. Edited and translated by Stephen J. Gendzier. New York: Harper Torchbooks.

——, ed. 1754. *Encyclopédie ou Dictionnaire Raisonné des Sciences, des Arts et des Métiers*, vol. 4. Paris: Briasson, David, Le Breton, and Durand.

——. 1750. "Prospectus" to the *Encyclopédie*. ARTFL Encyclopédie Project, Robert Morrissey, gen. ed., http://encyclopedie.uchicago.edu/node/174.

Dobrovolskaya, E. M. 2008. "Euler's Investigations in the Theory of Continued Fractions" [title translated from Russian]. In *Leonard Eĭler: K 300-letiiu so dnia rozhdenii / Leonhard Euler: 300th Anniversary*, ed. Vladimir N. Vasilyev, pp. 63–68. Saint Petersburg: Nestor-Historia. LEJA 8.

Dostrovsky, Sigalia. 1974. "Early Vibration Theory: Physics and Music in the Seventeenth Century." *AHES* 14, no. 3, pp. 169–218.

Duffy, Christopher. 1985. *Frederick the Great: A Military Life*. London: Routledge and Kegan Paul.

Dugac, Pierre. 1983. "Euler, d'Alembert et les fondements de l'analyse." In *Leonhard Euler, 1707–1783: Beiträge zu Leben und Werk*, ed. J. J. Burckhardt, E. A. Fellmann, and W. Habicht, pp. 171–185. Basel: Birkhäuser Verlag. LEJA 4.

Dugas, René. 1955. *A History of Mechanics.* Translated by J. R. Maddox. Neuchatel, Switzerland: Éditions du Griffon.

Dunham, William, 2007a. "Euler and the Fundamental Theorem of Algebra." In *The Genius of Euler: Reflections on His Life and Work*, ed. William Dunham, pp. 243–257. Washington, D.C.: Mathematical Association of America.

——, ed. 2007b. *The Genius of Euler: Reflections on His Life and Work.* Washington, D.C.: Mathematical Association of America.

——. 1999. *Euler: The Master of Us All.* Washington, D.C.: Mathematical Association of America.

Dvoichenko-Markov, Eufrosina. 1965. "The Russian Members of the American Academy of Sciences." *Proceedings of the American Philosophical Society* 109, no. 1, pp. 53–56.

Eckert, Michael. 2008. "Water-Art Problems at Sanssouci—Euler's Involvement in Practical Hydrodynamics on the Eve of Ideal Flow Theory." *Physica D: Nonlinear Phenomena* 237, pp. 1870–1877.

——. 2007. "Mathematics for the King: Was Euler an Impractical Theorist?" In *Euler Reconsidered*, ed. Roger Baker, pp. 11–39. Heber City, Utah: Kendrick.

——. 2002. "Euler and the Fountains of Sanssouci." *AHES* 56, no. 6, pp. 451–468.

Edwards, Harold M. 2007. "Euler and Quadratic Reciprocity." In *The Genius of Euler: Reflections on His Life and Work*, ed. William Dunham, pp. 233–243. Washington, D.C.: Mathematical Association of America.

——. 1977. *Fermat's Last Theorem.* New York: Springer.

Emerson, Roger L. 1990. "The Organization of Science and Its Pursuit in Early Modern Europe." In *Companion to the History of Modern Science*, ed. Robert C. Olby, G. M. Cantor, J. R. Christie, and M.J.S. Hodge, pp. 960–979. London: Routledge.

Erdős, Paul, and Underwood Dudley. 2007. "Some Remarks and Problems in Number Theory Related to the Work of Euler." In *The Genius of Euler: Reflections on His Life and Work*, ed. William Dunham, pp. 215–225. Washington, D.C.: Mathematical Association of America.

Ermolaeva, Natalia. 2008. "Les mathématiciens de Saint-Petersbourg et les problèmes cartographiques." *Archives Internationales d'Histoire des Sciences* 58, pp. 225–270.

Español González, Luis, and Emilio Fernández Moral. 2008. "Euler, Rey Pastor y la sumabilidad de series." *Quaderns d'Història de l'Enginyeria* 9, pp. 183–204.

Euler, Johann Albrecht. 2011. *Disquisitio de Causa Physica Electricitatis.* Charleston, S.C.: BiblioBazaar.

Euler, Karl. 1955. *Das Geschlecht Euler-Schölpi: Geschichte Einer Alten Familie.* Giessen, Germany: Wilhelm Schmitz Verlag.

Evans Despaux, Sloan, and Glen von Brummeln. 2003. "Abstracts." *HM* 32, pp. 241–267.

Fantet de Lagny, Thomas. 1692. "Nouvelle méthode de Mr. T. F. de Lagny pour l'approximation des racines cubiques." *Journal des Sçavans* 19, no. 17, pp. 200–202.

Fasanelli, Florence. 2007. "Images of Euler." In *Leonhard Euler: Life, Work and Legacy*, ed. Robert E. Bradley and C. Edward Sandifer, pp. 109–121. Elsevier Studies in the History and Philosophy of Mathematics 5. New York: Elsevier.

Fellmann, Emil A. 2007. *Leonhard Euler.* Translated by Erika and Walter Gautschi (Basel: Birkhäuser Verlag). The English edition has a bibliography on pp. 159–169.

——. 2003. "Christoph Jetzler und Leonhard Euler: Johann Jakob Burkhardt zur Vollendung seines 100. Lebensjahres am 13 Juli 2003 in Verehrung und Freundschaft zugeeignet." *NTM: Zeitschrift für Geschichte der Naturwissenschaften, Technik und Medizin* 11, pp. 145–154.

——. 1995. *Leonhard Euler.* Reinbeck bei Hamburg, Germany: Rowohlt Taschenbuch Verlag.

——. 1992. "*Non-Mathematica* im Briefwechsel Leonhard Eulers mit Johann Bernoulli." In *Amphora: Festschrift für Hans Wüssing zu seinem 65. Geburtstag*, ed. Sergei S. Demidov, Menso Folkerts, David E. Rowe, and Christoph J. Scriba, pp. 189–228. Basel: Birkhäuser Verlag.

——. 1983a. "Leonhard Euler—Ein Essay über Leben und Werk." In *Leonhard Euler, 1707–1783: Beiträge zu Leben und Werk*, ed. J. J. Burckhardt, E. A. Fellmann, and W. Habicht, pp. 13–99. Basel: Birkhäuser Verlag. LEJA 4.

——. 1983b. "Leonhard Eulers Stellung in der Geschichte der Optik." In *Leonhard Euler, 1707–1783: Beiträge zu Leben und Werk*, ed. J. J. Burckhardt, E. A. Fellmann, and W. Habicht, pp. 303–330. Basel: Birkhäuser Verlag. LEJA 4.

Ferlin, Fabrice. 2008. "Les lunettes achromatiques: en enjeu européen dans la deuxième moitié du 18e siècle." *Bollettino di Storia delle Scienze Matematiche* 28, pp. 221–237.

Ferraro, Giovanni. 2010. "Euler's Analytical Program." *Quaderns d'història de l'enginyeria* 11, pp. 185–208.

——. 2008a. "D'Alembert visto da Eulero." *Bollettino di Storia delle Scienze Matematiche* 28, pp. 273–291.

——. 2008b. "The Integral as an Anti-Differential: An Aspect of Euler's Attempt to Transform the Calculus into an Algebraic Calculus." *Quaderns d'Història de l'Enginyeria* 9, pp. 25–58.

——. 2008c. *The Rise and Development of the Theory of Series up to the Early 1820s*. New York: Springer.

——. 2007a. "Convergence and Formal Manipulation in the Theory of Series from 1730 to 1815." *HM* 34, pp. 62–88.

——. 2007b. "Euler's Treatises on Infinitesimal Analysis: *Introductio in analysin infinitorum, Institutiones calculi differentialis, Institutionum calculi integralis*." In *Euler Reconsidered*, ed. Roger Baker, pp. 39–102. Heber City, Utah: Kendrick.

——. 2004. "Differentials and Differential Coefficients in the Eulerian Foundation of the Calculus." *HM* 31, pp. 34–61.

——. 2000. "Functions, Functional Relations and the Laws of Continuity in Euler." *HM* 27, pp. 107–132.

——. 1998. "Some Aspects of Euler's Theory of Series: Inexplicable Functions and the Euler-Maclaurin Summation Formula." *HM* 25, pp. 290–317.

Ferreiro, Larrie D. 2007. *Ships and Science: The Birth of Naval Architecture in the Scientific Revolution*. Boston: The MIT Press.

Ferzola, Anthony P. 2007. "Euler and Differentials." In *The Genius of Euler: Reflections on His Life and Work*, ed. William Dunham, pp. 155–167. Washington, D.C.: Mathematical Association of America.

Figurovskij, N. A. 1958. "Aus der Geschichte wissenschaftlicher Begegnung und Zusammenarbeit deutscher und russischer Chemiker im 18. Jahrhundert." In *Die Deutsch-Russische Begegnung und Leonhard Euler*, ed. Eduard Winter, 49–63. Berlin: Akademie Verlag.

Finkel, B. F. 2007. "Leonhard Euler." In *The Genius of Euler: Reflections on His Life and Work*, ed. William Dunham, pp. 5–13. Washington, D.C.: Mathematical Association of America.

Fisher, Raymond H. 1977. *Bering's Voyages: Whither and Why*. Seattle: University of Washington Press.

Fitzgerald, Michael, and Ioan James. 2007. *The Mind of the Mathematician*. Baltimore, Md.: Johns Hopkins University Press.

Fletcher, Colin R. 1989. "Fermat's Theorem." *HM* 16, no. 2, pp. 149—154.

Folkerts, Menso, and Andreas Kühne. 2006. *Astronomy as a Model for the Sciences in Early Modern Times: Papers from the International Symposium, Munich, 10–12 March 2003*. Augsburg, Germany: Rauner.

Fontenelle, Bernard Le Bovier de. [1790] 2010. *Oeuvres choisies*. Reprint, edited by Georges-Bernard Depping. Oeuvres de Bernard de Fontenelle 2. Paris: Nabu Press.

——. [1686] 1990. *Entretiens sur la pluaralité des mondes*. Translated as *Conversations on the Plurality of Worlds* by H. A. Hargreaves (Berkeley: University of California Press).

——. 1727. "Éloge de Newton." *Histoire de l'Académie Royale des Sciences de Paris* (1727): pp. 151–172.

Formey, Jean-Henri-Samuel. 1786. "Discours prononcé dans l'assemblée publique du 29 Janvier 1784 pour célébrer l'anniversaire de la naissance du Roi." *N. Mém. de Berlin,* 1786 (for the year 1784), pp. 5–9.

Frangsmyr, Tore, J. L. Heilbron, and Robin E. Rider, eds. 1990. *The Quantifying Spirit of the Eighteenth Century.* Berkeley: University of California Press.

Fraser, Craig G. 2005. "Leonhard Euler's Book on the Calculus of Variations (1744)." In *Landmark Writings in Western Mathematics 1640–1940*, ed. Ivor Grattan-Guinness, pp. 168–180. Amsterdam: Elsevier Science.

——. 2003a. "The Calculus of Variations: A Historical Survey." In *A History of Analysis*, ed. H. N. Jahnke, 355–384. Providence, R.I.: American Mathematical Society.

——. 2003b. "History of Mathematics in the Eighteenth Century." In *The Cambridge History of Science*, vol. 4: *The Eighteenth Century*, ed. Roy Porter, 305–327. Cambridge: Cambridge University Press.

——. 1997a. "The Background to and Early Emergence of Euler's Analysis." In *Analysis and Synthesis in Mathematics*, ed. Michael Otte and Marco Panza, 47–78. Dordrecht, Netherlands: Kluwer Academic.

——. 1997b. *Calculus and Analytical Mechanics in the Age of Enlightenment.* Norfolk, England: Variorum.

——. 1994. "The Origins of Euler's Variational Calculus." *Archive for History of Exact Sciences* 47, pp. 103–141.

——. 1991. "Mathematical Techniques and Physical Conception in Euler's Investigation of the Elastica." *Centaurus* 34, pp. 211–246.

——. 1985. "J. L. Lagrange's Changing Approach to the Foundations of the Calculus of Variations." *AHES* 32, no. 2, pp. 151–191.

Gandt, François de. 2001. *Cirey dans la vie intellectuelle: La réception de Newton en France.* Oxford: Voltaire Foundation.

Garber, Daniel. 2000. "A Different Descartes: Descartes and the Programme for a Mathematical Physics in His Correspondence." In *Descartes' Natural Philosophy*, ed. Stephen Gaukroger, John Schuster, and John Sutton, 113–130. London: Routledge.

Gaukroger, Stephen, John Schuster, and John Sutton, eds. 2000. *Descartes' Natural Philosophy.* London: Routledge.

Gautschi, Walter. 2008. "Leonhard Euler: His Life, the Man, and His Works." *SIAM Review* 50, no. 1, pp. 3–33.

——. 2007. "Leonhard Eulers Umgang mit langsam konvergenten Reihen." *Elemente der Mathematik* 62, pp. 174–183.

Gehr, Sulamith, and Martin Mattmüller. 2010, January. "Leonhard Euler, seine Heimstadt und ihre Universität." Historische Seminar Basel. http://www.unigeschichte.unibas.ch/cms/upload/FaecherUndFakultaeten/Downloads/Gehr_u._Matt mueller_Euler_neu.pdf.

Gekker, R., and A. A. Euler. 2007. "Leonhard Euler's Family and Descendants" Translated by Robert Burns. In *Euler and Modern Science*, ed. N. N. Bogolyubov, G. K. Mikhaĭlov, and A. P. Yushkevich, pp. 397–417. Washington, D.C.: Mathematical Association of America. LEJA 7.

Gelfond, Aleksander O. 1983. "Über einige charakteristische Züge in den Ideen L. Eulers auf den Gebiet der mathematischen Analysis und in seiner 'Einführung in die Analysis des Unendlichen.'" In *Leonhard Euler, 1707–1783: Beiträge zu Leben und Werk*, ed. J. J. Burckhardt, E. A. Fellmann, and W. Habicht, pp. 99–111. Basel: Birkhäuser Verlag. LEJA 4.

George, Arthur L. 2003. *St Petersburg: Russia's Window to the Future. The First Three Centuries.* New York: Taylor.

Gerhardt, Carl Immanuel. 1898. "Über die vier Briefe von Leibniz, die Samuel König in dem Appel au public veröffentlichten hatte." *Sitzungsberichte der Königlich Preussischen Akadémie der Wissenschaften zu Berlin* 32, pp. 419–427.

Gertsman, E. V. 2007. "Euler and the History of a Certain Musical-Mathematical Idea." Translated by Robert Burns. In *Euler and Modern Science*, ed. N. N. Bogolyubov, G. K. Mikhaĭlov, and A. P. Yushkevich, pp. 335–349. Washington, D.C.: Mathematical Association of America. LEJA 7.

Gibbon, Edward. 1952. *The Decline and Fall of the Roman Empire*. Chicago: University of Chicago Press. Originally published in six volumes, 1776–89.

Gillings, R. J. 1954. "The So-Called Euler-Diderot Incident." *American Mathematical Monthly* 61, pp. 77–80.

Gillispie, Charles C. 1956. Review of Robert Shackelton's translation in 1955 of Bernard Le Bovier Fontenelle's *Entretiens sur la pluaralité des mondes*. *Isis* 47, no. 4, pp. 452–453.

Gillispie, Charles Coulston. [1980] 2008. *Science and Polity in France*. Princeton, N.J.: Princeton University Press.

——. 1997. *Pierre-Simon Laplace, 1749–1827: A Life in Exact Science*. Princeton, N.J.: Princeton University Press.

Giusti, Ernesto. 2003. "Clifford Truesdell, Historian of Mathematics." *Journal of Elasticity* 70, pp. 15–22.

Glaisher, J.W.L. 2007. "On the History of Euler's Constant." In *The Genius of Euler: Reflections on His Life and Work*, ed. William Dunham, pp. 147–153. Washington, D.C.: Mathematical Association of America.

Glasberg, Ronald. 2003. "Mathematics and Spiritual Interpretation: A Bridge to Genuine Interdisciplinarity." *Zygon* 38, pp. 277–294.

Godard, Roger. 2007. "The Euler Advection Equation." In *Euler at 300: An Appreciation*, ed. Robert E. Bradley, Lawrence A. D'Antonio, and C. Edward Sandifer, pp. 261–273. Washington, D.C: Mathematical Association of America.

Goldenbaum, Ursula. 2008. "Leonhard Eulers Schwierigkeiten mit der Freiheit der Gelehrtenrepublik." In *Kosmos und Zahl*, ed. Hartmut Hecht, Regina Mikosch, Ingo Schwarz, Harald Siebert, and Romy Werther, 105–121. Stuttgart: Franz Steiner Verlag.

——. 1999. "Die Bedeutung der öffentlichen Debatte über das Jugement der Berliner Akademie für die Wissenschaftsgeschichte. Eine kritische Sichtung hartnäckiger Verurteile." In *Pierre Louis Moreau de Maupertuis: Eine Bilanz nach 300 Jahren*, ed. Hartmut Hecht, 383–417. Berlin: Verlag Arno Spitz.

Golder, F. A. 1941. *Russian Expansion on the Pacific 1641–1850*. Cleveland: Arthur H. Clark.

Goldstein, Bernard R., and Giora Han. 2007. "Celestial Charts and Spherical Triangles: The Unifying Power of Symmetry." *Journal for the History of Astronomy* 38, pp. 1–14.

Goldstine, Herman H. 1980. *A History of the Calculus of Variations from the 17th through the 19th Century*. Berlin: Springer Verlag.

Golland, Louisa Ahrndt, and Ronald William Golland. 1993. "Euler's Troublesome Series: An Early Example of the Use of Trigonometric Series." *HM* 20, pp. 54–67.

González Redondo, Francisco A. 2007. "Constants, Units, Measures, and Dimensions in Leonard Euler's *Mechanics*, 1736–1765." In *Euler Reconsidered*, ed. Roger Baker, pp. 205–232. Heber City, Utah: Kendrick.

——. 2003. "La contribución de Leonard Euler a la matematización de la magnitudes y las leyes de la mecánica, 1736–1765." *Llull* 26, pp. 837–857.

González Urbaneja, Pedro Miguel. 2008. "Euler y la Geometria analitica." *Quaderns d'Història de l'Enginyeria* 9, pp. 83–116.

Goodman, Dena. 1994. *The Republic of Letters: A Cultural History of the French Enlightenment*. Ithaca, N.Y.: Cornell University Press.

Goss, Victor Godfrey Alan. 2009. "The History of the Planar Elastica: Insights into Mechanics and Scientific Method." *Science and Education* 18, pp. 1057–1082. An examination of Leonhard Euler's classic paper "De Curvis Elasticis."

Gossman, Lionel. 2000. *Basel in the Age of Burckhardt*. Chicago: University of Chicago Press.

Gossman, Lionel. 1994. "Basel." In Nicholas Bouvier, Gordon A. Craig, and Lionel Gossman, *Geneva, Zurich, Basel: History, Culture, and National Identity*, pp. 63–98. Princeton, NJ.: Princeton University Press.

Gouzévitch, Irina, and Dmitri Gouzévitch. 2009. "Introducing Mathematics, Building an Empire: Russia under Peter I." In *The Oxford Handbook of the History of Mathematics*, ed. Eleanor Robson and Jacqueline Stedall, pp. 353–375. Oxford: Oxford University Press.

Grant, Robert. 1842. *History of Physical Astronomy*. London: Henry G. Bohn.

Grasshof, Helmut. 1966. *Antioch Dmitrievič Kantemir und Westeuropa*. Berlin: Akademie Verlag.

Grattan-Guinness, Ivor. 2007. "On the Recognition of Euler among the French, 1790–1830." In *Leonhard Euler: Life, Work and Legacy*, ed. Robert E. Bradley and C. Edward Sandifer, pp. 441–459. Elsevier Studies in the History and Philosophy of Mathematics 5. New York: Elsevier.

——, ed. 2005. *Landmark Writings in Western Mathematics 1640–1940*. Amsterdam: Elsevier Science.

——. 1983. "Euler's Mathematics in French Science, 1795–1815." In *Leonhard Euler, 1707–1783: Beiträge zu Leben und Werk*, ed. J. J. Burckhardt, E. A. Fellmann, and W. Habicht, pp. 395–409. Basel: Birkhäuser Verlag. LEJA 4.

Grau, K. 2007. "Leonhard Euler and the Berlin Academy of Sciences." Translated by Robert Burns. In *Euler and Modern Science*, ed. N. N. Bogolyubov, G. K. Mikhaĭlov, and A. P. Yushkevich, pp. 75–87. Washington, D.C.: Mathematical Association of America. LEJA 7.

Gray, Jeremy. 2008. "A Short Life of Euler." *British Society for the History of Mathematics Bulletin* 23, pp. 1–12.

Grigor'ian, A. T., and V. S. Kirsanov. 2007. "*Letters to a German Princess* and Euler's Physics." Translated by Robert Burns. In *Euler and Modern Science*, ed. N. N. Bogolyubov, G. K. Mikhaĭlov, and A. P. Yushkevich, pp. 307–317. Washington, D.C.: Mathematical Association of America. LEJA 7.

Grigor'jan, Ašot T., and Vladimir S. Kirsanov. 1983. "Euler's Physics in Russia." In *Leonhard Euler, 1707–1783: Beiträge zu Leben und Werk*, ed. J. J. Burckhardt, E. A. Fellmann, and W. Habicht, pp. 385–395. Basel: Birkhäuser Verlag. LEJA 4.

Habermas, Jürgen. 1989. *The Structural Transformation of the Public Sphere: An Inquiry into a Category of Bourgeois Society*. Translated by Thomas Burger with the assistance of Frederick Lawrence. Cambridge, Mass.: Harvard University Press.

Habicht, Walter. 1983a. "Betrachtung zu Eulers Dioptrik." In *Leonhard Euler, 1707–1783: Beiträge zu Leben und Werk*, ed. J. J. Burckhardt, E. A. Fellmann, and W. Habicht, pp. 383–302. Basel: Birkhäuser Verlag. LEJA 4.

——. 1983b. "Einige grundlegende Themen in Leonhard Eulers Schiffstheorie." In *Leonhard Euler, 1707–1783: Beiträge zu Leben und Werk*, ed. J. J. Burckhardt, E. A. Fellmann, and W. Habicht, pp. 243–271. Basel: Birkhäuser Verlag. LEJA 4.

Hahn, Roger. 1971. *The Anatomy of a Scientific Institution: The Paris Academy of Sciences, 1666–1803*. Berkeley: University of California Press.

Hakfoort, Casper. 1995. *Optics in the Age of Euler: Conceptions of the Nature of Light, 1700–1795*. Cambridge: Cambridge University Press.

——. 1983. "Betrachtungen zu Eulers Dioptrik." In *Leonhard Euler, 1707–1783: Beiträge zu Leben und Werk*, ed. J. J. Burckhardt, E. A. Fellmann, and W. Habicht, pp. 283–303. Basel: Birkhäuser Verlag. LEJA 4.

——. 1982. "Nicolas Beguelin and His Search for a Crucial Experiment on the Nature of Light (1772)." *Annals of Science* 39, no. 3, pp. 297–310.

Hankins, Thomas L. 1986. "Review of Leonhard Euler, *OO*, IVA, 6." *Isis* 77, no. 4, pp. 716–717.

——. 1979. "Defense of Biography: The Use of Biography in the History of Science." *History of Science* 17, pp. 1–16.

——. 1970. *Jean d'Alembert: Science and the Enlightenment.* Oxford: Clarendon.

——. 1965. "Eighteenth-Century Attempts to Resolve the *Vis Viva* Controversy." *Isis* 56, no. 3, pp. 281–297.

Hans, Nicholas. 1963. *The Russian Tradition in Education.* London: Routledge and Kegan Paul.

Harman, P. M., and Alan E. Shapiro, eds. 1992. *The Investigation of Difficult Things: Essays on Newton and the History of Exact Sciences in Honour of D. T. Whiteside.* Cambridge: Cambridge University Press.

Harnack, Adolf. 1900. *Geschichte der Königlich Preussischen Akademie der Wissenschaften zu Berlin.* 3 vols. Berlin: Reichsdruckerei.

Hartkopf, Werner, and Gerhard Dunken. 1967. *Von der Brandenburgischen Sozietät der Wissenschaften zur Deutschen Akademie der Wissenschaften zu Berlin.* Berlin: Akademie Verlag.

Havil, Julian. 2003. *Gamma: Euler's Constant.* Princeton, N.J.: Princeton University Press.

Hecht, Hartmut, ed. 1999. *Pierre Louis Moreau de Maupertuis: Eine Bilanz nach 300 Jahren.* Berlin: Verlag Arno Spitz.

Hecht, Hartmut, Regina Mikosch, Ingo Schwarz, Harald Siebert, and Romy Werther, eds. 2008. *Kosmos und Zahl.* Stuttgart: Franz Steiner Verlag.

Heefer, Albrecht. 2007. "The Origin of the Problems in Euler's Algebra." *Bulletin of the Belgian Mathematical Society–Simon Stevin* 13, no. 5, pp. 949–952.

Heilbron, J. L. 1993. "Weighing Imponderables and Other Quantitative Sciences around 1800." *Historical Studies in the Physical and Biological Sciences* 24, no. 1, supplement.

——. 1990. "Introductory Essay." In *The Quantifying Spirit of the Eighteenth Century*, ed. Tore Frangsmyr, J. L. Heilbron, and Robin E. Rider, pp. 1–25. Berkeley: University of California Press.

Heine, George W., III. 2007. "Lambert, Euler, and Lagrange as Map Makers." In *Euler at 300: An Appreciation*, ed. Robert E. Bradley, Lawrence A. D'Antonio, and C. Edward Sandifer, pp. 281–295. Washington, D.C: Mathematical Association of America.

Henry, Philippe. 2007. *Leonhard Euler: "incomparable géomètre."* Geneva: Université de Genève.

Hepburn, Brian. 2010. "Euler, *Vis Viva,* and Equilibrium." *Studies in History and Philosophy of Science* 41, pp. 120–127.

——. 2008. "Equilibrium and Explanation in 18th-Century Mechanics." *Dissertation Abstracts International* A 68/09. http://d-scholarship.pitt.edu/8682/1/BHepburn2007.pdf.

Hermann, Jakob. 1728. "De mensura virium corporum." *Commentarii Academiae scientiarum Imperialis Petropolitanae 3, pp. 4–18.*

Herzog, Johann Wernhard. 1780. *Adumbratio eruditorum basiliensium. . . .* Basel: C. A. Serinus, 1750.

Heusler, Andreas. 1957. *Geschichte der Stadt Basel.* Basel: Frobenius.

Hine, Ellen McNiven. 1989. "Dortous de Mairan, the Cartonian." In *Studies on Voltaire and the Eighteenth Century*, vol. 266, pp. 163–179. Geneva: Institut et Musée Voltaire.

Hofer, G.E.H. 2008. "Introductory Words" [title translated from Russian]. In *Leonard Eiler: K 300-letiiu so dnia rozhdenii / Leonhard Euler: 300th Anniversary*, ed. Vladimir N. Vasilyev, p. 6. Saint Petersburg: Nestor-Historia. LEJA 8.

Hofmann, Joseph Ehrenfried. 1959. "Um Eulers erste Reihenstudien." In *Sammelband der zu Ehren des 250. Geburtstages Leonhard Eulers der Deutschen Akademie der*

Wissenschaften zu Berlin vorgelegten Abhandlungen, ed. Kurt Schröder, pp. 139–208. Berlin: Akademie Verlag. LEJA 3.

Hoffmann, Peter. 2007. "Leonhard Euler and Russia." In *Leonhard Euler: Life, Work and Legacy*, ed. Robert E. Bradley and C. Edward Sandifer, pp. 61–75. Elsevier Studies in the History and Philosophy of Mathematics 5. New York: Elsevier.

Home, Roderick W. 1988. "Leonhard Euler's 'Anti-Newtonian' Theory of Light." *Annals of Science* 45, pp. 521–533.

Hoogenboom, Hilde. 2007, July. "Catherine the Great, Academician Peter Simon Pallas, and Princess Dashkova: Dictionaries for Empire." Paper presented to the Twelfth International Enlightenment Congress, International Society of Eighteenth Century Studies, Montpellier, France.

Hopkins, Brian, and Robin Wilson. 2007a. "Euler's Science of Combinations." In *Leonhard Euler: Life, Work and Legacy*, ed. Robert E. Bradley and C. Edward Sandifer, pp. 395–409. Elsevier Studies in the History and Philosophy of Mathematics 5. New York: Elsevier.

——. 2007b. "The Truth about Königsberg." In *Leonhard Euler: Life, Work and Legacy*, ed. Robert E. Bradley and C. Edward Sandifer, pp. 409–421. Elsevier Studies in the History and Philosophy of Mathematics 5. New York: Elsevier.

——. 2007c. "The Truth about Königsberg." In *The Genius of Euler: Reflections on His Life and Work*, ed. William Dunham, pp. 263–273. Washington, D.C.: Mathematical Association of America.

Howson, A. G. 1975, 2007. "Addendum to: Euler and the Zeta Function." In *The Genius of Euler: Reflections on His Life and Work*, ed. William Dunham, pp. 133–135. Washington, D.C.: Mathematical Association of America.

Huber, Friedrich. 1959. *Daniel Bernoulli (1700–1782) als Physiologe und Statistiker.* Basel: Benno Schwabe.

Ignatieff, Yurie A., and Heike Willig. 1999. *Clifford Truesdell: Eine wissenschasfliche Biographies des Dichters, Mathematikers und Naturphilosophen.* Aachen, Germany: Shaker Verlag.

Ilic, Mirjana. 2008. "The Euler-Wettstein Correspondence." In *Leonard Eĭler: K 300-letiiu so dnia rozhdenii / Leonhard Euler: 300th Anniversary*, ed. Vladimir N. Vasilyev, pp. 274–279. Saint Petersburg: Nestor-Historia. LEJA 8.

Iliffe, Robert. 2012. "Servant of Two Masters? Fatio de Duiller, Isaac Newton and Christian Huygens." In *Newton and the Netherlands: How Isaac Newton Was Fashioned in the Netherlands*, ed. Erik Jorink and Ad Maas, pp. 67–91. Leiden, Netherlands: Leiden University Press.

——. 1993. "'Aplatisseur du monde et de Cassini': Maupertuis, Precision Measurement, and the Shape of the Earth in the 1730s." *History of Science* 3, pp. 335–375.

Ineichen, Robert. 2007. "The Contributions of Leonhard Euler to Actuarial Mathematics." In *Euler Reconsidered*, ed. Roger Baker, pp. 102–119. Heber City, Utah: Kendrick.

Jahnke, H. N., ed. 2003. *A History of Analysis.* Providence, R.I.: American Mathematical Society.

Jaki, Stanley L. 1977, September. "Lambert: Self-Taught Physicist." *Physics Today*, pp. 25–33.

James, Ioan. 2003. *Remarkable Mathematicians from Euler to von Neumann.* Cambridge: Cambridge University Press.

Jaquel, Roger. 1983. "Léonard Euler, son fils Johann Albrecht et leur ami Jean III Bernoulli." In *Leonhard Euler, 1707–1783: Beiträge zu Leben und Werk*, ed. J. J. Burckhardt, E. A. Fellmann, and W. Habicht, pp. 435–447. Basel: Birkhäuser Verlag. LEJA 4.

Jardine, Dick. 2007. "Taylor and Euler: Linking the Discrete and Continuous." In *Euler at 300: An Appreciation*, ed. Robert E. Bradley, Lawrence A. D'Antonio, and

C. Edward Sandifer, pp. 157–169. Washington, D.C: Mathematical Association of America.

Jia Xiaoyong and Li Yuewu. 2009. "The Calculus of Variations from Euler to Lagrange: As a Result of Formalized Improvement" [title translated from Chinese]. *Ziran Kexueshi Yánjiū* 28, pp. 312–325.

Jorink, Erik, and Ad Maas, eds. 2012. *Newton and the Netherlands: How Isaac Newton Was Fashioned in the Netherlands.* Leiden, Netherlands: Leiden University Press.

Jouve, Guillaume. 2008. "Le rôle de d'Alembert dans les débuts d'une étude programmatique des équations aux dérivées partielles (1760–1783)." *Bollettino di Storia della Scienze Matematiche* 28, pp. 167–181.

Yushkevich, Adolf P. 1983. "L. Euler's Unpublished Manuscript *Calculus Differentialis.*" In *Leonhard Euler, 1707–1783: Beiträge zu Leben und Werk*, ed. J. J. Burckhardt, E. A. Fellmann, and W. Habicht, pp. 161–171. Basel: Birkhäuser Verlag. LEJA 4.

——. 1963. "Leonhard Euler und die Universität Halle (Saale)." *Nova Acta Leopoldina* 27, pp. 367–378.

Južnic, Stanislav. 2008. "Euler and the Jesuits in Russia." *Quaderns d'Història de l'Enginyeria* 9, pp. 219–247.

Kant, Immanuel. 1999. *Correspondence.* Translated and edited by Arnulf Zweig. Cambridge: Cambridge University Press.

Katz, Eugene A. 2008. "Euler's Theorem Concerning Polyhedrons and Modern Ideas about Molecular Structure of Fullerenes and Fullerene-Like Nanostructures" [title translated from Russian]. In *Leonard Eĭler: K 300-letiiu so dnia rozhdenii / Leonhard Euler: 300th Anniversary*, ed. Vladimir N. Vasilyev, pp. 89–103. Saint Petersburg: Nestor-Historia. LEJA 8.

Katz, Victor J. 2007a. "Change of Variables in Multiple Integrals: Euler to Cartan." In *The Genius of Euler: Reflections on His Life and Work*, ed. William Dunham, pp. 185–197. Washington, D.C.: Mathematical Association of America.

——. 2007b. "Euler's Analysis Textbooks." In *Leonhard Euler: Life, Work and Legacy*, ed. Robert E. Bradley and C. Edward Sandifer, pp. 213–235. Elsevier Studies in the History and Philosophy of Mathematics 5. New York: Elsevier.

——. 1987. "The Calculus of Trigonometric Functions." *HM* 14, pp. 311–324.

Kemp, Jonathan, ed. 1963. *Diderot: Interpreter of Nature.* New York: International.

Kholchevnikov, Konstantin Vladislavovich. 2008. "Euler as Astronomer" [title translated from Russian]. In *Leonard Eĭler: K 300-letiiu so dnia rozhdenii / Leonhard Euler: 300th Anniversary*, ed. Vladimir N. Vasilyev, pp. 190–200. Saint Petersburg: Nestor-Historia. LEJA 8.

Khruschev, Sergey. 2008. *Orthogonal Polynomials and Continued Fractions from Euler's Point of View.* Cambridge: Cambridge University Press.

Kizilova, N. N. 2008. "L. Euler and the History of Biomechanics" [title translated from Russian]. In *Leonard Eĭler: K 300-letiiu so dnia rozhdenii / Leonhard Euler: 300th Anniversary*, ed. Vladimir N. Vasilyev, pp. 171–182. Saint Petersburg: Nestor-Historia. LEJA 8.

Kleinert, Andreas, and Martin Mattmüller. 2008. "Leonhardi Euleri Opera Omnia: A Centenary Project" [title translated from Russian]. In *Leonard Eĭler: K 300-letiiu so dnia rozhdenii / Leonhard Euler: 300th Anniversary*, ed. Vladimir N. Vasilyev, pp. 280–291. Saint Petersburg: Nestor-Historia. LEJA 8.

Kline, Morris. 2007. "Euler and Infinite Series." In *The Genius of Euler: Reflections on His Life and Work*, ed. William Dunham, pp. 101–113. Washington, D.C.: Mathematical Association of America.

Klyve, Dominic, and Lee Stemkoski. 2007a. "The Euler Archive: Giving Euler to the World." In *Euler at 300: An Appreciation*, ed. Robert E. Bradley, Lawrence A. D'Antonio, and C. Edward Sandifer, pp. 33–41. Washington, D.C: Mathematical Association of America.

——. 2007b. "Graeco-Latin Squares and a Mistaken Conjecture of Euler." In *The Genius of Euler: Reflections on His Life and Work*, ed. William Dunham, pp. 273–289. Washington, D.C.: Mathematical Association of America.

Knapp, Richard Gilbert. 1971. *Studies on Voltaire and the Eighteenth Century*, vol. 82: *The Fortunes of Pope's Essay on Man in 18th Century France*, ed. Theodore Besterman. Geneva: Institut et Musée Voltaire.

Knobloch, Eberhard. 2008. "Euler Transgressing Limits: The Infinite and Music Theory." *Quaderns d'Història de l'Enginyeria* 9, pp. 9–24.

——. 2007. "Euler's Mathematical Notebooks." Translated by Robert Burns. In *Euler and Modern Science*, ed. N. N. Bogolyubov, G. K. Mikhaĭlov, and A. P. Yushkevich, pp. 97–119. Washington, D.C.: Mathematical Association of America. LEJA 7.

——. 1992. "Eulers frühste Studie zum Dreikorper Problem." In *Amphora: Festschrift für Hans Wüssing zu seinem 65. Geburtstag*, ed. Sergei S. Demidov, Menso Folkerts, David E. Rowe, and Christoph J. Scriba, pp. 388–405. Basel: Birkhäuser Verlag.

——. 1991. "Leibniz and Euler: Problems and Solutions concerning Infinitesimal Geometry and Calculus." *Giornate di storia della matematica, Cetraro (Cosenza), Settembre 1988*, ed. Massimo Galluzzi, pp. 269–293. Rende, Italy: Editel.

——. 1990. "L'analogie et la pensée mathematique." In *Mathématique et philosophie de l'antiquité à l'age classique. Hommage à Jules Vuillemin*, ed. Roshdi Rashed, pp. 215–235. Paris: CNRS Éditions.

——. 1989. "Leonhard Eulers Mathematische Notizbücher." *Annals of Science* 46, pp. 277–302.

Koetsier, Teun. 2007. "Euler and Kinematics." In *Leonhard Euler: Life, Work and Legacy*, ed. Robert E. Bradley and C. Edward Sandifer, pp. 167–195. Elsevier Studies in the History and Philosophy of Mathematics 5. New York: Elsevier.

Koetsier, Teun, and Luc Bergmann, eds. 2005. *Mathematics and the Divine: A Historical Study.* Amsterdam: Elsevier.

Kopelevich, Judith, Kh. 2007. "Leonhard Euler, Active and Honored Member of the Petersburg Academy of Sciences." Translated by Robert Burns. In *Euler and Modern Science*, ed. N. N. Bogolyubov, G. K. Mikhaĭlov, and A. P. Yushkevich, pp. 39–53. Washington, D.C.: Mathematical Association of America. LEJA 7.

——. 1983. "Leonhard Euler und die Petersburger Akademie." In *Leonhard Euler, 1707–1783: Beiträge zu Leben und Werk*, ed. J. J. Burckhardt, E. A. Fellmann, and W. Habicht, pp. 373–385. Basel: Birkhäuser Verlag. LEJA 4.

——. 1977. *Osnovanie Petersburgkoj Akademii nauk.* Leningrad: Nauka.

——. 1966. "The Petersburg Astronomy Contest in 1751." *Soviet Astronomy* 9, no. 4, pp. 653–660.

Koser, Reinhold, and Hans Droysen, eds. 1965–68. *Briefwechsel Friedrichs des Grossen mit Voltaire.* Osnabrück, Germany: Otto Zeller. 3 vols.

Kowalenko, V., N. E. Frankel, M. L. Glasser, and T. Taucher. 1995. *Generalized Euler-Jacobi Inversion Formula and Asymptotics beyond All Orders.* Cambridge: Cambridge University Press.

Koyré, Alexandre. 1968. *Newtonian Studies.* Chicago: University of Chicago Press.

Krämer, Stefan. 2005. "Die Eulersche Konstante γ und verwandte Zahlen: Eine mathematische und historische Betractung." PhD diss., University of Gottingen.

Kreiszig, Erwin. 2007. "On the Calculus of Variations and Its Major Influences on the Mathematics of the First Half of Our Century." In *The Genius of Euler: Reflections on His Life and Work*, ed. William Dunham, pp. 209–215. Washington, D.C.: Mathematical Association of America.

Kusitcheva, Z. A. 2008. "Euler and Lambert: Logic Treatises" [title translated from Russian]. In *Leonard Eĭler: K 300-letiiu so dnia rozhdenii / Leonhard Euler: 300th Anniversary*, ed. Vladimir N. Vasilyev, pp. 128–136. Saint Petersburg: Nestor-Historia. LEJA 8.

Lagrange, Joseph-Louis. 1867–1892. *Oeuvres complètes*, vol. 13. Paris: Gauthier-Villars et Fils.

——. 1788. *Mécanique analytique*, vol. 1. Paris: Desaint.

Laitko, Hubert. 1987. *Wissenschaft in Berlin: Von den Anfängen bis sum Neubeginn nach 1945*. Berlin: Dietz Verlag.

Langton, Stacy G. 2007a. "Euler on Rigid Bodies." In *Leonhard Euler: Life, Work and Legacy*, ed. Robert E. Bradley and C. Edward Sandifer, pp. 195–213. Elsevier Studies in the History and Philosophy of Mathematics 5. New York: Elsevier.

——. 2007b. "The Quadrature of Lunes, from Hippocrates to Euler." In *Euler at 300: An Appreciation*, ed. Robert E. Bradley, Lawrence A. D'Antonio, and C. Edward Sandifer, pp. 53–63. Washington, D.C: Mathematical Association of America.

——. 2007c. "Some Combinatorics in Jacob Bernoulli's *Ars Conjectandi*." In *Euler at 300: An Appreciation*, ed. Robert E. Bradley, Lawrence A. D'Antonio, and C. Edward Sandifer, pp. 191–203. Washington, D.C: Mathematical Association of America.

Lathrop, Carolyn, and Lee Stemkoski. 2007. "Parallels in the Work of Leonhard Euler and Thomas Clausen." In *Euler at 300: An Appreciation*, ed. Robert E. Bradley, Lawrence A. D'Antonio, and C. Edward Sandifer, pp. 217–227. Washington, D.C: Mathematical Association of America.

Laudan, L. L. 1968. "The *Vis Viva* Controversy, A Post Mortem." *Isis* 59, no. 3, pp. 131–143.

Laugwitz, Detlef. 1983. "Die Nichtstandard-Analysis: Eine Wiederaufnahme der Ideen und Methoden von Leibniz und Euler." In *Leonhard Euler, 1707–1783: Beiträge zu Leben und Werk*, ed. J. J. Burckhardt, E. A. Fellmann, and W. Habicht, pp. 185–199.

Lauridsen, Peter. 1889. *Vitus Bering: The Discoverer of Bering Strait*. Translated by Julius E. Olson. Chicago: S. C. Griggs.

Lavrinenko, T. A. 2007. "Diophantine Equations in Euler's Works." Translated by Robert Burns. In *Euler and Modern Science*, ed. N. N. Bogolyubov, G. K. Mikhaĭlov, and A. P. Yushkevich, pp. 153–167. Washington, D.C.: Mathematical Association of America. LEJA 7.

Lehmann, Fritz. 1972. "Unter der Herrschaft der 'Gnädigen Herren' von Basel, 1522–1798." In *Riehen: Geschichte eines Dorfes*, ed. Albert Bruckner, pp. 267–318. Riehen, Switzerland: Verlag A. Schudel.

Leibniz, Gottfried Wilhelm. 2011. *Die mathematischen Zeitschriftenartikel*. Translated by Heinz-Jürgen Heß and Malte-Ludolf Babin. Zurich: Olms. This volume contains all of the mathematical papers of Leibniz from different journals.

Lemmermeyer, Franz. 2000. *Reciprocity Laws from Euler to Eisenstein*. Berlin: Springer Verlag.

Liebing, Heinz. 1961. *Zwischen Orthodoxie und Aufklärung: Das philosophische und theologische Denken Georg Bernhard Bilfingers*. Tübingen, Germany: J.C.B. Mohr.

Lincoln, W. Bruce. 2000. *Sunlight at Midnight: St. Petersburg and the Rise of Modern Russia*. New York: Basic Books.

Lipski, Alexander. 1959. "Some Aspects of Russia's Westernization during the Reign of Anna Ioannovna, 1730–1740." *American Slavic and Eastern European Review* 18, no. 1, pp. 1–11.

Liu Jianjun and Liu Qinying. 2003. "Euler's Achievements in Combinatorics" [title translated from Chinese]. *Ziran Kexueshi Yánjiū* 22, pp. 361–367.

Lundbaek, Knud. 1995. *T. S. Bayer (1694–1738): Pioneer Sinologist*. London: Taylor and Francis.

Lusa Montforte, Guillermo. 2008. "Congrés Internacional 300 Aniversari Leonhard Euler (1707–2007): Una presentación informal," *Quaderns d'Història de l'Enginyeria* 9, pp. 3–8.

Lützen, Jesper. 2007. "Euler's Vision of a General Partial Differential Calculus for a Generalized Kind of Function." In *The Genius of Euler: Reflections on His Life and Work*, ed. William Dunham, pp. 197–209. Washington, D.C.: Mathematical Association of America.

Mac Lane, Saunders. 1992. "The Protean Character of Mathematics." In *The Space of Mathematics: Philosophical, Epistemological, and Historical Explorations*, ed. Javier Echeverria, Andoni Ibarra, and Thomas Mormann, pp. 1–13. Berlin: Walter de Gruyter.

——. 1986. *Mathematics: Form and Function*. New York: Springer.

Maclaurin, Colin. 1724. "Demonstration des loix du choc des corps." *Recueil prix Paris* 1, pp. 1–24.

Magnan, André. 1986. *Dossier Voltaire en Prusse (1750–1753)*. Oxford: Voltaire Foundation.

Maindron, Ernest. 1881. *Les foundations de prix à l'Académie des sciences*. Paris: Gauthier-Villars.

Mallion, Roger. 2008. "A Contemporary Eulerian Walk over the Bridges of Kaliningrad." *British Society for the History of Mathematics Bulletin* 23, pp. 24–36.

Maltese, Giulio. 2003. "The Ancients' Inferno: The Slow and Tortuous Development of Newtonian Principles of Motion in the Eighteenth Century." In *Essays on the History of Mechanics: In Memory of Clifford Ambrose Truesdell and Edoardo Benvenuto*, ed. Antonio Becchi, Massimo Corradi, Federico Foce, and Orietta Pedemonte, pp. 199–221. Basel: Birkhäuser Verlag.

——. 2000. "On the Relativity of Motion in Leonhard Euler's Science." *AHES* 54, no. 4, 319–348.

——. 1992. "Taylor and John Bernoulli on the Vibrating String: Aspects of the Dynamics of Continuous Systems at the Beginning of the Eighteenth Century." *Physis* 29, pp. 703–744.

Malyh, A. E. 2008. "From the Combinatorial Legacy of Euler" [title translated from Russian]. In *Leonard Eĭler: K 300-letiiu so dnia rozhdenii / Leonhard Euler: 300th Anniversary*, ed. Vladimir N. Vasilyev, pp. 69–78. Saint Petersburg: Nestor-Historia. LEJA 8.

Mandryka, A. P. 2007. "The Significance of Euler's Research in Ballistics." Translated by Robert Burns. In *Euler and Modern Science*, ed. N. N. Bogolyubov, G. K. Mikhaĭlov, and A. P. Yushkevich, pp. 241–245. Washington, D.C.: Mathematical Association of America. LEJA 7.

Mann, Tony. 2011. "History of Mathematics and History of Science." *Isis* 102, no. 3, pp. 518–526.

Manuel, Frank E. 1963. *Isaac Newton, Historian*. Cambridge, Mass.: Harvard University Press.

Maor, Eli. 1994. *e: The Story of a Number*. Princeton, N.J.: Princeton University Press.

Marion, Charlie, and William Dunham. 2007. "A Response to "Bell's Conjecture" (a Poem)." In *The Genius of Euler: Reflections on His Life and Work*, ed. William Dunham, pp. 95–99. Washington, D.C.: Mathematical Association of America.

Marsden, Brian G. 1995. "Eighteenth- and Nineteenth-Century Development in the Theory and Practice of Orbit Determination." In *The General History of Astronomy*, ed. René Taton and Curtis Wilson, vol. 2, part B, pp. 181–191. Cambridge: Cambridge University Press.

Marston, Daniel. 2001. *The Seven Years' War*. Westminster, Md.: Osprey.

Marty, Martin E. 1983–96. *Modern American Religion*. 3 vols. Chicago: University of Chicago Press.

——. [1970] 1977. *Righteous Empire: The Protestant Experience in America*. New York: Harper Torchbooks.

Mason, Haydn T., ed. 1998. *The Darnton Debate: Books and Revolution in the Eighteenth Century*. Oxford: Voltaire Foundation.

Massa Esteva, Maria Rosa. 2008a. "Congrés Internacional 300 Aniversari Leonhard Euler (1707–2007)," *Quaderns d'Història de l'Enginyeria* 9, pp. 307–310.

——. 2008b. "Symbolic Language in Early Modern Mathematics: The *Algebra* of Pierre Hérigone (1540–1643)." *HM* 35, pp. 285–301.

Mattmüller, Martin. 2008. "The First Modern Mathematician? Euler's Influence on the Development of Scientific Style." In *Leonard Eĭler: K 300-letiiu so dnia rozhdenii /*

Leonhard Euler: 300th Anniversary, ed. Vladimir N. Vasilyev, pp. 37–50. Saint Petersburg: Nestor-Historia. LEJA 8.

Mattmüller, Martin, and Larisaa Ivanova Brylevskaya. 2008. "Euler's Opera Omnia" [title translated from Russian]. In *Leonard Eĭler: K 300-letiiu so dnia rozhdenii / Leonhard Euler: 300th Anniversary*, ed. Vladimir N. Vasilyev, pp. 292–300. Saint Petersburg: Nestor-Historia. LEJA 8.

Matvievskaya, Galina Pavlovna. 2008. "A Brief Survey of Euler's Unpublished Notebooks" [title translated from Russian]. In *Leonard Eĭler: K 300-letiiu so dnia rozhdenii / Leonhard Euler: 300th Anniversary*, ed. Vladimir N. Vasilyev, pp. 265–273. Saint Petersburg: Nestor-Historia. LEJA 8.

Matvievskaya, G. P. 2007. "On Euler's Surviving Manuscripts and Notebooks." Translated by Robert Burns. In *Euler and Modern Science*, ed. N. N. Bogolyubov, G. K. Mikhaĭlov, and A. P. Yushkevich, pp. 199–127. Washington, D.C.: Mathematical Association of America. LEJA 7.

Matvievskaya, G. P., and E. P. Ozhigova. 2007. "The Manuscript Materials of Euler on Number Theory." Translated by Robert Burns. In *Euler and Modern Science*, ed. N. N. Bogolyubov, G. K. Mikhaĭlov, and A. P. Yushkevich, pp. 127–137. Washington, D.C.: Mathematical Association of America. LEJA 7.

——. 1983. "Eulers Manuskripte zur Zahlentheorie." In *Leonhard Euler, 1707–1783: Beiträge zu Leben und Werk*, ed. J. J. Burckhardt, E. A. Fellmann, and W. Habicht, pp. 151–161. Basel: Birkhäuser Verlag. LEJA 4.

Mazière, Pierre. 1727. "Les loix du choc des corps à resort parfait ou imparfait, deduites d'une explication probable de la cause physique du ressort." *Recueil prix Paris* 2, pp. 1–57.

McClellan, James E., III. 1993. "L'Europe des academies." *Dix-Huitième Siècle* 25, pp. 153–165.

——. 1985. *Science Reorganized: The Scientific Societies in the Eighteenth Century*. New York: Columbia University Press.

McKinzie, Mark. 2007a. "Euler's Observations on Harmonic Progressions." In *Euler at 300: An Appreciation*, ed. Robert E. Bradley, Lawrence A. D'Antonio, and C. Edward Sandifer, pp. 131–143. Washington, D.C: Mathematical Association of America.

——. 2007b. "Origins of a Classic Formalist Argument: Power Series Expansions of the Logarithmic and Exponential Functions." In *Euler at 300: An Appreciation*, ed. Robert E. Bradley, Lawrence A. D'Antonio, and C. Edward Sandifer, pp. 143–157. Washington, D.C: Mathematical Association of America.

McNeill, William H. 1982. *The Pursuit of Power: Technology, Armed Force, and Society since A.D. 1000*. Chicago: University of Chicago Press.

Meli, Domenico Bertolini. 1993. "The Emergence of Reference Frames and the Transformation of Mechanics in the Enlightenment." *Historical Studies in the Physical and Biological Sciences* 23, no. 2, pp. 301–336.

Merton, Robert K. 2008. *On the Social Structure of Science*. Chicago: University of Chicago Press.

——. 1938. "Science, Technology, and Society in Seventeenth Century England." *Osiris* 4, 360–632.

Mervaud, Christiane. 1985. *Voltaire et Frédéric II: une dramaturgie des lumières, 1736–1778*. Oxford: Voltaire Foundation.

Mikhaĭlov, Gleb Konstantinovich. 2008a. "Leonhard Euler (on the 300th Anniversary of His Birthday)" [title translated from Russian]. In *Leonard Eĭler: K 300-letiiu so dnia rozhdenii / Leonhard Euler: 300th Anniversary*, ed. Vladimir N. Vasilyev, pp. 8–21. Saint Petersburg: Nestor-Historia. LEJA 8.

——. 2008b. "L. Euler and Rational Mechanics Establishment," [title translated from Russian]. In *Leonard Eĭler: K 300-letiiu so dnia rozhdenii / Leonhard Euler: 300th Anniversary*, ed. Vladimir N. Vasilyev, pp. 137–151. Saint Petersburg: Nestor-Historia. LEJA 8.

——. 2002. "General Introduction." In Daniel Bernoulli, *Werke*, vol. 5. Basel: Birkhaüser Verlag.
——. 1999. "The Origins of Hydraulics and Hydrodynamics in the Work of the Petersburg Academicians of the 18th Century." *Fluid Dynamics* 34, no. 6, pp. 787–800.
——. 1983. "Leonhard Euler und die Entwicklung der theoretischen Hydraulik im zweiten Viertel des 18. Jahrhunderts." In *Leonhard Euler, 1707–1783: Beiträge zu Leben und Werk*, ed. J. J. Burckhardt, E. A. Fellmann, and W. Habicht, pp. 229–243. Basel: Birkhäuser Verlag. LEJA 4.
——. 1959. "Notizen über die unveröffentlichten Manuskripte von Leonhard Euler." In *Sammelband der zu Ehren des 250. Geburtstages Leonhard Eulers der Deutschen Akademie der Wissenschaften zu Berlin vorgelegten Abhandlungen*, ed. Kurt Schröder, pp. 256–280. Berlin: Akademie Verlag. LEJA 3.
Mikhaĭlov, Gleb Konstantinovich, Patricia Radelet-de-Grave, and David Speiser, eds. 2002. *Daniel Bernoulli's Werke*, vol. 5: *Hydrodynamik II.* Basel: Birkhäuser Verlag.
Mikhaĭlov, Glebv Konstantinovich, and L. I. Sedov. 2007. "The Foundations of Mechanics and Hydrodynamics in Euler's Works." Translated by Robert Burns. In *Euler and Modern Science*, ed. N. N. Bogolyubov, G. K. Mikhaĭlov, and A. P. Yushkevich, pp. 167–183. Washington, D.C.: Mathematical Association of America. LEJA 7.
——. 1959. "Notizen über die unveröffentlichte Manuskripte von Leonhard Euler." In *Sammelband der zu Ehren des 250. Geburtstages Leonhard Eulers der Akademie der Wissenschaften der UdSSR vorgelegten Abhandlungen*, ed. Mikhail Alekseevich Lavrentiev, Adolph P. Yushkevich, and Ašot T. Grigor'ian, pp. 256–280. Moscow: Izdatel'stvo Akademii nauk. LEJA 2.
——. 1957. "Zur Übersiedlung Eulers nach Petersburg." *AN SSR, otdelenije technitscheskich nauk* 3, pp. 10–37.
Mills, Stella. 1985. "The Independent Derivations of Leonhard Euler and Colin Maclaurin of the Euler-Maclaurin Summation Formula." *AHES* 33, nos. 1–3, pp. 1–13.
——, ed. 1982. *The Collected Letters of Colin Maclaurin.* Cheshire, England: Shiva.
Mincenko, L. S. 1957. "Euler's Physics" [title translated from Russian]. *Proceedings of the Institute for the History of the Natural Sciences and Technology, Academy of Sciences USSR* [title translated from Russian] 19, pp. 221–270.
Mizler, Lorenz. 1752. *Musikaliche Bibliothek . . .*, vol. 3. Leipzig: Im Mizlerischen Verlag.
Montucla, Jean-Étienne. [1758] 1960. *Histoire des mathématiques.* 2 vols. Reprint, Paris: Albert Blanchard.
Morozov, Nikita Feodorovich, and Pjetr Eugenievich Tovstik. 2008. "Leonhard Euler and Modern Mechanics" [title translated from Russian]. In *Leonard Eĭler: K 300-letiiu so dnia rozhdenii / Leonhard Euler: 300th Anniversary*, ed. Vladimir N. Vasilyev, pp. 152–158. Saint Petersburg: Nestor-Historia. LEJA 8.
Nagel, Fritz. 2008. "Euler's Roots in Basel" [title translated from Russian]. In *Leonard Eĭler: K 300-letiiu so dnia rozhdenii / Leonhard Euler: 300th Anniversary*, ed. Vladimir N. Vasilyev, pp. 22–36. Saint Petersburg: Nestor-Historia. LEJA 8.
Nahin, Paul J. 2006. *Dr. Euler's Fabulous Formula: Cures Many Mathematical Ills.* Princeton, N.J.: Princeton University Press.
——. 1998. *An Imaginary Tale: The Story of* $\sqrt{-1}$. Princeton, N.J.: Princeton University Press.
Navarro Loidi, Juan Miguel. 2008. "El número *e* en los textos matemáticos españoles del siglo XVIII." *Quaderns d'Història de l'Enginyeria* 9, pp. 145–166.
Neumann, Olaf. 2007. "Cyclotomy: From Euler through Vandermonde to Gauss." In *Leonhard Euler: Life, Work and Legacy*, ed. Robert E. Bradley and C. Edward Sandifer, pp. 323–363. Elsevier Studies in the History and Philosophy of Mathematics 5. New York: Elsevier.
Nevskaia, Nina Ivanovna. 1984. *Peterburgkaia astronomcheskaia shkola*, vol. 18. Leningrad: Nauka.

Nevskaja, Nina I. 1983. "Euler als Astronom." In *Leonhard Euler, 1707–1783: Beiträge zu Leben und Werk*, ed. J. J. Burckhardt, E. A. Fellmann, and W. Habicht, pp. 363–373. Basel: Birkhäuser Verlag. LEJA 4.

Nevskaya, N. I. 2007. "New Evidence Concerning Euler's Development as an Astronomer and Historian of Science." Translated by Robert Burns. In *Euler and Modern Science*, ed. N. N. Bogolyubov, G. K. Mikhaĭlov, and A. P. Yushkevich, pp. 269–289. Washington, D.C.: Mathematical Association of America. LEJA 7.

Nevskaya, N. I., and K. V. Kholshenikov. 2007. "Euler and the Evolution of Celestial Mechanics." Translated by Robert Burns. In *Euler and Modern Science*, ed. N. N. Bogolyubov, G. K. Mikhaĭlov, and A. P. Yushkevich, pp. 263–269. Washington, D.C.: Mathematical Association of America. LEJA 7.

Newton, Isaac. 1999. *The Principia Mathematica: The Mathematical Principles of Natural Philosophy.* Edited by I. Bernard Cohen and Anne Whitman. Berkeley: University of California Press.

——. 1687. *Philosophiae Naturalis Principia Mathematica.* London: Royal Society, at the printer Joseph Streater.

Obolensky, Wladimir. 1986. *Reh oder Einhorn? Das Wappentier Leonhard Eulers. Jahrbuch 1986.* Basel: Schweizerische Gesellschaft für Familienforschung.

Okrepilov, V. V. 2008. "L. Euler and His Contribution to Metrology" (in Russian with English abstract). In *Leonard Eĭler: K 300-letiiu so dnia rozhdenii / Leonhard Euler: 300th Anniversary*, ed. Vladimir N. Vasilyev, pp. 212–220. Saint Petersburg: Nestor-Historia. LEJA 8. In Russian with English abstract.

Olby, Robert, G. M. Cantor, J. R. Christie, and M.J.S. Hodge, eds. 1990. *Companion to the History of Modern Science.* London: Routledge.

Oldfather, W. A., C. A. Ellis, and Donald M. Brown. 1933. "Leonhard Euler's Elastic Curves." *Isis* 20, no. 1, pp. 72–160.

Ortega, Romeo, Antonio Loría, Per Johan Nicklasson, and Hebertt Sira-Ramrez, 1998. *Passivity-Based Control of Euler-Lagrange Systems: Mechanical, Electrical, and Electromechanical Applications.* Berlin: Springer Verlag.

Otte, Michael, and Marco Panza, eds. 1997. *Analysis and Synthesis in Mathematics.* Dordrecht, Netherlands: Kluwer Academic.

Ozhigova, E. P. 2007. "The Part Played by the Petersburg Academy of Sciences (the Academy of Sciences of the USSR) in the Publication of Euler's Collected Works." In *Euler and Modern Science*, ed. N. N. Bogolyubov, G. K. Mikhaĭlov, and A. P. Yushkevich, pp. 53–75. Washington, D.C.: Mathematical Association of America. LEJA 7.

Panza, Marco. 2007. "Euler's *Introductio in analysin infinitorum* and the Program of Algebraic Analysis: Quantities, Functions and Numerical Partitions." In *Euler Reconsidered*, ed. Roger Baker, pp. 119–167. Heber City, Utah: Kendrick.

Pasquier, L.-Gustav du. [1927] 2008. *Léonard Euler et ses amis.* Paris: Hermann. Translated as *Leonard Euler and His Friends* by John S. D. Glaus (privately published, available through Amazon.com).

Paul, Charles B. 1970. "Jean-Philippe Rameau (1683–1764), the Musician as Philosophe." *Proceedings of the American Philosophical Society* 114, no. 2, pp. 140–54.

Pavlova, G. E. 1958. "Forgotten Testimony of Contemporaries concerning the Death of Leonhard Euler." In *Leonhard Euler*, ed. M. A. Lavrent'ev, pp. 605–608. Moscow: Akademiia nauk, SSSR.

Pedersen, Kurt Møller. 2008. "Leonhard Euler's Wave Theory of Light." *Perspectives on Science* 16, pp. 392–416.

Pekarskyi, Petr Petrovic. 1864. "Yekaterina II i Eylera (Catherine II and Euler)." *Zapiski Akademii nauk*, vol. 1, pt. 1, pp. 59–92.

Pengelley, David J. 2007. "Dances between Continuous and Discrete: Euler's Summation Formula." In *Euler at 300: An Appreciation*, ed. Robert E. Bradley, Lawrence A. D'Antonio, and C. Edward Sandifer, pp. 169–191. Washington, D.C: Mathematical Association of America.

Petrovich, Aleksandar. 2008. "The Meaning behind [the] Cartesian Wall (or How Euler Built St. Petersburg)." In *Leonard Eĭler: K 300-letiiu so dnia rozhdenii / Leonhard Euler: 300th Anniversary*, ed. Vladimir N. Vasilyev, pp. 234–240. Saint Petersburg: Nestor-Historia. LEJA 8.

Philidor, François André Danican. 1749. *L'Analyse des Echecs. . . .* London: Ludimus Effigie Belli. Over the next 125 years, this book went through seventy editions.

Pier, Jean-Paul. 2008. "Leonhard Euler and the Emergence of Harmonic Analysis." In *Leonard Eĭler: K 300-letiiu so dnia rozhdenii / Leonhard Euler: 300th Anniversary*, ed. Vladimir N. Vasilyev, pp. 57–63. Saint Petersburg: Nestor-Historia. LEJA 8.

Pla i Carrera, Josep. 2008. "El Tractat d'Euler De la Doctrina dels nombres, exposada en setze capítols. Una comparativa amb el Disquisitiones Arithmeticae de Gauss." *Quaderns d'Història de l'Enginyeria* 9, pp. 167–182.

Plofker, Kim. 2007. "Euler and Indian Astronomy." In *Leonhard Euler: Life, Work and Legacy*, ed. Robert E. Bradley and C. Edward Sandifer, pp. 147–167. Elsevier Studies in the History and Philosophy of Mathematics 5. New York: Elsevier.

Pocock, John Grevile Agard. 2005. "The Politics of Historiography." *Historical Research* 78, no. 1999, pp. 1–14.

Pólya, George. 2007. "Guessing and Proving." In *The Genius of Euler: Reflections on His Life and Work*, ed. William Dunham, pp. 257–263. Washington, D.C.: Mathematical Association of America.

Polyakhov, N. N. 2007. "Euler's Research in Mechanics during the First Petersburg Period." In *Euler and Modern Science*, ed. N. N. Bogolyubov, G. K. Mikhaĭlov, and A. P. Yushkevich, pp. 237–241. Washington, D.C.: Mathematical Association of America. LEJA 7.

Polyakhova, Tatjana S. 2008. "L. Euler and the Formation of Mathematical Education in Russia" [title translated from Russian]. In *Leonard Eĭler: K 300-letiiu so dnia rozhdenii / Leonhard Euler: 300th Anniversary*, ed. Vladimir N. Vasilyev, pp. 241–254. Saint Petersburg: Nestor-Historia. LEJA 8.

Popkin, Jeremy. 1998. "Robert Darnton's Alternative (to the) Enlightenment." In *The Darnton Debate: Books and Revolution in the Eighteenth Century*, ed. Haydn T. Mason, pp. 105–128. Oxford: Voltaire Foundation.

Porter, Roy, ed. 2003. *The Cambridge History of Science*, vol. 4: *The Eighteenth Century*. Cambridge: Cambridge University Press.

Porter, Roy, and Mikulaś Teich, eds. 1992. *The Scientific Revolution in National Context.* Cambridge: Cambridge University Press.

Posamentier, Alfred S. 2004. *Pi: A Biography of the World's Most Mysterious Number.* New York: Prometheus.

Preuss, Johann David Erdemann, ed. 1846–57. *Oeuvres de Fréderic le Grand.* 30 vols. Berlin: Impremerie Royale.

Principe, Lawrence M. 2013. *The Secrets of Alchemy.* Chicago: University of Chicago Press.

Pritchard, James. 1987. "From Shipwright to Naval Constructor: The Professionalization of 18th Century French Naval Shipbuilding." *Technology and Culture* 28, no. 1, pp. 1–25.

Pulte, Helmut. 1989. *Das Prinzip der kleinsten Wirkung und die Kraftkonzeptionen der rationalen Mechanik. Eine Untersuchung zur Grundlegungsproblematik bei Leonhard Euler, Pierre Louis Moreau de Maupertuis und Joseph Louis Lagrange*, Studia Leibnitiana, Sonderheft 19. Stuttgart: F. Steiner Verlag.

Radelet-de-Grave, Patricia. 2007. "Euler and Magnetism." In *Euler Reconsidered*, ed. Roger Baker, pp. 167–205. Heber City, Utah: Kendrick.

Raith, Michael. 1983. "Der Vater Paulus Euler. Beiträge zum Verständnis der geistigen Herkunft Leonhard Eulers." In *Leonhard Euler, 1707–1783: Beiträge zu Leben und Werk*, ed. J. J. Burckhardt, E. A. Fellmann, and W. Habicht, pp. 459–471. Basel: Birkhäuser Verlag. LEJA 4.

——. 1982. "Pietismus in Riehen." In *z'Rieche 1982, ein heimatliches Jahrbuch*, pp. 6–31. Riehen, Switzerland: Stiftung z'Rieche.

——. 1972. "Das kirchliche Leben seit der Reformation." In *Riehen: Geschichte eines Dorfes*, ed. Albert Bruckner, pp. 165–214. Riehen, Switzerland: Verlag A. Schudel.

Rashed, Roshdi, ed. 1990. *Mathématique et philosophie de l'antiquité à l'age classique. Hommage à Jules Vuillemin*. Paris: CNRS Éditions.

Raskin, N. M. 2007. "Euler and I. P. Kulibin." Translated by Robert Burns. In *Euler and Modern Science*, ed. N. N. Bogolyubov, G. K. Mikhaĭlov, and A. P. Yushkevich, pp. 317–334. Washington, D.C.: Mathematical Association of America. LEJA 7.

Recasens Gallart, Eduard. 2008. "Sobre un dels treballs d'Euler en geometria clàssica: l'E135." *Quaderns d'Història de l'Enginyeria* 9, pp. 205–218.

Reich, Karen. 2011. "Ein neues Blatt in Euler Lorbeerkranz, durch Carl Friedrich Gauβ eingeflochten." *Abhandlungen der Akademie der Wissenschaften zu Göttingen*, new series, 10, pp. 223–273.

——. 2007. "Euler's Contribution to Differential Geometry and its Reception." In *Leonhard Euler: Life, Work and Legacy*, ed. Robert E. Bradley and C. Edward Sandifer, pp. 479–503. Elsevier Studies in the History and Philosophy of Mathematics 5. New York: Elsevier.

——. 2005. "Gauss' geistige Vater: nicht nur 'summus Newton' sondern auch 'summus Euler.'" In *Wie der Blitz einschlägt, hat sich das Räthsel gelöst*, pp. 105–115. Göttingen, Germany: Göttingen Niedersächsische Staats- und Universitätsbibliothek.

Remmert, Volker R. 2009. "Antiquity, Nobility, and Utility: Picturing Early Modern Mathematical Sciences." In *The Oxford Handbook of the History of Mathematics*, ed. Eleanor Robson and Jacqueline Stedall, pp. 537–564. Oxford: Oxford University Press.

Richeson, David. 2008. *Euler's Gem: The Polyhedron Formula and the Birth of Topology*. Princeton, N.J.: Princeton University Press.

——. 2007. "The Polyhedral Formula." In *Leonhard Euler: Life, Work and Legacy*, ed. Robert E. Bradley and C. Edward Sandifer, pp. 421–440. Elsevier Studies in the History and Philosophy of Mathematics 5. New York: Elsevier.

Robins, Benjamin. 1739. *Remarks on Mr. Leonard Euler's Treatise Entitled Mechanics*. London: J. Nourse.

Robson, Eleanor, and Jacqueline Stedall, eds. 2009. *The Oxford Handbook of the History of Mathematics*. Oxford: Oxford University Press.

Romero, Angel E. 2007. "Physics and Analysis: Euler and the Search for Fundamental Principles of Mechanics." In *Euler Reconsidered*, ed. Roger Baker, pp. 232–281. Heber City, Utah: Kendrick.

Rouse, Hunter. 1978. *Elementary Mechanics of Fluids*. New York: Dover.

Rumyantsev, V. V. 2007. "Leonhard Euler and the Variational Principles of Mechanics." Translated by Robert Burns. In *Euler and Modern Science*, ed. N. N. Bogolyubov, G. K. Mikhaĭlov, and A. P. Yushkevich, pp. 183–213. Washington, D.C.: Mathematical Association of America. LEJA 7.

Saine, Thomas. 1987. "Who's Afraid of Christian Wolff?" In *Anticipations of the Enlightenment in England, France, and Germany*, ed. Alan C. Kors and Paul Korshin, pp. 102–133. Philadelphia: University of Pennsylvania Press.

Sandifer, C. Edward. 2008, February. "How Euler Did It: Fallible Euler." http://euler archive.maa.org/hedi/HEDI-2008-02.pdf.

——. 2007a. *The Early Mathematics of Leonhard Euler*. Washington, D.C: Mathematical Association of America.

——. 2007b. "Euler Rows the Boat." In *Euler at 300: An Appreciation*, ed. Robert E. Bradley, Lawrence A. D'Antonio, and C. Edward Sandifer, pp. 273–281. Washington, D.C: Mathematical Association of America.

——. 2007c. "Euler's Fourteen Problems." In *Euler at 300: An Appreciation*, ed. Robert E. Bradley, Lawrence A. D'Antonio, and C. Edward Sandifer, pp. 25–33. Washington, D.C: Mathematical Association of America.

——. 2007d. "Euler's Solution of the Basel Problem—The Longer Story." In *Euler at 300: An Appreciation*, ed. Robert E. Bradley, Lawrence A. D'Antonio, and C. Edward Sandifer, pp. 105–119. Washington, D.C: Mathematical Association of America.

——. 2007e. *How Euler Did It*. Washington, D.C: Mathematical Association of America.

——. 2007f, October. "How Euler Did It: Gamma the Constant." http://eulerarchive.maa.org/hedi/HEDI-2007-10.pdf.

——. 2007g, September. "How Euler Did It: Gamma the Function." http://eulerarchive.maa.org/hedi/HEDI-2007-09.pdf.

——. 2007h, November. "How Euler Did It: Inexplicable Functions." http://eulerarchive.maa.org/hedi/HEDI-2007-11.pdf.

——. 2007i, June. "How Euler Did It: Partial Fractions." http://eulerarchive.maa.org/hedi/HEDI-2007-06.pdf.

——. 2007j. "Some Facets of Euler's Work on Series." In *Leonhard Euler: Life, Work and Legacy*, ed. Robert E. Bradley and C. Edward Sandifer, pp. 279–303. Elsevier Studies in the History and Philosophy of Mathematics 5. New York: Elsevier.

——. 2006a, April. "How Euler Did It: Knight's Tour." http://eulerarchive.maa.org/hedi/HEDI-2006-04.pdf.

——. 2006b, May. "How Euler Did It: 19th Century Triangle Geometry." http://eulerarchive.maa.org/hedi/HEDI-2006-05.pdf.

——. 2006c, October. "How Euler Did It: How Euler Discovered America." http://eulerarchive.maa.org/hedi/HEDI-2006-10.pdf.

——. 2005a, November. "How Euler Did It: Amicable Numbers." http://eulerarchive.maa.org/hedi/HEDI-2005-11.pdf.

——. 2005b, April. "How Euler Did It: Euler and Pell." http://eulerarchive.maa.org/hedi/HEDI-2005-04.pdf.

——. 2005c, May. "How Euler Did It: The Euler Society." http://eulerarchive.maa.org/hedi/HEDI-2005-05.pdf.

——. 2005d, December. "How Euler Did It: Factors of Forms." http://eulerarchive.maa.org/hedi/HEDI-2005-12.pdf.

——. 2005e, October. "How Euler Did It: Philip Naudé's Problem." http://eulerarchive.maa.org/hedi/HEDI-2005-10.pdf.

——. 2004a, December. "How Euler Did It: A Mystery About the Laws of Cosines." http://eulerarchive.maa.org/hedi/HEDI-2004-12.pdf.

——. 2004b, February. "How Euler Did It: Propulsion of Ships." http://eulerarchive.maa.org/hedi/HEDI-2004-02.pdf.

——. 2004c, June. "How Euler Did It: V, E, and F, Part 1." http://eulerarchive.maa.org/hedi/HEDI-2004-06.pdf.

——. 2004d, July. "How Euler Did It: V, E, and F, Part 2." http://eulerarchive.maa.org/hedi/HEDI-2004-07.pdf.

——. 2004e, January. "How Euler Did It: Venn Diagrams." http://eulerarchive.maa.org/hedi/HEDI-2004-01.pdf.

Schafheitlin, Paul, Eugen Jahnke, and Carl Färber. 1907. "Preface." In *Festschrift zur Feier des 200. Geburtstages Leonhard Eulers*, ed. Berliner Mathematische Gesellschaft, Fasc. xxv. Berlin: Abhandlungen zur Geschichte der mathematischen Wissenschaften. LEJA 1. This volume contains four papers.

Scharlau, Winfried. 1983. "Eulers Beiträge zur *partitio numerorum* und zur Theorie der erzeugenden Funktionen." In *Leonhard Euler, 1707–1783: Beiträge zu Leben und Werk*, ed. J. J. Burckhardt, E. A. Fellmann, and W. Habicht, pp. 135–151. Basel: Birkhäuser Verlag. LEJA 4.

Schmidt, Alexandra. 2008. "Intersecting Arcs: The Equation of Time and the Equation of Noon." MAT thesis, Union Graduate College.

Schoenberg, Isaac J. 1983. "Euler's Contribution to Cardinal Spline Interpolation: The Exponential Euler Splines." In *Leonhard Euler, 1707–1783: Beiträge zu Leben und*

Werk, ed. J. J. Burckhardt, E. A. Fellmann, and W. Habicht, pp. 199–215. Basel: Birkhäuser Verlag. LEJA 4.

Schwab, Richard N. 1995. *Preliminary Discourse to the Encyclopedia of Diderot*. Chicago: University of Chicago Press.

Scriba, Christoph J. 2008. Book Review of Emil A. Fellmann, *Leonhard Euler*, 2007. *HM* 35, pp. 142–143.

Shabel, Lisa. 2003. *Mathematics in Kant's Critical Philosophy*. New York: Routledge.

Sheynin, Oscar B., comp. and trans. 2009. *Portraits: Leonhard Euler, Daniel Bernoulli, Johann-Heinrich Lambert*. Berlin: NG Verlag.

——. 2007. "Euler's Work in Probability and Statistics." In *Euler Reconsidered*, ed. Roger Baker, pp. 287–317. Heber City, Utah: Kendrick.

——. 1972. "On the Mathematical Treatment of Observations by L. Euler." *AHES* 9, no. 1, pp. 45–56.

Shimura, Goro. 1997. *Euler Products and Eisenstein Series*. Providence, R.I.: American Mathematical Society.

Smith, David Eugene. 1929. *A Source Book in Mathematics*. New York: McGraw-Hill.

Smith, Haywood C. 2010, July. "Mayer Earned the Lunar-Table Prize." Letter to the editor, *Physics Today*, p. 9.

Smith, Margarete G. 1993. "In Defence of an Eighteenth-Century Academician, Philosopher, and Journalist: Jean-Henri-Samuel Formey." In *Studies on Voltaire and the Eighteenth Century*, vol. 311, pp. 93–108. Geneva: Institut et Musée Voltaire.

Smith, Sydney. 1805. Review of Dieudonné Thiébault's *Souveniers. . . . Edinburgh Review* 7, pp. 218–244.

Snow, C. P. 1964. *The Two Cultures and a Second Look*. Cambridge: Cambridge University Press.

Spalt, Detlef D. 2011. "Welche Funktionbegriffe gab Leonhard Euler?" *HM* 38, pp. 485–505.

Speiser, David. 2003. "Clifford A. Truesdell's Contributions to the Euler and the Bernoulli Edition." *Journal of Elasticity* 70, pp. 39–53.

——. 1983. "Eulers Schriften zur Optik, zur Elektrizität und zum Magnetismus." In *Leonhard Euler, 1707–1783: Beiträge zu Leben und Werk*, ed. J. J. Burckhardt, E. A. Fellmann, and W. Habicht, pp. 215–229. Basel: Birkhäuser Verlag. LEJA 4.

Speziali, Pierre. 1983. "Léonard Euler et Gabriel Cramer." In *Leonhard Euler, 1707–1783: Beiträge zu Leben und Werk*, ed. J. J. Burckhardt, E. A. Fellmann, and W. Habicht, pp. 421–435. Basel: Birkhäuser Verlag. LEJA 4.

——. 1955. "Une correspondance inédite entre Clairaut et Cramer." *Revue d'histoire des sciences et de leurs applications* 8, no. 3, pp. 193–237.

Spiess, Otto. 1929. *Leonhard Euler*. Leipzig: Huber.

Stäckel, Paul. 1910. *Johann Albrecht Euler*. Zurich: Zürcher und Furrer.

Stähelin, Andreas. 1957. *Geschichte der Universität Basel 1632–1818 (Studien zur Geschichte der Wissenschaften in Basel)*, vols. 4–5. Basel: Broschiert. Published on the occasion of the five hundred year Jubilee of the University of Basel, 1460–1960.

Stén, Johan C.-E. 2014. *A Comet of the Enlightenment: Andres Johan Lexell's Life and Discoveries*. Vita Mathematica vol 17. Basel: Birkhäuser Verlag.

Steele, Brett D. 1994. "Muskets and Pendulums: Benjamin Robins, Leonhard Euler, and the Ballistics Revolution." *Technology and Culture* 35, no. 2, pp. 348–382.

Strachey, Lytton. 1997. "Voltaire and Frederick the Great." Edited by Geoffrey Sauer. http://books.eserver.org/nonfiction/nonfiction/strachey/voltaire-and-frederick.html.

——. 1922. *Books and Characters, French and English*. London: Chatto and Windus.

Struik, Dirk J. 1939–40. "A Story concerning Euler and Diderot." *Isis* 31, pp. 431–432.

Struik, D. J., ed. 1969. *A Source Book in Mathematics, 1200–1800*. Cambridge, Mass.: Harvard University Press.

Stuart, J. T. 1987. "Nonlinear Euler Partial Differential Equations: Singularities in Their Solution." In *Applied Mathematics, Fluid Mechanics, Astrophysics*. Cambridge, MA.: Harvard University Press.

Suay Belenguer, Juan Miguel. 2008. "Los Molinos y las Cometas de Mr. Euler le fils. Modelos matemáticos para las máquinas hidráulicas en el siglo XVIII." *Quaderns d'Història de l'Enginyeria* 9, pp. 117–144.

Suisky, Dieter. 2009. *Euler as Physicist.* Berlin: Springer Verlag.

——. 2007a. "Euler's Early Relativity Theory." In *Euler Reconsidered*, ed. Roger Baker, pp. 317–377. Heber City, Utah: Kendrick.

——. 2007b. "Euler's Mechanics as a Foundation of Quantum Mechanics." In *Leonhard Euler: Life, Work and Legacy*, ed. Robert E. Bradley and C. Edward Sandifer, pp. 503–527. Elsevier Studies in the History and Philosophy of Mathematics 5. New York: Elsevier.

Suzuki, Jeff. 2007. "Euler and Number Theory: A Study of Mathematical Invention." In *Leonhard Euler: Life, Work and Legacy*, ed. Robert E. Bradley and C. Edward Sandifer, pp. 363–385. Elsevier Studies in the History and Philosophy of Mathematics 5. New York: Elsevier.

Sved, Marta, and Dave Logothetti. 2007. "Discoveries (a Poem)." In *The Genius of Euler: Reflections on His Life and Work*, ed. William Dunham, pp. 91–93. Washington, D.C.: Mathematical Association of America.

Takase, Masahito. 2007. "Euler's Theory of Numbers." In *Euler Reconsidered*, ed. Roger Baker, pp. 377–422. Heber City, Utah: Kendrick.

Taton, René. 2008. "Les correspondants français d'Euler." *Archives Internationales d'histoire des Sciences* 58, pp. 133–149.

——. 1983. "Les Relations d'Euler avec Lagrange." In *Leonhard Euler, 1707–1783: Beiträge zu Leben und Werk*, ed. J. J. Burckhardt, E. A. Fellmann, and W. Habicht, pp. 409–421. Basel: Birkhäuser Verlag. LEJA 4.

Taton, René, and Curtis Wilson, eds. 1995. *The General History of Astronomy*, vol. 2: *Planetary Astronomy from the Renaissance to the Rise of Astrophysics.* Cambridge: Cambridge University Press.

Taylor, Ronald. 1997. *Berlin and Its Culture: A Historical Portrait.* New Haven, Conn.: Yale University Press.

Tent, M.B.W. 2009. *Leonhard Euler and the Bernoullis: Mathematicians from Basel.* Natick, Mass.: A. K. Peters.

Terrall, Mary. 2002. *The Man Who Flattened the Earth: Maupertuis and the Sciences in the Enlightenment.* Chicago: University of Chicago Press.

——. 1992. "Representing the Earth's Shape: The Polemics Surrounding Maupertuis's Expedition to Lapland." *Isis* 83, no. 4, pp. 218–237.

——. 1990. "The Culture of Science in Frederick the Great's Berlin." *History of Science* 28, pp. 333–364.

——. 1989. "Review of Leonhard Euler, *OO*, IVA 6." *HM* 16, pp. 96–97.

Tetzner, J. 1958. "Bücher deutscher Autoren in Prokopovičs Bibliothek." In *Die Deutsch-Russische Begegnung und Leonhard Euler*, ed. Eduard Winter, pp. 125–142. Berlin: Akademie Verlag.

Thiébault, Dieudonné. [1804–13] 2008. *Mes souvenirs de vingt ans de séjour à Berlin.* 2nd ed., 2 vols. Paris: F. Buisson.

Thiele, Rüdiger. 2007a. "Euler and the Calculus of Variations." In *Leonhard Euler: Life, Work and Legacy*, ed. Robert E. Bradley and C. Edward Sandifer, pp. 235–255. Elsevier Studies in the History and Philosophy of Mathematics 5. New York: Elsevier.

——. 2007b. "Leonhard Euler, the Decade 1750–1760." In *Euler at 300: An Appreciation*, ed. Robert E. Bradley, Lawrence A. D'Antonio, and C. Edward Sandifer, pp. 1–25. Washington, D.C: Mathematical Association of America.

——. 2007c. "The Rise of the Function Concept in Analysis." In *Euler Reconsidered*, ed. Roger Baker, pp. 422–462. Heber City, Utah: Kendrick.

——. 2007d. " . . . unsere Mathematiker können es mit denen aller Akademien aufnehmen: Leonhard Eulers Wirken an der Berliner Akademie als Mathematiker und Mechaniker (1741–1766)." *Wissenschaft und Fortschritt* 8, pp. 296–299.

——. 2007e. "What Is a Function?" In *Euler at 300: An Appreciation*, ed. Robert E. Bradley, Lawrence A. D'Antonio, and C. Edward Sandifer, pp. 63–85. Washington, D.C: Mathematical Association of America.

——. 1982. *Leonhard Euler.* Leipzig: B. G. Teubner.

Timerding, Heinrich Emil. 1919. "Kant und Euler." *Kant-Studien* 23, pp. 18–64.

Tjulina, Irina Alexandrova. 2008. "About Euler's Works in the Theory of the Hydraulic Turbine" [title translated from Russian]. In *Leonard Eĭler: K 300-letiiu so dnia rozhdenii / Leonhard Euler: 300th Anniversary*, ed. Vladimir N. Vasilyev, pp. 183–189. Saint Petersburg: Nestor-Historia. LEJA 8.

Todhunter, Isaac. [1873] 1962. *A History of the Mathematical Theories of Attraction and the Figure of the Earth.* 2 vols. London: Macmillan. Reprint, New York: Dover.

Truesdell, Clifford Ambrose. 1984a. *An Idiot's Fugitive Essays on Science: Methods, Criticism, Training, Circumstances.* New York: Springer.

——. 1984b. "Leonard Euler, Supreme Geometer." In *An Idiot's Fugitive Essays on Science: Methods, Criticism, Training, Circumstances*, pp. 337–380. New York: Springer.

——. 1982. "Euler's Contribution to the Theory of Ships and Mechanics: An Essay Review." *Centaurus* 26, no. 2, pp. 323–335.

——. 1972. Review of Charles Naux's *Histoire des logarithmes de Neper à Euler*. *Isis* 63, no. 3, pp. 443–444.

——. 1968. *Essays in the History of Mechanics.* Berlin: Springer Verlag.

——. 1954. "Rational Fluid Mechanics, 1687–1765." In Leonhard Euler. *OO* II.12, pp. xii–cxxv.

Tserlyuk-Askadskaya, S. S. 2007. "Euler's Music-Theoretical Manuscripts and the Formation of His Conception of the Theory of Music." Translated by Robert Burns. In *Euler and Modern Science*, ed. N. N. Bogolyubov, G. K. Mikhaĭlov, and A. P. Yushkevich, pp. 349–361. Washington, D.C.: Mathematical Association of America. LEJA 7.

Tuffet, Jacques, ed. 1967. *Histoire de Docteur Akakia et du natif de Saint-Malo* [by Voltaire], critical ed. Paris: A. G. Nizet.

Tweedie, Charles. 1922. *James Stirling.* Oxford: Oxford University Press.

Van Brummeln, Glen, and Michael Kinyon, eds. 2005. *Mathematics and the Historian's Craft: The Kenneth O. May Lectures.* New York: Springer.

Van der Waerden, Bartel L. 1983. "Eulers Herleitung des Drehimpulssatzes." In *Leonhard Euler, 1707–1783: Beiträge zu Leben und Werk*, ed. J. J. Burckhardt, E. A. Fellmann, and W. Habicht, pp. 271–283. Basel: Birkhäuser Verlag. LEJA 4.

Varadarajan, V. S. 2006. *Euler through Time: A New Look at Old Themes.* Providence, R.I.: American Mathematical Society.

Vasilyev, Vladimir N. 2008a. "Introductory Words" [title translated from Russian]. In *Leonard Eĭler: K 300-letiiu so dnia rozhdenii / Leonhard Euler: 300th Anniversary*, ed. Vladimir N. Vasilyev, p. 5. Saint Petersburg: Nestor-Historia. LEJA 8.

——. 2008b. "Euler in the Russian Internet of the Twenty-First Century" [title translated from Russian]. In *Leonard Eĭler: K 300-letiiu so dnia rozhdenii / Leonhard Euler: 300th Anniversary*, ed. Vladimir N. Vasilyev, pp. 301–315. Saint Petersburg: Nestor-Historia. LEJA 8.

Verdun, Andreas. 2012. "Leonhard Euler's Early Lunar Theories, 1725–1752." *AHES* 64, no. 3, pp. 235–303.

——. 2011. "Die (Wieder) Entdeckung von Eulers Mondtafeln." *NTM: Zeitschrift für Geschichte der Naturwissenschaften, Technik und Medizin* 19, no. 3, pp. 271–297.

——. 2004. "The Determination of the Solar Parallax from Transits of Venus in the 18th Century." *Archives des Sciences* 55, pp. 45–68.

——. 2003. "Leonhard Eulers Einfuhrung und Anwendung von Bezugssystem in Mechanik und Astronomie." *Elemente der Mathematik* 58, pp. 169–176.

Volk, Otto. 1983. "Eulers Beiträge zur Theorie der Bewegungen der Himmelskörper." In *Leonhard Euler, 1707–1783: Beiträge zu Leben und Werk*, ed. J. J. Burckhardt, E. A. Fellmann, and W. Habicht, pp. 345–363. Basel: Birkhäuser Verlag. LEJA 4.

Volkert, K. 1987. "History of Pathological Functions—On the Origins of Mathematical Methodology." *AHES* 37, no. 3, pp. 193–232.

Voltaire. [1752] 2005. *Histoire du Docteur Akakia et du Natif de Saint-Malo.* Translated as *The Story of Doctor Akakia and the Native from Saint Malo* by John S. D. Glaus (privately published).

——. [1752] 1879. *Histoire du Docteur Akakia et du Natif de Saint-Malo.* In *Oeuvres complètes de Voltaire*, vol. 23. Paris: Imprimerie de la Société littéraire-typographique.

——. [1732] 2007. *Lettres Philosophiques.* Edited by John Leigh and translated by Prudence Steiner as *Philosophical Letters*. Indianapolis: Hackett.

Vucinich, Alexander. 1963. *Science in Russian Culture: A History to 1866.* Stanford, Calif.: Stanford University Press.

Ward, Morgan. 2007. "A Mnemonic for Euler's Constant." In *The Genius of Euler: Reflections on His Life and Work*, ed. William Dunham, pp. 153–155. Washington, D.C.: Mathematical Association of America.

Waxell, Sven. 1952. *The American Expedition.* London: William Hodge.

Webster, Charles. 2008. *Paracelsus: Medicine, Magic, and Mission at the End of Time.* New Haven, Conn.: Yale University Press.

Weil, André. 2007. "Euler (Abridged)." In *The Genius of Euler: Reflections on His Life and Work*, ed. William Dunham, pp. 43–51. Washington, D.C.: Mathematical Association of America.

——. 1984. *Number Theory: An Approach through History from Hammurapi to Legendre.* Basel: Birkhäuser Verlag.

——. 1983. "L'oeuvre arithmétique d'Euler." In *Leonhard Euler, 1707–1783: Beiträge zu Leben und Werk*, ed. J. J. Burckhardt, E. A. Fellmann, and W. Habicht, pp. 111–135. Basel: Birkhäuser Verlag. LEJA 4.

Wepster, Steven A. 2009. *Between Theory and Observations: Tobias Mayer's Explorations of Lunar Motion, 1751–1755.* New York: Springer.

Werrett, Simon. 2010. "Reconfiguring Academic Expertise across Dynasties in Eighteenth-Century Russia." *Osiris* 25, no. 1. pp. 104–126.

Weyl, Hermann. 1968a. "David Hilbert, 1862–1943." In *Gesammelte Abhandlungen*, pp. 121–173. Berlin: Springer Verlag.

——. 1968b. *Gesammelte Abhandlungen.* Edited by K. Chandrasekharan. Berlin: Springer Verlag.

White, Homer S. 2007. "The Geometry of Leonhard Euler." In *Leonhard Euler: Life, Work and Legacy*, ed. Robert E. Bradley and C. Edward Sandifer, pp. 303–323. Elsevier Studies in the History and Philosophy of Mathematics 5. New York: Elsevier.

Whiteside, D. T. 1970. "The Mathematical Principles Underlying Newton's *Principia mathematica*." *Journal for the History of Astronomy* 1, pp. 116–138.

Wikipedia 2011. *Talento Excepcional: Prodigios em Calculos, John von Neumann, Carl Friedrich Gauss, Leonhard Euler, Stephen Hawking, Paul Erdős, William Rowan Hamilton, Memória eidética, Síndrome de savant, Kim Peek, Zerah Colburn, Rüdiger Gamm.* E-book. Memphis, Tenn.: General Books.

Wilson, Arthur M. 1972. *Diderot.* New York: Oxford University Press.

Wilson, Curtis. 2008a. "Euler's Combinatorial Mathematics." *British Society for the History of Mathematics Bulletin* 23, pp. 13–23.

——. 2008b. "The Nub of the Lunar Problem: From Euler to G. W. Hill." *Journal for the History of Astronomy* 39, pp. 453–468.

——. 2007. "Euler and Applications of Analytical Mathematics in Astronomy." In *Leonhard Euler: Life, Work and Legacy*, ed. Robert E. Bradley and C. Edward Sandifer, pp. 121–147. Elsevier Studies in the History and Philosophy of Mathematics 5. New York: Elsevier.

——. 2003. "Astronomy and Cosmology." In *The Cambridge History of Science*, vol. 4: *The Eighteenth Century*, ed. Roy Porter, pp. 328–353. Cambridge: Cambridge University Press.

——. 1992. "Euler on Action-at-a-Distance and Fundamental Equations in Continuum Mechanics." In *The Investigation of Difficult Things: Essays on Newton and the History of Exact Sciences in Honour of D. T. Whiteside*, ed. P. M. Harman and Alan E. Shapiro, 399–422. Cambridge: Cambridge University Press.

——. 1987. "D'Alembert versus Euler on the Precession of the Equinoxes and the Mechanics of Rigid Bodies." *AHES* 37, no. 3, pp. 233–273.

——. 1985. "The Great Inequality of Jupiter and Saturn: from Kepler to Laplace." *AHES* 33, nos. 1–3, pp. 15–290.

——. 1980a. "Perturbations and Solar Tables from Lacaille to Delambre: The Rapprochement of Observation and Theory," part 1. *AHES* 22, nos. 1–2, pp. 54–188.

——. 1980b. "Perturbations and Solar Tables from Lacaille to Delambre: The Rapprochement of Observation and Theory," part 2. *AHES* 22, no. 3, pp. 189–304.

Winter, Eduard, ed. 1958a. *Die Deutsch-Russische Begegnung und Leonhard Euler.* Berlin: Akademie Verlag.

——. 1958b. "Euler und die Begegnung der deutschen mit der russischen Aufklärung." In *Die Deutsch-Russische Begegnung und Leonhard Euler*, ed. Eduard Winter, pp. 1–18. Berlin: Akademie Verlag.

——, ed., 1957. *Die Registres der Berliner Akademie der Wissenschaften 1746–1766.* Berlin: Akademie Verlag.

Wolff, Christian. 1860. *Briefe von Christian Wolff aus den Jahren 1719–1753.* St. Petersburg: Eggers and Comp.

——. 1730. *Elementa Matheseos Universae*, vol. 1. Halle, Germany: Magdeburgicae.

——. 1728. "Principia dynamica," *Comm. Ac. Pb.* 1 (for 1726), pp. 217–236.

Woolf, Henry. 1959. *The Transits of Venus: A Study of Eighteenth-Century Science.* Princeton, N.J.: Princeton University Press.

Woronzoff-Dashkoff, A. 2008. *Dashkova: A Life of Influence and Exile.* Transactions of the American Philosophical Society 97, part 3. Philadelphia: American Philosophical Society.

Xambó Deschamps, Sebastian. 2008. "Euler and the Dynamics of Rigid Bodies." *Quaderns d'Història de l'Enginyeria* 9, pp. 279–303.

Yakovlev, Vadim Ivanovich. 2007. "Leonhard Euler and the Foundations of Mechanics." In *Euler Reconsidered*, ed. Roger Baker, pp. 462–474. Heber City, Utah: Kendrick.

Yuan Min and Jia Xiaoyong. 2008. "A Historical Survey: Why Did Lagrange Redefine the Complete Integral of a First-Order Partial Differential Equation?" [title translated from Chinese]. *Ziran Kexueshi Yánjiū* 27, pp. 485–497.

Yurkin, M. I., and S. A. Tolchelnikova. 2008. "Leonhard Euler and the Dynamics of Rigid Bodies" [title translated from Russian]. In *Leonard Eĭler: K 300-letiiu so dnia rozhdenii / Leonhard Euler: 300th Anniversary*, ed. Vladimir N. Vasilyev, pp. 201–211. Saint Petersburg: Nestor-Historia. LEJA 8.

Yushkevich, A. P. 2007. "Leonhard Euler: His Life and Work." Translated by Robert Burns. In *Euler and Modern Science*, ed. N. N. Bogolyubov, G. K. Mikhaĭlov, and A. P. Yushkevich, pp. 7–39. Washington, D.C.: Mathematical Association of America. LEJA 7.

Yushkevich, A. P., and R. Taton. 2007. "Leonhard Euler in Correspondence with Clairaut, D'Alembert, and Lagrange." Translated by Robert Burns. In *Euler and Modern Science*, ed. N. N. Bogolyubov, G. K. Mikhaĭlov, and A. P. Yushkevich, pp. 289–307. Washington, D.C.: Mathematical Association of America. LEJA 7.

Zacharov, A. S., and V. V. Nikolseva. 2008. "Leonhard Euler and the First Societies of Insurance in Russia with the Cause of Euler's Article in 1776" [title translated from Russian]. In *Leonard Eĭler: K 300-letiiu so dnia rozhdenii / Leonhard Euler: 300th Anniversary*, ed. Vladimir N. Vasilyev, pp. 104–114. Saint Petersburg: Nestor-Historia. LEJA 8.

Zaytsev, Evgeny. 2008. "Euler's Problem of Königsberg Bridges and Leibniz' *Geometria Situs*." *Archives Internationales d'histoire des Sciences* 58, pp. 151–170.
Zhang Sheng. 2007. "Euler and Euler Numbers" [title translated from Chinese]. *Nei Menguu Shifan Daxue Xuebao* 36, no. 6, pp. 778–779.
Zhelnina, Anna. 1999. "Euler in Saint Petersburg," in *Olympiad Work*.
Zubov, V. P. 1958. "Die Begegnung der deutschen mit der russischen Naturwissenschaft im 18. Jahrhundert." In *Die Deutsch-Russische Begegnung und Leonhard Euler*, ed. Eduard Winter, pp. 19–48. Berlin: Akademie Verlag.

Register of Principal Names

This register covers all principal persons mentioned in the body of the biography, but not in the preface or bibliography. The register employs the following abbreviations:

b.	born
Berlin AcS	Berlin Academy of Sciences, referring to the Royal Prussian Society of Sciences (1701–44); later developed into the Royal Prussian Academy of Sciences and Belle Lettres (1744–1810)
ca.	circa or about
CM	corresponding member
d.	died
fl.	flourished
FM	foreign member
FRS	fellow of the Royal Society of London
HM	honorary member
OM	ordinary member
Paris AcS	Royal Academy of Sciences, Paris
Pb AcS	Petersburg Imperial Academy of Sciences (1724–47); later renamed the Imperial Academy of Sciences and Arts (1747–1803)

Ackeret, Jakob (1898–1981). Swiss aeronautical engineer at the Eidgenössische Technische Hochschule, Zurich: 138.

Adams, John (1735–1826). Second president of the United States; supported founding the American Academy of Arts and Sciences, which was established by the Massachusetts legislature in 1780 and is now located in Cambridge, Massachusetts: 525.

Adodurov (Adadurov), Vasilii Yevdokimovich (1709–80). Russian mathematician, Russianist, and translator. Sole Russian HM of the early Pb AcS, adjunct for higher mathematics (1733–), removed (1741), HM (1778–). Curator of Moscow

University (1762–); translator of Euler's *Einleitung zur Rechen-Kunst: Commentationes ad physicam generalem pertinentes et Miscellanea*: 167.

Aepinus, Franz Ulrich Theodor Hoch (1724–1802). German and Russian natural philosopher and astronomer. OM (1755–57) of the Berlin AcS, director of the Berlin astronomical observatory, and then professor of physics and OM (1757–98) of the Pb AcS; head of the Russian cryptographic service (1764–97), published the initial mathematical theory of electricity and magnetism (1759): 387, 405, 411, 501.

Algarotti, Count Francesco (1712–64). Venetian polymath, poet, essayist, art critic and collector, and natural philosopher. Popularizer of Newton's science for women; HM of the Berlin AcS (1746–): 150, 181, 183, 314, 331, 417, 465.

Anhalt-Dessau, Leopold I, Prince of (1676–1747). German prince, married Anna Louise Föhse. Stern military disciplinarian known as the "Old Dessauer," he modernized the Prussian infantry and served under Frederick II: 417, 418.

Anna Ivanovna (1693–1740). Duchess of Courland (1711–30). Empress of imperial Russia (1730–40): 110, 139, 150, 171.

Anna Leopoldovna (1718–46). Grand Duchess of Russia, mother of Ivan VI: 171, 187.

Apollonius of Rhodes (ca. 295 B.C.–ca. 246 B.C.). Greek poet and grammarian, Alexandrian chief librarian, (perhaps 260 B.C.–247 B.C.), author of *Argonautica*: 144.

Archimedes of Syracuse (ca. 287 B.C.–212 B.C.). Mathematician, physicist, engineer, astronomer, and inventor in ancient Greece. One of the greatest mathematicians of antiquity, he studied buoyancy and the lever and developed the rigorous double method of exhaustion in what would later develop into calculus: 126, 492, 534.

Aristotle (384 B.C.–322 B.C.). Philosopher in ancient Greece. Contributed to methodology, physics, physical astronomy, meteorology, biology, and psychology: 23, 56, 153, 154, 226, 353, 354, 482.

Aristoxenus of Tarentum (fl. 335 B.C.). Peripatetic philosopher and student of Aristotle in ancient Greece. Initial authority on music theory in the classical world: 154.

August Wilhelm (1722–58). Crown prince of Prussia. Younger brother of Frederick II, designated heir: 252.

Avramov, Mikhail (fl. 1720–50). Russian director of the Saint Petersburg printing office, commanded by Peter I to print Christiaan Huygens's *Cosmotheoros*. He opposed the idea, but made thirty copies that were not circulated: 72.

Bach, Johann Sebastian (1685–1750). German classical composer, organist, harpsichordist, violist, and violinist. Member of the court orchestra in Weimar, director of music in Leipzig: 319, 487.

Bacon, Francis (Baron Verulam) (1561–1626). English author, lawyer, and statesman. Lord chancellor, historian, proponent of new sciences and of the inductive method; pioneer in English materialism: 56, 341, 397.

Bailly, Jean Sylvain (1736–93). French astronomer, mathematician, freemason, and political leader. OM (1763–) of the Paris AcS: 450.

Bärmann, Georg Friedrich (1717–69). German physician, natural philosopher, mathematician. Professor of mathematics at the University of Wittenberg and then the University of Leipzig; member of the German Society in Leipzig, specialist in algebra: 386, 388.

Battier, Reinhard (1724–79). Swiss physician, natural philosopher, and mathematician; cousin of Euler. OM of the Berlin AcS (1748–), and a teacher and OM (late 1748–1750) of the Paris AcS: 269, 307, 329.

Baudan, Charles de (d. 1756). Officer in the Russian military, 1729–41, married Mademoiselle Mirabel in 1742, named general quartermaster of the Russian Army from 1748: 187.

Bauer, General Fedor Villimovich (Friedrich Wilhelm) (1731–83). Swedish-born military engineer. Specialist in cartography and hydraulic engineering technology, head of the general staff of the Russian military; designed and built water supply and sewage systems in Moscow and Saint Petersburg: 509.

Baumgarten, Alexander Gottlieb (1714–62). German philosopher, Wolffian, and educator; professor at the University of Frankfurt an der Oder. He coined the term *aesthetics* and made it a distinct field of philosophy: 465.

Bayer, Gottlieb (Theophilus) Siegfried (1693–1738). Prussian sinologist, classical scholar, philologist, and natural historian. Professor of ancient Greek and Roman history and OM of the Pb AcS (1725–37): 49, 65, 110, 118.

Bayle, Pierre (1647–1706). French philosopher, writer, Huguenot refugee. Participant in the debate over faith and reason, author of *Historical and Critical Dictionary* (1695–97, 1702); forerunner of the Encyclopedists. Voltaire called him "the greatest master of the art of reasoning": 55, 56, 219, 341, 428.

Beausobre, Louis Isaac de (1730–83). German philosopher and political economist of French Huguenot descent. Member of the Berlin Consistory, OM of the Berlin AcS (1755). Adopted by Frederick II: 408, 439.

Beck, Johann Rudolf (1657–1726). Professor of logic (1695–1711) and then physics at the University of Basel: 32.

Beckenstein, Johann Simon (1684–1742). Baltic German jurist, heraldist, professor of jurisprudence and OM of the Pb AcS (1725–) who wrote on the use of weaponry: 51.

Béguelin, Nicolas de (1714–89). Swiss lawyer, mathematician, and writer, court master for Prince Frederick William II, Prussian subgovernor, commentator on Gottfried Wilhelm Leibniz, contributed to studies on probability and on meteorology, inspector of the French gymnasiums, OM and a director of the philosophy department of the Berlin AcS (1747–): 421, 510, 516, 518.

Belidor, Bernard Forest de (1698–1761). Spanish-born military and civil engineer in France. Professor at the French military school La Frére, inspector of artillery; wrote on engineering mechanics; mathematics, including integral calculus; artillery; ballistics. Author of the classic *L'architecture hydraulique* (1737–53, 4 vols.); FRS (1726–): 511.

Bell, Karl Joseph von (late eighteenth century). Prussian immigrant to Russia, major and chief quartermaster of the Russian Army. Became Russian noble; husband of Katharina Helene Bell): 523.

Bell, Katharina Helene (Elene), née Euler (1741–81). Leonhard Euler's daughter, wife of Karl Joseph von Bell: 189, 523.

Bering, Vitus Jonassen (1681–1741). Danish naval officer and explorer. In the Russian Navy (1704–) and captain (1730–), led the First and Second Kamchatka Expeditions (1725–30, 1733–43) that reached and studied what would later become named the Bering Strait; died there: 110, 111, 280.

Berkeley, George (1684–1753). Irish cleric and philosopher, bishop, critic of the foundations of calculus: 128, 267.

Berndisz (eighteenth century) Russian official: 15.

Bernoulli, Christoph (1782–1863). Swiss natural scientist and pedagogue: 15, 465.

Bernoulli, Daniel (1700–1782). Swiss natural philosopher, mathematician, and physician. OM and professor of physics (1725–31) and of mathematics (1731–33) at the Pb AcS; MD (1733), professor of anatomy and botany (1733–43), physiology (1743–50), and physics (1750–) at the University of Basel, dean of the medical faculty several times, and rector. Winner of ten Prix de Paris, FM (1733–) of the Pb AcS, of the Berlin AcS (1746–), of the Paris AcS (1748–), FM as FRS (1750–): 16, 25–27, 32, 37, 38, 49, 50, 54, 60–66, 68–70, 72, 77, 79, 83–86, 94, 96–98, 105, 109, 110, 113, 116, 123, 131, 145, 147, 152, 159, 160, 161, 162, 165, 167, 168, 172, 191, 195, 203, 205, 207, 211, 221, 226, 227, 230, 232, 235, 236, 245, 254, 255, 265, 277, 283, 285, 287, 292, 304, 305, 312, 329, 334, 335, 347, 356, 363, 380, 381, 386, 391, 392, 395, 399, 413, 430, 470, 480, 482, 494, 496, 506, 508, 511, 512, 518, 535.

Bernoulli, Jakob (Jacques) (1655–1705). Lecturer on experimental physics (1683–87), professor of mathematics (1687–) and rector (1701–) at the University of Basel; FM of Paris AcS (1699–) and Berlin AcS (1702–): 10, 23, 24, 49, 58, 77, 78, 96, 98, 100, 105, 121, 122, 133, 226, 492.

Bernoulli, Johann I (Jean I) (1667–1748). Swiss professor of mathematics and physics at the University of Groningen (1695–1705). Professor of mathematics at the University of Basel (1705–46), and several times dean of the philosophy faculty and rector. FM of the Paris AcS (1699–), Berlin AcS (1701–), Pb AcS (1725–), and the Institute of Bologna (1724–); FM as FRS (1712–): 10, 17–19, 21, 22, 24, 25, 28, 29, 31, 38, 56, 58, 60, 63, 73, 74, 76–78, 82, 87, 93, 95, 105, 122, 123, 126, 130, 132, 146, 150, 152, 170, 172, 173, 179, 183, 194, 195, 202, 226, 232, 235, 241, 260, 263, 285, 286, 312, 335, 347, 350, 351, 355, 366, 371, 392, 399, 431, 505, 514, 518, 535.

Bernoulli, Johann II (Jean II) (1710–90). Swiss mathematician. Professor of mathematics and eloquence at the University of Basel (1743–48), youngest son of Johann I Bernoulli, whom he succeeded as professor of mathematics at the University of Basel. FM of the Berlin AcS (1746) and the Paris AcS (1782): 82, 89, 172, 173, 245, 264, 278, 285, 329, 340, 343, 350, 351, 362, 363, 417, 518.

Bernoulli, Johann III (1744–1807). Swiss mathematician and astronomer, son of Johann II Bernoulli. Royal astronomer in Berlin known for his travels; OM of the Berlin AcS (1764–) and director of its mathematical department (1767–); FM of Pb AcS (1777–): 417, 437, 477, 478, 485, 494, 518–520.

Bernoulli, Nikolaus I (Niklaus) (1687–1759). Basel mathematician and painter. Professor of mathematics at the University of Padua (1716–19); professor of logic at the University of Basel (1722–) and several times dean of the law faculty and rector; FM of the Berlin AcS (1713–) and the Bologna Academy of Sciences (1724–); FM as FRS (1714–): 78.

Bernoulli, Nikolaus II (1695–1726). Swiss mathematician, son of Johann I Bernoulli. Professor of law at the University of Bern (1723–25); OM of the Pb AcS (1724–): 10, 25, 27, 49, 50, 61, 62, 66, 96, 129, 169, 482, 539.

Bertrand, Louis (1731–1812). Swiss mathematician and geologist, professor of mathematics at the Akademie Gauf in Geneva and its rector (1783–); FM of the Berlin AcS (1754–): 189, 393, 394, 411, 412, 414, 423.

Bevis, John (1695–1771). English physician and amateur astronomer. Discoverer of what is today known as the Crab Nebula; prepared the *Uranographia britannica* (1750); FRS (1765–) and later the society's foreign secretary: 315.

Bilfinger (Bülfinger), Georg Bernhardt (1693–1750). German natural philosopher and statesman. OM and professor of physics and mechanics (1725–30), then FM (1731–) at the Pb AcS; professor of philosophy and physics at the University of Tübingen (1731–): 31, 38, 48, 49, 51, 52, 54, 60–65, 70, 82, 83, 147, 195.

Biron (Biren or Bühren), Ernst Johann von (1690–1772). German statesman in Russia, duke of Courland, favorite of Anna Ivanovna, responsible for oppressive *Bironovschina* during her reign: 82, 90, 171.

Black, Joseph (1728–99). Scottish physician and chemist who rediscovered "fixed air" (carbon dioxide) and conducted research in modern quantitative chemistry. Professor of medicine at the University of Glasgow and later the University of Edinburgh; fellow of the Royal Society of Edinburgh (1783–): 527, 532.

Blaeu, Johan Williamson (1596–1673). Dutch cartographer and publisher of maps; son of Willem Janszoon Blaeu: 45, 71.

Blaeu, Willem Janszoon (1571–1638). Dutch cartographer and atlas maker, proprietor of notable cartographic publishing firm. With his son Johan Blaeu prepared the *Atlas novus* (first ed. 1635, 2 vols.) that supported Copernican astronomy. By 1662 it had grown to eleven or twelve volumes: 45, 71.

Blondel, Nicholas François (1618–86). Belgian soldier, military and civil engineer, and diplomat who wrote on fortifications and architecture. Associate geometer of the Paris AcS: 210.

Blumentrost, Laurentius (Lavrentii) (1692–1755). Royal physician to Peter I and Anna Ivanovna of Russia. First president of the Pb AcS (1725–33), chief physician of the Moscow military hospital, curator at the University of Moscow (1754–): 27, 38, 44–46, 48, 50, 53, 60, 62, 70, 84.

Blumler (eighteenth century). German telescope maker: 333.

Boerhaave, Hermann (1668–1738). Dutch physician, botanist, and chemist. *Praeceptor Europa* (teacher of Europe); reformed medical education as professor at the University of Leiden; FM of Paris AcS (1728–) and FM as FRS (1730–): 400.

Bolingbroke, Henry Saint John, First Viscount (1678–1751). English statesman, orator, political philosopher, writer, and Tory politician: 266.

Bonnet, Charles (1720–93). Swiss naturalist, biologist, and philosopher. Lawyer near Geneva, member of the Geneva government; conducted studies of insects; CM of the Paris AcS (1740–), FM as FRS (1743–); FM of the Royal Swedish Academy of Sciences (1753–) and the Royal Danish Academy of Sciences and Letters (1769–): 482.

Borcke, Kaspar Wilhelm von (1704–47). German diplomat. Cabinet minister in Berlin; scholar, translator of the works of William Shakespeare: 183, 195, 222, 244.

Borelli, Giovanni Alfonso (1608–79). Renaissance Italian physiologist, natural philosopher, and mathematician: 61.

Bouguer, Pierre (1698–1758). French geodesist, hydrographer, natural philosopher, astronomer, and expert on naval architecture. A pioneer in the field of photometry, professor of hydrography; OM of Paris AcS (1735–), member of the Peru Expedition, with Charles-Marie de la Condamine, that was sent by the Paris Academy in 1735 to measure an arc of meridian. Competitor of Euler, won the Prix de Paris three times (1727, 1729, and 1731) over Euler; FM as FRS (1750–): 31, 32, 141, 143, 161, 208, 217, 269, 303.

Bousquet, Marcus-Michael (fl. 1740s). Bookseller and publisher in Lausanne and Geneva: 202–204, 224, 277, 371.

Boyle, Robert (1627–91). Anglo-Irish natural philosopher and chemist. British court official; FM as FRS (1662–): 44, 176–178, 180–182, 185.

Bradley, James (1693–1762). English astronomer. Discovered the aberration of starlight (1725–28); Savilian Professor at the University of Oxford (1721–); British royal astronomer (1742–), FRS (1718); FM of the Berlin AcS (1746–); FM of the Bologna Academy of Sciences, the Paris AcS (1748–), and the Pb AcS (1754–): 208, 255, 294, 307.

Brahe, Tycho (1546–1601), Danish astronomer and high noble, made comprehensive astronomical and planetary observations, especially of Mars, known for their accuracy at Uraniborg on the island of Hven. Designed and built superior astronomical instruments, advocated heliocentric theory but not the Copernican system, accepted Habsburg Emperor Rudolph II's invitation to Prague in 1597, settled there as Imperial Mathematician in 1599, Johannes Kepler was his assistant, 1600–1601: 53.

Brand, Bernhard (1523–94). Basel lawyer and high guild master: 10.

Brandes, Karl Philipp (1720–76). German physician and chemist, FM and then OM of the Berlin AcS: 375, 419.

Brandmüller, Johann and Ludwig (d. 1751). Basel booksellers. Sent books to Euler to sell in Saint Petersburg: 213.

Braun, Joseph Adam (1712–68). German philosopher and grammarian. Professor of philosophy and OM of the Pb AcS (1748–), elected FM of the Berlin AcS (1761), but Frederick II would not confirm him: 255, 281.

Brauser, Benjamin (mid-eighteenth century). Teacher, mathematician in Halle and Berlin; correspondent of Euler (1746–47): 281.

Brevern, Karl von (1704–44). Diplomat and statesman, president of the Pb AcS (1740–41), conference minister for empress Elizabeth Petrovna: 174.
Brius, Iakov Vilimovich (Jacob Daniel Bruce) (1670–1735). Russian general field marshal, statesman, and partisan of the Enlightenment: 44, 72.
Brouncker, William, Second Viscount (1620–84). English mathematician. A founder and first president of the Royal Society of London (1663–77); FRS (1663–): 95, 134.
Brucker, Heinrich (mid-eighteenth century). Swiss pastor of Saint Peter's Church in Basel and Euler's uncle. Delivered funeral oration for Johann I Bernoulli: 285.
Brucker, Johann Georg (mid-eighteenth century). Teacher at the gymnasium of the Pb AcS (1735–37); later returned to Kiel: 162.
Brucker, Johann Heinrich (1636–1702). Pastor in Kilchberg, 1672–90, and Basel hospital vicar. Euler's maternal grandfather: 10.
Brucker-Faber, Maria Magdalena (Magdalene) (1652–1744). Mother of Katharina Euler: 14, 222.
Bruckner (Brucker), Isaac (1686–1762). Basel mechanic and geographer; master for mathematical instruments at the Pb AcS (1733–48): 39, 66, 301, 316.
Bruckner, Johann Georg (eighteenth century). Artist: 162.
Brumundt, Dietrich Siegvert (eighteenth century). Royal Danish mathematician in Copenhagen: 322.
Buffon, Georges-Louis Leclerc, Comte de (1707–88). French naturalist, mathematician, cosmologist, and encyclopedist. Intendant of the royal gardens in Paris, author of *Histoire naturelle* (1749–88, 366 vols.); OM of the Paris AcS (1734–) and French Academy (1753–); FM of the Berlin AcS and Pb AcS; FRS; and foreign member of the American Academy of Arts and Sciences (1782–): 150, 277, 515, 525.
Bülfinger, Georg. *See* Bilfinger, Georg Bernhardt.
Burckhardt, Bonifacius (1656–1708). Swiss pastor in Riehen: 11.
Burckhardt, Johannes (Johann Jakob) (1691–1743). Swiss pastor in Kleinhüningen, and (1732–) in Oltingen. Euler's tutor (1715–20): 15, 17, 19, 195.
Burja, Abel (mid-eighteenth century). Pastor and mathematician in Saint Petersburg. Was with Euler at his house on Euler's last two days: 530.
Büsching, Anton. Rector of the Gray Kloster Gymnasium in Berlin (1767–): 10.
Buxtorf, Johannes (1564–1629). Swiss Hebraist, a scholar ancestor in the Brucker line: 6.
Calvin, John (1509–64). French reformer and theologian: 9.
Capito, Wolfgang (1478–1541). German religious reformer: 32.
Carathéodory, Constantin (1873–1950). Greek mathematician. Contributed to measure theory, the theory of real functions, the calculus of variations, and axiomatic thermodynamics; professor of mathematics at the University of Göttingen (1913–19) and the University of Berlin (1919–20); visiting professor at the American Mathematical Society and Harvard University (1928–29); professor at the University of Munich (1924–38), OM (1919–1920) and FM (1920–) of the Berlin AcS: 225.
Carlyle, Thomas (1795–1881). Scottish philosopher, historian, essayist, satirist, and teacher. Known as the Sage of Chelsea; wrote the masterpiece *Friedrich II of Prussia* (1858–65; 6 vols.): 598.
Cartheuser, Johann Friedrich (1704–77). German physician, chemist, and naturalist. Professor of chemistry, pharmacy, and *materia medica* at the University of Frankfurt an der Oder; FM of the Berlin AcS (1758–): 422.
Cassini, Gian Domenico (1625–1712). Italian-French astronomer, geodesist, mathematician, and engineer. Professor at the University of Bologna; OM of Paris AcS (1668–), director of the Paris Observatory (1669–1712): 142, 144, 166, 208.
Cassini, Jacques II (1677–1756). French astronomer. Succeeded his father Gian Domenico Cassini as director of Paris Observatory (1712–); OM of the Paris AcS (1694–), FRS (1696–): 21, 143, 144, 161, 167, 184, 217, 229, 258, 292, 516.

Castillon, Frédéric Adolf Maximilian Gustav von (1747–1814). Son of Jean de Castillon; wrote his father's eulogy. Even without mathematics credentials, appointed over Johann Albrecht Euler as professor of mathematics and philosophy at the new knights' academy in Berlin (1766), OM of the Berlin AcS (1786–) and director of its philosophy department (1800–): 443.

Castillon, Jean de (Johann Francesco, Gian Francesco Mauro Melchoire Salvemini) (1708–91). Italian mathematician and astronomer. Translator of John Locke and Isaac Newton; editor, active early in Swiss cities; then professor of mathematics and philosophy at the University of Utrecht (1755–63); professor of mathematics at the artillery school in Berlin; FM as FRS; FM (1764–), OM (1764), and director of the mathematics department of the Berlin AcS (from 1787): 224, 301, 438, 439.

Cataldi, Pietro (1548–1626). Italian mathematician and astronomer, critic of Euclid's fifth postulate, teacher at the University of Bologna. Working mainly with Princess Ekaterina Romanovna Dashkova, he improved the finances, organization, and foreign contacts of the Pb AcS: 134.

Catherine I (1684–1727; r. 1725–27). Empress of Russia; successor to Peter I: 48, 68.

Catherine II (Catherine the Great; Yekaterina Alexeevnja 1729–96; r. 1762–96). Empress of Russia: 319, 426, 429, 433, 446, 447, 450–455, 458, 468, 471, 486, 487, 495, 498, 499, 508–511, 520, 526, 527, 529.

Catt, Henri Alexandre de (1725–95). Reader for Frederick II, OM of the Berlin AcS(1760–); recruited Euler for the Pb AcS: 367, 431, 439, 447, 448.

Cauchy, Augustin-Louis (1789–1857). French mathematician. Professor of mathematics at the École polytechnique (1815–30); OM of the Paris AcS (1816–). In exile (1830–38); made a baron (1835); FM of the Pb AcS (1832–), the Berlin AcS (1836–), and FM as FRS (1832–). Pioneer in mathematical analysis, general theory of convergence, and rigorous methods; a pioneer in the mathematical theory of elasticity: 392.

Celsius, Anders (1701–44). Swedish astronomer. Professor of astronomy at Uppsala University, built the Uppsala Astronomical Observatory, devised temperature unit of measurement that today bears his name: 144.

Chappe d'Auteroche, Jean-Baptiste (1722–69). French astronomer. Assistant astronomer at the Royal Observatory in Paris; saw the transit of Venus (1761); worked on longitude measurement; OM of the Paris AcS (1759–): 480.

Charles II (1630–85; reigned 1649–85). King of England, Scotland, and Ireland. Reign known as the Restoration; founder of the Royal Society of London: 42.

Charles XII (1682–1718; r. 1697–1718). King of Sweden. King during the Great Northern War against Peter I: 41.

Châtelet, Gabrielle Émilie, Marquise du (b. Gabrielle Émilie Le Tonnelier de Breteul) (1706–49). French mathematician, natural philosopher, and author. Mistress of Voltaire, translator of Isaac Newton's *Principia Mathematica* (published posthumously, 1759): 144, 147, 148, 150, 176, 179, 183, 197, 273, 276, 293, 299, 313, 357, 373.

Chauchot, M. (d. 1755). French naval engineer. Assistant constructor at Brest; won the Prix de Paris (1755): 406.

Chernyshev, Count Zachar' Grigorievich (1726–97). Russian field marshal and general admiral. Supported the coup by Catherine II (1762) and her reign; president of the war college: 425.

Chevalier, Vincent Jacques Louis (1771–1841). French optician. Designer of an achromatic microscope by following Euler (1825–26), and the first to employ it: 472.

Cicero, Marcus Tullius (106–43 B.C.). Roman orator, lawyer, statesman, politician, and scholar: 9, 223, 287, 534.

Clairaut, Alexis Claude (1713–65). French astronomer, physicist, and mathematician. Worked to advance lunar theory; adjunct (1731–38) and OM (1738–) of the Paris AcS; FM of the Berlin AcS (1744–) and Pb AcS (1754–); FM as FRS (1737–). Also

Darbès, Joseph Frédéric-Auguste (1747–1819). German painter who created portrait (oil on canvas) of Euler (1778): 518–520.

D'Argens, Jean Baptiste de Boyer, Marquis (1704–71). French writer and philosopher, Catholic aristocrat. Resided at the court of Frederick II; royal chamberlain from 1742, spread skeptical Enlightenment ideas; helped found the new Royal Prussian AcS; OM and director of the historical-philological or belles-lettres department of the Berlin AcS (1750–); friend of Euler: 188, 331, 419, 420.

Darwin, Charles Robert (1809–82). Eminent English naturalist. Devised the theory of evolution by natural selection: 19.

Dashkova, Princess Ekaterina Romanovna, née Vorontsova (1743–1810). Russian friend of Catherine II and often called Catherine the Little. Director of the Pb AcS, improved its finances and operations and had a new building constructed; escorted by Euler at her inauguration as director of the Pb AcS (1783–): 499, 515, 520, 526–528, 532, 534.

Dee, John (1527–1609). Welsh mathematician, natural philosopher, astronomer, navigator, astrologer, and occultist. Attached to the royal court and consultant to Queen Elizabeth I; supported search for the Northwest Passage, recommended adoption of the Gregorian calendar, and edited first English translation of Euclid's *Elements*: 247.

Delen, Charlotte, née Euler (1744–80). One of Euler's daughters: 435, 449, 523.

Delen (or Dehlen), Johann Jakob, Baron van (1743–86). Dutch noble, Prussian military officer, husband of Charlotte Delen. Traveled with the Euler family to Saint Petersburg (1766), and took his family to his estate in Hückelhoven near Aachen (1770): 433, 435, 449.

Delisle, Joseph-Nicolas (1688–1768). French astronomer. Student (1714–16), adjunct (1716–40), and OM (1741–) of the Paris AcS; professor of astronomy and OM of the Pb AcS (1726–47) and HM, 1748; FM as FRS (1724): 50–52, 56, 57, 62, 72, 80, 81, 84–87, 94, 101, 103, 110, 111, 115, 117, 118, 143, 161, 185, 198, 201, 208, 209, 216, 217, 222, 237, 253, 254, 256, 281, 282, 300, 305, 320, 331, 383, 446, 465, 479, 480, 517, 535.

Delisle de la Croyère, Louis (1688–1741). Brother of Joseph-Nicholas, assoc. prof. of astronomy at the Pb AcS (1727–): 110, 111, 169.

Descartes, René du Perron (1596–1650). Major French philosopher, mathematician, natural philosopher, and physiologist who worked in France, Holland, and Sweden. Inventor of analytical geometry: 1, 19, 20, 23, 53, 56, 57, 73, 81, 101, 113, 149, 155, 157, 271, 285, 325, 326, 328, 339, 354, 360, 428, 461–463, 468–470.

Deslandes, Pierre François. French organ builder based in Paris; taught François Thierry: 155.

Diderot, Dennis (1713–84). French *philosophe*, man of letters, art critic, and writer. Chief editor of the *Encyclopédie*, the principal multivolume work of the eighteenth-century Enlightenment. OM of the Paris AcS; HM of Pb AcS (1773–). Had dispute with Franz Ulrich Theodor Aepinas, but not Euler, in Saint Petersburg: x, 26, 239, 240, 340, 362, 374, 391, 438, 461, 468, 499, 527.

Dido (fl. 814 B.C.). Legendary founder and first queen of Carthage. She perhaps fell in love with the Trojan Aeneas and committed suicide when he left her: 76.

Dietrich, François (1731–1803). Pastor in Benken near Basel. Nephew of Johannes Dietrich, tutor to the Euler children: 385.

Dietrich, Johann (d. 1758). Swiss mechanist and instrument maker in Basel. Known for his research on magnets: 392, 413.

Diophantus of Alexandria (fl. 250). Hellenistic Greek mathematician, especially in algebra; author of *Arithmetica*: 101, 102, 104, 137.

Dirichlet, Peter Gustav Lejeune (1805–59). German mathematician. Greatly advanced number theory and mathematical analysis; professor of mathematics at the University of Berlin and Carl Friedrich Gauss's successor at the University of Göttingen: 498.

Dohna, Count Albrecht Christoph zu (1698–1752). Grand court chamberlain of the Prussian queen. Private secretary to Frederick II; HM of the Berlin AcS: 249.
Dolgorukij, Vladimir Sergeevič (1717–83). Russian prince and diplomat. Russian envoy to Prussia in Berlin (1762–83): 430, 442, 447.
Dollond, John (1706–61). English optician and mechanist. Improved optical instruments, reduced chromatic aberration in telescopes with two different lenses, disputed Euler on the method with crown and flint glass; optician to the king, proposed for FRS (1761): 278, 308, 349, 383, 412, 413, 415, 422, 428, 429, 480.
Domaschnev, Sergei Gerassimowitsch (1743–95). Director of the Pb AcS (1774–Jan. 1783): 508, 509, 511, 515, 516, 526.
Dorville, J. L. (fl. 1760s). Prussian state minister. Correspondent of Henri Alexandre de Catt and of Euler from 1765 to 1766: 447.
Dryden, John (1631–1700). English poet, dramatist, and literary critic: 76.
Dürer, Albrecht (1471–1528). German painter, engraver, printmaker, and graphic artist. Studied geometry and perspective, proportions, and magic squares: 521.
Duvernoy (Duvernois), Johann Georg (1691–1759). German anatomist, surgeon, and zoologist. MD, University of Basel (1710); professor and OM (1725–41) and FM (1741–) of the Pb AcS: 51, 61, 70.
Ehler, Karl Leonhard Gottlieb (1685–1753). German amateur mathematician, mayor of Danzig, correspondent of Euler: 130, 146.
Einstein, Albert (1879–1955). German American physicist and humanist; Nobel Prize laureate: 19.
Eisenstein, Ferdinand Gotthold Max (1823–52). German mathematician. Conducted major work in number theory; professor of mathematics at the University of Berlin; member of the Göttingen Academy and OM of the Berlin AcS (1852–): 498.
Elizabeth (Elizaveta) Petrovna (1709–62). Daughter of Peter I; grand duchess, then empress of Russia (1741–): 187, 253, 280, 410, 426, 427, 429, 430, 452.
Eller, Johann Theodor (1689–1760). German royal physician in Berlin, surgeon, natural philosopher, and chemist. Director of the Collegium medico-chirurgicum in Berlin (1725–); OM and director of the physics department of the Berlin AcS (1735–60): 197, 243, 246, 248, 249, 251, 262.
Elsner, Jakob (d. 1750). German pastor, philologist, and consistory official in Berlin: 243.
Eneström, Gustav (Gustaff) (1852–1923). Swedish mathematician, historian of mathematics, and bibliographer. He is responsible for the Eneström system that numbered Euler's writings: ix–x.
Erasmus, Desiderius (b. Gerrit Gerritszoon) (ca. 1467–1536). Dutch humanist and theologian: 6, 7, 15.
Errard, Jean (ca. 1554–1610). French mathematician and military engineer who, wrote on mathematical instruments. Called the Father of French Fortifications; precursor to Sébastien Le Prestre Vauban: 210.
Euclid (ca. 330 to ca. 260 B.C.). Leading mathematician of Greco-Roman antiquity, known for his *Elements* (13 vols.): 103, 271, 354, 472, 478.
Euler, Albertina Benedict (1766–1829). Second daughter of Johann Albrecht Euler. Married Nicholas Fuss, 1784: 496.
Euler, Anna Margaretha, née Hugelschoffer (d. 1750). Wife of Johann Heinrich Euler: 161.
Euler, Anna Maria, née Gassner (1642–1712). Leonhard Euler's paternal grandmother: 11.
Euler, Christoph (1743–1808). Son of Leonhard Euler. Physician in Berlin, especially to poor French Huguenots; in Saint Petersburg, colonel and then major general in the Russian artillery: 189, 415, 424, 427, 445, 447, 449, 452, 471, 484, 497.
Euler, Ertmuth Louise (April 13–August 9, 1749). Daughter of Leonhard Euler: 316.
Euler, Hans-Georg (Jörg) (1573–1663). Leonhard Euler's great-great grandfather: 8.
Euler, Hans-Georg (d. ca. 1595). One of four sons of Hans-Georg Euler (1573–1663); died soon after birth: 8.

Folkes, Martin (1690–1754). English antiquarian, numismatist, archaeologist, mathematician, and astronomer. FRS (1713–) and the Royal Society's president (1741–), FM of the Paris AcS (1742–) and Berlin AcS (1746–): 255, 269, 315.

Fontaine, Alexis (des Bertins) (1704–71). French mathematician and natural philosopher, royal notary. Adjunct (1733–42) and OM (1742–) geometer of the Paris AcS; FM of the Berlin AcS (1747–). Solved brachistochrone problem, conducted research in dynamics analogous to that of Pierre Maupertuis, who rejected it; contributed to the foundations of the calculus of several variables: 160, 169, 436, 475.

Fontenelle, Bernard le Bovier de (1657–1757). French man of science and author. Member of the French Academy; OM of the Paris AcS (1691–) and its standing secretary (1697–1740); helped secure the acceptance of Copernican astronomy: 43, 52, 63, 71, 86, 95, 99, 179, 195, 300, 356, 357, 417, 463.

Formey, Jean-Henri-Samuel (1711–97) Theologian, philosopher, and leading Huguenot (Calvinist) pastor in Berlin. OM (1744–), historiographer (1745–), and perpetual secretary (1748–) of the Berlin AcS; FM of the Pb AcS (1748–); FM as FRS (1750–), FM of the Göttingen Academy of Science and the Scholars' Society of Greifswald; partisan of the philosophy of Christian Wolff: 130, 170, 197, 218, 223, 243, 245, 247, 248, 262, 264, 265, 279, 284, 296, 299, 300, 309, 324, 342, 344, 371, 386, 403, 408, 419, 440, 456, 460, 476, 492, 495, 513, 533 , 534.

Fourier, (Jean Baptiste) Joseph (1768–1830). French mathematician and physicist; pioneered the Fourier series: 98.

Francheville, Joseph du Fresne de (1704–81). Invited by Frederick II to Berlin. OM of the Berlin AcS; coauthor with Voltaire of the sixth volume of *Le Siècle de Louis XIV* (1752): 331, 412.

Francke, August Hermann (1663–1727). German Lutheran leader, educator, biblical scholar, and Pietist pastor in Halle: 45.

Franklin, Benjamin (1706–90). American writer, diplomat, natural philosopher, and satirist. A signer of the Declaration of Independence, and printer in Philadelphia. First postmaster general of the United States, commissioner (ambassador) to France, and a founder and first president of the American Philosophical Society; FM as FRS, 1756, and of Pb AcS, 1789. A founder of the University of Pennsylvania; conducted major research on electricity and wrote *Experiments and Observations on Electricity* (1751): 382, 422, 525.

Frederick II (Friedrich II; Frederick the Great) (1712–86, reigned 1740–86). King of Prussia, talented military commander. Reorganized and strengthened the Berlin AcS (1744–45; president, 1764–); HM of the Pb AcS (1776–): xi, 3, 160, 164, 170, 172–174, 176, 178, 180, 184, 186, 191, 192, 195, 198, 215, 221–223, 237, 238, 242–244, 246, 266, 268, 282, 297, 308–310, 312, 318, 330, 331, 334, 340, 350,, 357, 358, 365, 367, 374, 375, 386, 401, 408, 409–412, 414, 415, 417–419, 419, 426–432, 439–441, 445, 448–450, 474, 511, 512, 515, 532.

Frederick William I (Friedrich Wilhelm I) (1688–1740). King of Prussia and elector of Brandenburg (1713–40). Built up the army, made economic reforms, and centralized the administration; father of Frederick II: 46, 177, 179.

Friederike Charlotte Leopoldine Louise of Brandenburg Schwedt (1745–1808). Daughter of Margrave Friedrich Heinrich. Princess and the last abbess of Herford Abbey. When she was a teenager Euler taught her natural philosophy, mainly via correspondence: 417, 418, 461.

Friedrich Heinrich (Frederick Henry) (1709–88). Margrave of Brandenburg-Schwedt and a Prussian prince. Took his daughters to meetings of the Berlin AcS (1748–65); invited Euler to play music and to teach his daughter Friederike Charlotte Leopoldine Louise: 417, 426, 460, 495, 523.

Frisi, Paolo (1728–84). Imperial Austrian and Italian mathematician, natural philosopher, and astronomer. Member of Barnabite religious order; conducted major

work on hydraulics, prepared an engineering workbook, and wrote commentaries on Galileo Galilei and Isaac Newton: 405.
Frisschmann, Martin (d. before 1582). Basel shoemaker: 8.
Froben, Johannes (ca. 1460–1527). Basel printer and publisher: 7.
Fuss, Nicholas (Niklaus, Nicolaus) (1755–1826). Mathematician from Basel. Euler's assistant, adjunct (1776–83), OM (1783–), and standing secretary (1793–) of the Pb AcS; FM of the Berlin AcS (1793–); of the Stockholm AcS, (1808–), the Munich AcS, and others; author of a major Euler eulogy at the Pb AcS: 12, 25, 115, 152, 163, 211, 219, 268, 488, 496, 497, 508, 511, 514, 516, 519, 528, 530, 532.
Fuss, Paul Heinrich von (1798–1855). Mathematician. Adjunct (1818–23), OM (1823–), and standing secretary (1826–) of the Pb AcS. Son of Nicholas Fuss: xi, 402, 519, 520.
Galileo Galilei (1564–1642). Italian physicist, natural philosopher, mathematician, astronomer, and methodologist. Supporter of Copernican astronomy, professor at the Universities of Pisa and Padua (1592–1610), then with an appointment as mathematician, philosopher, and engineer to the Medici in Florence. Member of the Lincean Academy: 20, 23, 97, 126, 155, 165, 210, 252, 268, 278, 406, 443, 532.
Gassendi, Pierre (1592–1655). French philosopher, geometer, natural philosopher, astronomer, and priest. Taught mathematics at the College Royal at Paris; attempted to moderate skepticism and make Epicurean atomism acceptable to Christianity: 428.
Gauss, Carl Friedrich (1777–1855). Leading German mathematician, physicist, and astronomer. Author of *Disquisitiones Arithmeticae* (1801) on number theory, one of the first books to include non-Euclidean geometry; professor and director of the observatory at the University of Göttingen (1807–), FM of the Berlin AcS (1810–), Pb AcS (1824–), and Paris AcS (1820–); and FM as FRS (1804–): ix, 14, 26, 78, 134, 204, 345, 392, 490, 506.
Geer, Charles de (1720–78). Swedish entomologist, naturalist, and wealthy industrialist who wrote eight volumes on insects. OM of the Royal Swedish Academy of Sciences (1739–), CM of the Paris AcS (1748–): 502.
Gellert, Christian Fürchtegott (1715–69). German poet, novelist, and moralist who wrote verse fables. Major representative of the German Enlightenment preceding Gotthold Ephraim Lessing; professor at the University of Leipzig (1751–): 420.
George I (George Louis) (1660–1727). Elector of Hanover (1698–). First Hanoverian king of Great Britain (r. 1714–27): 187.
Gerritszoon, Gerrit. *See* Erasmus, Desiderius.
Gibbon, Edward (1737–94). English rationalist historian and parliamentarian. Author of *The Decline and Fall of the Roman Empire* (1776–88, 6 vols.): 56.
Gleditsch, Johann Gottlieb (1714–86). Physician, and director of the Botanic Gardens in Berlin. OM of the Berlin AcS (1744–), and FM of the Pb AcS (1776–): 242, 281, 348, 419, 420, 431, 515.
Gmelin, Johann Georg (1709–55). German physician, botanist, chemist, naturalist, and geographer. Adjunct (1727–31), and then professor of chemistry and natural history and OM of the Pb AcS (1731–47); participant in the Second Kamchatka Expedition (1733–43). Professor of medicine at the University of Tübingen (1747–) and director of its botanic garden (1751–). FM of the Royal Swedish Academy of Sciences (1749–): 70, 81, 84, 110, 111, 331, 332, 368, 484, 494, 502.
Goethe, Johann Wolfgang von (1749–1832). Leading German writer, artist, politician, and literary celebrity well known for his two-part drama *Faust* (1808–32). Manager of a theater and government minister in Weimar: 469, 519.
Goldbach, Christian (1690–1764). German mathematician and diplomat. OM of the Pb AcS (1725–42), its secretary (1725–28), and then HM (1742–); active in the Collegium of the Foreign Affairs Council in Moscow (1742–) and privy counselor (1760–); correspondent with Euler on number theory: 50, 66, 69, 77, 79, 80, 89, 96,

101, 102, 104, 118, 132, 136, 160, 186, 187, 194, 202, 205–207, 212, 222, 233, 237, 244, 278, 310, 317, 321, 334, 335, 394, 395, 412, 416, 435, 438, 482, 492, 535.

Golovin, Mikhail Evseevič (1756–90). Russian mathematician. Adjunct for mathematics (1776–86) and HM (1786–) of the Pb AcS: 496, 497, 513, 514, 518.

Golovin, Nikolaj Fedorovich (1695–1745). Russian count, statesman, and admiral. President of the admiralty college (1733–) and later a senator: 214.

Golovkin, Count Aleksandr Gavrilovich (1688–1760). Russian diplomat and partisan of the Enlightenment. Envoy to Paris; Russian ambassador to Berlin (1711–27) and envoy to the Hague (1751–): 46, 47.

Gottsched, Johann Christoph (1700–1766). Major German author, critic, and playwright. Professor of poetry, logic, and metaphysics at the University of Leipzig; OM of the Berlin AcS (1729–); supporter of the Enlightenment: 151, 159, 251, 341.

Graffiny, Madame Françoise de (1695–1758). French writer and dramatist, author of *Lettres d'une Péruvienne* (1748). Important salon hostess; among those who attended her salon were Jean le Rond d'Alembert, Denis Diderot, Jean-Jacques Rousseau, and Voltaire: 183.

Grammatico, Nicaise (late seventeenth century–1736). Jesuit astronomer and teacher at the University of Freiburg. Died at Ratisbonne: 258.

Gravesande. *See* 'sGravesande, Willem Jacob.

Gregory (Gregorie), James (1638–1675). Scottish mathematician, astronomer, and optician in Edinburgh who improved the reflecting telescope: 95, 119, 120, 149.

Gregory (Gregorius) of Saint Vincent (1584–1667). Belgian Jesuit; pioneer of calculus and the origins of analytical geometry: 132.

Grimm, Peter (eighteenth century). Basler handyman for Euler in Saint Petersburg: 487.

Grischow, Augustin Nathanael (1726–60). German mathematician and astronomer. Director of the Berlin Observatory (1745–49); OM of the Berlin AcS (1749–); professor of optics at the Berlin Academy of Arts. Moved to Pb AcS as a student (1750–51), then OM and professor of astronomy (1751–) and conference secretary (1751–54): 216, 244, 304, 307, 325, 329, 349.

Gross, Christian Friedrich (ca. 1698–1742). German ethicist. Professor of moral philosophy and OM of the Pb AcS (1725–31); FM Pb AcS (1731–); house tutor for Andrei Ivanovich Ostermann (1728–). Diplomatic representative for Wolfenbüttel to the Russian court: 51, 65, 187.

Gsell, Dorothea Maria Henriette (1678–1743). Daughter of Swiss painter Maria-Sybilla Merian, second wife of Georg Gsell and mother of Salome Abigail Euler. Painter at the Pb AcS: 87, 88.

Gsell, Georg (1673–1740). Euler's father-in-law, from Saint Gallen. Artist at the Pb AcS (1726–31) and friend of Johann Daniel Schumacher: 87, 88.

Gsell, Katharina. *See* Euler, Katharina: 87, 88, 90.

Gsell, Marie Gertrud, née von Loen (early eighteenth century). Katharina Euler's mother: 513.

Gsell, Salome Abigail. *See* Euler, Salome Abigail: 513, 514.

Hadamard, Jacques (1865–1963). French mathematician who devised proof of prime number theorem: 93.

Haecks (late eighteenth century). Russian artillery captain who witnessed Euler's will in 1776: 514.

Hagemeister, Anna Charlotte Sophie (1734–1805). Wife of Johann Albrecht Euler: 424.

Haller, Albrecht von (1708–77). Swiss anatomist, physiologist, botanist, naturalist, and poet in Göttingen; MO, FM of the Berlin AcS (1749–) and of the Pb AcS (1776–): 33, 34, 51, 85, 394, 399–401.

Halley, Edmond (1656–1742). English astronomer, geophysicist, mathematician, and meteorologist. Second astronomer royal (1720–42), FRS (1678–1742) and secretary of the Royal Society of London (1713–21); FM of the Paris AcS (1729–).

Studied comets, including Halley's Comet, which is named for him: 34, 209, 305, 348–408, 428, 516.
Hamilton, William Rowan (1805–65). Irish physicist, mathematician, and astronomer. Discoverer of the algebra of quaternions and contributor in the field of optics. Professor of astronomy at Trinity College, Dublin; royal astronomer of Ireland at Dunsink Observatory; president of the Royal Irish Academy (1837–46): 2, 201.
Hardy, Godfrey Harold (1877–1947). English mathematician specializing in number theory and analysis. Savilian Professor of Geometry at the University of Oxford, visiting professor at Princeton University, and Sadlerian Professor at the University of Cambridge. President of the London Mathematical Society (1926–28, 1939–41) and FRS (1910–): 108.
Harrison, John (1693–1776). English carpenter, clockmaker, and horologist. Invented the first practical marine chronometer, making possible the accurate computation of longitude at sea; won the British parliamentary prize over Johann Tobias Mayer: 445, 466.
Haude, Ambrose (1690–1748). Bookseller and publisher who printed Euler's *Theoria motuum planetarum et cometarum* (1744). Codirector (with Johann Carl Spener) of the Haude and Spener publishing firm in Berlin: 229.
Hedlinger, Johann Karl (1691–1771). Swiss maker of medals who worked on heraldry. FM of the Berlin AcS (1744–): 192, 375.
Heinius, Johann Philipp (1668–1775). Philosopher. Rector of the gymnasium in Joachimsthal, director of the philosophy department of the Berlin AcS (1744–): 243, 249, 320, 344, 403, 439.
Heinsius, Gottfried Wilhelm (1709–69). German astronomer, geographer, and mathematician. OM, associate professor, then professor of astronomy at (1736–44) and FM (1744–) the Pb AcS; then a professor at the University of Leipzig (1744–): 117, 167, 169, 198, 208, 209, 213, 230, 234, 258, 305, 321, 368.
Helvétius, Claude Adrien (1715–71). French *philosophe*, controversialist, and *litterateur*. French farmer-general (1738–51), wrote *De l'esprit* (1758), which attacked morality and was burned. Invited to visit Berlin by Frederick II: 468.
Henninger, Johann Conrad (1696–1763). Private teacher at the academic gymnasium of the Pb AcS (1725–) and later for the prominent family of Prince Aleksandr Menshikov in Saint Petersburg. Private secretary for empress Anna Ivanovna (1740–); vice president of the Manufacture College (1752–60) and brother-in-law of Johann Daniel Schumacher: 213, 283.
Henzi, Samuel (d. 1749). Swiss member of the lower gentry and book dealer. Critical of aristocratic government, he was beheaded in Berne in 1749: 338, 343, 351, 352.
Hermann, Jakob (1668–1733). Basler mathematician and mechanician. Professor at the Universities of Padua (1707–13) and Frankfurt an der Oder (1713–24); OM (1725–31) and FM (1731–) of the Pb AcS, then professor of ethics at the University of Basel (1731–): 24, 35, 38, 48, 49, 51, 53, 54, 61, 62, 65–67, 69, 82, 83, 101, 105, 125, 126, 130, 146, 169, 211, 226, 312, 333, 337, 338, 343, 355, 373.
Herschel, (Frederick) William (Friedrich Wilhelm) (1738–1822). German-born British observational astronomer, composer, and technical expert. Discoverer of Uranus, its two major moons, two moons of Saturn, and infrared radiation; and improved the reflecting telescope. Director of the Bath Orchestra, and "King's astronomer" (not royal astronomer) (1782–); FRS (1782–): 531.
Hilbert, David (1862–1943). Major German mathematician. Professor of mathematics at Königsberg (1893–95). Professor at the Mathematical Institute at the University of Göttingen (1895–1930); advanced the axiomatization of geometry, consolidated number theory, contributed to the formalistic foundations of mathematics, and in 1900 proposed the famous twenty-three Paris problems. OM of the Berlin AcS and FM as FRS: 75, 202.
Hippocrates of Chios (ca. 470–ca. 410 B.C.). Merchant, geometer, and astronomer of ancient Greece who taught in Athens. A Pythagorean, he attempted to square the

circle through study of the quadrature of lunes (crescent-shaped figures); preceded Euclid in writing possibly the first work on the elements of geometry: 482, 483.
Hire, Philippe de la (1640–1718). French mathematician and observational astronomer: 101.
Hirschel, Abraham (mid-eighteenth century). Berlin banker who had a dispute with Voltaire: 358.
Hogarth, William (1697–1764). English artist, painter, engraver, social critic, and satirist; sergeant painter to the king (1757–): 315.
Huber, Johann Jakob (1733–78). Swiss astronomer in Basel. Student of Albrecht von Haller and physician; OM (1756–58), and FM (1758–) of the Berlin AcS: 85, 420.
Humbert, Abraham von (1689–1761). Engineer and major in the Prussian Army. OM of the Berlin AcS: 198, 215, 242, 420.
Hume, David (1711–76). Scottish philosopher, historian, economist, and essayist known for his philosophical empiricism. He accepted Isaac Newton's work as the model for the scientific method: 374.
Huygens, Christiaan (1629–95). Dutch mathematician, natural philosopher, astronomer, probabilist, and inventor of lenses and clocks. Proponent of the wave theory of light. FM as FRS (1663–), OM of the Paris AcS (1666–): 29, 34, 52, 60, 78, 97–99, 123, 126, 140, 150, 259, 357, 372, 443, 468.
Ibn al-Banna (1256–ca. 1321). Moroccan astronomer, mathematician, and astrologer; also a writer on linguistics, logic, and rhetoric, and translator of Euclid's *Elements* into Arabic: 271.
Ibn Yunus (950–1009). Egyptian Muslim astronomer and mathematician. Author of *Al-Zij al-Kabir al-Hakimi*, an original work of astronomical tables based on accurate observations, half of which survives today: 267.
Inochodzev (Inokhodtsev), Petr Borisovič (1742–1806). Adjunct (1768–79), associate professor of astronomy (1779–83), and professor of astronomy and OM of the Pb AcS (1783–); censor in Riga (1797–99): 457, 474, 513.
Ivan VI (1740–64). Infant emperor of Russia (r. 1740–41): 82, 171.
Jablonski, Daniel Ernst (1660–1741). Czech-German theologian and reformer who sought a union between Lutherans and Calvinists. A founder, secretary, vice president, and president (1733–41) of the Brandenburg Society of the Sciences (the old Berlin AcS): 177, 178.
Jacobi, Carl Gustav Jacob (1804–51). German mathematician and mathematical physicist. Professor at the University of Königsberg (1832–51); introduced research seminars in mathematics, promoted an edition of Euler; OM of the Berlin AcS: 347, 476.
James, William (fl. 1706). English mathematician: 289.
Jariges, Philippe Joseph de (1706–70). Prussian statesman and jurist. Director of the French high court in Berlin; OM (1731–), standing secretary (1733–48), and FM (1755–) of the Berlin AcS; worked with the privy state and war ministries (1755–): 186, 299, 344.
Jaucourt, Chevalier Louis de (1704–79). French author. Prolific contributor to the *Encyclopédie* and collaborator of Denis Diderot; wrote a biography of Gottfried Wilhelm Leibniz (1734); elected FM to the Berlin AcS, and FM as FRS (1764): 430.
Jetzler, Christoph (1734–91). Swiss mathematician and natural philosopher. Studied in Berlin with Euler; professor of mathematics and, after 1775, also physics at the Schaffhausen Collegium Humanitatis: 436.
Joncourt, Elie de (eighteenth century). French editor: 143.
Jordan, Charles Étienne (1700–1745). Prussian-born Huguenot historian and writer in the French language. Friend of Frederick II and official in his court; vice president of the Berlin AcS: 331, 367.
Juncker, Gottlieb Friedrich Wilhelm (1703–46). Swiss poet from Saint Gallen. Adjunct (1731–34), professor of politics and morals, then eloquence (1734–37), and FM of the Pb AcS (1737–): 88.
Jurin, James (1684–1750). English natural philosopher and physician in London. fellow (1717–) and secretary of the Royal Society of London (1721–27); editor of its

Kirch, Margarethe (eighteenth century). Sister of Christina Kirch; kept a weather diary for years: 293.
Kircher (Fuldensis), Athanasius (1601–80). Jesuit scholar distinguished in natural sciences and oriental studies. An inventor of a magic lantern: 155.
Kirilov, Ivan (ca. 1689–1737). Russian geographer and statistician; chief secretary of the Russian Senate: 110, 237.
Kleist, Ewald Georg von (1700–1748). German jurist, Lutheran cleric, administrator, and natural philosopher. Studied electricity and discovered the principle of the Kleistan (or Leiden) jar: 218.
Klingenstern (Klingenstierna), Samuel (1698–1765). Swedish mathematician and natural philosopher. Professor of geometry at Uppsala University (1728–) and physics (1756–58); FM as FRS (1730–); OM of the Royal Swedish Academy of Sciences (1739–), contributor to the invention of the achromatic telescope: 422.
Knowles, Charles (ca. 1697–1777). Rear admiral in the British Navy and briefly admiral in the Russian Navy (1770–74) including during the Seven Years' War: 504.
Knutzen, Martin (1713–51). German Wolffian philosopher with a basic knowledge of natural philosophy. Professor of logic and metaphysics at the University of Königsberg; teacher of Immanuel Kant: 198, 216, 217, 314.
Kohl, Johann (eighteenth century). German church historian at the Pb AcS: 51.
Köhler, David (fl. 1740–60). Financial officer at the Berlin AcS, calendar administrator, and friend of Euler. His embezzlement of funds left only part for the academy: 190, 199, 216, 299, 315, 439–441.
König, Johann Samuel (1712–57). German mathematician, philosopher, and lawyer. Professor at the war academy in the Hague, disputed the origins of the principle of least action with Pierre Maupertuis and Leonhard Euler. FM of the Berlin AcS (1749–52); librarian to Prince William IV of Orange in the Hague, returned his diploma from the Berlin AcS (1752): 18, 234, 281, 313, 318, 337–343, 350–354, 355, 356, 363, 369, 370, 373, 413, 533.
Korff, Johann Albrecht von (1697–1766). Russian diplomat and godfather of Euler's eldest son. President of the Pb AcS (1734–40), then HM; in diplomatic service (1740–): 82, 89, 109, 116, 117, 124, 174, 214.
Kotel'nikov, Semjon Kirillovič (1723–1806). Russian mathematician, student of Euler, and translator of the writings of Christian Wolff. Lectured on higher mathematics at the Pb AcS gymnasium (1785–96). Adjunct (1751–57), OM and professor of mathematics (1757–97), and HM and censor (1797–) of the Pb AcS; royal secretary (1768–ca. 1797): 368, 393, 394, 405, 454, 457, 484, 497, 503, 511, 526.
Koyré, Alexander (1892–1964). French historian of science, philosopher, and savant of the Russian origin. Studied under David Hilbert; expert on Galileo Galilei and the Scientific Revolution. Between 1955 and 1962, he spent half of each year at the Institute for Advanced Study by Princeton University; visiting professor at Harvard, Yale, and Johns Hopkins Universities and the Universities of Chicago and Wisconsin; associated (1958–) with what would later become the Centre Alexandre Koyré in Paris; member of the American Academy of Arts and Sciences: 465, 609.
Kozel'sky, Yakov Pavlovič (1728–94). Russian landowner and philosopher who taught mathematics and mechanics in the Russian Artillery Cadet Corps. Influenced by Mikhail Lomonosov and the French and German Enlightenments, he was a Voltairean and materialist who supported a social contract and opposed the ideas of Jean-Jacques Rousseau: 468.
Krafft, Georg Wolfgang (1701–54). German physicist and mathematician. Adjunct (1725–31), professor of physics and OM (1731–44), Pb AcS and also its conference secretary (1730–33) FM of the Pb AcS (1744–); professor of physics at the University of Tübingen (1744–); FM of the Berlin AcS (1745–): 51, 81, 84, 108, 109, 118, 171, 213, 234, 236, 253, 271, 321, 371, 474.

Krafft, Wolfgang Ludwig (1743–1814). Son of Georg Wolfgang Krafft. Adjunct (1768–71) and professor of physics and OM (1771–) of the Pb AcS: 470, 474, 489, 502, 508, 513, 532.

Kratzenstein, Christian Gottlieb 1723–95, German-born physician, natural philosopher, mechanist, and engineer. HM of the Pb AcS (1748–); HM and professor of experimental physics at the University of Copenhagen (1753–) and rector there four times: 254, 281, 282.

Kritter, Johann August (1721–98). German author, senator, and *camerarius* (chamberlain) in Göttingen: 512.

Kulibin, Ivan Petrovich (1735–1818). Self-taught Russian mechanic and inventor who was praised and visited by Euler and who made many mechanical tools and toys. Built a complete egg-shaped clock that played a cantata for Catherine II. Head of the workshops of the Pb AcS (1769–1801) and designer of a single-span bridge for the Neva River. Declined an offer of Russian nobility since it would require him to shave his beard and wear typical clothes: 485, 486, 506, 507, 511, 513.

Kurakin, Prince Alexander (early eighteenth century). Russian noble who owned the house that Euler purchased on his return to Saint Petersburg: 451.

Kurella (eighteenth century). German chemist in Berlin, son-in-law of Johann Heinrich Pott: 375.

Kushyar ibn Labban (971–1029). Persian astronomer, mathematician, and geographer who gave perhaps the first treatment of Hindu numerals and reckoning in Arabic arithmetic: 115.

Kutner, Samuel G. (eighteenth century). Engraver in Berlin: 519.

Lacaille, Nicolas Louis de (1713–62). French observational astronomer and geodesist. Abbé of Lacaille, professor of mathematics at Mazarin College in Paris, mapped stars in the southern sky. OM of the Paris AcS (1741–), FM of the Berlin AcS (1755–), and FM of the Pb AcS (1756–): 481.

La Croix, Cesar Marie de (1690–1747). French naval officer and architect. Worked on measuring cargo volumes and weights in ships; his study of these problems and ship hydrostatics and stability may have drawn Euler's interest: 504.

La Croix, Charles Eugene de (1727–1801). Secretary of state of the French fleet, marshal of France.

La Croyère, Louis de l'Isle de (before 1688–1741). Brother of Joseph Delisle. Associate professor of astronomy at Pb AcS (1727–), poor observer, died on the second Kamchatka expedition: 169.

Lagrange, Joseph-Louis, Comte de (b. Giuseppe Luigi Lagrangia) (1736–1813). Italian-French mathematician, physicist, and astronomer. Professor at the artillery school in Turin, significant contributor to analysis, number theory, and classical and celestial mechanics. FM (1756–) and then, on Euler's recommendation, OM and director of the mathematics department (1766–87) and again FM (1787–) of the Berlin AcS. FM of the Pb AcS (1776–); FM (1772–87) and OM (1787–) of the Paris AcS: 101, 127, 186, 260, 265, 347, 360, 392, 401, 402, 406–409, 413, 422, 423, 431, 450, 456, 458, 468, 475, 476, 484, 485, 492, 493, 496, 505, 506, 515, 516, 529.

Lalande, Joseph Jerôme le Français de (1732–1807). French astronomer, author, mathematician, and natural sciences researcher. Royal reader on mathematical questions and censor in Paris and professor of astronomy at the College de France. FM to the Berlin AcS (1751–), the Pb AcS (1764–), and the Royal Swedish Academy of Sciences; OM of the Paris AcS (1753–) and member of most large European science academies: 365, 369, 422, 475, 480.

Lambert, Johann Heinrich (1728–77). Swiss-German mathematician, natural philosopher, and astronomer who wrote proofs of pi and e as irrational based on continued fractions. OM of the Bavarian Academy of Sciences and of the Berlin AcS (elected 1764, approved by the king 1765): 134, 420, 421, 422, 427, 428, 435, 439, 441, 442, 467, 494, 517, 530.

La Mettrie, Julien Offray de (1709–51). French physician, philosopher, hedonist, architect, and materialist. Author of *L'Homme machine* (1747); court reader in Berlin (1748–), given refuge at the court of Frederick II; OM of the Berlin AcS (1748–): 297.

Landsberg, Hermann (1680–1746). Author of *Nouveaux plans et projets de fortification* (1731) and a later work on the same subject: 210.

Lange, Johann Joachim (1699–1765). German professor of mathematics at the University of Halle: 386, 387, 426.

Laplace, Pierre Simon (1749–1827). French mathematician, physicist, and astronomer. One of the most important scholars in these fields after Euler, referred to as the Newton of France. Demonstrated the stability of the solar system, and author of *Traité de mécanique celeste* (1799–1825, 5 vols.). Adjunct (1773–85) and OM (1785–) of Paris AcS: 260, 465, 481.

Laxmann, Erik G. (1737–96). Pioneer glass technologist who conducted research in Siberia and produced silicate glass: 484, 500.

Leadbetter, Charles (1688–1744). English mathematics teacher; writer on astronomy and practical mathematics; author of astronomical tables: 258.

Le Blond, Jean-Baptiste Alexandre (1679–1719). French architect and garden designer in Saint Petersburg: 41.

Legendre, Adrien-Marie (1752–1833). French mathematician and contributor to work on elliptic integrals. Adjunct (1783–85), associate (1785–95), and OM (1795–) of the Paris AcS: 134, 497.

Leibniz, Gottfried Wilhelm (1646–1716). German philosopher, mathematician, natural philosopher, historian, writer, logician, and researcher on language. Courtier as jurist-politician in Hannover (1676). FM of the Paris AcS (1669–) and FM as FRS (1676–); founder and first president of the Berlin AcS (1700), invented early phase of calculus: 1, 22, 28, 34, 42, 43, 45, 49, 50, 53–56, 60, 73, 78, 96, 97, 119, 121, 126, 176, 180, 197, 212, 216, 219, 226, 236, 243, 247, 262, 263, 259, 266, 285, 286, 298, 308, 313, 325, 326, 332, 337, 338, 342, 344, 350, 351, 353–356, 365, 366, 371, 373, 374, 392, 413, 428, 450, 458, 532.

Lemonnier, Pierre Charles (1715–99). French astronomer, OM of the Paris AcS: 291, 307.

Lessing, Gotthold Ephraim (1729–81). German writer, dramatist, critic, and publicist: 319, 403, 420, 437.

Leutmann, Johann Georg (1667–1736). German theologian, technologist, and natural philosopher known for mechanical and optical inventions. OM and professor of mechanics and optics at the Pb AcS (1725–33); worked at the mint in Moscow (1733–): 38, 49, 51.

Lexell, Anders Johan (Andrei Ivanovich) (1740–84). Finnish-Swedish mathematician, astronomer, and geographer. Professor of mathematics at Uppsala University; adjunct (1769–71) and professor of astronomy and OM (1771–) of the Pb AcS. Collaborator with and close friend of Euler, and his successor in the chair of mathematics at the Pb AcS: 471, 481, 482, 484, 489, 494, 502, 508, 513, 530, 531.

L'Hôpital, Guillaume-François-Antoine, Marquis de (1661–1704). French mathematician. OM (1693–99) and HM (1699–) of the Paris AcS: 24, 150, 287, 288.

Lieberkühn, Johann Nathanael (1711–56). German physician, anatomist, astronomer, and natural philosopher. OM of the Berlin AcS (1734–); FM as FRS (1734–): 189, 220, 382, 409.

Linnaeus, Carol (1707–78). Swedish botanist, physician, naturalist, and explorer who devised binomial nomenclature. Professor of medicine and botany at Uppsala University (1741–72); founder and member of the Royal Swedish Academy of Sciences (ca. 1738–) and FM of the Berlin AcS (1747–): 246.

Locke, John (1632–1704). English philosopher. Proponent of empiricism and opponent of authoritarianism: 21, 44, 57, 176, 178, 197, 262, 458, 465.

Lomonosov, Mikhail Vasil'evich (1711–65). Russian polymath, chemist, natural philosopher, and poet. Adjunct (1742–45) and professor of chemistry and OM (1745–)

of the Pb AcS; FM of the Royal Swedish Academy of Sciences (1760–) and the Bologna Academy of Sciences (1764–): 213, 236, 241, 281, 300, 302, 305, 314, 315, 336, 382, 402, 452, 455, 496, 535.
Louis XIV (1638–1715; r. 1643–1715). French monarch known as the Sun King; founder of the Paris AcS (1666): 42, 155, 176.
Louis XV (1710–1774; r. 1723–1774). French king: 197.
Lozeran du Fesc, Louis-Antoine de (1697–1755). French Jesuit. Shared with Euler the Prix de Paris on the nature of fire (1738); conducted research on cannons and fortifications (1740): 148.
Ludolff, Christian Friedrich (1707–63). German physician, natural philosopher, and professor at the Collegium Medico-chirurgicum. OM of the Berlin AcS (1738–): 218, 382, 419, 420, 503.
Luther, Martin (1483–1546). German theologian and seminal religious reformer: 9, 319.
Mac Lane, Saunders (1909–2005). American mathematician. Professor at Harvard University, Cornell University, and the University of Chicago; president of the American Mathematical Society and the Mathematical Association of America; vice president of the National Academy of Science (1949). Awarded the National Medal of Science (1989): xv.
Maclaurin, Colin (1698–1746). Scottish mathematician and natural philosopher. FRS (1719–) and professor of mathematics at the University of Edinburgh (1725–). He extended Isaac Newton's work in calculus in the two-volume *Treatise of Fluxions* (1742) and the Maclaurin series is named for him: 60, 96, 121, 132, 145, 165, 241, 292, 465, 478.
Mairan, Jean-Jacques d'Ortous de (1678–1771). French astronomer, natural philosopher, and chronobiologist. OM of the Paris AcS (1718–) and its perpetual secretary (1741–43 and 1746–); FM of the Pb AcS (1734–) and the Royal Swedish Academy of Sciences (1718–), and FM as FRS (1735–). Editor of the *Journal de Sçavans*: 147, 150, 179, 259.
Malebranche, Nicolas (1638–1715). French philosopher, Cartesian rationalist, and theologian: 44, 146, 259, 325, 360.
Malthus, Thomas Robert (1766–1834). English cleric, demographer, and economist. Professor of history and political economy at the East India Company College in Hertfordshire; conducted research on population growth and demography; FRS (1819–): 290.
Malygin, Stepan Gavrilovič (d. 1764). Lieutenant captain commander of the Russian fleet, research on the Arctic and author of *Navigation nach der Carte de Reduction*: 83.
Manfredi, Eustachio (1674–1739). Italian astronomer, mathematician, natural philosopher, geographer, geodesist, and poet. Professor at the University of Bologna (1698–), FM of the Berlin AcS and the Paris AcS (1726); FM as FRS (1729–): 216.
Manstein, Christoph (1711–57). Officer in the Prussian Army, general in the Russian Army: 171.
Mardefeld, Axel von (1691–1748). Baron, Prussian envoy to Russia (1728–40), and later an ambassador: 172, 174.
Marggraf, Andreas Sigismund (1709–82). German chemist, discoverer of beet sugar, pioneer of analytical chemistry, and last major German defender of the phlogiston theory. OM of the Berlin AcS (1738–), helped reorganize it (1740s), and directed its chemical laboratory and physical sciences department (1760–). FM of the Pb AcS (1776–): 197, 412, 414, 419, 420, 437, 439, 515.
Maria Leopoldine of Anhalt-Dessau (1716–82). Cousin of Frederick II and wife of Margrave Friedrich Heinrich of Brandenburg-Schwedt; mother of Friederike Charlotte Leopoldine Louise, whom Euler taught natural philosophy: 417.
Maria Theresa (1717–80, reigned 1742–1780). Archduchess, queen of Bohemia and Hungary, Holy Roman empress; last of the House of Habsburg in imperial Austria: 172, 184, 410, 417.

Marinoni, Giovanni Jacopo (1678–1755). Imperial Austrian astronomer, mathematician, and cartographer in Vienna. FM of the Berlin AcS and the Pb AcS (1746): 113, 246, 256.

Marpurg, Friedrich Wilhelm (1718–95). German music theorist, critic, composer, and celebrated violinist. Director of the Royal Prussian lotteries (1763–95); given the title of war councilor: 160.

Mariotte, Edme (1620–84). French physicist, plant physiologist, and priest: 61, 97.

Martini, Christian (1699–1739). Wolffian natural philosopher. Professor of physics (1725–26) and professor of logic and metaphysics (1726–29) at the Pb AcS: 49.

Maskelyne, Nevil (1732–1811). English royal astronomer and director of Greenwich Observatory (1765–1811) who improved time measurement and contributed to the science of navigation with new methods for finding longitude via studies of the moon. Anglican minister, fellow of Trinity College, Cambridge; FRS (1758–), receiving its Copley medal (1775): 379, 429–431.

Maslov, Aleksey Michailovich (mid-eighteenth century). Major general in the Russian Army during the Seven Years' War: 426.

Maupertuis, Pierre-Louis Moreau de (1698–1759). French mathematician, natural philosopher, and biologist. Adjunct (1723–31), OM (1731–35), and FM (1735) of the Paris AcS; FM of the Pb AcS (1738–); FM (1735–46), then OM and president of the Berlin AcS (1746–): 3, 143–145, 165, 172, 173, 178, 179, 181–184, 189, 209, 221–224, 242, 245, 246, 251, 252, 254, 260–262, 264, 265, 277–279, 283, 284, 289, 294, 296–300, 300, 304, 313–315, 318, 325–327, 334, 337–344, 350–357, 363–367, 370, 372–375, 386, 389, 391, 394, 401–403, 406–409, 413, 415–422, 430, 431, 535.

Maupertuis, René Moreau de (d. 1746). Father of Pierre Maupertuis: 243, 246.

Maurepas, Jean-Frédéric Phélypeaux, Comte de (1701–81). French statesman, freemason, royal adviser, and minister of the French marine. Skilled in military and naval strategy, he supported the geodesic mission of the Paris AcS (1734): 184.

Mayer, Friedrich Christoph (1697–1729). German mathematician and natural philosopher. Adjunct (1725–26) and associate professor (1726–) at the Pb AcS: 51, 66, 96.

Mayer, Johann Tobias (1723–62) German astronomer, mathematician, and geographer. Professor of mathematics and economics at the University of Göttingen (1750–) noted for lunar studies and a method for the determination of longitude at sea; prepared a catalog of stars, including double stars. Member of the Göttingen Academy of Sciences: 228, 274, 360, 375, 376, 378, 379, 391, 400, 412, 427, 429, 463, 481, 489, 535.

Melanchthon (b. Schwartzerdt), Philipp (1497–1560). German humanist and theologian who conducted the first systematic treatment of Wittenberg theology. Collaborator with Martin Luther; called a teacher of Germany for establishing and bettering public schools: 59, 179, 180.

Mendelssohn, Moses (1729–86). German Jewish philosopher, literary critic, Bible translator and commentator, and writer on aesthetics, political theory, and theology who authored works on Homer, Alexander Pope, Pierre Maupertuis, and Jean-Jacques Rousseau. A champion of religious toleration, he was known as the Jewish Luther; had a close association with Gotthold Ephraim Lessing. Awarded the 1763 prize of the Berlin AcS for "Über die Evidenz in metaphysischen Wissenschaften," an essay that applied mathematical proofs to metaphysics; Immanuel Kant came in second: 319, 403, 450.

Mengoli, Pietro (1625–86). Italian mathematician and cleric. Professor of mechanics at the University of Bologna: 78.

Menshikov, Prince Aleksandr Danilovich (1673–1729). Russian statesman, close friend of Peter I. Won the post of field marshal at Poltava (1709), and ruled during the reign of Catherine I: 44, 48, 68.

Menzel, Adolph Friedrich Erdmann von (1815–1905). German artist and realist painter. Works include four hundred drawings and wood-engraved prints, many portraying the reign and army of Frederick II: 330.

Merian, Johann Bernhard (1723–1807). Swiss philosopher. Professor of rhetoric at the University of Basel; OM (1750–) in the speculative philosophy department, librarian (1757–), director of the philosophy department (1771–97), and standing secretary (1797–1807) of the Berlin AcS: 329–331, 340, 353, 354, 369, 374, 403, 408, 413, 414, 419, 421, 431, 439, 442.

Merian, Maria Sybilla (1647–1717). German-born naturalist and scientific illustrator. Grandmother of Salome Abigail Euler: 513.

Mersenne, Marin (1588–1648). French theologian, natural philosopher, and mathematician. Cleric (minorit) in Nevers and Paris, superior of Place Royale monastery in Paris (1616–48). Major correspondent with a growing number of those engaged in the sciences across Europe and beyond. Best known for Mersenne primes $n = 2^p - 1$ in number theory: 20, 61, 80, 103, 150, 155, 158, 254, 350, 360.

Meslay, Jean-Baptiste Rouillé de (1656–1715). French printer; counselor at the Paris Parlement: 30.

Metzen, Frau. (mid- to late-eighteenth century) Formerly the housekeeper for lieutenant general Fedor Villimovich Bauer, she was the woman to whom Euler proposed in 1775. His family forced her to renounce the engagement: 509, 510.

Mirabel, Mademoiselle (eighteenth century). The woman from whom Euler purchased his Berlin house: 187.

Mizler, Lorenz (1711–78). German composer, mathematician, physician, historian, natural scientist, and Wolffian professor at the University of Leipzig who briefly studied music composition with Johann Sebastian Bach: 159, 160.

Moivre, Abraham de (1667–1754). French Huguenot–English mathematician and friend of Isaac Newton who conducted research on trigonometry, complex numbers, and probability. FRS (1697–): 79, 399.

Montecuccoli, Count Raimondo von (1609–80). Italian aristocrat. Field marshal in imperial Austrian service and master of fortifications and maneuvers: 211.

Montesquieu, Charles-Louis de Secondat, Baron de (1689–1755). French social commentator and political philosopher of the Enlightenment who opposed despotism. Author of *The Persian Letters* (1721) and *The Spirit of the Laws* (1748): 184, 239, 246.

Mozart, Wolfgang Amadeus (1756–91). Austrian composer from the Viennese classical school; one of the greatest composers in Western classical music: 14.

Müller, Gerhard Friedrich (1705–83). German historian, geographer, and pioneer in ethnography who conducted research in Siberia. Adjunct (1725–28) and OM and professor (1728–32), and conference secretary (1754–65) of the Pb AcS (reduced to adjunct, 1750–51; restored as OM, 1752–); FM as FRS (1730–). Participant in the Second Kamchatka Expedition (1733–43). FM of the Royal Swedish Academy of Sciences (1761–); director of the archives of the Collegium of Foreign Affairs in Moscow (1766–). Published an appendix to the popular science monthly *Sankt Peterburgskie Vedomosti* (1755–): 51, 71, 84, 110–112, 172, 174, 254, 265, 388, 389, 393, 404, 405, 409, 410, 411, 421, 425, 427, 430, 432, 433, 442, 451, 453, 485, 507.

Münnich, Count Burkhardt Christoph von (1683–1767). Russian count, statesman, and general field marshal of the Russian Army. Banished in 1741, returned in 1762: 87, 111, 171, 187.

Murr, Christoph Gottlieb von (1733–1811). German historian, polymath, bibliographer, and magistrate in Nuremberg: 495, 503.

Musschenbroek, Pieter (Petrus) van (1692–1761). Dutch mathematician, natural philosopher, and discoverer of the principle of the Leiden jar. Professor of mathematics at the Universities of Utrecht (1723–39) and Leiden (1739–); FM of the Pb AcS (1754–) the Berlin AcS (1746–) and the Royal Swedish AcS (1747–): 148, 149, 161, 218, 246, 356.

Napier, John (1550–1617). Scottish mathematician and theological writer known to have first come up with the concept of logarithms: 14.

Nartov, Andrej Konstantinovich (1693–1756). Russian natural philosopher, official, military engineer, and sculptor who worked in applied mechanics for Peter I. Leader of the academic Werkstätten and the academic chancery or administrative commission (1742–43): 67, 214.

Naudé, David (1720–94). German mathematician and astronomer at the Berlin AcS; son of Phillipp Naudé: 170, 216.

Naudé, Philipp, der Jüngere (the Younger) (1684–1745). Prussian mathematician. OM of the Berlin AcS (1711–), FM as FRS (1738–): 101, 170, 178, 190, 194, 205, 210, 216, 220, 345.

Navier, Claude-Louis-Marie-Henri (1785–1836). French engineer, physicist (in mechanics), and editor. Teacher of applied mechanics (1819–31) and then professor (1831–) at the École des Ponts et Chausées; technical consultant to the French state (1830–); OM of the Paris AcS (1824–): 312.

Neipperg, Count Wilhelm Reinhard von (1684–1774). Imperial Austrian general: 184.

Newton, Isaac (1642–1727). English natural philosopher (physicist), mathematician, astronomer, and alchemist known for his theory of gravitation. Inventor of calculus, disputed Gottfried Wilhelm Leibniz. Lucasian Professor at the University of Cambridge, author of the *Principia mathematica* (1687) and *Opticks* (1704). Master of the Royal Mint; fellow (1671–), and president of the Royal Society of London (1703–); FM of the Paris AcS (1699–): ix, 19, 22, 42, 50, 53, 54, 56, 61, 63, 64, 95, 98, 113, 115, 119, 121, 125, 126, 132, 142, 143, 145, 149, 153, 155, 161, 166, 176, 178, 201, 211, 218, 220, 228, 232, 247, 257, 259, 265, 272, 273, 275–278, 285, 287, 305, 307, 308, 320, 325, 326, 333, 336, 342, 344, 349, 357, 365, 377, 392, 428, 443, 462, 468, 469, 478, 504, 507, 532, 535.

Nörbel, Johann Jacob (d. 1758). Euler's brother-in-law: 162.

Nollet, Jean Antoine (also known as Abbé Nollet) (1700–1770). French clergyman and natural philosopher who conducted research on electricity and reportedly named the Leiden jar. Entered Paris AcS (1739) becoming OM (1758–). professor at the University of Paris, contributed to the *Mémoires* of the Paris AcS and the *Philosophical Transactions* of the Royal Society of London; FM as FRS (1734–).

Nostradamus (1503–66). French apothecary and reputed seer, published a famous collection of prophecies: 409.

Oechlitz, Christian Friedrich (b. 1723). German mathematician. Philosophy teacher at the University of Leipzig (1745–): 254, 281, 301.

Oecolampadius, Ulrich (Johannes Heussgen) (1482–1531). South German religious reformer, Christian humanist, and philologist: 6, 9.

Orlov, Count Grigorij Grigor'evich (1734–83). Russian general in chief and statesman. Lover of Catherine II, he fathered two of her children and supported her coup in 1762: 455.

Orlov, Count Vladimir Grigor'evich (1743–1831). Youngest of the famous Orlov brothers, named by Catherine II as director of the Pb AcS (1766–74), where he replaced Latin with German as the official language; later an HM there (1809–). Patronized such German men of science as Peter Simon Pallas: 455, 474, 480, 481, 487, 495, 496, 500, 501, 503, 508, 509, 534.

Ostermann, Andrei Ivanovich (b. Heinrich Johann Friedrich Ostermann) (1686–1747). German-born Russian count and statesman. Vice president of foreign affairs (1723–40); vice chancellor of all Russia (1740–41) and also cabinet minister of commerce, governor (1730–40); then exiled (1741). Pursued an imperial Austrian alliance: 171, 174, 187.

Paccassi, Johann Freiherr (Baron) von (fl. 1760–80). Astronomer, mathematician, and translator in Vienna: 229.

Pallas, Peter Simon (1741–1811). German zoologist, botanist, naturalist, and natural historian who made expeditions to central Russia and conducted research on native

Praskovea, Princess (early eighteenth century). The widowed sister-in-law of Peter I: 53, 528.

Prémontval, André Pierre Le Quay de (1716–64). Mathematician, philosopher, pedagogue, and defender of traditional marriage rather than common unofficial unions. OM of the Berlin AcS (1752): 344, 438.

Prokopovich, Feofan (Theophan) (1681–1736). Ukrainian statesman, theologian, and religious reformer. Bishop of Pskov, archbishop of Novgorod; helped Peter I in modernizing Russia and was a founder of the Pb AcS: 46, 82, 149.

Protassov, Aleksey Protas'evich (1724–96). Russian physician who conducted research in anatomy. Professor of anatomy and OM of the Pb AcS (1763–): 484.

Ptolemy, Claudius (ca. 100–ca. 170). Hellenistic astronomer, mathematician, and geographer in Alexandria. Author of the *Almagest*: 2, 114, 115, 154–156, 315, 339, 354.

Pufendorf, Samuel von (1632–94). German jurist, political philosopher, historian, economist, and statesman. Advocate of Cartesian philosophy who wrote on moral obligations and natural law. Taught at the Universities of Leipzig and Jena; made a baron in 1684: 180.

Pugachev, Yemel'yan Ivanovich (1740 or 1742–1775). Cossack pretender to the Russian throne who led the peasants' rebellion with the Cossacks against Catherine II (1773–75), wanting an end to serfdom: 456, 495.

Pythagoras of Samos (ca. 570–ca. 490 B.C.). Philosopher of ancient Greece who held a belief in a rational cosmology and who is credited with devising the Pythagorean Theorem on right triangles. Some of his followers pushed a scientific and mathematical direction of philosophy: 153, 156.

Quirini, Cardinal Angelo Maria (1680–1755). Italian scholar and Benedictine cardinal of the Roman Catholic Church who wrote on liturgy and the history of the Greek church and the papacy (Paul II). Bishop of Brescia, Cardinal, Prefect of the Vatican Library and then of the Index congregation (1747); FM of the Vienna Academy of Sciences (1747–) and the Berlin AcS (1748–) and the PAcS (1749–): 297.

Rachette, Jean-Dominique (1744–1809). French sculptor who designed models for the imperial porcelain factory in Russia. Made a bust of Euler: 534.

Rameau, Jean-Philippe (1683–1764). French composer and music theorist of the Baroque era, known for his harpsichord music and operas; created a synthesis of harmonic rules. Displaced Jean-Baptiste Lully as the chief composer of French opera: 153, 159, 360, 361, 394.

Ramus, Petrus (1515–72). French humanist, logician, educational reformer, and rhetorician who taught reformed Aristotelian logic: 179.

Raynal, Guillaume François, Abbé de (1711–96). French writer and propagandist, briefly a Jesuit, who helped set the climate for the French Revolution. Wrote a history of the East and West Indies, denounced European cruelty toward colonial peoples, and condemned the practice of slavery: 357, 358, 501.

Razumovskij, Count Kirill Gregor'evich (1728–1803). General field marshall, Hetman of Ukraine (1746–1798), president of the Pb AcS (1746–98), HM of the Berlin AcS (1748–): 189, 286, 300, 306, 321, 331, 333, 368, 376, 377, 404, 406, 433, 454.

Réamur, René Antoine Ferchault de (1683–1757). French naturalist and natural philosopher specializing in entomology. Developed what would become known as the Réamur temperature scale and Réamur porcelain. OM of Paris AcS (1708–); FM of the Berlin AcS and the Pb AcS (1737–); FM as FRS (1738–), and FM of the Royal Swedish Academy of Sciences (1748–): 313.

Redern, Count Sigismund Ehrenreich von (1719–89). Prussian court marshal and then chief marshal for Frederick II's mother; HM and curator of the Berlin AcS (1751–). He wrote articles on the perfection of optical instruments—especially telescopes, including the research of John Dollond—and late in his career also wrote on metaphysics and chemistry: 351, 456.

Reinbeck, Johann Gustav (1683–1741). Lutheran theologian. Curator in Berlin and provost at the University of Halle. Forerunner of Enlightenment theology in Prussia: 179, 180.

Reshevski, Alexey (1711–73). Juncker noble, occasional substitute for Vladimir Orlov as director of the Pb AcS (1775–73). Had good relations with Leonhard Euler and his son Johann Albrecht: 455, 486.

Reuchlin, Johannes (1455–1522). German humanist, political counselor, and scholar of Greek and Hebrew: 6.

Riccati, Jacopo (1676–1754). Italian mathematician, natural philosopher, and hydraulicist in Venice and Treviso: 50.

Richmann, Georg Wilhelm (1711–53). German natural philosopher. Adjunct (1740–41) and OM (1741–) of the Pb AcS and director of its physics laboratory; pioneer in the study of electricity. Killed when accidentally struck by lightning in an experiment similar to that of Benjamin Franklin: 235, 236, 282, 305, 336, 382, 411.

Riemann, Bernhard (1826–66). Major German mathematician credited with contributions to differential geometry that prepared the way for general relativity, as well as contributions to complex analysis and number theory. Professor at the University of Göttingen (1854–66) and head of its mathematical department (1859–); CM of the Berlin AcS and the Gesellschaft der Wissenschafter (Science Society) in Göttingen (1859–): 78, 136.

Ritter, Johann Jakob (1714–84). Swiss physician. MD from the University of Basel (1737); professor of anatomy, medicine, and botany at Franeker University (1747–50); physician in Silesia (1750–): 18, 19.

Robins, Benjamin (1707–51). English mathematician and military engineer, Newtonian, and critic of Euler. Conducted research on fortifications and ballistics and authored *New Principles of Gunnery* (1742). FRS (1727–): 128–130, 141, 211, 231, 308.

Rohault, Jacques (1618–72). Cartesian natural philosopher and mathematician: 23, 54.

Römer, Ole (1644–1710). Danish astronomer who demonstrated that the speed of light is finite. He spent nine years working at the Paris Observatory, professor of astronomy at the University of Copenhagen (1681–1705), cheif of Copenhagen police (1705–10): 155.

Roloff, Christian Ludwig (1726–1800). German botanist and author of *Index plantarum tem peregrimarum quam ostro nascentium*: 420.

Röse, Anton Ferdinand (fl. 1750s–60s). Danish editor and publisher in Rostock: 421.

Rosén von Rosenstein, Nils (1706–73). Swedish physician, specialist in applied anatomy. Opposed by Carl Linnaeus on medical treatments of illnesses. Professor of medicine at Uppsala University (1740–); first physician to the king of Sweden (1743–); member and president of the Royal Swedish Academy of Sciences: 511.

Rousseau, Jean-Jacques (1712–78). Genevan philosopher, writer, composer, music theorist, and botanist. Contributor in the fields of political philosophy, moral psychology, sociology, and education, and influenced Immanuel Kant: 153, 394, 501.

Rudolff, Christoph (1499–1545). German cossist (algebraist): 13, 14, 101.

Rumovskij, Stepan Jakolevich (1734–1812). Russian astronomer, mathematician, geographer, humanist, and student of Euler. Adjunct (1753–63); director of the geographical department (1763–67), professor of astronomy and OM (1767–), and vice president (1800–) of the Pb AcS; FM to the Royal Swedish Academy of Sciences (1761–): 235, 393, 405, 454, 457, 461, 480, 484, 497, 510.

Russell, Bertrand Arthur William (1872–1970). English philosopher, logician, epistemologist, and mathematician. FRS (1908–), Nobel Prize laureate (1950–): 19, 397.

Sack, August Friedrich Wilhelm (1703–86). German philosopher, theologian, and cleric. Court pastor in Berlin and pastor of the city's consistory; OM (1744–60) and veteran member (1760–) of the Berlin AcS: 249.

Salchow, Ulrich Christoph (1722–87). German physician and chemist. Professor of chemistry and OM of the Pb AcS (1753–60): 405.

Sanches/z, (Antonio Nuñes) Ribeiro (1699–1783). Portuguese physician, translator, encyclopedist, philosopher, pedagogue, and historian. First doctor of the Russian Army (1735–40); first physician to the Russian court (1740–47) and to Louis XV. FM of the Pb AcS (1747–48 and 1762–70), the Royal Medical Society in Paris (1747–), and the newly founded Royal Academy of Sciences in Lisbon. He discovered a new treatment for venereal disease: 471.

Scaliger, Joseph Justus (1540–1609). Dutch historian, textual critic, religious leader, and philologist who worked on chronology with the aid of astronomy and on comparative linguistics; expanded studies on classical history and taught at the University of Leiden: 118.

Schepin, Konstantin Ivanovich (1728–70). Russian physician and botanist. Professor at the chief hospital in Saint Petersburg: 426.

Schmettau, Count Samuel von (1684–1751). Prussian general field marshal and geographer, in Imperial Austrian and then Prussian military service (1717–41), field marshall (1744–51), curator of the Berlin AcS (1744–51): 199, 200, 216, 232, 301.

Schopenhauer, Arthur (1788–1860). German philosopher who stressed aesthetics. Lecturer at the University of Berlin (1825–32) and author of *Die Welt als Wille und Vorstellung* (2 vols., 1819): 469.

Schorndorf, Johann (d. 1769). Postmaster and friend of the Euler family in Basel: 283.

Schumacher, Johann Daniel (1690–1761). Alsatian bureaucrat and librarian. Officiated over the Pb AcS (1725–59, with one brief interruption); also managed its library and Kunstkamera: 44, 46, 70, 81, 84, 85, 89, 91, 110, 141, 175, 212, 213, 234–236, 238, 253, 254, 269, 281, 294, 314, 320, 327, 331, 332, 336, 368, 382, 392, 393, 404, 455.

Segner, Jan Andrej (Johann Andreas) von (1704–77). Hungarian-born natural philosopher and mathematician. Professor of natural science and mathematics at the University of Göttingen and the University of Halle (1754–); FM of the Berlin AcS (1746–) and the Pb AcS (1754–): 138, 312, 347, 375, 386, 387, 400, 426, 427, 435, 480, 485, 535.

'sGravesande, Willem Jacob (1688–1742). Dutch mathematician, natural philosopher, and lawyer who criticized the concept of perpetual motion. FM as FRS (1715–); professor at University of Leiden (1717–); offered but declined posts in the Pb AcS (1724) and the Berlin AcS (1737): 21, 45, 50, 85, 149, 339, 354.

Shafirov, Baron Peter Pavlovich (1670–1739). Russian statesman; coadjutor for Peter I: 53.

Short, James (1710–68). Leading British instrument maker and astronomer who created and built instruments for the Royal Society of London, including reflecting telescopes, in part to observe the transit of Venus in 1769, but he did not live long enough to do so. FRS (1736–): 415.

Sievers, Pieter de (eighteenth century). Danish marine lieutenant and later Russian captain (1712–19) and admiral (1719–40); vice president of the Russian Admirals' College (1727–40): 69.

Silberschlag, Johann Essais (1721–91). German Lutheran theologian and natural scientist. Pastor in Bergen; director of the Realschule in Berlin (1769–).FM (1760–86) and OM (1786–) of the Berlin AcS: 420.

Sloane, Hans (1660–1753). Irish physician, naturalist, collector, and a founder of the British Museum. FRS (1685–), editor of the *Philosophical Transactions* (1693–1713), and president of the Royal Society succeeding Isaac Newton (1727–41). FM of the Paris AcS (1709–),the Berlin AcS (1712–), and the Pb AcS (1734–): 149.

Snow, Charles Percy (1905–80). British physicist and novelist responsible for the two cultures theory: 149, 268, 346.

Sofronov, Mikhail Ivanovič (1729–60). Student of Euler. Adjunct in mathematics at the Pb AcS (1753–): 393.

Sophia-Dorothea of Hanover (1687–1757). German princess interested in art, literature, the sciences, and fashion who wrote on the improvement of scientific instruments, including the telescope. Queen consort in Prussia (1713–); HM of the Berlin AcS (1751–) and one of its curators: 187, 417.

Sophie Charlotte (1668–1705). Electress of Brandenburg (1688–); queen consort of Prussia (1701–5). Friend of Gottfried Wilhelm Leibniz, who was a guest at her palace at Lützenburg outside Berlin: 177, 385.

Spener, Johann Carl (d. 1756). Codirector (with Ambrose Haude) of the Haude and Spener publishing firm in Berlin: 251.

Spener, Philipp Jacob (1635–1705). German Christian theologian of the Holy Roman Empire who called for the reform of the Lutheran Church. A founder of the University of Halle (1694). Instrumental in the advancement of Pietism (though not a strict adherent) and publisher of the German "musical library": 46.

Spinoza, (Baruch) Benedict de (1632–77). Dutch Jewish philosopher: 45, 218.

Stähelin, Benedict (1695–1750). Basel physician and experimental physicist. CM of the Paris AcS (1727–); professor of physics (1727–) and rector (1737–) of the University of Basel. FM of several academies: 85.

Stahl, Georg Ernest (1660–1734). German physician and chemist who proposed the theory of phlogiston to explain combustion; the theory was dominant for most of the eighteenth century. Court physician at Weimar, professor of medicine at the University of Halle (1694–1715), and royal physician in Berlin (1715–): 69.

Stählin, Jacob (1709–85). German-born adjunct (1735–37), OM and professor of rhetoric and poetry (1737–), and conference secretary (1765–69) of the Pb AcS: 448, 452, 454, 457, 484, 528, 529.

Stapff, Gottfridus Magnus Maria (mid-eighteenth century). Freiburg theologian; participant in the Pb AcS annual prize competition in 1751: 322.

Stevin, Simon (1548–1620). Dutch mathematician, music theorist, engineer dealing with fortifications and navigation, and defender of Copernican astronomy. Known for devising decimal notation. Established an engineering school at the University of Leiden (1600); quartermaster general of the Dutch army (1604–): 14.

Stifel, Michael (1487–1567). German cossist (algebraist): 13.

Stirling, James (1692–1770). Scottish mathematician who contributed to the theory of infinite series and differential calculus. FRS (1726–53), FM of the Berlin AcS (1746–) and the Pb AcS: 95, 122, 132, 246, 255.

Stokes, George Gabriel, First Baronet (1819–1903). British physicist, mathematician, politician, and theologian who studied viscous liquids and was a pioneer of geodesy. Lucasian Professor of Mathematics at Cambridge University (1849–1903); FRS (1851–), and later the society's secretary and then its president: 312.

Stratico, Simone (eighteenth century). Italian translator: 508.

Suhm, Ulrich Friedrich von (1691–1740). Saxon count; friend of Frederick II and Prussian ambassador to Russia: 170.

Sully, Henri (early eighteenth century). Mathematician: 28.

Sulzer, Johann Georg (1720–79). Swiss Wolffian philosopher and aesthetician. Professor of mathematics at the gymnasium in Joachimsthal; OM of the Berlin AcS (1750–) and director of its philosophy department shortly before his death. Translator into German of David Hume's *An Enquiry Concerning the Principles of Morals,* (1755): 330, 344, 403, 439, 441, 445.

Süssmilch, Johann Peter (1707–67). German pastor, theologian, and statistician in Berlin. Pioneer in demography in Prussia; worked with a new concept of social statistics: 431, 474.

Tartaglia, Niccolò Fontana (1500–57). Italian mathematician, engineer, and surveyor, and the first to solve cubic equations algebraically. Debated Girolamo Cardano;

translated Euclid into Italian; advanced the field of ballistics with his theory of gunnery: 210.

Tatishchev, Vasilii Nikititch (1686–1750). Russian statesman, historian, geographer, and economic developer: 44, 47, 101, 116.

Taubert, Johann Kaspar (1717–71). Son-in-law of Johann Daniel Schumacher. Adjunct in history (1738–), librarian, and chancery official of the Pb AcS. Correspondent of Euler (1759–66); supervisor of the Kunstkammer (1761–67); business organizer: 433.

Taylor, Brook (1685–1731). English mathematician at the University of Cambridge, known for the Taylor Theorem and the Taylor Series in calculus. FRS (1712–) and the society's secretary (1714–18): 15, 24, 106, 132, 225.

Teplov, Grigorij Nikolaevich (1725–79). Russian senator, academic administrator, and adviser and assistant to Count Kirill Gregor'evich Razumovskij. Adjunct in botany (1742–47), secretary, assessor (1746–) of the academic chancery, and HM (1747–) of the Pb AcS. Correspondent of Euler and translator of the writings of Christian Wolff: 189, 253, 281, 286, 300, 306, 320, 332, 433, 434.

Thiébault, Dieudonné (1730–1807). French man of letters. Professor of French grammar at the military school and the Royal Court in Berlin (1765–84) and member of the Berlin AcS (1765–), for which he corrected the grammar in many publications. Wrote about Frederick II and his court before returning to France in 1784: 282, 419, 421, 441, 445, 450.

Thomasius, Christian (1655–1728). German jurist and philosopher. In Leipzig a central figure in the German Enlightenment who lectured in German; a founder of the University of Halle: 180.

Tindal, Matthew (1657–1733). English philosopher, theologian, proponent of Christian deism with reason foremost over faith, and defender of the Erastian view of the place of the state over the church: 266.

Tiresias. In Greek mythology, Tiresias was a blind Theban seer of Apollo who had great prophetic gifts. Among others, Homer, Euripides, and Ovid mention him. In later European literature, he was cited as a prophet: 534.

Totleben (Tottleben), Count Gottlob Heinrich (1710–73). General in the Russian Army, part of the brief Berlin occupation (1760); pardoned from death sentence by Catherine II (1763): 425.

Toussaint, François-Vincent (1715–72). French author, publisher, and translator who wrote the novel *Les Moeurs* (1748); OM of the Berlin AcS (1751–): 438.

Trediakovskii, Vasily Kirillovich (1703–69). Russian poet, prose writer, playwright, literary theorist, and translator of Bernhard Le Bovier de Fontenelle's *Adventures of Telemak*. Professor of eloquence at the Pb AcS (1745–): 118.

Trezzini, Domenico Andrea (ca. 1670–1734). Swiss architect famous for the Peter and Paul Fortress in Russia: 40, 41.

Truesdell, Clifford (1919–2000). American mathematician, natural philosopher, historian of science, and specialist on rational thermodynamics. Professor at Johns Hopkins University, editor of five volumes of Euler's *Opera omnia*: xii, 398.

Turgot, Anne-Robert Jacques, Baron de Laune (1727–1781). French economist, reformer, statesman, physiocrat, and (under Louis XVI) comptroller general of finance who ended *corvée* labor for peasants and supported religious tolerance: 431, 505, 507.

Van Schooten, Franz (1615–60). Dutch mathematician; popularizer of Cartesian geometry: 101.

Varignon, Pierre (1654–1722). French mathematician and researcher in graphical statics and mechanics. Professor in Paris at the Collège Mazaria (1688–) and the Collège Royale (1704–); OM of the Paris AcS (1688–); FM of the Berlin AcS (1713–); and FM as FRS (1718–): 24, 97, 287, 374.

Vauban, Sébastien Le Prestre, Marquis de (1633–1707). Leading French military engineer who transformed fortification and siege strategies: 211.

Vaucanson, Jacques de (1709–82). French engineer. Prolific inventor of automata (including an automated loom and the first robots), machines, and machine tools important to the Industrial Revolution: 181.

Venn, John (1834–1923). English mathematician, logician, philosopher, and Anglican priest skilled in building machines. Lecturer in moral science at the University of Cambridge and FRS (1897–). Introduced what would later become known as the Venn diagram: 158, 464, 467, 477.

Vermeulen, Georg Wilhelm (b. 1726). Cousin of Euler. Student at the Petersburg Academy's gymnasium (1738), afterward serving as an officer in the Prussian and Russian military: 282, 375.

Vermeulen, Karl Rudolf (b. 1728). Brother of Georg Wilhelm Vermeulen and cousin of Euler. Student at the Pb AcS gymnasium (ca. 1734), afterward serving in the Prussian military: 282.

Viereck, Adam Otto von (1684–1758). Prussian minister of state, privy councilor, and diplomat: 177, 198, 199.

Viète, François (1540–1603). French attorney and councilor to Henry III and Henry IV; independent mathematician with algebra as his major field: 101, 133, 483.

Vignoles, Alphonse des (1649–1744). French noble, savant, and historian. FM (1701–3), OM (1703–), director of the mathematics department (1721–), and vice president (1729–30) of the Berlin AcS: 178, 189, 215.

Vignon, P. (eighteenth century). French instrument maker: 81.

Virgil (Publius Vergilius Maro) (70–19 B.C.). Roman poet, author of the *Aeneid*: 22, 330.

Vlacq, Adriaan (ca. 1600–1667). Dutch publisher and mathematician who printed early tables of logarithms to fourteen decimal places: 94.

Voltaire (pen name of François Marie Arouet) (1694–1778). French writer, dramatist, *philosophe*, historian, public activist, and leader in the French Enlightenment. Adviser to Frederick II, FM as FRS (1743–), HM of the Berlin AcS (1746–), and FM of the Pb AcS (1746–): 26, 52, 144, 148, 149, 176–178, 187, 237, 239, 295, 297–299, 313, 322, 330, 351, 355–358, 363–366, 370, 372, 412, 417, 419, 431, 465, 527, 532.

Voroncov (Vorontsov), Count Michail Illarionovich (1714–67). Russian statesman and diplomat; high vice chancellor (1744–): 442, 446.

Wagner, Johann Wilhelm (1681–1745). German astronomer and mathematician. OM of the old Berlin AcS (1716–); professor of construction at the Painting Academy in Berlin (1730–); librarian (1736–), astronomer, and director (1740–) of the Berlin Observatory and the Berlin AcS: 185.

Wallis, John (1616–1703). English mathematician. Savilian Professor of Geometry at the University of Oxford (1649–): 50, 61, 78, 95, 101, 132–134, 287.

Waltz, Johann Theophil (1713–47). German astronomer and mathematician in Dresden. FM of the Berlin AcS (1745–): 254.

Waring, Edward (1734–98). English mathematician. Lucasian Professor at the University of Cambridge (1760–); FRS (1763–): 529.

Wegelin, Jakob Daniel (1721–91). Swiss historian, evangelical theologian, and philosopher. Professor of history at the Ritterakademie in Berlin, OM and archivist of the Berlin AcS (1765–): 5, 421, 447.

Weggersløff, Friedrich (1702–63). Danish naval officer: 141.

Weitbrecht, Josias (1702–47). German physiologist. Adjunct (1725–31) and professor and OM (1731–) at the Pb AcS: 51, 84, 235, 236, 282, 287.

Wenz (Wentz), Ludwig (1695–1772). Swiss astronomer and mathematician in Basel. Correspondent of Euler (1742–65): 234.

Wenzel, Baron Jakob Michael Johann Baptist von (1724–90). German ophthalmologist and skillful cataract surgeon who operated on Euler. Later moved to London where he founded the first ophthalmic hospital; oculist to King George III: 484, 488.

General Index